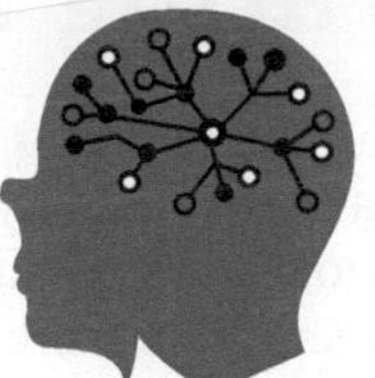

- **Dynamic Study Modules** help students study chapter topics and the language of MIS on their own by continuously assessing their knowledge application and performance in real time. These are available as graded assignments prior to class, and are accessible on smartphones, tablets, and computers.

- **Learning Catalytics™** is a student response tool that helps you generate class discussion, customize your lecture, and promote peer-to-peer learning based on real-time analytics. Learning Catalytics uses students' smartphones, tablets, or laptops to engage them in more interactive tasks.

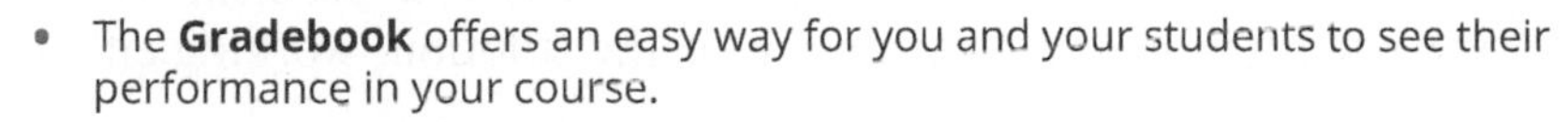

- The **Gradebook** offers an easy way for you and your students to see their performance in your course.

  Item Analysis lets you quickly see trends by analyzing details like the number of students who answered correctly/incorrectly, time on task, and more.

  And because it's correlated with the AACSB Standards, you can track students' progress toward outcomes that the organization has deemed important in preparing students to be leaders.

- **Pearson eTextbook** enhances learning—both in and out of the classroom. Students can take notes, highlight, and bookmark important content, or engage with interactive lecture and example videos that bring learning to life anytime, anywhere via MyLab or the app.

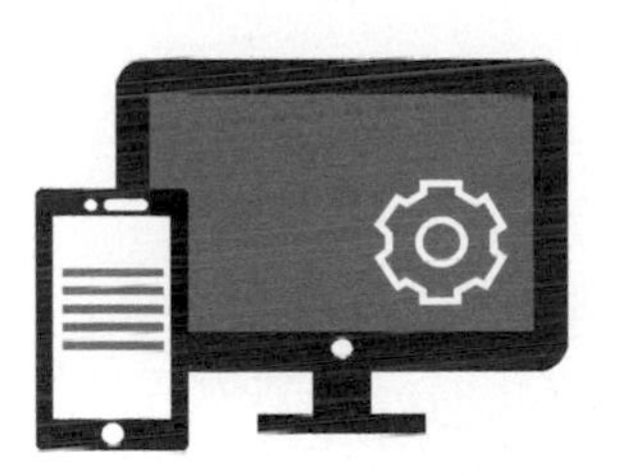

- **Accessibility (ADA)**—Pearson is working toward WCAG 2.0 Level AA and Section 508 standards, as expressed in the **Pearson Guidelines for Accessible Educational Web Media.** Moreover, our products support customers in meeting their obligation to comply with the Americans with Disabilities Act (ADA) by providing access to learning technology programs for users with disabilities.

  Please email our Accessibility Team at **disability.support@pearson.com** for the most up-to-date information.

- With **LMS Integration**, you can link your MyLab course from Blackboard Learn™, Brightspace® by D2L®, Canvas™, or Moodle®.

http://www.pearsonmylabandmastering.com

# Integrating Business with Technology

By completing the projects in this text, students will be able to demonstrate business knowledge, application software proficiency, and Internet skills. These projects can be used by instructors as learning assessment tools and by students as demonstrations of business, software, and problem-solving skills to future employers. Here are some of the skills and competencies students using this text will be able to demonstrate:

**Business Application skills:** Use of both business and software skills in real-world business applications. Demonstrates both business knowledge and proficiency in spreadsheet, database, and web page/blog creation tools.

**Internet skills:** Ability to use Internet tools to access information, conduct research, or perform online calculations and analysis.

**Analytical, writing and presentation skills:** Ability to research a specific topic, analyze a problem, think creatively, suggest a solution, and prepare a clear written or oral presentation of the solution, working either individually or with others in a group.

*** Dirt Bikes Running Case in MyLab MIS**

## Business Application Skills

| Business Skills | Software Skills | Chapter |
|---|---|---|
| **Finance and Accounting** | | |
| Payroll accounting | Spreadsheet formulas<br>SUM, AVERAGE functions<br>Absolute and relative addressing | Chapter 1 |
| Financial statement analysis | Spreadsheet formulas<br>Spreadsheet charts | Chapter 2* |
| Budgeting | Spreadsheet formulas<br>Spreadsheet charts | Chapter 2 |
| Pricing hardware and software | Spreadsheet formulas | Chapter 5 |
| Technology Rent vs. Buy Decision<br>Total Cost of Ownership (TCO) analysis | Spreadsheet formulas<br>SUM, VLOOKUP | Chapter 5* |
| Analyzing wireless services and costs | Spreadsheet formulas | Chapter 7 |
| Financial statement analysis | Spreadsheet formulas<br>Spreadsheet downloading and formatting | Chapter 10 |
| **Human Resources** | | |
| Analyzing security events | Spreadsheet sorting, data filtering | Chapter 8 |
| Employee training and skills tracking | Database design<br>Database querying and reporting | Chapter 12* |
| **Manufacturing and Production** | | |
| Inventory management | Database design, querying, and reporting | Chapter 6 |
| Analyzing supplier performance and pricing | Spreadsheet date functions<br>Data filtering, AVERAGE function | Chapter 9 |
| Bill of materials cost sensitivity analysis | Spreadsheet data tables<br>Spreadsheet formulas | Chapter 11* |
| **Sales and Marketing** | | |
| Sales trend analysis | Database querying and reporting | Chapter 3 |
| Blog creation and design | Blog creation tool | Chapter 4 |
| Marketing decisions | Spreadsheet pivot tables | Chapter 11 |

| | | |
|---|---|---|
| Customer reservation system | Database querying and reporting | Chapter 12 |
| Customer profiling | Database design<br>Database querying and reporting | Chapter 6* |

## Internet Skills

| | |
|---|---|
| Using online software tools for job hunting and career development | Chapter 1 |
| Using online interactive mapping software to plan efficient transportation routes | Chapter 2 |
| Researching product information<br>Evaluating websites for auto sales | Chapter 3 |
| Analyzing web browser privacy protection | Chapter 4 |
| Researching travel costs using online travel sites | Chapter 5 |
| Searching online databases for products and services | Chapter 6 |
| Using web search engines for business research | Chapter 7 |
| Researching and evaluating business outsourcing services | Chapter 8 |
| Researching and evaluating supply chain management services | Chapter 9 |
| Evaluating e-commerce hosting services | Chapter 10 |
| Using shopping bots to compare product price, features, and availability | Chapter 11 |
| Analyzing websites design | Chapter 12 |

## Analytical, Writing, and Presentation Skills*

| **Business Problem** | **Chapter** |
|---|---|
| Management analysis of a business | Chapter 1 |
| Value chain and competitive forces analysis<br>Business strategy formulation | Chapter 3 |
| Formulating a corporate privacy policy | Chapter 4 |
| Employee productivity analysis | Chapter 7 |
| Disaster recovery planning | Chapter 8 |
| Locating and evaluating suppliers | Chapter 9 |
| Developing an e-commerce strategy | Chapter 10 |

*Fifteenth Edition*

# Essentials of Management Information Systems

Kenneth C. Laudon
New York University

Jane P. Laudon
Azimuth Information Systems

Carol Guercio Traver
Azimuth Interactive, Inc.

Content Management: Ellen Thibault
Content Production: Purnima Narayanan
Product Management: Marcus Scherer
Product Marketing: Wayne Stevens
Rights and Permissions: Jenell Forschler

Please contact https://support.pearson.com/getsupport/s/ with any queries on this content

Cover Image by yurihoayda/123RF



**Library of Congress Cataloging-in-Publication Data**

Names: Laudon, Kenneth C., author. | Laudon, Jane P. (Jane Price), author. | Carol Guercio Traver, author.
Title: Essentials of management information systems / Kenneth C. Laudon, New York University, Jane P. Laudon, Azimuth Information Systems., Carol Guercio Traver Azimuth Interactive, Inc.
Description: Fifteenth edition. | Hoboken : Pearson Education, Inc., 2024. | Includes bibliographical references and index.
Identifiers: LCCN 2022035017 | ISBN 9780137946792 (paperback)
Subjects: LCSH: Management information systems.
Classification: LCC T58.6 .L3753 2022 | DDC 658.4/038—dc23/eng/20220824
LC record available at https://lccn.loc.gov/2022035017

1 2022

ISBN 10: 0-13-794679-1
ISBN 13: 978-0-13-794679-2

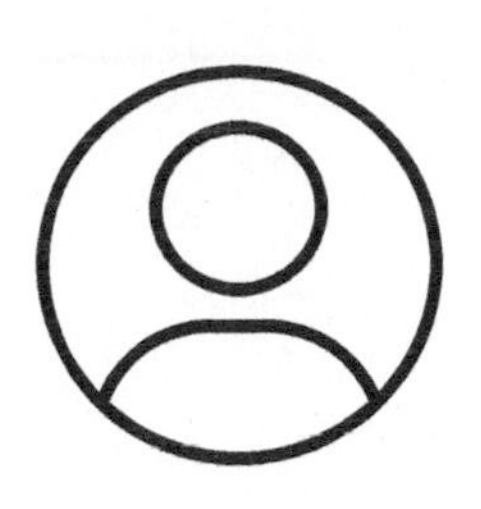

# Pearson's Commitment to Diversity, Equity, and Inclusion

**Pearson is dedicated to creating bias-free content that reflects the diversity, depth, and breadth of all learners' lived experiences.**

We embrace the many dimensions of diversity, including but not limited to race, ethnicity, gender, sex, sexual orientation, socioeconomic status, ability, age, and religious or political beliefs.

Education is a powerful force for equity and change in our world. It has the potential to deliver opportunities that improve lives and enable economic mobility. As we work with authors to create content for every product and service, we acknowledge our responsibility to demonstrate inclusivity and incorporate diverse scholarship so that everyone can achieve their potential through learning. As the world's leading learning company, we have a duty to help drive change and live up to our purpose to help more people create a better life for themselves and to create a better world.

**Our ambition is to purposefully contribute to a world where:**

- Everyone has an equitable and lifelong opportunity to succeed through learning.
- Our educational content accurately reflects the histories and lived experiences of the learners we serve.
- Our educational products and services are inclusive and represent the rich diversity of learners.
- Our educational content prompts deeper discussions with students and motivates them to expand their own learning (and worldview).

### Accessibility

We are also committed to providing products that are fully accessible to all learners. As per Pearson's guidelines for accessible educational Web media, we test and retest the capabilities of our products against the highest standards for every release, following the WCAG guidelines in developing new products for copyright year 2022 and beyond.

You can learn more about Pearson's commitment to accessibility at **https://www.pearson.com/us/accessibility.html**

### Contact Us

While we work hard to present unbiased, fully accessible content, we want to hear from you about any concerns or needs with this Pearson product so that we can investigate and address them.

Please contact us with concerns about any potential bias at **https://www.pearson.com/report-bias.html**

For accessibility-related issues, such as using assistive technology with Pearson products, alternative text requests, or accessibility documentation, email the Pearson Disability Support team at **disability.support@pearson.com**

# Brief Contents

# Contents

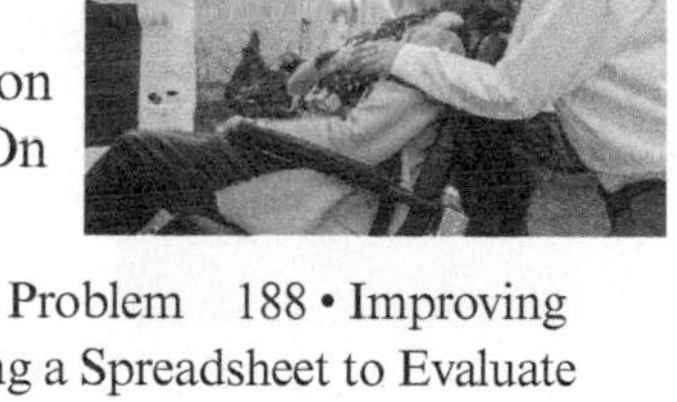

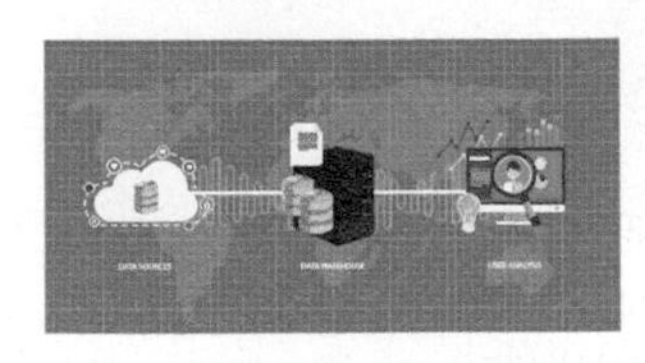



## 7 Telecommunications, the Internet, and Wireless Technology 230

RANSOMWARE

# Business Cases

Here are some of the business firms described in the cases in this book:

## Chapter 1: Business Information Systems in Your Career

Smart Shelves Reinvent the Retail Space
UPS Competes Globally with Information Technology
Will a Robot Steal Your Job?
Will the Covid-19 Pandemic Make Working from Home the New Normal?

## Chapter 2: Global E-Business and Collaboration

Microsoft Teams Helps Toyota Motor North America (TMNA) Do Even Better
Carbon Lighthouse Lights Up with the Internet of Things (IoT), Big Data, and Cloud Computing
Zoom: Quality Videoconferencing for Every Budget
How Much Does Technology Help Collaboration?

## Chapter 3: Achieving Competitive Advantage with Information Systems

Walmart's Supercenter Strategy
Customer Experience Management: A New Strategic Weapon
Signet Jewelers Sparkles with a Virtual Sales Process
Shipping Wars

## Chapter 4: Ethical and Social Issues in Information Systems

Apps That Track: A Double-Edged Sword
Section 230: Should the Law that "Created" Today's Internet Be Repealed or Revised?
Immersed in the Metaverse: What Will It Mean for the Future?
Facebook's Many Ethical Challenges

## Chapter 5: IT Infrastructure: Hardware and Software

QRyde Rides High with the Cloud
The Mobile Platform Comes to Healthcare
"Smart" Cities Become Smarter with Edge Computing
How Green Is the Cloud?

## Chapter 6: Foundations of Business Intelligence: Databases and Information Management

Better Data, Better Decisions for the State of Maine
New Cloud Database Tools Help Vodafone Fiji Make Better Decisions
Higher Data Quality Helps Vyaire Save Lives
Pursuing Sustainability with Blockchain

## Chapter 7: Telecommunications, the Internet, and Wireless Technology

The National Hockey League Scores with Wireless Technology and the Internet of Things (IoT)
Can Low Earth Orbit Satellite Internet Systems Solve the Digital Divide?
Monitoring Employees on Networks: Unethical or Good Business?
Google, Apple, and Meta Battle for Your Internet Experience

## Chapter 8: Securing Information Systems

Ransomware Everywhere
Race Against Time: The Scramble to Fix the Log4j Zero-Day Vulnerability
PayPal Ups Its Digital Resiliency
SolarWinds Hack Shines a Light on Software Supply Chain Attacks

## Chapter 9: Achieving Operational Excellence and Customer Intimacy: Enterprise Applications

Lenzing Sustainably Balances Supply and Demand
MillerKnoll Uses Salesforce.com to Transform Its Business Strategy
Versum's ERP Transformation
The Coronavirus Pandemic Disrupts Supply Chains Around the World

## Chapter 10: E-commerce: Digital Markets, Digital Goods

Chewy Blends B2C and B2B E-commerce
Lemonade Disrupts the Insurance Industry with "Insurtech"
Engaging "Socially" with Customers on TikTok
Uber Discovers That Becoming the Uber of Everything Is Not So Easy

## Chapter 11: Improving Decision Making and Managing Artificial Intelligence

Big Data Analytics: A New Way to Fight Wildfires
What Happened to Watson Health?
Do You Know Who Is Using Your Face?
Should an Algorithm Make Our Decisions?

## Chapter 12: Making the Business Case for Information Systems and Managing Projects

Angostura Builds a Mobile Sales System
McAfee Turns to Automated Software Testing
Sauder Woodworking Gets ERP Implementation Right
JEDI and JWCC: A Cloud of Controversy

# Preface

## New To This Edition

*Essentials of Management Information Systems,* 15th edition has been thoroughly updated to cover the latest industry and technology changes that impact the course and to provide new interactive learning opportunities.

### MyLab MIS

The goal of *Essentials of Management Information Systems* is to provide students and instructors with an authoritative, up-to-date, interactive, and engaging introduction to the MIS field. MyLab MIS for *Essentials of Management Information Systems* is an extension of this goal in an interactive digital environment.

MyLab is the teaching and learning platform that empowers you to reach every student. By combining trusted author content with digital tools and a flexible platform, MyLab personalizes the learning experience and improves results for each student.

MyLab MIS features videos, animations, and interactive quizzes to foster student comprehension of concepts, theories, and issues. The MyLab MIS environment reflects the new learning styles of students, which are more social, interactive, and usable on digital devices such as smartphones and tablets.

### WHAT'S INCLUDED

- **Pearson eTextbook** – Enhances learning – both in and out of the classroom. Students can highlight, take notes, and review key vocabulary all in one place, even when offline. Seamlessly integrated interactivities and Figure Videos bring concepts to life via MyLab or the app.
- **Figure Videos** – Have author Ken Laudon walk students through important concepts in each chapter (23 total) using a contemporary animation platform. Available not only in the Pearson eTextbook that lives in MyLab MIS but can also be purchased with a standalone eTextbook.
- **New Video Cases** – A brand new collection of video cases (one per chapter) draws from Pearson's extensive library of business and technology video clips. The cases cover key concepts and experiences in the MIS world, illustrating how real-world businesses and managers are using information technology and systems. Each case is accompanied by 3 auto-graded true/false or multiple choice questions and 2 manually graded essay questions. Video cases are listed at the beginning of each chapter.
- **MIS Simulations** – Foster critical decision making skills with these interactive exercises that allow students to play the role of a manager and make business decisions.
- **Chapter Warm Ups, Chapter Quizzes** – These objective-based quizzes evaluate comprehension.
- **Excel & Access Activities** provided inside MyLab MIS support classes covering Office tools. In addition, Hands-On MIS Projects from the book are available.
- **Running Case** on Dirt Bikes USA provides additional hands-on projects for each chapter.
- **Dynamic Study Modules** help students study chapter topics and the language of MIS on their own by continuously assessing their knowledge application and performance in real time. These are available as graded assignments prior to class, and are accessible on smartphones, tablets, and computers.
- **Learning Catalytics** is a student response tool that helps you generate class discussion, customize your lecture, and promote peer-to-peer learning based on real-time analytics. Learning Catalytics uses students' devices to engage them in more interactive tasks.

## ENHANCED STAND-ALONE PEARSON eTEXTBOOK

*Essentials of Management Information Systems* is also available as a stand-alone eTextbook which extends the learning experience, anytime and anywhere: The mobile app lets students use their eTextbook whenever they have a moment in their day, on Android and iPhone mobile phones and tablets. Offline access ensures students never miss a chance to learn. The eTextbook engages students with compelling media: Videos and animations written and produced by the authors bring key concepts to life, helping students place what they are reading into context. Other features include highlights that allow educators to share information directly with students within their eTextbook, and analytics that let educators gain insight into how students use their eTextbook, allowing them to plan more effective instruction.

Both the MyLab MIS and eTextbook platforms provide an affordable, simple-to-use mobile reading experience that lets instructors and students extend learning beyond class time.

## NEW AND UPDATED TOPICS

The 15th edition features all new opening, closing, and "Spotlight on" cases as well as new in-text examples of organizations using IT applications. There are 4 cases per chapter. The text, figures, tables, and cases have been updated through July 2022 with the latest sources from industry and MIS research. New topics and coverage include:

- **Cloud computing, Big Data, and the Internet of Things (IoT):** We have added more coverage of these topics throughout the text and learning package because of their importance in the MIS world. Chapter 1 now contains an introduction to these topics, followed by in-depth coverage in Chapters 2, 5, 6, 7, and 11. Case studies on Big Data can be found in Chapters 1, 4, 6, 7, and 11. There are case studies on IoT in Chapters 1, 2, 5, 6, and 7.

  We have updated and expanded coverage of cloud computing in Chapter 5 (IT Infrastructure) with more detail on types of cloud services, private and public clouds, hybrid clouds, and managing cloud services. Cloud computing is also covered in Chapter 6 (databases in the cloud), Chapter 8 (cloud security), Chapter 9 (cloud-based CRM and ERP), Chapter 10 (e-commerce), and Chapter 12 (cloud software services). There are case studies on cloud computing in Chapters 2, 5, 6, and 12.
- **Sustainability and ESG:** We have added new coverage of how information systems promote sustainability and Environmental, Social, and Governance (ESG) goals. Chapter 1 now includes ESG leadership as a major objective of information systems. Case studies on sustainability can be found in Chapters 2, 5, 6, 9, and 11.
- **Updated and expanded coverage of artificial intelligence (AI):** Chapter 11 has been rewritten to include new expanded coverage of machine learning, "deep learning," natural language systems, computer vision systems, and robotics, reflecting the surging interest in business uses of AI and "intelligent" techniques.
- **System impacts of the coronavirus pandemic:** Up-to-date coverage of the impact of the coronavirus pandemic on business uses of information systems. Two "Spotlight on" cases (Chapters 2 and 6) and Chapters 1 and 9 ending cases cover topics such as working remotely, supply chain disruptions, and rethinking global supply chains.
- **Expanded coverage of blockchain,** including a new Chapter 6 case study on blockchain and sustainability. New and/or expanded coverage of the following topics:
  - Digital resiliency
  - Cryptocurrencies
  - Metaverse
  - Customer experience management
  - Low-code and no-code development
  - Automated testing
  - Windows 11
  - JavaScript
  - Zero trust

The Laudon text, MyLab MIS, and eTextbook provide the most up-to-date and comprehensive overview of information systems used by business firms today. After using this learning package, we expect students will be able to participate in, and even lead, management discussions of information systems for their firms and understand how to use information technology in their jobs to achieve bottom-line business results. Regardless of whether students are accounting, finance, management, operations management, marketing, or information systems majors, the knowledge and information in this book will be valuable throughout their business careers.

## NEW VIDEO CASES

The Video Cases are all new to this edition and based on Pearson's collection of business and technology video clips. They are available in the eTextbook and MyLab MIS.

| Chapter | Video |
|---|---|
| 1. Business Information Systems in Your Career | The New HQ Is in the Cloud: Salesforce President |
| 2. Global E-business and Collaboration | How Slack Is Preparing for the Future of Work |
| 3. Achieving Competitive Advantage with Information Systems | Celonis Tops $11 Billion Valuation with New Round of Funding |
| 4. Ethical and Social Issues in Information Systems | Australia Passes Law Forcing Tech Giants to Pay for News |
| 5. IT Infrastructure: Hardware and Software | IBM Expands Cloud to Daimler |
| 6. Foundations of Business Intelligence: Databases and Information Management | Stitch Fix CEO Sees Business of Personalization as Key to Success; Stitch Fix President on Booming Growth Amid Pandemic |
| 7. Telecommunications, the Internet, and Wireless Technology | Nokia CEO Suri Sees 5G Market Maturity in 2021 |
| 8. Securing Information Systems | Fastly Internet Outage Exposes Vulnerability of Major Websites; Ransomware Is a Worldwide Problem: Palo Alto Networks |
| 9. Achieving Operational Excellence and Customer Intimacy: Enterprise Applications | Software Startup Freshworks Not in a Rush to Raise Capital |
| 10. E-commerce: Digital Markets, Digital Goods | Shopify Earnings Soar Even as Economies Reopen; Shopify is Writing the Future of Commerce, Says President |
| 11. Improving Decision Making and Managing Artificial Intelligence | Predictive Tech Can Save $3B-$4B A Year: Tom Siebel |
| 12. Making the Business Case for Information Systems and Managing Projects | Software Design Is "Really on a Tear," Figma CEO Says |

# Solving Teaching and Learning Challenges

MyLab MIS is the teaching and learning platform that empowers you to reach every student. By combining trusted authors' content with digital tools and a flexible platform, MyLab MIS personalizes the learning experience and improves results for each student. And with MIS Sims and Excel and Access Activities, students understand how MIS concepts will help them succeed in their future careers.

MyLab MIS and the Pearson eTextbook offer unique digital interactive features that hold student attention spans longer and make learning more effective, including 23 Figure Videos that walk students through key concepts in each chapter, a collection of online video cases, and interactive quizzes. All of this is available anytime, anywhere, on any digital device. The result is a comprehensive learning environment that will heighten student engagement and learning in the MIS course.

The Laudon learning package is more current, real-world, and authoritative than competitors. Laudon *Essentials of MIS* 15e, MyLab MIS, and Pearson eTextbook help students understand MIS concepts and issues through extensive use of real-world company examples, a wide variety of text and video cases based on real-world organizations, and numerous line art illustrations, interactive animations, and hands-on software projects.

The Laudons are known for their outstanding real-world case studies, which describe how well-known business firms are using IT to solve problems and achieve objectives. Students are often asked to analyze the business problem and propose alternative solutions. The Laudons also provide hands-on MIS software and management decision-making problems in each chapter that are based on real-world companies and business scenarios.

The Laudon text and learning package now has a very strong career focus, which incentivizes students to learn by showing exactly how each chapter will help them prepare for future jobs. In addition to Career Opportunities, MyLab MIS features Career Resources, including how to incorporate MIS knowledge into resumes, cover letters, and job interviews.

## THE CORE TEXT

The Core text provides an overview of fundamental MIS concepts using an integrated framework for describing and analyzing information systems. This framework shows information systems composed of people, organization, and technology elements and is reinforced in student projects and case studies. The Core text consists of 12 chapters with hands-on projects covering the most essential topics in MIS. An important part of the Core text are the Video Cases, a case study for each chapter built around one or two videos available on the Pearson Clips platform. Videos are keyed to the topics of each chapter.

### Chapter Organization

Each chapter contains the following elements:

- A list of Learning Objectives
- Lists of the Case Studies and Video Cases for each chapter
- A chapter opening case describing a real-world organization to establish the theme and importance of the chapter
- A diagram analyzing the opening case in terms of the people, organization, and technology model used throughout the text
- Two "Spotlight on" case studies with Case Study Questions

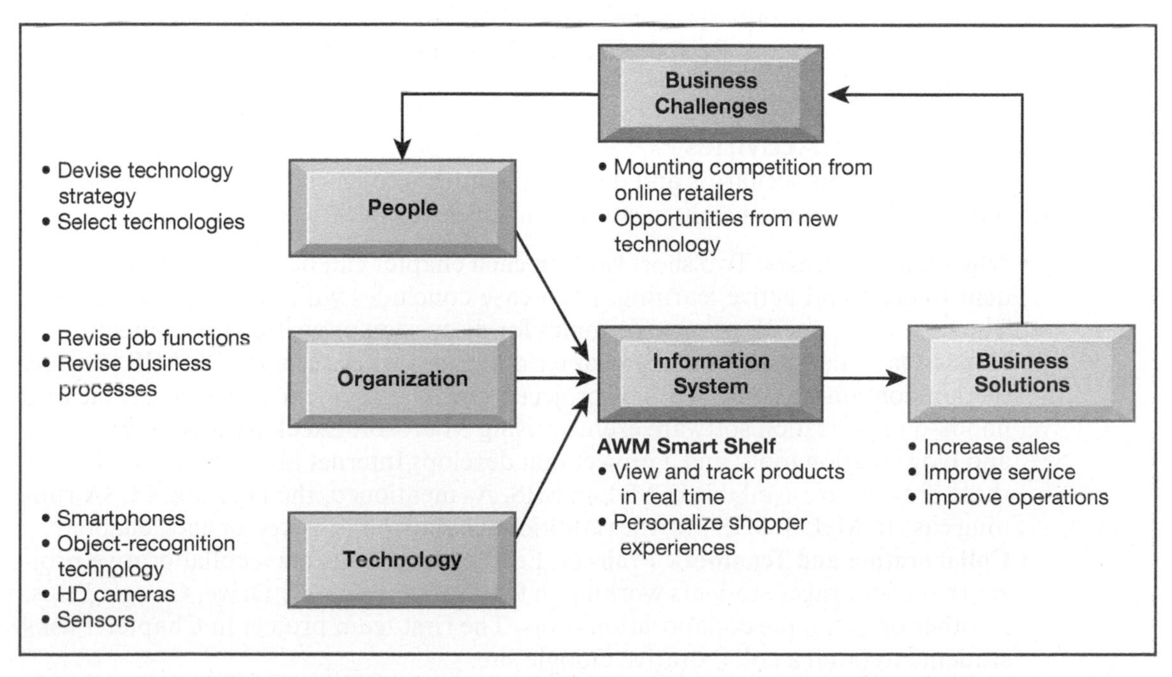


A diagram accompanying each chapter-opening case graphically illustrates how people, organization, and technology elements work together to create an information system solution to the business challenges discussed in the case.

- A Career Opportunities section showing students how to use the text for job hunting and career preparation
- A Review Summary keyed to the Student Learning Objectives
- A list of Key Terms that students can use to review concepts
- Review questions for students to test their comprehension of chapter material
- Discussion questions raised by the broader themes of the chapter
- A series of Hands-on MIS Projects consisting of one Management Decision Problem, a hands-on application software project, and a project to develop Internet skills
- A Collaboration and Teamwork Project to develop teamwork and presentation skills with options for using open source collaboration tools
- A chapter-ending case study for students to apply chapter concepts
- Chapter references

### Student Learning-Focused

Student Learning Objectives are organized to focus student attention. Each major chapter section is based on a Learning Objective and each chapter concludes with a Review Summary and Review Questions organized around these Learning Objectives.

## KEY FEATURES

We have enhanced the text to make it more interactive, leading edge, and appealing to both students and instructors. The features and learning tools are described in the following sections.

### Business-Driven with Real-World Business Cases and Examples

The text helps students see the direct connection between information systems and business performance. It describes the main business objectives driving the use of information systems and technologies in corporations all over the world: operational excellence, new products and services, customer and supplier intimacy, improved decision making, competitive advantage, survival and promoting ESG leadership. In-text examples and case studies show students how specific companies use information systems to achieve these objectives.

We use current (2022) examples from business and public organizations throughout the text to illustrate the important concepts in each chapter. Most of the case studies describe companies or organizations that are familiar to students, such as Uber, Facebook, Walmart, Amazon, PayPal, Chewy, Coca Cola, the National Hockey League, and Zoom.

### Hands-On Text Activities

Real-world business scenarios and data help students to learn firsthand what MIS is all about. These projects heighten student involvement in this exciting subject.

- **"Spotlight on" Cases.** Two short cases in each chapter can be used to stimulate student interest and active learning. Each case concludes with case study questions. The case study questions provide topics for discussion or written assignments.
- **Hands-On MIS Projects.** Every chapter concludes with a Hands-On MIS Projects section containing three types of projects: one Management Decision Problem; a hands-on application software exercise using Microsoft Excel, Access, or web page and blog creation tools; and a project that develops Internet business skills. Files for these projects are available in MyLab MIS. As mentioned, the Dirt Bikes USA running case in MyLab MIS provides additional hands-on projects for each chapter.
- **Collaboration and Teamwork Projects.** Each chapter features a collaborative project that encourages students working in teams to use Google Drive, Google Docs, or other open source collaboration tools. The first team project in Chapter 1 asks students to build a collaborative Google site.

## SPOTLIGHT ON: ORGANIZATIONS

## Carbon Lighthouse Lights Up with the Internet of Things (IoT), Big Data, and Cloud Computing

According to the US Environmental Protection Agency (EPA), more carbon emissions come from operating commercial, residential, and commercial properties than from transportation. Emissions from cars, trucks, and aircraft account for only **29 percent of carbon emissions, while 68 percent** comes from generation of electric power, (manufacturing) industry, commercial and residential property, and agriculture.

San Francisco–based Carbon Lighthouse is tackling this problem head-on by focusing on reducing carbon emissions produced by commercial and industrial buildings. The company offers energy savings services that make it profitable for building owners to reduce their energy consumption using their existing equipment. Carbon Lighthouse has reduced more than 260,000 metric tons of emissions, equivalent to the energy produced by 18 power plants, while providing **$250 million in savings to its clients.**

How is this possible? Many businesses complain that it costs more money to be environmentally responsible. This does not have to be so. Acting on findings from advanced data analytics, commercial and building owners can reduce carbon emissions while producing cost savings.

Carbon Lighthouse uses a software platform called Carbon Lighthouse Unified Engineering System (CLUES) to analyze more than 100 million square feet of clients' commercial and industrial real estate and 5 billion data points. CLUES collects the building data in real time as it is generated. It then develops models that lead to insights about how to reduce the client's carbon emissions and how much money will be saved. Carbon Lighthouse uses Amazon's cloud services to run its computers and store data such as building addresses and square footage, along with time-series data such as the data collected on-site at buildings.

When a new client contracts for Carbon Lighthouse services, the company sends a sensor kit to start new data streaming from the client better. Internet of Things (IoT) sensors, Big Data, cloud computing, and powerful data analysis software rather than expensive capital upgrades allow Carbon Lighthouse to unlock hidden returns in existing building mechanical systems and translate energy efficiency into tangible long-term savings.

One of Carbon Lighthouse's clients is L&B Realty Advisors, a Dallas-based real estate investment advisor with $9 billion under management. L&B has over 50 years' experience acquiring, managing, and selling real estate for its clients. For L&B, sustainability is a key consideration when evaluating and managing investments. L&B worked with Carbon Lighthouse on its One Biscayne Tower property. The building was already rated as highly energy-efficient, with engineering staff using a dashboard to track overall utility consumption.

Carbon Lighthouse first established a data stream that fed detailed heating, ventilation, and air conditioning (HVAC); lighting; and occupancy data into the CLUES platform. The company then worked with the L&B property and engineering teams to assess One Biscayne Tower's existing building systems and look for energy savings and operational improvements. CLUES analysis of the data stream identified and quantified new HVAC control optimizations as well as lighting retrofits that property and building engineering teams had long sought to implement. (A lighting retrofit is an upgrade to light fixtures or lamps, increasing energy efficiency.)

The annual expense savings from these energy conservation measures for One Biscayne Tower are financially guaranteed by Carbon Lighthouse. And as CLUES collects more data, its algorithms become more accurate and provide additional zone isolation and central plant measures for One Biscayne Tower, generating more energy efficiency and cost savings than projected.

The building engineering and property manager knew what lighting fixtures and aesthetic output they wanted, but they did not have the

Each chapter contains two Spotlight on cases on People, Organizations, or Technology, using real-world companies to illustrate chapter concepts and issues.

## CASE STUDY QUESTIONS

1. Identify the problem described in this case study. Is it a people problem, an organizational problem, or a technology problem? Explain your answer.
2. What role have the IoT, Big Data analytics, and cloud computing played in developing a solution for this problem?
3. Describe Carbon Lighthouse's problem-solving methodology for reducing both carbon emissions and costs.

Case Study Questions encourage students to apply chapter concepts to real-world companies in class discussions, student presentations, or writing assignments.

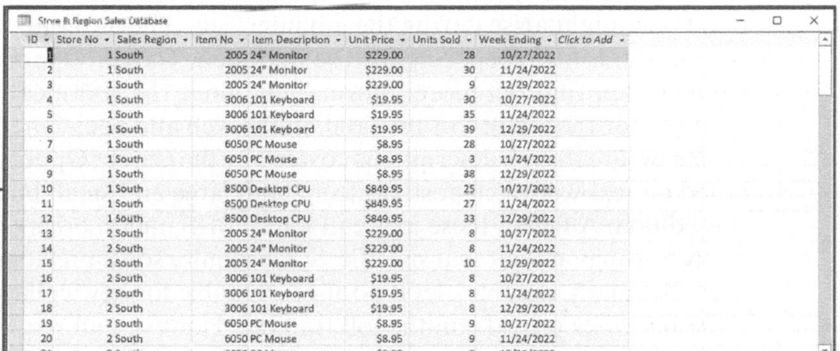

Store & Region Sales Database

| ID | Store No | Sales Region | Item No | Item Description | Unit Price | Units Sold | Week Ending | Click to Add |
|---|---|---|---|---|---|---|---|---|
| 1 | 1 | South | 2005 | 24" Monitor | $229.00 | 28 | 10/27/2022 | |
| 2 | 1 | South | 2005 | 24" Monitor | $229.00 | 30 | 11/24/2022 | |
| 3 | 1 | South | 2005 | 24" Monitor | $229.00 | 9 | 12/29/2022 | |
| 4 | 1 | South | 3006 | 101 Keyboard | $19.95 | 30 | 10/27/2022 | |
| 5 | 1 | South | 3006 | 101 Keyboard | $19.95 | 35 | 11/24/2022 | |
| 6 | 1 | South | 3006 | 101 Keyboard | $19.95 | 39 | 12/29/2022 | |
| 7 | 1 | South | 6050 | PC Mouse | $8.95 | 28 | 10/27/2022 | |
| 8 | 1 | South | 6050 | PC Mouse | $8.95 | 3 | 11/24/2022 | |
| 9 | 1 | South | 6050 | PC Mouse | $8.95 | 38 | 12/29/2022 | |
| 10 | 1 | South | 8500 | Desktop CPU | $849.95 | 25 | 10/27/2022 | |
| 11 | 1 | South | 8500 | Desktop CPU | $849.95 | 27 | 11/24/2022 | |
| 12 | 1 | South | 8500 | Desktop CPU | $849.95 | 33 | 12/29/2022 | |
| 13 | 2 | South | 2005 | 24" Monitor | $229.00 | 8 | 10/27/2022 | |
| 14 | 2 | South | 2005 | 24" Monitor | $229.00 | 8 | 11/24/2022 | |
| 15 | 2 | South | 2005 | 24" Monitor | $229.00 | 10 | 12/29/2022 | |
| 16 | 2 | South | 3006 | 101 Keyboard | $19.95 | 8 | 10/27/2022 | |
| 17 | 2 | South | 3006 | 101 Keyboard | $19.95 | 8 | 11/24/2022 | |
| 18 | 2 | South | 3006 | 101 Keyboard | $19.95 | 8 | 12/29/2022 | |
| 19 | 2 | South | 6050 | PC Mouse | $8.95 | 9 | 10/27/2022 | |
| 20 | 2 | South | 6050 | PC Mouse | $8.95 | 9 | 11/24/2022 | |
| 21 | 2 | South | 6050 | PC Mouse | $8.95 | 8 | 12/29/2022 | |

Record: 1 of 96 Search

Students can practice using software in real-world settings for achieving operational excellence and enhancing decision making.

Each chapter features a project to develop Internet skills for accessing information, conducting research, and performing online calculations and analysis.

Chapter 3: Achieving Competitive Advantage with Information Systems 111

**IMPROVING DECISION MAKING: USING WEB TOOLS TO CONFIGURE AND PRICE AN AUTOMOBILE**

Software skills: Internet-based software
Business skills: Researching product information and pricing

**3-10** In this exercise, you will use software at car-selling websites to find product information about a car of your choice and then use that information to make an important purchase decision. You will also evaluate two of these sites as selling tools.

You are interested in purchasing a lightly used, low-mileage Honda CR-V (or some other car of your choice). Go to the Carsdirect.com website, and begin your investigation. Research the various used Honda CR-V models for sale; choose one you prefer in terms of price, features, and safety ratings. Then visit the Carvana.com website. Compare the information on pricing and available inventory at Carvana's website with that of Carsdirect for the Honda CR-V. Try to locate the lowest price for the car you want in inventory along with financing and vehicle pickup or delivery. Compare the buying experience at both websites. Which website do you prefer for this type of purchase? Explain your answer.

## Developing Career Skills

For students to succeed in a rapidly changing job market, they should be aware of their career options and how to go about developing a variety of skills. With MyLab MIS and *Essentials of Management Information Systems 15e*, we focus on these skills in the following ways.

### CAREER OPPORTUNITIES AND RESOURCES

Every student who reads this text wants to know: How will this book help my career? The Career Opportunities feature shows how to use this book and MyLab MIS as tools for job-hunting and career-building. Job interviewers will typically ask about why you want the job, along with your ability to communicate, multitask, work in a team, show leadership, solve problems, and meet goals. These are general skills and behaviors you'll need to succeed in any job, and you should be prepared to provide examples from your course work and job experiences that demonstrate these skills. But there are also business knowledge and professional skills that employers will ask you about. Career Opportunities will show you how to use what you have learned in this text to demonstrate these skills.

The Career Opportunities section, identified by this icon is the last major section of each chapter under the heading "Understand how MIS can help your career". There you will find a description of an entry-level job for a recent college graduate based on a real-world job description from major online job sites related to the topics covered in that chapter. The name of the company offering the job and its location have been changed. Each chapter's job posting describes the required educational background and specific job skills, and suggests some of the business-related questions that might arise during the job interview. The authors provide tips for answering the questions and preparing for the interview. Career Opportunities also show where students can find out more information about the technical and business knowledge required for the job in this text and on the web and social media.

Below are the job descriptions covered by the Career Opportunities sections. They are based on real-world job postings from both large and small businesses. Many are new to this edition. A few of these jobs call for an MIS major, others for MIS course work, but many postings are not that specific. Some require some previous internship or job experience, but many are entry-level positions suitable for new college graduates, and some of these positions provide on-the-job training. However, all require knowledge of business information systems and applications and the ability to work in a digital environment.

| Chapter | Career Opportunity Job Description |
| --- | --- |
| 1. Business Information Systems in Your Career | Client Support Assistant |
| 2. Global E-business and Collaboration | Customer Success Analyst |
| 3. Achieving Competitive Advantage with Information Systems | Entry Level Business Development Representative |
| 4. Ethical and Social Issues in Information Systems | Junior Privacy Analyst |
| 5. IT Infrastructure: Hardware and Software | Coordinating Product Manager |
| 6. Foundations of Business Intelligence: Databases and Information Management | Entry Level Data Analyst |
| 7. Telecommunications, the Internet, and Wireless Technology | Web Developer |
| 8. Securing Information Systems | Identity and Access Management Support Specialist |
| 9. Achieving Operational Excellence and Customer Intimacy: Enterprise Applications | Supply Chain Analyst |
| 10. E-commerce: Digital Markets, Digital Goods | Junior E-Commerce Associate |
| 11. Improving Decision Making and Managing Artificial Intelligence | Sales Coordinator |
| 12. Making the Business Case for Information Systems and Managing Projects | Junior Business Systems Analyst |

Students can use Career Opportunities to shape their resumes and career plans as well as to prepare for interviews. For instructors, Career Opportunities are potential projects for student research and in-class discussion.

In MyLab MIS we have provided additional Career Resources, including job-hunting guides and instructions on how to build a Digital Portfolio demonstrating the business knowledge, application software proficiency, and Internet skills acquired from using the text. The portfolio can be included in a resume or job application or used as a learning assessment tool for instructors.

## Instructor Teaching Resources

| Supplements available to instructors at www.pearson.com/laudon | Features of the Supplement |
| --- | --- |
| Instructor's Manual | • Chapter-by-chapter summaries<br>• Examples and activities not in the main book<br>• Teaching outlines<br>• Teaching tips<br>• Solutions to all questions and problems in the book |
| Test Bank<br>authored by Professor Kenneth Laudon, New York University | The authors have worked closely with skilled test item writers to ensure that higher-level cognitive skills are tested. Test bank multiple-choice questions include questions on content but also include many questions that require analysis, synthesis, and evaluation skills.<br>**AACSB Assessment Guidelines**<br>As a part of its accreditation activities, the AACSB has developed an Assurance of Learning Program designed to ensure that schools do in fact teach students what they promise. Schools are required to state a clear mission, develop a coherent business program, identify student learning objectives, and then prove that students do in fact achieve the objectives.<br>We have attempted in this book to support AACSB efforts to encourage assessment-based education. The end papers of this edition identify student learning objectives and anticipated outcomes for our Hands-On MIS projects. The authors will provide custom advice on how to use this text in colleges with different missions and assessment needs. Please email the authors or contact your local Pearson representative for contact information. |

| Supplements available to instructors at www.pearson.com/laudon | Features of the Supplement |
|---|---|
| Computerized TestGen | TestGen allows instructors to:<br>• Customize, save, and generate classroom tests<br>• Edit, add, or delete questions from the Test Item Files<br>• Analyze test results<br>• Organize a database of tests and student results |
| PowerPoints<br>authored by Professor Kenneth Laudon, New York University | The authors have prepared a comprehensive collection of PowerPoint slides for each chapter to be used in your lectures. Many of these slides are the same as used by Ken Laudon in his MIS classes and executive education presentations. Each of the slides is annotated with teaching suggestions for asking students questions, developing in-class lists that illustrate key concepts, and recommending other firms as examples in addition to those provided in the text. The annotations are like an Instructor's Manual built into the slides and make it easier to teach the course effectively.<br>PowerPoints meet accessibility standards for students with disabilities. Features include but are not limited to:<br>• Keyboard and Screen Reader access<br>• Alternative text for images<br>• High color contrast between background and foreground colors |

# About the Authors

**Kenneth C. Laudon** has been a Professor of Information Systems at New York University's Stern School of Business. He holds a B.A. in Economics from Stanford and a Ph.D. from Columbia University. He has authored twelve books dealing with electronic commerce, information systems, organizations, and society. Professor Laudon has also written more than forty articles concerned with the social, organizational, and management impacts of information systems, privacy, ethics, and multimedia technology.

At NYU's Stern School of Business, Ken Laudon has taught courses on Managing the Digital Firm, Information Technology and Corporate Strategy, Professional Responsibility (Ethics), and Electronic Commerce and Digital Markets.

**Jane Price Laudon** is a management consultant in the information systems area and the author of seven books. Her special interests include systems analysis, data management, MIS auditing, software evaluation, and teaching business professionals how to design and use information systems.

Jane received her Ph.D. from Columbia University, her M.A. from Harvard University, and her B.A. from Barnard College. She has taught at Columbia University and the New York University Stern School of Business. She maintains a lifelong interest in languages and civilizations of Asia.

**Carol Guercio Traver** is a graduate of Yale Law School and Vassar College. She has had many years of experience representing major corporations, as well as small and medium sized businesses, as an attorney with a leading international law firm, with specific expertise in technology law, Internet law, privacy law, intellectual property law, and general corporate law. Carol is the co-author of *E-commerce: business.technology.society* (Pearson), as well as several other texts on information technology, and has been the lead project manager/editor on a number of technology-related projects. Carol is the co-founder and president of Azimuth Interactive, one of the first edtech firms and a provider of digital media and publisher services for the education industry.

## Acknowledgments

The production of any book involves valued contributions from a number of persons. We would like to thank all of our editors for encouragement, insight, and strong support for many years. We thank our editors, Ellen Thibault, Jenifer Niles, and Content Producer, Purnima Narayanan, for their roles in managing the project. Thanks also to Product Manager Marcus Scherer for his contributions and to Gowthaman Sadhanandham and his Integra team for their production work.

We want to thank our supplement authors for their work, including the following MyLab MIS content contributors: Robert J. Mills, Utah State University; Chris Parent, Rivier University; Maureen Steddin; Roberta Roth, University of Northern Iowa; Gipsi Sera, Indiana University; and John Hupp, Columbus State University. We are indebted to Todd Traver for his help with security topics, to Erica Laudon for her contributions to Career Opportunities, and to Megan Miller for her help during production. We thank Christopher Traver for his help with database topics and software projects.

Special thanks to Professor Mark Gillenson, Fogelman College of Business and Economics, University of Memphis for his contributions to the discussion of agile development and testing.

We also want to especially thank all our reviewers whose suggestions helped improve our texts. Reviewers for recent editions include:

Abdullah Albizri, *Montclair State University*
Robert M. Benavides, *Collin College*
Gordon Bloom, *Virginia Commonwealth University*
Brett Cabradillia, *Coastal Carolina Community College*
Qiyang Chen, *Montclair State University*
Amita Chin, *Virginia Commonwealth University*
Lynn Collen, *St. Cloud State University*
Reet Cronk, *Harding University*
Uldarico Rex Dumdum, *Marywood University*
Mahmoud Elhussini, *Montclair State University*
Anne Formalarie, *Plymouth State University*
Sue Furnas, *Collin College*
Scott Hamerink, *Oakland University*
Terry Howard, *University of Detroit Mercy*
Dae Youp Kang, *University of Memphis*
Rajkumar Kempaiah, *College of Mount Saint Vincent*
Channa J. Kumarage, *Saint Cloud State University*
Weiqi Li, *University of Michigan-Flint*
Liu Liu, *Old Dominion University*
Susan Mahon, *Collin College*
Robert Morphew, *Texas Woman's University*
John Newman, *Coppin State University*
Jose Ng, *Montclair State University*
Richard Peterson, *Montclair State University*
Robin Poston, *University of Memphis*
Dr. Michael Raisinghani, *Texas Woman's University*
Patricia Ryan, *Southeast Missouri State University*
Ethné Swartz, *Montclair State University*
Amir Talaei-Khoei, *University of Nevada Reno*
Paulus Van Vliet, *University of Nebraska at Omaha*

# Essentials of Management Information Systems

PART I



# Information Systems in the Digital Age

Part I introduces the major themes and the problem-solving approaches that are used throughout this book. While surveying the role of information systems in today's businesses, this part examines a series of major questions: What is an information system? Why are information systems so essential in businesses today? What are the different types of information systems that can be found in organizations? How can information systems help businesses become more competitive? What do I need to know about information systems to succeed in my business career? What ethical and social issues do widespread use of information systems raise?

CHAPTER

# 1 Business Information Systems in Your Career

## LEARNING OBJECTIVES

After completing this chapter, you will be able to:

1-1 Understand why information systems are essential for running and managing a business.

1-2 Define an information system, explain how it works, and identify its people, organizational, and technology components.

1-3 Apply a four-step method for business problem solving to solve information system-related problems.

1-4 Describe the information systems skills and knowledge that are essential for business careers.

1-5 Understand how MIS can help your career.

## CHAPTER CASES

- Smart Shelves Reinvent the Retail Space
- UPS Competes Globally with Information Technology
- Will a Robot Steal Your Job?
- Will the Covid-19 Pandemic Make Working from Home the New Normal?

**MyLab MIS**

- **Video Case:** The New HQ Is in the Cloud: Salesforce President
- Discussion Questions: 1-5, 1-6, 1-7
- Hands-on MIS Projects: 1-8, 1-9, 1-10, 1-11
- eText with Figure Videos

# SMART SHELVES REINVENT THE RETAIL SPACE

Although there has been an upsurge in online shopping, physical retail stores are not going away. Some traditional retailers are fighting back by using innovative information technologies such as computer vision, facial recognition, artificial intelligence, Big Data analysis, and the Internet of Things (IoT) to provide new ways to entice people into physical stores as well as enhance their in-store experiences. Retailers are also employing many of the same technologies to make the process used to fulfill online orders for curbside pickup or home delivery more efficient.

Shelves have become more than just a surface for storing and displaying objects. New systems for "smart" shelves use proximity sensors, 3D cameras, microphones, RFID tags and readers, and weight sensors to enable interactions between shoppers in physical stores and the shelves they're standing in front of. These systems can create a highly personalized shopping experience that fundamentally improves the way shoppers move inside physical stores.

For instance, retailers such as Walmart, ShopRite, and Stop & Shop and brands such as Pepsi, Hasbro, and Hershey are using Smart Shelf by AWM to replicate the benefits of the online experience in physical retail environments. Smart Shelf is powered by an integrated set of physical products, including super-wide-angle low-light HD cameras and LED displays, coupled with AWM's Automated Inventory Intelligence, Product Mapper, Content Management, and Retail Data Engine (RDE) software solutions. Smart Shelf can be implemented in a wide range of store sizes and formats, from micromarkets to convenience stores to larger-format retailers.

Smart Shelf enables retailers to provide real-time on-shelf marketing to consumers, with up-to-the-minute advertising and pricing. The shelves' LED digital displays use dynamic video and animation to create an immersive environment, drawing shoppers toward them. As the shopper moves nearer, proximity sensors shift the display to content-specific promotions and then, as the shopper gets even closer, change the display again to show price tags and product information directly beneath the product. Retailers can choose the specific distances that trigger the shifts in display.

© Moncpoly1919/Shutterstock

Smart Shelf is also able to personalize shoppers' experiences when they are in stores based on the items they pick up, even if they don't purchase them. For example, if a customer picks up a box of cookies and then puts it back, the retailer can use the system to offer a discount on the shelf beneath that item the next time the shopper encounters it in the store. When retailers connect Smart Shelf to their mobile apps, they can help shoppers locate products themselves through their mobile devices. According to AWM, Smart Shelf has helped some retailers increase sales by more than 30 percent.

AWM also offers a consumer behavior tracking application it calls a Demographic Engine, which includes facial recognition technology, that triggers delivery of content based on demographic criteria like age, gender, and ethnicity. To protect consumer privacy, the system does not collect images or tie data to personally identifiable information. Nonetheless, some remain alarmed about the privacy implications of such software.

Smart Shelf also helps retailers improve operational efficiencies. Retailers deploying Smart Shelf can view and track their products in real-time, highlighting specific shelves that need restocking. It also can be employed by stores to make the process of fulfilling online orders more efficient, directing pickers to items even if the products have been moved from their proper location. The system also helps retailers track how long pickers take to retrieve items. According to AWM Chief Executive Kevin Howard, it can cut stores' fulfillment costs by 60 percent. However, labor experts note that the system raises concerns about employee monitoring.

The global market for smart shelves such as those offered by AWM is expected to grow from $1.8 billion in 2020 to more than $7 billion by 2026, as physical retailers strive to offer a better shopping experience and compete more effectively with online retailers. In the process, technology is redefining the role of the shelf in retail marketing.

Sources: Smartshelf.com, accessed June 7, 2022; Caroline Hayes, "The Doors Never Close on Digital Shopping," Eandt.theiet.org, August 16, 2021; Sapna Maheshwari, "It's Hard Work to Make Ordering Groceries Online So Easy," *New York Times*, June 4, 2021; Marketsandmarkets.com, "Smart Shelves Market by Component, Application, and Region—Global Forecast to 2026," Marketsandmarkets.com, March 2021; Brian Wreckler, "The Definitive Guide to Interactive Retail Smart Shelves," Perchinteractive.con, January 11, 2021; Jared Council, "Retailers Hope In-Store Tech Will Keep Shoppers in Stores," *Wall Street Journal*, January 15, 2020.

The companies and technologies described here show how essential information systems are today. Today, retail stores are struggling to stay alive and relevant as more shoppers gravitate to online shopping and the Internet. One solution is to use leading-edge innovative information technology to provide new ways of drawing buyers into physical stores and making the in-store buying experience more efficient, safe, and pleasant. The information flows that drive these reimagined retail businesses have become much more digital, making use of mobile tools and object-recognition technology.

The chapter-opening diagram calls attention to important points raised by this case and this chapter. To compete more effectively against online retailers and take advantage of new technology solutions, brick-and-mortar retail stores are using innovative systems based on artificial intelligence, Big Data analysis, computer vision and facial recognition technology, IoT sensors, and smartphones. The use of leading-edge digital technologies to drive business operations and management decisions is a key topic today in the management information systems (MIS) world and will be discussed throughout this text. At the same time, the use of some of these technologies raises ethical concerns, in this case, about consumer privacy and employee monitoring, topics that also will be further discussed throughout the text.

It is also important to note that deploying information technology has changed the way retailers using AWM Smart Shelf run their businesses. To effectively use new digital tools, these companies had to redesign jobs and procedures for gathering, inputting, and accessing information. These changes had to be carefully planned to make sure they enhanced efficiency, service, and profitability.

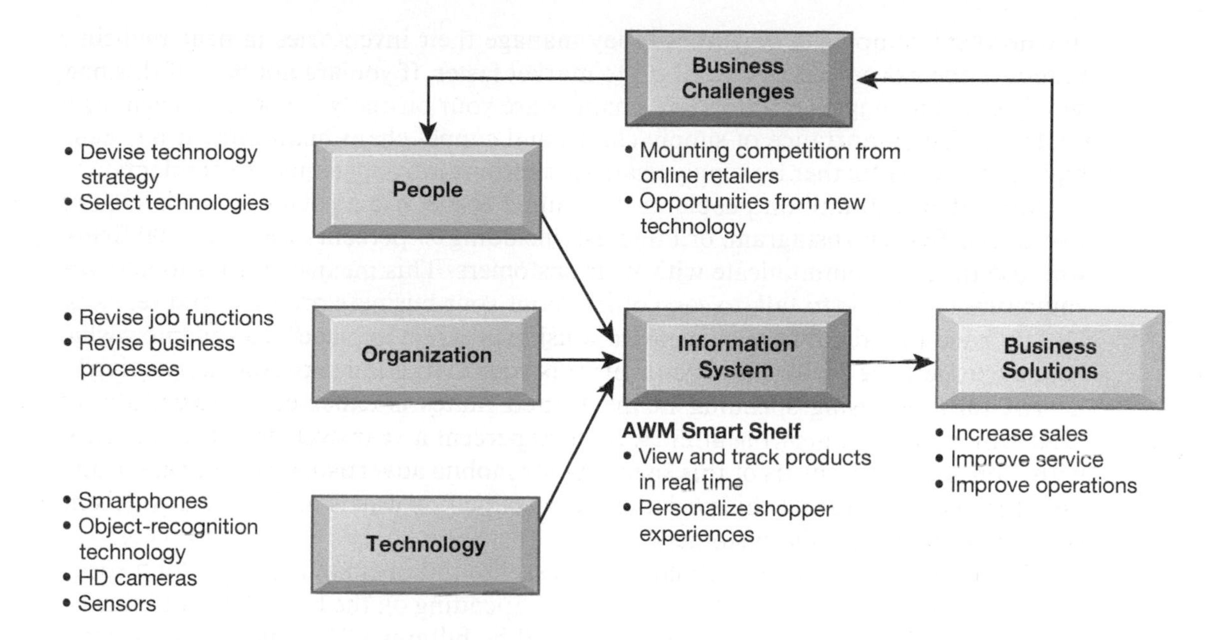


Here are some questions to think about: How do AWM's systems change retail operations? How do they improve the customer experience? What ethical concerns do they raise?

## 1-1 Understand why information systems are essential for running and managing a business.

Over the past two decades, tremendous changes have taken place in the way business is conducted, driven in large part by investments in information technology and information systems. In 2021, global spending on information technology (IT) and IT services topped $4 trillion. In addition, businesses worldwide were expected to spend an additional $900 billion on business and management consulting and information technology services, much of which involves redesigning firms' business operations to take advantage of these new technologies (Gartner, 2021; Statista, 2021).

### HOW INFORMATION SYSTEMS ARE TRANSFORMING BUSINESS

You can see the results of this spending around you every day. Smartphones, tablet computers, email, and online videoconferencing have all become essential tools of business. In 2021, more than 170 million businesses worldwide had registered .com or .net Internet sites. More than 210 million people in the United States bought something online, with almost 180 million using a mobile device to do so, and 255 million used a search engine, with 230 million of these searchers using their mobile devices. What this means is that if you and your business aren't connected to the Internet and mobile apps, chances are you are not being as effective as you could be (VeriSign, 2022; Insider Intelligence/eMarketer, 2022a).

United Parcel Service (UPS) moves more than 25 million packages a day worldwide; FedEx moves more than 18.5 million, with 6.5 million of those overnight. The growth of e-commerce has had a significant impact on shipping volume, particularly in the wake of the pandemic. For instance, Amazon accounted for about 12 percent of UPS's revenue in 2021. Businesses are attempting to sense and respond to rapidly changing customer demand, reduce inventories to the lowest possible levels, and achieve higher levels of operational efficiency. Supply chains have become faster paced, with companies of all sizes depending on the delivery of just-in-time inventory to help them compete. Companies today manage their inventories in near real time to reduce their overhead costs and get to market faster. If you are not part of this new supply chain management economy, chances are your business is not as efficient as it could be. The importance of supply chains and supply chain management has been highlighted even further by supply chain disruptions introduced by the pandemic.

More than 220 million people in the United States use a social network such as Facebook, Twitter, Instagram, or Pinterest, including 97 percent of *Fortune* 500 firms, who use them to communicate with their customers. This means your customers are empowered and able to talk to each other about your business products and services. Do you have a solid online customer relationship program in place? Do you know what your customers are saying about your firm? Is your marketing department listening?

Digital advertising spending in the United States is expected to reach almost $250 billion in 2022, growing at more than 20 percent a year over the past five years, with more than two-thirds of this spending for mobile advertising. In contrast, traditional advertising during this period either declined or was flat. Is your advertising reaching web and mobile customers?

Various federal laws require many businesses to retain email messages for a specified period, ranging from two to seven years, depending on the law. These and similar laws, as well as all of the data being generated by billions of Internet-linked sensors comprising the Internet of Things (IoT), online consumer data, and social media data, are spurring the explosive growth of digital information known as Big Data. At the same time, the retention of all this data has created privacy concerns and has led to the passage of laws about how such data must be safeguarded. Does your compliance department meet the minimal requirements for storing financial, health, and occupational information? If it doesn't, your entire business may be at risk (Insider Intelligence/eMarketer, 2022b, 2022c; Porteous, 2021; TitanHQ, 2021).

Briefly, it's a constantly evolving world of doing business, one that will greatly affect your future career. Along with the changes in business come changes in jobs and careers. No matter whether you are a finance, accounting, management, marketing, operations management, or information systems major, how you work, where you work, and how well you are compensated will all be affected by business information systems. The purpose of this book is to help you understand and benefit from today's business realities and opportunities.

## KEY CHALLENGES IN MANAGEMENT INFORMATION SYSTEMS

What makes management information systems the most exciting topic in business today is the continual change in technology, management use of the technology, and the impact on business success. New start-up firms arrive in traditional industries using the latest technologies and business models. These changes present challenges to all business managers who need to decide how to adapt their firm to new developments. What are the benefits and costs of these new developments in hardware, software, and business practice?

Table 1.1 summarizes the major challenges in business uses of information systems. These challenges confront all managers, not just information systems professionals. These challenges will appear throughout the book in many chapters, so it might be a good idea to take some time now to discuss them with your professor and classmates.

## GLOBALIZATION CHALLENGES AND OPPORTUNITIES: A FLATTENED WORLD

Prior to 1500 CE, there was no truly global economic system of trade that connected all the continents, although there were active regional trade markets. After the sixteenth century, a global trading system began to emerge based on advances in navigation and ship technology. The world trade that ensued after these developments has brought the peoples and cultures of the world much closer together. The Industrial Revolution was a worldwide phenomenon energized by expansion of trade among

**TABLE 1.1**

Key Challenges in MIS

| Change | Management Challenge |
| --- | --- |
| ***Technology*** | |
| Cloud computing platform emerges as a major business area of innovation. | Companies are increasingly turning to a flexible collection of computers on the Internet to perform tasks traditionally performed at corporate data centers. Major business applications can be delivered online as an Internet service (software as a service [SaaS]). What are the costs and benefits of cloud computing and how much of the firm's IT infrastructure should be moved to cloud providers? |
| Big Data and the Internet of Things (IoT) | Businesses look for insights in huge volumes of data from web traffic, email messages, social media content, and Internet-connected devices (sensors). More powerful data analytics and interactive dashboards can provide real-time performance information to managers to enhance decision making. Does your firm have the ability to analyze and use Big Data and analytics? How can you use IoT to provide better products and services? |
| Artificial intelligence (AI) | Computer programs can find patterns in large databases that can help managers understand their business and provide better products. Where could your company use AI, and where can you find the expertise? What benefits can you expect? How much will it cost? |
| The mobile platform | Business and personal computing is increasingly moving to smartphones, tablet computers, car infotainment systems, and wearable devices. Thousands of applications are now available to support collaboration, coordination of work, communication with colleagues and customers, and online purchases on mobile devices. More than 90 percent of Internet users access the web with mobile devices. Is your firm making the best use of mobile capabilities for its employees and customers? How could your firm improve? What are the costs and benefits? |
| ***Management and People*** | |
| Return on investment (ROI) | Although firms spend millions on information systems and services, they typically have little understanding of how much benefit they receive. How can your firm measure and understand the benefit it is receiving from IS/IT expenditures? Are there alternative sources of these services that would cost less? |
| Online collaboration and social networking | Millions of business professionals use Google Apps, Google Drive, Microsoft 365, Yammer, Zoom, and IBM Connections to support blogs, project management, online meetings, personal profiles, and online communities. Is your firm making a coordinated effort to use new technologies to improve coordination, collaboration, and knowledge sharing? Which of the many alternatives should it be using? |
| ***Organizations*** | |
| Security and privacy | Security lapses and protecting customer privacy are major public issues that affect all businesses. How does your company know its data are secure? How much does it spend on security now? What privacy policies does your firm have in place? How should the firm expand its privacy protections as new laws emerge? |
| Social business | Businesses use social network platforms, including Facebook, Twitter, Instagram, and internal corporate social tools, to deepen interactions with employees, customers, and suppliers. What use is your company making of social business tools? Where should it go from there? Is your company getting real value from these platforms? |
| Remote work (telework) surges | The Internet, cloud computing, smartphones, and tablet computers make it possible for growing numbers of people to work away from the traditional office. This ability proved critical during the height of the Covid-19 pandemic, when more than 70 percent of workers worked from home. Going forward, what will your firm's policies and technologies be for remote work, and what are the risks for productivity? |

nations, making nations both competitors and collaborators in business. The Internet has greatly heightened the competitive tensions among nations as global trade expands and strengthened the benefits that flow from trade and also created significant dislocations in labor markets.

In 2005, journalist Thomas Friedman wrote an influential book declaring that the world was now flat, by which he meant that the Internet and global communications had greatly expanded the opportunities for people to communicate with one another and reduced the economic and cultural advantages of developed countries. The United States and European countries were in a fight for their economic lives, according to Friedman, competing for jobs, markets, resources, and even ideas with highly educated, motivated populations in developing countries (Friedman, 2007). This globalization presents you and your business with both challenges and opportunities.

A growing percentage of the economy of the United States and other developed industrial countries in Europe and Asia depends on imports and exports. In 2021, 23 percent of the US economy resulted from foreign trade of goods and services, both imports and exports. In certain countries in Europe and Asia, the number exceeds 50 percent (Organisation for Economic Co-operation and Development [OECD], 2022).

It's not just goods that move across borders. So too do jobs, some of them high-level jobs that pay well and require a college degree. Between 1991 and 2019, the United States lost a net 3.5 million manufacturing jobs to offshore, low-wage producers, so manufacturing is now a much smaller part of US employment than it once was. US multinational companies employ more than 14 million people outside the United States, many of them in service jobs in information technology, customer call centers, human resources, financial services, consulting, engineering, architecture, and even medical services, such as radiology (Amadeo, 2021; Bureau of Economic Analysis, 2021; Rose, 2021).

On the plus side, the US economy continues to create millions of new jobs, especially in 2021 as the country began to emerge from the pandemic. From 2020 to 2030, employment is projected to grow by almost 12 million jobs. Employment in information systems and the other service occupations listed previously has rapidly expanded in sheer numbers, wages, productivity, and quality of work. Outsourcing has actually accelerated the development of new information systems in the United States and worldwide by reducing the cost of building and maintaining them. In 2021, job openings in information systems and technologies far exceeded the supply of applicants (Bureau of Labor Statistics, 2021a).

The challenge for you as a business student is to develop high-level skills through education and on-the-job experience that cannot be outsourced. The challenge for your business is to avoid markets for goods and services that can be produced offshore much less expensively. The opportunities are equally immense. You can learn how to profit from the lower costs available in world markets and the chance to serve a marketplace with billions of customers. You have the opportunity to develop higher-level and more profitable products and services. Throughout this book, you will find examples of companies and individuals who either failed or succeeded in using information systems to adapt to this new global environment.

What does globalization have to do with management information systems? The answer is simple: everything. The emergence of the Internet into a full-blown international communications system has drastically reduced the costs of operating and transacting on a global scale. Communication between a factory floor in Shanghai and a distribution center in Sioux Falls, South Dakota, is now instant and virtually free. Customers now can shop in a worldwide marketplace, obtaining price and quality information reliably 24 hours a day. Firms producing goods and services on a global scale achieve extraordinary cost reductions by finding low-cost suppliers and managing production facilities in other countries. Internet service firms such as Google and eBay can replicate their business models and services in multiple countries without having to redesign their expensive, fixed-cost information systems infrastructure.

## BUSINESS DRIVERS OF INFORMATION SYSTEMS

What makes information systems so essential today? Why are businesses investing so much in information systems and technologies? They do so to achieve seven important business objectives: operational excellence; new products, services, and business models; customer and supplier intimacy; improved decision making; competitive advantage; and survival. In addition, businesses today are increasingly being pressured by various stakeholders, such as employees, customers, governmental authorities, and investors, to show leadership in pursuing broader environmental, social, and governance (ESG) goals. ESG is becoming a strategic business imperative, and information systems are central in enabling companies to meet their ESG goals.

### Operational Excellence

Businesses continuously seek to improve the efficiency of their operations to achieve higher profitability. Information systems and technologies are some of the most important tools available to managers for achieving higher levels of efficiency and productivity in business operations, especially when coupled with changes in business practices and management behavior.

Walmart, the largest retailer on earth, exemplifies the power of information systems coupled with sophisticated business practices and supportive management to achieve world-class operational efficiency. In fiscal 2021, Walmart generated $559 billion in sales—in large part because of its Retail Link and Global Replenishment System, which digitally links its suppliers to every one of Walmart's stores worldwide. As soon as a customer purchases an item, the supplier monitoring the item knows to ship a replacement to the shelf. Walmart is the most efficient retail store in its industry.

Amazon, the largest online retailer on earth, which is expected to generate an estimated $650 billion in retail e-commerce sales worldwide in 2022, invested an astounding $56 billion in technology and content in 2021 so that it can continue to enhance the customer experience and improve process efficiency while operating at an ever-increasing scale (Amazon.com Inc., 2022; Insider Intelligence/eMarketer, 2022d).

### New Products, Services, and Business Models

Information systems and technologies are a major enabling tool for firms to create new products and services as well as entirely new business models. A **business model** describes how a company produces, delivers, and sells a product or service to create wealth. Today's music industry is vastly different from the industry a decade ago. Apple transformed an old business model of music distribution based on vinyl records, tapes, and CDs into an online, legal download distribution model based on its own operating system and iTunes store. Apple has prospered from a continuing stream of innovations, including the iPod, iTunes and Apple Music music services, iPhone, and iPad.

### Customer and Supplier Intimacy

When a business really knows its customers and serves them well, the way they want to be served, its customers generally respond by returning and purchasing more. This raises revenues and profits. Likewise with suppliers: the more a business engages its suppliers, the better the suppliers can provide vital inputs. This lowers the business's costs. How really to know your customers, or suppliers, is a central problem for businesses with millions of offline and online customers.

High-end hotel chains such as Ritz Carlton exemplify the use of information systems and technologies to achieve customer intimacy. Every time a new prospect shares their information or a guest checks into a Ritz hotel, their information is stored in a cloud-based relationship management system. When that guest returns to the hotel or logs into their online account, the system cross-references their data profile and tailors that customer's guest experience to their exact preferences (such as room temperature, check-in and check out times, and room location) and history. This highly personalized experience makes the customer feel valued, included, and part of the Ritz brand. The hotels also analyze their customer data to identify their best customers and develop individualized marketing campaigns based on customers' preferences.

Hong Kong-headquartered TAL Apparel, one of the world's largest contract apparel manufacturers, exemplifies the use of information systems to enable supplier intimacy. TAL Apparel produces shirts, blouses, knits, pants, outerwear, and suits for many of the most famous garment brands, including Brooks Brothers, Bonobos, and L.L.Bean. TAL manufactures one out of every six dress shirts sold in the United States. The company also helps its clients manage their supply chains. Every time its clients sell a dress shirt, for example, the record of the sale appears immediately on TAL's computers in Hong Kong. TAL runs the numbers through a computer model it developed and decides how many replacement shirts to make and in what styles, colors, and sizes. TAL then sends the shirts to the retailer. TAL's systems reduce inventory costs and ensure that what customers want is actually on retailers' shelves.

### Improved Decision Making

Many business managers operate in an information fog bank, never really having the right information at the right time to make an informed decision. Instead, managers rely on forecasts, best guesses, and luck. The result is over- or underproduction of goods and services, misallocation of resources, and poor response times. These poor outcomes raise costs and lose customers. Information systems and technologies have now made it possible for managers to use real-time data from the marketplace when making decisions.

For instance, Coca-Cola Bottling Company Consolidated (CCBCC) uses a series of digital dashboards that display metrics like delivery operations, budget, and profitability, consolidating data from hundreds of disparate sources. Leadership dashboards focus on strategy, business growth, and profitability. Sales teams can now access mobile dashboards on iPads in the field, increasing timeliness to sell more product. Field sales dashboards allow sales teams to see a full overview of customer portfolios or track sales quotas.

### Competitive Advantage

When firms achieve one or more of these business objectives—operational excellence; new products, services, and business models; customer/supplier intimacy; and improved decision making—chances are they have already achieved a competitive advantage. Doing things better than your competitors, charging less for superior products, and responding to customers and suppliers in real time all add up to higher sales and higher profits that your competitors cannot match. Apple, Walmart, and UPS are industry leaders because they know how to use information systems for this purpose.

### Survival

Business firms also invest in information systems and technologies because they are necessities of doing business. Sometimes these necessities are driven by industry-level changes. For instance, after Citibank introduced the first automated teller machines (ATMs) in the New York region to attract customers through higher service levels, its competitors rushed to provide ATMs to their customers to keep up with Citibank. Today, virtually all banks in the United States have regional ATMs and link to national and international ATM networks, such as CIRRUS. Providing ATM services to retail banking customers is simply a requirement of being in and surviving in the retail banking business.

Many federal and state statutes and regulations create a legal duty for companies and their employees to retain records, including digital records. For instance, the Toxic Substances Control Act, which regulates the exposure of US workers to more than 75,000 toxic chemicals, requires firms to retain records on employee exposure for 30 years. The Sarbanes–Oxley Act, which was intended to improve the accountability of public firms and their auditors, requires public companies to retain audit working papers and records, including all email messages, for five years. The Dodd–Frank Act requires financial service firms to expand their public reporting greatly on derivatives and other financial instruments. Firms turn to information systems and technologies to provide the capability to respond to information retention and reporting requirements.

### Environmental, Social, and Governance (ESG) Leadership

Over the past decade or so, there has been a growing awareness of the important role that businesses play in, and the impact that they have, on society. With that growing awareness has come pressure from employees, customers, investors, and governmental authorities with respect to business conduct in three primary areas: the environment, social responsibility, and corporate governance, often referred to as ESG.

Environmental criteria encompass such aspects as energy use, carbon footprint, sustainability, recycling practices, pollution, and natural resource conservation. Social criteria focus on whether the company is a good corporate citizen and social actor in terms of its employees, suppliers, customers, and the community. Governance criteria relate primarily to the actions of the company's leadership, executive pay, internal controls, and shareholder rights. ESG also includes a company's actions with respect to diversity, equity, and inclusion (DEI). Many aspects of ESG are intertwined with one another.

Showing leadership in these areas can add business value by facilitating growth, reducing costs, minimizing regulatory and legal issues, increasing employee productivity, and optimizing assets and investments. For instance, many consumers are no longer willing to support companies that don't prioritize ESG. Many investors are now basing investment decision on an assessment of a company's ESG performance.

Information systems are central in enabling companies to achieve ESG goals. For instance, a major industrial company advised by consulting firm PWC has elevated sustainability to being a strategic priority. One step it recently took to achieve its goal of zero net emissions by 2050 was to implement a cloud-based ERP system for its entire supply chain, which is helping its suppliers to track, report, and reduce their carbon impact. This was an important step because the bulk of the company's carbon footprint is in its supply chain, not within the company itself. We discuss other instances of information systems being used for ESG purposes throughout the text.

## 1-2 Define an information system, explain how it works, and identify its people, organizational, and technology components.

So far we've used *information systems and technologies* informally without defining the terms. **Information technology (IT)** consists of all the hardware and software that a firm needs to use to achieve its business objectives. This includes not only computers, storage technology, and mobile devices but also software, such as the Windows or Linux operating systems, the Microsoft Office desktop productivity suite, and the many thousands of computer programs that can be found in a typical large firm. Information systems are more complex and can be understood best by looking at them from both a technology and a business perspective.

### WHAT IS AN INFORMATION SYSTEM?

An **information system (IS)** can be defined technically as a set of interrelated components that collect (or retrieve), process, store, and distribute information to support decision making, coordinating, and control in an organization. In addition, information systems may also help managers and workers analyze problems, visualize complex subjects, and create new products.

Information systems contain information about significant people, places, and things within the organization or in the environment surrounding it. By **information**,

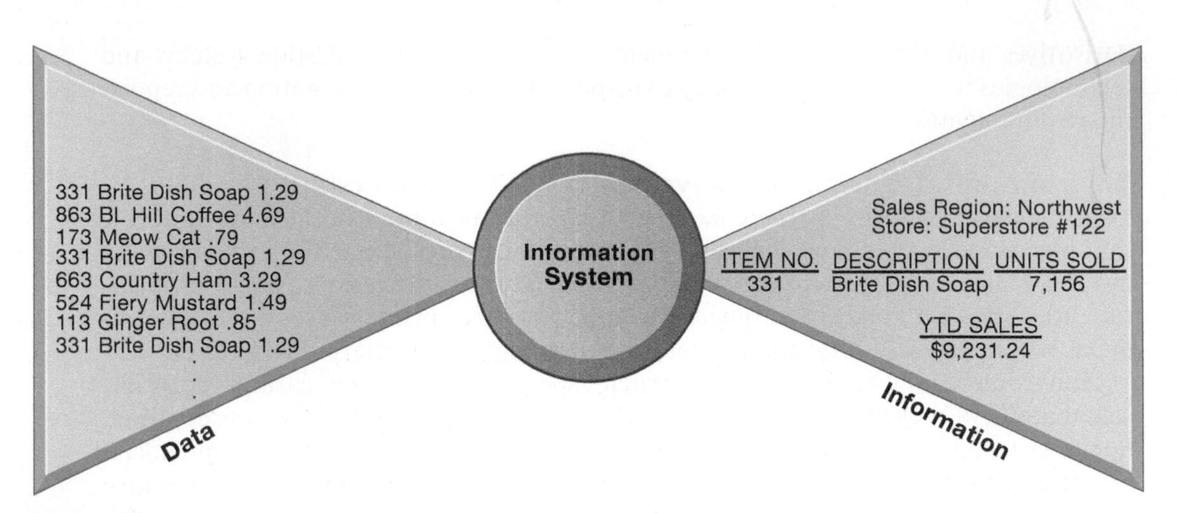


**Figure 1.1**
Data and Information
*Raw data from a supermarket checkout counter can be processed and organized to produce meaningful information, such as the total unit sales of dish detergent or the total sales revenue from dish detergent for a specific store or sales territory.*

we mean data that have been shaped into a form that is meaningful and useful to human beings. **Data**, in contrast, are streams of raw facts representing events occurring in organizations or the physical environment before they have been organized and arranged into a form that people can understand and use.

A brief example contrasting information and data may prove useful. Examine Figure 1.1, which provides an example of an information system used by a supermarket. Supermarket checkout counters scan bar codes on products. The bar codes contain data about the product, such as a numerical identifier, name, and price ("331 Brite Dish Soap, 1.29), shown in the left portion of the diagram. An information system enables these pieces of data to be totaled and analyzed to provide meaningful information, displayed in the right portion of the diagram, such as the total number of bottles of dish detergent sold at a particular store, which brands of dish detergent were selling the most rapidly at that store or sales territory, or the total amount spent on that brand of dish detergent at that store or sales region. View the Figure 1.1 video in the eText for an animated and more detailed discussion of this figure.

Three activities in an information system produce the information that organizations need to make decisions, control operations, analyze problems, and create new products or services. These activities are input, processing, and output. Examine Figure 1.2, which illustrates how these activities work together within an information system. **Input** captures or collects raw data from within the organization or from its external environment. **Processing** converts this raw input into a meaningful form by classifying, arranging, and performing calculations on it. **Output** transfers the processed information to the people who will use it or to the activities for which it will be used. Information systems also provide **feedback**, which is output that is returned to appropriate members of the organization to help them evaluate or correct the input stage. Figure 1.2 also illustrates that an organization's information systems do not operate within a vacuum: they are impacted by various elements in the environment that surround them, such as the organization's suppliers, customers, competitors, and stockholders as well as governmental agencies that regulate the organization. View the Figure 1.2 video in the eText for an animated and more detailed discussion of this figure.

In the AWM Smart Shelf system, input includes the digital results from scanning on-shelf products and store shelf identification codes, along with scanned images of items customers have selected for purchase. A computer stores and processes these data to keep track of the items on each shelf, the items each customer purchases or examines, and a history of the customer's purchases and items of interest. The system then determines what items on what shelves need restocking and what items should be

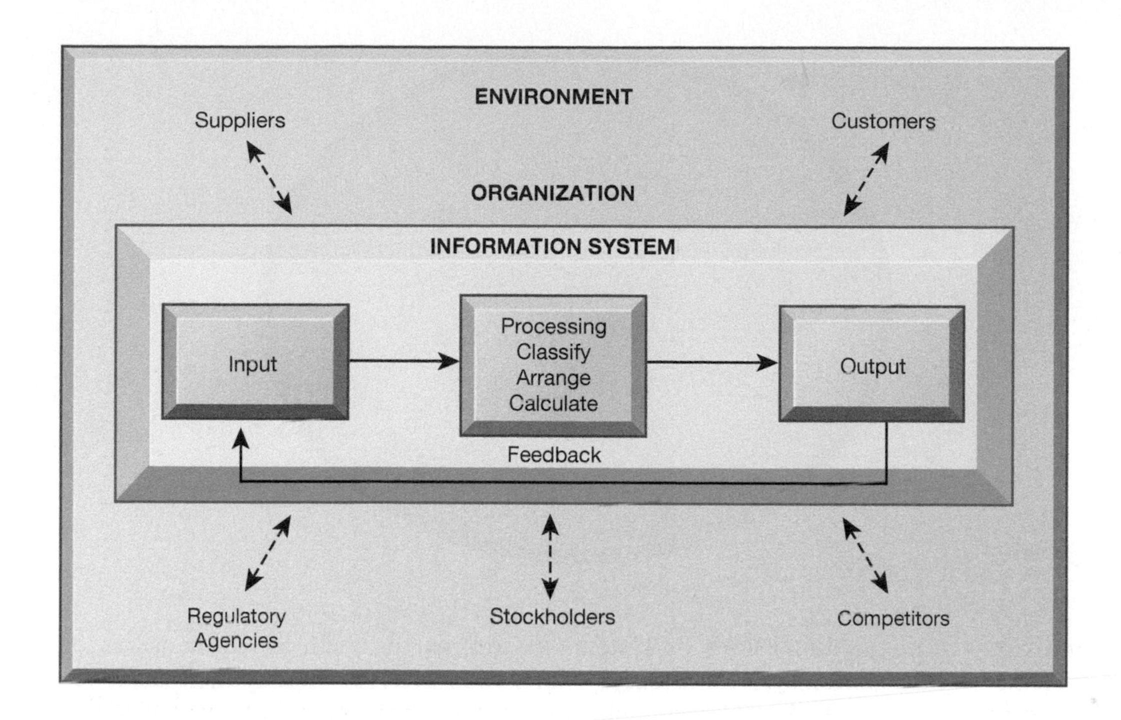


**Figure 1.2**
Functions of an Information System
*An information system contains information about an organization and its surrounding environment. Three basic activities—input, processing, and output—produce the information organizations need. Feedback is output returned to appropriate people or activities in the organization to evaluate and refine the input. Environmental actors, such as customers, suppliers, competitors, stockholders, and regulatory agencies, interact with the organization and its information systems.*

recommended to each customer. The system provides meaningful information such as all the items sold in a particular store or on a particular store shelf on a specific day, what items have been purchased by a specific customer, and which items need restocking.

Although computer-based information systems use computer technology to process raw data into meaningful information, there is a sharp distinction between a computer and a computer program and an information system. Computers and related software programs are the technical foundation, the tools and materials, of modern information systems. Computers provide the equipment for storing and processing information. Computer programs, or software, are sets of operating instructions that direct and control computer processing. Knowing how computers and computer programs work is important in designing solutions to organizational problems, but computers are only part of an information system.

A house is an appropriate analogy. Houses are built with hammers, nails, and wood, but these alone do not make a house. The architecture, design, setting, landscaping, and all of the decisions that lead to the creation of these features are part of the house and are crucial for solving the problem of putting a roof over one's head. Computers and programs are the hammer, nails, and lumber of computer-based information systems, but alone they cannot produce the information a particular organization needs. To understand information systems, you must understand the problems they are designed to solve, their architectural and design elements, and the organizational processes that lead to these solutions.

## IT ISN'T SIMPLY TECHNOLOGY: THE ROLE OF PEOPLE AND ORGANIZATIONS

To understand information systems fully, you will need to be aware of the broader organizational, people, and information technology dimensions of systems and their power to provide solutions to challenges and problems in the business environment. Examine Figure 1.3, which highlights the interconnected nature of these dimensions, showing that organizations, technology, and people are all interrelated components underlying the effective use of information systems. We refer to this broader understanding of information systems, which encompasses an understanding of the people

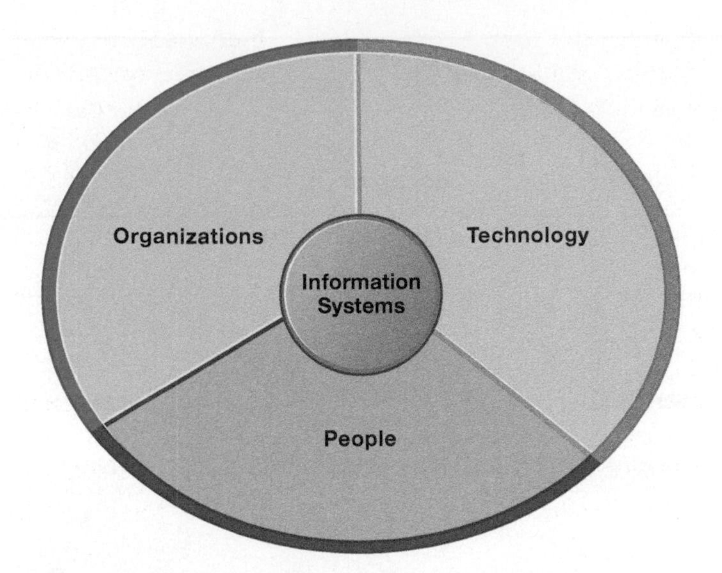


**Figure 1.3**
Information Systems Are More Than Computers
*Using information systems effectively requires an understanding of the organization, people, and information technology shaping the systems. An information system provides a solution to important business problems or challenges facing the firm.*

and organizational dimensions of systems as well as the technical dimensions of systems, as **information systems literacy**. Information systems literacy includes a behavioral as well as a technical approach to studying information systems. **Computer literacy**, in contrast, focuses primarily on knowledge of information technology.

The field of **management information systems (MIS)** tries to achieve this broader information systems literacy. MIS deals with behavioral issues as well as technical issues surrounding the development, use, and impact of information systems that managers and employees in the firm use.

## DIMENSIONS OF INFORMATION SYSTEMS

Let's examine each of the dimensions of information systems—organizations, people, and information technology.

### Organizations

Information systems are an integral part of organizations, and although we tend to think about information technology changing organizations and business firms, it is, in fact, a two-way street. The history and culture of business firms also affect how the technology is used and how it should be used. To understand how a specific business firm uses information systems, you need to know something about the structure, history, and culture of the company.

Organizations have a structure that is composed of different levels and specialties. Their structures reveal a clear-cut division of labor. A business firm is organized as a hierarchy, or a pyramid structure, of rising authority and responsibility. The upper levels of the hierarchy consist of managerial, professional, and technical employees, whereas the lower levels consist of operational personnel. Experts are employed and trained for different business functions, such as sales and marketing, manufacturing and production, finance and accounting, and human resources. The firm builds information systems to serve these different specialties and levels of the firm. Chapter 2 provides more detail on these business functions and organizational levels and the ways in which information systems support them.

An organization accomplishes and coordinates work through this structured hierarchy and through its **business processes**, which are logically related tasks and behaviors for accomplishing work. Developing a new product, fulfilling an order, and hiring a new employee are examples of business processes.

Most organizations' business processes include formal rules that have been developed over a long time for accomplishing tasks. These rules guide employees in a variety of procedures, from writing an invoice to responding to customer complaints. Some of these business processes have been written down, but others are informal work practices, such as a requirement to return telephone calls from co-workers or customers that are not formally documented. Information systems automate many business processes. For instance, how a customer receives credit or how a customer is billed is often determined by an information system that incorporates a set of formal business processes.

Each organization has a unique **culture**, or fundamental set of assumptions, values, and ways of doing things, that has been accepted by most of its members. Parts of an organization's culture can always be found embedded in its information systems. For instance, the United Parcel Service's concern with placing service to the customer first is an aspect of its organizational culture that can be found in the company's package tracking systems.

Different levels and specialties in an organization create different interests and points of view. These views often conflict. Conflict is the basis for organizational politics. Information systems come out of this cauldron of differing perspectives, conflicts, compromises, and agreements that are a natural part of all organizations.

## People

A business is only as good as the people who work there and run it. Likewise with information systems, they are useless without skilled people to build and maintain them or people who can understand how to use the information in a system to achieve business objectives.

For instance, a call center that provides help to customers by using an advanced customer relationship management system (described in later chapters) is useless if employees are not adequately trained to deal with customers, find solutions to their problems, and leave the customer feeling that the company cares for them. Similarly, employee attitudes about their jobs, employers, or technology can have a powerful effect on their abilities to use information systems productively.

Business firms require many kinds of skills and people, including managers as well as rank-and-file employees. The job of managers is to make sense out of the many situations organizations face, make decisions, and formulate action plans to solve organizational problems. Managers perceive business challenges in the environment, they set the organizational strategy for responding to those challenges, and they allocate the human and financial resources to coordinate the work and achieve success. Throughout, they must exercise responsible leadership.

However, managers must do more than manage what already exists. They must also create new products and services and even re-create the organization from time to time. A substantial part of management responsibility is creative work driven by new knowledge and information. Information technology can play a powerful role in helping managers develop novel solutions to a broad range of problems.

As you will learn throughout this text, technology is relatively inexpensive today, but people are very expensive. Because people are the only ones capable of business problem solving and converting information technology into useful business solutions, we spend considerable effort in this text looking at the people dimension of information systems.

## Technology

Information technology is one of many tools managers use to cope with change and complexity. **Computer hardware** is the physical equipment used for input, processing, and output activities in an information system. It consists of the following: computers of various sizes and shapes; various input, output, and storage devices; and networking devices that link computers.

**Computer software** consists of the detailed, preprogrammed instructions that control and coordinate the computer hardware components in an information system. Chapter 5 describes the contemporary software and hardware platforms firms use today in greater detail.

**Data management technology** consists of the software governing the organization of data on physical storage media. More detail on data organization and access methods can be found in Chapter 6.

**Networking and telecommunications technology**, consisting of both physical devices and software, links the various pieces of hardware and transfers data from one physical location to another. Computers and communications equipment can be connected in networks for sharing voice, data, images, sound, and video. A **network** links two or more computers to share data or resources such as a printer.

The world's largest and most widely used network is the **Internet**, a global network of networks that uses universal standards (described in Chapter 7) to connect millions of networks in more than 230 countries around the world.

The Internet provides a universal technology platform on which to build new products, services, strategies, and business models. This same technology platform has internal uses, providing the connectivity to link different systems and networks within the firm. Internal corporate networks based on Internet technology are called **intranets**. Private intranets extended to authorized users outside the organization are called **extranets**, and firms use such networks to coordinate their activities with other firms for making purchases, collaborating on design, and performing other interorganizational work. For most business firms today, using Internet technology is a business necessity and a competitive advantage.

The **World Wide Web** (today, normally referred to as just the "web") is a service the Internet provides that uses universally accepted standards for storing, retrieving, formatting, and displaying information in a page format on the Internet. Web pages contain text, graphics, animations, sound, and video and are linked to other web pages. By clicking highlighted words or buttons on a web page, you can link to related pages to find additional information and links to other locations on the web. The web can serve as the foundation for new kinds of information systems such as UPS's web-based package tracking system.

All these technologies, along with the people required to run and manage them, represent resources that can be shared throughout the organization and constitute the firm's **information technology (IT) infrastructure**. The IT infrastructure provides the foundation, or *platform*, on which the firm can build its specific information systems. Each organization must carefully design and manage its IT infrastructure so that it has the set of technology services it needs for the work it wants to accomplish with information systems. Part II of this text examines each major technology component of information technology infrastructure and shows how they all work together to create the technology platform for the organization.

Among all these technologies, three developments are especially important today. They are the Internet of Things (IoT), Big Data, and cloud computing.

The **Internet of Things (IoT)** describes a network of physical objects—"things"—that are embedded with sensors, software, and other technologies for the purpose of connecting and exchanging data with other devices and systems over the Internet. These devices range from ordinary household objects to sophisticated industrial tools. The number of connected IoT devices is projected to reach 30.9 billion by 2025 (Vailshery, 2021). Digitally linking these devices gives them enough intelligence to make simple decisions and remember particular patterns and routines to be carried out without any human involvement.

These connected IoT devices as well as data from websites and other sources generate many millions of data points that can be analyzed to create predictions or unearth patterns of behavior. Such large quantities of structured and unstructured data provide useful information but are often too vast or complex to process using traditional methods and are called **Big Data**.

**Cloud computing** is a centralized system for storing, managing, and processing data across the Internet using a network of remote computing centers. Organizations and individuals can store, analyze, and access their data from these cloud computing

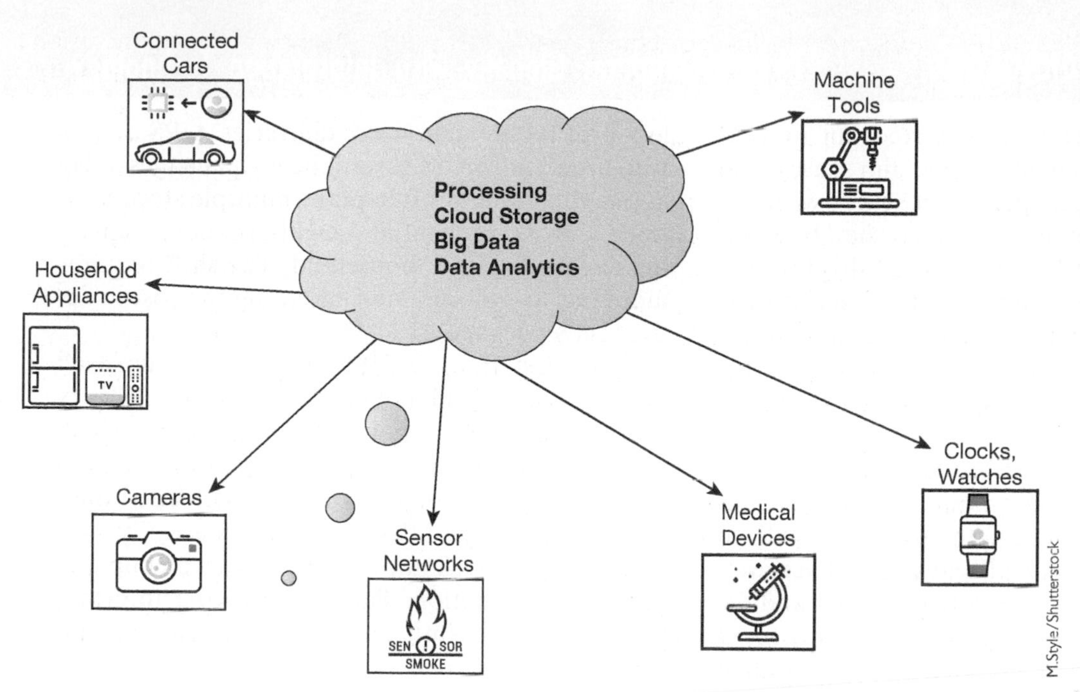


**Figure 1.4**
Cloud Computing, Big Data, and the Internet of Things (IoT)
*Using remote cloud computing centers has made it possible to link large numbers of many different kinds of devices via the Internet so that they can exchange data. Cloud computing has also enabled organizations to create more robust IT infrastructures for managing and analyzing Big Data.*

centers (rather than on-site) using an Internet-connected device such as a desktop, laptop, or mobile device. We provide many examples of how organizations are using these technologies throughout this text.

Examine Figure 1.4, which illustrates the relationship between the IoT, Big Data, and cloud computing. It shows different kinds of IoT devices, including consumer appliances, sensors, medical devices, and connected cars, which generate millions and trillions of data points: Big Data. The devices transmit these data via the Internet to a remote cloud computing center for storage, processing, and analysis. The results might include instructions to shut down a computer-controlled machine tool, to replace a defective part, to monitor a person's heartbeat, or to raise the heating temperature in a house when the weather turns cold. Cloud computing centers have the scalability and capacity to process Big Data efficiently.

The Spotlight on Technology case describes some of the typical technologies used in information systems today. UPS invests heavily in information systems technology to make its business more efficient and customer oriented. It uses an array of information technologies, including bar code scanning systems, wireless networks, large mainframe computers, handheld computers, the Internet, and many pieces of software for tracking packages, calculating fees, maintaining customer accounts, and managing logistics. As you read this case, try to identify the problem this company was facing, what alternative solutions were available to management, and how well the chosen solution worked.

Let's identify the organization, people, and technology elements in the UPS package tracking system we have just described. The organization element anchors the package tracking system in UPS's sales and production functions (the main product of UPS is a service—package delivery). It specifies the required procedures for identifying packages with both sender and recipient information, taking inventory, tracking the packages en route, and providing package status reports for UPS customers and customer service representatives.

The system must also provide information to satisfy the needs of managers and workers. UPS drivers need to be trained in both package pickup and delivery procedures and in how to use the package tracking system so that they can work efficiently and effectively. UPS customers may need some training to use UPS in-house package tracking software or the UPS website or mobile apps.

## UPS Competes Globally with Information Technology

United Parcel Service (UPS) started out in 1907 in a closet-sized basement office. Jim Casey and Claude Ryan—two teenagers from Seattle with two bicycles and one phone—promised the "best service and lowest rates." UPS has used this formula successfully for more than a century to become the world's largest ground and air package-delivery company. It's a global enterprise with almost 535,000 employees, more than 125,000 delivery vehicles, and 595 aircraft.

Today UPS delivers more than 6 billion packages and documents a year in more than 220 countries and territories. The firm has been able to maintain leadership in small-package delivery services despite stiff competition from FedEx and the US Postal Service by investing heavily in advanced information technology. UPS has spent more than $3 billion over the past five years to maintain a high level of customer service while keeping costs low and streamlining its overall operations. Its Global Advanced Technology Group operates a Robotics AI Lab (RAIL) and a rapid prototyping lab to help UPS quickly identify, evaluate, and implement new technologies to meet customer needs.

It all starts with the scannable bar-coded label attached to a package, which contains detailed information about the sender, the destination, and when the package should arrive. Customers can download and print their own labels using special software provided by UPS or by accessing the UPS website or mobile app. Before the package is even picked up, information from the "smart" label is transmitted to one of UPS's computer centers in Mahwah, New Jersey, or Alpharetta, Georgia, and sent to the distribution center nearest its final destination.

Dispatchers at this center download the label data and use special routing software called ORION, which is based on advanced algorithms, artificial intelligence, and machine learning, to create the most efficient delivery route for each driver that considers traffic, weather conditions, and the location of each stop. Each UPS driver makes an average of 135 stops per day. In a network with 66,000 routes in the United States alone, shaving even one mile off each driver's daily route translates into big savings in time, fuel consumption, miles driven, and carbon emissions—as much as $400 million per year.

These savings are critical as UPS tries to boost earnings growth as more of its business shifts to less-profitable e-commerce deliveries. UPS drivers who used to drop off several heavy packages a day at one retailer now often make multiple stops scattered across residential neighborhoods, delivering one package per household. The shift requires more fuel and more time, increasing the cost to deliver each package.

The first thing a UPS driver picks up each day is a handheld computer called a Delivery Information Acquisition Device (DIAD), which can access a wireless cell phone network. As soon as drivers log on, their route is downloaded onto the handheld. The DIAD also automatically captures customers' signatures along with pickup and delivery information. Package tracking information is then transmitted to UPS's computer network for storage and processing. From there, the information can be accessed worldwide to provide proof of delivery to customers or to respond to customer queries. It usually takes less than 60 seconds from the time a driver presses "complete" on the DIAD for the new information to be available on the web.

Through its automated package tracking system, UPS can monitor and even reroute packages throughout the delivery process. At various points along the route from sender to receiver, bar code devices scan shipping information on the package label and feed data about the progress of the package into the central computer. Customer service representatives are able to check the status of any package from desktop computers linked to the central computers and respond immediately to inquiries from customers. UPS customers can also access this information from the company's website using their own computers or mobile device. UPS now has a mobile website and mobile apps for iOS and Android devices.

Anyone with a package to ship can access the UPS website or apps to track packages, check delivery routes, calculate shipping rates, determine time in transit, print labels, and schedule a pickup. The data collected are transmitted to the UPS central computer and then back to the customer after processing. UPS typically gets almost 375 million tracking requests per day and at peak times more than 1 billion. UPS also provides tools that enable customers, such as Cisco Systems, to embed UPS functions, such as tracking and cost calculations, into their own websites so that they can track shipments directly.

UPS is now leveraging its decades of expertise managing its own global delivery network to manage logistics and supply chain activities for other companies. It created a UPS Supply Chain Solutions division that provides a complete bundle of standardized services to subscribing companies at a fraction of what it would cost to build their own systems and infrastructure. These services include supply chain design and management, freight forwarding, customs brokerage, mail services, multimodal transportation, and financial services in addition to logistics services.

UPS technology and business services are helpful to businesses of all sizes, including small startups. Cookie DŌ is a New York–based company that creates and sells safe-to-eat raw cookie dough. UPS made it possible for Cookie DŌ to ship its product safely and securely throughout the United States and Canada using a temperature-controlled insulated container that includes a small digital thermometer. Cookie DŌ can monitor the temperature of the package until it reaches the customer's doorstep. Cookie DŌ also turned to UPS for practical help with marketing technology.

Green Circle Salons is dedicated to recycling, recovering, and repurposing hard-to-recycle products used by beauty salons and spas, such as foils, aersosol cans, color tubes, and hair clippings. UPS, Green Circle's exclusive shipping provider, provides eco-friendly carbon-neutral shipping options and enables Green Circle to track all shipments and quantify their environmental impact.

Sources: UPS, "Company Facts," About.ups.com, accessed August 4, 2022; UPS, "Testing Ideas in the Real World," About.ups.com, August 25, 2021; UPS, "UPS Fact Sheet," About.ups.com, August 24, 2021; UPS, "Making Beauty Sustainable," About.ups.com, April 23, 2021; UPS, "From Her Mom's Kitchen to Small Business Success," About.ups.com, March 24, 2021; Paul Sawers, "UPS Will Now Use Dynamic Routing to Get Parcels to You on Time," Venturebeat.com, January 29, 2020.

## CASE STUDY QUESTIONS

1. What are the inputs, processing, and outputs of UPS's package tracking system?
2. What technologies are used by UPS? How are these technologies related to UPS's business strategy?
3. What strategic business objectives do UPS's information systems address?
4. What would happen if UPS's information systems were not available?

UPS's management is responsible for monitoring service levels and costs and for promoting the company's strategy of combining low cost and superior service. Management decided to use automation to increase the ease of sending a package via UPS and of checking its delivery status, thereby reducing delivery costs and increasing sales revenues.

The technology supporting this system consists of handheld computers, bar code scanners, wired and wireless communications networks, desktop computers, UPS's central computer, storage technology for the package delivery data, UPS in-house package tracking software, and software to access the web. The result is an information system solution to the business challenge of providing a high level of service with low prices in the face of mounting competition.

## 1-3 Apply a four-step method for business problem solving to solve information system-related problems.

Our approach to understanding information systems is to consider information systems and technologies as solutions to a variety of business challenges and problems. We refer to this as a problem-solving approach. Businesses face many challenges and problems, and information systems are one major way of solving these problems. All the cases in this book illustrate how a company used information systems to solve a specific problem.

The problem-solving approach has direct relevance to your future career. Your future employers will hire you because you can solve business problems and achieve business objectives. Your knowledge of how information systems contribute to problem solving will be very helpful to both you and your employers.

## THE PROBLEM-SOLVING APPROACH

At first glance, problem solving in daily life seems to be perfectly straightforward: a machine breaks down, parts and oil spill all over the floor, and, obviously, somebody has to do something about it. So, of course, you find a tool around the shop and start repairing the machine. After a cleanup and proper inspection of other parts, you start the machine, and production resumes.

No doubt, some problems in business are this straightforward, but few problems are this simple in the real world of business. In real-world business firms, a number of major factors are simultaneously involved in problems. These major factors can usefully be grouped into three categories: *organization, technology,* and *people.* In other words, a whole set of problems is usually involved.

## A MODEL OF THE PROBLEM-SOLVING PROCESS

There is a simple model of problem solving that you can use to help you understand and solve business problems by using information systems. You can think of business problem solving as a four-step process. Examine Figure 1.5, which illustrates each of these steps, beginning with problem identification, then solution design and solution evaluation and choice, and culminating with implementation. Once the solution has been implemented, feedback on how well it is working is provided to enable the problem solvers to make any necessary adjustments. This may require revisiting earlier steps in the problem-solving process. Let's now take a more detailed look at each step.

### Problem Identification

The first step in the problem-solving process is to understand what kind of problem exists. Contrary to popular beliefs, problems are not like basketballs on a court simply waiting to be picked up by some objective problem solver. Before problems can be solved, there must be agreement in a business that a problem exists, about what the

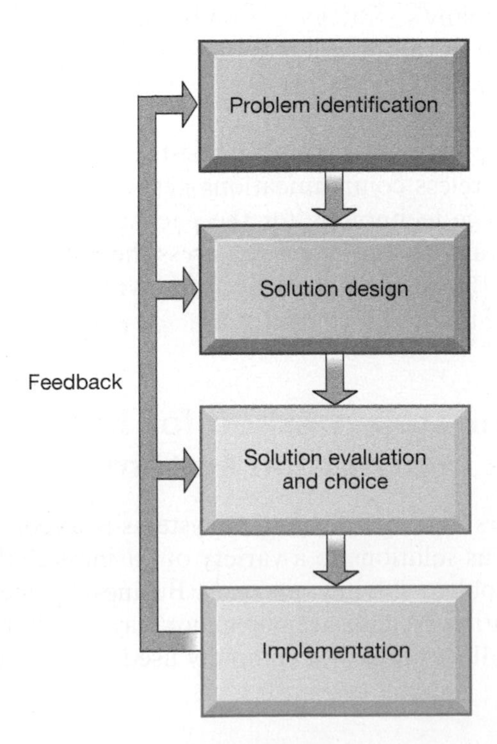


**Figure 1.5**
Problem Solving Is a Continuous Four-Step Process
*During implementation and thereafter, the outcome must be continually measured, and the information about how well the solution is working is fed back to the problem solvers. In this way, the identification of the problem can change over time, solutions can be changed, and new choices can be made, all based on experience.*

problem is, about its causes, and about what can be done about it, given the limited resources of the organization. Problems have to be properly defined by people in an organization before they can be solved.

For instance, what at first glance what might seem like a problem with employees not adequately responding to customers in a timely and accurate manner might in reality be a result of an older, out-of-date information system for keeping track of customers; or it might be a combination of both poor employee incentives for treating customers well and an outdated system. Once you understand this critical fact, you can start to solve problems creatively. Finding answers to these questions will require fact gathering, interviews with people involved in the problem, and analysis of documents and data.

In this text, we emphasize three different and typical dimensions of business problems: organizations, technology, and people (see Table 1.2). Typical organizational problems include poor business processes (usually inherited from the past), unsupportive culture, political infighting, and changes in the organization's surrounding environment. Typical technology problems include insufficient or aging hardware, outdated software, inadequate database capacity, insufficient network capacity, and the incompatibility of old systems with new technology. Typical people problems involve employee training, difficulties of evaluating performance, legal and regulatory compliance, ergonomics, poor or indecisive management, and employee support and participation. When you begin to analyze a business problem, you will find these dimensions are helpful guides to understanding the kind of problem with which you are working.

## Solution Design

The second step is to design solutions to the problem(s) you have identified. As it turns out, there are usually a great many solutions to any given problem, and the choice of solution often reflects the differing perspectives of people in an

**TABLE 1.2**

Dimensions of Business Problems

| Dimension | Description |
|---|---|
| Organizational dimensions | Outdated business processes |
| | Unsupportive culture and attitudes |
| | Political conflict |
| | Turbulent business environment, change |
| | Complexity of task |
| | Inadequate resources |
| Technology dimensions | Insufficient or aging hardware |
| | Outdated software |
| | Inadequate database capacity |
| | Insufficient network capacity |
| | Incompatibility of old systems with new technology |
| | Rapid technological change; failure to adopt new technology |
| People dimensions | Lack of employee training |
| | Difficulties of evaluating performance |
| | Legal and regulatory compliance |
| | Work environment |
| | Lack of employee support and participation |
| | Indecisive management |
| | Poor management |
| | Wrong incentives |

organization. You should try to consider as many solutions as possible so that you can understand the range of possible solutions. Some solutions emphasize technology; others focus on change in the organization and people aspects of the problem. As you will find throughout the text, most successful solutions result from an integrated approach in which changes in organization and people accompany new technologies.

### Solution Evaluation and Choice

Choosing the best solution for your business firm is the next step in the process. Some of the factors to consider when trying to find the best single solution are the cost of the solution, the feasibility of the solution for your business given existing resources and skills, and the length of time required to build and implement the solution. Also very important at this point are the attitudes and support of your employees and managers. A solution that does not have the support of all the major interests in the business can quickly turn into a disaster.

### Implementation

The best solution is one that can be implemented. Implementation of an information system solution involves building the solution and introducing it into the organization. This includes purchasing or building the software and hardware—the technology part of the equation. The software must be tested in a realistic business setting; then employees need to be trained, and documentation about how to use the new system needs to be written.

You will definitely need to think about change management. **Change management** refers to the many techniques used to bring about successful change in a business. Nearly all information systems require changes in the firm's business processes and, therefore, changes in what hundreds or even thousands of employees do every day. You will have to design new, more efficient business processes and then figure out how to encourage employees to adapt to these new ways of doing business. This may require meeting sessions to introduce the change to groups of employees, new training modules to bring employees quickly up to speed on the new information systems and processes, and, finally, some kind of rewards or incentives to encourage people to support the changes enthusiastically.

Implementation also includes the measurement of outcomes. After a solution has been implemented, it must be evaluated to determine how well it is working and whether any additional changes are required to meet the original objectives. This information is fed back to the problem solvers. In this way, the identification of the problem can change over time, solutions can be changed, and new choices made, all based on experience.

### Problem Solving: A Process, Not An Event

It is easy to fall into the trap of thinking about problem solving as an event that is over at some point, like a relay race or a baseball game. Often in the real world, this does not happen. Sometimes the chosen solution does not work, and new solutions are required.

For instance, the US National Aeronautics and Space Administration (NASA) spent more than $1 billion to fix a problem with shedding foam on the space shuttle. Experience proved the initial solution did not work. More often, the chosen solution partially works but needs a lot of continuous changes to fit the situation well. Sometimes, the nature of the problem changes in a way that makes the initial solution ineffective. For instance, hackers create new variations on computer viruses that require continually evolving antivirus programs to hold them in check. For all these reasons, problem solving is a continuous process rather than a single event.

## THE ROLE OF CRITICAL THINKING IN PROBLEM SOLVING

It is amazingly easy to accept someone else's definition of a problem or to adopt the opinions of some authoritative group that has objectively analyzed the problem and offers quick solutions. You should try to resist this tendency to accept existing definitions of any problem. It is essential for you to try to maintain some distance from any specific solution until you are sure you have properly identified the problem, developed understanding, and analyzed alternatives. Otherwise, you may leap off in the wrong direction, solve the wrong problem, and waste resources. You will have to engage in some critical-thinking exercises.

**Critical thinking** can be briefly defined as the sustained suspension of judgment with an awareness of multiple perspectives and alternatives. It involves at least four elements:

- Maintaining doubt and suspending judgment
- Being aware of different perspectives
- Testing alternatives and letting experience guide
- Being aware of organizational and personal limitations

Simply following a rote pattern of decision making, or a model, does not guarantee a correct solution. The best protection against incorrect results is to engage in critical thinking throughout the problem-solving process.

First, maintain doubt and suspend judgment. Perhaps the most frequent error in problem solving is to arrive prematurely at a judgment about the nature of the problem. By doubting all solutions at first and refusing to rush to a judgment, you create the necessary mental conditions to take a fresh, creative look at problems, and you keep open the chance to make a creative contribution.

Second, recognize that all interesting business problems have many dimensions and that the same problem can be viewed from different perspectives. In this text, we have emphasized the usefulness of three perspectives on business problems: technology, organizations, and people. Within each of these broad perspectives are many subperspectives, or views. The *technology perspective*, for instance, includes a consideration of all the components in the firm's IT infrastructure and the way they work together. The *organization perspective* includes a consideration of a firm's business processes, structure, culture, and politics. The *people perspective* includes consideration of the firm's management as well as employees as individuals and their interrelationships in workgroups.

You will have to decide for yourself which major perspectives are useful for viewing a given problem. The ultimate criterion here is usefulness: Does adopting a certain perspective tell you something more about the problem that is useful for solving the problem? If not, reject that perspective as not meaningful in this situation and look for other perspectives.

The third element of critical thinking involves testing alternatives, or modeling solutions to problems, letting experience be the guide. Not all contingencies can be known in advance, and much can be learned through experience. Therefore, experiment, gather data, and reassess the problem periodically.

## THE CONNECTIONS AMONG BUSINESS OBJECTIVES, PROBLEMS, AND SOLUTIONS

Now let's make the connection between business information systems and the problem-solving approach. At the beginning of this chapter, we identified seven business objectives of information systems: operational excellence; new products, services, and business models; customer/supplier intimacy; improved decision making; strategic advantage; survival; and meeting ESG goals. When firms cannot achieve these objectives, they become challenges or problems that receive attention.

Managers and employees who are aware of these challenges often turn to information systems as one of the solutions or the entire solution.

Review the diagram at the beginning of this chapter. The diagram shows how the AWM Smart Shelf system helps solve the business problem of brick-and-mortar stores losing market share to online retailers. This system provides a solution that takes advantage of opportunities provided by new object-recognition and wireless digital technology. The AWM Smart Shelf digitally enables key business processes for inventory management, sales, and marketing, helping retailers improve their overall business performance. The diagram also illustrates how people, technology, and organizational elements work together to create the system.

Each chapter of this text begins with a diagram similar to this one to help you analyze the chapter-opening case. You can use this diagram as a starting point for analyzing any information system or information system problem you encounter.

## 1-4 Describe the information systems skills and knowledge that are essential for business careers.

Looking out to 2030, the US economy will create almost 12 million new jobs, and millions more will will open up as their occupants retire. Employment in the leisure and hospitality sectors is projected to grow the fastest, driven largely by recovery from the pandemic, with the fastest-growing number of jobs in healthcare services. Strong employment growth is also expected in professional, business, and scientific services industries as well as management, scientific and technical consulting services. More than 33 percent of the jobs cataloged by the Bureau of Labor Statistics require either a bachelor's degree or at least some postsecondary education (Bureau of Labor Statistics, 2021a, 2021b).

What this means is that US business firms are looking for candidates who have a broad range of problem-solving skills—the ability to read, write, and present ideas—as well as the technical skills required for specific tasks. Regardless of your major or your future occupation, information systems and technologies will play an important and expanding role in your day-to-day work and your career. Your career opportunities, and your compensation, will in part depend on your ability to help business firms use information systems to achieve their objectives.

### HOW INFORMATION SYSTEMS WILL AFFECT BUSINESS CAREERS

In the following sections, we describe how specific occupations will be affected by information systems and what skills you should be building to benefit from the current and future labor market based on the research of the Bureau of Labor Statistics (Bureau of Labor Statistics, 2021c).

#### Accounting

There are about 1.4 million accountants and auditors in the US labor force today, and the field is expected to expand by 7 percent by the year 2030, adding almost 100,000 new jobs. The growth in accounting is driven in part by new accounting laws for public companies, greater scrutiny of public and private firms by government tax auditors, and a growing demand for management and operational advice.

Accountants rely heavily on information systems to summarize transactions, create financial records, organize data, and perform financial analysis. Because of new public laws, accountants require an intimate knowledge of databases, reporting systems, and networks to trace financial transactions. Because so many transactions are occurring over the Internet, accountants need to understand online transaction and reporting systems and how systems are used to achieve management accounting functions in an online and mobile business environment.

## Finance

If you include financial analysts, brokers and financial services sales representatives, loan officers, financial examiners, budget analysts, financial advisors, and financial managers, there are currently about 2.8 million workers employed in finance-related occupations. These financial occupations are expected to add about 220,000 new jobs between 2020 and 2030.

Financial managers play important roles in planning, organizing, and implementing information system strategies for their firms. Financial managers work directly with a firm's board of directors and senior management to ensure that investments in information systems help achieve corporate goals and high returns. The relationship between information systems and the practice of modern financial management and services is so strong that many advise finance majors to co-major in information systems (and vice versa).

## Marketing

No field has undergone more technology-driven change in the past five years than marketing and advertising. Digital advertising is the fastest-growing form of advertising, reaching more than $210 billion in 2021. Product branding and customer communication are moving online at a fast pace.

There are about 1.4 million public relations managers and specialists, marketing research analysts, and marketing and sales managers in the US labor force. Occupations related to marketing are growing faster than average and overall are expected to add more than 235,000 jobs by 2030. There are also millions of employees in marketing-related occupations (art, design, entertainment, sports, and media) and more than 10 million employees in sales. Marketing and advertising managers and specialists deal with large databases of customer behavior both online and offline in the process of creating brands and selling products and services. They develop reports on product performance, retrieve feedback from customers, and manage product development. These managers need an understanding of how enterprise-wide systems for product management, sales force management, and customer relationship management are used to develop products that consumers want, to manage the customer relationship, and to manage an increasingly mobile sales force.

## Operations Management In Services and Manufacturing

The growing size and complexity of modern industrial production and the emergence of huge global service companies have created a growing demand for employees who can coordinate and optimize the resources required to produce goods and services.

© fizkees/Shutterstock

*The job of management requires extensive use of information systems to support decision making and monitor the performance of the firm.*

Operations management as a discipline is directly relevant to three occupational categories: industrial production managers, administrative service managers, and operations analysts.

Production managers, administrative service managers, and operations analysts will be employing information systems and technologies every day to accomplish their jobs, with extensive use of database and analytical software.

### Management

Management is the largest single group in the US labor force with almost 9 million members, not including an additional 900,000 management analysts and consultants. Overall, the management corps in the United States is expected to expand at an average pace of 9 percent, adding about 900,000 new jobs by 2030. The Bureau of Labor Statistics tracks more than 20 types of managers, all the way from chief executive officer to human resource managers, production managers, project managers, lodging managers, medical managers, and community service managers.

Arguably, it would be impossible to manage business firms today, even very small firms, without the extensive use of information systems. Nearly all US managers use information systems and technologies every day to accomplish their jobs, from desktop productivity tools to mobile applications coordinating the entire enterprise. Managers today manage through a variety of information technologies without which it would be impossible to control and lead the firm.

### Information Systems

The information systems field is one of the most dynamic and fast changing of all the business professions because information technologies are among the most important tools for achieving business firms' key objectives. The explosive growth of business information systems has generated a growing demand for information systems employees and managers who work with other business professionals to design and develop new hardware and software systems to serve the needs of business.

There are more than 5 million people employed in occupations related to computers and information technology, ranging from information systems managers (more than 480,000) to computer systems analysts (more than 600,000), network and computer systems adminstrators (more than 350,000), software developers (more than 1.8 million) to web developers and digital designers (about 200,000), among others. These occupations are expected to add around 680,000 new jobs by 2030, driven by a greater emphasis on cloud computing, the collection and storage of Big Data and concerns about information security.

### Outsourcing and Offshoring

The Internet has created new opportunities for outsourcing many information systems jobs, along with many other service sector and manufacturing jobs. There are two kinds of outsourcing: outsourcing to domestic US firms and offshore outsourcing to low-wage countries such as India, China, and eastern European countries. Even this distinction blurs as US service providers develop global outsourcing centers offshore.

The most common and successful offshore outsourcing projects involve production programming and system maintenance programming work, along with call center work related to customer relationship management systems. However, inflation in Indian and Chinese wages for technology work, coupled with the additional management costs incurred in outsourcing projects, is leading to a counter movement of some IT jobs back to the United States. Moreover, although routine technical IS jobs such as software maintenance can be outsourced easily, all the management and organizational tasks required in systems development—including business process design, user interface design, and supply chain management—often remain in the United States.

Innovative new products, services, and systems are rarely outsourced either domestically or globally. Software outsourcing of routine IS work to low-wage countries lowers the cost of building and maintaining systems in the United States and other high-wage countries. As systems become less expensive, more are built. The net result is that offshore outsourcing likely increases demand domestically for employment in a wide variety of IS positions.

Given all these factors in the IT labor market, on what kinds of skills should information system majors focus? Following is a list of general skills we believe will optimize employment opportunities:

- An in-depth knowledge of how business firms can use new and emerging hardware and software tools to make them more efficient and effective, enhance customer and supplier intimacy, improve decision making, achieve competitive advantage, and ensure firm survival. This includes an understanding of artificial intelligence, cloud computing, business analytics, business intelligence, databases, system implementation, and mobile application development.
- An ability to take a leadership role in the design and implementation of new information systems, work with other business professionals to ensure systems meet business objectives, and work with cloud computing services and software firms providing new system solutions.

### Information Technology and The Future of Jobs

Few topics have recently captured the attention of journalists, academics, and the general public more than the impact of artificial intelligence, and automation in general, on employment in the United States and worldwide. This general concern is not new but has recurred with every advance in information technology since the 1950s. The Spotlight on People case illustrates some of the differing views on how new information technologies like robotics and other forms of automation may impact employment in coming years.

## INFORMATION SYSTEMS AND BUSINESS CAREERS: WRAP-UP

Looking back at the information system skills and knowledge required for specific careers, there are some common themes that apply to many different occupations. No matter your intended career, you will need to:

- Understand how information systems and technologies can help firms achieve business objectives.
- Develop skills in analysis of information and helping firms understand and make sense of their data.
- Understand how companies can use information systems to respond to ever-evolving changes in the ethical, social, and legal environment surrounding the business.
- Be able to communicate effectively, both verbally and in writing, and collaborate with others as part of a team.

## HOW THIS BOOK PREPARES YOU FOR THE FUTURE

This book is explicitly designed to prepare you for your future business career. It provides you with the necessary knowledge and foundational concepts for understanding the role of information systems in business organizations. You will be able to use this knowledge to identify opportunities for increasing the effectiveness of a business. You will learn how to use information systems to improve operations, create new products and services, improve decision making, increase customer intimacy, and promote competitive advantage.

## Will a Robot Steal Your Job?

More than a century ago, a Czech playwright coined the term "robot" in a fictional story about artificial human beings. In the past few decades, advances in information technology, particularly in the field of artificial intelligence, have made robots an everyday reality. Robots can be found performing jobs in many different types of workplaces. Today's robots include not only physical robots, such as Boston Dynamic's dog-like Spot, an agile mobile robot that can be used for a variety of purposes, but also software robots, sometimes referred to as robotic process automation (RPA) or just software "bots." Software bots can emulate human actions in interacting with digital systems, such as identifying and extracting data and understanding what is on a screen, but do so faster and in a more consistent manner than a person might.

While many businesses champion the use of robots and other forms of automation as engines of higher productivity, operational efficiency, improved profit margins, safer workplaces, and job creation, others are concerned that these technologies are stealing jobs away from the people who most need them. This controversy is not new: tension around the use of automation has been a recurring theme since the first Industrial Revolution in the eighteenth century. However, the Covid-19 pandemic has ushered in a new host of variables, accelerating the progress of automation as well as potentially deepening its impact.

Industrial robots are primarily used in manufacturing and for logistics, such as in warehouses. Many are fully autonomous and able to complete their tasks without any human intervention. A cobot (short for "collaborative robot") works collaboratively alongside humans. Worldwide, the use of industrial robots in factories has nearly doubled over the past five years, according to the International Federation of Robotics. Sales in the past two years have particularly boomed, in part due to businesses attempting to adjust to changes wrought by the Covid-19 pandemic, such as changes in buying habits. For example, Texas grocery chain H-E-B has installed several automated microfulfillment systems from robotics company AutoStore to support its curbside pickup and delivery services. Giant stacks of cubes about 6 inches apart are packed with bins filled with products. It would not be possible for humans to reach the bins. Instead, a robot travels across the top of the cube, digging out bins and delivering them to stations where workers assemble the orders. Currently, AutoStore has installed similar systems at more than 750 sites across 40 countries.

Robots are also being called upon to help with labor shortages that have followed in the wake of the pandemic. Kenco, a third-party logistics provider, is deploying self-driving robots and testing autonomous tractors that can tow pallets as well as remotely operated warehouse forklifts that can be operated from anywhere in the world. Software bots are being employed to handle customer service call center requests, office tasks such as processing payroll data and expense reports, and even complex tasks such as reviewing legal documents. Applied Materials, which supplies equipment, services, and software to the semiconductor industry, is attempting to stretch strained resources by using more than 250 bots to automate financial accounting and other workplace processes. According to research firm International Data Corp, up to 40 percent of companies worldwide have increased their use of software bots and other forms of automation in response to the pandemic.

Research is mixed on whether robots and other forms of automation result in the permanent replacement of workers. Some have found that robots and other forms of automation are reducing the demand for workers, weighing down wages, and pushing workers into low-paying parts of the economy. For instance, one study by economists Daron Acemoglu of MIT and Pascual Restrepo of Boston University found that for every robot per thousand workers, up to six workers lost their jobs and wages fell as much as 0.75 percent. Acemoglu and Restrepo found little employment increase in other occupations to offset job losses in manufacturing. Acemoglu and Restrepo noted that a specific local economy, such as Detroit, could be especially hard-hit, although nationally the effects of robots are smaller because jobs were created in other places. The new jobs created by technology are not necessarily in the places losing jobs, such as the Rust Belt. Those forced out of a job by robots generally do not have the skills or mobility to assume the new jobs created by automation. They are forced to compete with other workers for whatever jobs are left, which increases labor supply and depresses wages.

However, another study by Dixon, Hong, and Wu that examined data from Canadian businesses over a five-year period found that investments in robotics resulted in an increase in total firm employment, in part because enhanced productivity led to increased demand for the firm's products, requiring more employees. The research also found that those investments were primarily motivated by a desire to improve product and service quality rather than to reduce labor costs.

A 2021 report by McKinsey Global Institute on the post-Covid-19 labor force predicted that more than 45 million US workers could be displaced by automation by 2030, but it is important to note that this number is not the same as actual jobs lost. The report further notes that it expects that 90 percent of those workers would transition to jobs within the same occupational category. Likewise, the World Economic Forum predicted that while technology may eliminate 85 million jobs by 2025, it would at the same time create 97 million new ones, a net addition of 12 million.

Although the research may appear contradictory, what is mostly undisputed is that robots and other forms of automation are likely to displace workers in certain types of jobs in certain industries in the short term. In addition, the effects of automation are not equally distributed and may contribute to economic inequality. Researchers from the Brookings Institution found that advances in automation will disproportionately affect Black and Latinx workers, who are overrepresented on a percentage basis in the job occupations that have the highest risk of being automated in the next two decades: those that are characterized as "middle-skilled," which often require specialization in routine tasks. In contrast, high-skill jobs that involve nonroutine cognitive tasks and low-paying service-sector jobs in hotels, restaurants, and nursing homes are much less likely to be automated. Much of this service work is difficult to automate, and employers have less incentive to replace low-wage workers with machines.

How can you "future-proof" your career to deal with the potential impact of automation? Experts in the field offer some guidance. The World Economic Foundation reports that the top skills employers will be seeking include critical thinking and problem solving. Daniel Zhao, a senior economist at Glassdoor, recommends that workers build skills that are complementary to technology and sharpen up soft skills that are hard for machines to replace. Mark Muro, senior fellow and policy director at the Brookings Institution, suggests that workers should seek work that has a creative, face-to-face, human element and notes that personalized work, teamwork, and creative work will last. Ravin Jesuthasan, a global transformation leader at Mercer, believes that there will always be demand for work requiring judgment, creativity, and empathy—in short, the things that make people human.

Sources: Kenco, "Labor Shortage Impacts: Mitigation Through Innovation," Blog.kencogroup.com, accessed June 7, 2022; IFR, "Robot Density Nearly Doubled Globally," Ifr.org, December 14, 2021; Allison Prang, "Companies Order Record Number of Robots amid Labor Shortage," *Wall Street Journal*, November 11, 2021; Ashley Nunes, "Automation Doesn't Just Create or Destroy Jobs—It Transforms Them," *Harvard Business Review*, November 2, 2021; Angus Loten, "Workplace Automation Bots Gain Clout amid Covid-19 Pandemic," *Wall Street Journal*, September 22, 2021; World Economic Forum, "Robots and Your Job: How Automation Is Changing the Workplace," Weforum.org, June 24, 2021; Jennifer Smith, "Warehouses Look to Robots to Fill Labor Gaps, Speed Deliveries," *Wall Street Journal*, May 24, 2021; Jay Dixon et al, "The Robot Revolution: Managerial Consequences for Firms," *Management Science*, March 31, 2021; Kevin Carey, "Do Not Be Alarmed by Wild Predictions of Robots Taking Everyone's Jobs," Slate.com; March 31, 2021; Kristen Broady et al, "Race and Jobs at Risk of Being Automated in the Age of Covid-19," Brookings.edu, March 4, 2021; McKinsey Global Institute, "The Postpandemic Economy: The Future of Work After Covid-19," February 2021; Angus Loten, "Software Bots Multiply to Cope with 'Stretched' Resources, *Wall Street Journal*, January 25, 2021; Christopher Mims, "On the 100th Anniversary of 'Robot', They're Finally Taking Over," *Wall Street Journal*, January 23, 2021; Jennifer Smith, "Smaller Is Big in New E-commerce Warehouses," *Wall Street Journal*, November 8, 2020; Daron Acemoglu and Pascual Restrepo, "Robots and Jobs: Evidence from US Labor Markets," *Journal of Political Economy*, 2020.

## CASE STUDY QUESTIONS

1. How does automating jobs pose an ethical dilemma? Who are the stakeholders? Identify the options that can be taken and the potential consequences of each.
2. If you were the owner of a factory deciding on whether to acquire robots to perform certain tasks, what people, organization, and technology factors would you consider?
3. How has the Covid-19 pandemic impacted the process of automation?
4. What types of work and workers are likely to be among the most affected by the increasing use of robots and other forms of automation?

Equally important, this book develops your ability to use information systems to solve problems that you will encounter on the job. You will learn how to analyze and define a business problem and how to design an appropriate information system solution. You will deepen your critical-thinking and problem-solving skills. The following features of the text and the accompanying learning package reinforce this problem-solving and career orientation.

### A Framework for Describing and Analyzing Information Systems

The text provides you with a framework for analyzing and solving problems by examining the people, organizational, and technology components of information systems. This framework is used repeatedly throughout the text to help you understand information systems in business and analyze information systems problems.

### A Four-Step Model for Problem Solving

The text provides you with a four-step method for solving business problems, which we introduced in this chapter. You will learn how to identify a business problem, design alternative solutions, choose the correct solution, and implement the solution. You will be asked to use this problem-solving method to solve the case studies in each chapter. Chapter 12 will show you how to use this approach to design and build new information systems and determine their business value.

### Hands-On MIS Projects for Stimulating Critical Thinking and Problem Solving

Each chapter concludes with a series of hands-on MIS projects to sharpen your critical-thinking and problem-solving skills. These projects include a Management Decision Problem, hands-on application software problems, and projects for building Internet skills. For each of these projects, we identify both the business skills and the software skills required for the solution.

### Career Resources

To make sure you know how the text is directly useful in your future business career, we've added a full set of career resources to help you with career development and job hunting.

**Career Opportunities Feature** To show you how this book can help you find a job and build your career, we have added a "Career Opportunities" feature, identified by this icon, to each chapter. The last major section of each chapter, titled "Understand how MIS can help your career" presents a description of an entry-level job for a recent college graduate based on a real-world job description. The job requirements are related to the topics covered in that chapter. The job description shows the required educational background and skills, lists business-related questions that might arise during the job interview, and provides author tips for answering the questions and preparing for the interview. Students and instructors can find more detail about how to use this feature in the Preface and in MyLab MIS.

**Digital Portfolio** MyLab MIS includes a template for preparing a structured digital portfolio to demonstrate the business knowledge, application software skills, Internet skills, and analytical skills you have acquired in this course. You can include this portfolio in your résumé or job applications. Your professors can also use the portfolio to assess the skills you have learned.

**Additional Career Resources** A Career Resources section in MyLab MIS shows you how to integrate what you have learned in this course in your résumé, cover letter, and job interview to improve your chances for success in the job market.

## 1-5 Understand how MIS can help your career.

Here is how Chapter 1 can help you find an entry-level job as a client support assistant.

### THE COMPANY

Alpha Financial Analytics Data Services is a cloud-based business intelligence software company serving the financial services industry with offices in New York City, Atlanta, Los Angeles, and Chicago. Alpha Financial is looking to fill a remote entry-level client support position for its customer success team, whose mission is to build customer loyalty and satisfaction. The company has 1,600 employees around the country, many of whom now work either fully remotely or on a hybrid basis.

### POSITION DESCRIPTION

The client support assistant will be part of the firm's customer success team. The customer success team combines a thorough understanding of finance and technology with specific expertise in Alpha Financial Analytics Data Services software and assists clients in a variety of ways. The company provides on-the-job training in its software and client support methods. Job responsibilities include:

- Supporting Alpha Financial Analytics Data Services applications used by the client.
- Helping the team create custom models and screens for the client.
- Training clients via videoconference such as Zoom and via webinars.
- Providing expert consultation to clients via email, videoconference, and telephone.

### JOB REQUIREMENTS

- Recent college graduate, preferably with one to two years of experience in a client support position. Applicants with backgrounds in finance, MIS, economics, accounting, business administration, and mathematics are preferred.
- Knowledge of, or interest in learning about, the financial services industry
- Sound working knowledge of spreadsheet and videoconference software
- Very strong communication and interpersonal skills
- Ability to work independently in a remote environment

### INTERVIEW QUESTIONS

1. What kind of work have you done, if any, with clients? Can you give examples of how you provided client service or support?
2. What is your proficiency level with spreadsheet software such as Microsoft Excel? What work have you done with Excel spreadsheets? Can you show examples of your work?
3. How will current trends in the financial services industry impact Alpha Financial's business model and client base?
4. Discuss your experiences, both positive and negative, using videoconference software such as Zoom in either an educational setting or business setting.
5. Can you give us an example of a finance-related problem or other business problem that you helped solve? Did you do any writing and analysis? Can you provide examples?

### AUTHOR TIPS

1. Use the web to learn about the industry the company serves—in this case, financial services.
2. Use the web to research the company, its products, and the tools and services it offers customers. Additionally, examine the company's social media channels, such as LinkedIn and Facebook, for trends and themes.
3. Inquire exactly how you would be using spreadsheets for this job. Provide examples of how you used spreadsheets to solve problems in the classroom or for a job assignment.
4. Bring examples of your writing (including some from your Digital Portfolio described in MyLab MIS) demonstrating your analytical skills and project experience. Be prepared to discuss how you helped customers solve a business problem or the business problem solving you did for your courses.

## Review Summary

1-1 **Understand why information systems are essential for running and managing a business.** Information systems are a foundation for conducting business today. In many industries, survival and even existence is difficult without extensive use of information technology. Businesses use information systems to achieve seven major objectives: operational excellence; new products, services, and business models; customer/supplier intimacy; improved decision making; competitive advantage; day-to-day survival; and meeting ESG goals.

1-2 **Define an information system, explain how it works, and identify its people, organizational, and technology components.** From a technical perspective, an information system collects, stores, and disseminates information from an organization's environment and internal operations to support organizational functions and decision making, communication, coordination, control, analysis, and visualization. Information systems transform raw data into useful information through three basic activities: input, processing, and output. From a business perspective, an information system provides a solution to a problem or challenge facing a firm and represents a combination of people, organization, and technology elements.

The people dimension of information systems involves issues such as training, job attitudes, and management behavior. The technology dimension consists of computer hardware, software, data management technology, and networking/telecommunications technology, including the Internet. The Internet of Things (IoT), Big Data, and cloud computing are especially important technologies. The organizational dimension of information systems involves issues such as the organization's hierarchy, functional specialties, business processes, culture, and political interest groups.

1-3 **Apply a four-step method for business problem solving to solve information system-related problems.** Problem identification involves understanding what kind of problem is being presented and identifying people, organizational, and technology factors. Solution design involves designing several alternative solutions to the problem that has been identified. Evaluation and choice entail selecting the best solution, taking into account its cost and the available resources and skills in the business. Implementation of an information system solution entails purchasing or building hardware and software, testing the software, providing employees with training and documentation, managing change as the system is introduced into the organization, and measuring the outcome. Problem solving requires critical thinking in which one suspends judgment to consider multiple perspectives and alternatives.

**1-4 Describe the information system skills and knowledge that are essential for business careers.** Business careers in accounting, finance, marketing, operations management, and management all rely heavily on information systems. The information systems field is one of the most dynamic and fast-growing because information technologies are among the most important tools for achieving business firms' key objectives. Although the Internet has created new opportunities for outsourcing many information system jobs, it has also increased demand domestically for employment in a wide variety of IS positions. The most important skills for IS majors to focus on developing include an in-depth knowledge of new and emerging hardware and software tools, as well as the ability to take a leadership role in the design and implementation of new information systems. No matter their intended career, all students need to understand how information systems and technologies help firms achieve business objectives and respond to changes in the ethical, social, and legal environment; develop data analytic skills; and be able to communicate effectively and collaborate with others.

## Key Terms

Big Data, 16
Business model, 9
Business processes, 14
Change management, 22
Cloud computing, 16
Computer hardware, 15
Computer literacy, 14
Computer software, 15
Critical thinking, 23
Culture, 15
Data, 12
Data management technology, 16
Extranets, 16
Feedback, 12
Information, 11
Information system (IS), 11
Information systems literacy, 14
Information technology (IT), 11
Information technology (IT) infrastructure, 16
Input, 12
Internet, 16
Internet of Things (IoT), 16
Intranets, 16
Management information systems (MIS), 14
Network, 16
Networking and telecommunications technology, 16
Output, 12
Processing, 12
World Wide Web, 16

## Review Questions

**1-1** Understand why information systems are essential for running and managing a business.

- List and describe the seven reasons information systems are so important for business today.
- Describe the challenges and opportunities of globalization.

**1-2** Define an information system, explain how it works, and identify its people, organization, and technology components.

- List and describe the organizational, people, and technology dimensions of information systems.
- Define an information system and describe the activities it performs.
- Distinguish between data and information and between information systems literacy and computer literacy.
- Explain how the Internet and the World Wide Web are related to the other technology components of information systems.
- Which three recent developments in information technology are especially important today? How are these developments interrelated?

**1-3** Apply a four-step method for business problem solving to solve information system-related problems.

- List and describe each of the four steps for solving business problems.
- Give some examples of people, organizational, and technology problems found in businesses.
- Describe the relationship of critical thinking to problem solving.
- Describe the role of information systems in business problem solving.

**1-4** Describe the information system skills and knowledge that are essential for business careers.

- Describe the role of information systems in careers in accounting, finance, marketing, management, and operations management and explain how careers in information systems have been affected by new technologies and outsourcing.
- List and describe the information system skills and knowledge that are essential for all business careers.

**MyLab MIS**
To complete these problems, go to EOC Discussion Questions in **MyLab MIS**.

## Discussion Questions

**1-5** MyLab MIS What are the implications of globalization when you have to look for a job? What can you do to prepare yourself for competing in a globalized business environment? How would knowledge of information systems help you compete?

**1-6** MyLab MIS If you were setting up the website for a new minor league baseball team, what people, organizational, and technology issues might you encounter?

**1-7** MyLab MIS Identify some of the people, organizational, and technology issues that UPS had to address when creating its successful information systems.

## Hands-On MIS Projects

The projects in this section give you hands-on experience in analyzing a visitor management problem, using spreadsheet software to improve payroll reporting operations, and using Internet software for researching job requirements. Visit MyLab MIS to access this chapter's Hands-On MIS Projects.

### MANAGEMENT DECISION PROBLEM

**1-8** The Brooklyn Navy Yard (BNY) is a shipyard and industrial complex located in Brooklyn, New York. It encompasses a 300-acre waterfront site for more than 450 businesses employing more than 11,000 people and generating more than $2.5 billion per year in economic impact for New York City. The BNY needs to keep track of its thousands of visitors and nearly 10,000 employees per day, but it had been unable to do so efficiently and safely. Its internally hosted web portal for managing visitors was slow and not very user-friendly and was unable to scale to handle the many thousands more visitors projected for the future. Most BNY tenants were not pre-registering their visitors or delivery drivers, causing long lines and traffic at the security gates. Security checks were performed manually by checking IDs against a printed watch list. BNY was unable to create a welcoming experience for visitors and employees. Use the four-step problem-solving method introduced in this chapter to identify BNY's problem and analyze its business impact. What business functions were affected? Is it a people problem, an organizational problem, or a technology problem? Suggest an information system solution and identify some implementation challenges that the solution would have to address.

### ACHIEVING OPERATIONAL EXCELLENCE: CREATING AN EXECUTIVE PAYROLL REGISTER WITH SPREADSHEET SOFTWARE

Spreadsheet software skills: Formulas; absolute and relative addressing; worksheet formatting; SUM; AVERAGE
Business skills: Payroll accounting

**1-9** In this project, you will create a payroll register for the eight senior executives of a machine tool company with 8,500 employees. All employees are paid through the firm's automated payroll system except for the eight members of the company's executive steering committee, including the CEO. The payroll system for the steering committee is produced separately on a spreadsheet because it contains highly confidential information about senior executives' salaries. This group of executives is paid monthly.

In MyLab MIS you will find the EMIS15Chap1 Question File, which displays the names of the executive steering committee members, their Social Security numbers, and annual salaries. Develop a worksheet that creates a payroll register report for these employees for the first month of the year (January 2023). The worksheet should automatically calculate monthly gross pay, year-to-date gross pay, net pay, and all deductions. These deductions include federal income tax withholding (25 percent of gross pay), state income tax withholding (5 percent of gross pay), FICA (Social Security) (6.2 percent of gross pay), Medicare (1.45 percent of gross pay), group health insurance ($600 per month), and monthly contributions to a profit-sharing plan (5 percent of gross pay). It should also provide totals for each of these categories for the current pay period. Create an "Assumptions" section for all deductions and other variables in the upper left of the worksheet so you can easily make changes in deductions and formulas using the addressing function of the spreadsheet software. Listing variables also allows all assumptions to be clearly visible and reported. Use formulas to calculate monthly gross pay, year-to-date gross pay, all deductions, and net pay. Be sure these formulas reference the appropriate cells in the "Assumptions" section of your worksheet (e.g., $A$56) rather than actual values (e.g., 20 percent). Use the SUM function to provide totals for gross pay, net pay, and each deduction category. Use the AVERAGE function to calculate the average annual salary paid to these top eight executives.

### IMPROVING DECISION MAKING: USING THE INTERNET TO LOCATE JOBS REQUIRING INFORMATION SYSTEMS KNOWLEDGE

Software skills: Internet-based software
Business skills: Job searching

**1-10** Visit a job-posting website such as Monster.com or Indeed.com. Spend some time at the site examining jobs for accounting, finance, sales, marketing, and human resources. Find two or three descriptions of jobs that require some information systems knowledge. What information systems knowledge do these jobs require? What do you need to do to prepare for these jobs? Write a one- to two-page report summarizing your findings.

## Collaboration and Teamwork Project

### SELECTING TEAM COLLABORATION TOOLS

**1-11** Form a team with three or four classmates and review the capabilities of Google Drive and Google Sites for your team collaboration work. Compare the capabilities of these two tools for storing team documents, project announcements, source materials, work assignments, illustrations, presentations, and web pages of interest. Learn how each works with Google Docs. Explain why Google Drive or Google Sites is more appropriate for your team. If possible, use Google Docs to brainstorm and develop a presentation of your findings for the class. Organize and store your presentation by using the Google tool you have selected.

## BUSINESS PROBLEM-SOLVING CASE

### WILL THE COVID-19 PANDEMIC MAKE WORKING FROM HOME THE NEW NORMAL?

As Covid-19 spread around the globe in 2020 and 2021, companies large and small started to make changes to the way they work, shuttering their offices and requiring most or all of their employees to work remotely from their homes. Here are just a few examples:

- Tax Analysts (TA), a small nonprofit publisher in Falls Church, Virginia, closed its offices in mid-March 2020, supplying laptops to employees who didn't have one so that the whole company could work from home.
- Many large law firms, including Reed Smith, Baker McKenzie, and Nixon Peabody, closed their offices and required work at home. The law firms emphasized that they could continue to serve clients despite office closings and remote work.
- George Washington University Law School in Washington, D.C., was among the many colleges and universities around the world that opted to cancel traditional classes in favor of online learning.
- Elementary and high schools across the nation moved their classes online. For the rest of 2020 through mid-2021, many classrooms went totally virtual.
- Twitter and Facebook quickly said they would embrace remote work long-term. Some companies even vowed to give up their physical office spaces entirely.

At the pandemic's peak in early May 2020, 52 percent of people employed in the United States reported always working from home, and another 18 percent reported sometimes working from home, for a total of 70 percent, according to a Gallup Poll survey. Prior to the pandemic, the percentage of the US workforce regularly working from home was in the single digits, with only about 4 percent working from home at least half the time. However, the trend of working from home had been slowly gaining momentum, thanks to advances in information technology for remote work and changes in corporate work culture. The coronavirus pandemic may mark a tipping point.

It's likely that many people who started working from home for the first time during the pandemic will continue to do so in the future. Pandemic-related health guidelines about distancing have required some workplaces to expand to accommodate all their employees or to have a significant percentage of employees work permanently from home.

A variety of information technologies have made this shift to remote work possible, including widespread availability of broadband high-speed Internet connections, laptop computers, tablets, smartphones, email, messaging, and videoconferencing tools. As companies shift their work from face-to-face to remote, videoconferencing is becoming the new normal for meetings. Although less than ideal for face-to-face interactions, people are adapting to having good conversations, sharing critical information, generating new ideas, reaching consensus, and making decisions quickly on videoconference platforms.

There are now many powerful and affordable videoconferencing options, including Skype, Skype for Business, Microsoft Teams, Amazon Chime, Verizon's BlueJeans, Cisco's WebEx, LogMeIn's GoToMeeting, and Google Meet. Some businesspeople are using the same tools they do in their personal communications, such as FaceTime, which now supports group video chat with up to 32 people, and Facebook Messenger.

Videoconference software such as WebEx and BlueJeans is designed for more corporate uses. Other software such as Microsoft's Skype and Zoom is more consumer-friendly and easier to set up, with free or low-cost versions suitable for smaller businesses.

Skype works for video chats, calls, and instant messaging and can handle up to 100 people in a single video call. Skype allows calls to be recorded in case someone misses a meeting. Skype also has file-sharing capabilities, caller ID, voicemail, a split-view mode to keep conversations separate, and screen share on mobile devices.

Up to 1,000 users can participate in a single Zoom video call, and 49 videos can appear on the screen at once. Zoom includes collaboration tools like simultaneous screen-sharing and co-annotation and the ability to record meetings and generate transcripts. Users can adjust meeting times, select multiple hosts, and communicate via chat if microphones and cameras are turned off.

There are definite benefits to remote work: lower overhead, more flexible schedules, reductions in employee commuting time and attrition rates, and increases in productivity. Executives at many companies were amazed at how well their workers performed remotely. According to Global Workplace Analytics, a typical company saves about $11,000 per half-time telecommuter per year.

In May 2020, Twitter was the first major US company to publicly announce a permanent work-from-home policy. Other employers have followed suit in making remote work permanent. Twitter anticipates that half of its employees will permanently work from home and that distributed work will give employees more autonomy and freedom, which it believes improves morale, employee retention, and productivity. Outdoor apparel company REI, Facebook, and e-commerce platform Shopify have all announced some measures making work from home the new norm.

However, working remotely also poses challenges. Constant connectedness causes many people to experience "Zoom fatigue": tiredness, worry, or burnout associated with the overuse of virtual communication platforms particularly videoconferencing (on any platform, not just Zoom).

In a video call, minds are together, but bodies are not. INSEAD associate professor Gianpiero Petriglier has suggested that Zoom fatigue comes about because people need to pay more attention to nonverbal cues such as pitch and tone of voice, facial expressions, and body language. This requires the mind to work much harder than it would need to in a face-to-face setting. Participants use high levels of cognitive energy to recognize nonverbal cues, which are difficult to visualize in a virtual environment. Moreover, although in face-to-face communication silence gives natural rhythm to a conversation, in a video call, silence can generate anxiety. In addition, according to Marissa Shuffler, an associate professor at Clemson University, people have a greater awareness of being watched when on camera and can feel a greater sense of self-awareness when seeing their own image. When you're on a videoconference, you know everyone is looking at you; you are on stage, and for many people, this creates social pressure and feeling like you need to perform.

The decision to allow many or all employees to work remotely may also negatively impact innovation and creativity, with great ideas at work often born from daily in-person interactions. Apple founder Steve Jobs asserted that creativity comes from spontaneous meetings and random discussions. The ideas for Facebook and Google were generated in a college dorm room and a garage near Stanford University. Although email and text messaging are very useful, they are not as effective tools for communication compared to the information exchange and personal connection of face-to-face conversations. Remote work inhibits the creativity and innovative thinking that take place when people interact with each other face-to-face, and videoconferencing is only a partial solution. Studies have found that people working together in the same room tend to solve problems more quickly than remote collaborators and that team cohesion suffers when members work remotely.

Internet access is another issue. Not everyone in the United States has access to the Internet at home. About 85 percent of US adults have high-speed broadband Internet service at home. However, according to a Pew Research Center study, racial minorities, older adults, rural residents, and people with lower levels of education and income are less likely to have in-home broadband service. In addition, one in five US adults accesses the Internet only through their smartphone.

Full-time employees are four times more likely to have remote work options than part-time employees. According to Global Workplace Analytics, a typical remote worker is college-educated, is at least 45 years old, and earns an annual salary of $58,000 while working for a company with more than 100 employees.

An October 2021 study by the Gallup Poll found 45 percent of full-time employees working partly or fully remotely in September and that nine in 10 remote workers want to maintain remote work to some degree. One way to marry the benefits of remote work with the efficiency of face-to-face meetings is to adopt a hybrid work program, where employees work part of the week at the office and the rest off-premises. For example, Google will let employees spend two days per week wherever they work best. Morgan Stanley's Chief Executive Officer James Gorman sees office-working at "not 100% but not zero percent" of total hours. Both workers and managers say that two to three days a week of working from home is ideal, according to Chiara Criscuolo, who researches productivity for the Organisation for Economic Co-operation and Development (OECD). After that, professional relationships can suffer. It seems a near-certainty that there will be substantially more remote work going forward, and this will change the nature of work and the way work gets done.

Sources: Global Workplace Analytics, "Latest Work-at-Home/Telecommuting/Remote Work Statistics," Globalworkplaceanalytics.com, accessed June 7, 2022; Danielle Abril, "Four-day Weeks and the Freedom to Move Anywhere: Companies Are Rewriting the Future of Work (Again)," *Washington Post*, January 20, 2022; Daniel Akst, "Zoom, Slack, Google Hangouts and More: The Hidden Risks of Remote Work," *Wall Street Journal*, October 29, 2021; Lydia Saad and Ben Wigert, "Remote Work Persisting and Trending Permanent," Gallup News, News.gallup.com, October 13, 2021; Global Workplace Analytics, "U.S. Employers Stand to Save Over $500B a Year with a Combination of In-Office/Remote Work Strategies," Gobalworkplaceanalytics.com, January 12, 2021; Chiara Criscuolo, "Productivity and Business Dynamics Through the Len of COVID-19: the Shock, Risks, and Opportunities", OECD, Ecb.europa.eu, 2021; Christopher Mims "The Work-from-Home Shift Shocked Companies—Now They're Learning Its Lessons," *Wall Street Journal*, July 25, 2020; Carolyn Reinach Wolf, "Virtual Platforms Are Helpful Tools but Can Add to Our Stress," *Psychology Today*, May 14, 2020; Manyu Jiang, "The Reason Zoom Calls Drain Your Energy," Bbc.com, April 22, 2020; Rani Molla, "This Is the End of the Office as We Know It," Vox.com, April 14 2020; Cate Pye, "Coronavirus: What Does the 'New Normal' Mean for How We Work?," *Computer Weekly*, April 3, 2020; Kevin Roose, "Sorry, but Working from Home Is Overrated," *New York Times*, March 10, 2020.

## CASE STUDY QUESTIONS:

**1-12** Define the problem described in this case. What were the people, organization, and technology issues raised by this problem?

**1-13** Identify the information technologies used to provide a solution to this problem. Was this a successful solution? Why or why not?

**1-14** Will working from home become the dominant way of working in the future? Why or why not?

## Chapter 1 References

Amadeo, Kimberly. "How Outsourcing Jobs Affects the US Economy." Thebalance.com (February 20, 2021).

Amazon.com Inc. "Annual Report on Form 10-K for the Fiscal Year ended December 31, 2021." Sec.gov (February 4, 2022).

Brynjolfsson, Erik. "VII Pillars of IT Productivity." *Optimize* (May 2005).

Bureau of Economic Analysis. "Activities of U.S. Multinational Enterprises, 2019." Bea.gov (November 12, 2021).

Bureau of Labor Statistics, U.S. Department of Labor. "Employment Projections: 2020–2030." Bls.gov (September 8, 2021a).

Bureau of Labor Statistics, U.S. Department of Labor. "Table 5.4 Education and Training Assignments by Detailed Occupation, 2020." Bls.gov (September 8, 2021b).

Bureau of Labor Statistics, U.S. Department of Labor. *Occupational Outlook Handbook*. Bls.gov (September 15, 2021c).

"Coca Cola Bottling Company Empowers the Enterprise with Tableau Mobile Dashboards to Drive Bottom Line." Tableau.com (accessed June 7, 2022).

Insider Intelligence/eMarketer. "US Digital Buyers and Penetration." (February 2022); "US Mobile Buyers and Penetration." (February 2022); "US Search Users and Penetration." (February 2022); "US Mobile Phone Search Users." (February 2022a).

Insider Intelligence/eMarketer. "US Social Network Users and Penetration." (April 2022b).

Insider Intelligence/eMarketer. "US Digital Ad Spending." (March 2022); "US Mobile Ad Spending." (March 2022); "US Traditional Ad Spending." (March 2022). (2022c)

Insider Intelligence/eMarketer. "Worldwide Amazon Retail Ecommerce Sales" (June 2022d).

FedEx. "Company Structure & Facts." Fedex.com (accessed June 7, 2022).

Friedman, Thomas. *The World Is Flat*. New York: Picador, 2007.

Gartner Inc. "Gartner Forecasts Worldwide IT Spending to Exceed $4 Trillion in 2022." Gartner.com (October 20, 2021).

Gassmann, Peter, Casey Herman, and Colm Kelly. "Are You Ready for the ESG Revolution." Pwc.com (June 14, 2021).

Georgetown University Center on Education and the Workforce. *Workplace Basics: The Competencies Employers Want*. (2020)

Henisz, Witold, Tim Koller, and Robin Nutall. "Five Ways That ESG Creates Value." Mckinsey.com (2019).

KPMG. "The ESG Imperative for Technology Companies." Home.Kpmg.com (2020).

Laudon, Kenneth C. *Computers and Bureaucratic Reform*. New York: Wiley, 1974.

McKenna, Nick. "Explaining the Relationship Between IoT, Big Data, and the Cloud."Mckennaconsultants.com (May 5, 2021).

Organisation for Economic Co-operation and Development (OECD). "Trade in Goods and Services." Data.oecd.org (accessed June 7, 2022).

"Personalize Your Customer Journey Like the Ritz Carlton." Hospitalityinsights.ehl.edu (accessed June 7, 2022).

Porteus, Chris. "97% of Fortune 500 Companies Rely on Social Media. Here's How You Should Use It for Maximum Impact." Entrepeneur.com (March 18, 2021).

Rose, Stephen J. "Do Not Blame Trade for the Decline in Manufacturing Jobs." Center for Strategic & International Studies Csis.org (October 2021).

Statista Research Department. "Market Size of the Management Consulting Services Industry in 2020 with a Forecast for 2021 and 2025." Statista.com (June 10, 2021).

TitanHQ. "Email Retention Laws in the United States." Spamtitan.com (February 22, 2021).

United Parcel Services, Inc. "Annual Report on Form 10-K for the Fiscal Year Ended December 31, 2021." Sec.gov (February 22, 2022).

Vailshery, Lionel. "IoT and Non-IoT Connections Worldwide 2010–2025." Statista.com (March 8, 2021).

Verisign. "The Domain Name Industry Brief." Verisign.com (April 15, 2022).

CHAPTER

2

# Global E-business and Collaboration

## LEARNING OBJECTIVES

After completing this chapter, you will be able to:

- **2-1** Identify the major features of a business that are important for understanding the role of information systems.
- **2-2** Explain how information systems serve different management groups in a business and how systems that link the enterprise improve organizational performance.
- **2-3** Understand why systems for collaboration, social business, and knowledge management are so important and the technologies they use.
- **2-4** Describe the role of the information systems function in a business.
- **2-5** Understand how MIS can help your career.

## CHAPTER CASES

- Microsoft Teams Helps Toyota Motor North America (TMNA) Do Even Better
- Carbon Lighthouse Lights Up with the Internet of Things (IoT), Big Data, and Cloud Computing
- Zoom: Quality Videoconferencing for Every Budget
- How Much Does Technology Help Collaboration?

**MyLab MIS**

- **Video Case:** How Slack Is Preparing for the Future of Work
- Discussion Questions: 2-5, 2-6, 2-7
- Hands-On MIS Projects: 2-8, 2-9, 2-10, 2-11
- eText with Figure Videos

## MICROSOFT TEAMS HELPS TOYOTA MOTOR NORTH AMERICA (TMNA) DO EVEN BETTER

Toyota Motor North America (TMNA) oversees all operations of the Toyota Motor Corporation in Canada, Mexico, and the United States. Its operations include research and development, manufacturing, sales, marketing, and after-sales and corporate functions. The company is headquartered in Plano, Texas, with 14 sales regions, 14 manufacturing plants, 1,500 dealerships, 36,000 direct employees, and over 30 million vehicles built in the United States.

Toyota is known for its reliable vehicles and its manufacturing principles and management philosophy of continuous improvement and respect for people. The Toyota Way always looks for ways to improve operations, innovate, and collaborate. New digital tools are empowering employees to work more efficiently by promoting teamwork, gradual improvement of everyday business processes, and increased engagement.

When the Covid-19 pandemic shut TMNA's factories and forced office workers to work from home, TMNA used Microsoft Teams to keep the company running in a safe work environment. Microsoft Teams is a business communication platform developed by Microsoft as part of the Microsoft 365 family of products. Teams has capabilities for workspace chat, audio and videoconferencing, screen sharing, online meetings, file storage, and communication through the public switched telephone network. Teams have channels, which are conversation boards for teammates. All team members can view and add to different conversations for which they are authorized. Chat can take place between teams, groups, and individuals.

Adopting Teams helped TMNA create a more uniform and secure work environment, replacing ad hoc videoconferencing tools such as Zoom that had been adopted unofficially by different groups in the organization with a single software platform. TMNA's implementation of Teams included Teams Phone, which enables employees to make, receive, and transfer phone calls from landlines and mobile devices over the public switched telephone network, all within Teams. IT support employees are able to contact the right people to respond to problems more quickly. Previously they struggled to reach colleagues using a global address book.

© Fizkes/Shutterstock

After a large number of office-based employees started working remotely, TMNA used Teams for all of its meeting and collaboration activities. For example, a global meeting on enterprise resource planning (ERP) used Teams live events to bring together Toyota representatives from six continents to share presentations and collaborate on best practices. Before adopting Teams, it would have taken TMNA months to schedule a meeting with representatives from every continent, and the cost would be very high.

Employees throughout the company used Teams to work together virtually and drive productivity. Before the Covid-19 pandemic, high-level managers inspected manufacturing plants in person. After the onset of the pandemic, inspections went virtual. Senior VPs and executives have daily Teams meetings with plant floor members where they can actually see what's taking place on the factory floor. Each plant has a cart with an iPad and external headset that is moved around. TMNA implemented a new process in a plant using Teams meetings to show the process in action for the entire executive team, which felt the Teams presentation was more efficient and cost-effective than an in-person meeting.

TMNA's corporate culture values employee wellness and uses Teams for this purpose. During the pandemic shutdown, managers used Teams to connect with employees and ensure everyone was doing well working from home. Meeting experiences in Teams have lessened the need for business travel, which reduces stress as well as costs. Teams has empowered employees to support each other as they solve problems and produce more value for the firm.

Sources: "Driving Employee Engagement with Microsoft Teams Accelerates Business at Toyota Motor North America," Customers.microsoft.com, accessed July 7, 2022; "Toyota Motor North America Migrates 40,000 to Microsoft Teams in Three Weeks," Sitetechservices.com, November 30, 2021.

---

TMNA's experience illustrates how much organizations today rely on information systems to improve their performance and remain competitive. It also shows how much systems supporting collaboration and teamwork make a difference in an organization's ability to innovate, execute, grow, profit, and survive.

The chapter-opening diagram calls attention to important points raised by this case and this chapter. TNMA is known for its focus on quality, efficiency, continuous improvement, and employee well-being. Corporate culture encourages everyone in the company to do their best and to do better every day. The challenge was to continue to operate a very large multinational company following these principles when the Covid-19 pandemic forced factories and offices to shut down worldwide. This prevented employees and managers in this far-flung company from freely sharing information and performing job tasks, impacting production, sales, and earnings.

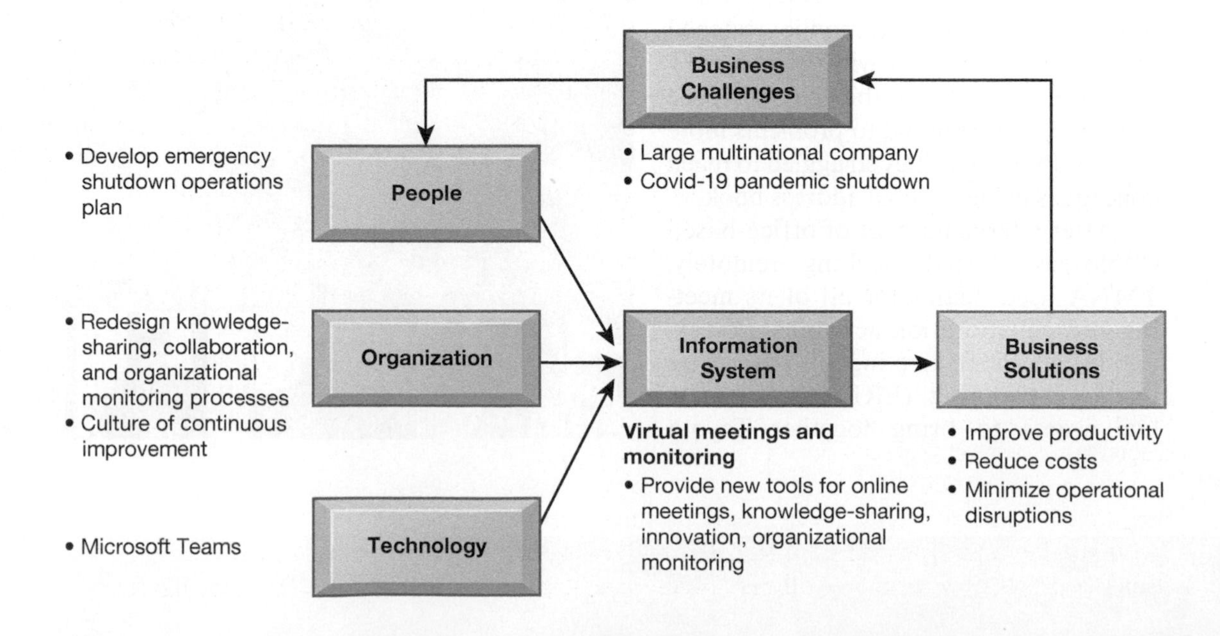

TMNA management found that the best solution was to deploy new technology to help the entire workforce to operate remotely. The company adopted Microsoft Teams communication and collaboration tools to increase employee collaboration and engagement and monitor operations from afar. Virtual meeting and messaging tools actively engaged employees and enabled them to obtain more knowledge from colleagues and managers. There is now more effective sharing of employee knowledge.

New technology alone would not have solved TMNA's problem. To make the solution effective, TMNA had to change its business processes for knowledge dissemination and collaborative work, and the new technology made these changes possible.

Here are some questions to think about: How did Microsoft Teams make it possible for TMNA workers to work remotely? How effective was Teams in promoting collaboration and employee engagement? How did using Microsoft Teams reinforce corporate culture and change the way work was performed at TMNA?

## 2-1 Identify the major features of a business that are important for understanding the role of information systems.

A **business** is a formal organization whose aim is to produce products or provide services for a profit—that is, to sell products or services at a price greater than the costs of production. Customers are willing to pay this price because they believe they receive a value greater than or equal to the sale price. Business firms purchase inputs and resources from the larger environment (suppliers who are often other firms). Employees of the business firm transform these inputs by adding value to them in the production process.

There are, of course, nonprofit firms and government agencies that are complex formal organizations that produce services and products but do not operate to generate a profit. Nevertheless, even these organizations consume resources from their environments, add value to these inputs, and deliver their outputs to constituents and customers. In general, the information systems found in government and nonprofit organizations are remarkably similar to those found in private industry.

### ORGANIZING A BUSINESS: BASIC BUSINESS FUNCTIONS

Imagine you want to set up your own business. Simply deciding to go into business is the most important decision, but next is the question of what product or service to produce (and hopefully sell). The decision of what to produce is called a *strategic choice* because it determines your likely customers, the kinds of employees you will need, the production methods and facilities required, the marketing themes, and many other choices.

Examine Figure 2.1, which illustrates the core business functions surrounding the creation and sale of a product or service. First, you need to develop a manufacturing and/or production division—an arrangement of people, physical infrastructure, and business processes (procedures) that will produce the product or service. Second, you need a sales and marketing group to attract customers, sell the product or service, and maintain a relationship with the customer. Third, once you generate sales, you will need a finance and accounting group to keep track of financial transactions, such as orders, invoices, disbursements, and payroll. In addition, this group will seek out sources of credit and finance. Finally, you will need a group of people to focus on recruiting, hiring, training, and retaining employees.

Note that if you are an entrepreneur or your business is small, with only a few employees, you would not need and probably could not afford all these separate groups of people. Instead, in a small firm, you would be performing all these functions yourself or with a few others. In any event, even in small firms, the four basic

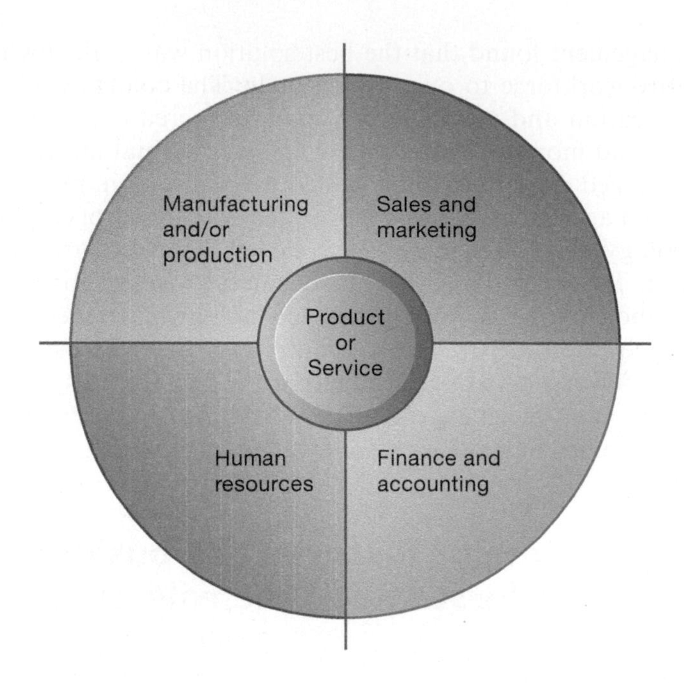


**Figure 2.1**
The Four Major Functions of a Business
*Every business, regardless of its size, must perform four functions to succeed. It must produce the product or service, market and sell the product or service, keep track of accounting and financial transactions, and perform basic human resources tasks such as hiring and retaining employees.*

functions of a firm are required. Larger firms often will have separate departments for each function: manufacturing and/or production, sales and marketing, finance and accounting, and human resources.

Figure 2.1 is also useful for thinking about the basic entities that make up a business. The five basic entities in a business with which it must deal are: suppliers, customers, employees, invoices/payments, and, of course, products and services. A business must manage and monitor many other components, but these are the basic ones at the foundation of any business.

## BUSINESS PROCESSES

Once you identify the basic business functions and entities for your business, your next job is to describe exactly how you want your employees to perform these functions. What specific tasks do you want your sales personnel to perform, in what order, and on what schedule? What steps do you want production employees to follow as they transform raw resources into finished products? How will customer orders be fulfilled? How will vendor bills be paid?

The actual steps and tasks that describe how work is organized in a business are called **business processes**. A business process is a logically related set of activities that defines how specific business tasks are performed. Business processes also refer to the unique ways in which work, information, and knowledge are coordinated in a specific organization.

Every business can be seen as a collection of business processes. Some of these processes are part of larger, encompassing processes. Many business processes are tied to a specific functional area. For example, the sales and marketing function would be responsible for identifying customers, and the human resources function would be responsible for hiring employees. Table 2.1 describes some typical business processes for each of the functional areas of business.

Other business processes cross many functional areas and require coordination across departments. Examine Figure 2.2, which illustrates the seemingly simple business process of fulfilling a customer order for a physical product. When the sales department receives a customer order, it generates and then submits a sales order to accounting to ensure that the customer can pay for the order by either a credit verification or a request for immediate payment prior to shipping. Once the customer credit

**TABLE 2.1**
Examples of Functional Business Processes

| Functional Area | Business Process |
| --- | --- |
| Manufacturing and production | Assembling the product<br>Checking for quality<br>Producing bills of materials |
| Sales and marketing | Identifying customers<br>Making customers aware of the product<br>Selling the product |
| Finance and accounting | Paying creditors<br>Creating financial statements<br>Managing cash accounts |
| Human resources | Hiring employees<br>Evaluating employees' job performance<br>Enrolling employees in benefits plans |

is established, the production department has to pull the product from inventory or produce the product. Next, the product needs to be shipped (which may require working with a logistics firm such as UPS or FedEx). The accounting department then generates a bill or invoice and sends a notice to the customer, indicating that the product has shipped. Sales has to be notified of the shipment and prepare to support the customer by answering calls or fulfilling warranty claims. View the Figure 2.2 video in the eText for an animated and more detailed discussion of this figure.

What at first appears to be a simple process—fulfilling an order—turns out to be a very complicated series of business processes that require the close coordination of major functional groups in a firm. Moreover, to perform all these steps efficiently in the order fulfillment process requires the rapid flow of a great deal of information within the firm, with business partners such as delivery firms, and with the customer. The particular order fulfillment process we have just described is not only *cross-functional*, it is also *interorganizational* because it includes interactions with delivery firms and customers who are outside the boundaries of the organization. Ordering raw materials or components from suppliers would be another interorganizational business process.

To a large extent, the efficiency of a business firm depends on how well its internal and interorganizational business processes are designed and coordinated.

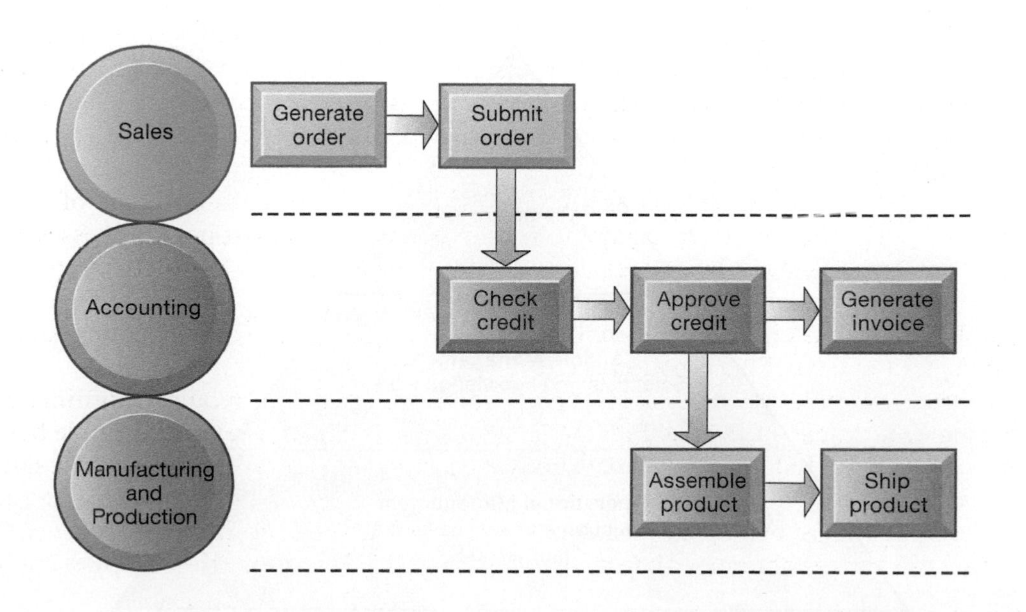


**Figure 2.2**
The Order Fulfillment Process
*Fulfilling a customer order involves a complex set of steps that requires the close coordination of the sales, accounting, and manufacturing functions.*

A company's business processes can be a source of competitive strength if they enable the company to innovate or to execute better than its rivals. Business processes can also be liabilities if they are based on outdated ways of working that impede organizational responsiveness and efficiency.

### How Information Technology Enhances Business Processes

Exactly how do information systems enhance business processes? Information systems automate many steps in business processes that were formerly performed manually, such as checking a client's credit or generating an invoice and shipping order. Today, however, information technology can do much more. New technology can actually change the flow of information, making it possible for many more people to access and share information, replacing sequential steps with tasks that can be performed simultaneously and eliminating delays in decision making. It can even transform the way the business works and drive new business models. Ordering a book online from Amazon.com and streaming a song from Apple Music are new business processes based on new business models that are inconceivable without information technology.

That's why it's so important to pay close attention to business processes, both in your information systems course and in your future career. By analyzing business processes, you can achieve a clear understanding of how a business actually works. Moreover, by conducting a business process analysis, you will also begin to understand how to change the business to make it more efficient or effective. Throughout this book, we examine business processes with a view to understanding how they might be changed, or replaced, by using information technology to achieve greater efficiency, innovation, and customer service. Chapter 3 discusses the business impact of using information technology to redesign business processes.

## MANAGING A BUSINESS AND FIRM HIERARCHIES

Each business function has its own goals and processes, and they obviously need to cooperate for the whole business to succeed. Business firms, like all organizations, achieve coordination by hiring managers whose responsibility is to ensure that all the various parts of an organization work together. Firms coordinate the work of employees in various divisions by developing a hierarchy in which authority (responsibility and accountability) is concentrated at the top.

Examine Figure 2.3, which illustrates the hierarchy of management in the form of a pyramid. **Senior management**, which makes long-range strategic decisions about

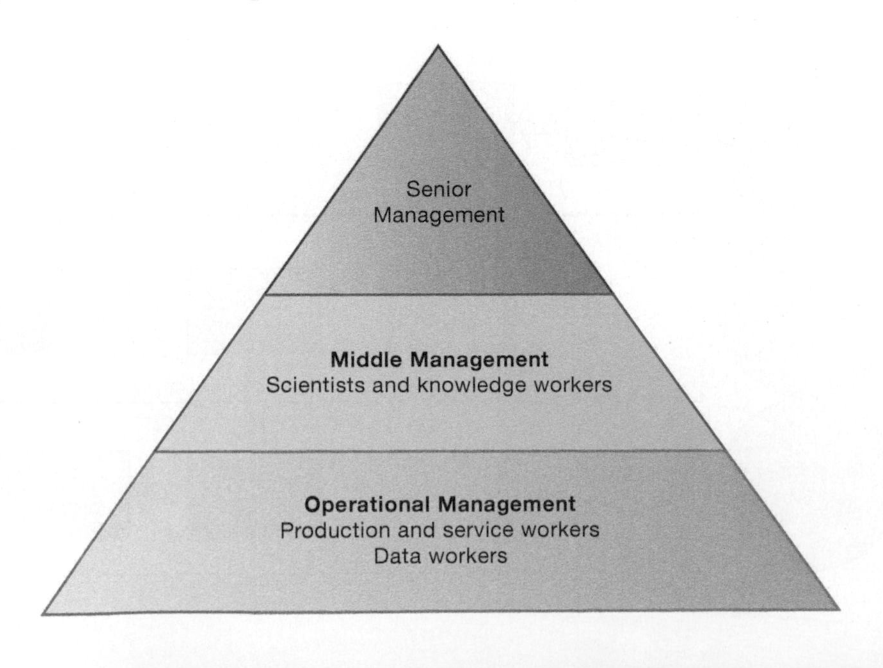


**Figure 2.3**
Levels in a Firm
*Business organizations are hierarchies consisting of three principal levels: senior management, middle management, and operational management. Information systems serve each of these levels. Scientists and knowledge workers often work with middle management.*

products and services and ensures financial performance of the firm, is at the top of the pyramid. **Middle management**, which carries out the programs and plans of senior management, forms the middle layer. **Knowledge workers**, such as engineers, scientists, or architects, design products or services and create new knowledge for the firm, and frequently work with middle management. **Operational management**, which is responsible for monitoring the daily activities of the business, comprises the base of the pyramid. Operational management typically oversees **data workers**, such as bookkeepers or clerks, who assist with administrative work at all levels of the firm, and **production or service workers**, who actually produce the product or deliver the service.

Each of these groups has different needs for information given their different responsibilities. Senior managers need summary information that can quickly inform them about the overall performance of the firm, such as gross sales revenues, sales by product group and region, and overall profitability. Middle managers need more specific information about the results of specific functional areas and departments of the firm such as sales contacts by the sales force, production statistics for specific factories or product lines, employment levels and costs, and sales revenues for each month or even each day. Operational managers need transaction-level information such as the number of parts in inventory each day or the number of hours logged on Tuesday by each employee. Knowledge workers may need access to external scientific databases or internal databases with organizational knowledge. Finally, production workers need access to information from production machines, and service workers need access to customer records to take orders and answer questions from customers.

## THE BUSINESS ENVIRONMENT

So far, we have talked about business as if it operated in a vacuum, but nothing could be further from the truth. In fact, business firms depend heavily on their environments to supply capital, labor, customers, new technology, services and products, stable markets and legal systems, and general educational resources. Even a pizza parlor cannot survive long without a supportive environment that delivers the cheese, tomato sauce, and flour!

Examine Figure 2.4, which depicts the key actors in the environment of every business. In their immediate environment, firms need to track and share information

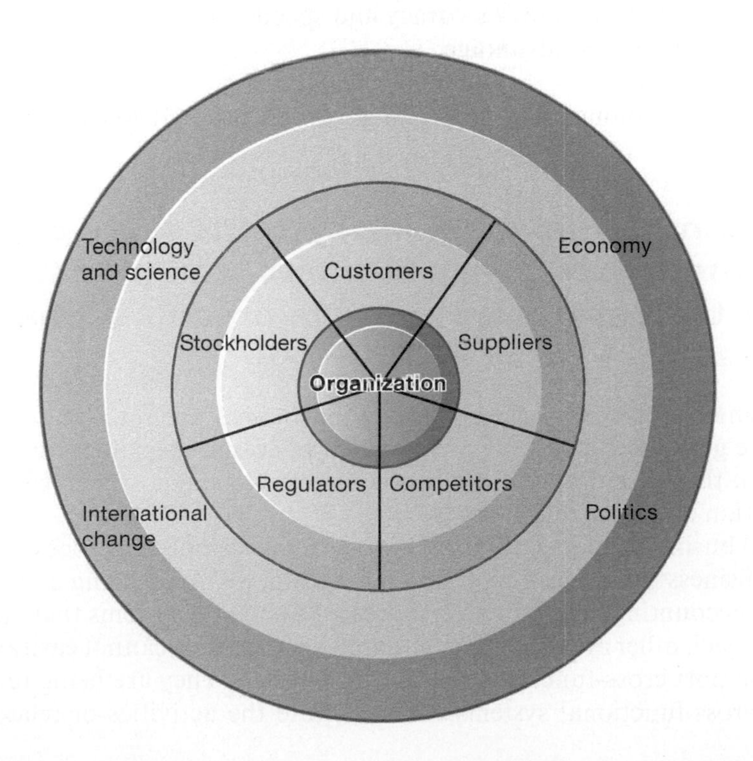


**Figure 2.4**
**The Business Environment**
*To be successful, an organization must constantly monitor and respond to—or even anticipate—developments in its environment. A firm's environment includes specific groups with which the business must deal directly, such as customers, suppliers, competitors, regulators, and stockholders, as well as the broader general environment, including socioeconomic trends, political conditions, technological innovations, and global events.*

with suppliers, customers, stockholders, and regulators as well as monitor and respond to the activities of their competitors. But just doing so is not enough. To stay in business, a firm must also monitor and respond to changes in its broader environment. For instance, a firm must monitor and respond to political shifts, changes in the overall economy (such as changes in labor rates and price inflation), new technologies and scientific developments, and changes in the global business environment (such as foreign exchange rates).

Business environments are constantly changing; new developments in technology, politics, customer preferences, and regulations happen all the time. In general, when businesses fail, it is often because they failed to respond adequately to changes in their environments.

Changes in technology, such as the Internet, are forcing entire industries and leading firms to change their business models or suffer failure. Apple's iTunes and other online music services have made the music industry's traditional business model based on distributing music on CDs obsolete. Traditional cameras with film have been largely supplanted by digital photography, and digital cameras themselves have lost ground to iPhones and other mobile devices with cameras.

### THE ROLE OF INFORMATION SYSTEMS IN A BUSINESS

From the brief review of business functions, entities, and environments, you can see the critical role that information plays in the life of a business. Up until the mid-1950s, firms managed all this information and information flow with paper records. Since then, more and more business information, and the flow of information among key business actors in the environment, has been moved from manual to digital systems.

Businesses invest in information systems as a way to manage their internal production functions and cope with the demands of key actors in their environments. Specifically, as we noted in Chapter 1, firms invest in information systems for the following business objectives:

- To achieve operational excellence (productivity, efficiency, agility)
- To develop new products and services
- To attain customer intimacy and service (continuous marketing, sales, and service; customization and personalization)
- To improve decision making (accuracy and speed)
- To achieve competitive advantage
- To ensure survival
- To promote environmental, social, and governance (ESG) goals

## 2-2 Explain how information systems serve different management groups in a business, and how systems that link the enterprise improve organizational performance.

Now it is time to look more closely at how businesses use information systems to achieve these goals. Because there are different interests, specialties, and levels in an organization, there are different kinds of systems. No single system can provide all the information an organization needs.

A typical business organization will have systems supporting processes for each of the major business functions—sales and marketing, manufacturing and production, finance and accounting, and human resources. Functional systems that operate independently of each other are becoming outdated because they cannot easily share information to support cross-functional business processes. They are being replaced with large-scale cross-functional systems that integrate the activities of related business processes and organizational units. We describe these integrated cross-functional applications later in this section.

A typical firm will also have different systems supporting the decision-making needs of each of the main management groups described earlier. Operational management, middle management, and senior management each use a specific type of system to support the decisions they must make to run the company. Let's look at these systems and the types of decisions they support.

## SYSTEMS FOR DIFFERENT MANAGEMENT GROUPS

A business firm has systems to support decision making and work activities at different levels of the organization. They include transaction processing systems and systems for business intelligence.

### Transaction Processing Systems

Operational managers need systems that keep track of the elementary activities and transactions of the organization, such as sales, receipts, cash deposits, payroll, credit decisions, and the flow of materials in a factory. **Transaction processing systems (TPS)** provide this kind of information. A transaction processing system is a computerized system that performs and records the daily routine transactions necessary to conduct business, such as sales order entry, hotel reservations, payroll, employee record keeping, and shipping.

The principal purpose of systems at this level is to answer routine questions and to track the flow of transactions through the organization. How many parts are in inventory? What happened to Mr. Garcia's payment? To answer these kinds of questions, information generally must be easily available, current, and accurate.

At the operational level, tasks, resources, and goals are predefined and highly structured. The decision to grant credit to a customer, for instance, is made by a lower-level supervisor according to predefined criteria. All that must be determined is whether the customer meets the criteria.

Managers need TPS to monitor the status of internal operations and the firm's relations with the external environment. TPS are also major producers of information for the other systems and business functions. Examine Figure 2.5, which illustrates a TPS for payroll processing. A payroll system keeps track of money paid to employees. An employee time sheet with employee data such as the employee's name, identification number, and number of hours worked per week represents a single transaction for this system. Once this transaction is input in the payroll system, it updates the system's master employee file (or database—see Chapter 6) that permanently maintains employee payroll data for the organization. The data in the system can typically be accessed and queried online and, in addition to being used to generate employee paychecks, can be combined in different ways to create reports of interest to management and government agencies. For instance, payroll data can be used to supply employee payment history data for insurance, pension, and other benefits calculations to the firm's human resources function and employee payment data to government agencies such as the US Internal Revenue Service and Social Security Administration. The payroll system, along with other accounting TPS, also supplies data to the company's general ledger system, which is responsible for maintaining records of the firm's income and expenses and for producing reports such as income statements and balance sheets. View the Figure 2.5 video in the eText for an animated and more detailed discussion of this figure.

TPS failure for even a short period of time can have a significant negative impact on a firm as well as on other firms linked to it. For example, in May 2021, travelers arriving at airports across the globe were greeted with long lines and flight delays due to a system outage at Sabre, a technology company that provides airline reservation and operational systems, leaving airlines with no means to check in passengers or dispatch flights.

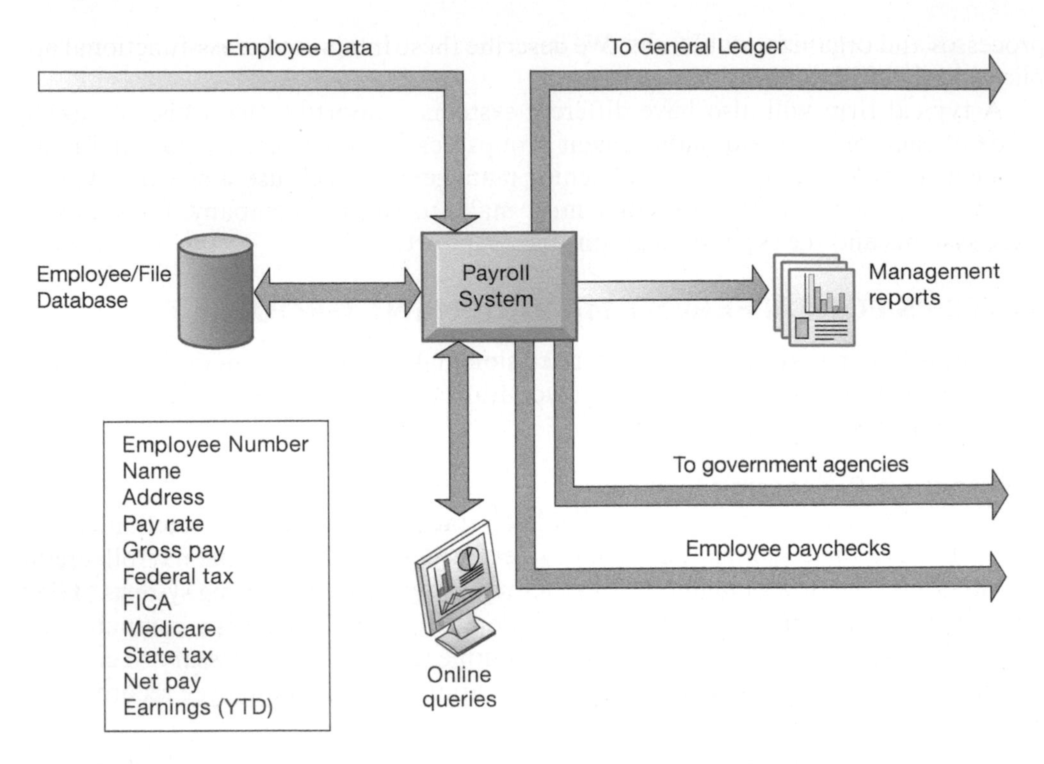


**Figure 2.5**
A Payroll TPS
*A TPS for payroll processing captures employee payment transaction data (such as a timecard). System outputs include online and hard copy reports for management and employee paychecks.*

## Systems for Business Intelligence

Firms also have business intelligence systems that focus on delivering information to support management decision making. **Business intelligence** is a contemporary term for data and software tools for organizing, analyzing, and providing access to data to help managers and other enterprise users make more informed decisions. Business intelligence addresses the decision-making needs of all levels of management. This section provides a brief introduction to business intelligence. You'll learn more about this topic in Chapters 6 and 11.

Business intelligence systems for middle management help with monitoring, controlling, decision-making, and administrative activities. In Chapter 1, we defined management information systems as the study of information systems in business and management. The term **management information systems (MIS)** also designates a specific category of information systems serving middle management. MIS provide middle managers with reports about the organization's current performance. Managers use this information to monitor and control the business and predict future performance.

MIS summarize and report on the company's basic operations using data supplied by transaction processing systems. The basic transaction data from TPS are compressed and usually presented in reports that are produced on a regular schedule. Today, many of these reports are delivered online. Examine Figure 2.6, which shows how a typical MIS obtains transaction-level order, production, and accounting data from order processing, manufacturing resource planning, and general ledger transaction processing systems, respectively. The MIS transforms this data into MIS files such as sales data, unit production cost data, product change data, and expense data that managers can access via online displays and dashboards and use to generate various reports.

Next, examine Figure 2.7, which shows a sample report from the system illustrated in Figure 2.6. MIS typically provide answers to routine questions that have been specified in advance and have a predefined procedure for answering them. In this instance, the MIS report compares total annual sales figures for specific products (carpet cleaner and room freshener) to planned targets. These systems generally are

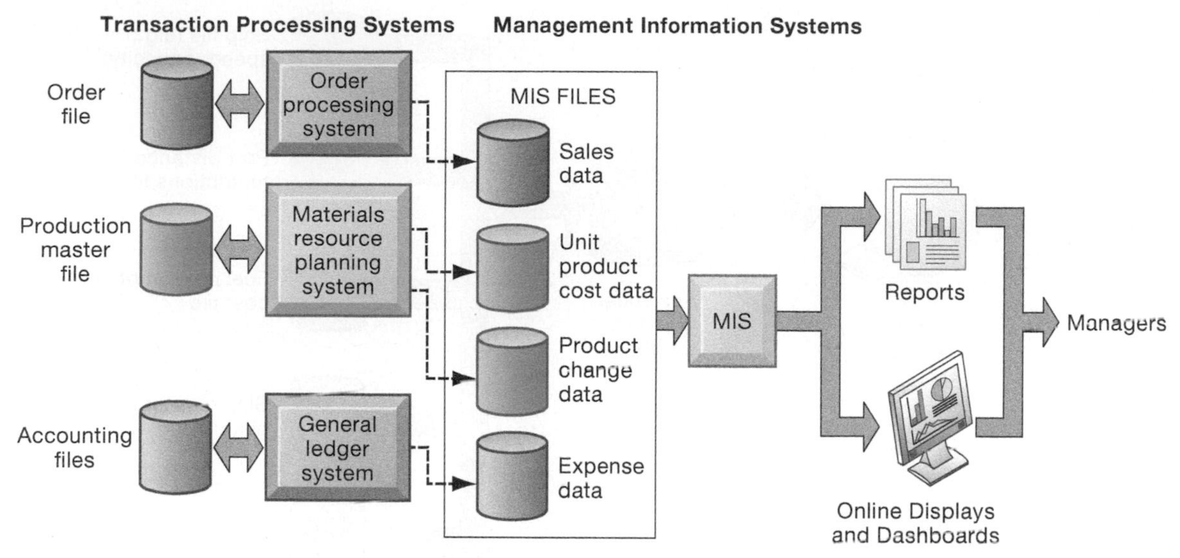


**Figure 2.6**
How Management Information Systems Obtain Their Data from the Organization's TPS
*In the system illustrated by this diagram, three TPS supply summarized transaction data to the MIS reporting system at the end of the time period. Managers gain access to the organizational data through the MIS, which provides them with the appropriate reports.*

not flexible and have little analytical capability. Most MIS use simple routines, such as summaries and comparisons, as opposed to sophisticated mathematical models or statistical techniques.

Other types of business intelligence systems support more nonroutine decision making. **Decision-support systems (DSS)** focus on problems that are unique and rapidly changing, for which the procedure for arriving at a solution may not be fully predefined in advance. They try to answer questions such as these: What would be the impact on production schedules if we were to double sales in the month of December? What would happen to our return on investment if a factory schedule were delayed for six months?

Although DSS use internal information from TPS and MIS, they often bring in information from external sources, such as current stock prices or product prices of competitors. Super-user managers and business analysts who want to use sophisticated analytics and models to analyze data employ these systems.

Consolidated Consumer Products Corporation Sales by Product and Sales Region: 2022

| PRODUCT CODE | PRODUCT DESCRIPTION | SALES REGION | ACTUAL SALES | PLANNED | ACTUAL versus PLANNED |
|---|---|---|---|---|---|
| 4469 | Carpet Cleaner | Northeast | 4,066,700 | 4,800,000 | 0.85 |
| | | South | 3,778,112 | 3,750,000 | 1.01 |
| | | Midwest | 4,867,001 | 4,600,000 | 1.06 |
| | | West | 4,003,440 | 4,400,000 | 0.91 |
| | TOTAL | | 16,715,253 | 17,550,000 | 0.95 |
| 5674 | Room Freshener | Northeast | 3,676,700 | 3,900,000 | 0.94 |
| | | South | 5,608,112 | 4,700,000 | 1.19 |
| | | Midwest | 4,711,001 | 4,200,000 | 1.12 |
| | | West | 4,563,440 | 4,900,000 | 0.93 |
| | TOTAL | | 18,559,253 | 17,700,000 | 1.05 |

**Figure 2.7**
Sample MIS Report
*This report, showing summarized annual sales data, was produced by the MIS in Figure 2.6.*

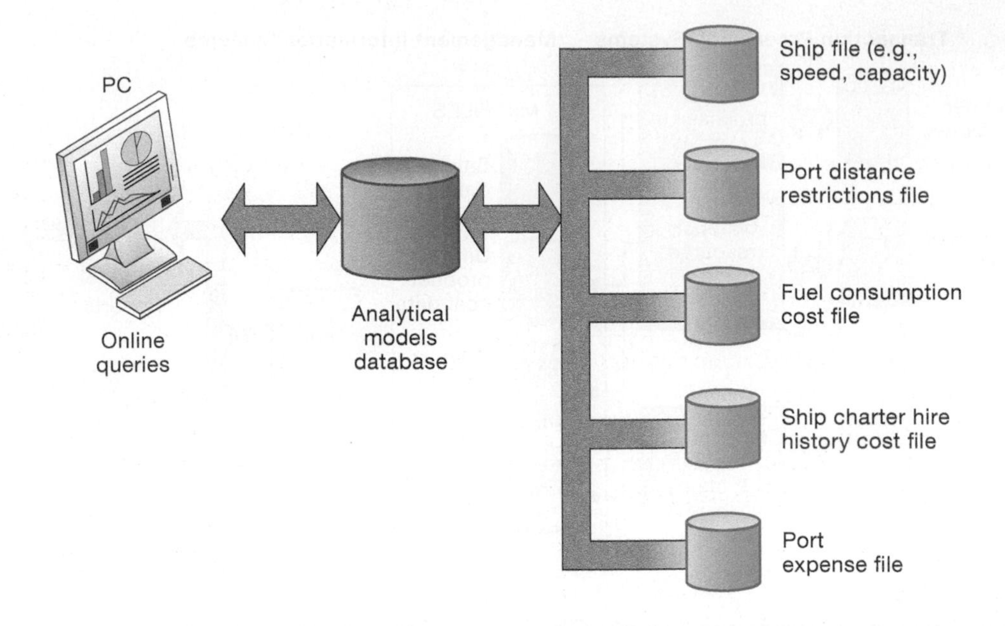


**Figure 2.8**
Voyage-Estimating Decision-Support System
*This DSS operates on a powerful PC. Managers who must develop bids on shipping contracts use it daily.*

Examine Figure 2.8, which illustrates an interesting, highly-specific and powerful voyage-estimating DSS used by a large global shipping company that primarily carries bulk cargoes of coal, oil, ores, and finished products for its parent company. The firm owns some vessels, charters others, and bids for shipping contracts in the open market to carry general cargo. The voyage-estimating DSS calculates financial and technical voyage details. Financial calculations include ship/time costs (fuel, labor, capital), freight rates for various types of cargo, and port expenses. Technical details include myriad factors, such as ship cargo capacity, speed, port distances, fuel and water consumption, and loading patterns (location of cargo for different ports).

The system can answer questions such as the following: Given a customer delivery schedule and an offered freight rate, which vessel should be assigned at what rate to maximize profits? What is the optimal speed at which a particular vessel can optimize its profit and still meet its delivery schedule? What is the optimal loading pattern for a ship bound for the US West Coast from Malaysia? The system operates on a powerful desktop personal computer, providing a system of menus that makes it easy for users to enter data or obtain information.

The voyage-estimating DSS we have just described draws heavily on models. Other business intelligence systems are more data-driven, focusing instead on extracting useful information from Big Data, including the Internet of Things (IoT). For example, major ski resort companies such as Alterra Mountain Company and Vail Resorts collect and store large amounts of customer data from call centers, lift tickets, lodging and dining reservations, ski schools, and ski equipment rental stores. They use special software to analyze these data to determine the value, revenue potential, and loyalty of each customer to help managers make better decisions about how to target their marketing programs.

The Spotlight on Organizations case provides another example. Carbon Lighthouse offers services that use Big Data, the Internet of Things (IoT), and cloud computing to analyze data from large commercial building heating and cooling systems to improve decisions about how to reduce energy consumption and lower costs. Users of these services are able to reduce carbon emissions using existing equipment while saving on energy costs and realizing ESG goals.

Business intelligence systems also address the decision-making needs of senior management. Senior managers need systems that focus on strategic issues and long-term trends, both in the firm and in the external environment. They are concerned with questions such as: What will employment levels be in five years? What are the long-term industry cost trends? What products should we be making in five years?

## SPOTLIGHT ON: ORGANIZATIONS

## Carbon Lighthouse Lights Up with the Internet of Things (IoT), Big Data, and Cloud Computing

According to the US Environmental Protection Agency (EPA), more carbon emissions come from operating commercial, residential, and industrial properties than from transportation. Emissions from cars, trucks, and aircraft account for only 29 percent of carbon emissions, while 68 percent comes from generation of electric power, (manufacturing) industry, commercial and residential property, and agriculture.

San Francisco-based Carbon Lighthouse is tackling this problem head-on by focusing on reducing carbon emissions produced by commercial and industrial buildings. The company offers energy savings services that make it profitable for building owners to reduce their energy consumption using their existing equipment. Carbon Lighthouse has reduced more than 260,000 metric tons of emissions, equivalent to the energy produced by 18 power plants, while providing $250 million in savings to its clients.

How is this possible? Many businesses complain that it costs more money to be environmentally responsible. This does not have to be so. Acting on findings from advanced data analytics, commercial and building owners can reduce carbon emissions while producing cost savings.

Carbon Lighthouse uses a software platform called Carbon Lighthouse Unified Engineering System (CLUES) to analyze more than 100 million square feet of clients' commercial and industrial real estate and 5 billion data points. CLUES collects the building data in real time as it is generated. It then develops models that lead to insights about how to reduce the client's carbon emissions and how much money will be saved. Carbon Lighthouse uses Amazon's cloud services to run its computers and store data such as building addresses and square footage, along with time-series data such as the data collected on-site at buildings.

When a new client contracts for Carbon Lighthouse services, the company sends a sensor kit to start new data streaming from the client's buildings. The recommendations generated are more complex than just turning off lights or lowering the heat. Findings are more on the order of directing control systems so specific valves automatically open and shut at a specific time or when building temperature reaches a specific level. These are not major changes such as replacing windows or boilers; the recommendations focus instead on making existing equipment work a little better. Internet of Things (IoT) sensors, Big Data, cloud computing, and powerful data analysis software rather than expensive capital upgrades allow Carbon Lighthouse to unlock hidden returns in existing building mechanical systems and translate energy efficiency into tangible long-term savings.

One of Carbon Lighthouse's clients is L&B Realty Advisors, a Dallas-based real estate investment advisor with $9 billion under management. L&B has over 50 years' experience acquiring, managing, and selling real estate for its clients. For L&B, sustainability is a key consideration when evaluating and managing investments. L&B worked with Carbon Lighthouse on its One Biscayne Tower property. The building was already rated as highly energy-efficient, with engineering staff using a dashboard to track overall utility consumption.

Carbon Lighthouse first established a data stream that fed detailed heating, ventilation, and air conditioning (HVAC); lighting; and occupancy data into the CLUES platform. The company then worked with the L&B property and engineering teams to assess One Biscayne Tower's existing building systems and look for energy savings and operational improvements. CLUES analysis of the data stream identified and quantified new HVAC control optimizations as well as lighting retrofits that property and building engineering teams had long sought to implement. (A lighting retrofit is an upgrade to light fixtures or lamps, increasing energy efficiency.)

The annual expense savings from these energy conservation measures for One Biscayne Tower are financially guaranteed by Carbon Lighthouse. And as CLUES collects more data, its algorithms become more accurate and provide additional zone isolation and central plant measures for One Biscayne Tower, generating more energy efficiency and cost savings than projected.

The building engineering and property manager knew what lighting fixtures and aesthetic output they wanted, but they did not have the lighting material wattage details, detailed verification checklists, and financial savings guarantee for almost one dozen lighting retrofit configurations. Carbon Lighthouse was able to apply a precise financial analysis for different product configurations backed with a guarantee of financial results. When the retrofits were complete, CLUES and the data stream confirmed the energy savings from the retrofits were actually realized.

CLUES and the data stream also found new ways to optimize the central plant, including improved control of existing HVAC equipment based on building demand and outdoor air conditions and ability to isolate energy zones after hours and on unoccupied floors. CLUES evaluated over 700 individual zone locations requiring a different amount of air-conditioning at any one time, examining several hundred different ways of controlling air pressure to produce cost savings without affecting tenant comfort. Without Carbon Lighthouse, the building management system (BMS), which controls and monitors a building's ventilation, lighting, power systems, fire systems, and security systems, would have implemented optimization measures at most of the 700 locations using off-the-shelf programming that would not have been as finely tuned as CLUES. This would have eliminated any savings.

Sources: "L&B Realty Advisors," Carbonlighthouse.com, accessed July 5, 2022; Eric Avidon, "AWS Analytics Helps Vendor Fight Climate Change," Searchbusinessanalytics.com, June 9, 2021.

## CASE STUDY QUESTIONS

1. Identify the problem described in this case study. Is it a people problem, an organizational problem, or a technology problem? Explain your answer.
2. What role have the IoT, Big Data analytics, and cloud computing played in developing a solution for this problem?
3. Describe Carbon Lighthouse's problem-solving methodology for reducing both carbon emissions and costs.

**Executive support systems (ESS)** help senior management make these decisions. They address nonroutine decisions requiring judgment, evaluation, and insight because there is no agreed-on procedure for arriving at a solution. ESS present graphs and data from many sources through an interface that is easy for senior managers to use. Often the information is delivered to senior executives through a **portal**, which uses a web interface to present integrated personalized business content.

ESS are designed to incorporate data about external events such as new tax laws or competitors, but they also draw summarized information from internal MIS and DSS. They filter, compress, and track critical data, displaying the data of greatest importance to senior managers. Increasingly, such systems include business intelligence analytics for analyzing trends, forecasting, and drilling down to data at greater levels of detail.

For example, the chief operating officer (COO) and plant managers at Valero, the world's largest independent petroleum refiner, use a Refining Dashboard to display real-time data related to plant and equipment reliability, inventory management, safety, and energy consumption. With the displayed information, management can review the performance of each Valero refinery in the United States and Canada in terms of how each plant is performing compared to the production plan of the firm. The headquarters group can drill down from executive level to refinery level and individual system-operator level displays of performance. Valero's Refining Dashboard is an example of a **digital dashboard**, which displays on a single screen graphs and charts of key performance indicators for managing a company. Digital dashboards are becoming an increasingly popular tool for management decision makers.

© Andrey_Popov/Shutterstock

*A digital dashboard delivers comprehensive and accurate information for decision making, often using a single screen. The graphical overview of key performance indicators helps managers quickly spot areas that need attention.*

## SYSTEMS FOR LINKING THE ENTERPRISE

Reviewing all the types of systems we have just described, you might wonder how a business can manage all the information in these differing systems. You might also wonder how costly it is to maintain so many systems. You might also wonder how all these systems can share information and how managers and employees can coordinate their work. In fact, these are all important questions for businesses today.

### Enterprise Applications

Getting the different kinds of systems in a company to work together has proven a major challenge. Typically, companies are built both through normal organic growth and through acquisition of smaller firms. Over time, companies end up with a collection of systems, most of them older, and face the challenge of getting them all to talk with one another and work together as one corporate system. There are several solutions to this problem.

One solution is to implement **enterprise applications**, which are systems that span functional areas, focus on executing business processes across the business firm, and include all levels of management. Enterprise applications help businesses become more flexible and productive by coordinating their business processes more closely and integrating groups of processes so they focus on efficient management of resources and customer service.

Examine Figure 2.9, which illustrates the interrelationship that the four major types of enterprise applications—enterprise systems, supply chain management systems, customer relationship management systems, and knowledge management systems—have with the business functions and processes of a firm. Each of these enterprise applications integrates a related set of functions and business processes to enhance the performance of the organization as a whole. Figure 2.9 shows that the architecture for these enterprise applications encompasses processes spanning the entire organization and, in some cases, extending beyond the organization to customers, suppliers, and other key business partners.

**Enterprise Systems** Firms use **enterprise systems**, also known as *enterprise resource planning (ERP)* systems, to integrate business processes in manufacturing and production, finance and accounting, sales and marketing, and human resources into a

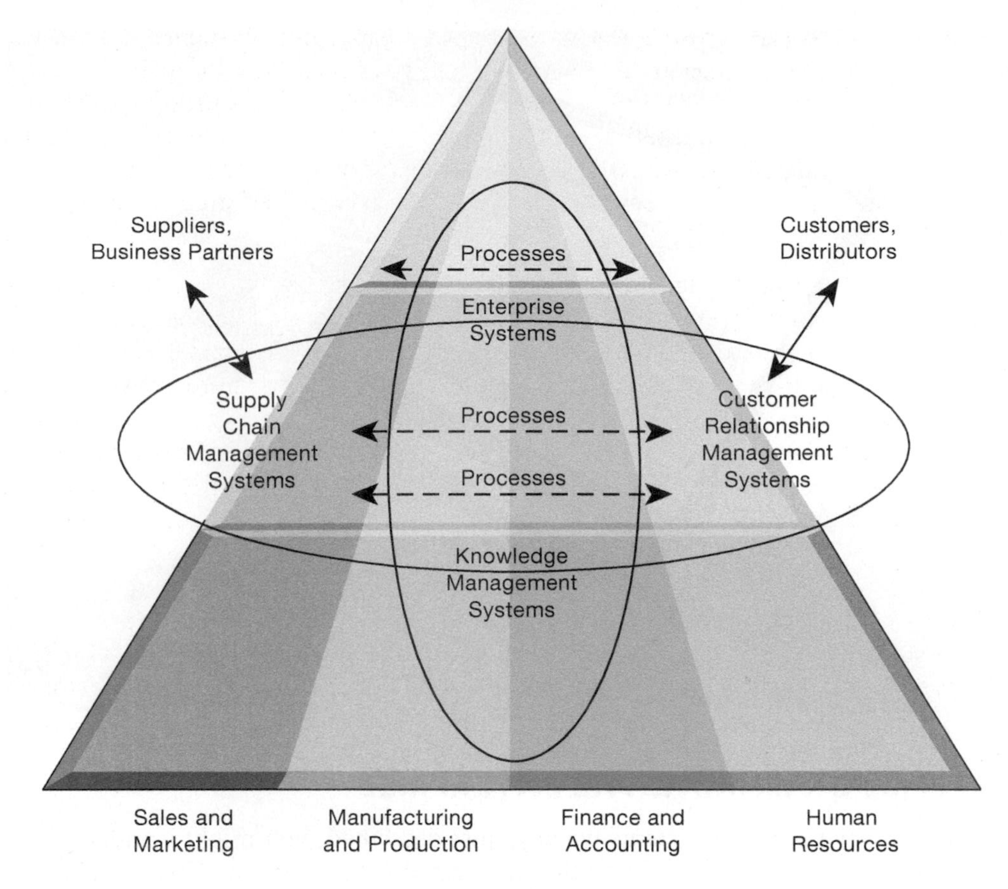


**Figure 2.9**
Enterprise Application Architecture
*Enterprise applications automate processes that span multiple business functions and organizational levels and may extend outside the organization.*

single software system. Information that was previously fragmented in many systems is stored in a single comprehensive data repository where it can be used by many parts of the business.

For example, when a customer places an order, the order data flow automatically to other parts of the company that they affect. The order transaction triggers the warehouse to pick the ordered products and schedule shipment. The warehouse informs the factory to replenish whatever has been depleted. The accounting department is notified to send the customer an invoice. Customer service representatives track the progress of the order through every step to inform customers about the status of their orders. Managers can use firmwide information to make more precise and timely decisions about daily operations and longer-term planning.

**Supply Chain Management Systems** Firms use **supply chain management (SCM) systems** to help manage relationships with their suppliers. These systems help suppliers, purchasing firms, distributors, and logistics companies share information about orders, production, inventory levels, and delivery of products and services so that they can source, produce, and deliver goods and services efficiently. The ultimate objective is to get the right number of their products from their source to their point of consumption in the shortest time and at the lowest cost. These systems increase firm profitability by lowering the costs of moving and making products and by enabling managers to make better decisions about how to organize and schedule sourcing, production, and distribution.

Supply chain management systems are one type of **interorganizational system** because they automate the flow of information across organizational boundaries. You will find examples of other types of interorganizational information systems throughout this text because such systems make it possible for firms to link to their customers and to outsource their work to other companies.

**Customer Relationship Management Systems** Firms use **customer relationship management (CRM) systems** to help manage their relationships with their customers. CRM systems provide information to coordinate all the business processes that deal with customers in sales, marketing, and service to optimize revenue, customer satisfaction, and customer retention. This information helps firms identify, attract, and retain the most profitable customers; provide better service to existing customers; and increase sales.

**Knowledge Management Systems** Some firms perform better than others do because they have better knowledge about how to create, produce, and deliver products and services. **Knowledge management systems (KMS)** collect all relevant knowledge and experience in the firm and make it available wherever and whenever it is needed to improve business processes and management decisions. They also link the firm to external sources of knowledge. We examine enterprise systems and systems for supply chain management and customer relationship management in greater detail in Chapter 9. We discuss knowledge management along with systems for collaboration in Section 2-3.

### Intranets and Extranets

Enterprise applications create deep-seated changes in the way the firm conducts its business, offering many opportunities to integrate important business data into a single system. They are often costly and difficult to implement. Intranets and extranets deserve mention here as alternative tools for increasing integration and expediting the flow of information within the firm and with customers and suppliers.

Intranets are simply internal company websites that are accessible only by employees. The term *intranet* refers to an internal network in contrast to the Internet, which is a public network linking organizations and other external networks. Intranets use the same technologies and techniques as the larger Internet, and they often are simply a private access area in a larger company website. Extranets are company websites that are accessible to authorized vendors and suppliers and often used to coordinate the movement of supplies to the firm's production apparatus.

For example, Cubic Telecom, which provides globally connected software and analytics services for the Internet of Things (IoT) market, implemented an intranet based on the Workvivo employee communications platform to facilitate communication with its employees. The company is headquartered in Dublin and has 180 employees across the globe. Cubic Telecom's management was dissatisfied with email as a communication tool because it discouraged conversation, and management couldn't always tell if it was getting through to people. The intranet solution can replace email and reach everyone equally, while improving engagement (Workvivo, 2022).

We describe the technology for intranets and extranets in more detail in Chapter 7.

## E-BUSINESS, E-COMMERCE, AND E-GOVERNMENT

The systems and technologies we have just described are transforming firms' relationships with customers, employees, suppliers, and logistic partners into digital relationships by using networks and the Internet. So much business is now enabled by or based on digital networks that we use the terms *e-business* and *e-commerce* frequently throughout this text.

**E-business**, or **electronic business**, refers to the use of digital technology and the Internet to execute the major business processes in the enterprise. E-business includes activities for the internal management of the firm and for coordination with suppliers and other business partners. It also includes **e-commerce**, or **electronic commerce**. E-commerce is the part of e-business that deals with buying and selling goods and services via the Internet. It also encompasses activities supporting those market transactions, such as advertising, marketing, customer support, security, delivery, and payment.

The technologies associated with e-business have also brought about similar changes in the public sector. Governments on all levels are using Internet technology to deliver information and services to citizens, employees, and businesses with which they work. **E-government** refers to the application of the Internet and networking technologies to digitally enable government and public sector agencies' relationships with citizens, businesses, and other arms of government. In addition to improving delivery of government services, e-government can empower citizens by giving them easier access to information and the ability to network with other citizens.

## 2-3 Understand why systems for collaboration, social business, and knowledge management are so important and the technologies they use.

With all these systems and information, you might wonder how it is possible to make sense of them. How do people working in firms pull it all together, work toward common goals, and coordinate plans and actions? In addition to the types of systems we have just described, businesses need special systems to support collaboration and teamwork.

### WHAT IS COLLABORATION?

**Collaboration** is working with others to achieve shared and explicit goals. Collaboration focuses on task or mission accomplishment and usually takes place in a business or other organization and between businesses. You collaborate with a colleague in Tokyo who has expertise on a topic about which you know nothing. You collaborate with many colleagues in publishing a company blog. If you're in a law firm, you collaborate with accountants in an accounting firm in servicing the needs of a client with tax problems.

Collaboration can be short-lived, lasting a few minutes, or longer term, depending on the nature of the task and the relationship among participants. Collaboration can be one-to-one or many-to-many.

Employees may collaborate in informal groups that are not a formal part of the business firm's organizational structure, or they may be organized into formal teams. **Teams** have a specific mission that someone in the business assigned to them. Team members need to collaborate on the accomplishment of specific tasks and collectively achieve the team mission. The team mission might be to "win the game" or "increase online sales by 10 percent." Teams are often short-lived, depending on the problems they tackle, and the length of time needed to find a solution and accomplish the mission.

Collaboration and teamwork are more important today than ever for a variety of reasons.

- *Changing nature of work.* The nature of work has changed from factory manufacturing and pre-computer office work where each stage in the production process occurred independently of one another and was coordinated by supervisors. Work was organized into silos. Within a silo, work passed from one machine tool station to another, from one desktop to another, until the finished product was completed. Today, jobs require much closer coordination and interaction among the parties involved in producing the service or product. Even in factories, workers today often work in production groups, or pods.
- *Growth of professional work.* "Interaction" jobs tend to be professional jobs in the service sector that require close coordination and collaboration. Professional jobs require substantial education and the sharing of information and opinions to get work done. Each actor on the job brings specialized expertise to the problem, and all the actors need to take one another into account in order to accomplish the job.

- *Changing organization of the firm.* For most of the industrial age, managers organized work in a hierarchical fashion. Orders came down the hierarchy, and responses moved back up the hierarchy. Today, work is organized into groups and teams, and the members are expected to develop their own methods for accomplishing the task. Senior managers observe and measure results but are much less likely to issue detailed orders or operating procedures. In part, this is because expertise and decision-making power have been pushed down in organizations.
- *Changing scope of the firm.* The work of the firm has changed from a single location to multiple locations—offices or factories throughout a region, a nation, or even around the globe. For instance, Henry Ford developed the first mass-production automobile plant at a single Dearborn, Michigan, factory. In 2022, Ford employed 183,000 people at 61 plants and 93 office locations worldwide. With this kind of global presence, the need for close coordination of design, production, marketing, distribution, and service obviously takes on new importance and scale. Large global companies need to have teams working on a global basis.
- *Emphasis on innovation.* Although we tend to attribute innovations in business and science to great individuals, these great individuals are most likely working with a team of brilliant colleagues. Think of Bill Gates and Steve Jobs (founders of Microsoft and Apple), both of whom are highly regarded innovators and both of whom built strong collaborative teams to nurture and support innovation in their firms. Their initial innovations derived from close collaboration with colleagues and partners. Innovation, in other words, is a group and social process, and most innovations derive from collaboration among individuals in a lab, a business, or government agencies. Strong collaborative practices and technologies are believed to increase the rate and quality of innovation.
- *Changing culture of work and business.* It is widely believed that diverse teams produce better outputs faster than individuals working on their own. Popular notions of the crowd ("crowdsourcing" and the "wisdom of crowds") also provide cultural support for collaboration and teamwork.

## WHAT IS SOCIAL BUSINESS?

Many firms today enhance collaboration by embracing **social business**—the use of social network platforms, including Facebook, Twitter, and internal corporate social tools—to engage their employees, customers, and suppliers. These tools enable workers to set up profiles, form groups, and "follow" each other's status updates. The goal of social business is to deepen interactions with groups inside and outside the firm to expedite and enhance information sharing, innovation, and decision making.

A key word in social business is *conversations.* Customers, suppliers, employees, managers, and even oversight agencies continually have conversations about firms, often without the knowledge of the firm or its key actors (employees and managers).

Supporters of social business argue that if firms could tune in to these conversations, they would strengthen their bonds with consumers, suppliers, and employees, increasing their emotional involvement in the firm.

All of this requires a great deal of information transparency. People need to share opinions and facts with others quite directly, without intervention from executives or others. Employees get to know directly what customers and other employees think, suppliers will learn very directly the opinions of supply chain partners, and even managers presumably will learn more directly from their employees how well they are doing. Nearly everyone involved in the creation of value will know much more about everyone else.

If such an environment could be created, it is likely to drive operational efficiencies, spur innovation, and accelerate decision making. If product designers can learn directly about how their products are doing in the market in real time, based on consumer feedback, they can speed up the redesign process. If employees can use social

**TABLE 2.2**

Applications of Social Business

| Application | Description |
|---|---|
| Social networks | Connect through personal and business profiles |
| Crowdsourcing | Harness collective knowledge to generate new ideas and solutions |
| Shared workspaces | Coordinate projects and tasks; co-create content |
| Blogs and wikis | Publish and rapidly access knowledge; discuss opinions and experiences |
| Social commerce | Share opinions about purchasing on social platforms |
| File sharing | Upload, share, and comment on photos, videos, audio, text documents |
| Social marketing | Use social media to interact with customers; derive customer insights |
| Communities | Discuss topics in open forums; share expertise |

connections inside and outside the company to capture new knowledge and insights, they will be able to work more efficiently and solve more business problems.

Table 2.2 describes important applications of social business inside and outside the firm. This chapter focuses on enterprise social business—its internal corporate uses. Chapters 7 and 10 describe social business applications relating to customers and suppliers outside the company.

## BUSINESS BENEFITS OF COLLABORATION AND SOCIAL BUSINESS

There is a general belief among both business and academic communities that the more a business firm is "collaborative," the more successful it will be, and that collaboration within and among firms is more essential than in the past. *MIT Sloan Management Review*'s research found that a focus on collaboration is central to how digitally advanced companies create business value and establish competitive advantage (Kiron, 2017).

Table 2.3 summarizes some of the benefits of collaboration and social business that have been identified.

## BUILDING A COLLABORATIVE CULTURE AND BUSINESS PROCESSES

However, not all collaboration efforts succeed. A joint study of 1,100 companies by the Institute for Corporate Productivity and Rob Cross, professor of Global Business at Babson College, found that companies that promoted collaborative work were more than five times more likely to be high-performing. But it also found that among the companies studied, two-thirds of which claimed collaboration to be an explicit organizational value, only a minority were able to achieve good results. Collaboration won't take place spontaneously in a business firm, especially in the absence of supportive culture or business processes, such as how information flows throughout the organization, how decisions are made, the physical design of the workplace, and how employees are rewarded (Samdahl, 2017). Collaboration tools and technologies must be appropriate for the tasks at hand (see the chapter-ending case study).

Review Table 2.3. Then examine Figure 2.10, which introduces the elements that impact the quality of collaboration within an organization and the organization's ability to use collaboration to enhance firm performance. These elements include the

**TABLE 2.3**
Business Benefits of Collaboration and Social Business

| Benefit | Rationale |
|---|---|
| Productivity | People interacting and working together can capture expert knowledge and solve problems more rapidly than the same number of people working in isolation from one another. There will be fewer errors. |
| Quality | People working collaboratively can communicate errors and corrective actions faster than if they work in isolation. Collaborative and social technologies help reduce time delays in design and production. |
| Innovation | People working collaboratively can come up with more innovative ideas for products, services, and administration than the same number working in isolation from one another. There are advantages to diversity and the "wisdom of crowds." |
| Customer service | People working together using collaboration and social tools can solve customer complaints and issues faster and more effectively than if they were working in isolation from one another. |
| Financial performance (profitability, sales, and sales growth) | As a result of all of the above, collaborative firms have the potential for superior sales, sales growth, and financial performance. |

culture and business processes of the firm and the breadth with which it employs collaboration as well as the collaboration and social business technologies used by the firm. We explore each of these elements in further detail in the following sections.

With a collaborative organizational culture and business processes, senior managers are responsible for achieving results but rely on teams of employees to achieve and implement the results. Policies, products, designs, processes, and systems are much more dependent on teams at all levels of the organization to devise, to create, and to build. Teams are rewarded for their performance, and

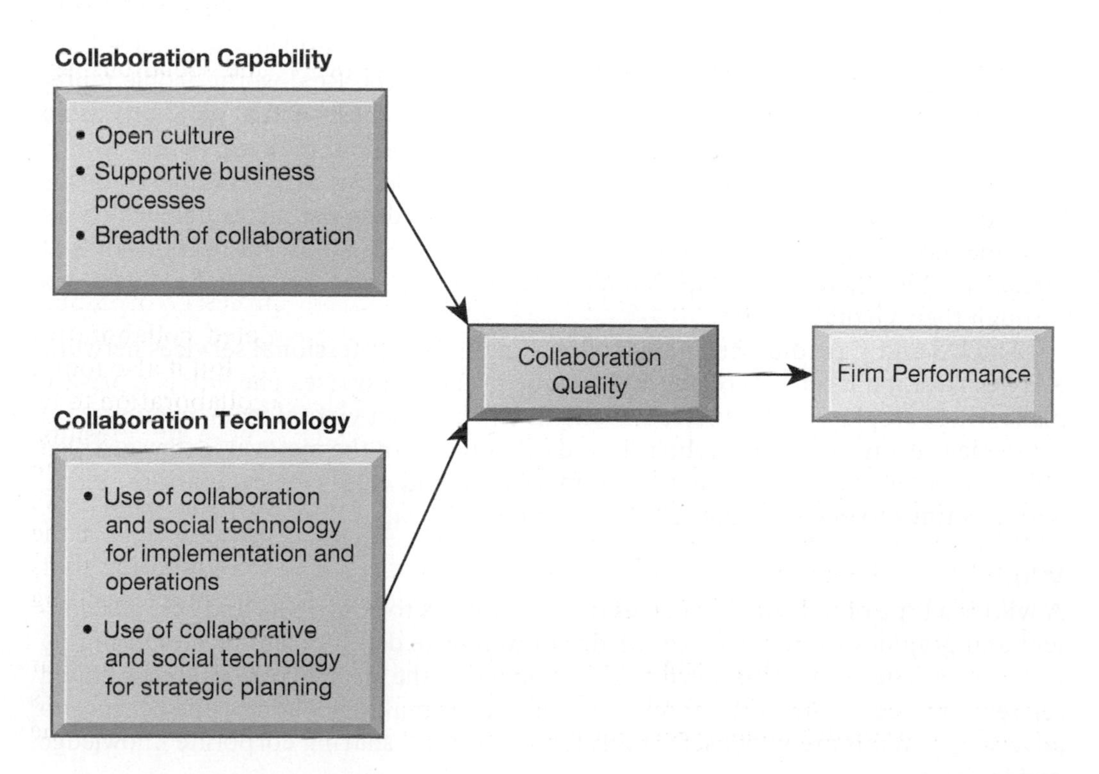


**Figure 2.10** Requirements for Collaboration
*Successful collaboration requires appropriate business processes and organizational culture along with appropriate collaboration technology.*

individuals are rewarded for their performance in a team. The function of middle managers is to build the teams, coordinate their work, and monitor their performance. The business culture and business processes are more "social." In a collaborative culture, senior management establishes collaboration and teamwork as vital to the organization.

## TOOLS AND TECHNOLOGIES FOR COLLABORATION AND SOCIAL BUSINESS

A collaborative, team-oriented culture won't produce benefits without information systems in place to enable collaboration, group decision making, and social business. Currently there are hundreds of tools designed to deal with the fact that, in order to succeed in our jobs, we are all much more dependent on one another, our fellow employees, customers, suppliers, and managers. Some of these tools are expensive, but others are available online for free (or with premium versions for a modest fee). Let's look more closely at some of these tools.

### Email and Instant Messaging (IM)

Email and instant messaging (including text messaging) are major communication and collaboration tools for interaction jobs. Their software operates on computers and mobile devices and includes features for sharing files as well as transmitting messages. Many instant messaging systems allow users to engage in real-time conversations with multiple participants simultaneously. In recent years, email use has declined, with messaging and social media becoming preferred channels of communication. Slack is an instant messaging platform for business that provides a single place for messaging, tools, and files. It has capabilities for integrated file sharing, video and phone calls, screen sharing directly through its platform, and the ability to invite others from outside the organization to collaborate. Teams can work in dedicated spaces called channels that bring together the right people and information. You can find out more about Slack in the Video Case for this chapter.

### Virtual Worlds

**Virtual worlds** are online, 3-D environments populated by residents who have built graphical representations of themselves known as avatars. Real-world people represented by avatars meet, interact, and exchange ideas at these virtual locations using gestures, chat box conversations, and voice communication, making virtual worlds potential online communication and collaboration tools. An early example of this is Second Life. Today, interest in virtual worlds is focused on the metaverse, which we describe more fully in Chapters 4 and 7. The metaverse is a nascent network of 3-D virtual worlds where people can interact, do business, and forge social connections through their virtual avatars.

The US and Canadian businesses of multinational professional services network KPMG have launched a "collaboration hub" in the metaverse. The hub is a private place where employees, partners, and clients can conduct virtual team meetings and share ideas using virtual whiteboards and other tools. In the metaverse they are also able to conduct training sessions, job interviews, and employee evaluations, along with meetings to demonstrate KPMG's services and capabilities (Strack, 2022).

### Wikis

A wiki is a type of website that makes it easy for users to contribute and edit text content and graphics without any knowledge of web page development or programming techniques. The most well-known wiki is Wikipedia, the largest collaboratively edited reference project in the world. It relies on volunteers, makes no money, and accepts no advertising. Wikis are very useful tools for storing and sharing corporate knowledge and insights.

### Collaboration and Social Business Platforms

There are now suites of software products providing multifunction platforms for collaboration and social business among teams of employees who work together from many different locations. The most widely used are Internet-based audioconferencing and videoconferencing systems; cloud collaboration services such as Google's online services and tools; corporate collaboration systems such as Microsoft SharePoint; and enterprise social networking tools such as Salesforce Chatter, Microsoft Yammer, Meta's Workplace, and IBM Connections.

**Virtual Meeting Systems** To reduce travel expenses and help people in different locations to meet and collaborate, many companies, both large and small, have adopted videoconferencing and web conferencing technologies. More companies are using these systems to enable their employees to work remotely as well as in the office.

A videoconference allows individuals at two or more locations to communicate simultaneously through two-way video and audio transmissions. High-end videoconferencing systems feature **telepresence** technology, an integrated audio and visual environment that allows people to give the appearance of being present at a location other than their true physical location. Free or low-cost Internet-based systems such as Microsoft Teams and Amazon Chime are lower quality but improving, and they are very useful for smaller companies. Apple's FaceTime is appropriate for one-to-one or small group videoconferencing. Zoom has emerged as a powerful videoconferencing tool with many features of high-end systems but is much less expensive and easier to use. Some of these tools are available on mobile devices (see the Spotlight on Technology case).

**Cloud Collaboration Services** Google offers many online tools and services, and some are suitable for collaboration. They include Google Drive, Google Docs, Google Workspace, and Google Sites. Most are free of charge.

Google Drive is file storage and synchronization service for cloud storage, file sharing, and collaborative editing. Such web-based online file-sharing services allow users to upload files to secure online storage sites from which the files can be shared with others. Microsoft OneDrive and Dropbox are other leading cloud storage services. They feature both free and paid services, depending on the amount of storage space and administration required. Users are able to synchronize their files stored online with their local PCs and other devices, with options for making the files private or public and for sharing them with designated contacts.

Google Drive and Microsoft OneDrive are integrated with tools for document creation and sharing. OneDrive provides online storage for Microsoft 365 documents and other files and works with Microsoft 365 apps, both installed and on the web. It can share with Facebook as well. Google Drive is integrated with Google Docs, Sheets, and Slides (often called Google Docs), a suite of productivity applications that offer collaborative editing on documents, spreadsheets, and presentations. Google's cloud-based productivity suite for businesses, called Google Workspace, also works with Google Drive. Google Sites allows users to quickly create online team-oriented websites where multiple people can collaborate and share files.

**Microsoft SharePoint** Microsoft SharePoint is a collaboration and document management service that helps organizations share and manage content, knowledge, and applications with on-premise and cloud versions. SharePoint has a web-based interface and close integration with productivity tools such as Microsoft 365. SharePoint software makes it possible for employees to share their documents and collaborate on projects using Microsoft documents as the foundation.

SharePoint can be used to host internal websites that organize and store information in one central workspace to enable teams to coordinate work activities, collaborate on and publish documents, maintain task lists, implement workflows, and share information via wikis and blogs. Users are able to control versions of

## Zoom: Quality Videoconferencing for Every Budget

When it comes to collaboration and managing employees from afar, videoconferencing has become a very popular tool for organizations of all sizes. In the past, high-quality videoconferencing was limited to the very largest companies that could afford dedicated videoconference rooms and expensive networking technologies and software for this purpose. Today, videoconferencing has been democratized. There are now high-quality, low-cost videoconferencing tools for organizations of all sizes, and some of these tools can be used on mobile devices.

Zoom became the online meeting tool of choice for classrooms and offices all over the world that had to operate remotely from people's homes during the 2020 coronavirus shutdown and thereafter. Zoom is a cloud platform for online video and audioconferencing, collaboration, chat, screen sharing, and webinars across mobile devices, desktops, telephones, and room systems. Zoom allows you to conduct a video meeting or webinar directly from your computer, mobile device, or Zoom-configured conference room. Zoom software is available in a number of different versions and prices. There is a free Basic version for short personal meetings with a maximum of 100 participants and more full-featured business versions ranging from $149.90 to $249.90 per year per license. Zoom can accommodate up to 1,000 video participants, 10,000 view-only attendees, and as many as 49 HD videos on-screen simultaneously.

The two core pieces of Zoom technology are Zoom Meetings and Zoom Rooms. Zoom Meetings supports online, video-hosted meetings that are free to join for users with Zoom videoconferencing software installed on their computer or mobile device. Participants can join Zoom Meetings from different locations regardless of whether they're joining the meeting from home, in a conference room in the office, or on their mobile device. Zoom Rooms consist of hardware and software technology to power physical conference rooms. Zoom Meetings can be activated with just a tap of a button on a tablet. All of the video and audio is integrated into the conference room equipment and calendar systems.

Butler Health System (BHS) based in Butler, Pennsylvania, used Zoom to expand its telehealth services during the Covid-19 pandemic. BHS had been using some telehealth services to increase access to care in rural communities. However, these services had been limited and required patients to travel to a specified location for an appointment in order to have their visits covered by insurance. The pandemic forced BHS to close its facilities. In order to have its network of over 200 healthcare providers continue serving patients, BHS needed to bring telehealth into patients' homes within days.

Zoom for Healthcare provided a solution, enabling providers to offer much of the clinical experience without having to see patients in person. (Zoom for Healthcare is Zoom's secure, web-based virtual care videoconferencing platform.) It enabled BHS to start offering expanded telehealth appointments to patients in their homes within three days. Using Zoom's Scheduling Privilege feature, office staff can schedule appointments for multiple providers at the same time. Staff also use Zoom's Waiting Room feature to simulate an urgent care setting. Patients with acute symptoms who can't wait for a scheduled appointment are given a meeting link to a physician's personal meeting ID and placed in a "virtual waiting room" queue. The patients are then admitted one by one into the physicians' virtual office for a private telehealth consultation.

Before the pandemic, BHS had only a handful of telehealth visits each week. Zoom made it possible for the organization to stage thousands. And by removing long travel times and other barriers to treatment in rural areas, BHS has improved access to high-quality care for the people who need it the most. Virtual visits in patients' houses have also reduced appointment cancellations and no-shows.

Blue Yonder (formerly JDA Software), a leading vendor of supply chain management solutions, used Zoom to create a communications platform that could support both its employees working from home and its customers around the globe during the Covid-19 pandemic and thereafter. Blue Yonder is headquartered in Scottsdale, Arizona, with more than 5,500 associates in regional offices around the world. Over 3,000 companies in the manufacturing, distribution, transportation, retail, and services industries depend on Blue Yonder software to coordinate the movement and delivery of goods from source to customer. Blue Yonder's legacy communications solution was not up to the task. It was unreliable and provided poor connectivity in regions of the world that don't have great reception.

Blue Yonder implemented Zoom across the organization in February 2020, right before the pandemic was in full force. Zoom helped Blue Yonder transition to a remote work environment and support customers facing pandemic-induced demand spikes, transportation disruptions, and labor shortages. Tools such as Zoom Meetings, Zoom Rooms, and Zoom Webinar enabled the company to maintain business continuity and communication with employees and customers. Blue Yonder used Zoom to train employees in new products and product updates and for monthly meetings with the CEO and leadership to update them on the company and customers. Blue Yonder's marketing team uses Zoom to stage webinars for customers and sales prospects about successful customer experiences with Zoom tools.

Sources: "From the Hospital into the Home: How Butler Health System Scaled Telehealth During Covid-19" and "How Blue Yonder Shifted to Work from Home & Supported Critical Supply Chains During Covid-19," Zoom.us.com, accessed February 1, 2022; Danielle Abril, "Four-Day Weeks and the Freedom to Move Anywhere: Companies Are Rewriting the Future of Work (Again)," *Washington Post*, January 20, 2022.

## CASE STUDY QUESTIONS

1. How is videoconferencing related to the business models and business strategies of the organizations described in this case?
2. Describe the specific ways in which Zoom helped each of the organizations in this case improve their operations and decision making.
3. If you were a small or medium-sized business, what people, organization, and technology criteria would you use to determine whether to use Zoom videoconferencing?

documents and document security. Because SharePoint stores and organizes information in one place, users can find relevant information quickly and efficiently while working closely together on tasks, projects, and documents. Enterprise search tools help locate people, expertise, and content. SharePoint now features social tools.

**Enterprise Social Networking Tools** The tools we have just described include capabilities for supporting social business, but there are also more specialized social tools for this purpose, such as Salesforce Chatter, Microsoft Yammer, Meta's Workplace, and IBM Connections. Enterprise social networking tools create business value by connecting the members of an organization through profiles, updates, and notifications similar to Facebook features but tailored to internal corporate uses. Table 2.4 provides more detail about these internal social capabilities.

### Checklist for Managers: Evaluating and Selecting Collaboration and Social Software Tools

With so many collaboration tools and services available, how do you choose the right collaboration technology for your firm? You need a framework for understanding just what problems these tools are designed to solve. Examine Figure 2.11, which illustrates a time/space collaboration and social tool matrix developed by a number of collaborative-work scholars that can be helpful in addressing this issue.

The time/space matrix focuses on two dimensions of the collaboration problem: time and space (place or location). For instance, you need to collaborate with people in different time zones, and you cannot all meet at the same time. Midnight in New York is 10:30 a.m. in Mumbai, so this makes it difficult to have a videoconference (the people in New York are too tired). Time is clearly an obstacle to collaboration on a global scale.

Place (location) also inhibits collaboration in large global or even national and regional firms. Assembling people for a physical meeting is made difficult by the

**TABLE 2.4**
Enterprise Social Networking Software Capabilities

| Social Software Capability | Description |
| --- | --- |
| Profiles | Ability to set up member profiles describing who individuals are, educational background, and interests. Includes work-related associations and expertise (skills, projects, teams). |
| Content sharing | Share, store, and manage content including documents, presentations, images, and videos. |
| Feeds and notifications | Real-time information streams, status updates, and announcements from designated individuals and groups. |
| Groups and team workspaces | Establish groups to share information, collaborate on documents, and work on projects with the ability to set up private and public groups and to archive conversations to preserve team knowledge. |
| Tagging and social bookmarking | Indicate preferences for specific pieces of content, similar to the Facebook Like button. Tagging lets people add keywords to identify content they like. |
| Permissions and privacy | Ability to make sure private information stays within the right circles, as determined by the nature of relationships. In enterprise social networks, there is a need to establish who in the company has permission to see what information. |

physical dispersion of distributed firms (firms with more than one location), the cost of travel, and the time limitations of managers.

The collaboration and social technologies we have just described are ways of overcoming the limitations of time and space. Using this time/space framework will help you choose the most appropriate collaboration and teamwork tools for your firm. Note that some tools are applicable in more than one time/place scenario. For example, Internet collaboration suites such as Microsoft Teams have capabilities for both synchronous (instant messaging, meeting tools) and asynchronous (email, wikis, document editing) interactions.

Here's a "to-do" list to get started. If you follow these six steps, you should be led to investing in the correct collaboration software for your firm at a price you can afford and within your risk tolerance.

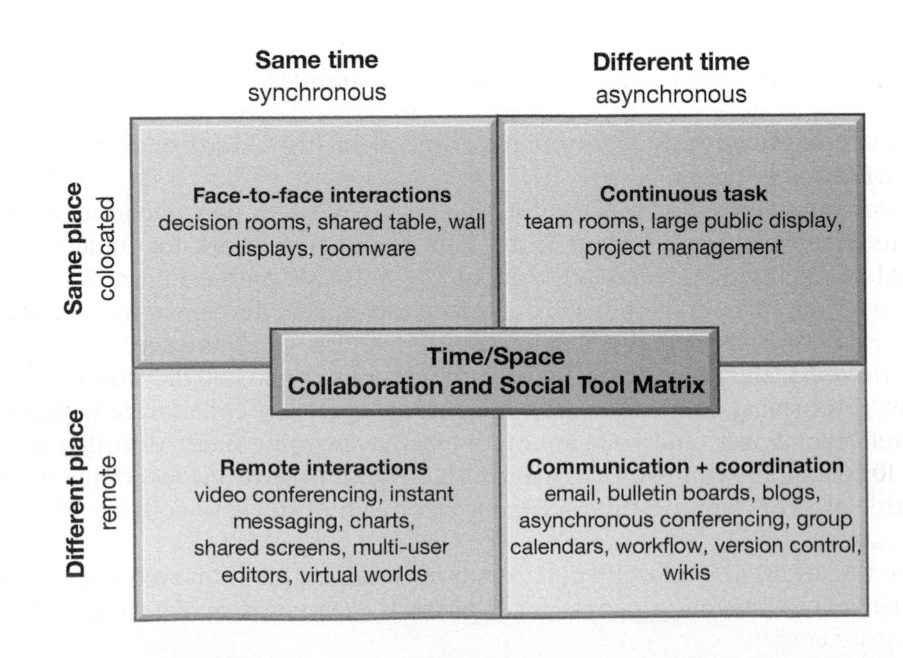


**Figure 2.11**
The Time/Space Collaboration and Social Tool Matrix
*Collaboration and social technologies can be classified in terms of whether they support interactions at the same or different times or places and whether these interactions are remote or colocated.*

1. What are the collaboration challenges facing the firm in terms of time and space? Locate your firm in the time/space matrix. Your firm can occupy more than one cell in the matrix. Different collaboration tools will be needed for each situation.
2. Within each cell of the matrix where your firm faces challenges, exactly what kinds of solutions are available? Make a list of vendor products.
3. Analyze each of the products in terms of its cost and benefits to your firm. Be sure to include the costs of training in your cost estimates and the costs of involving the information systems department, if needed.
4. Identify the risks to security and vulnerability involved with each of the products. Is your firm willing to put proprietary information into the hands of external service providers over the Internet? Is your firm willing to expose its important operations to systems controlled by other firms? What are the financial risks facing your vendors? Will they be here in three to five years? What would be the cost of making a switch to another vendor in the event the vendor firm fails?
5. Seek the help of potential users to identify implementation and training issues. Some of these tools are easier to use than others.
6. Make your selection of candidate tools and invite the vendors to make presentations.

In addition to selecting the right digital tools for communication and collaboration, organizations need to make sure the tools actually support productive ways of working. Access to technology alone often can't fix poorly designed business processes and entrenched organizational behaviors (Dodge et al., 2018).

## SYSTEMS FOR KNOWLEDGE MANAGEMENT

Collaboration and knowledge-sharing will be less effective and efficient if the requisite knowledge for collaboration, decision-making, and ongoing operations is not available. Knowledge management systems, which we introduced earlier in this chapter, are a specific capability for addressing this need.

Some knowledge exists within the firm in the form of **structured knowledge**, consisting of formal text documents (reports and presentations). According to experts, however, at least 80 percent of an organization's business content is semistructured or unstructured—information in folders, messages, memos, proposals, emails, graphics, chat room exchanges, slide presentations, and even videos created in different formats and stored in many locations. In still other cases, there is no formal or digital information of any kind, and the knowledge resides in the heads of employees. Much of this knowledge is **tacit knowledge** and is rarely written down. Enterprise-wide knowledge management systems feature capabilities for storing and searching for all of these different types of knowledge and for locating employee expertise within the firm.

**Enterprise content management (ECM) systems** help organizations manage structured, semistructured, and unstructured types of information. They have capabilities for knowledge capture, storage, retrieval, distribution, and preservation. Examine Figure 2.12, which illustrates how an ECM system functions. Such systems include a unified corporate repository of documents, reports, presentations, and best practices as well as capabilities for collecting and organizing **semistructured knowledge** such as email. Major ECM systems also enable users to access external sources of information, such as news feeds and research, and to communicate by email, chat/instant messaging, discussion groups, and videoconferencing. They have started to incorporate blogs, wikis, and other enterprise social networking tools.

A key problem in managing knowledge is the creation of an appropriate classification scheme to organize information into meaningful categories. Once the categories for classifying knowledge have been created, each knowledge object needs to be tagged, or classified, so that it can be easily retrieved. ECM systems have capabilities for tagging, interfacing with corporate data repositories where the documents are stored, and creating an enterprise portal environment for employees to use when searching for corporate knowledge. OpenText, IBM, and Oracle are leading vendors of enterprise content management software.

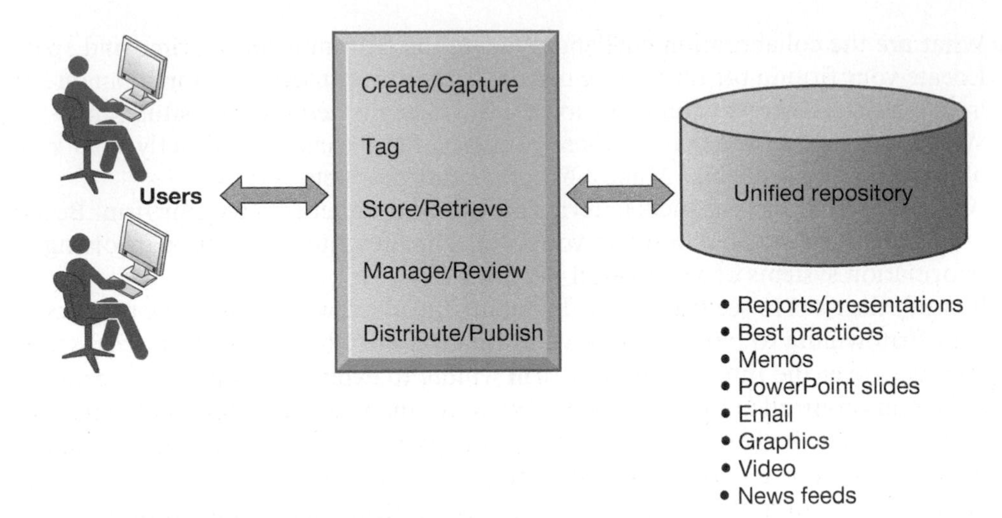


**Figure 2.12**
An Enterprise Content Management System
*An enterprise content management system has capabilities for classifying, organizing, and managing structured, semistructured, and unstructured knowledge and making it available throughout the enterprise.*

### Locating and Sharing Expertise

Some of the knowledge that businesses need is not in the form of a digital document but instead is tacit knowledge that resides in the memory of individual experts in the firm. Contemporary enterprise content management systems, along with the systems for collaboration and social business, have capabilities for locating experts and tapping their knowledge. These include online directories of corporate experts and their profiles, with details about their job experience, projects, publications, and educational degrees, and repositories of expert-generated content. Specialized search tools make it easier for employees to find the appropriate expert in a company.

## 2-4 Describe the role of the information systems function in a business.

We've seen that businesses need information systems to operate today and that they use many kinds of systems, but who is responsible for running these systems? End users manage their systems from a business standpoint, but managing the technology requires a special information systems function. In all but the smallest of firms, the **information systems department** is the formal organizational unit responsible for information technology services. The information systems department is responsible for maintaining the hardware, software, data storage, and networks that comprise the firm's IT infrastructure. We describe IT infrastructure in detail in Chapter 5.

### THE INFORMATION SYSTEMS DEPARTMENT

The information systems department consists of specialists such as programmers, systems analysts, and information systems managers. **Programmers** are highly trained technical specialists who write the software instructions for computers. **Systems analysts** constitute the principal liaisons between the information systems groups and the rest of the organization. It is the systems analyst's job to translate business problems and requirements into information requirements and systems. **Information systems managers** are leaders of teams of programmers and analysts, project managers, physical facility managers, telecommunications managers, or database specialists. They are also managers of computer operations and data entry staff. External specialists, such as hardware vendors and manufacturers, software firms, and consultants, also frequently participate in the day-to-day operations and long-term planning of information systems.

In many companies, the information systems department is headed by a **chief information officer (CIO)**. The CIO is a senior manager who oversees the use of information technology in the firm. Today's CIOs are expected to have a strong business background, as well as information systems expertise, and to play a leadership role in integrating technology with the firm's business strategy. Large firms today also have positions for a chief security officer, chief knowledge officer, chief data officer, and chief privacy officer, all of whom work closely with the CIO.

The **chief security officer (CSO)** is in charge of information systems security for the firm and is responsible for enforcing the firm's information security policy (see Chapter 8). (Where information systems security is separated from physical security, this position is sometimes called the chief information security officer [CISO]). The CSO is responsible for educating and training users and information systems specialists about security, keeping management aware of security threats and breakdowns, and maintaining the tools and policies chosen to implement security.

Information systems security and the need to safeguard personal data have become so important that companies collecting vast quantities of personal data have established positions for a **chief privacy officer (CPO)**. The CPO is responsible for ensuring that the company complies with data privacy laws and regulations.

The **chief knowledge officer (CKO)** is responsible for the firm's knowledge management program. The CKO helps design programs and systems to find new sources of knowledge or to make better use of existing knowledge in organizational and management processes.

The **chief data officer (CDO)** is responsible for enterprise-wide governance and usage of information to maximize the value of the organization's data. The CDO ensures that the firm is collecting appropriate data to serve its needs, deploying appropriate technologies for analyzing the data, and using the results to support business decisions. This position arose to deal with the massive amounts of data that organizations are now generating and collecting (see Chapter 6).

**End users** are representatives of departments outside the information systems group for whom applications are developed. These users are playing an increasingly large role in the design and development of information systems.

In the early years of computing, the information systems group was composed mostly of programmers who performed highly specialized but limited technical functions. Today, a growing proportion of staff members are systems analysts and network specialists, with the information systems department acting as a powerful change agent in the organization. The information systems department suggests new business strategies and new information-based products and services and coordinates both the development of the technology and the planned changes in the organization.

## INFORMATION SYSTEMS SERVICES

Services the information systems department provides include the following:

- Computing services connect employees, customers, and suppliers into a coherent digital environment, including mainframes (in very large companies), server computers, desktop and laptop computers, and mobile devices.
- Telecommunications services provide data, voice, and video connectivity to employees, customers, and suppliers.
- Data management services store and manage corporate data and provide capabilities for analyzing the data.
- Application software services provide development and support services for the firm's business systems.
- IT management services plan and develop the infrastructure, coordinate with the business units for IT services, manage accounting for the IT expenditure, and provide project management services.
- IT standards services provide the firm and its business units with policies that determine not only which information technology will be used but when and how it is used.

- IT educational services provide training in system use to employees and IT specialists.
- IT research and development services provide the firm with research on potential future information systems projects and investments that could help the firm differentiate itself in the marketplace.

In the past, firms generally built their own software and managed their own computing facilities. As our discussion of collaboration systems has shown, many firms are turning to external vendors and Internet-based services to provide these services (see also Chapters 5 and 12) and are using their information systems departments to manage these service providers.

## 2-5 Understand how MIS can help your career.

Here is how this chapter and this book can help you find a job as a customer success analyst.

### THE COMPANY

CareerKNOWLEDGE is a 50-person employment service company based in Ogden, Utah, that specializes in recruiting and preparing recent college graduates for paid internships with corporations. Its paid internship program includes coaching/mentoring, on-the-job training, industry knowledge transfer, and career networking. The company is looking for a customer success analyst who will be part of its Customer Success Team leading customer implementation, monitoring usage, and maximizing retention of the company's proprietary workforce development learning management system.

### POSITION DESCRIPTION

This job will consist of responding to and assisting the support desk, assisting in new system implementations, providing ongoing training, conducting testing and quality assurance, and monitoring usage and feedback. Job responsibilities include:

- Monitoring and resolving customer support issues and concerns
- Analyzing information to identify trends, diagnose problem areas, and coordinate teams to create long-term effective solutions
- Supporting organization technology training efforts including facilitating and building online support tools such as documentation, videos, and other information content
- Consulting with customers on implementation and usage of the workforce development learning management system
- Working with the marketing director and graphic designer on visual design
- Proactively following up with customers to ensure a positive user experience

### JOB REQUIREMENTS

- Excellent communication skills
- Experience in software customer success
- Proficiency in time management and working on multiple projects
- Knowledge of Microsoft 365 and Google Workspace
- Experience with learning management systems such as Blackboard, Moodle, or Desire2Learn desirable
- Knowledge of support desk software such as Freshdesk, Zendesk, Hubspot, or equivalent desirable
- Bachelor's degree preferred
- One or more years of job experience desirable

## INTERVIEW QUESTIONS

1. What do you know about learning management systems? Have you ever worked with a learning management system?
2. What do you know about customer support systems? Have you ever worked with support desk software tools? If so, what system did you use?
3. What can you do with Microsoft 365 and Google Workspace? Have you ever taken courses in Excel, Access, Google Sheets, or Google Slides?
4. Do you have any specific experience working in customer service? What exactly was your role?
5. Can you give an example of a client service challenge you had to face? How did you approach this challenge? What were the key takeaways from this specific client service challenge that you will use going forward?

## AUTHOR TIPS

1. Review the section of this chapter on enterprise applications and the Chapter 9 discussion of customer relationship management.
2. Use the web and LinkedIn to research the company and its learning management and career coaching products and services as well as the way the company operates. Think about what the company needs to do in order to operate its client service support systems.
3. Provide examples of your work with Microsoft 365 and Google Workspace demonstrating your proficiency with the software.
4. Bring samples of your writing demonstrating communication skills.

# Review Summary

**2-1 Identify the major features of a business that are important for understanding the role of information systems.** A business is a formal, complex organization producing products or services for a profit. Businesses have specialized functions such as finance and accounting, human resources, manufacturing and production, and sales and marketing. Business organizations are arranged hierarchically into levels of management. A business process is a logically related set of activities that define how specific business tasks are performed. Business firms must monitor and respond to their surrounding environments.

**2-2 Explain how information systems serve different management groups in a business, and how systems that link the enterprise improve organizational performance.** Systems serving operational management are transaction processing systems (TPS), such as payroll or order processing, that track the flow of the daily routine transactions necessary to conduct business. Business intelligence systems serve multiple levels of management and help employees make more informed decisions. Management information systems (MIS) and decision-support systems (DSS) support middle management. Most MIS reports condense information from TPS and are not highly analytical. DSS support management decisions that are unique and rapidly changing, using advanced analytical models and data analysis capabilities. Executive support systems (ESS) support senior management by providing data that are often in the form of graphs and charts delivered in portals and dashboards using many sources of internal and external information.

Enterprise applications are designed to coordinate multiple functions and business processes. Enterprise systems integrate the key internal business processes of a firm into a single software system to improve coordination and decision making. Supply chain management (SCM) systems help the firm manage its relationship with suppliers to optimize the planning, sourcing, virtual manufacturing, and delivery of products and services. Customer relationship management (CRM) systems coordinate the business processes surrounding the firm's customers. Knowledge management systems (KMS) enable firms to optimize the creation, sharing, and distribution of knowledge. Intranets and extranets are private corporate networks based on Internet technology. Extranets make portions of private corporate intranets available to outsiders.

2-3 **Understand why systems for collaboration, social business, and knowledge management are so important and the technologies they use**. Collaboration means working with others to achieve shared and explicit goals. Social business is the use of internal and external social network platforms to engage employees, customers, and suppliers, and it can enhance collaborative work. Collaboration and social business have become increasingly important in business because of globalization, the decentralization of decision making, and growth in jobs where interaction is the primary value-adding activity. Collaboration and social business enhance innovation, productivity, quality, and customer service. Tools for collaboration and social business include email and instant messaging, wikis, virtual meeting systems (such as Zoom and Microsoft Teams),virtual worlds, cloud-based file-sharing services, corporate collaboration platforms such as Microsoft SharePoint, and enterprise social networking tools such as Chatter, Yammer, Workplace, and IBM Connections. Knowledge management systems such as enterprise content management systems help organizations collect, store, and distribute the requisite knowledge for collaboration, decision-making, and ongoing operations.

2-4 **Describe the role of the information systems function in a business** The information systems department is the formal organizational unit responsible for information technology services. It is responsible for maintaining the hardware, software, data storage, and networks that comprise the firm's IT infrastructure. The department consists of specialists, such as programmers, systems analysts, and information systems managers, and is often headed by a chief information officer (CIO).

## Key Terms

## Review Questions

**2-1** Identify the major features of a business that are important for understanding the role of information systems.

- Define a business and describe the major business functions.
- Define business processes and describe the role they play in organizations.
- Identify and describe the different levels in a business firm and their information needs.
- Explain why environments are important for understanding a business.

**2-2** Explain how information systems serve different management groups in a business, and how systems that link the enterprise improve organizational performance.

- Define business intelligence systems.
- Describe the characteristics of transaction processing systems (TPS) and the role they play in a business.
- Describe the characteristics of management information systems (MIS), decision support systems (DSS), and executive support systems (ESS) and explain how each type of system helps managers make decisions.
- Explain how enterprise applications improve organizational performance.
- Define enterprise systems, supply chain management (SCM) systems, customer relationship management (CRM) systems, and knowledge management systems (KMS) and describe their business benefits.
- Explain how intranets and extranets help firms improve business performance.

**2-3** Understand why systems for collaboration, social business, and knowledge management are so important and the technologies they use.

- Define collaboration and social business and explain why they have become so important in business today.
- List and describe the business benefits of collaboration and social business.
- Describe a supportive organizational culture for collaboration.
- List and describe the various types of collaboration and social business tools.
- Describe how knowledge management systems help organizations make better use of their structured, semistructured, and unstructured knowledge and also their tacit knowledge assets.

**2-4** Describe the role of the information systems function in a business.

- Describe how the information systems function supports a business.
- Compare the roles programmers, systems analysts, information systems managers, the chief information officer (CIO), chief security officer (CSO), chief data officer (CDO), chief privacy officer (CPO), and chief knowledge officer (CKO) play.

## Discussion Questions

**2-5** **MyLab MIS** How could information systems be used to support the order fulfillment process illustrated in Figure 2.2? What are the most important pieces of information these systems should capture? Explain your answer.

**2-6** **MyLab MIS** Identify the steps that are performed in the process of selecting and checking out a book from your college library and the information that flows among these activities. Diagram the process. Are there any ways this process could be adjusted to improve the performance of your library? Diagram the improved process.

**2-7** **MyLab MIS** Use the Time/Space Collaboration and Social Tool Matrix to classify the collaboration and social technologies Toyota Motor North America uses.

**MyLab MIS**
To complete these problems, go to EOC Discussion Questions in **MyLab MIS**.

# Hands-On MIS Projects

The projects in this section give you hands-on experience analyzing opportunities to improve business processes with new information system applications, using a spreadsheet to develop a town budget, and using Internet software to plan efficient transportation routes. Visit MyLab MIS to access this chapter's Hands-On MIS Projects.

## MANAGEMENT DECISION PROBLEM

**2-8** The US Census, conducted every 10 years, counts every resident in the United States. The results of the census are used to determine the number of seats each state has in the US House of Representatives, Electoral College votes, and an estimated $1.5 trillion each year in federal funding for the states for health care, education, infrastructure improvements, and more.

In the past, all census data were collected by paper surveys, transported to the US Census Bureau, and input manually into the Bureau's information systems. The 2020 census was primarily conducted online, with about 60 percent of the US population of 330 million filling out a form on their Internet-linked smartphones, tablets, or personal computers. The Bureau also collected census responses over the phone and on paper forms. An unprecedented amount of data had to be collected, stored, secured, and interpreted. What are some of the people, organizational, and technology issues that the Census Bureau had to address in modernizing a massive, important paper-based system?

## IMPROVING DECISION MAKING: USING A SPREADSHEET TO CREATE A TOWN BUDGET

Software skills: Spreadsheet charts, formulas, assumptions
Business skills: Preparing a town budget

**2-9** In this exercise, you will learn how to use spreadsheet software to improve management decisions about a budget and budget projections, in this case for a town. The town of ClearSkies with a population of 9,042, is located in Oklahoma. Each November the town manager works with the mayor and the board of trustees to develop the town's budget for the forthcoming year. The town's main source of revenue is the local property tax, but it also receives some aid from the state government and some revenue from miscellaneous licenses and fees.

The town leaders limit tax increases to 3 percent or less. They fear revenues may be declining because the state wants to cut the state aid it provides to local governments. Anticipating continued cutbacks in state aid, the town manager would like to develop preliminary budgets for the next two years to see if they can be supported by anticipated revenues. If planned expenditures exceed revenues, the town manager, the mayor, and the trustees must develop an alternative budget that is balanced.

In MyLab MIS, you fill find the spreadsheet file EMIS15Chap2 Question File, showing receipts and disbursements for the town in 2023, along with additional instructions. The 2023 budget can be used as the basis for projecting the town budget for the next two years. In projecting the next two years' budgets, the town manager wants to use the following assumptions: She expects the state will reduce its aid to the town by 6 percent each year. Historically, expenditures for employee salaries and benefits have been rising 5 percent annually. She expects all other expenditures to rise at a rate of 6 percent annually and miscellaneous receipts to rise 4 percent annually. Tax increases should be 3 percent per year. The town's expenditure for debt service to pay off previous loans will remain constant. Can the town balance its budget if it keeps its promise not to raise taxes beyond 3 percent or go further into debt?

The town manager wants to highlight the portion of salary expenditures represented by the town court and the best way to do that is visually in a pie chart. Create a pie chart showing the various categories of salary disbursements for 2023.

## ACHIEVING OPERATIONAL EXCELLENCE: USING INTERNET SOFTWARE TO PLAN EFFICIENT TRANSPORTATION ROUTES

**2-10** In this exercise, you will use Google Maps to map out transportation routes for a business and select the most efficient route.

You have just started working as a dispatcher for Cross-Country Transport, a new trucking and delivery service based in Cleveland, Ohio. Your first assignment is to plan a delivery of office equipment and furniture from Elkhart, Indiana (at the corner of E. Indiana Ave. and Prairie Street), to Hagerstown, Maryland (corner of Eastern Blvd. N. and Potomac Ave.). To guide your trucker, you need to know the most efficient route between the two cities. Use Google Maps to find the route that is the shortest distance between the two cities and also to find the route that takes the least time. Compare the results. Which route should Cross-Country use?

## COLLABORATION AND TEAMWORK PROJECT

### Identifying Management Decisions and Systems

**2-11** With a team of three or four other students, search business publications on the web to find a description of a manager of a corporation. Gather information about what the manager does and the role the manager plays in the company. Identify the organizational level and business function where this manager works. Make a list of the kinds of decisions this manager has to make and the kind of information that manager would need for those decisions. Suggest how information systems could supply this information. If possible, use Google Docs and Google Drive or Google Sites to brainstorm, organize, and develop a presentation of your findings for the class.

## BUSINESS PROBLEM-SOLVING CASE

### HOW MUCH DOES TECHNOLOGY HELP COLLABORATION?

During the spring of 2020, there was a huge shift in the way people worked, as the Covid-19 pandemic shuttered offices and factories around the world. Instead of meeting face-to-face, many organizations turned to virtual meeting systems such as Zoom and Microsoft Teams so that employees could perform their jobs at home. The low cost, ease of use, and widespread availability of virtual meeting software helped employees work and collaborate remotely and gave managers a tool for managing employees from afar.

The popularity of collaborative technologies throughout the pandemic has accelerated the move toward virtual work by allowing teams to collaborate even when physical offices are closed. The market for collaborative technologies continues to grow and is predicted to reach $50.7 billion by 2025. However, as this move to work virtually takes root in the workplace, organizations need to pay closer attention to the risks and rewards of collaborative technologies. Is virtual meeting technology good or bad for employees and managers, and how much does the technology help or hinder the collaboration process?

Because many companies went completely remote in 2020, the number of meetings per day has increased. Upwork's recent "Future Workforce Pulse Report" predicted that 36.2 million Americans will be working remotely, a near 90 percent increase from pre-pandemic levels. It looks like virtual meetings—and lots of them—are here to stay.

Even before the Covid pandemic, 71 percent of managers thought meetings were costly and unproductive. Studies show that spending too much time in meetings negatively affects people's psychological, physical, and mental well-being. Having too many meetings diverts workers during their most productive hours, interrupts people's train of thought, and detracts from effective collaboration. Regardless of company size, both the number of collaborative demands and the diversity of them are overwhelming people. This is a serious problem.

In addition to the problems caused by the frequency and dynamics of virtual meetings, there are problems caused by the nature of the technology itself. Recent research has found that videoconferencing technology can actually make it harder for people to collaborate. Dr. Anita Williams Wooley, associate professor of organizational behavior and theory at Carnegie Mellon's Tepper School of Business, compared how well two-person teams of unacquainted workers in different locations using videoconferencing and similar teams communicating only with audio tools were able to perform a collaborative task. Wooley and her fellow researchers found that the teams communicating via only audio means scored higher than those using videoconferencing. These teams were more successful in synchronizing their vocal cues and speaking turns. Wooley believes communicating via video can draw attention away from audio cues in many cases, reducing capacity to carry out the work itself.

Meeting virtually creates more complexity for both meeting leaders and attendees. People aren't used to the unnatural lack of nonverbal cues, prolonged eye contact, or overload of faces to process on the screen. Seeing our own faces as we talk or listen, and awareness of how we appear online is stressful. The amount of effort required to process all of these stimuli while simultaneously thinking and communicating is fatiguing. The term "Zoom fatigue" is now part of our vocabulary. Constant connectedness and overuse of virtual communication platforms, particularly videoconferencing, cause many people to experience feelings of tiredness, anxiety, or burnout.

During the summer of 2020, doctoral students Katie Kavanagh, Nicole Voss, and Liana Kreamer, along with Professor Steven G. Rogelberg, all from the University of North Carolina at Charlotte, collected data from 150 employees from a range of industries in the United States and Europe to learn about their experiences in virtual meetings. The vast majority of those surveyed reported feeling fatigued and drained during and after their virtual meetings—more so than with in-person meetings. In addition to feeling exhausted from the number and scheduling of meetings, survey respondents varied widely in their explanations for why this was so. Different people had different virtual meeting preferences. Some employees reported, for example, that virtual meetings are more fatiguing than in-person meetings because there are fewer social cues, while others appreciated the detached virtual meeting style. Managers need to take all of these factors into account when using virtual meeting technology.

Virtual collaboration also introduces risks of isolation, exclusion, surveillance, and self-censorship. Heavy reliance on virtual meeting technology in the manager–employee relationship can increase isolation between managers and their employees. This isolation becomes amplified when there is an upsurge in videoconferencing fatigue or when technology cannot adequately replace less formal social interactions in remote work environments. Diminishing opportunities for spontaneous face-to-face social interaction widen the manager–employee gap.

Virtual meeting technology may also hinder some employees' ability to fully participate in organizational life. There may be generational and personality differences in facility with technology and work behaviors that affect attitudes toward virtual meetings and levels of participation. The virtual meeting platform will showcase people who actively participate in virtual team activities while unintentionally excluding people perceived as not participating or those who lack the technology skills to do so. Technology has the power to reshape "in" and "out" groups in the work environment.

Collaboration technologies can be used by companies and managers to monitor employees (see the Chapter 7 Spotlight on People case). Many collaboration platforms track user activity and duration of time spent on them. The data can tell managers about individual user engagement and can potentially be used to make judgments about employee productivity, even though this is not the technology's main purpose. For instance, Microsoft Teams's capability to provide detailed automated user analytics can be used to show how people are working together. It is important to inform employees exactly what data are collected by the technologies they use, how they are collected, and for what purpose. Research shows that negative perceptions of managerial surveillance lower trust in the manager–employee relationship, adversely affecting employee performance and job satisfaction.

Collaboration technologies create a sense of openness where managers and employees can easily text, call, see each other, and freely share information. However, this openness may also make some people feel they will be judged if they say the wrong thing. An employee may hold back on sharing their thoughts in a recorded online meeting if they're not confident about sounding intelligent. Collaboration technologies can create conditions for self-censorship, even if the technology is perceived to be user-friendly. This may limit opportunities to build closer connections with managers and other employees.

Although each organization has its own specific needs, recommendations for making virtual meetings more effective and less tiring include the following:

- Cancel unnecessary meetings and make necessary meetings shorter.
- Use breakout rooms for problem solving, discussions, and social interactions.
- Use asynchronous meetings, such as by creating a shared Google Doc for employees to contribute to throughout the day.
- Build in breaks during long meetings and in between back-to-back meetings.
- Moderate and facilitate virtual meetings more actively, moving topics along when needed and ensuring that everyone has an opportunity to contribute.
- Turn off "self-view," if possible, on your meeting platform and make camera use optional for some meetings.
- Implement meeting-free blocks of time or days.

This last recommendation is especially important. Although building trust and achieving team cohesion rely on frequent, quality interactions, increasing virtual meetings is no longer the best way to accomplish this. As a result, many organizations, including Meta, Asana, and Highfive have adopted no-meeting days, during which people work according to their own rhythms and collaborate with others at a pace and on a schedule that is not forced.

Professor Ben Laker, University of Reading Henley Business School, along with human resource management professors Vijay Pereira, Pawan Budhwar, and Ashish Malik, surveyed 76 companies, with more than 1,000 employees each and operations in more than 50 countries. These companies had introduced from one to five no-meeting days per week (including one-on-one meetings) during 2021. Laker and his colleagues examined data comparing employee stress levels before and after a reduction in meetings and assessed the impact on productivity, collaboration, and engagement.

Nearly half (47 percent) of the companies Laker and his team studied reduced meetings by 40 percent by introducing two no-meeting days per week. Thirty-five percent instituted three no-meeting days, and 11 percent implemented four. The remaining 7 percent eradicated meetings entirely.

The subsequent impact of introducing meeting-free days was profound. Regardless of the number of meeting-free days instituted, autonomy, communication, engagement, and satisfaction all improved, resulting in decreased stress and a rise in productivity. The best results were achieved at companies that had three meeting-free days per week. Eliminating 60 percent of meetings, equivalent to three days per week, increased cooperation by 55 percent and lowered the risk of stress by 57 percent. Workers replaced meetings with better ways of connecting one-on-one, at a pace suitable for them.

Sources: Ben Laker, Vijay Pereira, Pawan Budhwar, and Ashish Malik, "The Surprising Impact of Meeting-Free Days," *MIT Sloan Management Review* January 18, 2022; Lebene Soga, Yemisi Bolade-Ogunfodun, and Ben Laker, "Design Your Work Environment to Manage Unintended Tech Consequences," *MIT Sloan Management Review* September 20, 2021; Rob Cross, "Easing the Invisible Burdens of Collaboration," *MIT Sloan Management Review* September 14, 2021; Heidi Mitchell, "How Videoconferences Make It Harder for Employees to Collaborate," *Wall Street Journal* June 10, 2021; Katie Kavanagh, Nicole Vrees, Liana Kreamer, and Steven G. Rogelberg, "How to Combat Virtual Meeting Fatigue," *MIT Sloan Management Review* March 30, 2021; Paul Leonardi, "Picking the Right Approach to Digital Collaboration," *MIT Sloan Management Review* January 27, 2021.

## CASE STUDY QUESTIONS

**2-12** How do virtual meeting technologies enhance collaboration?

**2-13** How do virtual meeting technologies hinder collaboration? Identify the people, organization, and technology factors that contribute to this problem.

**2-14** What can organizations do to improve the effectiveness of collaboration technology?

**2-15** If a midsized company wants to implement virtual meeting technology to support remote work and collaboration, what issues should be considered?

## Chapter 2 References

Banker, Rajiv D., Nan Hu, Paul A. Pavlou, and Jerry Luftman. "CIO Reporting Structure, Strategic Positioning, and Firm Performance." *MIS Quarterly* 35, No. 2 (June 2011).

Bughin, Jacques, Michael Chui, and Martin Harrysson. "How Social Tools Can Reshape the Organization." McKinsey Global Institute (May 2016).

Bureau of Labor Statistics. "Occupational Outlook Handbook 2020–2021." Bls.gov, accessed March 4, 2022.

Colony, George F. "CIOs and the Future of IT." *MIT Sloan Management Review* 59, No. 3 (Spring 2018).

Cross, Rob. "Easing the Invisible Burdens of Collaboration." *MIT Sloan Management Review* (September 14, 2021).

Cummings, Jeff, and Alan Dennis. "Virtual First Impressions Matter: The Effect of Enterprise Social Networking Sites on Impression Formation in Virtual Teams." *MIS Quarterly* 42, No. 3 (September 2018).

De Vreede, Gert-Jan, and Robert O. Briggs. "A Program of Collaboration Engineering Research and Practice: Contributions, Insights, and Future Directions." *Journal of Management Information Systems* 36, No. 1 (2019).

Dodge, Sheila, Don Kieffer, and Nelson P. Repenning. "Breaking Logjams in Knowledge Work." *MIT Sloan Management Review* (Fall 2018).

Feng, Cong, Pankaj C., Patel, and Scott Fay. "The Value of the Structural Power of the Chief Information Officer in Enhancing Forward-Looking Firm Performance." *Journal of Management Information Systems* 38 No. 3 (2021).

Harvard Business Review Analytic Services. "Collaboration Technology Boosts Organizations." Insight Enterprises Inc. (February 13, 2017).

Hill, N. Sharon, and Kathryn M. Bartol. "Five Ways to Improve Communication in Virtual Teams." *MIT Sloan Management Review* 60, No. 1 (Fall 2018).

Kane, Gerald C., Rich Nanda, Anh Phillips, and Jonathan Copulsky. "Redesigning the Post-Pandemic Workplace. *MIT Sloan Management Review* 62 No. 3 (Spring 2021).

Kiron, David. "Why Your Company Needs More Collaboration." *MIT Sloan Management Review* 59, No. 1 (Fall 2017).

Leonardi, Paul. "Picking the Right Approach to Digital Collaboration." *MIT Sloan Management Review* 62 No. 3 (Spring 2021).

Leonardi, Paul, and Tsedal Neely. "What Managers Need to Know About Social Tools." *Harvard Business Review* (November–December 2017).

"Looking to Replace Email with a Solution to Increase Engagement, Cubic Telecom Chose Workvivo." Workvivo.com, accessed July 3, 2022.

Magni, Massimo, and Likoebe Maruping. "Unleashing Innovation with Collaboration Platforms." *MIT Sloan Management Review Frontiers* (April 30, 2019).

McKinsey & Company. "Transforming the Business through Social Tools" (2015).

McKinsey Global Institute. "The Social Economy: Unlocking Value and Productivity through Social Technologies." McKinsey & Company (July 2012).

Reck, Fabian, and Alexander Fliaster. "Four Profiles of Successful Digital Executives," *MIT Sloan Management Review Frontiers* (April 10, 2019).

Samdahl, Erik. "Top Employers Are 5.5X More Likely to Reward Collaboration." I4cps.com (June 22, 2017).

Soga, Lebene, Yemisi Bolade-Ogunfodun, and Ben Laker. "Design Your Work Environment to Manage Unintended Tech Consequences," *MIT Sloan Management Review* (September 20, 2021).

Srivastava, Shirish C., and Shalini Chandra. "Social Presence in Virtual World Collaboration: An Uncertainty Reduction Perspective Using a Mixed Methods Approach." *MIS Quarterly* 42, No. 3 (September 2018).

Strack, Ben. "KPMG Kicks Off Metaverse Collaboration Hub." *Blockworks* (June 22, 2022).

Subramani, Mari, Mihir Wagle, Gautam Ray, and Aolk Gupta. "Capability Development through Just-in-Time Access to Knowledge in Document Repositories: A Longitudinal Examination of Technical Problem Solving." *MIS Quarterly* 45 No. 3 (September 2021).

Zaffron, Steve, and Gregory Unruh. "Your Organization Is a Network of Conversations." *MIT Sloan Management Review Frontiers* (July 10, 2018).

CHAPTER 3

# Achieving Competitive Advantage with Information Systems

## LEARNING OBJECTIVES

After completing this chapter, you will be able to:

3-1 Demonstrate how Porter's competitive forces model, the value chain model, synergies, core competencies, and network-based strategies help companies use information systems for competitive advantage.

3-2 Describe how information systems help businesses compete globally.

3-3 Describe how information systems help businesses compete using quality and design.

3-4 Explain the role of business process management (BPM) in enhancing competitiveness.

3-5 Understand how MIS can help your career.

## CHAPTER CASES

- Walmart's Supercenter Strategy
- Customer Experience Management: A New Strategic Weapon
- Signet Jewelers Sparkles with a Virtual Sales Process
- Shipping Wars

**MyLab MIS**

- **Video Case:** Celonis Tops $11 Billion Valuation with New Round of Funding
- Discussion Questions: 3-5, 3-6, 3-7
- Hands-on MIS Projects: 3-8, 3-9, 3-10, 3-11
- eText with Figure Videos

## WALMART'S SUPERCENTER STRATEGY

**Walmart** is the dominant retail chain in the United States but has spent years struggling to catch up with e-commerce leader Amazon, investing billions to grow online and improve its stores. Although its e-commerce sales have grown significantly, reaching more than $70 billion in 2021, Walmart's e-commerce operations are not yet profitable.

In December 2019 CEO Doug McMillon pointed out that Walmart wasn't going to win by building an unprofitable e-commerce operation or other standalone ventures. Instead, the company would use its giant supercenters—webs of businesses all working together to attract shoppers and generate profits. More than half of Walmart's growth in US e-commerce sales came from expanding online grocery pickup or delivery service run out of stores. Supercenters would be a way to further monetize Walmart's geographic footprint.

Walmart's supercenters are sprawling stores of up to 180,000 square feet, offering 100,000 products, including groceries, clothing, and televisions. But the supercenters are also places where customers can fill medical prescriptions, transfer money, have their hair cut, or get fitted for eyeglasses. The stores are the bases for same-day pickup and delivery, digital entertainment, financial services, and one-stop shops for healthcare, offering primary care, urgent care, diagnostics, x-rays, behavioral health, and dental care. In the future, Walmart could also use the supercenters to provide warehouse and shipping capacity to third parties selling through Walmart, allowing more companies to easily sell their products on Walmart.com. For an increasing number of customers, Walmart will be viewed as a service.

Walmart's strategy is to have all parts of the business interacting to drive profitable growth. For instance, Walmart is now using the data it has amassed on hundreds of millions of customers to sell online advertising to brands such as Tide and Kellogg's. Walmart Connect is an advertising network that helps brands target online ads based on Walmart shopper data. Walmart is not selling individual shopper data but uses anonymized data to help companies target ads more precisely.

Walmart already has a number of data centers and uses edge computing technology, in which data are processed physically close to where they are being collected, which enables the data to be processed more rapidly by cloud computing. It plans to take edge computing to the next level, spreading out edge computing centers among Walmart retail locations for use by autonomous vehicles and other systems that need to

process large amounts of data quickly. Walmart could also rent out data processing power to local customers, bypassing traditional remote cloud computing providers.

Some experts warn, however, that Walmart's supercenter strategy might prove tricky to implement successfully. Some supercenter capabilities could require expensive and complicated work, with no assurance of substantial payoff. And even if Walmart's supercenter initiatives do well, they risk distracting Walmart from its core retail operations, which matter much more.

Sources: Dan Berthiaume, "Walmart Has Big Plans for Its Walmart Connect Ad Platform," Chainstoreage.com, March 30, 2022; Sai Balasubramanian, "Walmart May Soon Become the Largest Primary Care Provider in the Country, *Forbes*, February 22, 2022; James Melton, "Walmart's Ecommerce Sales Grew 11% in Fiscal 2022," Digitalcommerce360.com, February 17, 2022 "Walmart to Open 4,000 Healthcare 'Supercenters' by 2029 that Include 'Comprehensive' Clinical Laboratory Services," *Dark Daily*, May 3, 2021; Mike Troy, "Walmart Disrupting Grocery Again," *Progressive Grocer*, March 16, 2021; James Brumle, "Walmart's Supercenter Plans Are a Double-Edged Sword," *The Motley Fool*, January 1, 2020; and Sarah Nassauer, "Walmart's Secret Weapon to Fight Off Amazon: The Supercenter," *Wall Street Journal*, December 21, 2019.

---

Walmart's supercenter strategy illustrates some of the ways that information systems help businesses compete, as well as the challenges of finding the right business strategy and how to use technology in that strategy. Retailing today is an extremely crowded and competitive playing field, both online and in physical, brick-and-mortar stores. Even though Walmart is the world's leading retailer, it has powerful competitors, such as Amazon, and Walmart is searching for ways to keep growing its business. Customers are increasingly doing more retail shopping online, but Walmart's foray into e-commerce has proved costly. Walmart needs a business strategy that includes e-commerce but that makes better use of its competitive strengths—its wide footprint of physical store locations.

The chapter-opening diagram calls attention to important points raised by this case and this chapter. Walmart's business model is based on a low-cost retail leadership strategy, but management now wants to broaden this strategy to include a greater focus on customer niche (targeted advertising), more emphasis on creating shopping spaces with more customer conveniences, and new services such as healthcare and edge computing centers. Walmart is using information technology both as support for the operations of its new supercenters and also as a service for sale (edge computing).

Here are some questions to think about: What are the components of Walmart's supercenter business strategy? How much does technology support that strategy? Explain your answers.

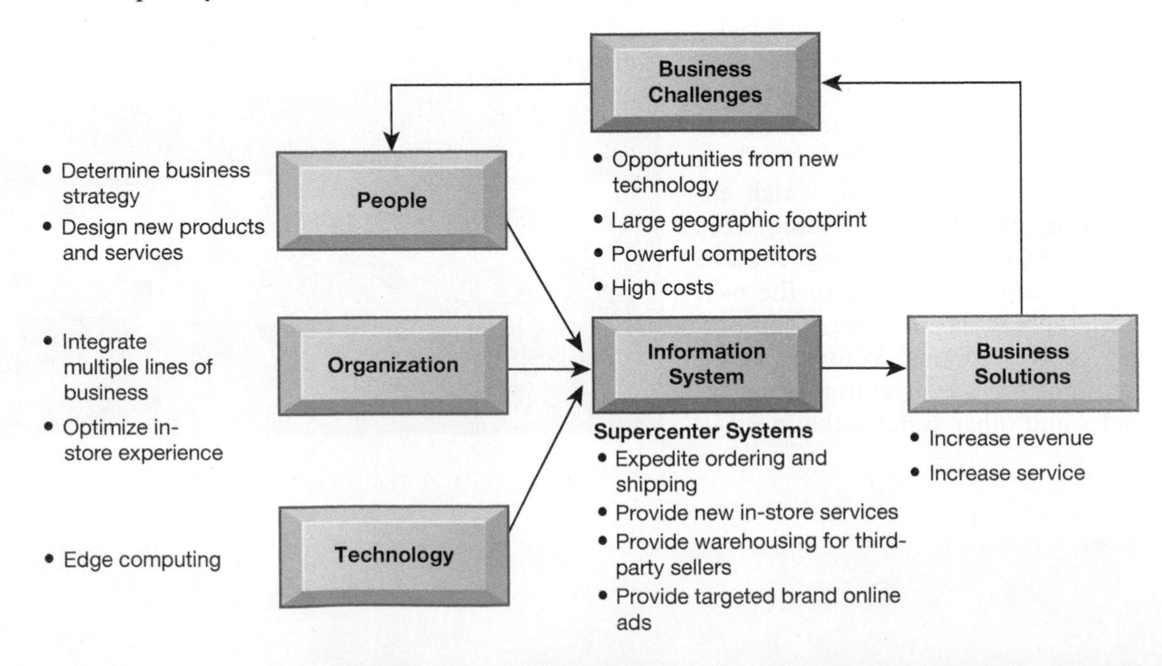

# 3-1 Demonstrate how Porter's competitive forces model, the value chain model, synergies, core competencies, and network-based strategies help companies use information systems for competitive advantage.

In almost every industry you examine, you will find that some firms do better than most others do. There's almost always a standout firm. In online retail, Amazon is the leader; in offline retail, Walmart, the largest retailer on earth, has been the leader. In search engine services, Google is considered the leader.

Firms that do better than others are said to have a competitive advantage. Either they have access to special resources that others do not or they use commonly available resources more efficiently—usually because of superior knowledge and information assets. In any event, more successful firms do better in terms of revenue growth, profitability, or productivity growth (efficiency), all of which ultimately translate into higher stock market valuations than their competitors.

But why do some firms do better than others, and how do they achieve competitive advantage? How can you analyze a business and identify its strategic advantages? How can you develop a strategic advantage for your own business? How do information systems contribute to strategic advantages? One answer to these questions is Michael Porter's competitive forces model.

## PORTER'S COMPETITIVE FORCES MODEL

Arguably, the most widely used model for understanding competitive advantage is Michael Porter's **competitive forces model**. This model provides a general view of the firm, its competitors, and the firm's environment. Recall that in Chapter 2 we described the importance of a firm's environment and the dependence of firms on their environments. Porter's model is all about the firm's general business environment.

Examine Figure 3.1, which illustrates Porter's model. In this model, five competitive forces shape the fate of the firm: competitors that directly compete with the firm and also four forces—new market entrants, substitute products and services, customers, and suppliers—within the firm's industry environment. View the Figure 3.1 video in the eText for an animated and more detailed discussion of this figure. We also further examine each of these forces in the following sections.

### Traditional Competitors

Firms compete with other firms that offer similar products and services by developing new, more efficient ways to produce their products, by introducing new products and services, and by developing their brands to attract customers and impose switching costs on their customers.

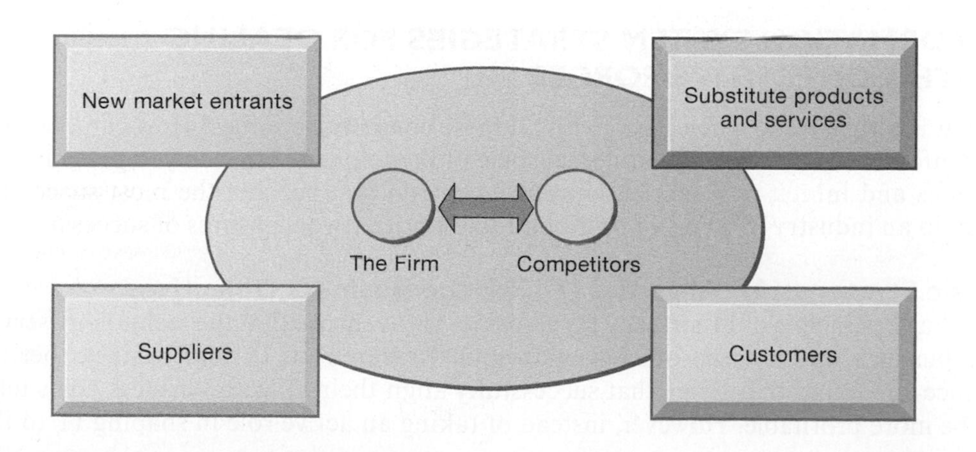


**Figure 3.1**
Porter's Competitive Forces Model
*In Porter's competitive forces model, the strategic position of the firm and its strategies are determined not only by competition with its traditional, direct competitors but also by four forces in the industry's environment: new market entrants, substitute products and services, customers, and suppliers.*

### New Market Entrants

In a free economy with mobile labor and financial resources, new companies are always entering the marketplace. In some industries, there are very low barriers to entry, whereas in other industries, entry is very difficult. For instance, it is fairly easy to start a pizza business or just about any small retail business, but it is much more expensive and difficult to enter the computer chip business, which has very high capital costs and requires significant expertise and knowledge that is hard to obtain. New companies have several possible advantages: They are not locked into old plants and equipment; they often hire younger workers, who are less expensive and perhaps more innovative; they are not encumbered by old, worn-out brand names; and they are more hungry (more highly motivated) than traditional occupants of an industry. These advantages are also their weaknesses: They may need to depend on outside financing for new plants and equipment, which can be expensive; they have a less-experienced workforce; and they have little brand recognition.

### Substitute Products and Services

In just about every industry, there are substitutes that your customers might use if your prices become too high. New technologies create new substitutes all the time. Ethanol can substitute for gasoline in cars; vegetable oil for diesel fuel in trucks; and wind, solar, and hydropower for coal, oil, and gas electricity generation. Likewise, Internet telephone service has substituted for traditional telephone service, and streaming Internet music services have substituted for CDs, music stores, and digital download services like iTunes. The more substitute products and services in your industry, the less you can control pricing and the lower your profit margins.

### Customers

A profitable company depends in large measure on its ability to attract and retain customers (while denying this ability to competitors) and charge high prices. The power of customers grows if they can easily switch to a competitor's products and services or if they can force a business and its competitors to compete on price alone in a transparent marketplace, where there is little product differentiation, and all prices are known instantly (such as on the Internet). For instance, the Internet has enabled students (customers) to find multiple suppliers of just about any current college textbook, giving them significant power over college bookstores, which had previously been the only place such books could be purchased.

### Suppliers

The market power of suppliers can have a significant impact on firm profits, especially when the firm cannot raise prices as quickly as suppliers can. The more suppliers a firm has, the greater the control it can exercise over those suppliers in terms of price, quality, and delivery schedules. For instance, manufacturers of laptop PCs usually have multiple competing suppliers of key components, such as keyboards, hard drives, and display screens.

## INFORMATION SYSTEM STRATEGIES FOR DEALING WITH COMPETITIVE FORCES

What is a firm to do when faced with all these competitive forces? How can the firm use information systems to counteract some of these forces? How do you prevent substitutes and inhibit new market entrants? How do you become the most successful firm in an industry in terms of profit and share price (two measures of success)?

### Basic Strategy 101: Align the IT with the Business Objectives

The basic principle of IT strategy for a business is to ensure that the technology serves the business and not the other way around. Research on IT and business performance has found that firms that successfully align their IT with business goals tend to be more profitable. However, instead of taking an active role in shaping IT to the enterprise, many businesses ignore IT, claim not to understand IT, and tolerate failure in the IT area as just a nuisance to work around. Such firms pay a hefty price in poor performance. Successful firms and managers understand what IT can do and how it works, take an active role in shaping its use, and measure its impact on revenues and profits (Hind, Leidner, and Preston, 2019; Luftman, 2003; Weill and Aral, 2005).

So, how do you as a manager achieve this alignment of IT with business? In the following sections, we discuss some basic ways to do this, but here's a summary:

- Identify your business strategy and goals.
- Break these strategic goals down into concrete activities and processes.
- Identify how you will measure progress toward the business goals (e.g., by using metrics).
- Ask yourself, "How can information technology help me achieve progress toward our business goals, and how will it improve our business processes and activities?"
- Measure actual performance. Let the numbers speak.

Let's see how this works out in practice. There are four generic strategies, each of which often is enabled by using information technology and systems: low-cost leadership, product differentiation, focus on market niche, and strengthening customer and supplier intimacy.

### Low-Cost Leadership

Use information systems to achieve the lowest operational costs and the lowest prices. The classic example is Walmart. By keeping prices low and shelves well stocked using a legendary inventory replenishment system, Walmart became the leading retail business in the United States. Point-of-sale terminals record the bar code of each item passing the checkout counter and send a purchase transaction directly to a central computer at Walmart headquarters. The computer collects the orders from all Walmart stores and transmits the orders to suppliers. Suppliers can also access Walmart's sales and inventory data by using web technology.

Because the system replenishes inventory at lightning speed, Walmart does not need to spend much money on maintaining large inventories of goods in its own warehouses. The system also enables Walmart to adjust the items stocked in its stores to meet customer demands. By using systems to keep operating costs low, Walmart can charge less for its products than competitors do yet reap higher profits.

Walmart's continuous replenishment system is also an example of an **efficient customer response system**. An efficient customer response system directly links consumer behavior to distribution and to production and supply chains. Walmart's continuous replenishment system provides such an efficient customer response.

*Supermarkets and large retail stores such as Walmart use sales data captured at the checkout counter to determine which items have sold and need to be reordered. Walmart's continuous replenishment system transmits orders to restock directly to its suppliers. The system enables Walmart to keep costs low while fine-tuning its merchandise availability to meet customer demands.*

## Product Differentiation

Use information systems for **product differentiation**, by creating new products and services or by enhancing customer convenience in using your existing products and services. For instance, Google continuously introduces new and unique digital services, such as its Google Pay mobile payment service, and improvements in Google Photos and Google Assistant. Apple has continued to differentiate its mobile devices with nearly annual introductions of new iPhone and iPad models. Crayola is creating new digital products and services such as Crayola Color Camera, which allows children to turn their favorite photos into digital coloring book pages.

Manufacturers and retailers are using information systems to create products and services that are customized and personalized to fit the precise specifications of individual customers. For example, Baume, an offshoot of the luxury brand Baume et Mercier, uses recycled, upcycled, and sustainable materials to craft distinctive watches. With Baume's mobile-optimized website, which is built on Salesforce Commerce Cloud, customers can pick and mix different watch elements, including dial style, casing finish, strap material, and engraved messages. Customers can choose from 2,160 variations to create their own unique product. This ability to offer individually tailored products or services using the same production resources as mass production is called **mass customization**.

More and more companies are differentiating their products not just by the features of the products themselves but also by the entire experience of buying and using the product. This is called the "customer experience," and **customer experience management** as a competitive strategy is described more fully in the Spotlight on People case.

Table 3.1 lists a number of companies that have developed IT-based products and services that other firms have found difficult to copy, and which, as a result, provide a competitive advantage.

## Focus on Market Niche

Use information systems to enable a specific market focus and to serve this narrow target market better than competitors do. Information systems support this strategy by producing and analyzing data for finely tuned sales and marketing techniques. Information systems enable companies to analyze customer buying patterns, tastes, and preferences closely so that they can efficiently pitch advertising and marketing campaigns to smaller and smaller target markets.

The data come from a range of sources—credit card transactions, demographic data, purchase data from checkout counter scanners at supermarkets and retail stores, and data collected when people access and interact with websites. Sophisticated software tools find patterns in these large pools of data and infer rules from them that can be used to guide decision making. Analysis of such data drives one-to-one marketing by which personal messages can be created based on individualized preferences. A data-driven strategy is more likely to succeed if the data offer high and lasting value, lead to improvements that can't be easily imitated, or generate insights that can be quickly utilized (Hagiu and Wright, 2020).

**TABLE 3.1**

IT-Enabled New Products and Services Providing Competitive Advantage

| | |
|---|---|
| One-click shopping: Amazon | For many years, Amazon held a patent on one-click shopping that it licensed to other online retailers. |
| Online music: Apple Music | Apple sells access to an online library of 50 million songs. |
| Sneaker customization: Nike | Nike sells customized sneakers through its NIKE BY YOU program on its website. Customers can select the type of shoe and the shoe's colors, material, outsoles, and laces. The sneakers take only about three weeks to reach the customer. |
| Online person-to-person payment: PayPal.com | PayPal enables the transfer of money between individual bank accounts and between bank accounts and credit card accounts. |

## Customer Experience Management: A New Strategic Weapon

More and more companies are finding new strategic opportunities by looking at how customers interact with their products. They are finding that customers don't want an emotionless, straightforward transaction—they want a good experience, too. Customer experience is emerging as a major brand differentiator and an increasingly powerful driver of business success. Traditional advertising is becoming more difficult; IT provides new ways to cater to customers; and social media can amplify bad experiences.

Customer experience is distinguished from customer service in that the former encompasses the entire customer journey—all the interactions, including awareness, discovery, cultivation, advocacy, purchases, and service, between a customer and a company throughout their entire business relationship.

Businesses are finding that the experience of the product goes beyond the product alone: It's everything concerning the product. How does the product feel? How well does it work? According to Donald Chestnut, Mastercard's chief experience officer, the experience of the product is more important than the product itself.

The chief experience officer (CXO) is a new position that many companies have created to manage their customer experiences. Hundreds of large companies in the United States, Canada, and United Kingdom surveyed by Gartner Inc. consultants employ a chief experience officer or equivalent position. Restaurant chain TGI Fridays, Citizens Financial Group, TripAdvisor, Volkswagen, and Under Armour are among the companies that now have a CXO.

Companies that implement a successful customer experience strategy achieve higher customer satisfaction, increased revenues, and reduced customer turnover. Customers are willing to pay higher prices (higher by as much as 13 percent) if they have had a great customer experience. Customers are willing to pay companies with high customer experience ratings 140 percent more and remain loyal for up to six years. A survey by Bloomberg Businessweek found that delivering a great customer experience has become a top strategic objective.

The best customer experiences result when a company is able to create an emotional connection with a customer. For example, when a customer was late returning a pair of shoes to online retailer Zappos because her mother had passed away, Zappos, which is known for providing outstanding customer service and embracing change, paid for the return shipping and had a courier pick up the returned shoes at no cost. The next day the customer received a bouquet of flowers delivered to her home with a note expressing condolences from the Zappos customer service team.

Amazon is another leader in customer experience management, and its business strategy focuses on improving all aspects of customers' interactions with the company's pricing, payment, checkout, and shipping components. Business processes focus on ways of making the customer experience more pleasant. An example is Amazon Prime, a paid subscription service that gives users access to services that would otherwise be unavailable or cost extra, such as free two-day delivery, music streaming, and video. Amazon's shipping times were too long, so Amazon created its own Prime shipping network. Checking out was too difficult, so Amazon introduced one-click ordering for Prime customers. Customers weren't loyal to the Amazon brand—they looked for the best price wherever it was available. The Amazon Prime subscription service decreases price sensitivity because customers feel like they are missing out if they don't take advantage of the fast shipping they have already paid for.

Netflix's subscribers indicated that they wanted a service to highlight coming attractions, so the company tested a large preview box at the top of the screen. However, although customers had asked for this capability, they didn't use it, and the box was actually making it harder for users to access programs they wanted to view right away. Netflix solved this problem by moving the coming attractions box to a section that viewers could navigate to if they were interested in those attractions.

Netflix made another change to let binge watchers skip the opening credits of its serialized TV shows. Long opening-credit sequences make sense when TV show episodes arrive a week apart and people watch other shows in between. However, binge viewing has become more popular, causing Netflix to re-examine the customer experience for this type of viewer. The company found that the old opening-credit format degraded the experience for people watching several episodes of the same show in a row. Netflix allowed some members to use a skip button, and they loved it.

The Covid-19 pandemic forced retail and hospitality companies to respond quickly to sudden changes in consumer preferences and needs. Health and safety became customers' top concerns, and companies had to create new digital customer experiences that would be as personal and engaging as face-to-face interactions.With customer health concerns in mind, the Coca-Cola Company redesigned the customer experience with its Freestyle drink dispensers. Coca-Cola operates more than 50,000 Freestyle drink dispensers in restaurants, convenience stores, cinemas, and other locations. Coca-Cola Freestyle is a touch screen soda fountain that features more than 100 different Coca-Cola drink products as well as customised flavors. The machine allows users to select from mixtures of flavors of Coca-Cola-branded products, which are then individually dispensed.

At the beginning of the Covid-19 pandemic, Coca-Cola Freestyle conducted research on consumer feelings about interacting with the machines. The company found that 50 percent of soda fountain customers worried about touching a Freestyle machine. These findings prompted Coca-Cola to offer its customers a contactless Freestyle experience. Using Amazon Web Services, the company quickly developed a Mobile Pour capability where consumers choose and have drinks dispensed from Freestyle machines in near real time using their smartphones and without having to create an account or download an app.

Sources: Zoe Hillenmeyer, "Powering the Future of Retail and Hospitality," Partners.wsj.com, accessed March 4, 2022; "You're Not Just Binge Watching Netflix. You're Having an 'Experience,'" *Wall Street Journal*, February 21, 2020; and Steven MacDonald, "7 Ways to Create a Great Customer Experience Strategy," Superoffice.com, February 4, 2020.

## CASE STUDY QUESTIONS

1. What is customer experience management? How can it contribute to competitive advantage?
2. How does information technology support customer experience management? Give some examples.
3. How did information technology and customer experience management change operations and decision making at the organizations described in this case?

For example, the Charlotte Hornets NBA basketball team maintains detailed records on its millions of fans that include real-time data on buying tickets online and at games, purchasing team gear in-store and online, and eating and drinking at games. The team combines these fan records to create a single profile for each fan. The Hornets use these profiles to build a more personal relationship with fans, which can lead to a better fan experience, more targeted marketing, and more sales. If a fan's profile shows that the person spends $5,000 per year on games and food, always buys nachos in the second quarter of every game, or purchased a LaMelo Ball jersey, that fan might receive free nachos at one game or an invitation to a meet and greet with LaMelo Ball. Contemporary customer relationship management (CRM) systems feature analytical capabilities for this type of intensive data analysis (see Chapters 9 and 11).

### Strengthen Customer and Supplier Intimacy

Use information systems to tighten linkages with suppliers and to develop intimacy with customers. Toyota, Ford, and other automobile manufacturers have information systems that give their suppliers direct access to their production schedules, enabling suppliers to decide how and when to ship supplies to the plants where cars are assembled. This allows suppliers more lead time in producing goods. On the customer side, Amazon.com keeps track of user preferences for all the products it sells, and can recommend products purchased by others to its customers. Strong linkages to customers and suppliers increase **switching costs** (the cost of switching from a product or service to a competitor's) and loyalty to your firm.

Table 3.2 summarizes the competitive strategies we have just described.

**TABLE 3.2**

Four Basic Competitive Strategies

| Strategy | Description | Example |
| --- | --- | --- |
| Low-cost leadership | Use information systems to produce products and services at a lower price than competitors do while enhancing quality and level of service. | Walmart |
| Product differentiation | Use information systems to differentiate products and provide new services and products. | Uber, Nike, Apple, Baume |
| Focus on market niche | Use information systems to enable a focused strategy on a single market niche; specialize. | Charlotte Hornets, Harrah's |
| Customer and supplier intimacy | Use information systems to develop strong ties and loyalty with customers and suppliers. | Toyota Corporation, Amazon |

As shown by the cases throughout this book, successfully using information systems to achieve a competitive advantage requires a precise coordination of technology, organizations, and people. Indeed, as many have noted with regard to Walmart, Apple, and Amazon, the ability to implement information systems successfully is not equally distributed, and some firms are much better at it than others are.

## THE INTERNET'S IMPACT ON COMPETITIVE ADVANTAGE

Because of the Internet, the traditional competitive forces are still at work, but competitive rivalry has become much more intense (Porter, 2001). Internet technology is based on universal standards that any company can use, making it easier for rivals to compete on price alone and for new competitors to enter the market. Because information is available to everyone, the Internet raises the bargaining power of customers, who can quickly find the lowest-cost provider on the web. Profits have often been dampened as a result of increased competition. Table 3.3 summarizes some of the potentially negative impacts of the Internet on business firms that Porter has identified.

The Internet has nearly destroyed some industries and has severely threatened others. For instance, the printed encyclopedia industry and the travel agency industry have been nearly decimated by the availability of substitutes on the Internet. Likewise, the Internet has had a significant impact on the retail, music, book, retail brokerage, software, and telecommunications industries. However, the Internet has also created

**TABLE 3.3**

Impact of the Internet on Competitive Forces and Industry Structure

| Competitive Force | Impact of the Internet |
| --- | --- |
| Substitute products or services | Enables new substitutes to emerge with new approaches to meeting needs and performing functions |
| Customers' bargaining power | Shifts bargaining power to customers because of the availability of global price and product information |
| Suppliers' bargaining power | Tends to raise bargaining power over suppliers in procuring products and services; however, suppliers can benefit from reduced barriers to entry and from the elimination of distributors and other intermediaries standing between them and their users |
| Threat of new entrants | Reduces barriers to entry such as the need for a sales force, access to channels, and physical assets; provides a technology for driving business processes that makes other things easier to do |
| Positioning and rivalry among existing competitors | Widens the geographic market, increasing the number of competitors and reducing differences among competitors; makes it more difficult to sustain operational advantages; puts pressure on businesses to compete on price |

entirely new markets; formed the basis for thousands of new products, services, and business models; and provided new opportunities for building brands with very large and loyal customer bases. Amazon, eBay, Apple, YouTube, Facebook, Uber, and Google are examples of companies that have successfully leveraged Internet technology for business success. In this sense, the Internet is transforming entire industries, forcing firms to change how they do business.

### Smart Products and the Internet of Things

The growing use of Internet of Things (IoT)–linked sensors in industrial and consumer products is an excellent example of how the Internet is changing competition within industries and enabling the creation of new products and services. Nike, Under Armour, and other sports and fitness companies are pouring money into wearable health trackers and fitness equipment that use sensors to report users' activities to remote computing centers where the data can be analyzed. John Deere tractors are loaded with field radar, GPS transceivers, and hundreds of sensors keeping track of the equipment. GE created a new business out of helping its aircraft and wind turbine clients improve operations by examining the data generated by the many thousands of sensors in the equipment. The result is what's referred to as smart products—products that are a part of a larger set of information services sold by firms (Porter and Heppelmann, 2014; Gandhi and Gervet, 2016).

Shoes, clothing, watches (think Apple Watch), water bottles, and even toothbrushes have been redesigned to incorporate sensors and metering devices connected to the Internet so that performance can be monitored and analyzed. Homes increasingly use smart devices such as smart thermostats, smart electrical meters, smart security systems, and smart lighting systems. Smart products offer new functionality, greater reliability, and more intense use of products. They expand opportunities for product and service differentiation while providing opportunities for improving both the product and the customer experience. When you buy a wearable digital health product, you get not only the product itself but also the host of cloud-based services available from the manufacturer. Smart products generally inhibit new entrants to a market simply because existing customers are enmeshed in the dominant firm's software environment. Finally, smart products may decrease the power of suppliers of industrial components if, as many believe, the physical product becomes less important than the software and hardware that make it function.

## THE BUSINESS VALUE CHAIN MODEL

Although the Porter model is helpful for identifying competitive forces and suggesting generic strategies, it is not specific about what exactly to do, and it does not provide a methodology to follow for achieving competitive advantages. If your goal is to achieve operational excellence, where do you start? Here's where the business value chain model is helpful.

The **value chain model** highlights specific activities in the business where competitive strategies can best be applied (Porter, 1985) and where information systems are most likely to have a strategic impact. This model identifies specific, critical advantage points at which a firm can use information technology most effectively to enhance its competitive position. The value chain model views the firm as a series or chain of basic activities that add a margin of value to a firm's products or services. These activities can be categorized as either primary activities or support activities.

**Primary activities** are most directly related to the production and distribution of a firm's products and services, which create value for the customer. Primary activities include inbound logistics, operations, outbound logistics, sales and marketing, and service. Inbound logistics includes receiving and storing materials for distribution to production. Operations transforms inputs into finished products. Outbound logistics entails storing and distributing finished products. Sales and marketing includes promoting and selling the firm's products. The service activity includes maintenance and repair of the firm's goods and services.

**Support activities** make the delivery of the primary activities possible and consist of organization infrastructure (administration and management), human resources (employee recruiting, hiring, and training), technology (improving products and the production process), and procurement (purchasing input).

Examine Figure 3.2, which illustrates the participants in an industry value chain (suppliers' suppliers, suppliers, the firm, distributors, and customers), with particular attention given to various systems for primary activities (inbound logistics, operations, sales and marketing, service, and outbound logistics) and support activities (administration and management, human resources, technology, and procurement) that add value to a firm's products and services. For example, an automated warehousing system (a form of sourcing and procurement system) adds value by creating more efficient inbound logistics (a primary activity). Similarly, a computerized purchasing system adds value by creating a more efficient procurement process (a support activity).

You can ask at each stage of the value chain, "How can we use information systems to improve operational efficiency and to improve customer and supplier intimacy?" This will force you to examine critically how you perform value-adding activities at each stage and how the business processes might be improved. For example, value chain analysis indicates that Walmart, described in the chapter-opening case, should use technology to improve its processes for sales, marketing, and customer service, and for differentiating its products.

You can also begin to ask how information systems can be used to improve the relationship with customers and with suppliers who lie outside the firm's value chain but who belong to the firm's extended value chain, where they are absolutely critical to your success. Here, supply chain management systems that coordinate the flow of resources into your firm, and customer relationship management systems that coordinate your sales and support employees with customers, are two of the most common system applications that result from a business value chain analysis. We discuss these enterprise applications in detail in Chapter 9.

Using the business value chain model will also encourage you to consider benchmarking your business processes against your competitors' or those of others in related industries and identifying industry best practices. **Benchmarking** involves comparing the efficiency and effectiveness of your business processes against strict standards and then measuring performance against those standards. Industry **best practices** are usually identified by consulting companies, research organizations, government agencies, and industry associations as the most successful solutions or problem-solving methods for consistently and effectively achieving a business objective.

Once you have analyzed the various stages in the value chain at your business, you can come up with a list of potential information systems applications. Then, when you have a list of candidate applications, you can decide which to develop first. By making improvements to your own business value chain that your competitors might miss, you can achieve competitive advantage by attaining operational excellence, lowering costs, improving profit margins, and forging a closer relationship with customers and suppliers. If your competitors are making similar improvements, then at least you will not be at a competitive disadvantage—the worst of all situations!

### Extending the Value Chain: the Value Web

Figure 3.2 shows that a firm's value chain is linked to the value chains of its suppliers, distributors, and customers. After all, the performance of most firms depends not only on what goes on inside a firm but also on how well the firm coordinates with direct and indirect suppliers, delivery firms (logistics partners, such as FedEx or UPS), and, of course, customers.

How can information systems be used to achieve strategic advantage at the industry level? By working with other firms, industry participants can use information

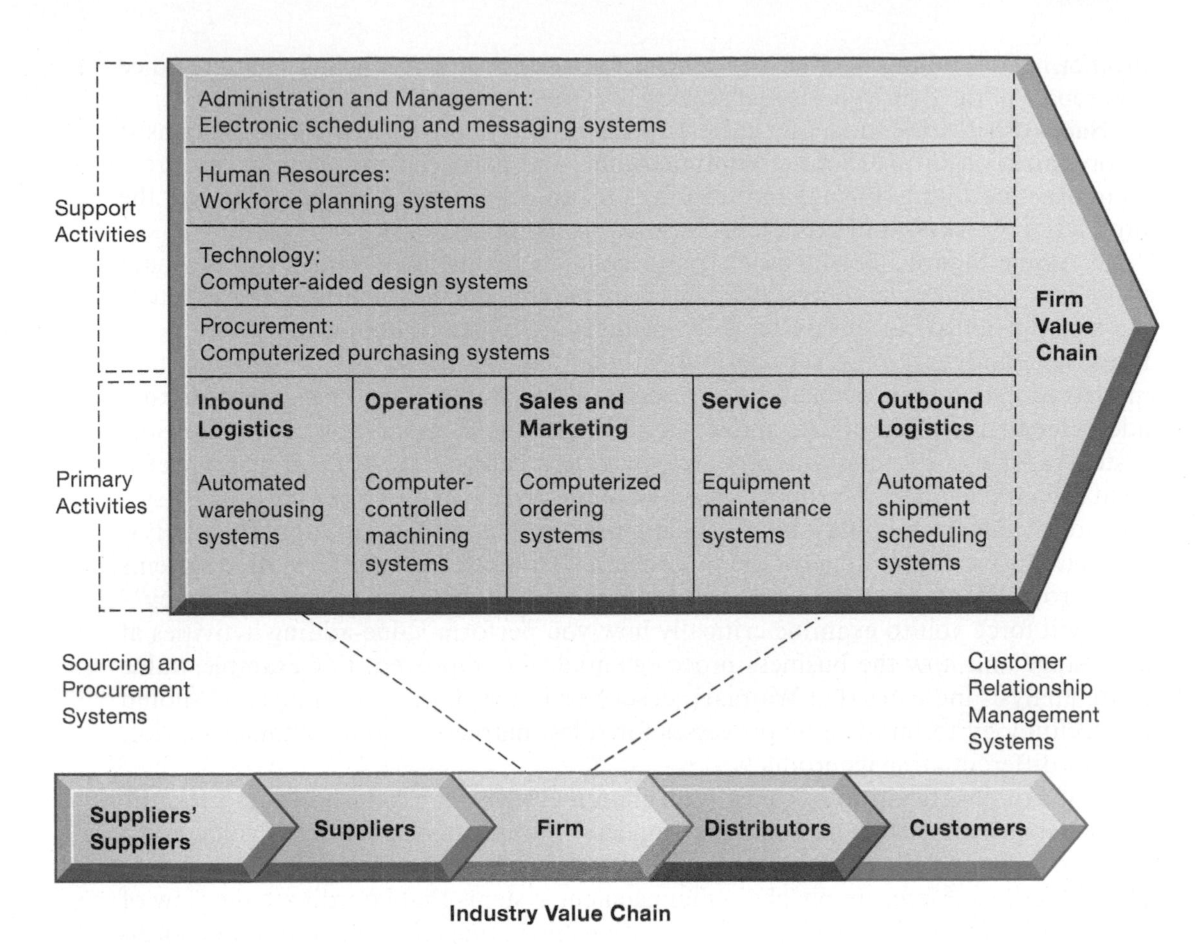


**Figure 3.2**
The Value Chain Model
*This figure provides examples of systems for both primary and support activities of a firm and of its value partners that would add a margin of value to the firm's products or services.*

technology to develop industry-wide standards for exchanging information or business transactions electronically that force all market participants to subscribe to similar standards. Such efforts increase efficiency, making product substitution less likely and perhaps raising entry costs—thus discouraging new entrants. Moreover, industry members can build industry-wide, IT-supported consortia, symposia, and communications networks to coordinate activities concerning government agencies, foreign competition, and competing industries.

Looking at the industry value chain encourages you to think about how to use information systems to link up more efficiently with your suppliers, strategic partners, and customers. Strategic advantage derives from your ability to relate your value chain to the value chains of other partners in the process. For instance, if you were Amazon.com, you would want to build systems that:

- make it easy for suppliers to display goods and open stores on the Amazon site.
- make it easy for customers to search for goods.
- make it easy for customers to order and pay for goods.
- track and coordinate the shipment of goods to customers.

In fact, this is exactly what Amazon has done to make it one of the most satisfying online retail shopping sites.

Internet technology has made it possible to create highly synchronized industry value chains called value webs. A **value web** is a collection of independent firms that use information technology to coordinate their value chains to produce a product or service for a market collectively. Examine Figure 3.3, which illustrates a value web as a networked system that operates in a less linear fashion than the traditional value chain

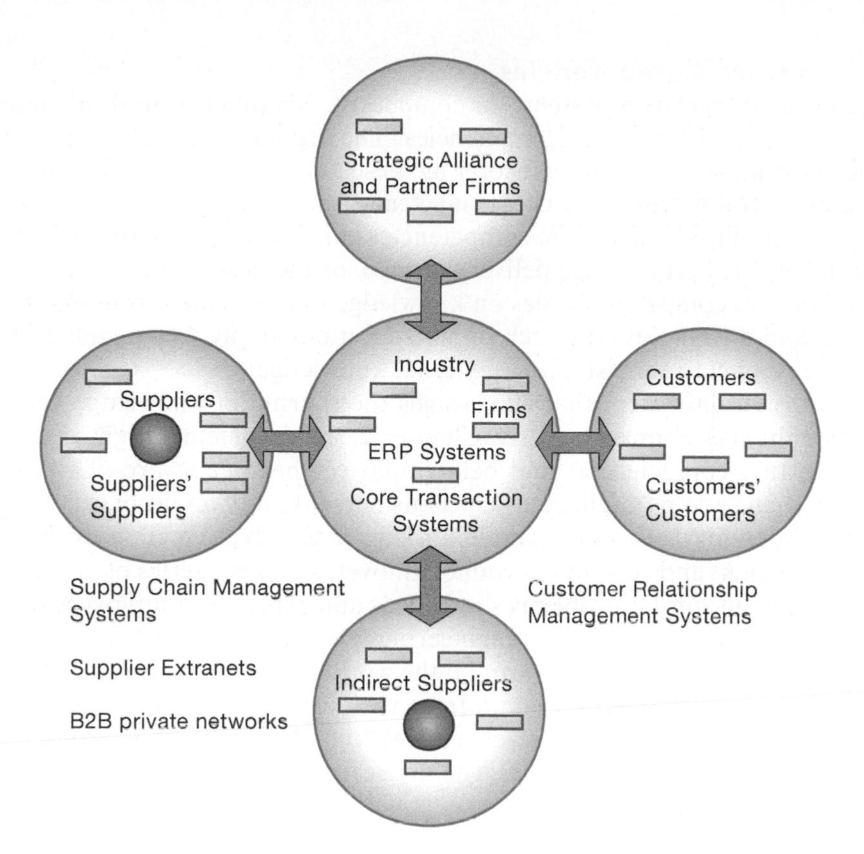


**Figure 3.3**
The Value Web
*The value web is a networked system that can synchronize the value chains of business partners within an industry to respond rapidly to changes in supply and demand.*

illustrated by Figure 3.2. A value web uses various systems such as core TPS, ERP, CRM, SCM, extranets, and B2B private networks (described in Chapter 10) to synchronize the business processes of customers, suppliers (both direct and indirect), and strategic alliance and trading partner firms among different companies in an industry or in related industries. These value webs are more customer-driven, flexible, and adaptive to changes in supply and demand. Relationships can be bundled or unbundled in response to changing market conditions. Firms will accelerate time to market and to customers by optimizing their value web relationships to make quick decisions on who can deliver the required products or services at the right price and location.

## SYNERGIES, CORE COMPETENCIES, AND NETWORK-BASED STRATEGIES

A large corporation is typically a collection of businesses. Often, the firm is organized financially as a collection of strategic business units, and the returns to the firm are directly tied to the performance of all the units. For instance, General Electric—a leading multinational conglomerate—is a collection of aviation, healthcare, power, and energy management businesses called *business units*. Information systems can improve the overall performance of these business units by promoting communication, synergies, and core competencies among the units.

### Synergies

Synergies develop when the output of some units can be used as inputs to other units, or two organizations can pool markets and expertise, and when these relationships lower costs and generate profits. Information technology in these synergy situations can help tie together the operations of disparate business units so that they can act as a whole. For example, acquiring MGM in March 2022 enabled Amazon to bolster its position in the streaming entertainment industry, giving its Prime Video streaming service an additional 4,000 films and 17,000 TV shows to offer. Information systems helped Amazon and MGM leverage their content, consolidate their operations, lower their costs, and increase their cross-marketing of entertainment products.

## Enhancing Core Competencies

Another use of information systems for competitive advantage is to think about ways that systems can enhance core competencies. The argument is that the performance of all business units can increase insofar as these business units develop, or create, a central core of competencies. A **core competency** is an activity for which a firm is an industry best-in-class leader. Core competencies may involve being the best miniature parts designer, the best package delivery service, or the best thin-film manufacturer. In general, a core competency relies on knowledge that is gained from many years of experience and a first-class research organization or, simply, key people who follow the literature and stay abreast of new external knowledge.

Any information system that encourages the sharing of knowledge across business units enhances competency (see Chapter 2). Such systems might encourage or enhance existing competencies and help employees become aware of new external knowledge; such systems might also help a business take advantage of existing competencies in related markets. For example, Procter & Gamble (P&G), a world leader in brand management and consumer product innovation, uses a series of systems to help people working on similar problems share ideas and expertise. Employees working in research and development (R&D), engineering, purchasing, marketing, legal affairs, and business information systems around the world have online access to documents, reports, charts, videos, and other data from various sources. These systems also enable users to locate subject matter experts within the company as well as external research scientists and entrepreneurs who are searching for new, innovative products worldwide.

## Network-Based Strategies

Internet and networking technology have spawned strategies that take advantage of firms' abilities to create networks or to network with each other. Network-based strategies include the use of a virtual company model, network economics, and business ecosystems.

A **virtual company** uses networks to link people, assets, and ideas, enabling it to ally with other companies to create and distribute products and services without being limited by traditional organizational boundaries or physical locations. The virtual company model is useful when a company finds it cheaper to acquire products, services, or capabilities from an external vendor or when it needs to move quickly to exploit new market opportunities and lacks the time and resources to respond on its own.

Fashion companies such as GUESS, Ann Taylor, Levi Strauss, and Reebok enlist Hong Kong–based Li & Fung to manage product design, raw material sourcing, manufacturing, quality assurance, and shipping for their garments. Li & Fung does not own any fabric, factories, or machines, outsourcing all of its work to a network of more than 15,000 suppliers in 40 countries. Customers place orders to Li & Fung over its private extranet. Li & Fung then sends instructions to appropriate raw material suppliers and factories where the clothing is produced. The Li & Fung extranet tracks the entire production process for each order. Working as a virtual company keeps Li & Fung flexible and adaptable so that it can design and produce its clients' products quickly to keep pace with rapidly changing fashion trends.

Business models based on a network may help firms strategically by allowing firms to take advantage of **network economics**. In traditional economics—the economics of factories and agriculture—production experiences diminishing returns. The more any given resource is applied to production, the lower the marginal gain in output, until a point is reached where the additional inputs produce no additional outputs. This is the law of diminishing returns, and it is one foundation of modern economics.

In some situations, the law of diminishing returns does not work. For instance, in a network, the marginal costs of adding another participant are about zero, whereas the marginal gain is much larger. The larger the number of subscribers in a telephone system or the Internet, the greater the value to all participants because each user can interact with more people. It is no more expensive to operate a television station with 1,000 subscribers than it is to operate a television station with 10 million subscribers. The value of a community of people grows as the number of participants increases, whereas the cost of adding new members is inconsequential. This is referred to as a "network effect."

From this network economics perspective, information technology can be strategically useful: Firms can use the Internet to build communities of like-minded customers who want to share their experiences. EBay, the giant online auction and retail site, is an example. This business is based on a network of millions of users and has built an online community by using the Internet. The more people offering products on eBay, the more valuable eBay is to everyone because more products are listed, and more competition among suppliers lowers prices. Network economics also provide strategic benefits to commercial software vendors such as Microsoft. The value of its software and complementary software products increases as more people use them, and there is a larger installed base to justify continued use of the product and vendor support.

**Business Ecosystems and Platforms** Instead of participating in a single industry, some of today's firms participate in industry sets—collections of industries that provide related services and products that deliver value to the customer. **Business ecosystem** is another term for these loosely coupled but interdependent networks of suppliers, distributors, outsourcing firms, transportation service firms, and technology manufacturers. Information technology plays an important role in enabling a dense network of interactions among the participating firms.

Business ecosystems typically have one or a few keystone firms that dominate the ecosystem and create the **platforms** used by other niche firms. For instance, both Microsoft and Apple provide platforms composed of information systems, technologies, and services that thousands of other firms in different industries use to enhance their own capabilities. Facebook is a platform used by billions of people and millions of businesses to interact and share information as well as to buy, market, and sell numerous products and services. More firms are trying to use information systems to develop into keystone firms by building IT-based platforms that other firms can use. Alternatively, firms should consider how information systems will enable them to become profitable participants in the larger ecosystems created by keystone firms.

Digital platforms have reorganized markets and industries, redefining the rules of business strategy and value creation. Dominant online platform companies such as Amazon and Apple are among the world's most valuable companies. Inspired by their success, an enormous number of platform startups have been set up. Some have prospered, but many have not, and firms hoping to use another company's platform to compete should understand the power dynamics and risks of platform business models (Cutolo, Hargadon, and Kenney, 2021; Hagiu and Wright, 2021; Pidun, Reeves, and Wesselink, 2021).

## DISRUPTIVE TECHNOLOGIES: RIDING THE WAVE

Sometimes a technology and its resulting business innovation come along to change the business landscape and environment radically. These innovations are loosely called *disruptive* (Christensen, 2003). In some cases, **disruptive technologies** are substitute products that perform as well as or better than anything currently produced. The automobile substituted for the horse-drawn carriage, the Apple iPod for portable CD players, digital photography for process film photography, and on-demand services like Uber for traditional taxi service. In these cases, entire industries are put out of business or significantly challenged.

In other cases, disruptive technologies simply extend the market, usually with less functionality and much less cost than existing products. Eventually they turn into

**TABLE 3.4**

Disruptive Technologies: Winners and Losers

| Technology | Description | Winners and Losers |
|---|---|---|
| Microprocessor chips | Thousands and eventually millions of transistors on a silicon chip | Microprocessor firms (Intel, Texas Instruments) win; transistor firms (GE) decline. |
| Personal computers | Small, inexpensive, but fully functional desktop computers | PC manufacturers (HP, Apple, IBM) and chip manufacturers (Intel) prosper; mainframe (IBM) and minicomputer (DEC) firms lose. |
| Digital photography | Using charge-coupled device (CCD) image sensor chips to record images | CCD manufacturers and smartphone companies win; manufacturers of film products lose. |
| World Wide Web | A global database of digital files and pages instantly available on the Internet | E-commerce, online stores benefit; small retailers and shopping malls lose. |
| Internet music, video, TV services | Repositories of downloadable music, video, TV on the web | Owners of Internet platforms, telecommunications providers owning Internet backbone (AT&T, Verizon), Internet service providers win; content owners and physical retailers (Tower Records, Blockbuster) lose. |
| PageRank algorithm | A method for ranking web pages in terms of their popularity to supplement web search using key terms | Google is the winner (it owned the patent); traditional keyword search engines (Alta Vista) lose. |
| Software as web service | Using the Internet to provide remote access to online software | Online software services companies (Salesforce.com) win; traditional boxed software companies (Microsoft, SAP, Oracle) lose. |

low-cost competitors for whatever was sold before. Disk drives are an example: Small hard-disk drives used in PCs extended the market for computer disk drives by offering inexpensive digital storage for files on small computers.

Some firms create these technologies and ride the wave to profits, whereas others learn quickly and adapt their business; still others are obliterated because their products, services, and business models become obsolete. There are also cases when no firms benefit, and all gains go to consumers (i.e., firms fail to capture any profits). Table 3.4 provides examples of some disruptive technologies.

Disruptive technologies are tricky. Firms that invent disruptive technologies as first movers do not always benefit. The MITS Altair 8800 is widely regarded as the first PC, but its inventors did not take advantage of their first-mover status. Second movers, so-called fast followers, such as IBM and Microsoft, reaped the rewards. Citibank's ATMs revolutionized retail banking, but other banks copied them. Now all banks use ATMs, and the benefits go mostly to the consumers.

## 3-2 Describe how information systems help businesses compete globally.

Look closely at your jeans or sneakers. Even if they have a US label, they may have been designed in California but stitched together in Hong Kong or Guatemala, using materials from China or India. Call Microsoft Support or Verizon Support, and chances are good you will be speaking to a customer service representative located in India.

For instance, examine Figure 3.4, which illustrates the path to market for an iPhone. The iPhone is designed by Apple engineers in the United States, includes more than 200 high-tech components sourced from around the world, and is assembled primarily

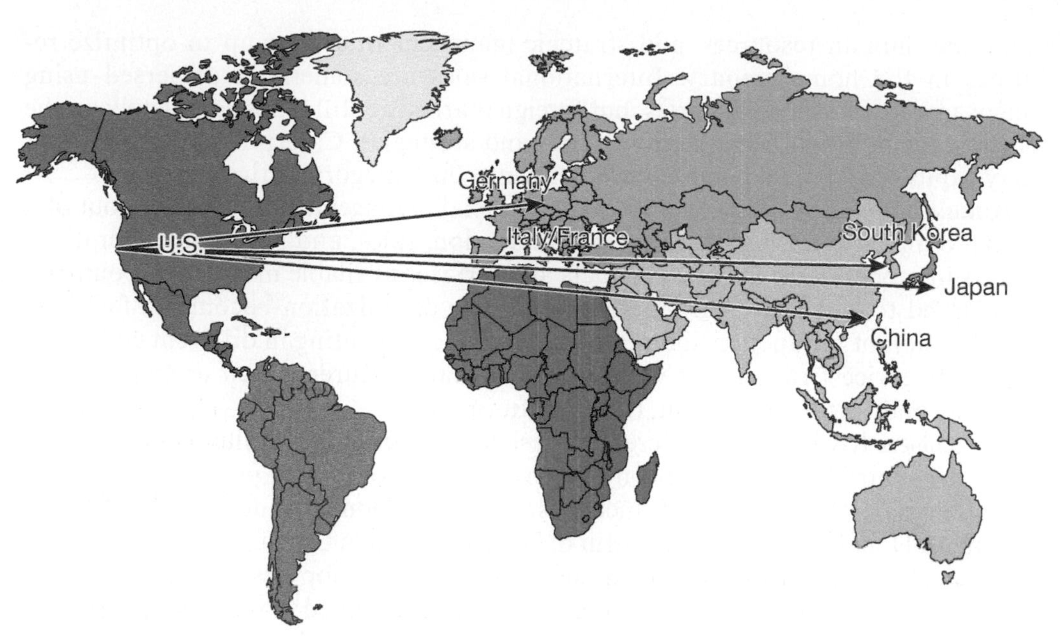


**Figure 3.4**
Apple iPhone's Global Supply Chain
*Apple designs the iPhone in the United States and relies on suppliers in the United States, Germany, Italy, France, Japan, and South Korea for parts. Much of the final assembly occurs in China.*

in China. Companies in Taiwan, South Korea, Japan, France, Italy, Germany, and the United States provide components such as the case, camera, processor, accelerator, gyroscope, electronic compass, power management chip, touch screen controller, and high-definition display screen. Taiwan-based Hon Hai Precision Industry, better known as Foxconn, is in charge of manufacturing and assembly.

Firms pursuing a global strategy benefit from economies of scale and resource cost reduction (usually wage cost reduction). For example, Apple spread design, sourcing, and production for its iPhone over multiple countries overseas to reduce tariffs and labor costs.

## THE INTERNET AND GLOBALIZATION

Huge multinational firms, such as General Electric, General Motors, Toyota, and IBM, used to dominate competition on a global scale. These large firms could afford huge investments in factories, warehouses, and distribution centers in foreign countries and proprietary networks and systems that could operate on a global scale. The emergence of the Internet into a full-blown international communications system has drastically reduced the costs of operating on a global scale, expanding the possibilities for large companies and simultaneously creating many opportunities for small and medium-sized firms.

The global Internet, along with internal information systems, puts manufacturing firms in nearly instant contact with their suppliers. Internet telephony (see Chapter 7) permits millions of customer service calls to US companies to be answered in India and Jamaica just as easily and cheaply as if the help desk were in New Jersey or California. Likewise, the Internet makes it possible to move very large computer files with hundreds of graphics or complex industrial designs across the globe in seconds.

## GLOBAL BUSINESS AND SYSTEM STRATEGIES

There are four main ways of organizing businesses internationally: domestic exporter, multinational, franchiser, and transnational, each with different patterns of organizational structure or governance. In each type of global business organization, business functions may be centralized (in the home country), decentralized (to local foreign units), and coordinated (all units participate as equals).

The **domestic exporter** strategy is characterized by heavy centralization of corporate activities in the home country of origin. Production, finance/accounting, sales/

marketing, human resources, and strategic management are set up to optimize resources in the home country. International sales are sometimes dispersed using agency agreements or subsidiaries, but foreign marketing still relies completely on the domestic home base for marketing themes and strategies. Caterpillar Inc. and other heavy capital equipment manufacturers fall into this category of firm.

A **multinational** strategy concentrates financial management and control out of a central home base while decentralizing production, sales, and marketing operations to units in other countries. The products and services available in different countries are adapted to suit local market conditions. The organization becomes a far-flung confederation of production and marketing facilities operating in different countries. Financial service firms, along with a host of manufacturers, such as Ford Motor Company and Intel Corporation, fit this pattern.

**Franchisers** have the product created, designed, financed, and initially produced in the home country but rely heavily on foreign personnel for further production, marketing, and human resources. Food franchisers, such as McDonald's and Starbucks, fit this pattern. McDonald's created a new form of fast-food chain in the United States and continues to rely primarily on the United States for inspiration for new products, strategic management, and financing. However, units outside the United States typically turn to local production for many items, as well as local employees and local marketing.

Transnational firms have no single national headquarters but instead have many regional headquarters and perhaps a world headquarters. In a **transnational** strategy, nearly all the value-adding activities are managed from a global perspective without reference to national borders, optimizing sources of supply and demand wherever they appear and taking advantage of any local competitive advantages. Such firms have common global business processes. There is a strong central management core of decision making but considerable dispersal of power and financial muscle throughout the global divisions.

Nestlé S.A., the largest food and beverage company in the world, is one of the world's most globalized companies, with more than 270,000 employees at 354 factories in 79 countries. Nestlé launched a $2.4 billion initiative to adopt a single set of business processes and systems for procurement, distribution, and sales management using SAP enterprise software. All of Nestlé's worldwide business units use the same processes and systems for making sales commitments, establishing factory production schedules, billing customers, compiling management reports, and reporting financial results. Nestlé has learned how to operate as a single unit on a global scale. Unilever and Eli Lilly are examples of other companies pursuing a transnational strategy.

## GLOBAL SYSTEM CONFIGURATION

There are four types of systems configurations for global business organizations. *Centralized systems* are those in which systems development and operation occur totally at the domestic home base. *Duplicated systems* are those in which development occurs at the home base, but operations are handed over to autonomous units in foreign locations. *Decentralized systems* are those in which each foreign unit designs its own, unique solutions and systems. *Networked systems* are those in which systems development and operations occur in an integrated and coordinated fashion across all units.

Examine Figure 3.5, which depicts the four types of global business strategies in terms of the system configurations that each type employs. For instance, domestic exporters tend to have highly centralized systems in which a single domestic systems development staff develops worldwide applications. However, some domestic exporters also use decentralized systems in local marketing regions. Multinationals typically use a decentralized system in which they allow foreign units to devise their own systems solutions based on local needs with few, if any, applications in common with headquarters. However, they often use networked systems for financial reporting and

| SYSTEM CONFIGURATION | STRATEGY | | | |
|---|---|---|---|---|
| | Domestic Exporter | Multinational | Franchiser | Transnational |
| Centralized | X | | | |
| Duplicated | | | X | |
| Decentralized | x | X | x | |
| Networked | | x | | X |

**Figure 3.5**
Global Business Organization and Systems Configurations
*The large Xs show the dominant patterns, and the small Xs show the emerging patterns. For instance, domestic exporters rely predominantly on centralized systems, but there is continual pressure and some development of decentralized systems in local marketing regions.*

some telecommunications applications. Franchisers typically develop a single system, usually at the home base, and then replicate it around the world. Each unit, no matter where it is located, typically has identical applications. However, they may also have some decentralized systems to deal with local needs. Firms such as Nestlé, which is organized along transnational lines, use networked systems that span multiple countries and support global business processes using powerful telecommunications networks and a shared management culture that transcends local cultural barriers.

## 3-3 Describe how information systems help businesses compete using quality and design.

Quality has developed from a business buzzword into a serious goal for many companies. Quality is a form of differentiation. Companies with reputations for high quality, such as BMW or Nordstrom, can charge premium prices for their products and services. Information systems have a major contribution to make in this drive for quality. In the services industries in particular, superior information systems and services generally enable quality strategies.

### WHAT IS QUALITY?

**Quality** can be defined from both producer and customer perspectives. From the perspective of the producer, quality signifies conformance to specifications or the absence of variation from those specifications. The specifications for a smartphone might include one stating that the phone should continue to function even if it is dropped from a four-foot height onto a wooden floor.

A customer definition of quality is much broader: First, customers are concerned with the quality of the physical product—its durability, safety, ease of use, and installation. Second, customers are concerned with the quality of service, by which they mean the accuracy and truthfulness of advertising, responsiveness to warranties, and ongoing product support. Finally, customer concepts of quality include psychological aspects: the company's knowledge of its products, the courtesy and sensitivity of sales and support staff, and the reputation of the product. Today, as the quality movement in business progresses, the definition of quality is increasingly from the perspective of the customer.

Many companies have embraced the concept of **total quality management (TQM)**. TQM makes quality the responsibility of all people and functions within an organization. TQM holds that the achievement of quality control is an end in itself. Everyone is expected to contribute to the overall improvement of quality—the engineer who prevents design errors, the production worker who spots defects, the sales representative who presents the product properly to potential customers, and even the administrative assistant who avoids typing mistakes. TQM derives from quality management concepts that US quality experts such as W. Edwards Deming and Joseph Juran developed, but the Japanese popularized it.

Another quality concept that is widely implemented today is Six Sigma, which Amazon uses to reduce errors in order fulfillment. **Six Sigma** is a specific measure of quality, representing 3.4 defects per million opportunities. Most companies cannot achieve this level of quality but use Six Sigma as a goal to implement a set of methodologies and techniques for improving quality and reducing costs. Studies have repeatedly shown that the earlier in the business cycle a problem is eliminated, the less the problem costs the company. Thus, quality improvements not only raise the level of product and service quality but also lower costs.

## HOW INFORMATION SYSTEMS IMPROVE QUALITY

Let's examine some of the ways companies face the challenge of improving quality to see how information systems can be part of the process.

### Reduce Cycle Time and Simplify the Production Process

Studies have shown that one of the best ways to reduce quality problems is to reduce **cycle time**, which refers to the total elapsed time from the beginning of a process to its end. Shorter cycle times mean that problems are caught earlier in the process, often before the production of a defective product is completed, saving some of the hidden production costs. Finally, finding ways to reduce cycle time often means finding ways to simplify production steps. The fewer steps there are in a process, the less time and opportunity there are for an error to occur. Information systems help eliminate steps in a process and critical time delays.

ArcScan is a small medical device manufacturer based in Golden, Colorado, that builds high-resolution, ultrasonic imaging devices for the eye. Many parties and activities are involved in approving parts, components, and procedures. ArcScan's quality control system was heavily manual and inefficient and could not handle the upsurge in paperwork as the company expanded. For example, one of the company's subcontractors hadn't been following all of its quality control procedures: Some corrective actions weren't taken, and there was no paper trail to document what was going on. This was unacceptable for a medical device company and required extra work and rework. ArcSan was able to solve this problem by implementing Qualio, a cloud-based quality management system for the life sciences that automates much of the product approval process. Qualio creates a digital paper trail that records every action that is taken. If a user doesn't take a required action within three days, the user will receive an email reminder to do so, and the quality engineer will be sent a copy of that email. With all the necessary paperwork accurate and complete, ArcScan is in a stronger position to meet the requirements of regulatory agencies.

### Benchmark

Companies achieve quality by using benchmarking to set standards for products, services, and other activities and then measuring performance against those standards. Companies may use external industry standards, standards other companies set, internally developed standards, or some combination of the three. L.L. Bean, the Freeport, Maine, clothing company, used benchmarking to achieve an order-shipping accuracy of 99.9 percent. Its old batch order fulfillment system could not handle the surging volume and variety of items to be shipped. After studying German and Scandinavian companies that have leading-edge order fulfillment operations, L.L. Bean carefully redesigned its order fulfillment process and information systems so that orders could be processed as soon as they were received and then shipped within 24 hours.

### Use Customer Demands to Improve Products and Services

Improving customer service, and making customer service the number-one priority, will improve the quality of the product itself. Delta Airlines decided to focus on its customers and installed a customer care system at its airport gates. For each flight, the airplane seating chart, reservations, check-in information, and boarding data are linked in a central database. Airline personnel can track which passengers are on board regardless of where they checked in and can use this information to help passengers reach their destination quickly, even if delays cause them to miss connecting flights.

### Improve Design Quality and Precision

Computer-aided design software has made a major contribution to quality improvements in many companies, from producers of automobiles to producers of razor blades. A **computer-aided design (CAD) system** automates the creation and revision of designs by using computers and sophisticated graphics software. The software enables users to create a digital model of a part, a product, or a structure and make changes to the design on the computer without having to build physical prototypes.

For example, Ford Motor Company used a computer simulation that came up with the most efficient design possible for an engine cylinder. Engineers altered that design to account for manufacturing constraints and tested the revised design on the computer, using models with decades of data on material properties and engine performance. Ford then created the physical mold to make a real part that could be bolted onto an engine for further testing. The entire process took days instead of months and cost thousands instead of millions of dollars.

CAD systems can supply data for **3-D printing**, also known as additive manufacturing, which uses machines to make solid objects, layer by layer, from specifications in a digital file. Unlike traditional techniques, in which objects are cut or drilled from molds, resulting in some wasted materials, 3-D printing lets workers model an object on a computer and print it out with plastic, metal, or composite materials. 3-D printing is currently being used for prototyping, custom manufacturing, and fashioning items with small production runs. Today's 3-D printers can handle materials including plastic, titanium, and human cartilage and produce fully functional components for specialized applications, including batteries, transistors, prosthetic devices, and LEDs. Large-scale 3-D printers designed specifically for printing concrete can pour foundations, build walls onsite, and create modular concrete sections that are later assembled on the job site. 3-D printing services are now available in retail stores such as Best Buy and Home Depot or on the web.

### Improve Production Precision and Tighten Production Tolerances

For many products, quality can be enhanced by making the production process more precise, thereby decreasing the amount of variation from one part to another. CAD software often produces design specifications for tooling and manufacturing

© RAGMA IMAGES/Shutterstock

*Computer-aided design (CAD) systems improve the quality and precision of product design by performng most of the design and testing work on the computer. Shown here is a 3-D CAD model of a tank design for the oil and gas industry.*

processes, saving additional time and money while producing a manufacturing process with far fewer problems. The user of this software can design a more precise production system, a system with tighter tolerances, than could ever be done manually.

## 3-4 Explain the role of business process management (BPM) in enhancing competitiveness.

Technology alone is often not enough to make organizations more competitive, efficient, or quality-oriented. The organization itself needs to be changed to take advantage of the power of information technology. These changes may require minor adjustments in work activities, but, often, entire business processes will need to be redesigned. Business process management (BPM) addresses these needs.

### WHAT IS BUSINESS PROCESS MANAGEMENT?

**Business process management (BPM)** is an approach to business that aims to improve business processes continuously. BPM uses a variety of tools and methodologies to understand existing processes, design new processes, and optimize those processes. BPM is never concluded because continuous improvement requires continual change. Companies practicing business process management need to go through the following steps:

1. **Identify processes for change:** One of the most important strategic decisions that a firm can make is not deciding how to use computers to improve business processes but, rather, understanding which business processes need improvement. When systems are used to strengthen the wrong business model or business processes, the business can become more efficient at doing what it should not do. As a result, the firm becomes vulnerable to competitors who may have discovered the right business model. Considerable time and cost may also be spent improving business processes that have little impact on overall firm performance and revenue. Managers need to determine which business processes are the most important and how improving these processes will help business performance.
2. **Analyze existing processes:** Existing business processes should be modeled and documented, noting inputs, outputs, resources, and the sequence of activities. The process design team identifies redundant steps, paper-intensive tasks, bottlenecks, and other inefficiencies.

   Examine Figure 3.6, which illustrates the as-is process for purchasing a book from a physical bookstore. A customer visits a physical bookstore and searches its shelves for a book. If the customer finds the book, the customer takes it to the checkout counter and pays for it by credit card, cash, or check. If the customer cannot locate the book, the customer must ask a bookstore clerk to search the shelves or check the bookstore's inventory records to see whether the book is in stock. If the clerk finds the book, the customer purchases it and leaves. If the book is not available locally, the clerk inquires about ordering it for the customer, either from the bookstore's warehouse or from the book's distributor or publisher. Once the ordered book arrives at the bookstore, a bookstore employee telephones the customer with this information. The customer would have to go to the bookstore again to pick up the book and pay for it. If the bookstore cannot order the book for the customer, the customer would have to try another bookstore. You can see that this process has many steps and might require the customer to make multiple trips to the bookstore.
3. **Design the new process:** Once the existing process is mapped and measured in terms of time and cost, the process design team will try to improve the process by designing a new one. A new, streamlined to-be process will be documented and modeled for comparison with the old process.

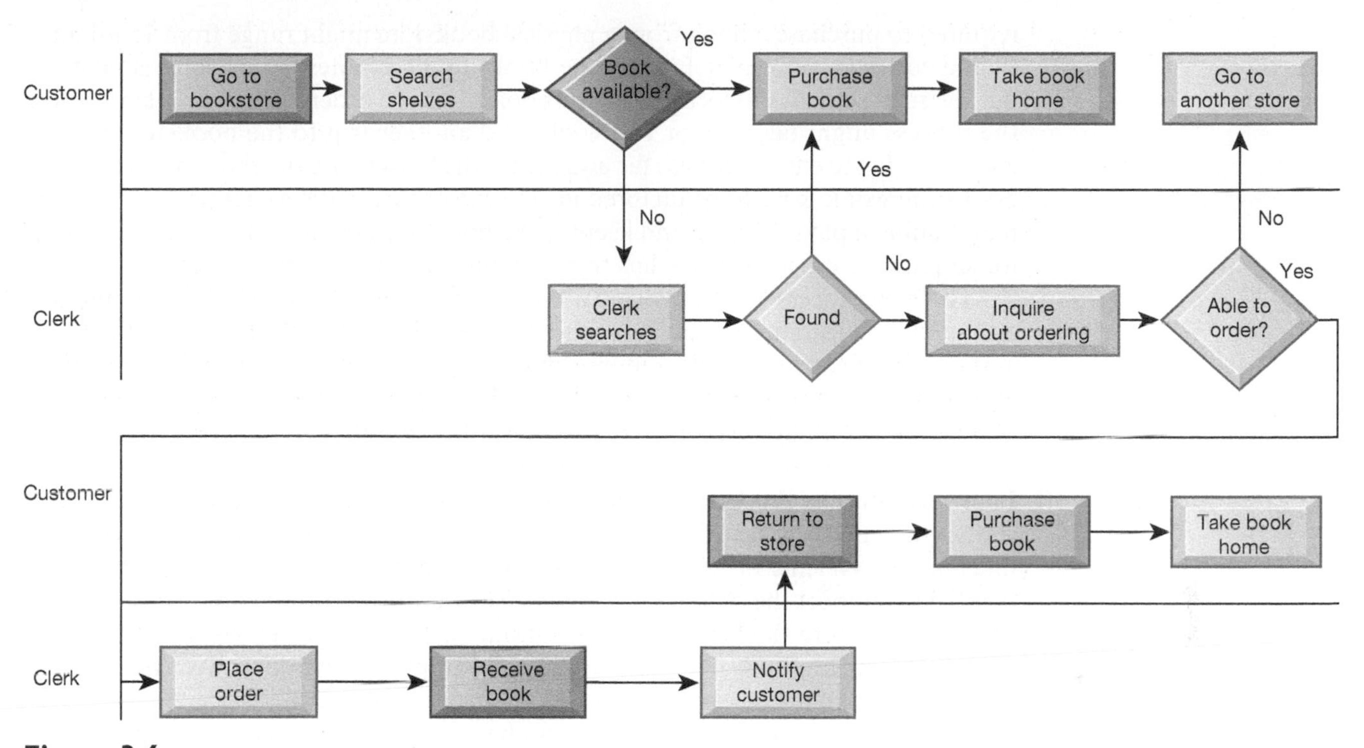


**Figure 3.6**
As-Is Business Process for Purchasing a Book from a Physical Bookstore
*Purchasing a book from a physical bookstore requires both the seller and the customer to perform many steps.*

Examine Figure 3.7, which illustrates how the book purchasing process can be redesigned by taking advantage of the Internet and e-commerce technology. The customer accesses an online bookstore. The customer searches the bookstore's online catalog for the book the customer wants. If the book is available, the customer orders the book online, supplying credit card and shipping address information, and the book is delivered to the customer's home. If the online bookstore does not carry the book, the customer selects another online bookstore and searches for the book again. This process has far fewer steps than for purchasing the book in a physical bookstore, requires much less effort on the part of the customer, and requires fewer sales staff for customer service. The new process is therefore much more efficient and timesaving.

The new process design needs to be justified by showing how much it reduces time and cost or enhances customer service and value. Management first measures the time and cost of the existing process as a baseline. In our example, the time

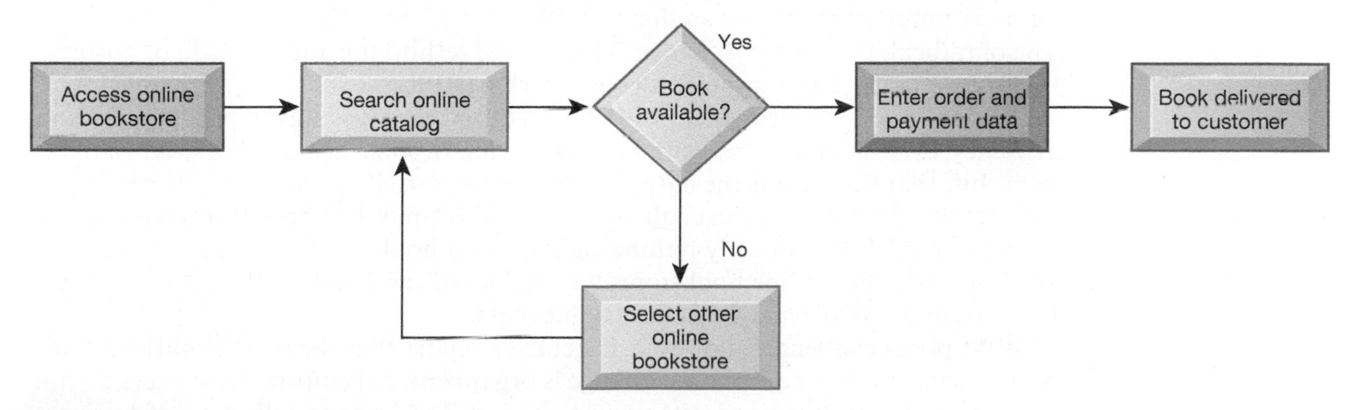


**Figure 3.7**
Redesigned Process for Purchasing a Book Online
*Using Internet technology makes it possible to redesign the process for purchasing a book so that it has only a few steps and consumes fewer resources.*

required to purchase a book from a physical bookstore might range from 15 minutes (if the customer immediately finds the book) to 30 minutes if the book is in stock but the sales staff has to locate it. If the book has to be ordered from another source, the process might take one or two weeks and another trip to the bookstore for the customer. If the customer lives far away from the bookstore, the time to travel to the bookstore would have to be factored in. The bookstore will have to pay the costs for maintaining a physical store and keeping the book in stock, for sales staff onsite, and for shipment costs if the book has to be obtained from another location.

The new process for purchasing a book online might take only several minutes, although the customer might have to wait several days or weeks to receive the book and may have to pay a small shipping charge. Nevertheless, the customer saves time and money by not having to travel to the bookstore or make additional visits to pick up the book. Booksellers' costs are lower because they do not have to pay for a physical store location or for local inventory.

4. **Implement the new process:** After the new process has been thoroughly modeled and analyzed, it must be translated into a new set of procedures and work rules. New information systems or enhancements to existing systems may have to be implemented to support the redesigned process. The new process and supporting systems are rolled out into the business organization. As the business starts using this process, problems are uncovered and addressed. Employees working with the process may recommend improvements.
5. **Continuous measurement:** After a process has been implemented and optimized, it needs to be measured continually. Why? Processes may deteriorate over time as employees fall back on old methods, or the processes may lose their effectiveness if the business experiences other changes.

Many software tools are available to facilitate various aspects of BPM. These tools help businesses identify and document processes requiring improvement, create models of improved processes, capture and enforce business rules for performing processes, and integrate existing systems to support new or redesigned processes. BPM software tools also provide analytics for verifying that process performance has been improved and for measuring the impact of process changes on key business performance indicators.

The Spotlight on Organizations case illustrates how a company can benefit from new technology and business processes. As you read this case, learn how changing business processes and their underlying technology improved Signet Jewelers' business performance.

## Business Process Reengineering

Many business process improvements are incremental and ongoing, but occasionally, more radical change is required. Our example of a physical bookstore redesigning the book purchasing process so that it can be carried out online is an example of this type of radical, far-reaching change. This radical rethinking and redesign of business processes is called **business process reengineering (BPR)**.

When properly implemented, BPR can lead to dramatic gains in productivity and efficiency, even changing the way a business is run. In some instances, it drives a paradigm shift that transforms the nature of the business itself. This actually happened in book retailing when Amazon challenged traditional, physical bookstores with its online retail model. By radically rethinking the way a book can be purchased and sold, Amazon and other online bookstores have achieved remarkable efficiencies, cost reductions, and a whole new way of doing business.

BPM poses challenges, however. Executives report that the most significant barrier to successful business process change is organizational culture. Employees do not like unfamiliar routines and often resist change. This is especially true of business process reengineering projects because the organizational changes are so far-reaching. Managing change is neither simple nor intuitive, and companies committed to extensive process improvement need a good change management strategy (see Chapter 12).

## SPOTLIGHT ON: ORGANIZATIONS Signet Jewelers Sparkles with a Virtual Sales Process

The average price of an engagement ring in the United States is around $3,000, and purchasing a ring or other diamond jewelry has traditionally been a very high-touch, in-person process. A customer visits a physical jewelry store and is greeted by a salesperson, who pulls out pieces of interest from a display case for the customer to examine and perhaps try on. The salesperson might need to educate the customer about how to select a diamond. If a customer chooses to purchase a particular ring in stock, the customer might have to wait one or two weeks for the ring to be properly sized, perhaps a month or two if the ring has to be custom-made. Once the ring is ready, the buyer completes payment, and the ring can be picked up in the store or shipped to the buyer's home.

If purchasing diamond jewelry is such an in-person experience, how could jewelers operate when they were forced to close because of the Covid-19 pandemic? Many couldn't. But Signet Jewelers, the world's largest diamond retailer, with flagship brands including Zales, Kay Jewelers, Jared, Diamonds Direct, Rocksbox, H.Samuel, and Ernest Jones, could. The company compensated for the shuttering of its 2,500 North American stores in the spring of 2020 by creating a virtual jewelry sales service. It deployed information technology to transform the jewelry sales process so that customers would be able to research and purchase engagement rings and other fine jewelry online from their homes.

Signet introduced virtual consultations in which customers can communicate with a sales consultant via text, voice, or video to ask questions. Potential buyers can use computers and smartphones to book one-hour, virtual appointments with a consultant by visiting the website of a Signet brand such as Zales.

Signet has conducted many hundreds of thousands of these sessions. It enhanced its photographic techniques so that buyers can enlarge the photo of a diamond 40 times its real size to see the diamond's cut or check for imperfections. It enhanced its personalized design capabilities so that buyers can order unique items. Signet also added visual search and collaborative browsing (cobrowsing) tools. Much like buyers who go into a jewelry store with a picture of a necklace or ring they like, users can upload a picture to find that product or something similar online. Cobrowsing technology allows sales or service agents to access a customer's browser and navigate it in real time. With joint navigation of a web page, agents can interact and collaborate online with customers, facilitating customer support or sales assistance. After making a selection, the customer can complete the purchase online, and the product will be shipped to the customer's home.

Signet retrofitted the iPads used in its stores with technology that senses when consultants are logging in from their home network, automatically linking them to the company's network via a secure connection. Using this secure connection, staff can select clienteling, endless aisle, and other applications that enable consultants to access their client "books" so that the staff can email or text personalized offers. Clienteling is a technique used by retail sales associates to establish long-term relationships with customers based on data about customers' preferences, behaviors, and prior purchases and provides a more personal shopping experience. Endless aisle is a marketing and sales technique by which retailers are able to provide in-store customers with the opportunity to order products that are either not normally sold in the store or that are currently out of stock. To help consultants show off merchandise without overwhelming customers with outreach, Signet revised its appointment application to support virtual appointments and disabled automated message reminders for jewelry cleaning and inspections.

Signet used cloud software from Amazon Web Services (AWS) along with data analysis software as the technology platform for its virtual sales process. Signet is able to add computing resources on the fly to handle spikes in network traffic during the peak holiday sales period. It also can draw on these computing resources for data-intensive machine-learning algorithms that support strategic decision making, such as determining which stores to reopen first or improving digital marketing capabilities.

It was a bit of a challenge for Signet to get its 2,000+ sales consultants accustomed to showing off merchandise in stores to replicate the same experience virtually in clients' homes. Signet had already started moving into e-commerce sales, and its virtual sales during the pandemic strengthened the company's digital sales channel. Despite the pandemic, Signet's 2020 holiday sales were the best in nine years, and sales from all channels have continued to grow. The company's business strategy has shifted from TV-centric marketing and real estate expansion to an omnichannel marketing mix and more e-commerce retail sales. For fiscal year 2022, Signet reported e-commerce sales of $556 million, up 8.7% from fiscal year 2021 and up 85.4% from fiscal year 2020.

As Signet's stores reopened, the virtual showcasing and contactless sales processes were so popular that Signet added more than 700 full-time digital consultants to continue the virtual shopping consultations. Jewelry consultants working in stores continued to use virtual consultations with customers. Shoppers who make a purchase can buy online and pick up their purchase in the store or use curbside delivery.

Signet's successful virtual sales effort has inspired further experimentation with digital sales technology. The company is now piloting electronic "e-tags" that use Bluetooth wireless networking and the Internet of Things (IoT) to track how customers interact with its store merchandise. The e-tags put all the relevant information about a particular product at customers' fingertips, reducing the need for paper signs and labels. By using a smartphone to scan an e-tag inside or on top of a jewelry box, a customer will be able to see real-time pricing, for example, or how many times the product has been picked up in a store or when the product was last cleaned.

Sources: Signetjewelers.com, accessed March 16, 2022; Clint Boulton, "Signet Adds Virtual Facet to Jewelry Sales in Pandemic Play," *CIO*, May 10, 2021; and Kenneth Hein, "How Diamond Retailer Signet Jewelers Got Its E-commerce Channels to Sparkle," *The Drum*, March 21, 2021.

## CASE STUDY QUESTIONS

1. Describe Signet's traditional business strategy and business model. How did they change when the company embraced e-commerce and digital technology?
2. Compare the in-store process of buying a diamond ring with Signet's virtual sales process. Diagram or list the steps in each process.
3. What people, organization, and technology issues had to be addressed when Signet designed its virtual sales process?
4. What are the advantages and disadvantages of selling items such as high-end jewelry virtually?

## 3-5 Understand how MIS can help your career.

Here is how this chapter and this book can help you find a job as an entry-level business development representative.

### THE COMPANY

Founded in 1999, Austin Texas-based SuperiorCRM.com is a global leader in customer relationship management (CRM) systems. The company's CRM platform has various tools and technologies that are used by companies of all sizes across many industries to acquire, keep, and increase their customer base. SuperiorCRM.com is looking for an entry-level business development representative.

### POSITION DESCRIPTION

The business development representative will help the company's sales team meet aggressive growth targets. The company provides an introductory, two-week training program in addition to ongoing, on-the-job training classes that focus on having high-level conversations with senior executives, assisting with strategic account research, learning how to write a pitch deck, and developing value propositions. Job responsibilities include:

- Researching, identifying, and generating new business opportunities
- Qualifying, building, and maintaining a sales pipeline while maintaining a steady volume of outbound calls, emails, and social media posts

### JOB REQUIREMENTS

- Bachelor's degree
- Some sales experience (software sales a plus)
- Experience communicating via email and telephone with new business leads
- Familiarity with attaining consistent metrics and goals
- Ability to work remotely
- Exceptional communication, interpersonal, and problem-solving skills
- Ability to quickly learn to anticipate and prepare for objections during sales process

### INTERVIEW QUESTIONS

1. Do you have any work experience in business development in the software industry? Exactly what did you do in this job?
2. What do you know about our CRM platform? Have you ever used the platform? If so, what did you do with it?
3. Have you ever worked with a sales pipeline before? What exactly did you do?
4. What experience do you have working with internal and/or external customers? Can you describe the work you did with the customers?
5. Can you give an example of a sales-related problem or another business problem that you helped solve?

### AUTHOR TIPS

1. Review this chapter's discussion of competitive forces, product differentiation, and the business value chain model. Also review the section on CRM in Chapter 9.
2. Review the company's social media profiles on LinkedIn, Twitter, and Facebook. Does the company focus on consistent themes across these social media channels, and if so, what are those themes? Be prepared to discuss the kinds of business challenges and opportunities facing this company.

## Review Summary

3-1 **Demonstrate how Porter's competitive forces model, the value chain model, synergies, core competencies, and network-based strategies help companies use information systems for competitive advantage.** In Porter's competitive forces model, the strategic position of the firm and its strategies are determined by competition with its traditional direct competitors. They are also greatly affected by new market entrants, substitute products and services, suppliers, and customers. Information systems help companies compete by allowing them to maintain low costs, differentiate products or services, focus on market niche, strengthen ties with customers and suppliers, and increase barriers to market entry with high levels of operational excellence. Information systems are most successful when the technology is aligned with business objectives.

The value chain model highlights the specific activities in the business where competitive strategies and information systems will have the greatest impact. The model views the firm as a series of primary and support activities that add value to the firm's products or services. Primary activities are directly related to production and distribution, whereas support activities make the delivery of primary activities possible. A firm's value chain can be linked to the value chains of its suppliers, distributors, and customers. A value web consists of information systems that enhance competitiveness at the industry level by promoting the use of standards and industry-wide consortia and by enabling businesses to work more efficiently with their value partners.

Information systems allow a business to achieve additional efficiencies or enhanced services by tying together the operations of disparate business units. Information systems help businesses use their core competencies by promoting the sharing of knowledge across business units. Information systems facilitate business models based on large networks of users or subscribers that take advantage of network economics. A virtual company strategy uses networks to link to other firms so that a company can use the capabilities of other companies to build, market, and distribute products and services. Firms can also redefine their businesses to become niche players or keystone firms in platform-based business ecosystems where multiple industries work together to deliver value to the customer. Disruptive technologies provide strategic opportunities, although first movers do not necessarily obtain long-term benefit.

3-2 **Describe how information systems help businesses compete globally.** Information systems and the Internet help companies operate internationally by facilitating coordination of geographically dispersed units of the company and communication with faraway customers and suppliers. There are four main strategies for organizing businesses internationally: domestic exporter, multinational, franchiser, and transnational. There are four types of systems configuration for global business organizations: centralized, duplicated, decentralized, and networked.

3-3 **Describe how information systems help businesses compete using quality and design.** Information systems can enhance quality by reducing product development cycle time and simplifying the production process, facilitating benchmarking, enabling the improvement of customer service, and increasing quality and precision in design and production.

3-4 **Explain the role of business process management (BPM) in enhancing competitiveness.** Organizations often have to change their business processes to execute their business strategies successfully. BPM combines and streamlines the steps in a business process to eliminate repetitive and redundant work and to achieve dramatic improvements in quality, service, and speed. BPM is most effective when it is used to strengthen a good business model and when it strengthens processes that have a major impact on firm performance. Business process reengineering (BPR) involves a radical rethinking and redesign of business processes and when properly implemented can lead to dramatic gains in productivity and efficiency.

## Key Terms

3-D printing, 101
Benchmarking, 91
Best practices, 91
Business ecosystem, 95
Business process management (BPM), 102
Business process reengineering (BPR), 104
Competitive forces model, 83
Computer-aided design (CAD) system, 101
Core competency, 94
Customer experience management, 86
Cycle time, 100
Disruptive technologies, 95
Domestic exporter, 97
Efficient customer response system, 85
Franchiser, 98
Mass customization, 86
Multinational, 98
Network economics, 94
Platforms, 95
Primary activities, 90
Product differentiation, 86
Quality, 99
Six Sigma, 100
Support activities, 91
Switching costs, 88
Total quality management (TQM), 99
Transnational, 98
Value chain model, 90
Value web, 92
Virtual company, 94

## Review Questions

**3-1** Demonstrate how Porter's competitive forces model, the value chain model, synergies, core competencies, and network-based strategies help companies use information systems for competitive advantage.

- Define Porter's competitive forces model, and explain how it works.
- List and describe four competitive strategies enabled by information systems that firms can pursue.
- Describe how information systems can support each of these competitive strategies, and give examples.
- Explain why aligning IT with business objectives is essential for strategic use of systems.
- Define and describe the value chain model.
- Explain how the value chain model can be used to identify opportunities for information systems.
- Define the value web, and show how it is related to the value chain.
- Describe how the Internet has changed competitive forces and competitive advantage.
- Explain how information systems promote synergies and core competencies that enhance competitive advantage.
- Explain how businesses benefit by using network economics.
- Define and describe a virtual company and the benefits of pursuing a virtual company strategy.
- Define and describe a business ecosystem and how it can provide competitive advantage.
- Explain how disruptive technologies create strategic opportunities.

**3-2** Describe how information systems help businesses compete globally.

- Describe how globalization has increased opportunities for businesses.
- List and describe the four main ways of organizing a business internationally and the types of systems configuration for global business organizations.

**3-3** Describe how information systems help businesses compete using quality and design.

- Define quality, and compare the producer and consumer definitions of quality.
- Describe the various ways in which information systems can improve quality.

**3-4** Explain the role of business process management (BPM) in enhancing competitiveness.

- Define BPM, and explain how it helps firms become more competitive.
- Distinguish between BPM and business process reengineering (BPR).
- List and describe the steps companies should take to make sure BPM is successful.

## Discussion Questions

**3-5** MyLab MIS It has been said that there is no such thing as a sustainable competitive advantage. Do you agree? Why or why not?

**3-6** MyLab MIS What are some of the issues to consider in determining whether the Internet would provide your business with a competitive advantage?

**3-7** MyLab MIS It has been said that the advantages that leading-edge retailers such as Walmart have over competitors isn't technology—it's their management. Do you agree? Why or why not?

**MyLab MIS**
To complete these problems, go to EOC Discussion Questions in **MyLab MIS**.

## Hands-On MIS Projects

The projects in this section give you hands-on experience analyzing the impact of business processes on business performance, using a database to improve sales performance, and using web tools to configure and price an automobile. Visit MyLab MIS to access this chapter's Hands-On MIS Projects.

### MANAGEMENT DECISION PROBLEM

**3-8** Entrust Corporation, formerly Datacard Group, produces hardware and software for establishing trusted identities and the issuance of personalized financial cards, passports, driver's licenses, identification cards, and mobile devices. Entrust is one of the world's leading suppliers of payment cards. As a part of doing business with Entrust, customers—which include financial institutions and credit card bureaus, government agencies, and major corporations—supply the Shakopee, Minnesota-based company with large volumes of business data. Entrust had a predominantly manual process for distributing pricing for its secure identification and card issuance products on a quarterly basis to about 250 partners by emailing a price book of multiple spreadsheets. Creating the book involved hours of manual input by people in marketing and sales. The process of changing a price was complex, time-consuming, and people-dependent. Partners who received the price books had to manually type them into their own systems and check the data—a process that could take up to two days. Assess the impact of this process on Entrust's business performance, and suggest some people, organization, and technology issues that would need to be addressed by a solution

### IMPROVING DECISION MAKING: USING A DATABASE TO ANALYZE SALES PERFORMANCE

Software skills: Database querying and reporting; database design
Business skills: Sales trend s analysis

**3-9** In this exercise, you will start out with raw transactional sales data and then use Microsoft Access database software to develop queries that help managers make better decisions about product pricing, sales promotions, and inventory replenishment. In MyLab MIS, you can find a Store and Regional Sales Database developed in Microsoft Access. The database contains raw data on weekly store sales of computer equipment in various sales regions. The database includes fields for store identification number, sales region, item number, item description, unit price, units sold, and the weekly sales period when the sales were made. Use Access to develop some queries to make this information more useful for running the business. This information includes the following:

- All 24-inch monitor sales by Sales Region for Week Ending 11/24/2022, sorted within region from highest to lowest number of units sold
- All 24-inch monitor sales by store for the East Sales Region, sorted by Store No. and Week Ending in ascending order.
- All sales of all products for the East Sales Region, sorted from highest to lowest total number of units sold by item number

Then describe how the information in these queries could be used for management decision making.

## IMPROVING DECISION MAKING: USING WEB TOOLS TO CONFIGURE AND PRICE AN AUTOMOBILE

Software skills: Internet-based software
Business skills: Researching product information and pricing

**3-10** In this exercise, you will use software at car-selling websites to find product information about a car of your choice and then use that information to make an important purchase decision. You will also evaluate two of these sites as selling tools.

You are interested in purchasing a lightly used, low-mileage Honda CR-V (or some other car of your choice). Go to the Carsdirect.com website, and begin your investigation. Research the various used Honda CR-V models for sale; choose one you prefer in terms of price, features, and safety ratings. Then visit the Carvana.com website. Compare the information on pricing and available inventory at Carvana's website with that of Carsdirect for the Honda CR-V. Try to locate the lowest price for the car you want in inventory along with financing and vehicle pickup or delivery. Compare the buying experience at both websites. Which website do you prefer for this type of purchase? Explain your answer.

## COLLABORATION AND TEAMWORK PROJECT

Identifying Opportunities for Strategic Information Systems

**3-11** With your team of three or four students, select a company described in the *Wall Street Journal, Fortune, Forbes*, or another business publication available on the web. Visit the company's website to find additional information about that company and to see how the firm is using the web. On the basis of this information, analyze the business. Include a description of the organization's features, such as important business processes, culture, structure, and environment, as well as its business strategy. Suggest strategic information systems appropriate for that particular business, including those based on Internet technology, if appropriate. If possible, use Google Docs and Google Drive or Google Sites to brainstorm, organize, and develop a presentation of your findings for the class.

## BUSINESS PROBLEM-SOLVING CASE

### SHIPPING WARS

Shipping and delivery have been vital to the success of e-commerce, both for retailers and for the shipping companies themselves. FedEx, UPS, and the United States Postal Service (USPS) have earned many billions of dollars handling the massive number of products ordered from Amazon and other e-commerce companies. Convenient and seamless online ordering and shipping processes, along with free or low-cost delivery or two-day delivery, are a source of competitive advantage for online merchants over traditional brick-and-mortar retailers.

Especially important in logistics is the “last mile,” which refers to the last step in a delivery that takes the package to the customer's doorstep. Instead of using the USPS, FedEx, or UPS for the last mile, Amazon has been building its own network of warehouses, trucks, airlines, and delivery vans. Amazon Air is a cargo airline operating exclusively to transport Amazon packages. By 2022, Amazon Air had 97 cargo aircraft operating out of 42 air gateways in the United States. Amazon additionally expanded its airport hub operations, building a $1.5 billion hub at Cincinnati/Northern Kentucky International Airport. Amazon also has operations at Fort Worth Alliance Airport and Chicago Rockford International Airport.

Amazon's taking over the “last mile” is draining billions of dollars of business away from the USPS, UPS, and FedEx. Amazon eventually wants to perform 80 to 90 percent of last-mile deliveries but will use the other delivery services for the hardest-to-reach places. Amazon uses UPS for deliveries to areas far outside Amazon's current shipping network, such as a remote farm in northern Canada. Amazon uses the USPS for customers who are within Amazon's delivery network but are less efficient and profitable to serve. Amazon packages assigned to the USPS for delivery are the ones that would be more expensive for Amazon to ship itself.

Amazon wants to upstage these shippers but not replace them entirely. It especially wants more control over logistics in order to guarantee that Amazon Prime members get their two-day shipping on time and that it has the capability to handle very large sales volumes during the holidays or bad weather periods. Amazon will also save on costs. According to Morgan Stanley, Amazon saves $2 to $4 per package, amounting to $2 billion annually, when it uses its own fleet. Additionally, having control over the entire shipping process makes it possible for Amazon to provide a better customer experience: It is easier to track lost packages and respond immediately to customer inquiries if Amazon does not have to work through another shipper. Amazon's shipping policies have been a principal driver of its rapid retail growth.

Amazon now delivers more than 70 percent of its own packages and is hoping to eventually deliver 80 to 90 percent of them. Using its own fulfillment and logistics network has enabled Amazon to shift from a two-day-delivery model to one-and even same-day delivery. By early 2022 Amazon had surpassed both FedEx and UPS as the largest US package delivery service.

Amazon also sees opportunities for turning logistics into a profitable business of its own. The company has started to sell its logistics services to outside companies that do not sell through Amazon. These services include Fulfilled By Amazon, or FBA, services for orders not made on Amazon.com (which is why some orders from eBay, Walmart, and others might be delivered to your door in Amazon packaging). Amazon Freight uses a network of more than 30,000 Amazon trailers and carriers to move other companies' full truckload freight. Amazon's algorithms also allow sellers to take advantage of LTL (less than load) truck space at discounted rates while enabling Amazon to make money on otherwise-wasted space.

When Amazon announced one-day shipping for Prime members in April 2019, Fed Ex canceled its express delivery contract with Amazon. FedEx's management believes FedEx doesn't really need Amazon to flourish because Amazon accounted for less than 1.3 percent of FedEx's $70 billion in consolidated annual revenues and had been one of FedEx's least profitable customers on a margin basis. Management also believed that working with Amazon was cannibalizing FedEx's own business. The direction FedEx has chosen calls for improving its ground delivery service and establishing new partnerships with other retailers and brands to serve the broader e-commerce market. FedEx thinks it can overtake Amazon and become the fastest, most cost-efficient e-commerce delivery service.

For example, in June 2019 FedEx and Dollar General announced a strategic alliance to offer new, convenient access to FedEx drop-off and pickup services at thousands of Dollar General stores. The effort is designed to increase access to FedEx for all customers, particularly those living in rural communities where Dollar General has a large footprint. FedEx and Dollar General began rolling out the service in more than 1,500 Dollar General stores in late summer 2019 and expected to be in more than 8,000 Dollar General stores by the end of 2020. The Dollar General alliance expanded the FedEx Retail Convenience Network to more than 62,000 retail locations. That move put more than 90 percent of Americans within 5 miles of a FedEx-hold retail location. Customers will be able to drop off pre-packaged and pre-labeled FedEx Express or FedEx Ground shipments at Dollar General stores and pick up packages sent to their neighborhood Dollar General stores. FedEx has similar arrangements with Walgreens, Target, and other retail chains. Using convenient drop-off points reduces the number of expensive, door-to-door deliveries FedEx has to make.

In addition to consolidating delivery locations, FedEx is hoping to attract more commercial shipping customers by offering logistics management services that many companies can't develop on their own. These services make better use of the Big Data that FedEx collects on shipments and customers to provide customers with more real-time information that will improve how FedEx moves goods. For example, a service called Surround, which features advanced monitoring of time-sensitive priority shipments, provides FedEx customers with early warnings of shipment delays caused by weather, traffic, or other events. A medical device maker might learn from FedEx Surround that its shipment of a heart-surgery kit to a hospital will be delayed because of a snowstorm. For such a high-priority shipment, FedEx could re-route the shipment to ensure it arrives on time. FedEx Surround was deployed in December 2020 and has helped FedEx monitor and manage the increase in priority shipment volume attributed to the Covid-19 pandemic. Surround will not cost FedEx customers extra and also handles data from FedEx scanner and sensor devices.

FedEx Surround is the result of a partnership with Microsoft that is behind other initiatives to help FedEx position itself as a logistics alternative to Amazon for small and large retailers. Microsoft provides cloud computing services and sophisticated analytics, including artificial intelligence, to build software that digitizes and tracks key parts of a customer's business. In January 2022 FedEx and Microsoft announced a joint effort to offer a cross-platform "logistics as a service" for retailers, merchants, and brands. FedEx and Microsoft will try to generate new insights from the 17 million packages that pass through the FedEx network each day, which will help brands deliver improved customer experiences.

Microsoft and FedEx are working on a data integration project that will combine data insights from FedEx with Microsoft Dynamics 365 Intelligent Order Management to help brands improve fulfillment, shipping, and servicing customer orders while easily integrating with their existing e-commerce platforms. Microsoft Dynamics 365 Intelligent Order Management is designed for complex environments where there are many internal and external systems and partners for supply chain processes. Intelligent Order Management enables customers to choose the best systems and partners, to integrate them into a single system, and to orchestrate order flows across different platforms and apps. This cross-platform approach makes it possible for brands to offer faster, more effectives delivery, communications of nearly real-time delivery status, and frictionless returns that will improve brands' overall e–commerce customer experience.

FedEx has 50 years of delivery experience, and UPS has been in the delivery business for more than a century. Although Amazon has been in business for a much shorter time period, it has much greater resources to leverage for fulfillment and delivery services. Extremely profitable sectors and subsidiaries (like Amazon Web Services, which is responsible for the majority of Amazon's operating income), enable Amazon to outspend or undercut competitors.

UPS invests heavily in IT and offers an array of technology-based shipping and logistics services for large and small businesses, some of which are described in the Chapter 1 Spotlight on Technology case. It, too, has powerful Big Data analytics tools to capture and analyze customer, operational, and planning data to track the real-time status of every package as it moves across the company's shipping network, feeding other systems for better planning and management of the network. Leading-edge technology investments help UPS compete effectively against FedEx and the USPS, and UPS does some shipping and packaging work for Amazon, but it is generally more expensive to ship via UPS than via Amazon.

What about the USPS? It still is the top US package delivery service, making 40 percent of all parcel deliveries in 2021. (Amazon accounts for 21 percent of US parcel shipments, UPS for 24 percent, and FedEx for 16 percent.) However, the USPS was designed primarily for delivering paper letters by human carriers who walk door-to-door, and it has more of a public service orientation than the private carriers. There are more than 35,000 different Post Office locations, each of which has to be staffed to transact with individual customers buying stamps, sending letters, and sending packages, each of which must be measured for shape and size. Compare that to making a purchase online on Amazon, where every single activity from the time the customer clicks "buy now" until the package arrives at the customer's door is tracked precisely using advanced information systems and analytics. The entire process is automated and optimized to create highly efficient warehousing and delivery services. Amazon's increasingly end-to-end control of a package's journey from warehouse to doorstep means that consumers get more precise estimates of delivery after they've placed an order,

Thanks to email and the Internet, there are fewer letters to deliver, so the USPS cannot recover the cost of the mail delivery person who drops off mail to your house each day. Even though the USPS's shipping and packages revenue has been increasing, the USPS continues to operate at a loss ($4.9 billion in fiscal year 2021). In March 2022 Congress passed legislation to return the USPS to solvency, but the measure did not call for fundamental changes in the USPS's technology platform or business model.

Which company will win the retail shipping wars? The outcome could determine the future direction of the entire e-commerce retail industry.

Sources: "Intelligent Order Management Overview," Docs.microsoft.com, accessed March 19, 2022; Thor Olavsrud, "Supply Chain Analytics: 3 Success Stories," *CIO*, February 22, 2022; "FedEx and Microsoft Announce New Cross-Platform Logistics Solution for E-commerce," News.microsoft.com, January 22, 2022; Jalen Small, "As Amazon's Business Grows, USPS Struggles to Fulfill Mission," *Newsweek*, December 9, 2021; Katie Schoolov, "Amazon Is Now Shipping Cargo for Outside Customers in Its Latest Move to Compete with FedEx and UPS, Cnbc.com, September 7, 2021; Erica Pandey, "Amazon Is Now a Bigger Shipper in the U.S. than FedEx," *Axios.com*, October 21, 2021; Bloomberg, "FedEx Has a Plan to Battle Amazon Shipping, CEO Says," *Bloomberg.com*, January 30, 2020; and Greg Petro, "Amazon versus FedEx: The Retail Shipping Wars," *Forbes*, June 28, 2019.

## CASE STUDY QUESTIONS

**3-12** Why is shipping so important for e-commerce? Explain your answer.

**3-13** Compare the shipping strategies of Amazon, FedEx, UPS, and the USPS. How are they related to each company's business model?

**3-14** Will FedEx succeed in its push into ground shipping? Will Amazon become a major logistics service supplier? Why or why not?

## Chapter 3 References

"ArcScan Helps Improve Diagnostic, Patient, and Surgical Outcomes." Qualio.com, accessed March 18, 2022.

"Baume Brings Customisable and Sustainable Watches to More People with Salesforce." Salesforce.com, accessed March 10, 2022.

Birkinshaw, Julian. "How Incumbents Survive and Thrive." *Harvard Business Review* (January–February 2022).

Bossert, Oliver, and Driek Desmet. "The Platform Play: How to Operate Like a Tech Company." McKinsey & Company (February 2019).

Christensen, Clayton. *Competitive Advantage: The Revolutionary Book That Will Change the Way You Do Business* (New York: HarperCollins, 2003).

Christensen, Clayton M., Michael E. Raynor, and Rory McDonald. "What Is Disruptive Innovation?" *Harvard Business Review* (December 2015).

Cutolo, Donato, Andrew Hargadon, and Martin Kenney. "Competing on Platforms." *MIT Sloan Management Review* (Spring 2021).

Desai, Veeral, Tim Fountaine, and Kayvaun Rowshankish." A Better Way to Put Your Data to Work." *Harvard Business Review* (July–August 2022).

El Sawy, Omar A. *Redesigning Enterprise Processes for E-Business* (New York: McGraw-Hill, 2001).

Gandhi, Suketo, and Eric Gervet. "Now That Your Products Can Talk, What Will They Tell You?" Special Collection: Getting Product Development Right. *MIT Sloan Management Review* (Spring 2016).

Hagiu, Andrei, and Julian Wright. "Don't Let Platforms Commoditize Your Business." *Harvard Business Review* (May–June 2021).

__________. "When Data Creates Competitive Advantage." *Harvard Business Review* (January–February 2020).

Hind, Benbiya, Dorothy E. Leidner, and David Preston. "Information Systems Alignment." *MIS Quarterly Research Curations* (March 14, 2019).

Iansiti, Marco, and Roy Levien. "Strategy as Ecology." *Harvard Business Review* (March 2004).

Kapoor, Rohit. "How Data Is Humanzing Customer Exeriences." *MIT Sloan Management Review* (March 23, 2022).

Kim, Keongtae, Jeongsik "Jay" Lee, and Anandasivam Gopal. "Soft but Strong: Software-Based Innovation and Product Differentiation in the IT Hardware Industry." *MIS Quarterly* 46, No. 2 (June 2022).

Li, He, Chen Zhang, and William J. Kettinger. "Digital Platform Ecosystem Dynamics: The Roles of Product Scope, Innovation, and Collaborative Network Centrality," *MIS Quarterly* 46, No. 2 (June 2022).

Luftman, Jerry. *Competing in the Information Age: Align in the Sand*, 2nd ed. (New York: Oxford University Press, 2003).

Miric, Milan, Margherita Pagani, and Omar A. El Sawy. "When and Who Do Platform Companies Acquire? Understanding the Role of Acquisitions in the Growth of Platform Companies." *MIS Quarterly* 45, No. 4 (December 2021).

Nambisan, Satish, and Yadong Luo. "Think Globally, Act Locally." *MIT Sloan Management Review* 63, No. 3 (Spring 2022).

Parker, Geoffrey, Marshall Van Alstyne, and Xiaoyue Jiang. "Platform Ecosystems: How Developers Invert the Firm." *MIS Quarterly* 41, No. 1 (March 2017).

Pidun, Ulrich, Martin Reeves, and Edzard Wesselink. "How Healthy Is Your Business Ecosystem?" *MIT Sloan Management Review* 62, No. 3 (Spring 2021).

Porter, Michael. *Competitive Advantage* (New York: Free Press, 1985).

———. "Strategy and the Internet." *Harvard Business Review* (March 2001).

Porter, Michael E., and James E. Heppelmann. "How Smart, Connected Products Are Transforming Competition." *Harvard Business Review* (November 2014).

Rahmati, Pouya, Ali Tafti, J. Christopher Westland, and Cesar Hidalgo. "When All Products Are Digital: Complexity and Intangible Value in the Ecosystem of Digitizing Firms." *MIS Quarterly* 445, No. 3a (September 2021).

Roca, Jaime Bonnin, Parth Vaishnav, Joana Mendonça, and M. Granger Morgan. "Getting Past the Hype about 3-D Printing." *MIT Sloan Management Review* 58, No. 3 (Spring 2017).

Ross, Jeanne W., Ina M. Sebastian, and Cynthia M. Beath. "How to Develop a Great Digital Strategy." *MIT Sloan Management Review* 58, No. 2 (Winter 2017).

Sabherwal, Rajiv, Sanjiv Sabherwal, Taha Havakhor, and Zach Steelman, "How Does Strategic Alignment Affect Firm Performance? The Roles of Information Technology Investment and Environmental Uncertainty." *MIS Quarterly* 43, No. 2 (June 2019).

Shapiro, Carl, and Hal R. Varian. *Information Rules* (Boston: Harvard Business School Press, 1999).

Siggelkow, Nicolai, and Christian Terwiesch. "The Age of Continuous Connection." *Harvard Business Review* (May–June 2019).

Song, Peijian, Ling Xue, Arun Rai, and Cheng Zhang. "The Ecosystem of Software Platform: A Study of Asymmetric Cross-Side Network Effects and Platform Governance." *MIS Quarterly* 42, No. 1 (March 2018).

Van Alstyne, Marshall W., Geoffrey G. Parer, and Sangeet Paul Choudary. "Pipelines, Platforms, and the New Rules of Strategy." *Harvard Business Review* (April 2016).

Weill, Peter, and Sinan Aral. "IT Savvy Pays Off: How Top Performers Match IT Portfolios and Organizational Practices," *MIT CISR* (May 2005).

Weill, Peter, and Stephen L. Weorner. "Thriving in an Increasingly Digital Ecosystem." *MIT Sloan Management Review* 56, No. 4 (Summer 2014).

Zhu, Feng, and Marco Iansiti. "Why Some Platforms Thrive and Others Don't." *Harvard Business Review* (January–February 2019).

CHAPTER

4

# Ethical and Social Issues in Information Systems

## LEARNING OBJECTIVES

After completing this chapter, you will be able to:

4-1 Identify the ethical, social, and political issues raised by information systems.

4-2 Describe principles for conduct that can be used to guide ethical decisions.

4-3 Explain how contemporary information systems technologies and the Internet pose challenges to the protection of individual privacy and intellectual property.

4-4 Discuss the issues that contemporary information systems raise with respect to system quality, accountability and control, and the quality of everyday life.

4-5 Understand how MIS can help your career.

## CHAPTER CASES

- Apps That Track: A Double-Edged Sword
- Section 230: Should the Law that "Created" Today's Internet Be Repealed or Revised?
- Immersed in the Metaverse: What Will It Mean for the Future?
- Facebook's Many Ethical Challenges

**MyLab MIS**

- **Video Case:** Australia Passes Law Forcing Tech Giants to Pay for News
- Discussion Questions: 4-5, 4-6, 4-7
- Hands-On MIS Projects: 4-8, 4-9, 4-10, 4-11
- eText with Figure Videos

## APPS THAT TRACK: A DOUBLE-EDGED SWORD

How often do you check your local weather forecast? If you're like many people, you might do so multiple times a day, probably by opening a weather app on a mobile device. But did you stop to think about the fact that your weather app is likely also recording your exact physical location each time you check the weather? Few realize that this simple act can open the door to the widespread sale of that data to third parties, who can then use it for purposes far removed from the function of the app that originally gathered it. And it's not just weather apps that collect and sell this kind of data. Location data form the basis for a multibillion-dollar industry, populated by largely unregulated companies that operate behind the scenes. For instance, one such company, Mobilewalla, has said that it has data from 75,000 mobile apps and 1.6 billion devices across more than 35 countries.

To figure out where you are, mobile devices use data from a variety of technologies embedded within the device, such as GPS sensors, Bluetooth, and connections to local Wi-Fi and cellular networks. Mobile apps often make use of a third-party SDK (software development kit) such as Foursquare's Pilgrim SDK, a free software tool that enables an app to collect data, or may be paid by an SDK developer to include its code within an app.

Apple's iOS 14.5 update, released at the end of April 2021, aims to make privacy issues related to apps more visible. If an app uses an SDK, the developer must describe what data the SDK collects and how the data may be used. Apps must now ask for your permission before tracking your activity (including your physical location). Google instituted similar policies in February 2022. However, many users are so conditioned to clicking "Yes" that they don't stop to consider the ramifications. In addition, some app developers employ "dark patterns," a design tactic that prompts you to make a choice detrimental to your own interests, for example, by prompting you to enable location tracking while simultaneously suggesting that the app might not work as intended if location tracking is not enabled. As a case in point, contrary to what many people assume, weather apps do not actually "need" to take constant readings of location data to provide a local weather forecast, but these apps do not necessarily make that fact apparent.

Once you agree to tracking, the app will typically refer you to a privacy policy that details how that data can be used. Research has shown that few users ever actually read such policies or understand their ramifications. Companies that sell user data often justify that doing so provides them with a necessary revenue stream. For example, AccuWeather's privacy policy notes that its ability to provide its app for free is supported by its sale of user information to third parties. Companies that sell user data also note that the information collected is usually anonymized (although studies have shown that it can be relatively easy to tie such data to a particular person). Companies that buy data may also be doing so for a variety of legitimate reasons such as for analytics, fraud detection, and advertising and marketing purposes.

© Olesky Mark/Shutterstock

You might not object to your data being used to provide you with relevant advertising. However, it's what happens to that data once it is in the hands of a data broker that is of concern. Currently, there are few laws or regulations in the United States that restrict what a data broker can do with that data. Once it has been collected, it can be sold over and over again. For instance, location data collected by Mobilewalla was ultimately acquired by a company named Venntel, which then sold the data to various US government agencies such as the IRS, DEA, FBI, DHS, and ICE. Although the US Supreme Court has held that the government needs a warrant to compel companies such as cellphone carriers to produce location data, no such restriction applies to the government's purchase of such data from data brokers. The agencies identified as using the data have defended the practice, saying it is widely available commercially and aids them in fulfilling their responsibilities to the public.

Highlighting this issue, in March 2022 researchers discovered that a Panamanian company reportedly linked to a US-based defense contractor had paid developers to incorporate an SDK that violated Google's rules on collecting user data. App developers were told that the SDK would collect non-personal data on behalf of ISPs and financial service and energy companies. However, in reality, the SDK harvested data about users' precise locations, email addresses, phone numbers, and nearby computers and mobile devices. The SDK was included in a number of consumer apps, such as weather apps, QR code scanners, and highway-radar detectors, as well as religious prayer apps, that together were downloaded on at least 60 million Android devices. Google has banned the SDK and removed the apps from the Google Play store, but that action doesn't restrict the SDK's ability to continue to gather data from apps that are already installed.

Sources: Bryan Tau and Robert McMillan, "Apps with Hidden Data-Harvesting Software Are Banned by Google," *Wall Street Journal*, April 5, 2022; Womble Bond Dickinson, "State Laws Shift Geolocation's Spot on the Privacy Map," Jdsupra.com, March 9, 2022; Alfred Ng and Jon Keegan, "Who Is Policing the Location Data Industry," Themarkup.org, February 24, 2022; Frederic Lardinois, "Google Wants to Bring Its Privacy Sandbox to Android," Techcrunch.com, February 16, 2022; Malwarebytes Lab, "Google Sued Over Deceptive Location Tracking," Blog.malwarebytes.com, January 26, 2022; Bryon Tau, "How Cellphone Data Collected for Advertising Landed at U.S. Government Agencies," *Wall Street Journal*, November 18, 2022; Jon Keegan and Alfred Ng, "There's a Multibillion-Dollar Market for Your Phone's Location Data," The markup.com, September 30, 2021; Thorin Klosowski, "We Checked 250 iPhone Apps—This Is How They're Tracking You," *New York Times*, May 6, 2021; Elizabeth Goitein, "The Government Can't Seize Your Digital Data, Except by Buying It," *Washington Post*, April 26, 2021.

The challenges that location tracking poses to privacy, described in the chapter-opening case, show that technology can be a double-edged sword. It can be the source of benefits, including the ability to use mobile devices to access maps and driving directions or local weather and business news. At the same time, digital technology creates new opportunities for invading privacy and using information that could cause harm.

The chapter-opening diagram calls attention to important points that this case and chapter raise. Advances in mobile device technologies; mobile communications; and data management and analytics have created opportunities for organizations to track the locations of individuals who are using mobile devices at any given time. The organizations described in the opening case are benefiting by collecting and using location-based data to monitor user behavior so that this information can be sold for marketing and other purposes. The use of location data is also taking benefits away from individuals because such use can invade their privacy. Individuals might be subject to job discrimination or special scrutiny based on behavior patterns revealed by location tracking. There are very few privacy protections for location tracking.

This case also illustrates an ethical dilemma because it shows two sets of interests at work: the interests of organizations that want to use location data to enhance profit and the interests of those who fervently believe that businesses and public organizations should not use personal data because doing so invades privacy or may harm

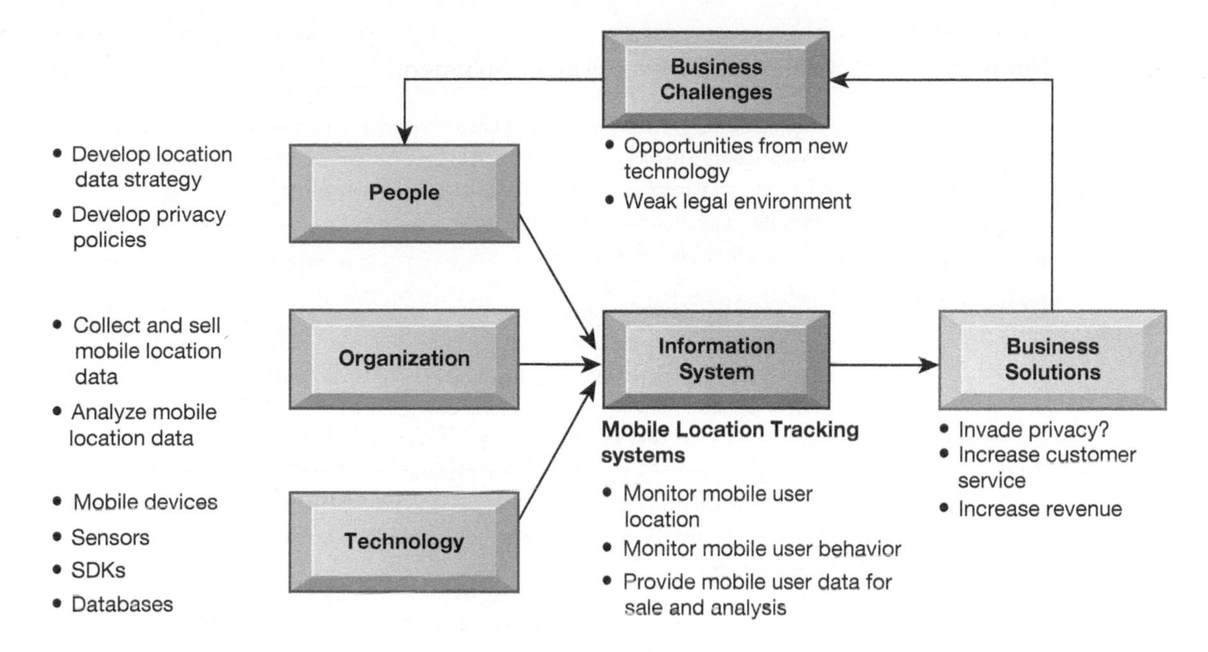


individuals. The case further illustrates the importance of being sensitive to both the positive and negative impacts of information systems and the importance of learning how to resolve ethical dilemmas involving information systems.

Here are some questions to think about: Does analyzing location data create an ethical dilemma? Why or why not? Should there be new privacy laws to protect personal data collected from mobile device users? Why or why not?

## 4-1 Identify the ethical, social, and political issues raised by information systems.

Organizations and their employees face numerous ethical challenges in conducting business. Table 4.1 provides a small sample of recent cases involving failures in ethical business judgment. These lapses in judgment occurred across a broad spectrum of industries.

In today's environment, employees who are convicted of violating the law face the risk of time in prison. US federal sentencing guidelines mandate that federal judges impose stiff sentences based on the monetary value of the crime, the presence of a conspiracy to prevent discovery of the crime, the use of structured financial transactions to hide the crime, and failure to cooperate with prosecutors.

Although business firms in the past often paid for the legal defense of employees enmeshed in civil charges and criminal investigations, firms are now encouraged to cooperate with prosecutors to reduce charges against the firm itself. More than ever, as a manager or employee, you will have to decide for yourself what constitutes proper legal and ethical conduct.

Failures in ethical business judgment are not usually masterminded by employees of information systems departments, but information systems are often instrumental in many of these frauds. In many cases, the perpetrators of these crimes artfully used information systems to bury their actions from public scrutiny.

We deal with the issue of information systems controls in Chapter 8. In this chapter, we will talk about the ethical dimensions of these and other actions based on the use of information systems.

**Ethics** refers to the principles of right and wrong that individuals, acting as free moral agents, use to make choices to guide their behavior. Information systems raise ethical questions for both individuals and societies because these systems create opportunities

**TABLE 4.1**

Recent Examples of Failures in Ethical Business Judgment

| Company | Failure in Ethical Business Judgment |
|---|---|
| Robinhood | Fintech brokerage firm Robinhood was fined a record $70 million in 2021 by the Financial Industry Regulatory Authority (FINRA) for systemic supervisory failures in several critical parts of its business, including distributing false and misleading information to customers of the firm, as well as systems outages that prevented customers from placing timely trades. |
| Volkswagen AG | Installed "defeat-device" emissions software on more than 500,000 diesel cars in the United States and on roughly 10.5 million more worldwide that allowed the cars to meet US emissions standards during regulatory testing while actually spewing unlawful levels of pollutants into the air in real-world driving. Penalties paid internationally have totaled $35 billion, with $9.5 billion going to US car owners. Volkswagen sued top company executives for being complicit in the fraud and obtained settlements totalling $365 million. Criminal charges were also brought against a number of VW executives, including US-based Oliver Schmidt, who was sentenced to seven years in prison and fined $400,000. |
| Wells Fargo | Wells Fargo employees admitted to opening millions of false accounts, manipulating terms of mortgages, and forcing auto loan customers to purchase unneeded insurance. Wells Fargo was fined $2.5 billion by the US government. |
| Takata Corporation | Takata executives admitted they covered up faulty airbags used in millions of cars over many years. Three executives were indicted on criminal charges, and Takata was fined $1 billion. Takata filed for bankruptcy in 2017. |

for intense social change and, thus, threaten existing distributions of power, money, rights, and obligations. Like other technologies, such as steam engines, electricity, and the telephone, information technology can be used to achieve social progress, but it can also be used to commit crimes and threaten cherished social values. The development of information technology will produce benefits for many and costs for others.

Ethical issues in information systems have been given additional urgency by the rise of the Internet and e-commerce. Internet and digital technologies make it easier than ever to assemble, integrate, and distribute information, unleashing concerns about the appropriate use of customer information, the protection of personal privacy, and the protection of intellectual property.

Other pressing ethical issues that information systems raise include setting standards to safeguard system quality that protect the safety of the individual and society, establishing accountability for the consequences of information systems, and preserving values and institutions considered essential to the quality of life in today's information society. When using information systems, it is essential to ask, "What is the ethical and socially responsible course of action?"

## A MODEL FOR THINKING ABOUT ETHICAL, SOCIAL, AND POLITICAL ISSUES

Ethical, social, and political issues are closely linked. The ethical dilemma you may face as someone who is involved in the use of information systems by your organization typically is reflected in social and political debate. Examine Figure 4.1, which illustrates one way to think about these relationships. Imagine society as a more or less calm pond on a summer day, a delicate ecosystem in partial equilibrium with individuals and with social and political institutions. Individuals know how to act in this pond because social institutions (family, education, organizations) have developed well-honed rules of behavior, and these rules are supported by laws developed in the political sector that prescribe behavior and promise sanctions for violations. Now toss a rock into the center of the pond. What happens? Ripples, of course.

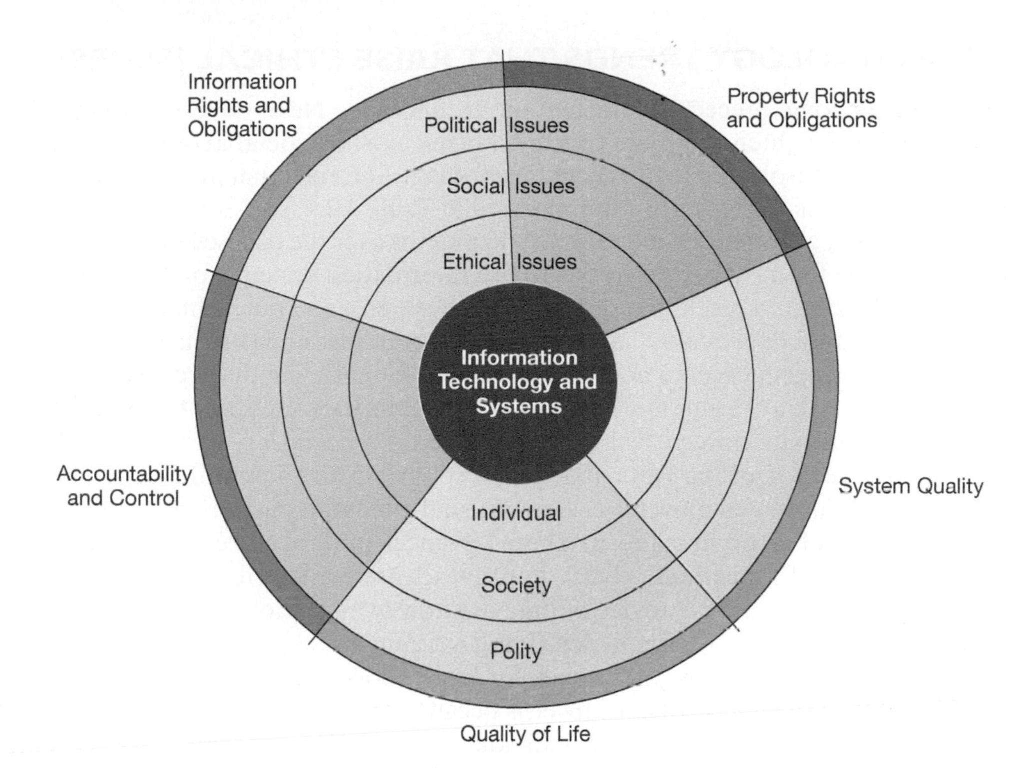


**Figure 4.1**
**The Relationship among Ethical, Social, and Political Issues in an Information Society**
*The introduction of new information technology has a ripple effect, raising new ethical, social, and political issues that must be dealt with on the individual, social, and political levels. These issues have five moral dimensions: information rights and obligations, property rights and obligations, system quality, quality of life, and accountability and control.*

Imagine instead that the disturbing force is a powerful shock of new information technology and systems hitting a society more or less at rest. Suddenly, individual actors are confronted with new situations often not covered by the old rules. Social institutions cannot respond overnight to these ripples—it may take years to develop etiquette, expectations, social responsibility, politically correct attitudes, or approved rules. Political institutions also require time to develop new laws and often require the demonstration of real harm before they act. In the meantime, you may have to act. You may be forced to act in a legal gray area.

We can use this model to illustrate the dynamics that connect ethical, social, and political issues. This model is also useful for identifying the main moral dimensions of the information society, which cut across various levels of action—individual, social, and political. View the Figure 4.1 video in the eText for an animated and more detailed discussion of this model.

## FIVE MORAL DIMENSIONS OF THE INFORMATION AGE

The major ethical, social, and political issues that information systems raise include the following moral dimensions.

- *Information rights and obligations* What **information rights** do individuals and organizations possess with respect to themselves? What can these rights protect?
- *Property rights and obligations* How will traditional intellectual property rights be protected in a digital society in which tracing and accounting for ownership are difficult while ignoring such property rights is so easy?
- *System quality* What standards of data quality and system quality should we demand to protect individual rights and the safety of society?
- *Accountability and control* Who can and will be held accountable and liable for the harm done to individual and collective information and property rights?
- *Quality of life* What values should be preserved in an information- and knowledge-based society? Which institutions should we protect from violation? Which cultural values and practices does the new information technology support?

We explore these moral dimensions in detail in Sections 4-3 and 4-4.

## KEY TECHNOLOGY TRENDS THAT RAISE ETHICAL ISSUES

Ethical issues long preceded information technology. Nevertheless, information technology has heightened ethical concerns, taxed existing social arrangements, and made some laws obsolete or insufficient. Technological trends that are responsible for many of these ethical stresses are summarized in Table 4.2.

The doubling of computing power every 18 months over the past several decades has made it possible for most organizations to use information systems for their core production processes. As a result, our dependence on systems and our vulnerability to system errors and poor data quality have increased. Social rules and laws have not yet fully adjusted to this dependence. In addition, standards for ensuring the accuracy and reliability of information systems (see Chapter 8) are not universally accepted or enforced.

Advances in data storage techniques and rapidly declining storage costs have enabled the collection of Big Data and been responsible for the proliferation of databases on individuals—employees, customers, and potential customers—maintained by private and public organizations. These advances in data storage have made the routine violation of individual privacy both inexpensive and effective. Data storage systems for terabytes and petabytes of data are now available onsite or as online services for firms of all sizes to use in identifying customers.

Advances in data analysis techniques for large pools of data are another technological trend that heightens ethical concerns because organizations can find out highly detailed personal information about individuals. With contemporary data management tools (see Chapter 6), companies can assemble and combine myriad pieces of data about you stored in information systems much more easily than they could in the past.

Think of all the ways you generate digital information about yourself—browsing online; using mobile apps; making online and in-person retail and service purchases; interacting with the financial services industry; participating in social networks; making mobile phone calls; using a smart TV, smart speaker, and other smart home devices; driving an Internet-enabled car; as well as interacting with local, state, and federal government institutions (including courts and the police). Each one of these actions generates digital data. Put together and mined properly, this information can reveal not only your credit information but also your tastes, your associations, what you read and watch, your political interests, your driving habits, and more.

Companies purchase personal information from a variety of sources to help them more finely target their marketing campaigns. Chapters 6 and 11 describe how

**TABLE 4.2**

Technology Trends that Raise Ethical Issues

| Trend | Impact |
|---|---|
| Computing power doubles every 18 months. | More organizations depend on computer systems for critical operations and become more vulnerable to system failures. |
| Data storage costs rapidly decline. | The steep reduction in cost enables the collection and storage of Big Data. Organizations can easily maintain detailed databases on individuals. There are no limits on the data collected about you. |
| Data analysis advances | Companies can analyze vast quantities of data gathered on individuals to develop detailed profiles of individual behavior. Large-scale population surveillance is enabled. |
| Networking advances | The cost of making data accessible from anywhere falls exponentially. Access to data becomes more difficult to control. |
| Mobile devices proliferate. | Mobile devices may be tracked without user consent or knowledge. The always-on device becomes a tether. |
| Artificial intelligence (AI) | Increased reliance on various forms of AI in decision making substitutes data-driven calculations for human judgment. |

© Cast Of Thousands/Shutterstock

*Credit card purchases can make personal information available to a wide variety of organizations. Advances in information technology thus facilitate the invasion of privacy.*

companies can analyze large pools of data from multiple sources to rapidly identify buying patterns of customers and make individualized recommendations. The use of computers to combine data from multiple sources and create digital dossiers of detailed information on individuals is called **profiling**.

For example, Google Marketing Platform tracks the activities of visitors of participating websites and uses this information to create a profile of each online visitor, adding more detail to the profile as the visitor accesses an associated Google Marketing Platform site. Over time, Google Marketing Platform can create a detailed dossier of a person's spending and online behavioral habits that is sold to companies to help them target their online ads more precisely. Advertisers can combine online consumer information with offline consumer information such as credit card purchases at stores.

LexisNexis Risk Solutions gathers data from police, criminal, and motor vehicle records; credit and employment histories; current and previous addresses; professional licenses; and insurance claims to assemble and maintain dossiers on almost every adult in the United States. The company sells this personal information to businesses and government agencies. Demand for personal data is so enormous that data broker businesses, such as LexisNexis Risk Solutions, Acxiom, CoreLogic, Datalogix, and DataRaker, are flourishing. The three major credit monitoring services, TransUnion, Equifax, and Experian, also maintain huge databases of personal data.

A data analysis technology called **nonobvious relationship awareness (NORA)** has given both the government and the private sector even more powerful profiling capabilities. Examine Figure 4.2, which illustrates how NORA works. NORA can take information about people from many disparate sources, such as "watch" lists, incident and arrest systems, consumer transaction systems, telephone records, and human resources systems, and correlate relationships to find obscure connections that might help identify criminals or terrorists. NORA technology processes the data and extracts information as the data are being generated so that, for example, it can—before a person boards an airplane—instantly discover that the person at the airline ticket counter has the same phone number as a known terrorist. NORA is also often used by the casino industry as an anti-fraud technology. Although NORA technology is considered a valuable security tool, it does have privacy implications because it can provide such a detailed picture of the activities and associations of a particular individual.

Advances in networking, including the Internet, have greatly reduced the costs of accessing large quantities of data and have enabled the mining of large pools of data using desktop/laptop computers, mobile devices, and cloud servers, permitting an invasion of privacy on a scale and with a precision heretofore unimaginable.

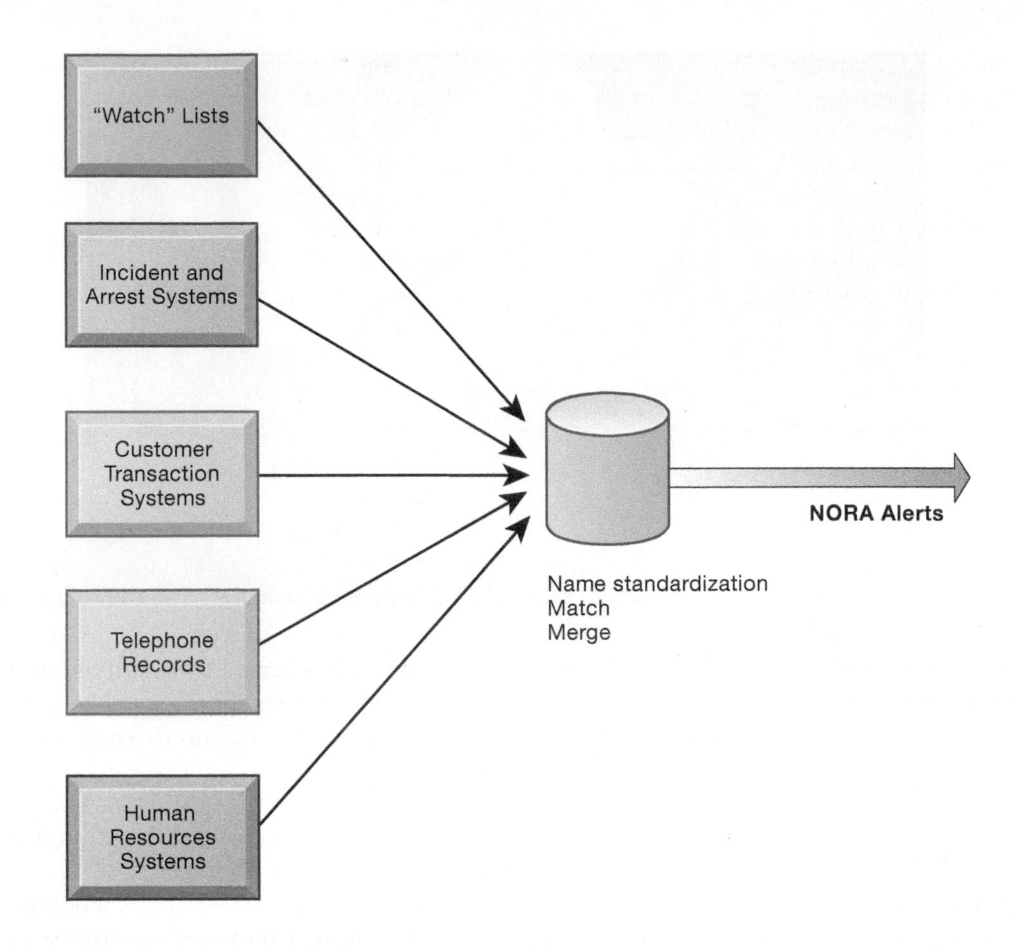


**Figure 4.2**
Nonobvious Relationship Awareness (NORA)
*NORA technology can take information about people from disparate sources and find obscure, nonobvious relationships. It might discover, for example, that an applicant for a job at a casino shares a telephone number with a known criminal and then issue an alert to the hiring manager.*

The proliferation of mobile devices, such as smartphones, that users carry everywhere means that users can potentially be tracked without their knowledge or consent.

Finally, companies and individuals are increasingly relying on AI technologies as a substitute for human judgment in decision making. However, doing so raises serious ethical issues. Human judgment relies not only on reasoning but also on other human capabilities, such as empathy, and also typically takes into account factors that inform moral and ethical behavior. AI, on the other hand, may make decisions without regard to consequences that might violate ethical and moral standards. Another issue with AI is that it reduces multidimensional reality into a more simplistic representation of the world, one that appears on the surface to be precise and undeniably correct, thus encouraging acceptance of its decisions without questioning them. We discuss this issue further in Chapter 11 (Moser, den Hond, and Lindebaum, 2022).

## 4-2 Describe principles for conduct that can be used to guide ethical decisions.

Ethics is a concern of humans who have freedom of choice. Ethics is about individual choice: When faced with alternative courses of action, what is the correct moral choice? What are the main features of ethical choice?

### BASIC CONCEPTS: RESPONSIBILITY, ACCOUNTABILITY, LIABILITY, AND DUE PROCESS

Ethical choices are decisions made by individuals who are responsible for the consequences of their actions. **Responsibility** is a key element of ethical action. Responsibility means that you accept the potential costs, duties, and obligations for the decisions you make. **Accountability** is a feature of systems and social institutions; it means that mechanisms are in place to determine who took action and who is responsible. Systems and institutions in which it is impossible to find out who took what action are inherently incapable of ethical analysis or ethical action. **Liability** extends the concept of responsibility further to the area of laws. Liability is a feature of political systems in which a body of laws is in place that permits individuals to recover the damages done to them by other actors, systems, or organizations. **Due process** is a related feature of law-governed societies and is a process in which laws are known and understood, and ability exists to appeal to higher authorities to ensure that the laws are applied correctly.

These basic concepts form the underpinning of an ethical analysis of information systems and those who manage them. First, information technologies are filtered through social institutions, organizations, and individuals. Systems do not have impacts by themselves. Whatever information system effects exist are products of institutional, organizational, and individual actions and behaviors. Second, responsibility for the consequences of technology falls clearly on the institutions, organizations, and individuals who choose to use the technology. Using information technology in a socially responsible manner means that you can and will be held accountable for the consequences of your actions. Third, in an ethical political society, individuals and others can recover damages done to them through a set of laws characterized by due process.

## ETHICAL ANALYSIS

When confronted with a situation that seems to present ethical issues, how should you analyze it? The following five-step process should help:

1. *Identify and describe the facts clearly.* Find out who did what to whom and where, when, and how. In many instances, you will be surprised at the errors in the initially reported facts, and often you will find that simply getting the facts straight helps define the solution. It also helps to get the opposing parties involved in an ethical dilemma to agree on the facts.
2. *Define the conflict or dilemma and identify the higher-order values involved.* Ethical, social, and political issues always reference higher values. The parties to a dispute all claim to be pursuing higher values (e.g., freedom, privacy, protection of property, or the free enterprise system). Typically, an ethical issue involves a dilemma: two diametrically opposed courses of action that support worthwhile values. For example, the chapter-opening case study illustrates two competing values: organizational efficiency and profit versus respect for individual privacy.
3. *Identify the stakeholders.* Every ethical, social, and political issue has stakeholders: players in the game who have an interest in the outcome, who have invested in the situation, and who usually have vocal opinions. Find out the identity of these groups and what they want. This will be useful later when designing a solution.
4. *Identify the options that you can reasonably take.* You may find that none of the options satisfy all the interests involved but that some options do a better job than others. Sometimes arriving at a good or ethical solution may not always be a balancing of consequences to stakeholders.
5. *Identify the potential consequences of your options.* Some options may be ethically correct but disastrous from other points of view. Other options may work in one instance but not in similar instances. Always ask yourself, "What if I choose this option consistently over time?"

## ETHICAL PRINCIPLES

Once your analysis is complete, what ethical principles or rules should you use to make a decision? What higher-order values should inform your judgment? Although you are the only one who can decide which among many ethical principles you will follow, and how you will prioritize them, it is helpful to consider some ethical principles with deep roots in many cultures that have survived throughout recorded history:

1. Do unto others as you would have them do unto you (the **Golden Rule**). Putting yourself in the place of others, and thinking of yourself as the object of the decision, can help you think about fairness in decision making.
2. If an action is not right for everyone to take, it is not right for anyone (**Immanuel Kant's categorical imperative**). Ask yourself, "If everyone did this, could the organization, or society, survive?"
3. If an action cannot be taken repeatedly, it is not right to take at all. This is the **slippery slope rule**: An action may bring about a small change now that is acceptable, but if it is repeated, it would bring unacceptable changes in the long run. In the vernacular, it might be stated as "once started down a slippery path, you may not be able to stop."
4. Take the action that achieves the higher or greater value (**Utilitarian principle**). This rule assumes you can prioritize values in a rank order and understand the consequences of various courses of action.
5. Take the action that produces the least harm or the least potential cost (**risk aversion principle**). Some actions have extremely high failure costs of very low probability (e.g., building a nuclear generating facility in an urban area) or extremely high failure costs of moderate probability (speeding and automobile accidents). Avoid actions that have extremely high failure costs; focus on reducing the probability of accidents occurring.
6. Assume that virtually all tangible and intangible objects are owned by someone else unless there is a specific declaration otherwise. (This is the **ethical no-free-lunch rule**.) If something someone else has created is useful to you, it has value, and you should assume the creator wants compensation for this work.

Actions that do not easily pass these rules deserve close attention and a great deal of caution. The appearance of unethical behavior may do as much harm to you and your company as actual unethical behavior.

## PROFESSIONAL CODES OF CONDUCT

When groups of people claim to be professionals, they take on special rights and obligations because of their special claims to knowledge, wisdom, and respect. Professional codes of conduct are promulgated by associations of professionals such as the American Medical Association (AMA), the American Bar Association (ABA), the Association of Information Technology Professionals (AITP), and the Association for Computing Machinery (ACM). These professional groups take responsibility for the partial regulation of their professions by determining entrance qualifications and competence. Codes of ethics are promises by professions to regulate themselves in the general interest of society. For example, avoiding harm to others, honoring property rights (including intellectual property), and respecting privacy are among the General Moral Imperatives of the ACM's Code of Ethics and Professional Conduct.

## SOME REAL-WORLD ETHICAL DILEMMAS

Information systems have created new ethical dilemmas in which one set of interests is pitted against another. For example, many companies use voice recognition software to reduce the size of their customer support staff by enabling computers to recognize a customer's responses to a series of automated questions. Many companies monitor what their employees are doing on the Internet to prevent them from wasting company resources on nonbusiness activities (see the Chapter 7 Spotlight on People case).

In each instance, you can find competing values at work, with groups lined up on either side of a debate. A company may argue, for example, that it has the right to use information systems to increase productivity and reduce the size of its workforce to lower costs and stay in business. Employees displaced by information systems may argue that employers have some responsibility for their welfare. Business owners might feel obligated to monitor employee email and Internet use to minimize drains on productivity. Employees might believe they should be able to use the Internet for short personal tasks. A close analysis of the facts can sometimes produce solutions that give each side "half a loaf." Try to apply some of the principles of ethical analysis described to each of these cases. What is the right thing to do?

## 4-3 Explain how contemporary information systems technologies and the Internet pose challenges to the protection of individual privacy and intellectual property.

In this section, we take a closer look at information rights and obligations and property rights and obligations, two of the moral dimensions of information systems first described in Figure 4.1. For each, we identify ethical, social, and political issues and use real-world examples to illustrate the values involved, the stakeholders, and the options chosen.

### INFORMATION RIGHTS: PRIVACY AND FREEDOM IN THE INTERNET AGE

**Privacy** is the claim of individuals to be left alone, free from surveillance or interference from other individuals or organizations, including the state. Claims to privacy are also involved at the workplace. Millions of employees are subject to digital and other forms of high-tech surveillance. Information technology and systems threaten individual claims to privacy by making the invasion of privacy cheap, profitable, and effective.

The claim to privacy is protected in the United States Constitution in a variety of ways and also through various statutes. The same holds true for many other countries. In the United States, the claim to privacy is protected primarily by the First Amendment's guarantees of freedom of speech and association, the Fourth Amendment's protections against unreasonable search and seizure of one's personal documents or home, and the Fifth and Fourteenth Amendments' guarantee of due process.

Table 4.3 describes the major US federal statutes that set forth the conditions for handling information about individuals in such areas as credit reporting, education, financial records, newspaper records, and electronic and digital communications. The Privacy Act of 1974 was one of the first, and arguably, the most important of these laws, regulating the federal government's collection, use, and disclosure of information.

Most US and European privacy law is based on a set of principles called **Fair Information Practices (FIP)**, first set forth in a 1973 report by a federal government advisory committee (U.S. Department of Health, Education, and Welfare, 1973). FIP is a set of principles governing the collection and use of information about individuals. FIP principles are based on the notion of a mutuality of interest between the record holder and the individual. The individual has an interest in engaging in a transaction, and the record keeper—usually a business or government agency—requires information about the individual to support the transaction. After information is gathered, the individual maintains an interest in the record, and the record may not be used to

**TABLE 4.3**

Federal Privacy Laws in the United States

| General Federal Privacy Laws | Privacy Laws Affecting Private Institutions |
|---|---|
| Freedom of Information Act | Fair Credit Reporting Act |
| Privacy Act<br>Privacy Protection Act | Family Educational Rights and Privacy Act |
| Electronic Communications Privacy Act | Right to Financial Privacy Act |
| Computer Matching and Privacy Protection Act | |
| Computer Security Act | Cable Communications Policy Act |
| Federal Managers Financial Integrity Act | |
| Driver's Privacy Protection Act | Video Privacy Protection Act<br>Telephone Consumer Protection Act (TCPA) |
| E-Government Act<br>USA Freedom Act | Health Insurance Portability and Accountability Act (HIPAA)<br>Children's Online Privacy Protection Act (COPPA)<br>Financial Services Modernization Act (Gramm-Leach-Bliley Act) |

support other activities without the individual's consent. In 1998, the Federal Trade Commission (FTC) restated and extended the original FIP to provide guidelines for protecting online privacy and updated them further in 2010 to take into account new privacy-invading technology. Table 4.4 describes the FTC's FIP principles.

The FTC's FIP principles have been used throughout the years as guidelines to drive changes in privacy legislation. For instance, the Children's Online Privacy Protection Act (COPPA) requires websites and mobile apps to obtain parental permission before collecting information on children under the age of 13. The FTC has also added three principles to its framework for privacy. Firms should adopt privacy by design, building products and services that protect privacy; firms should increase the transparency of their data practices; and firms should require consumer consent and provide clear options to opt out of data-collection schemes. The FTC has extended its privacy policies to address behavioral targeting, smartphone tracking, the Internet of Things (IoT), and mobile health apps (Federal Trade Commission, 2021).

**TABLE 4.4**

Federal Trade Commission's Fair Information Practice Principles

| Principle | Description |
|---|---|
| Notice/awareness (core principle) | Websites must disclose their information practices before collecting data. Includes identification of collector; uses of data; other recipients of data; nature of collection (active/inactive); voluntary or required status; consequences of refusal; and steps taken to protect confidentiality, integrity, and quality of the data. |
| Choice/consent (core principle) | A process enabling choice must be in place, allowing consumers to choose how their information will be used for purposes other than supporting the transaction, including internal use and transfer to third parties. |
| Access/participation | Consumers should be able to review and contest the accuracy and completeness of data collected about them in a timely, inexpensive process. |
| Security | Data collectors must take responsible steps to ensure that consumer information is accurate and secure from unauthorized use. |
| Enforcement | A mechanism must be in place to enforce FIP principles. This can involve self-regulation, legislation giving consumers legal remedies for violations, or federal statutes and regulations. |

Public opinion polls show an ongoing distrust of online marketers. Although there are many studies of privacy issues, there has been no significant federal legislation in recent years. A recent survey by the Pew Research Center found that almost 80 percent of those surveyed were concerned about their online privacy, most believe they have lost control of their personal information online, and more than 85 percent have taken steps to protect their information online.

Privacy protections have also been added to laws deregulating financial services and safeguarding the maintenance and transmission of health information about individuals. The Gramm-Leach-Bliley Act, which repeals earlier restrictions on affiliations among banks, securities firms, and insurance companies, includes some privacy protection for consumers of financial services. All financial institutions are required to disclose their policies and practices for protecting the privacy of nonpublic personal information and to allow customers to opt out of information-sharing arrangements with nonaffiliated third parties.

The Health Insurance Portability and Accountability Act (HIPAA) includes privacy protection for medical records. The law gives patients access to their personal medical records that healthcare providers, hospitals, and health insurers maintain and the right to authorize how protected information about themselves can be used or disclosed. Doctors, hospitals, and other healthcare providers must limit the disclosure of personal information about patients to the minimum amount necessary to achieve a given purpose.

### The European General Data Protection Regulation

In 2018 the European Commission implemented the European Union's **General Data Protection Regulation (GDPR)**, which is arguably the most important privacy legislation in the last 20 years since the FTC's FIP principles. It applies to all firms and organizations that collect, store, or process personal information of EU citizens, and these protections apply worldwide regardless of where the processing takes place (European Commission, 2018).

The GDPR is an updated framework for protecting personally identifiable information (PII) and replaces an earlier Data Protection Directive. In Europe, privacy protection is historically much stronger than it is in the United States. In the United States, there is no federal agency charged with enforcing privacy laws, and there is no single privacy statute governing use of PII by private organizations. Instead, privacy laws are piecemeal, sector by sector (e.g., medical privacy, educational privacy, and financial privacy laws). These are enforced by the FTC, through self-regulation by businesses, and by individuals, who must sue agencies or companies in court to recover damages. This is expensive and rarely done.

In the EU, data protection laws are comprehensive, apply to all organizations, and are enforced by data protection agencies in each country to pursue complaints brought by citizens and to actively enforce privacy laws. The GDPR protects a wide variety of PII: basic identity information such as name, address, and ID numbers; web data such as location, IP address, cookie data, and RFID tags; health and genetic data; mobile phone number; driver's license and passport number; biometric and facial data; racial and ethnic data; political opinions; and sexual orientation.

The main objective of the GDPR is to strengthen the rights of citizens to their own personal information and to strengthen oversight of firms to ensure they implement these individual rights. A second thrust was to harmonize conflicting data protection standards among the EU member nations and create a single EU agency to implement and enforce the regulation. And third, to enforce these conditions worldwide for all organizations that operate in the EU, or process data pertaining to EU citizens, regardless of where the organization is located.

For individuals, the GDPR requires organizations to allow consumers to access all their personal information without charge within one month; delete personal data (the right to be forgotten); ensure data portability so consumers are not locked into a particular service; and guarantee the right to sue providers for damages or abuse of PII, including class action lawsuits.

Organizational requirements have been strengthened. Organizations must have a data protection officer that reports to senior management. Explicit consent must be obtained from users before the organization can collect data (positive opt-in), eliminating default opt-in processes. Organizations must publish the rationale for data collection and the length of time data are retained, and report any breaches or hacks within 72 hours. Organizations must also build privacy protections into all new systems (privacy by design); limit targeting and retargeting of individuals to audience-level, anonymized data (rather than targeting based on intimate, personal profiles), and limit the collection of personal data to only that which is needed to support a task or a transaction, and then delete it shortly thereafter. Abuse of PII can be fined up to $20 million or 4 percent of the organization's global revenue, whichever is greater. Finally, the EU will enforce the GDPR requirements with non-EU countries like the United States using intergovernmental privacy shield agreements that ensure that EU data processed in non-EU nations meets GDPR standards. Privacy shield agreements are a more enforceable version of earlier **safe harbor** agreements. A safe harbor is a private self-regulating policy and enforcement mechanism that meets the objectives of government regulators and legislation but does not involve government regulation or enforcement.

The GDPR is aimed at Meta (owner of Facebook, Instagram, and WhatsApp), Google, and other online ad-based businesses that track and collect personal data from users in order to create comprehensive digital profiles and target these persons with ads. Google and Facebook are both extremely popular in Europe and dominate their markets, but at the same time they are widely criticized for invading privacy and not protecting PII.

### Internet Challenges to Privacy

Internet technology poses new challenges for the protection of individual privacy. It enables companies to track web searches that users conduct, the websites and web pages visited, the online content a person has accessed, and what items that person has inspected or purchased online. Mobile apps can also track users. This monitoring and tracking occurs in the background without the visitor's knowledge. It is conducted not just by individual websites (known as first-party tracking) but by advertising networks such as Microsoft Advertising, Google's Marketing Platform, and Meta's Audience Network that are capable of tracking personal browsing behavior across thousands of websites (known as third-party tracking). In the past, website publishers and the online advertising industry have defended tracking of individuals across the web because doing so allows more relevant ads to be targeted to users, and this helps pay for the cost of providing online content. The commercial demand for this personal information is virtually insatiable. However, these practices also impinge on individual privacy.

**Cookies** are one method used to monitor and track online users. Examine Figure 4.3, which illustrates how cookies work. When a user visits a website, the website's web server places a small text file (a "cookie") on the user's computer or mobile device. Cookies identify the visitor's web browser software, as well as other information, and track visits to the website. When the visitor returns to a site that has stored a cookie, the website software searches the visitor's computer or mobile device, finds the cookie, and knows what that person has done in the past. It may also update the cookie, depending on the activity during the visit. In this way, the site can customize its content for each visitor's interests. For example, if you purchase a book on Amazon.com and return later from the same browser, the site will welcome you by name and recommend other books of interest based on your past purchases. Cookies also make "quick checkout" options possible by allowing a site to keep track of users as they add items to a shopping cart. This type of cookie is known as a first-party

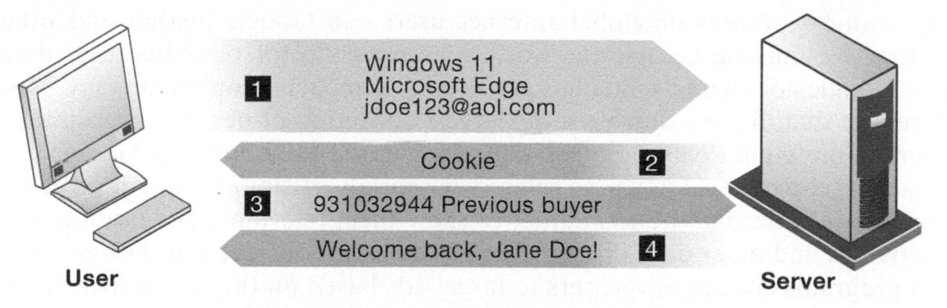


1. The web server reads the user's web browser and determines the operating system, browser name, version number, Internet address, and other information.
2. The server transmits a tiny text file with user identification information called a cookie, which the user's browser receives and stores on the user's computer.
3. When the user returns to the website, the server requests the contents of any cookie it deposited previously in the user's computer.
4. The web server reads the cookie, identifies the visitor, and calls up data on the user.

**Figure 4.3**
How First-Party Cookies Identify Web Visitors
*Cookies are placed by a website on a visitor's computer. When the visitor returns to that website, the web server requests the ID number from the cookie and uses it to access the data stored by that server on that visitor. The website can then use these data to display personalized information.*

cookie. First-party cookies are typically considered to be "good" cookies because they help enhance the user experience.

Websites using cookie technology cannot directly obtain visitors' names and addresses. However, if a person has registered at a site, that information can be combined with cookie data to identify the visitor. Website owners can also combine the data they have gathered from cookies and other website monitoring tools with personal data from other sources, such as offline data collected from surveys or paper catalog purchases, to develop very detailed profiles of their visitors. First-party cookies can be blocked and/or deleted by users, although the majority of users do not do so.

Third-party cookies operate similarly to first-party cookies but enable advertising platforms to track user behavior across websites and devices. In the past, third-party cookies were supported by all the major web browsers, but that has begun to change. Apple's Safari and Mozilla's Firefox browsers now block third-party cookies by default, and Google has announced that it will follow suit beginning in 2024. We discuss other technology-based efforts to enhance privacy in the following section on technological solutions.

In addition to cookies, there are other tools for surveillance of Internet users. A **web beacon**, also called a *web bug* (or simply tracking file), is a tiny image that keeps a record of users' online clickstreams. They report these data back to whoever owns the tracking file, which can be invisibly embedded in an email message or web page to monitor the behavior of the user visiting the website or receiving the email. Web beacons are placed on popular websites by third-party firms who pay the websites a fee for access to their audience. So how common is web tracking? In a path-breaking series of articles in the *Wall Street Journal*, researchers examined the tracking files on 50 of the most popular US websites. What they found revealed a very widespread surveillance system: On the 50 sites, they discovered 3,180 tracking files installed on visitor computers. Only one site, Wikipedia, had no tracking files. Two-thirds of the tracking files came from 131 companies whose primary business is identifying and tracking Internet users to create consumer profiles that can be sold to advertising firms looking for specific types of customers. The biggest trackers were Google, Microsoft, and Quantcast, all of whom are in the business of selling ads to advertising firms and marketers. A follow-up study found tracking on the 50-most-popular sites had risen nearly fivefold because of the growth of online ad auctions where advertisers buy the data about users' web-browsing behaviors.

**Adware** can secretly install itself on an Internet user's computer by piggybacking on larger applications. Once installed, adware calls out to websites to send ads and other unsolicited material to the user. A more malicious version of adware, known as **spyware** can also track the user's browsing habits. More information is available about intrusive software in Chapter 8.

More than 90 percent of global Internet users use Google Search and other Google services, making Google the world's largest collector of online user data. Whatever Google does with its data has an enormous impact on online privacy. Most experts believe that Google possesses the largest collection of personal information in the world—more data on more people than any government agency. Google uses behavioral targeting (also known as interest-based advertising) to target individuals as they move from one site to another to show them relevant ads based on their search activities and other data that Google has collected about them. For instance, one of its programs enables advertisers to target ads based on the search histories of Google users, along with any other information the user submits to Google such as age, demographics, region, and web activities (such as blogging). Google's AdSense program enables Google to help advertisers select keywords and design ads for various market segments based on search histories such as helping a clothing website create and test ads targeted at teenage females. Google displays targeted ads on YouTube and Google mobile applications, and its Google Marketing Platform ad network serves targeted ads.

The United States has allowed businesses to gather transaction information generated in the marketplace and then use that information for other marketing purposes without obtaining the **informed consent** of the individual whose information is being used. These firms argue that when users agree to the sites' terms of service, they are also agreeing to allow the site to collect information about their online activities. An **opt-out** model of informed consent permits the collection of personal information until the consumer specifically requests the data not to be collected. Privacy advocates would like to see wider use of an **opt-in** model of informed consent in which a business is prohibited from collecting any personal information unless the consumer specifically takes action to approve information collection and use. Here, the default option is no collection of user information. Many websites now employ "cookie banners"—a pop-up window that allows users to accept or reject cookies from the site. Although intended as a form of opt-in informed consent, most privacy advocates feel that cookie banners are ineffective, as few people actually read the disclosures about what data are being collected before clicking the Accept button.

The online industry has preferred self-regulation to privacy legislation for protecting consumers. Members of the online advertising industry have created an industry association called the Network Advertising Initiative (NAI) to develop privacy policies to help consumers opt out of advertising network programs and to provide consumers redress from abuses. Individual firms such as Apple, Google, and Microsoft have also launched privacy initiatives in an effort to address public concern about tracking people online.

In general, most businesses do little to protect the privacy of their customers, and consumers do not do as much as they should to protect themselves. For websites and apps that depend on advertising for support, most revenue derives from selling access to customer information. The vast majority of people claim to be concerned about online privacy, but most do not read the privacy statements on websites. In addition, website privacy policies are often ambiguous about key terms and too complicated for the average consumer to understand (Laudon and Traver, 2024). What firms call a privacy policy is in fact a data use policy. The concept of privacy is associated with consumer rights, which many firms do not wish to recognize. A data use policy simply tells customers how the information will be used and does not mention rights.

### Technological Solutions

In addition to legislation, a number of technological solutions have been developed to help protect user privacy. Table 4.5 lists some of the most common tools that are available. For the most part, technical solutions thus far have failed to provide effective protection for online privacy, in part because most of them require users to be proactive in implementing them.

**TABLE 4.5**

Technological Protections for Online Privacy

| Technology | Protection |
|---|---|
| Apple App Tracking Transparency (ATT) | Requires any app that wants to track user activity and share it with other apps or websites to ask user for permission |
| Apple Intelligent Tracking Prevention (ITP) for Safari web browser | Monitors and disables cross-site tracking cookies and blocks trackers' ability to identify users by IP address |
| Google Privacy Sandbox | System now being tested by Google to replace cookie-based, targeted advertising in the Google Chrome web browser by 2024 |
| Differential privacy software | Reduces the ability to merge different files and de-anonymize consumer data |
| Privacy default browsers | Eliminates tracking cookies and prevents IP tracking |
| Message encryption apps | Encrypts text and other data transferred using smartphones |
| Spyware blockers | Detects and removes spyware, adware, keyloggers, and other malware |
| Pop-up and ad blockers | Prevents calls to ad servers; restricts downloading of images at user request |
| Secure email | Email and document encryption |
| Anonymous remailers | Enhanced privacy protection for email |
| Anonymous surfing | Enhanced privacy protection for web browsing |
| Cookie managers | Blocks third-party cookies |
| Public key encryption | Enables encryption of email and documents |

## PROPERTY RIGHTS: INTELLECTUAL PROPERTY

Contemporary information systems have severely challenged existing laws and social practices that protect **intellectual property**. Intellectual property refers to products of the mind created by individuals or corporations. Information technology has made it difficult to protect intellectual property because digital information can be so easily copied or distributed. Intellectual property is subject to a variety of protections under four legal traditions: copyright, patents, trademarks, and trade secrets.

### Copyright

**Copyright** protects creators of intellectual property from having their work copied by others for any purpose for a certain period of time, depending on several factors. As a general rule, copyright protection lasts for the life of the author, plus an additional 70 years after the author's death. For corporate-owned works, copyright protection generally lasts for 95 years after initial creation. Copyright protects literary, dramatic, musical, and artistic works, such as novels, poetry, songs, plays, maps, drawings, and artwork. The intent behind copyright laws has been to encourage creativity and authorship by ensuring that creative people receive the financial and other benefits of their work. Most industrial nations have their own copyright laws, and there are several international conventions and bilateral agreements through which nations coordinate and enforce their copyright laws.

In the mid-1960s, the Copyright Office began registering software programs, and in 1980, Congress passed the Computer Software Copyright Act, which provides protection for software program code and copies of the original code sold in commerce; it sets forth the rights of the purchaser to use the software, while the creator retains legal title.

Copyright protects against copying entire programs or their parts. Damages and relief are readily obtained for infringement. The drawback to copyright protection is that the underlying ideas behind a work are not protected; only their manifestation in a work are protected. A competitor can use your software, understand how it works, and build new software that follows the same concepts without infringing on a copyright.

Look-and-feel copyright infringement lawsuits are precisely about the distinction between an idea and its expression. For instance, in the early 1990s, Apple sued Microsoft and Hewlett-Packard, claiming that the defendants copied the expression of overlapping windows in Apple's Macintosh interface. The defendants countered that the idea of overlapping windows can be expressed only in a single way and, therefore, was not protectable. When ideas and their expression merge, the expression cannot be copyrighted. Ultimately, the courts agreed with the defendants.

In general, courts follow the reasoning of a 1989 case—*Brown Bag Software v. Symantec Corp*—in which the court dissected the elements of software alleged to be infringing. The court found that similar concepts, functions, general functional features (e.g., drop-down menus), and colors are not protectable by copyright law (*Brown Bag Software v. Symantec Corp.*, 1992).

### Patents

A **patent** grants the owner an exclusive monopoly on the ideas behind an invention for a certain period of time, typically 20 years. The congressional intent behind patent law was to ensure that inventors of new machines, devices, or methods receive the full financial and other rewards of their labor and yet to make widespread use of the invention possible by providing detailed diagrams for those wishing to use the idea under license from the patent's owner. The granting of a patent is determined by the United States Patent and Trademark Office and relies on court rulings.

The key concepts in patent law are originality, novelty, and invention. The Patent Office did not accept applications for software patents routinely until a 1981 Supreme Court decision held that computer programs could be part of a patentable process.

The strength of patent protection is that it grants a monopoly on the underlying concepts and ideas of software. The difficulty is passing stringent criteria of nonobviousness (e.g., the work must reflect some special understanding and contribution), originality, and novelty as well as years of waiting to receive protection.

### Trademarks

**Trademarks** are the marks, symbols, and images used to distinguish products in the marketplace. Trademark laws protect consumers by ensuring that they receive what they paid for. These laws also protect the investments that firms have made to bring products to market. Typical trademark infringement violations occur when one firm appropriates or pirates the marks of a competing firm. Infringement also occurs when firms dilute the value of another firm's marks by weakening the connection between a mark and the product. For instance, if a search engine firm copies the trademarked Google icon, colors, and images, it would be infringing on Google's trademarks. It would also be diluting the connection between the Google search service and its trademarks, potentially creating confusion in the marketplace.

### Trade Secrets

Any intellectual work product—a formula, device, pattern, or compilation of data—used for a business purpose can be classified as a **trade secret**, provided that it is not based on information in the public domain. Protections for trade secrets vary from state to state. In general, trade secret laws grant a monopoly on the ideas behind a work product, but the monopoly can be very tenuous.

Software that contains novel or unique elements, procedures, or compilations can be considered a trade secret. Trade secret law protects the actual ideas in a work product, not only their manifestation. To make a claim of trade secret infringement, the creator or owner must take care to bind employees and customers with nondisclosure agreements and prevent the secret from falling into the public domain.

The limitation of trade secret protection is that although virtually all software programs of any complexity contain unique elements of some sort, it is difficult to prevent the ideas in the work from falling into the public domain when the software is widely distributed.

### Challenges to Intellectual Property Rights

Contemporary information technologies, especially software, pose severe challenges to existing methods of protecting intellectual property and, therefore, create significant ethical, social, and political issues. Digital media differ from books, periodicals, and other media in terms of the former's ease of replication; ease of transmission; ease of alteration; ease of theft because of compactness; and difficulties in establishing uniqueness.

The proliferation of digital networks, including the Internet, has made it even more difficult to protect intellectual property. Before the widespread use of digital networks, copies of software, books, magazine articles, or films had to be stored on physical media such as paper, computer disks, or videotape, creating some hurdles to distribution. Using digital networks, information can be more widely reproduced and distributed. A Global Software Survey conducted by International Data Corporation and BSA/The Software Alliance reported that 37 percent of the software installed on personal computers was unlicensed. Software piracy remains an ongoing issue, with the number of visits to software piracy sites worldwide increasing to 3.2 billion in 2021 (BSA/The Software Alliance, 2018; Statista Research Department, 2021).

The Internet was designed to transmit information, including copyrighted information, freely around the world. You can easily copy and distribute virtually anything to millions of people worldwide even if they are using different types of computer systems. Information can also be illicitly copied from one place and distributed throughout other systems and networks even though these parties do not willingly participate in the infringement.

Individuals have been illegally copying and distributing digitized music files on the Internet for several decades. File-sharing services such as Napster and, later, Grokster, Kazaa, Morpheus, Megaupload, and The Pirate Bay sprang up in the early years of the web to enable users to locate and swap digital music and video files, including those protected by copyright. Illegal file sharing became so widespread that it threatened the viability of the music recording industry and, at one point, consumed 20 percent of Internet bandwidth. The recording industry won several legal battles for shutting these services down, but it has not been able to halt illegal file sharing entirely. The motion picture and cable television industries continue to wage similar battles. Several European nations have worked with US authorities to shut down illegal sharing sites, with mixed results.

However, as legitimate online music stores such as iTunes and streaming services such as Spotify, Apple Music, Amazon Prime Music, and Pandora have expanded, illegal file sharing has significantly declined. The Apple iTunes Store legitimized paying for music and entertainment and created a closed environment from which music and videos could not be easily copied and widely distributed. Amazon's Kindle also protects the rights of publishers and writers because its books cannot be copied to the Internet and distributed. Music and video streaming services also inhibit piracy because the streams cannot be easily recorded on separate devices. Since 2015, music industry revenues have more than doubled after a precipitous drop of more than 50 percent between 1999 and 2014.

The **Digital Millennium Copyright Act (DMCA)** also provides some copyright protection. The DMCA implemented a World Intellectual Property Organization Treaty that makes it illegal to circumvent technology-based protections of copyrighted materials. Internet service providers (ISPs) are required to take down sites of copyright infringers they are hosting when the ISPs are notified of the infringement. Microsoft and other major software and information content firms are represented by the Software and Information Industry Association (SIIA), which lobbies for new laws and enforcement of existing laws to protect intellectual property around the world. The SIIA runs an antipiracy hotline for individuals to report piracy activities, offers educational programs to help organizations combat software piracy, and has published guidelines for employee use of software.

## 4-4 Discuss the issues that contemporary information systems raise with respect to system quality, accountability and control, and the quality of everyday life.

In this section, we take a closer look at the issues that new information technologies and systems raise with respect to system quality, accountability and control, and the quality of everyday life, which, along with information rights and obligations and property rights and obligations, are part of the moral dimensions of information systems first described in Figure 4.1. Given the centrality of information systems to the functioning of businesses and society, what is an acceptable level of system quality? If a person is injured by a machine controlled, in part, by software, who should be held accountable and, therefore, held liable? Should a social network be held accountable and liable for the posting of offensive material? What kinds of acts should be considered computer crimes? How should individuals, society, and governments deal with the potentially negative impacts of new information technologies on the quality of everyday life?

### SYSTEM QUALITY: DATA QUALITY AND SYSTEM ERRORS

In 2021, many of the largest cloud providers experienced significant outages, which took out sites and services throughout the United States and Europe. For example, in December 2021, a seven-hour Amazon Web Services (AWS) outage related to application programming interfaces took down a wide array of Amazon services, such as Alexa, Ring cameras, and Prime Video, as well as a host of third-party websites and apps, such as Netflix, Disney+, Ticketmaster, payment app Venmo, and stock trading app Robinhood. In November 2021, Google Cloud experienced an outage because of a network configuration glitch that took down Snapchat, Spotify, and Home Depot. In October, Facebook, Instagram, WhatsApp, and Facebook Messenger all went down for seven hours when Facebook engineers accidently made a configuration change to the backbone routers that coordinate network traffic among its data centers. The Facebook outage also impacted enterprises that use Facebook to authenticate users: When the service went down, users were no longer able to log in.

Despite these high-profile incidents, outages at cloud computing services are infrequent. But, as more and more firms rely on cloud providers and centralize their data and operations with a small group of cloud providers, these outages have called into question the reliability and quality of cloud services. Are these outages acceptable?

The debate raises a related but independent moral dimension: What is an acceptable, technologically feasible level of system quality? At what point should system managers say, "Stop testing, we've done all we can to perfect this software. Ship it!" Individuals and organizations may be held responsible for avoidable and foreseeable consequences, which they have a duty to recognize and correct. The gray area is that some system errors are foreseeable and correctable only at very great expense, an expense so great that pursuing this level of perfection is not feasible economically—no one could afford the product!

For example, although software companies try to debug their products before releasing them to the marketplace, they knowingly ship buggy products because the time and cost of fixing all the minor errors would prevent these products from ever being released. What if the product was not offered on the marketplace? Would social welfare as a whole falter and perhaps even decline? Carrying this further, just what is the responsibility of a producer of computer services—should the producer withdraw the product that can never be perfect, warn the user, or forget about the risk (let the buyer beware)?

Three principal sources of poor system performance are (1) software bugs and errors, (2) hardware or facility failures caused by natural or other causes, and (3) poor input data quality. Zero defects in software code of any complexity cannot be achieved, and the seriousness of the remaining bugs cannot be estimated (Choi, 2020). Hence, there is a technological barrier to perfect software, and users must be aware of the potential for catastrophic failure. The software industry has not yet arrived at testing standards that would result in software of acceptable but imperfect performance.

Although software bugs and facility catastrophes are likely to be widely reported in the press, by far the most common source of business system failure is data quality (see Chapter 6). Few companies routinely measure the quality of their data, but individual organizations report data error rates ranging from 0.5 to 30 percent.

## ACCOUNTABILITY AND CONTROL: SOFTWARE AND INTERNET LIABILITY ISSUES

Who, if anyone, is liable if a business or person is harmed by software that is defective? Poorly designed software can result in injury or economic loss. For instance, it has been reported that up to 60 percent of data breaches are caused by unpatched software vulnerabilities. Should the company that developed the software be held legally liable for any harm that results from using the software? What about the individual software developers? The main concern is: What level of defects, if any, should be considered ordinary and customary and, therefore, acceptable?

Liability for software-related problems is a complicated issue from a legal standpoint. If someone suffers a personal injury or property damage because of a defective physical product, the seller, distributor, and/or manufacturer of that product can normally be held liable, under the concept of strict product liability, whether any of those parties were negligent or not. Whether software is a "product" that is subject to strict product liability remains unsettled. Courts that have grappled with the question have generally found that software is not a product. This finding is important because it is typically very difficult to prove negligence, which is an alternative legal standard that might provide a basis for a lawsuit. However, when software is an essential part of a machine, such as a car, plane, or robot, it is more likely to be considered a product and potentially be subject to strict product liability laws or other types of liability. For example, Boeing has admitted to full responsibility for the 2019 crash of its 737 Max in Ethiopia because of a faulty software system.

Software vendors and software service providers typically use contracts and end-user license agreements to limit any potential liability they may have for defective software. Unfortunately, very few users actually read such agreements before using the software and, even when they do read them, most users do not understand that the agreements effectively exempt software vendors from almost all liability. Courts have typically upheld such agreements, severely impacting the ability of software users to recover for any damages they may have suffered.

The Internet has introduced additional questions surrounding the issue of liability. Should a website or social network such as Facebook or Twitter be held liable and accountable for the content a user posts, when that content contains false, offensive, or potentially harmful material, or should it be held harmless from any liability with respect to that content? The Spotlight on Organizations case further examines this issue.

## SPOTLIGHT ON: ORGANIZATIONS

## Section 230: Should the Law that "Created" Today's Internet Be Repealed or Revised?

In 1996, the Internet and web as we know it today did not yet really exist. There was no Facebook, Twitter, Google, YouTube, or Wikipedia. Only about 20 million US adults had access to the Internet, and they spent fewer than 30 minutes a month online. AOL was the world's largest ISP, along with Prodigy and Compuserve. But there were a few features that would be recognizable today, including bulletin boards (forums) and blogs (although they were not yet called that). The US Congress at that time was very concerned about online pornography and in that year, passed the Communications Decency Act, a law designed primarily to regulate obscenity and indecency on the Internet. Most of the law was quickly found to be unconstitutional by the Supreme Court, but one provision survived: Section 230. Congress had no inkling at the time that Section 230 would later come to be viewed as making the Internet of today possible.

Section 230 has two major provisions: Section 230(c)(1) and Section 230(c)(2). Section 230(c)(1) provides that online companies that enable the posting of user content should not be treated as a publisher or speaker of that content and, therefore, should not be held legally responsible (liable) for that content. This section of the law likens such services to the telephone system or bookstores, which are not responsible for unlawful content that they may play a part in distributing. Congress believed that such a provision would give needed support for what was, at the time, a nascent industry. Without this provision, early Internet companies might have had to spend too much time and resources moderating content to avoid being subjected to potential liability.

Section 230(c)(2) provides that online companies shall not be held liable if they choose, in good faith, to remove user content they deem offensive. This provision resulted from a lawsuit involving user content on a Prodigy bulletin board that a company alleged was defamatory. A New York state court found Prodigy liable because Prodigy exercised editorial control over bulletin board content, at times removing content it believed objectionable. The court case presented Internet firms with a difficult choice—either moderate content (but be subject to the risk of potential legal liability) or don't moderate any content (but run the risk of objectionable content that might harm their business interests). Section 230(c)(2) aimed to remove that concern.

In the years that followed the passage of Section 230, courts interpreted it as creating broad immunity with respect to any claims based on user content, even if the platform knew the content was unlawful. Doing so enabled business models based on user content, such as social media, to flourish. The law has been widely characterized as laying the foundation for the Internet we know today.

However, Section 230 has become a lightening rod in recent years, particularly in relation to social media platforms such as Facebook and Twitter that host a flood of divisive content, including what many believe to be harmful disinformation. Adding fuel to the fire, in October 2021, the *Wall Street Journal* published an exposé based on a trove of documents leaked by Frances Haugen, a former Facebook product manager, revealing that Facebook was aware of the harm the company's platforms can inflict, including on teenagers' mental health.

In the current, heightened political climate, Section 230 has been attacked from both sides. Many Republicans have called for its repeal or reform because they believe that Big Tech companies are biased against conservative viewpoints and that, enabled by Section 230(c)(2), they act to censor certain voices. These critics want to prevent Big Tech platforms from being able to unilaterally decide what content they will remove, label, or restrict. Many Democrats also want reform, but for a different reason. They contend that Section 230(c)(1), by granting Internet platforms legal immunity with respect to user content, reduces their incentive to proactively remove content that causes social harm, especially when that content can be economically valuable to them. Seemingly no one is happy with Section 230 anymore, except the technology companies it was originally designed to protect. Most of these companies want Section 230 to remain as it has always been, with some companies viewing any kind of reform effort as an existential threat to the Internet.

Some Section 230 critics focus on the amplification caused by algorithms and suggest removing liability protection when social media algorithms promote harmful content. However, it is unclear whether it would be possible to design an algorithm that amplifies only "good" content, nor is it clear as to how "good" versus "bad" content should be determined. Others have suggested that a platform's safe harbor protections be tied to the use of reasonable content moderation policies, although this raises First Amendment issues. Some lawmakers have proposed eliminating Section 230 protections for "market-dominant" Internet platforms or have said that eligibility for Section 230 safe harbors should be based on political neutrality. Still others warn that any Section 230 reform itself may have unintended and harmful side effects, pointing to a previous amendment aimed at fighting online sex trafficking that reportedly actually made prosecuting such crimes more difficult. The Internet Society warns that Section 230 impacts not just platforms such as Facebook and Google but also underlying infrastructure providers, such as cloud providers, that move content and data from one place to another.

Given the polarized views about Section 230 (according to Pew Research, 56 percent of US adults say people should not be able to sue social media companies for content that users post on their platforms, while 46 percent say people should be able to do so), it's unclear if enough bipartisan support can be developed to either repeal or amend Section 230 anytime soon. However, if and when that time comes, just as the initial passage of Section 230 helped create the Internet of today, its repeal or amendment will shape the Internet of the future.

Sources: Julie Pattison-Gordon, "Can Section 230 Reform Advocates Learn from Past Mistakes," Govtech.com, March 15, 2022; Marguerite Reardon, "Section 230: How It Shields Facebook and Why Congress Wants Changes," Cnet.com, October 6, 2021; Michael D. Smith and Marshall Van Alstyne, "It's Time to Update Section 230," *Harvard Business Review*, August 12, 2021; Colleen McClain, " 56% of Americans Oppose the Right to Sue Social Media Companies for What Users Post," Pewresearch.org, July 1, 2021; Congressional Research Service, "Section 230: An Overview," Crsreports.congress.gov, April 7, 2021; Adi Robertson, "What Will Changing Section 230 Mean for the Internet," Theverge.com, February 26, 2021; Stephen Engleberg, "Twenty-Six Words Created the Internet. What Will It Take to Save It," Propublica.com, February 9, 2021; Farhad Manjoo, "Jurassic Web," Slate.com, February 24, 2009.

## CASE STUDY QUESTIONS

1. What are the ethical, social, and political issues that Internet platforms face with respect to user content?
2. Do you think Section 230 should be repealed or amended? Why or why not?
3. In what ways does the controversy over Section 230 illustrate the difficulties of creating laws governing the use of new digital technologies?

### Computer Crime and Abuse

New technologies, including computers, create new opportunities for committing crime by creating new, valuable items to steal, new ways to steal them, and new ways to harm others. **Computer crime** can be generally defined as the commission of illegal acts by using a computer or against a computer system. Simply accessing a computer system without authorization or with intent to do harm, even accidentally, is now a federal crime. The most frequent types of incidents comprise a greatest hits list of cybercrime: malware, phishing, network interruption, spyware, and denial-of-service attacks. The true cost of all computer crime is unknown, but it is estimated to be in the billions of dollars. You can find a more detailed discussion of computer crime in Chapter 8.

**Computer abuse** is the commission of acts involving a computer that may not be illegal but are considered unethical. The popularity of the Internet, email, and mobile devices has turned one form of computer abuse—spamming—into a serious problem for both individuals and businesses. Originally, **spam** was junk email that an organization or individual sent to a mass audience who had expressed no interest in the product or service being marketed. Today, spam typically tries to entice users to select malicious links or may market fraudulent deals and services, outright scams, pornography, illegal or counterfeit drugs, or other objectionable products. Most spam originates from Russian- or Eastern European-based bot networks, which consist of thousands of captured PCs that can initiate and relay spam messages.

Spam costs for businesses are high because of the computing and network resources and the time required to deal with billions of unwanted email messages. In 2021, the percentage of all email that is spam worldwide was estimated to be around 45 percent (Johnson, 2021).

Identity and financial-theft cybercriminals are also targeting mobile devices with spam. Mobile phone spam usually comes in the form of SMS text messages, but increasingly, users are receiving spam in their Facebook Feed and Messenger app as well.

ISPs and individuals can combat spam by using spam filtering software to block suspicious email before it enters a recipient's email inbox. However, spam filters may block legitimate messages. Spammers know how to skirt filters by continually changing their email accounts, by incorporating spam messages in images, by embedding spam in email attachments and digital greeting cards, and by using other people's computers that have been hijacked by botnets (see Chapter 8). Many spam messages are sent from one country but hosted by another country.

Some countries have passed laws to outlaw spamming or restrict its use. For instance, spamming is more tightly regulated in Europe than in the United States. Unsolicited commercial messaging is banned in the European Union. Digital marketing can be targeted only to people who have given prior consent.

In the United States, the CAN-SPAM Act does not outlaw spamming but does ban deceptive email practices by requiring commercial email messages to display accurate subject lines, identify the true senders, and offer recipients an easy way to remove their names from email lists. It also prohibits the use of fake return addresses. Although a few people have been prosecuted under this law, it has had a negligible impact on spamming, in large part because of the Internet's exceptionally poor security and the use of offshore servers and botnets.

## QUALITY OF LIFE: EQUITY, ACCESS, BOUNDARIES, AND HEALTH

The negative social and personal costs of introducing information technologies and systems are beginning to mount along with the power of the technology. Many of these negative social consequences are not violations of individual rights or property rights. Nevertheless, they can be extremely harmful to individuals, societies, and political institutions. Information technologies potentially can destroy valuable elements of our culture and society even while they bring us benefits. They can also create serious personal health risks. If there is a balance of good and bad consequences of using information systems, whom do we hold responsible for the bad consequences? In this section, we briefly examine some of the negative social and personal consequences of systems, considering individual, social, and political responses.

### Big Tech: Concentrating Economic and Political Power

In the United States, Amazon accounts for almost 40 percent of all e-commerce retail sales, with more than 80 percent of all book, music, and video sales occurring on its platform. Google accounts for more than 87 percent of online searches. More than 80 percent of social network users use Facebook. More than 60 percent of all digital ad revenue goes to Google, Meta, and Amazon, collectively. More than 95 percent of online video viewers use YouTube (owned by Google) and almost 65 percent use Amazon. In the office, Microsoft dominates, with around 75 percent of the world's desktop computers using Windows operating system software and Microsoft software products. Apple accounts for more than 55 percent of the US market in smartphones, while phones with an Android operating system (developed by Google) account for the remaining 45 percent. In today's world of "Big Tech" firms, oligopolies and monopolies dominate. The wealth created by these firms inevitably translates into political influence: These same firms have amassed an army of lobbyists in Washington, DC and state capitals to ensure that legislation, or legislative inquiries, that might affect their market and tax concerns reflect their interests. In 2021, Google, Amazon, Meta, and Apple spent nearly $55 million to support their lobbying efforts in Washington (Birnbaum, 2022).

Concentrations of market power are not new in the United States or Europe. Beginning in 1890 with the Sherman Antitrust Act in the United States, and continuing through the 1960s, monopolies have been considered threats to competition and to smaller start-up businesses, generally restraining free trade. Monopolies typically achieve their size by purchasing smaller competitors, crushing competitors by developing similar products, or engaging in predatory pricing by dropping prices drastically for short periods of time to force smaller firms out of business. Big Tech firms have a well-documented history of these behaviors. But antitrust thinking changed in the 1970s to a different standard of harm: consumer welfare. In this view, bigness per se was not a danger, nor was anticompetitive behavior. Instead, price and consumer welfare became paramount. As long as consumers were not forced to pay higher prices, market power was not important and was not considered a social or economic harm. In this view, because the offerings of Meta, Google, and Amazon are either free or very low cost, there can be no harm.

Critics point out that consumer welfare is harmed in ways other than price, namely, by preventing new, innovative companies from market access or from surviving long enough to prosper as independent firms. Complaints and lawsuits originated by small start-up firms alleging anticompetitive and unfair practices, and concerns about the abuse of personal privacy by Big Tech firms, have led to a torrent of public criticism, governmental investigations and regulatory actions, and lawsuits. In Europe, the European Union, which has been at the forefront of regulating Big Tech companies, has agreed on a new Digital Markets Act (DMA), which will prevent "gatekeeper" platforms (defined on the basis of market value and which include Alphabet [owner of Google and YouTube], Meta, Amazon, Apple, and Microsoft) from exploiting their interlocking products and services to lock in their users and squash potential rivals. Once officially enacted, the law is expected to have a far-reaching impact on the operation of app stores, online advertising, e-commerce, and messaging services. US critics of Big Tech hope the passage of the DMA will provide a roadmap for the passage of similar legislation in the United States (Satariano, 2022).

### Equity and Access: Increasing Racial and Social Class Cleavages

Does everyone have an equal opportunity to participate in the digital age? Will the social, economic, and cultural gaps that exist in the United States and other societies be reduced by information systems technology? Or will the cleavages be increased?

These questions have not yet been fully answered because the impact of digital technologies on various groups in society has not been thoroughly studied. What is known is that information, knowledge, computers, and access to these resources through educational institutions and public libraries are inequitably distributed along racial, ethnic, and social class lines, as are many other information resources. Several studies have found that low-income groups in the United States are less likely to have computers or online Internet access even though computer ownership and Internet access have soared over the past decade. Although the gap in computer access is narrowing, higher-income families in each racial or ethnic group are still more likely to have home computers and broadband Internet access than are lower-income families in the same group. Moreover, the children of higher-income families are far more likely to use their Internet access to pursue educational goals, whereas lower-income children are much more likely to spend time on entertainment and games. This is called the "time-wasting" gap.

Left uncorrected, this **digital divide** could lead to a society of information haves, who are computer literate and skilled, versus a large group of information have-nots, who are computer illiterate and unskilled. Narrowing this digital divide by making digital information services—including broadband Internet service—available to as many people as possible, just as basic telephone service is now, has become an important priority for federal, state, and local governments in the United States.

### Employment: Trickle-Down Technology and Reengineering Job Loss

The reengineering of work is typically hailed in the information systems community as a major benefit of new information technology. It is much less frequently noted that the redesign of business processes and automation have also caused millions of people to lose their jobs. On the other hand, some economists believe that new technologies create as many or more jobs than they destroy. For instance, the growth of e-commerce has led to a decline in retail sales jobs but an increase in jobs for warehouse workers, supervisors, and delivery work. These economists also believe that bright, educated workers who are displaced by technology will move to better jobs in fast-growth industries. Review the Spotlight on People case, *Will a Robot Steal Your Job?*, in Chapter 1 for a further discussion of the impact of information technology on jobs.

### Rapidity of Change: Reduced Response Time to Competition

Information systems have helped to create much more efficient national and international markets. Today's rapidly moving global marketplace has reduced the normal social buffers that permitted businesses many years ago to adjust to competition. Time-based competition has an ugly side: The business you work for may not have enough time to respond to global competitors and may be wiped out in a year, along with your job. We run the risk of developing a just-in-time society with just-in-time jobs and just-in-time workplaces, families, and vacations. One impact of on-demand services firms such as Uber is to create just-in-time jobs with no benefits or insurance for employees.

### Maintaining Boundaries: Family, Work, and Leisure

Although the idea of ubiquitous computing, remote work and telecommuting, nomad computing, mobile computing, and the do-anything-anywhere computing environment is alluring to many, it also raises a number of issues. One impact of such an environment is that it weakens the traditional boundaries that separate work from family and leisure.

The advent of information systems, coupled with the growth of knowledge-work occupations, means that more and more people are working when traditionally they would have been interacting with family and friends. The work umbrella now extends far beyond the eight-hour day into commuting time, vacation time, and leisure time. The explosive growth and use of mobile devices have only heightened the sense of many employees that they are never away from work.

Even leisure time spent on the computer threatens these close social relationships. Extensive Internet and mobile phone use, even for entertainment or recreational purposes, takes people away from their family and friends. Among middle school and teenage children, it can lead to harmful antisocial behavior, such as cyberbullying.

Weakening these institutions poses clear-cut risks. Family and friends historically have provided powerful support mechanisms for individuals, and they act as balance points in a society by preserving private life and providing a place for people to collect their thoughts, think in ways contrary to their employer, and dream.

© Drazen Zigic/Shutterstock

*Although some people enjoy the convenience of working at home, the do-anything-anywhere computing environment can blur the traditional boundaries between work and family time.*

## Physical, Mental, and Cognitive Health Risks

Information technology can pose a number of physical, mental, and cognitive health risks. For instance, a common occupational disease today is **repetitive stress injury (RSI)**. RSI occurs when muscle groups are forced through repetitive actions that are often made with high-impact loads (such as playing tennis) or tens of thousands of repetitions under low-impact loads (such as working at a computer keyboard). The incidence of RSI is estimated to affect as much as one-third of the labor force and accounts for one-third of all disability cases.

The single-largest source of RSI is computer keyboards. The most common kind of computer-related RSI is **carpal tunnel syndrome (CTS)**, in which pressure on the median nerve through the wrist's bony structure, called a carpal tunnel, produces pain. The pressure is caused by the constant repetition of keystrokes. Symptoms of CTS include numbness, shooting pain, inability to grasp objects, and tingling. Millions of workers have been diagnosed with CTS. In the United States and most other developed countries, CTS affects an estimated 1 to 3 percent of the general population in any given year (Sevy and Varacallo, 2022).

However, RSI is avoidable. Designing workstations for a neutral wrist position (for example, using a wrist rest to support the wrist), proper monitor stands, and footrests all contribute to proper posture and reduced RSI. Ergonomically correct keyboards are also an option. These measures should be supported by frequent rest breaks and rotation of employees to different jobs.

© Orathai Mayoeh/Shutterstock

*Repetitive stress injury (RSI) is a leading occupational disease today. The single-largest cause of RSI is computer keyboard work.*

RSI is not the only occupational illness computers cause. Back and neck pain, leg stress, and foot pain also result from poor ergonomic designs of workstations. **Computer vision syndrome (CVS)** refers to any eyestrain condition related to display screen use with desktop computers, laptops, e-readers, smartphones, and handheld video game players. CVS affects about 90 percent of people who spend three hours or more per day using a display screen. Its symptoms, which are usually temporary, include headaches, blurred vision, and dry and irritated eyes.

Much has also been written about the challenges that the Internet, mobile apps, and social media pose for mental health, particularly that of children and young adults. The impact of the increase in screen time and time spent with digital devices during the Covid-19 pandemic has further highlighted this issue. Potential negative impacts include increased exposure to negative online social interactions (cyberbullying, feelings of missing out, and social comparisons); reduced physical activity; sleep disruption; isolation from in-person contact; and the possibility of Internet/social media addiction. The topic remains controversial, however, with one recent study of 430,000 UK and US adolescents finding little evidence for an increased association between adolescents' technology engagement and mental health problems over the past 30 years (APS, 2021).

Information technology also may be harming our cognitive functions or at least changing how we think and solve problems. Although the Internet has made it much easier for people to access, create, and use information, some experts believe that it is also preventing people from focusing and thinking clearly on their own. They argue that excessive use of computers and smartphones reduces intelligence. Some scholars believe that exposure to computers encourages looking up answers rather than engaging in real problem solving. People, in this view, don't learn much while surfing the web or answering email when compared to listening, drawing, arguing, looking, and exploring. In addition, increased reliance on various forms of artificial intelligence (see Chapter 11) may be impacting our power to make decisions based on our own judgments (Moser, den Hond, and Lindebaum, 2022; Carr, 2015).

The coming metaverse may also present new physical and mental health challenges. The metaverse is a visual, three-dimensional (3-D) virtual reality that some believe will be at the center of the future web. It will move the Internet beyond 2-D screens and toward an immersive 3-D experience. The Spotlight on Technology case explores the risks that the metaverse may pose.

Informaton technology has become part of our lives—personally as well as socially, culturally, and politically. It is unlikely that the issues and our choices will become easier as information technology continues to transform our world. The growth of the Internet and the information economy suggests that all the ethical and social issues we have described will be heightened as we move further into the first fully digital century.

## 4-5 Understand how MIS can help your career.

Here is how Chapter 4 and this book can help you find a job as a junior privacy analyst.

### THE COMPANY

LetsTravel is a US-based online travel reservation service for consumers and businesses, with headquarters in San Francisco, California, and operations throughout the United States and Europe. LetsTravel has an open entry-level position for a junior privacy analyst. LetsTravel collects a variety of consumer data, both directly as a part of its reservation service and also in connection with its online advertising efforts.

## SPOTLIGHT ON: TECHNOLOGY

## Immersed in the Metaverse: What Will It Mean for the Future?

In October 2021, Facebook made a surprising announcement. It was changing its name to Meta to highlight what it believes will be the future of not only Facebook but also the Internet as a whole: a digital reality known as the metaverse. The metaverse has its roots in current technologies such as virtual reality (VR), augmented reality (AR), and avatar-based virtual worlds and envisions a 3-D virtual reality in which users can connect, socialize, collaborate, and transact. Although Facebook's announcement has kicked off a wellspring of hype, buzz, and interest, it also raises some serious questions about the potential impacts the metaverse may have on our lives.

Mark Zuckerberg, Facebook's founder and CEO, believes that the metaverse is the ultimate expression of social technology. He characterizes it as an "embodied Internet," where instead of just viewing content in 2-D, you experience being "in it," just as if you were physically present. Zuckerberg likens a fully realized metaverse to a teleportation device. Although VR, AR, and metaverse-like experiences are currently being used primarily for gaming and advertising, Zuckerberg and others believe that the metaverse ultimately will create a profound change in the ways people experience online life and work.

The world is already grappling with a host of ethical, social, political, and legal issues related to current online practices and content. In a metaverse, those issues and risks are likely to be exacerbated, and new ones will likely be created.

Some of the simplest risks to describe are the physical ones. Meta's current health and safety warnings for its Oculus VR system provide a long list of potential physical risks, ranging from seizures to dizziness, nausea, vomiting, and visual issues. It notes that symptoms can persist for hours after use. It also warns that frightening, violent, or anxiety-provoking content can cause a user's body to react as if the content were real. For instance, one researcher recounted the experience of her avatar getting punched in the face. Although she "knew" her body was safe in her office, she said her mind and body registered the punches as real. With many companies working on technology such as haptic gloves, which will incorporate touch as an additional sensation in an immersive reality, physical harm in the metaverse is likely to feel even more "real."

Psychological impacts are potentially more complex. Immersion is a more potent experience than merely observing and interacting with 2-D content through a flat screen. It is possible that, similar to the impact of compulsive social media use, participating in an intensely immersive world will make some people prefer it to "real" life and replace behaviors that are healthy and supportive to mental health, such as appropriate exercise, engagement in real-life relationships, healthy sleep, and time spent outside.

The potential impacts on children and adolescents are particularly concerning. Research has shown a myriad of negative effects of social media on children and adolescents, from bullying and harassment to self-esteem and body image issues. These negative effects are likely to be just as prevalent, if not more so, in the metaverse. For example, one researcher teamed up with the Sesame Street Workshop to use Grover as an avatar. Her findings suggest that young children are more likely to comply with commands from a VR character than from the same character in 2-D. The researcher noted that when watching on television, children often would hesitate, look at the researcher, and ask what they were supposed to do. But in VR, the children just did whatever Grover told them to do, without stopping or looking around.

Exposure to harmful content on these platforms is another issue. For instance, a study by the Center for Countering Digital Hate found that minors are regularly exposed to graphic sexual content, racist and violent language, bullying, and other forms of harassment on various games, such as VRChat (accessed via Meta's Oculus VR platform) even though Meta has policies prohibiting these sorts of negative behaviors. Re-creations of the 2019 Christchurch, New Zealand, shooting aimed at children have been found multiple times on the Roblox game platform despite efforts by Roblox to prevent such content from appearing.

The potential impact of the metaverse on privacy and government surveillance is another concern. The privacy issues are particularly acute because metaverse companies may track and retain user biometric data, as well as data about their actual actions, and ultimately may be able to learn how users uniquely think and act. Tellingly, while Meta has stopped using facial recognition technologies on its Facebook platform, it continues to do so on its Oculus VR platform. This information would be a treasure trove for both advertisers and governments.

Some believe that the concerns about the potential impact of the metaverse are overblown. These critics point to a long history of alarmist predictions when a new technology is introduced, with such predictions usually proving to be unfounded. Others point to the beneficial applications already developed for VR and AR that put those technologies' power to blur the line between the real and the virtual to good use, such as VR applications for treating phobias, nightmares, and PTSD; for education and the arts; and for building empathy, diversity, and inclusion, among others. Zuckerberg has said that safety, privacy, and ethics will be front and center in the metaverse's development and that since a full-fledged metaverse is still several years away, regulators and policymakers will have ample time to implement safeguards.

But how realistic is it to expect that industry self-regulation or even laws and regulations will be able to adequately address the issues likely to be posed by the metaverse? Safety policies cannot be simply brought over from existing social media. Instead, specific policies for immersive environments based on how the technology interacts with our brains will be needed. Even these policies will be very difficult to monitor and enforce. Further, Meta's current business model is built on trying to engage user attention for as long as possible, making it more likely than not that its version of the metaverse will also evolve into something that focuses on hooking both kids' and adults' attention. Critics don't trust Meta to create a metaverse that protects people if that protection conflicts with its business goals of maximizing profit, especially as it grapples with declining use of its Facebook platform.

Sources: "The Promise and Perils of the Metaverse," Mckinsey.com, March 29, 2022; Tom Huddleston, "'This Is Creating More Loneliness': The Metaverse Could Be a Serious Problem for Kids, Experts Say," Cnbc.com, January 31, 2022; Gadjo Sevilla, "Tech Companies Reveal Blueprint for a Safer Metaverse," Emarketer.com, January 20, 2022; Cathy Li and Farah Lalani, "How to Address Digital Safety in the Metaverse," Weforum.com, January 14, 2022; Benoit Morenne, "The Metaverse's Effect on Mental Health: Trivial or Troubling?" *Wall Street Journal*, January 9, 2022; Greenberg Glusker LLP, "Law in the Metaverse," Jdsupra.com, December 27, 2021; Alexandra Levine, "The Health Concerns that Hang Over the Metaverse," Politico.com, November 17, 2021; Catherine Buni, "If Social Media Can Be Unsafe for Kids, What Happens in VR?" Slate.com, October 11, 2021; Casey Newton, "Mark in the Metaverse," Theverge.com, July 22, 2021.

## CASE STUDY QUESTIONS

**1.** What are some of the issues that the metaverse poses for quality of life?

**2.** What ethical dilemmas do companies who want to create the metaverse face?

**3.** Do you think industry self-regulation will be an effective method for safeguarding the metaverse? Why or why not?

## POSITION DESCRIPTION

The junior privacy analyst will help ensure compliance with relevant European and US federal and state privacy laws and regulations. Job responsibilities include:

- Acting as the first point of contact for internal and external privacy-related questions.
- Proactively informing and involving internal stakeholders (such as Customer Service, Legal, Technology, Finance).
- Supporting various departments to make sure processes and controls are aligned with privacy business strategies as well as supporting requests from data subjects and authorities.
- Working as part of the team in creating documentation and encouraging and facilitating privacy best practices.
- Providing support on privacy-related projects, improvements, and innovations, including assisting in the development and implementation of privacy processes.

### JOB REQUIREMENTS

- Bachelor's degree in liberal arts or business
- General knowledge of various privacy-related laws and regulations
- Critical thinking skills
- Strong communication and organizational skills
- Knowledge of the travel industry is beneficial but not essential.
- Ability to work remotely if required

### INTERVIEW QUESTIONS

1. What background or job experience do you have in the privacy protection field?
2. What do you know about various privacy laws that the company may need to comply with, such as the California Privacy Rights Act, the General Data Protection Regulation, and other privacy laws?
3. What do you know about consumer privacy protection practices?
4. If you were asked to improve privacy protection for an organization, how would you proceed?
5. Have you ever dealt with a problem involving privacy protection? What role did you play in its solution?

### AUTHOR TIPS

1. Review this chapter, giving special attention to the sections dealing with information systems and privacy.
2. Use the web to find out more about the California Privacy Rights Act, the General Data Protection Regulation, other potentially applicable privacy laws, and privacy protection best practices and procedures.
3. If you do not have any hands-on experience in the privacy area, explain what you do know about privacy and why it is so important to protect sensitive personal data, and indicate that you would be very interested in learning more and doing privacy-related work.

## Review Summary

4-1 **Identify the ethical, social, and political issues raised by information systems.** Information technology is introducing changes for which laws and rules of acceptable conduct have not yet been developed. Increasing computing power, storage, data analysis, and networking capabilities—including the Internet—expand the reach of individual and organizational actions and magnify their impacts. The increasing use of mobile devices and artificial intelligence also raises ethical issues. The ease and anonymity with which information is now communicated, copied, and manipulated in online environments pose new challenges to the protection of privacy and intellectual property. The main ethical, social, and political issues that information systems raise center on information rights and obligations, property rights and obligations, system quality, accountability and control, and quality of life.

4-2 **Describe principles for conduct that can be used to guide ethical decisions.** Basic concepts underlying ethical decisions include responsibility, accountability, liability, and due process. When confronted with an ethical issue, a five-step ethical analysis process that involves identifying and describing the facts, defining the conflict or dilemma and identifying the higher-order values involved, identifying the stakeholders, identifying the options that can be reasonably taken, and identifying the potential consequences of the options should be used. Ethical principles that should be followed in conjunction with an ethical analysis include the Golden Rule, Immanuel Kant's categorical imperative, the slippery slope rule, the utilitarian principle, the risk aversion principle, and the ethical no-free-lunch rule. These principles should be used in conjunction with an ethical analysis.

4-3 **Explain how contemporary information systems technologies and the Internet pose challenges to the protection of individual privacy and intellectual property.** Contemporary information systems enable companies to gather personal data about individuals from many sources easily and analyze these data to create detailed digital profiles about individuals and their behaviors. Data flowing over the Internet can be monitored at many points. Cookies and other monitoring tools closely track the activities of website visitors and mobile app users. Not all websites and mobile apps have strong privacy protection policies, and they do not always allow for informed consent regarding the use of personal information. Traditional copyright laws are insufficient to protect against software piracy because digital material can be copied so easily and transmitted to many locations simultaneously over the Internet.

4-4 **Discuss the issues that contemporary information systems raise with respect to system quality, accountability and control, and the quality of everyday life.** New information technologies are raising questions about what constitutes acceptable levels of system quality, given the centrality of information systems to the functioning of businesses and society. Who, if anyone, is liable if a business or person is harmed by software that is defective? Should a website or social network be held accountable for the content that a user posts? Widespread use of computers increases opportunities for computer crime and computer abuse: How should such occurrences be handled by the legal system? Although information systems have been sources of efficiency and wealth, they also have some negative social consequences that impact the quality of life by, for example, enabling the concentration of economic and politicial power. Not everyone has an equal opportunity to participate in the digital economy, and this digital divide may be exacerbating socioeconomic disparities among different ethnic groups and social classes. Jobs can be lost when computers replace workers or when tasks become unnecessary in reengineered business processes. Information systems and technology can also create issues with maintaining appropriate boundaries separating family, leisure and work, as well as create physical, mental, and cognitive health risks.

## Key Terms

Accountability, 125
Adware, 131
Carpal tunnel syndrome (CTS), 143
Computer abuse, 139
Computer crime, 139
Computer vision syndrome (CVS), 144
Cookies, 130
Copyright, 133
Digital divide, 142
Digital Millennium Copyright Act (DMCA), 136
Due process, 125
Ethical no-free-lunch rule, 126
Ethics, 119
Fair Information Practices (FIP), 127
General Data Protection Regulation (GDPR), 129
Golden Rule, 126
Immanuel Kant's categorical imperative, 126
Information rights, 121
Informed consent, 132
Intellectual property, 133
Liability, 125
Nonobvious relationship awareness (NORA), 123
Opt-in, 132
Opt-out, 132
Patent, 134
Privacy, 127
Profiling, 123
Repetitive stress injury (RSI), 143
Responsibility, 124
Risk aversion principle, 126
Safe harbor, 130
Slippery slope rule, 126
Spam, 139
Spyware, 131
Trade secret, 134
Trademarks, 134
Utilitarian principle, 126
Web beacons, 131

## Review Questions

**4-1** Identify the ethical, social, and political issues raised by information systems.
- Explain how ethical, social, and political issues are connected, and give some examples.
- List and describe the key technological trends that heighten ethical concerns.
- Differentiate among responsibility, accountability, and liability.

**4-2** Describe principles for conduct that can be used to guide ethical decisions.
- List and describe the five steps in an ethical analysis.
- Identify and describe six ethical principles.

**4-3** Explain how contemporary information systems technologies and the Internet pose challenges to the protection of individual privacy and intellectual property.
- Define privacy and Fair Information Practices.
- Explain how the Internet challenges the protection of individual privacy and intellectual property.
- Explain how informed consent, legislation, industry self-regulation, and technology tools help protect the individual privacy of Internet users.
- List and define four types of law that protect intellectual property rights.

**4-4** Discuss the issues that contemporary information systems raise with respect to system quality, accountability and control, and the quality of everyday life.
- Explain why it is so difficult to hold software vendors and service providers liable for failure or injury.
- List and describe the principal causes of system quality problems.
- Name and describe four quality-of-life impacts of computers and information systems.
- Define and describe computer vision syndrome and repetitive stress injury (RSI), and explain their relationship to information technology.

## Discussion Questions

**4-5** MyLab MIS Should cloud service providers be held liable for economic injuries suffered when their systems fail?

**4-6** MyLab MIS Should companies be responsible for unemployment that their information systems cause? Why or why not?

**4-7** MyLab MIS Discuss the pros and cons of allowing companies to amass personal data for behavioral targeting.

**MyLab MIS**
To complete these problems, go to EOC Discussion Questions in **MyLab MIS**.

## Hands-On MIS Projects

The projects in this section give you hands-on experience in analyzing the privacy implications of using online data brokers, developing a corporate policy for employee web usage, using blog creation tools to create a simple blog, and analyzing web browser privacy. Visit MyLab MIS to access this chapter's Hands-On MIS Projects.

### MANAGEMENT DECISION PROBLEM

**4-8** InfoFree's website is linked to massive databases that consolidate personal data on millions of people. Users can purchase marketing lists of consumers broken down by location, age, gender, income level, home value, home ownership, and length of residence. Do data brokers such as InfoFree raise privacy issues? Why or why not? If your name and other personal information were in this database, what limitations on access would you want in order to preserve your privacy? Consider the following data users: government agencies, your employer, private business firms, other individuals.

### ACHIEVING OPERATIONAL EXCELLENCE: CREATING A SIMPLE BLOG

Software skills: Blog creation
Business skills: Blog and web page design

**4-9** In this project, you'll learn how to build a simple blog of your own design using the online blog creation software available at Blogger.com or WordPress.com. Pick a sport, hobby, or topic of interest as the theme of your blog. Name the blog, give it a title, and choose a template for the blog. Post at least four entries to the blog, adding a label for each posting. Edit your posts if necessary. Upload an image, such as a photo from your computer or the web, to your blog. Add capabilities for other registered users, such as team members, to comment on your blog. Briefly describe how your blog could be useful to a company selling products or services related to the theme of your blog. List the tools available in Blogger that would make your blog more useful for business, and describe the business uses of each. Save your blog, and show it to your instructor.

## IMPROVING DECISION MAKING: ANALYZING WEB BROWSER PRIVACY

Software Skills: Web browser software
Business Skills: Analyzing web browser privacy protection features

**4-10** This project will help develop your Internet skills for using the privacy protection features of leading web browser software.

Examine the privacy protection features and settings for two leading web browsers such as Microsoft Edge, Mozilla Firefox, or Google Chrome. Make a table comparing the features of two of these browsers in terms of functions provided and ease of use.

- How do these privacy protection features protect individuals?
- How do these privacy protection features affect what businesses can do on the Internet?
- Which browser does the better job of protecting privacy? Why?

## COLLABORATION AND TEAMWORK PROJECT

Developing a Corporate Code of Ethics

**4-11** With three or four of your classmates, develop a corporate ethics code on privacy that addresses both employee privacy and the privacy of customers and users of the corporate website. Be sure to consider email privacy and employer monitoring of worksites as well as corporate use of information about employees concerning their off-the-job behavior (e.g., lifestyle, marital arrangements, and so forth). If possible, use Google Docs and Google Drive or Google Sites to brainstorm, organize, and develop a presentation of your findings for the class.

## BUSINESS PROBLEM-SOLVING CASE

### FACEBOOK'S MANY ETHICAL CHALLENGES

Meta's Facebook is the most "popular" social network in the world, based on the number of its users. At the same time, it is also one of the most reviled. This seemingly contradictory situation stems in large part from Facebook's business practices and how it has responded to a variety of ethical challenges throughout the years.

Facebook's stated mission is to give people the power to build community and bring the world closer together. It provides a platform that enables people to connect and share their opinions, photos, and videos with audiences ranging from just friends and family to the public at large, to find communities, and to grow businesses. As it has grown, its platform has also become a source of information for its users about what is happening in the world.

But from just about its earliest days, Facebook has been hounded by controversy over its business practices. When Facebook originally launched in 2004, it promised its users nearly complete control over who could see their personal profile and information. But that did not last long. Shortly after it launched, Facebook took its first steps toward monetizing the platform to help cover its costs. It quickly turned to online advertising to generate revenue. What the online advertising industry prizes is data— specifically, personal data about people as well as data that enables companies to predict how people will think and act. It soon became clear that Facebook was sitting on a trove of such data, setting the stage for one of Facebook's most enduring ethical challenges: the privacy of its users versus Facebook's desire to generate revenue based on that data. In that battle, the desire to maximize its revenue has usually won.

Facebook's Beacon program was an early indication of this challenge. Introduced in 2007, the Beacon program broadcast users' activities on participating websites, such as their purchases, to their friends' News Feeds, without user permission. Advertisers on Facebook could then run ads next to those posts. Facebook was blistered with criticism from privacy advocates, and more than 50,000 users signed a petition objecting to the program. Facebook initially said that users would accept Beacon once they had time to get used to it. But Facebook was wrong, and it ultimately was forced to withdraw the Beacon program.

Fast forward to 2022. Facebook's revenues and profits have grown each year, and in 2021, its various platforms produced more than $115 billion in total revenue and a profit of almost $57 billion. A look back over the past 15 years shows that Facebook has been repeatedly embroiled in conflicts over user privacy. It has been slapped by the Federal Trade Commission more than once: first in 2012, when it agreed to stop deceiving users about their ability to control their personal data and to stop sharing data with third parties without informing users and then again in 2019, when it was fined a record-breaking $5 billion for clearly and intentionally violating the 2012 settlement. The 2019 settlement followed closely on the heels of what has become known as the Cambridge Analytica scandal. A UK researcher had obtained permission from Facebook to use the platform for a personality quiz app. Unbeknownst to users, the app collected data not only from those who downloaded it but also from their friends and their friends' friends. Ultimately, it created a database on more than 87 million Facebook users, which was then sold to political consulting and data analytics firm Cambridge Analytica, which used the data in an effort to influence the 2016 US presidential election. Not long thereafter, a *New York Times* exposé revealed that Facebook had data-sharing partnerships with at least 60 device makers as well as selected app developers, which enabled those third parties to gain access not only to data about Facebook users but also to personal data about their Facebook friends, without their consent.

Facebook has also been repeatedly sued for privacy-related violations and has paid out millions to settle those lawsuits, including $650 million in 2021 to settle claims it violated an Illinois privacy law by storing facial recognition data without first obtaining users' consent and $90 million in 2022 to settle decade-long litigation over its use of tracking cookies to track Facebook users' online behavior even after they had logged off Facebook. It has also continually run afoul of the European Union's General Data Protection Regulation, which mandates more stringent privacy protections than the United States does.

Given Facebook's checkered history of privacy violations, why do so many people remain Facebook users and continue to share sensitive details of their lives on Facebook? Often it's because users do not realize all the data that are being collected about them and the purposes for which that data may be used. Many of Facebook's features and services are enabled by default, and a study by Siegel+Gale found that Facebook's privacy policy is more difficult to comprehend than government notices or typical bank credit card agreements, which are notoriously dense.

But issues surrounding user privacy are not Facebook's only ethical challenge: Within the last several years, there has been an increasing focus on Facebook's role in providing a platform for misinformation and other harmful content. For instance, in 2021, the *Wall Street Journal* published an exposé based on a review of Facebook documents including research reports, online employee discussions, and internal presentations. According to the *Wall Street Journal*, Facebook knew that its platforms caused harm but did little, if anything, about it. For example, Facebook researchers knew that its Instagram platform could be harmful to mental health, particularly for teenage girls, but in public statements, Facebook consistently downplayed its negative effect on teens. Facebook employees warned that its platforms were being used to incite ethnic violence and to facilitate criminal activity such as human trafficking in certain countries, but Facebook did little to respond to those concerns on a systemic basis. Facebook also knew that a change to its News Feed algorithm, which it hoped would increase user engagement, rewarded posts oriented toward outrage and sensationalism, making the Facebook environment more and more toxic. A former Facebook employee known as the Facebook whistleblower painted a picture of a company unable to control the "monster" it had created. For instance, Facebook's internal documents revealed Facebook's efforts to use artificial intelligence to keep its platforms free from offensive or dangerous content have fallen far short of its expectations.

Facebook's business practices with respect to its competitors have also come under fire. The FTC has sued Facebook for attempting to maintain its dominant position by engaging in a series of anti-competitive actions that violated US antitrust law. For example, the FTC pointed to Facebook's acquisitions of rivals Instagram and WhatsApp in 2014 as predatory moves meant to crush the competition. The FTC also argued that Facebook harmed rival app developers by blocking them from accessing data and tools on its platform. Facebook has faced similar charges from various state attorneys-general and the European Union.

Facebook, not surprisingly, generally denies most of the claims that its detractors allege, although at times it has apologized for some of its missteps. But Facebook's future may rely on how it responds to its ethical challenges going forward. Over the last several years, there has been an increasing focus on ethical business practices and a growing awareness that good ethics is good business. The ESG (environmental, social, and governance) movement, which focuses on a company's performance beyond just its profit and loss, is gaining increased traction and becoming a strategic business imperative. Facebook says it is taking a number of steps to respond to these challenges and has published a long list of environmental, social, and governance resources explaining how it is doing so. However, given that Facebook's business model relies almost entirely on the largely unfettered use of users' personal information, and given its past history and business practices, it is unclear whether Facebook can be trusted.

Sources: "Meta Platforms Inc. Form 10-K for the fiscal year ended December 31, 2021," Sec.gov, February 3, 2022; "States Ask U.S. Court to Reinstate Facebook Antitrust Lawsuit," Reuters.com, January 14, 2022; "The Facebook Files," *Wall Street Journal*, January 12, 2022; "FTC Alleges Facebook Resorted to Illegal Buy-or-Bury Scheme to Crush Competition after String of Failed Attempts to Innovate," Ftc.gov, August 19, 2021; "A Timeline of Trouble: Facebook's Privacy Record and Regulatory Fines," Guild.com, August 4, 2021; "Meta Investor Relations: Environmental Social Governance Resources," Investor.fb.com, accessed April 18, 2020; Federal Trade Commission, "In the Matter of Facebook, a Corporation," Ftc.gov, July 24, 2019; John D. McKinnon, Emily Glazer, Deepa Seetharaman, and Jeff Horwitz, "Facebook Worries Emails Could Show Zuckerberg Knew of Questionable Privacy Practices," *New York Times*, June 12, 2019; Deepa Seetharaman and Kirsten Grind, "Facebook Gave Some Companies Access to Additional Data about Users' Friends," *Wall Street Journal*, June 8, 2018; Cecilia Kang and Sheera Frenkel, "Facebook Says Cambridge Analytica Harvested Data of Up to 87 Million Users," *New York Times*, April 24, 2018; Eduardo Porter, "The Facebook Fallacy: Privacy Is Up to You," *New York Times*, April 24, 2018,

## CASE STUDY QUESTIONS

**4-12** Choose an ethical dilemma discussed in this case, and perform an ethical analysis of it.

**4-13** Will Facebook be able to have a successful business model without invading users' privacy? Explain your answer.

**4-14** Investigate the ESG movement. How might it impact Facebook's future?

**4-15** Would you want to work for Facebook? Why or why not?

# Chapter 4 References


Adjerid, Idris, Eyal Peer, and Alessandro Acquisti. "Beyond the Privacy Paradox: Objective versus Relative Risk in Privacy Decision Making." *MIS Quarterly* 42, No. 2 (June 2018).

APS (Association for Pyschological Science). "Little to No Link Between Adolescents' Mental Health Problems and Digital Technology." Psychologicalscience.org (June 28, 2021).

*Bilski v. Kappos*, 561 US (2010).

Birnbaum, Emily. "Tech Spent Big on Lobbying Last Year." Politico.com (January 24, 2022).

*Brown Bag Software vs. Symantec Corp*, 960 F2D 1465 (Ninth Circuit, 1992).

BSA/The Software Alliance. "BSA Global Software Survey 2018." (June 2018).

Carr, Nicholas. *The Glass Cage: How Our Computers Are Changing Us* (W.W. Norton & Company, 2015).

Choi, Bryan. "Software as a Profession." *Harvard Journal of Law & Technology* 33, No. 2 (Spring 2020).

European Commission. "2018 Reform of EU Data Protection Rules." Ec.europa.eu (2018).

European Parliament. "Directive 2009/136/EC of the European Parliament and of the Council of November 25, 2009." European Parliament (2009).

Federal Trade Commission. "FTC Report to Congress on Privacy and Security." (September 13, 2021).

Gopal, Ram D., Hooman Hidaji, Raymond A. Patterson, Erik Rolland, and Dmitry Zhdanov. "How Much to Share with Third Parties? User Privacy Concerns and Website Dilemmas." *MIS Quarterly* 42, No. 1 (March 2018).

Johnson, Joseph, "Spam: Share of Global Email Traffic 2014–2021." Statista.com (July 20, 2021).

Kim, Antino, Atanu Lahiri, Debabrata Dey, and Gerald C. Kane, "'Just Enough' Piracy Can Be a Good Thing." *MIT Sloan Management Review* 61, No. 1 (Fall 2019).

Korolov, Maria, and Alex Korolov. "Top 10 Outages of 2021." Networkworld.com (January 31, 2022).

Laudon, Kenneth C. *Dossier Society: Value Choices in the Design of National Information Systems* (New York: Columbia University Press, 1986).

Laudon, Kenneth C., and Carol Guercio Traver. *E-commerce 2023: business.technology.society*, 17th ed. (Pearson Education, 2024).

Moser, Christine, Frank den Hond, and Dirk Lindebaum, "What Humans Lose When We Let AI Decide." *MIT Sloan Management Review* (February 7, 2022).

Nocera, Joe. "How Cookie Banners Backfired." *New York Times* (January 29, 2022).

Nunes, Ashley. "Automation Doesn't Just Create or Destroy Jobs—It Transforms Them." *Harvard Business Review* (November 2, 2021).

Pew Research Center (Brooke Auxier). "How Americans See Digital Privacy Issues amid the Covid-19 Outbreak." Pewresearch.org (May 4, 2020).

Pew Research Center (Brooke Auxier et al.). "Americans and Privacy: Concerned, Confused, and Feeling Lack of Control Over Their Personal Information." Pewresearch.org (November 15, 2019).

RIAA. "RIAA Releases 2021 Year-End Music Industry Revenue Report." (2022).

Satariano, Adam. "E.U. Takes Aim at Big Tech's Power with Landmark Digital Act." *New York Times* (March 24, 2022).

Sevy, Justin, and Matthew Varacallo. "Carpal Tunnel Syndrome." *StatPearls* (StatPearls Publishing LLC, 2022).

Sivan, Liron, Michael D. Smith, and Rahul Telang. "Do Search Engines Influence Media Piracy? Evidence from a Randomized Study." *MIS Quarterly* 43 No. 4 (December 2019).

Statista Research Department. "Visits to Software Piracy Sites Worldwide 2020–2021, per Quarter." Statista.com (November 12, 2021).

U.S. Department of Health, Education, and Welfare. *Records, Computers, and the Rights of Citizens* (Cambridge: MIT Press, 1973).

U.S. Sentencing Commission. "Sentencing Commission Toughens Requirements for Corporate Compliance Programs." (April 13, 2004).

Venkatesh, Viswath, Tracy Ann Sykes, Frank K. Y. Chan, James Y. L. Thong, and Paul Jen-Hwa Hu. "Children's Internet Addiction, Family-to-Work Conflict, and Job Outcomes: A Study of Parent–Child Dyads." *MIS Quarterly* 43 No. 3 (September 2019).

Wessel, Max, and Nicole Holmer. "A Crisis of Ethics in Technology Innovation." *MIT Sloan Management Review* 61 No. 3 (Spring 2020).

World Economic Forum. "Robots and Your Job: How Automation Is Changing the Workplace." Weforum.org (June 24, 2021).

________. "The Future of Jobs Report." Weforum.org (October 2020).

PART II



# Information Technology Infrastructure

Part II provides the technical knowledge foundation for understanding information systems by examining hardware, software, databases, networking technologies, and tools and techniques for security and control. This part answers questions such as these: What technologies and tools do businesses today need in order to accomplish their work? What do I need to know about these technologies to make sure they enhance the performance of my firm? How are these technologies likely to change in the future? What technologies and procedures are required to ensure that systems are reliable and secure?

CHAPTER

5

# IT Infrastructure: Hardware and Software

## LEARNING OBJECTIVES

After completing this chapter, you will be able to:

5-1 Identify the components of IT infrastructure.

5-2 Describe the major computer hardware, data storage, input, and output technologies used in business and major hardware trends.

5-3 Describe the major types of computer software used in business and major software trends.

5-4 Identify the principal issues in managing hardware and software technology.

5-5 Understand how MIS can help your career.

## CHAPTER CASES

- QRyde Rides High with the Cloud
- The Mobile Platform Comes to Healthcare
- "Smart" Cities Become Smarter with Edge Computing
- How Green Is the Cloud?

**MyLab MIS**

- **Video Case:** IBM Expands Cloud to Daimler
- Discussion Questions: 5-5, 5-6, 5-7
- Hands-On MIS Projects: 5-8, 5-9, 5-10, 5-11
- eText

## QRYDE RIDES HIGH WITH THE CLOUD

Himanshu Bhatnagar, QRyde's founder and CEO, has devoted his career to helping people with physical or mental challenges obtain access to safe, affordable, and reliable transportation. He wants people in these traditionally underserved communities to have multiple transportation options so they can obtain the medical care they need, attend school, or travel to critical appointments. QRyde has helped make his dream come true.

QRyde is a first in class and best of breed cloud-based transportation management system. It enables independent transportation companies and government transit systems to optimize current operations using technology such as mobile scheduling, automated dispatching, and artificial intelligence (AI)-based route optimization to help those organizations provide services more efficiently and economically. Educational institutions, healthcare companies, and public transit agencies use QRyde's shared ride scheduling platform for ride booking, ride cost-sharing, and bidding management services to provide mobility options to people in need.

For example, Penn Highlands Healthcare, which serves residents of Northwestern/Central and Southwestern Pennsylvania, uses QRyde to coordinate rides with regional transportation providers in a service area covering 39 counties. The QRyde technology enables residents, facilities, and caregivers to identify available seats with each transportation provider, coordinate schedules and costs, and determine the transportation service that best fits their needs. The QRyde mobile app and web portal provide even more ways to request a ride for all residents through local transportation providers and transportation network companies.

QRyde is very data-and computing-intensive. with substantial logistics and data integration requirements. The system originally ran in on-premises data centers and colocation facilities. (*Colocation* is the practice of housing privately-owned servers and networking equipment in a third-party data center. The organization that owns this equipment is responsible for its management and upkeep.) However, the business continued to grow, to the point where it was coordinating 100,000 trips per day for more than 700 organizations. It needed a more powerful, flexible IT infrastructure to do this work.

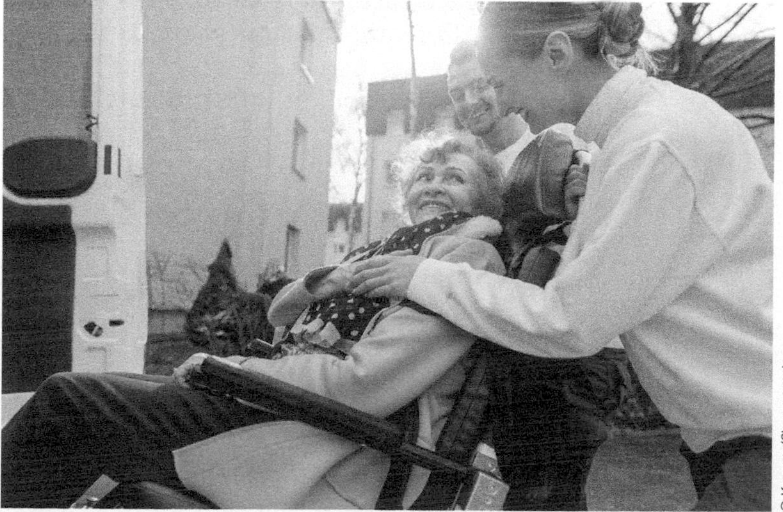
© Kzenon/Shutterstock

Oracle Cloud Infrastructure (OCI) provided the best solution. OCI is a platform of cloud services that enable organizations to build and run a wide range of applications in a highly-available, remote high-performance environment. These services include computing services, data storage and management services, networking services, developer services, application services, and security and governance services. Organizations adopting OCI can increase their use of these services as they grow, and pay for only the level of service that they need. Bhatnagar selected Oracle because of its expertise in data management services and its understanding of how to optimize the data. QRyde stores and secures data while active and at rest.

With QRyde Cloud all stakeholders of transportation systems can perform their entire business functions on one portal. These functions include client profile management, ride booking, scheduling, dispatching, ride-sharing, billing, electronic fare cards, and contract management on the same platform using text, voice and browser-based user interfaces and experiences. QRyde Cloud removes the high-cost-of-technology barrier for communities and providers and its cost-per-ride pricing model beats has helped organizations save hundreds of thousands of dollars for optimization of transportation systems.

Sources: QRyde.com, accessed April 14, 2022; Phhealthcare.org, accessed April 16, 2022; Justine Kavanaugh-Brown, "To Get Rides for People in Need, QRyde Taps OCI for Better Disaster Recovery, Security, and Scale," Blogs.oracle.com, November 5, 2021; Oracle. "Oracle Cloud Infrastructure Platform Overview." September 2021.

---

The experience of QRyde illustrates the importance of information technology infrastructure in running a business today. The right technology at the right price will improve organizational performance. QRyde was saddled with outdated hardware and software platforms that could no longer meet its computing requirements as the business grew. This prevented the company from operating efficiently and effectively and from providing its customers with the services they needed.

The chapter-opening case diagram calls attention to important points raised by this case and this chapter: Using cloud computing for its IT infrastructure enables QRyde to increase its workload as it expands and to provide new services to the people and organizations that need them. QRyde is also able to focus more on growing the business and serving customers because it delegated the operation and management of its IT infrastructure to a cloud service provider. The company works with state-of-the-art information technology tools, pays for only the computing capacity it uses on an as-needed basis, and does not have to make extensive and costly up-front IT investments.

Here are some questions to think about: How did using cloud computing help QRyde serve its customers better? What were the business benefits for QRyde of using a cloud computing infrastructure?

## 5-1 Identify the components of IT infrastructure.

If you want to know why businesses worldwide spent about $4.5 trillion in 2022 on computing and information systems, just consider what it would take for you personally to set up a business or manage a business today. Businesses require a wide variety of computing equipment, software, and communications capabilities simply to operate and solve basic business problems.

Do your employees travel or do some work from home? You will want to equip them with laptop computers, tablets, or smartphones. If you are employed by a medium or large business, you will also need larger server computers, perhaps an entire data center or server farm with hundreds or even thousands of servers. A **data center** is a facility that houses computer systems and associated components such as telecommunications, storage, security systems, and backup power supplies.

You will also need plenty of software. Each computer will require an operating system and a wide range of application software capable of dealing with spreadsheets,

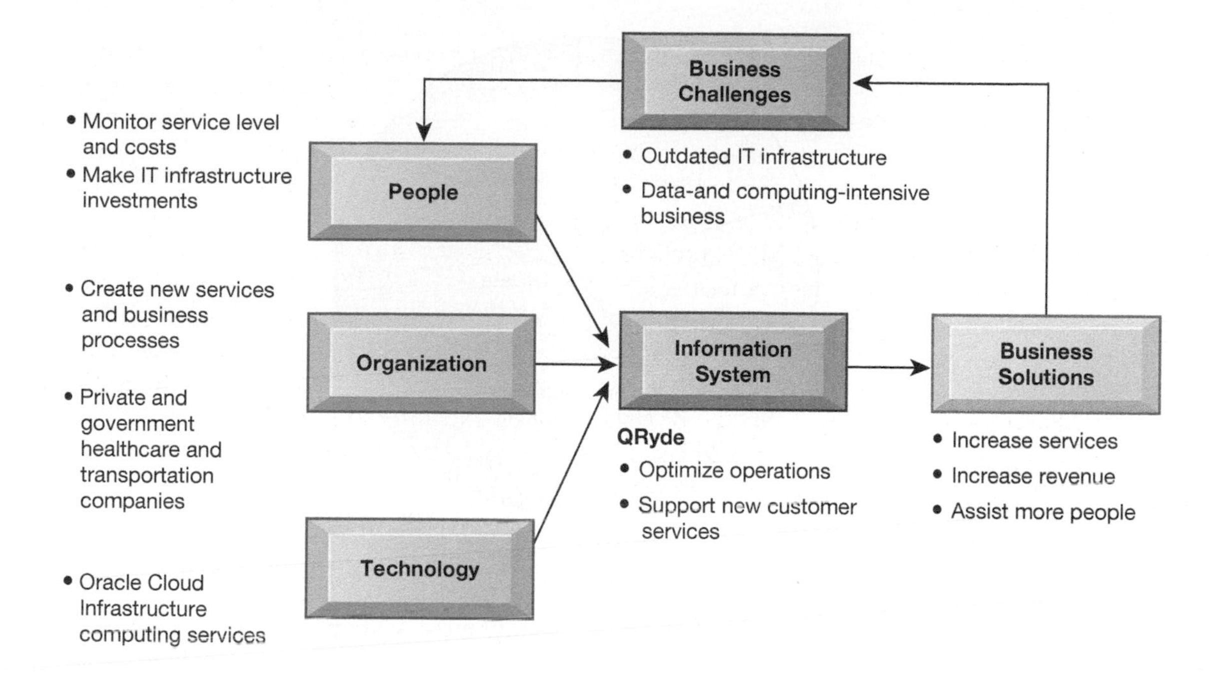


documents, and data files. Unless you are a single-person business, you will most likely want to have a network to link all the people in your business and perhaps your customers and suppliers. In fact, you will probably want several networks: a local area network connecting employees in your office and another providing remote access capabilities so that employees can share email and computer files while they are out of the office. You will also want all your employees to have access to landline phone systems, mobile phone networks, and the Internet. Finally, to make all this equipment and software work harmoniously, you will need the services of trained people to help you run and manage this technology.

All the elements we have just described combine to make up the firm's *information technology (IT) infrastructure*, which we first defined in Chapter 1. A firm's IT infrastructure provides the foundation, or platform, for supporting all the information systems in the business.

## IT INFRASTRUCTURE COMPONENTS

Examine Figure 5.1, which identifies the five major components of today's IT infrastructure: computer hardware, computer software, data management technology, telecommunications and networking technology, and technology services. These components must be coordinated with each other. We describe each of these in more detail in the following sections.

### Computer Hardware

Computer hardware consists of technology for computer processing, data storage, input, and output. This component includes large mainframes, servers, desktop and laptop computers, and mobile devices for accessing corporate data and the Internet. It also includes equipment for gathering and inputting data, physical media for storing the data, and devices for delivering the processed information as output.

### Computer Software

Computer software includes both system software and application software. **System software** manages the resources and activities of the computer. **Application software** directs the computer to complete a specific task, such as processing an order or

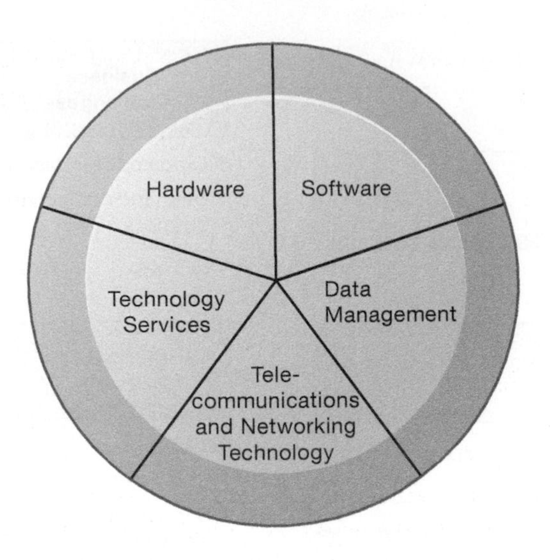


**Figure 5.1**
IT Infrastructure Components
*A firm's IT infrastructure is composed of hardware, software, data management technology, telecommunications and networking technology, and technology services.*

generating a mailing list, for an end user. Today, most system and application software is no longer custom programmed but rather is purchased from outside vendors. We describe these types of software in detail in Section 5-3.

### Data Management Technology

In addition to physical media for storing the firm's data, businesses need specialized software to organize the data and make them available to business users. **Data management software** organizes, manages, and processes business data concerned with inventory, customers, and vendors. Chapter 6 describes data management software in detail.

### Telecommunications and Networking Technology

Telecommunications and networking technology provides data, voice, and video connectivity to employees, customers, and suppliers. It includes technology for running a company's internal networks, telecommunications/telephone services, and technology for running websites and linking to other computer systems through the Internet. Chapter 7 provides an in-depth description of these technologies.

### Technology Services

Businesses need people to run and manage the infrastructure components we have just described and to train employees in how to use these technologies for their work. Chapter 2 described the role of the information systems department, which is the firm's internal business unit set up for this purpose. Today, many businesses supplement their in-house information systems staff with external technology consultants to provide expertise that is not available internally. When businesses need to make major system changes or implement an entirely new IT infrastructure, they typically turn to external consultants to help them with systems integration.

Systems integration means ensuring that the new infrastructure works with the firm's legacy systems and that the new elements of the infrastructure work with one another. **Legacy systems** are generally older transaction processing systems created for older computers that continue in use to prevent the high cost of replacing or redesigning them.

Thousands of technology vendors supply IT infrastructure components and services, and an equally large number of ways of putting them together exists. This chapter discusses the hardware and software components of infrastructure that you will need to run a business. Chapter 6 describes the data management component, and Chapter 7 is devoted to the networking and telecommunications technology component. Chapter 8 deals with hardware and software for ensuring that information systems are reliable and secure.

## 5-2 Describe the major computer hardware, data storage, input, and output technologies used in business and major hardware trends.

Business firms face many challenges and problems that computers and information systems can solve. To be efficient, firms need to match the right computer hardware to the nature of the business challenge, neither overspending nor underspending for the technology.

### TYPES OF COMPUTERS

Computers come in an array of sizes with differing capabilities for processing information, from the smallest mobile devices to the largest mainframes and supercomputers. If you're working alone or with a few other people in a small business, you'll probably be using a desktop or laptop **personal computer (PC)**. You will likely have a mobile device, such as an iPhone, iPad, or Android mobile device, with substantial computing capability. If you're doing advanced design or engineering work requiring powerful graphics or computational capabilities, you might use a **workstation**, which fits on a desktop but has more powerful mathematical and graphics-processing capabilities than a PC.

If your business has a number of networked computers or maintains a website, it will need a **server**. Server computers are specifically optimized to support a computer network, enabling users to share files, software, peripheral devices (such as printers), or other network resources.

Servers provide the hardware platform for e-commerce. By adding special software, servers can be customized to deliver web pages, process purchase and sale transactions, or exchange data with systems inside the company. You will sometimes find many servers linked to provide all the processing needs of large companies. If your company has to process millions of financial transactions or customer records, you will need multiple servers or a single large mainframe to do this work.

Mainframe computers first appeared in the mid-1960s, and large banks, insurance companies, stock brokerages, airline reservation systems, and government agencies still use them to keep track of hundreds of thousands or even millions of records and transactions. A **mainframe** is a large-capacity, high-performance computer that can process large amounts of data very rapidly. IBM has repurposed its mainframe systems so that they can be used as giant servers for large-scale enterprise networks and corporate websites. A single IBM mainframe can replace thousands of smaller Windows-based servers. A **supercomputer** is a specially designed and more sophisticated computer that is used for tasks requiring extremely rapid and complex calculations with thousands of variables, millions of measurements, and thousands of equations. Supercomputers traditionally have been used in engineering analysis of structures, scientific exploration and simulations, and military work such as classified weapons research.

If you are a long-term weather forecaster, such as the National Oceanic and Atmospheric Administration (NOAA) or the National Hurricane Center, and your challenge is to predict the movement of weather systems based on hundreds of thousands of measurements and thousands of equations, you would want access to a supercomputer or a distributed network of computers called a grid.

**Grid computing** involves connecting geographically remote computers into a single network and combining the computational power of all computers on the grid. Grid computing takes advantage of the fact that most computers in the United States use their central processing units on average about 25 percent of the time, leaving 75 percent of their capacity available for other tasks. By using the combined power of thousands of PCs and other computers networked together, the grid can solve complicated problems at supercomputer speeds but at far lower cost.

For example, GM and Ford use grids to simulate crash tests, saving millions in junked cars and speeding time to market. Pratt and Whitney test aircraft engine designs on a grid. Biotech firms such as Aventis, GlaxoSmithKline, and Pfizer use grid computing to accelerate their research.

### Computer Networks and Client/Server Computing

Unless you work at a small business and use a stand-alone computer, you'll be using networked computers for most processing tasks. The use of multiple computers linked by a communications network for processing is called **distributed processing**. **Centralized processing**, in which all processing is accomplished by one large central computer, is much less common.

One widely used form of distributed processing is **client/server computing**. Computing on the Internet uses the client/server model (see Chapter 7). Client/server computing splits processing between clients and servers. Both are on the network, but each machine is assigned the functions it is best suited to perform. Examine Figure 5.2, which illustrates the client/server computing concept implemented in the form of a simple client/server network consisting of a client computer networked to a server computer, with processing split between the two types of machines (known as *two-tiered client/server architecture*). The **client** is the user point of entry for the required function and is normally a desktop or laptop computer. It could also be a mobile device. The user generally interacts directly only with the client portion of the application. The server provides the client with services. Servers store and process shared data and perform such functions as managing printers, providing backup storage, handling security, and providing network resources. Whereas simple client/ server networks such as that depicted by Figure 5.2 can be found in small businesses, most corporations have more complex, multitiered networks (often called **N-tier client/server architectures**) in which the work of the entire network is balanced over several levels of servers, depending on the kind of service being requested.

Examine Figure 5.3, which illustrates this more complex form of client/server network. Here, the client's request for services goes first to a **web server**, which serves a web page to the client in response to that request. Web server software is responsible for locating and managing stored web pages. If the client requests access to a corporate system (to see a product list or price information, for instance), the request is passed along to an **application server**. Application server software handles all application operations between a user and an organization's back-end business systems. The application server may reside on the same computer as the web server or on its own dedicated computer. Chapters 6 and 7 provide more detail about other pieces of software that are used in multitiered client/server architectures for e-commerce and e-business.

**Figure 5.2**
Client/Server Computing
*In client/server computing, computer processing is split between client machines and server machines linked by a network. Users interface with the client machines.*

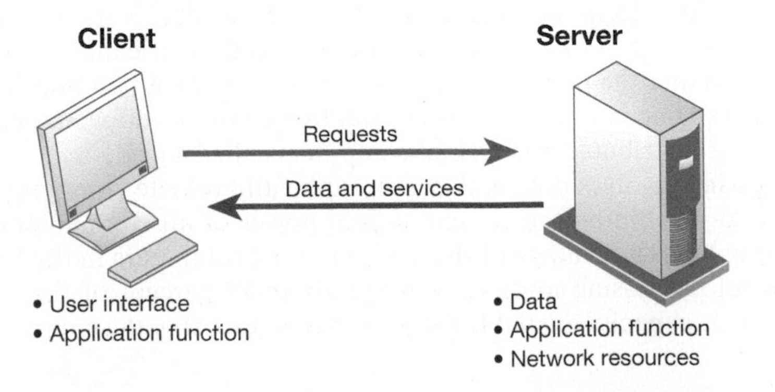

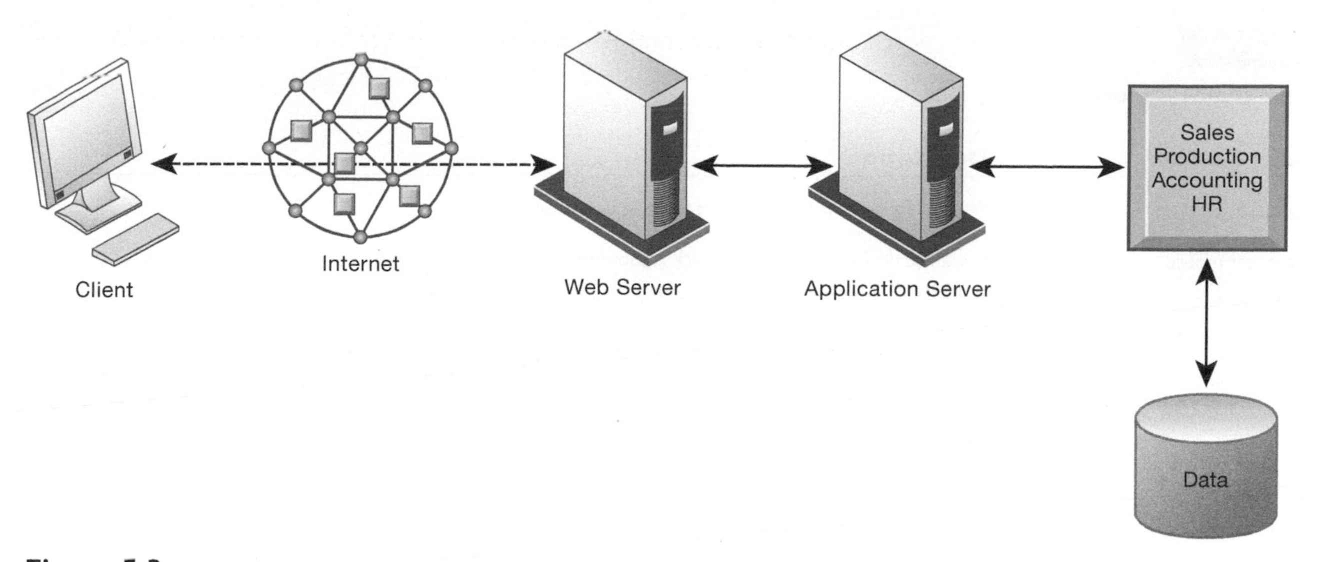


**Figure 5.3**
A Multitiered Client/Server Network (N-Tier)
*In a multitiered client/server network, client requests for service are handled by different levels of servers.*

## STORAGE, INPUT, AND OUTPUT TECHNOLOGY

In addition to hardware for processing data, you will need technologies for data storage and input and output. Storage and input and output devices are called *peripheral devices* because they are outside the main computer system unit.

### Secondary Storage Technology

E-commerce and e-business, and regulations such as the Sarbanes-Oxley Act and the Dodd-Frank Act, have made storage a strategic technology. The amount of data that companies now need to store is doubling every 12 to 18 months. Traditional storage technologies include magnetic disks, optical drives, magnetic tape, and storage networks, which connect multiple storage devices on a separate high-speed network dedicated to storage.

Magnetic disk drives are still used for secondary storage in large and midrange computers and in some PCs. However, data storage in many PCs, as well as in smartphones and tablets, is based on **solid state drives (SSDs)**, which use an array of semiconductors organized as a very fast internal disk drive. Portable USB flash drives use similar technology for external storage. Optical drives (CD, DVD, and Blu-ray) use laser beaming technology to store very large quantities of data, including sound and images, in compact form. Firms also use cloud computing services for large-scale data storage. We discuss cloud computing later in this chapter.

### Input and Output Devices

Humans interact with computer systems largely through input and output devices. **Input devices** gather data and convert them into digital form for use by the computer, whereas **output devices** display data after they have been processed. Table 5.1 describes the principal input and output devices.

## CONTEMPORARY HARDWARE TRENDS

The vastly increasing power of computer hardware and networking technology has dramatically changed how businesses organize their computing power, putting more of this power on networks and mobile devices. Let's look at seven hardware trends: the mobile platform, consumerization of IT and BYOD, virtualization, cloud computing, edge computing, green computing, and high-performance computing and processors.

**TABLE 5.1**

Input and Output Devices

| Input Device | Description |
|---|---|
| Keyboard | Principal method of data entry for text and numerical data. |
| Computer mouse | Handheld device with point-and-click capabilities for controlling a cursor's position on a computer display screen and selecting commands. Trackballs and touch pads often are used in place of the mouse as pointing devices on laptop PCs. |
| Touch screen | Device that allows users to interact with a computer by touching the surface of a sensitized display screen. Used in kiosks in airports, retail stores, and restaurants and in multitouch devices such as the iPhone, iPad, and multitouch PCs. |
| Optical character recognition | Device that can translate specially designed marks, characters, and codes into digital form. The most widely used optical code is the bar code. |
| Magnetic ink character recognition (MICR) | Technology used primarily in check processing for the banking industry. Characters on the bottom of a check identify the bank, checking account, and check number and are preprinted using special magnetic ink for translation into digital form for the computer. |
| Pen-based input | Handwriting-recognition devices that convert into digital form the motion made by an electronic stylus pressing on a touch-sensitive tablet screen. |
| Digital scanner | Device that translates images, such as pictures or documents, into digital form. |
| Audio input | Devices that convert voice, music, or other sounds into digital form for processing by the computer. |
| Sensors | Devices that collect data directly from the environment for input into a computer system. For instance, farmers can use sensors to monitor the moisture of the soil in their fields. |

| Output Device | Description |
|---|---|
| Display | Often a flat-panel (LCD) display screen for text, graphics, and video. |
| Printers | Devices that produce a printed hard copy of information output. They include impact printers (such as dot matrix printers) and nonimpact printers (such as laser, inkjet, and thermal transfer printers). |
| Audio output | Output devices that convert digital output data back into intelligible speech, music, or other sounds. |

## The Mobile Platform

A mobile computing platform has emerged as an alternative to PCs and larger computers. Mobile devices such as the iPhone and Android smartphones have taken on many functions of PCs, including transmitting data, surfing the web, transmitting email and instant messages, displaying digital content, and exchanging data with internal corporate systems. The mobile platform also includes small, lightweight subnotebooks such as the Google Chromebook, **tablet computers** such as the iPad, car infotainment systems, wearable devices, and digital e-book readers such as Amazon's Kindle.

Smartphones and tablet computers are becoming the primary means of accessing the Internet. These devices are increasingly used for business computing as well as for consumer applications. For example, PepsiCo managers use iPads customized to monitor their teams' performance; pull up pricing, planograms, and contracts; and coordinate deliveries with merchandising. The Spotlight on People case shows some of the ways mobile devices are improving healthcare.

Wearable computing devices are a recent addition to the mobile platform. These include smartwatches, smart glasses, smart badges, and activity trackers. Wearable computing devices have both consumer and business uses. They improve productivity by delivering information to workers without requiring them to interrupt their tasks, which in turn empowers workers to make more informed decisions more quickly.

## The Mobile Platform Comes to Healthcare

Today there are many job functions for both rank-and-file employees and their managers that can be performed using smartphones and tablets (including the iPad, iPhone, and Android mobile devices), and this is true for healthcare as well. Mobile technology is spreading to core work functions for doctors, nurses, and other healthcare professionals, and it is also powering applications that patients can use themselves.

There are more than 350,000 medical, health, and fitness apps in major app stores, and two-thirds of the largest US hospitals offer their own proprietary mobile health apps. Ninety percent of physicians use smartphones at work to access electronic health records, communicate with their team members, reference information, or manage their schedules. According to the Boston Technology Corporation, 74 percent of patients say using wearables and other mHealth tools helps them cope with and manage their conditions. Mobile telehealth visits have exploded because of the Covid-19 pandemic, but there are many other innovative applications that increase patient engagement and well-being.

Geisinger Medical Center (GMC) in Danville, Pennsylvania, is an example of how the mobile platform can improve many facets of healthcare. Geisinger has long been known for its innovative approach to patient care, which is reflected in its extensive use of mobile technology.

Dr. Jonathan Slotkin is a neurosurgeon and medical director of Geisinger in Motion, which is responsible for Geisinger's digital initiatives. According to Slotkin, Geisinger believes that the patient needs to be the center of things and that mobile IT applications help make that possible.

Mobile devices such as the iPad and iPhone help patients learn about their conditions and treatments and play a big role in patient care. GMC inpatients waiting for a procedure to begin or waiting to talk with their doctor can learn more about their conditions or their test results even before they see their physician. They can call up their lab results on a hospital-issued iPad available in many areas of the hospital. Patients also can upload their medical records to their smartphones. GMC has found that patients who use the iPads during their stay are more knowledgeable about their conditions.

Children in the hospital who are heading to surgery can download their favorite games onto the iPads. This provides a distraction that can relieve the stress of facing surgery or tests. Young patients entertaining themselves with games or shows on the tablets before surgery often require less sedation and pain medication, which promotes faster recovery.

Geisinger's hospital systems team delivers around 100 apps through a service portal that patients can download; these apps include games, news, weather, and social media. About 30 or 40 of those apps are for pediatric patients. Patients are allowed to supply their own Apple ID for some of the apps such as Minecraft. When a patient is discharged, the patient's apps are erased and the iPad is digitally and physically cleaned.

At Geisinger, children in the hospital for prolonged treatment don't have to miss school: A robot with an iPad attached on top travels from class to class, enabling the hospitalized students to see their friends and to see each other. The iPad raises when the patients raise their hand. Danville Middle School parents and students said the program was not just about keeping up with classes—it also reinforced connections between students and teachers.

Because of language barriers or physical impairments, many people cannot access the healthcare they need. When non-English-speaking patients lack interpretation services, patient engagement rates drastically decline. Studies show that patients who are engaged with their care are more likely to follow healthcare plans, to attend more appointments, and to fill more prescriptions. They are also less likely to be readmitted to the hospital within 30 days of discharge.

Tablets such as iPads can also help bridge the communication gap between healthcare professionals and people with different abilities: The tablets enable physicians to utilize video interpretation services in sign language as well as spoken language. This capability is especially helpful when treating people with hearing impairments.

The iPad and iPhone have a Translate app that addresses this problem. The Translate app can translate text, voice, and conversations between any two of the languages it supports. You can download languages to translate entirely on the device, even without an Internet connection. Geisinger uses the iPad for language translation services, as does the Harris Health System, which serves the residents of Harris County, Texas.

Smartphones and tablet applications enable healthcare workers to move from room to room without having to log out of the system they are using or access another computer to enter or extract information. It is much more efficient to be able to access a patient's full medical profile than to carry around a stack of papers. With the necessary data at their fingertips, doctors and nurses can focus more easily on the patient.

Sources: Geisinger.org, accessed April 24, 2022; "Empowering Patients at Geisinger," Apple.com, accessed April 24, 2022; Joe Sylvester, "IPad Links Geisinger Patients with Information," Dailyitem.com, accessed April 23, 2022; "What Can the New iPad Do for Healthcare?" Healthworkscollective.com, accessed April 22, 2022.

## CASE STUDY QUESTIONS

1. What kinds of applications are described here? What business functions do they support? How do they improve operational efficiency and decision making?
2. Identify the problems that healthcare organizations in this case study solved by using mobile digital devices.
3. How did the mobile platform change patient care at the organizations described in this case study?

### Consumerization of IT and BYOD

The popularity, ease of use, and rich array of useful applications for smartphones and tablet computers have created a groundswell of interest in allowing employees to use their personal mobile devices in the workplace, a phenomenon popularly called *bring your own device* (*BYOD*). **BYOD** is one aspect of the **consumerization of IT**, in which new information technology that first emerges in the consumer market spreads into business organizations. Consumerization of IT includes not only mobile personal devices but also business uses of software services that originated in the consumer marketplace, such as Google search, Gmail, Dropbox, and even Facebook and Twitter.

Consumerization of IT is forcing businesses to rethink the way they obtain and manage information technology equipment and services. Historically, at least in large firms, the IT department controlled selection and management of the firm's hardware and software. This ensured that information systems were protected and served the purposes of the firm. Today, employees and business departments are playing a much larger role in technology selection, in many cases demanding that employees be able to use their own personal mobile devices to access the corporate network. Although consumer technologies provide new tools to foster creativity, collaboration, and productivity, they are more difficult for firms to manage and control.

### Virtualization

**Virtualization** is the process of presenting a set of computing resources (such as computing power or data storage) so that they can all be accessed in ways that are not restricted by physical configuration or geographic location. Virtualization enables a single physical resource (such as a server or a storage device) to appear to the user as multiple logical resources. For example, a server or mainframe can be configured to run many instances of an operating system (or different operating systems) so that it acts like many different machines. Each virtual server "looks" like a real physical server to software programs, and multiple virtual servers can run in parallel on a single machine. VMware is a major virtualization software vendor for Windows and Linux servers.

Server virtualization is a common method of reducing technology costs by providing the ability to host multiple systems on a single physical machine. Most servers run at just 15 to 20 percent of their capacity, and virtualization can boost server utilization rates to 70 percent or higher. Higher utilization rates translate into fewer computers required to process the same amount of work, reduced data center space needed to house machines, and lower energy usage. Virtualization also facilitates centralization and consolidation of hardware administration.

Virtualization also enables multiple physical resources (such as storage devices or servers) to appear as a single logical resource, as in **software-defined storage (SDS)**, which separates the software for managing data storage from storage hardware. Using software, firms can pool and arrange multiple storage infrastructure resources and efficiently allocate them to meet specific application needs. SDS enables firms to replace expensive storage hardware with lower-cost commodity hardware and cloud storage hardware. By doing so, there is less under- or overutilization of storage resources.

## Cloud Computing

**Cloud computing** is a model of computing in which computer processing, storage, software, and other services are provided as a shared pool of virtualized resources over a network, primarily the Internet. These clouds of computing resources can be accessed on an as-needed basis from any connected device and location. Cloud computing is the fastest-growing form of computing, with worldwide cloud computing expenditures projected to exceed $832 billion by 2025 (Sumina, 2022).

Consumers use cloud services such as Apple's iCloud or Google Drive to store documents, photos, videos, and email. Cloud-based software powers social networks, streamed video and music, and online games. Private and public organizations are turning to cloud services to replace internally run data centers and software applications. Amazon, Google, IBM, Oracle, and Microsoft operate huge, scalable cloud computing centers offering computing power, data storage, and other IT services to firms that want to maintain their IT infrastructures remotely. Google, Microsoft, SAP, Oracle, Salesforce.com and many other companies sell software applications as cloud services delivered over the Internet.

The US National Institute of Standards and Technology (NIST) defines cloud computing as having the following essential characteristics:

- **On-demand self-service:** Consumers can obtain computing capabilities such as server time or network storage as needed, automatically, and on their own.
- **Ubiquitous network access:** Cloud resources can be accessed using standard network and Internet devices, including mobile platforms.
- **Location-independent resource pooling:** Computing resources are pooled to serve multiple users, with different virtual resources dynamically assigned according to user demand. The user generally does not know where the computing resources are located.
- **Rapid elasticity:** Computing resources can be rapidly provisioned, increased, or decreased to meet changing user demand.
- **Measured service:** Charges for cloud resources are based on the number of resources actually used.

Examine Figure 5.4, which illustrates the cloud computing concept and identifies the types of services that it can provide. Many different types of devices, including laptops, desktops, tablets, smartphones, and servers, can access these services remotely. Internet of Things (IoT) sensor devices can also link to cloud services.

Cloud computing consists of three types of services:

- **Infrastructure as a service (IaaS):** Customers use processing, storage, networking, and other computing resources from cloud service providers to run their information systems. For example, Amazon Web Services (AWS) uses the spare capacity of its IT infrastructure to provide a broadly based cloud environment selling IT infrastructure services. Examine Figure 5.5, which shows some of the major services AWS offers, including its Elastic Compute Cloud (EC2) service that provides secure, resizable computing capacity in the cloud for customer applications,

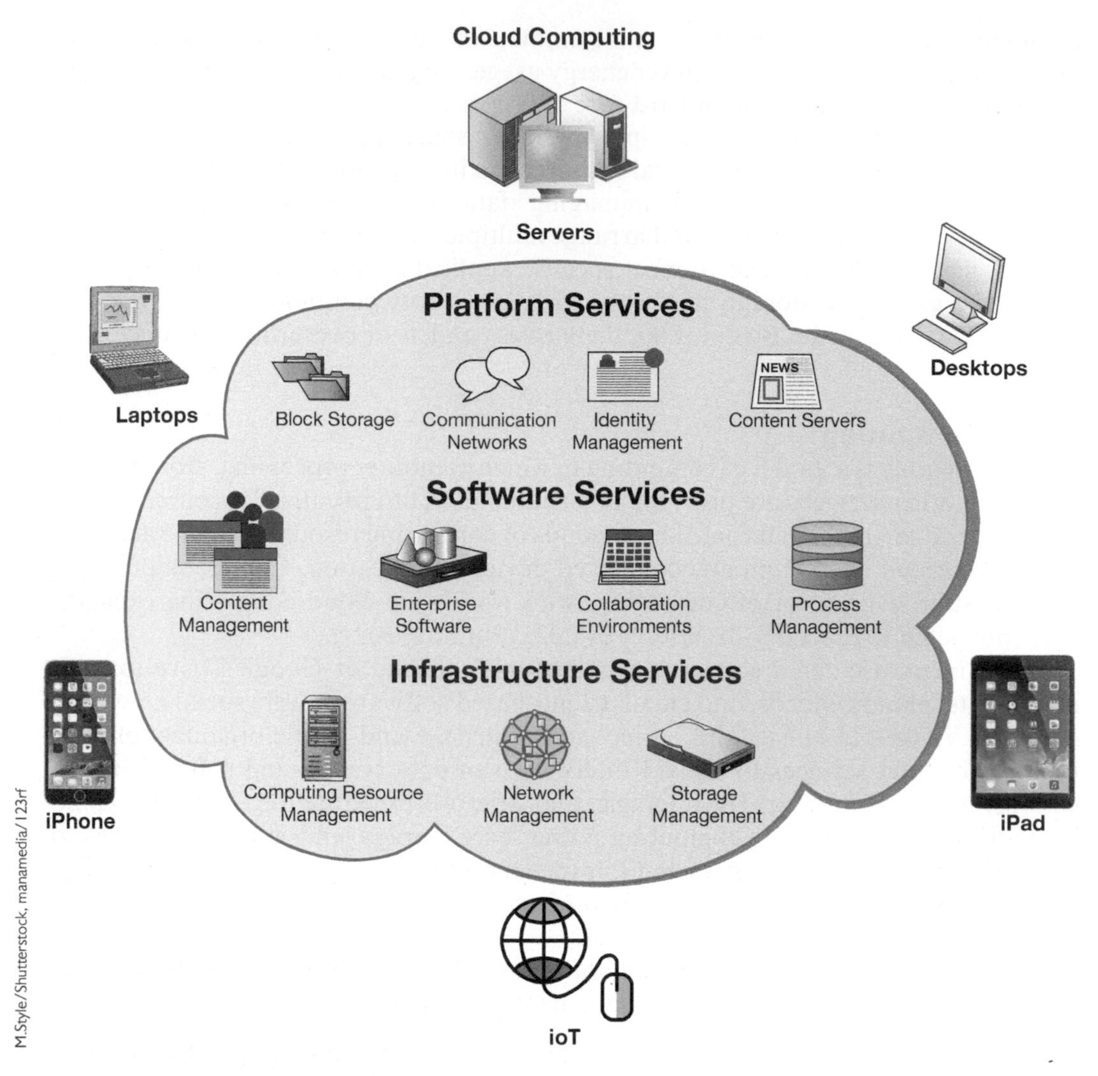


**Figure 5.4**
Cloud Computing Platform
*In cloud computing, hardware and software capabilities are a pool of virtualized resources provided over a network, often the Internet. Businesses and employees have access to applications and IT infrastructure anywhere and at any time.*

its Simple Storage Service (S3) for storing customers' data, its CloudFront content-delivery service, and a variety of networking, database, security, management, application, and analytics services, among others. Users pay only for the amount of services they actually use.

- **Software as a service (SaaS):** Customers use software hosted by the vendor on the vendor's cloud infrastructure and delivered as a service over a network. Leading **software as a service (SaaS)** examples are Microsoft 365, which provides common business applications online, and Salesforce.com, which licenses various software services (described more fully below) over the Internet. Both charge users an annual subscription fee. Users access these applications from a web browser, and the data and software are maintained on the providers' remote servers.
- **Platform as a service (PaaS):** Customers use infrastructure and programming tools supported by the cloud service provider to develop their own applications. For example, Microsoft offers PaaS tools and services for software development and testing among its Azure cloud services. Another example is Salesforce.com's Salesforce Platform.

Chapter 2 discussed Google Docs and Microsoft 365 software services for desktop productivity and collaboration. These are among the most popular software services for consumers as well as businesses. Salesforce.com is a leading SaaS provider for business, providing customer relationship management (CRM) and many other application software solutions. For instance, its Sales Cloud and Service Cloud offer applications for improving sales and customer service. A Marketing Cloud enables companies to engage in digital marketing interactions with customers through

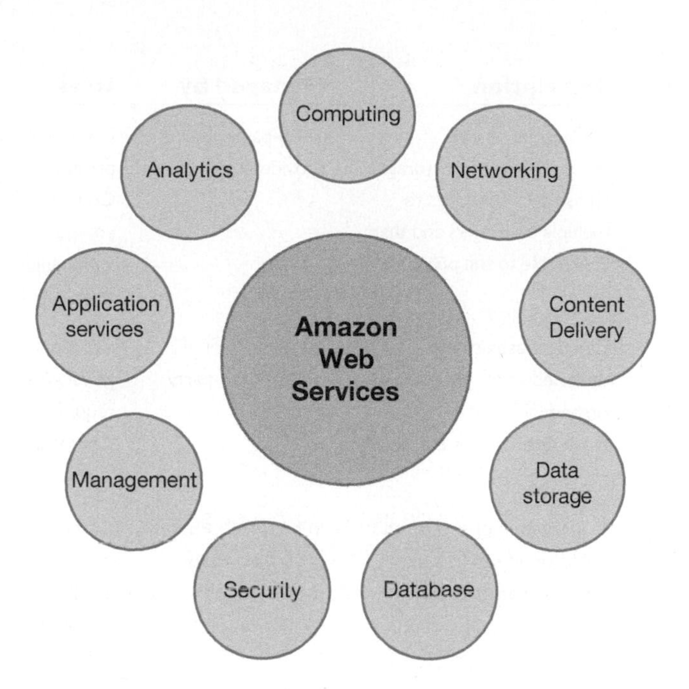


**Figure 5.5**
Major Amazon Web Services
*Amazon Web Services (AWS) is a collection of web services that Amazon provides to users of its cloud platform. AWS now offers hundreds of services and is the largest provider of public cloud-based services.*

email, mobile, social, web, and connected products. Salesforce.com also provides a Community Cloud platform for online collaboration and engagement, an Analytics Cloud platform to enable users to visualize and analyze data on many different kinds of devices without having to toggle among applications, and a Health Cloud for patient care management. A new IoT Cloud transmits data about product usage and customer behavior generated by Internet-connected devices to the Salesforce platform to improve customer relationship management.

Salesforce.com is also a leading example of platform as a service (PaaS). Its Salesforce Platform gives users the ability to develop, launch, and manage applications without having to deal with the infrastructure required for creating new software. The Salesforce Platform provides a set of development tools and IT services that enable users to build new applications and run them in the cloud on Salesforce.com's data center infrastructure. Salesforce.com also lists software from other independent developers on its AppExchange, an online marketplace for third-party applications that run on the Salesforce Platform.

A cloud can be public or private. A **public cloud** is owned and maintained by a cloud service provider, such as Amazon Web Services, and made available to the general public or industry group. Public cloud services are often used for websites with public information and product descriptions, one-time large computing projects, new application development and testing, and consumer services such as online storage of data, music, and photos. Google Drive, Dropbox, and Apple iCloud are leading examples of these public cloud services.

A **private cloud** is operated solely for an organization. It might be managed by the organization or a third party and hosted either internally or externally. Like public clouds, private clouds can allocate storage, computing power, or other resources seamlessly to provide computing resources on an as-needed basis. Companies that want flexible IT resources and a cloud service model while retaining control over their own IT infrastructure are gravitating toward these private clouds.

Because organizations using cloud platforms do not own the infrastructure, they do not have to make large investments in their own hardware and software. Instead, they purchase their computing services from remote providers and pay only for the amount of computing power they actually use (utility computing) or are billed on a monthly or annual subscription basis. The term **on-demand computing** has also been used to describe such services.

Cloud computing has some drawbacks. Unless users make provisions for storing their data locally, the responsibility for data storage and control is in the hands of the

**TABLE 5.2**

Cloud Computing Models Compared

| Type of Cloud | Description | Managed by | Uses |
|---|---|---|---|
| Public cloud | Third-party service offering computing, storage, and software services to multiple customers and that is available to the public | Third-party service providers | Companies without major privacy concerns<br>Companies seeking pay-as-you-go IT services<br>Companies lacking IT resources and expertise |
| Private cloud | Cloud infrastructure operated solely for a single organization and hosted either internally or externally | In-house IT or private third-party host | Companies with stringent privacy and security requirements<br>Companies that must have control over data sovereignty |
| Hybrid cloud | Combination of private and public cloud services that remain separate entities | In-house IT, private host, third-party providers | Companies requiring some in-house control of IT that are also willing to assign part of their IT infrastructures to a public cloud |

provider. Some companies worry about the security risks related to entrusting their critical data and systems to an outside vendor that also works with other companies. Companies expect their systems to be available 24/7 and do not want to suffer any loss of business capability if cloud infrastructures malfunction. Nevertheless, the trend is for companies to shift more of their computer processing and storage to some form of cloud infrastructure. Start-ups and small companies that do not have ample IT resources or budgets will find public cloud services especially helpful.

Large firms are most likely to adopt a **hybrid cloud** computing model in which they use their own infrastructure for their most essential, core activities and adopt public cloud computing for less critical systems or for additional processing capacity during peak business periods. Table 5.2 compares the three cloud computing models. Cloud computing will gradually shift firms from having a fixed infrastructure capacity toward a more flexible infrastructure—some of it owned by the firm and some of it rented from giant data centers owned by cloud vendors.

## Edge Computing

Having all the laptops, smartphones, tablets, wireless sensor networks, and local on-premise servers used in cloud computing systems interacting with a single, central public cloud data center to process all their data can be inefficient and costly. **Edge computing** is a method of optimizing cloud computing systems by performing some processing on a set of linked servers at the edge of the network, near the source of the data. This reduces the amount of data flowing back and forth between local computers and other devices and the central cloud data center and any delays in the transmission and processing of the data. This is helpful for situations where delays of milliseconds can be untenable, such as in financial services, manufacturing, or online gaming. For example, Network Next, which specializes in making the Internet better for gaming, uses edge computing to reduce delays in data transmission and response time for real-time, online multiplayer games. The Spotlight on Technology case provides more examples of edge computing applications.

## Green Computing

By curbing hardware proliferation and power consumption, virtualization has become one of the principal technologies for promoting green computing. **Green computing** or **green IT** refers to practices and technologies for designing,

## "Smart" Cities Become Smarter with Edge Computing

Cities and towns are becoming "smarter." A "smart" city is one that uses information and communication technology to improve operational efficiency, share information with the public. and provide a better quality of government service to citizens. Smart cities are proliferating around the world and use cloud computing, sensors, and Big Data to make them smarter.

Edge computing is another key technology for smart cities. Edge computing focuses on bringing computing as close to the source of data as possible in order to reduce transmission delays and the volume of information needed for operations and services as well as to help municipal officials at all levels make more informed decisions. Fewer processes run in a centralized cloud. Moving those processes to locations near the network's edge, such as on a user's computer, an IoT device, or an edge server, minimizes the amount of long-distance communication that needs to happen between a client and a server.

Edge computing allows organizations to take the power of the cloud all the way to the network edge, especially to areas where they have not been able to use it before. Agencies can perform data analytics and processing and gain insights at the edge before routing the data back to centralized data centers for further analysis. This capability is especially useful for smart city applications.

There are compelling reasons for smart cities to move data analytics to the edge, where the data are generated and captured, rather than sending everything to faraway corporate and cloud data centers. Many smart city applications require Big Data and data analysis that takes place in near real time as the data are generated. There isn't time to send data to a distant data center for analysis or to store it in the cloud when it might be needed for immediate purposes. Such considerations justify analyzing many types of data at the edge, where the sensors, cameras, or other devices are located and where intelligent systems can take immediate actions based on the results of data analytics. A high percentage of smart cities and government agencies are actively using or exploring edge computing and IoT.

There are many examples of edge and IoT solutions for cities spanning the range of municipal operations, from public safety and security to smart utility metering, traffic management, and parking. For example, the City of St. Petersburg, Florida, uses IoT and edge computing to capture and analyze data on its most dangerous intersection. It installed smart light poles, safety cameras, and environmental sensors and expanded Wi-Fi coverage at the intersection, which enabled the use of data-intensive video analytics and artificial intelligence software. Driver behaviors and activities can now be monitored and tracked automatically and in near real time 24/7/365. Traffic data visualization and insights from external sources are now displayed on a dashboard in a web browser. The system can also correlate multiple datasets to identify near-accidents, types of vehicles, illegal U-turns, jaywalking, and the time it takes pedestrians to cross the street. Correlations with environmental data enable the city to answer questions such as whether pedestrians cross the street more quickly when it rains. In addition, traffic engineers have a better understanding of "when, how, and what" occurs in the intersection.

Prior to 2019, the City of Syracuse street lighting infrastructure was owned and operated by an electric utility. The City had little visibility when lights were out, and residents had to notify the utility of outages, resulting in more services for affluent neighborhoods. Repairs took weeks or months. The lights were energy inefficient and costly to maintain. In 2019, the City required their streetlights to operate as a municipal service. The City replaced all streetlights with LED lights connected over a Wi-Fi network. Lights are now monitored using an automated, real-time, central dashboard. Outages can be resolved within three days, and lights are fixed within 24 hours of notification. The new lighting infrastructure has dramatically improved service and also provides that service more equitably. The system has also provided more visibility in other infrastructure such as underground power feeds and conduit.

The City of Lima, Ohio, has more than 80 railroad crossings. Vehicles must wait for crossing trains, and rail companies do not publish train schedules because of safety, security, and competitive reasons. In the past, unplanned gridlock for unknown durations inconvenienced citizens, visitors, local businesses, and first responders several times throughout the day. An Intel-sponsored study by Juniper Research found that gridlock costs drivers up to 70 hours per year and that an IoT-enabled intelligent traffic system, safer roads, and frictionless toll and parking payment could save drivers from spending 60 unproductive hours in their cars per year.

The city worked with US Ignite, DriveOhio, and networking solutions provider Spectrum Enterprise to develop a solution that helps mitigate the traffic delays by capturing data about train operations, using predictive analytics, and providing train metrics on a visualization platform. (DriveOhio is an Ohio Department of Transportation initiative to organize and accelerate smart vehicle and connected vehicle projects in the State of Ohio. US Ignite provides technology research to communities on smart city development.) The solution features an IoT system that has 15 multifaceted sensors. Data from the sensors are fed into a local edge computing processing unit that uses artificial intelligence and computer vision technology to detect a train crossing and capture the train's speed, length, and direction of travel. The captured data are sent to a cloud platform via cellular connectivity. The system then uses predictive algorithms to process the sensor data and display in real time the current status of the intersection as well as to predict outcomes, such as when the train will arrive at the next crossing and how long the crossing will be blocked.

Sources: "IDC Smart Cities North America Awards," Idc.com, accessed April 16, 2022; Kirsten Billhardt, "Smart Cities Drive Progress with Analytics at the Edge," *CIO*, January 29, 2020; Matt Parnofiello, "How Smart Cities Are Deploying Edge Computing," *State Tech*, December 2, 2019; Juniper Research, "Smart Cities–What's in It for Citizens?" (2018).

## CASE STUDY QUESTIONS

**1.** What problems did the organizations described in this case address by using edge computing, Big Data, and IoT?

**2.** What people, organization, and technology issues should a municipality address when building a smart city application?

**3.** How did information technology improve operations and decision making for the organizations described in this case?

manufacturing, using, and disposing of computers, servers, and associated devices such as monitors, printers, storage devices, and networking and communications equipment in order to minimize impact on the environment.

Reducing computer power consumption has been a very high green priority. A corporate data center can easily consume more than 100 times more power than a standard office building consumes. All this additional power consumption has a negative impact on the environment and on corporate operating costs. Data centers are now being designed with energy efficiency in mind, using state-of-the-art air-cooling techniques, energy-efficient equipment, virtualization, and other energy-saving practices. Organizations are achieving additional efficiencies by migrating from on-premise data centers to cloud computing. Large companies such as Microsoft, Google, Facebook (Meta), and Apple are starting to reduce their carbon footprints with clean energy-powered data centers, including extensive use of wind and hydropower. The chapter-ending case study provides more details on this topic.

### High-Performance Computing and Processors

As computing devices shrink in size and take on ever-larger workloads, information technologies for increasing processing power and reducing the amount of power consumed have come to the fore. These technologies include quantum computing, nanotechnology, and high-performance power-saving processors.

**Quantum Computing** **Quantum computing** uses the principles of quantum mechanics—the science of how matter and light behave at atomic and subatomic levels—to represent data and perform operations on that data. While conventional computers handle bits of data either as 0 or 1 but not both, quantum computing can process bits as 0, 1, or both simultaneously. A quantum computer gains enormous processing power through this ability to be in multiple states at once, allowing it to solve some scientific and business problems millions of times more quickly than can be done today. IBM has made quantum computing tools available to the general public through the IBM Cloud. Microsoft's Azure cloud platform also gives companies access to quantum resources, and Google provides a quantum computing service to researchers. Volkswagen and Lockheed Martin are other companies experimenting with quantum technology.

**Nanotechnology** Over the years, microprocessor manufacturers have been able to increase processing power exponentially while shrinking chip size by finding ways to pack more transistors into less space. They are now turning to nanotechnology to shrink the size of transistors to the width of several atoms. **Nanotechnology** uses individual atoms and molecules to create computer chips that are thousands of times smaller than traditional technologies permit. Long, microscopic strands of carbon atoms, called carbon nanotubes, can be fashioned into tiny transistors that provide twice the processing power of silicon chips while generating much less heat. Nanotechnology applications offer more efficient performance, thus conserving power and increasing battery life for smaller, portable digital devices. In addition to greater processing power, nanotechnology in computers is providing advanced forms of memory storage.

**High-Performance and Power-Saving Processors** Another way to reduce power requirements and hardware sprawl is to use more efficient and power-saving processors. Contemporary microprocessors now feature multiple processor cores (which perform the reading and execution of computer instructions) on a single chip. A **multicore processor** is an integrated circuit to which two or more processor cores have been attached for enhanced performance, reduced power consumption, and more efficient, simultaneous processing of multiple tasks. This technology enables two or more processing engines with reduced power requirements and heat dissipation to perform tasks faster than a resource-hungry chip with a single processing core. You'll find PCs with dual-core, quad-core, six-core, and eight-core processors and servers with 8-, 16-, and 32-core processors. Some of today's processors are also designed to handle intensive data-processing requirements for machine learning applications (AI processors), which we discuss in Chapter 11.

Chip manufacturers are working on microprocessors that minimize power consumption, which is essential for prolonging battery life in small mobile digital devices. Highly power-efficient microprocessors, such as the A14, A15, and A16 Bionic processors found in Apple's iPhone and iPad and Intel's Atom processor, are used in lightweight smartphones and tablets, intelligent cars, and healthcare devices. The A16

*Nanotubes are tiny tubes about 10,000 times thinner than a human hair. They consist of rolled-up sheets of carbon hexagons and have potential uses as minuscule wires or in ultrasmall digital devices and are very powerful conductors of electrical current.*

Bionic has two high-performance cores and four high-efficiency cores. Apple claims the A16 is about 40 percent faster than the competition, and its efficiency cores use one-third of the power compared to other efficiency cores on the market.

## 5-3 Describe the major types of computer software used in business and major software trends.

To use computer hardware, you need software, which provides the detailed instructions that direct the computer's work. Examine Figure 5.6, which illustrates the relationship among system software, application software, and users as a set of nested boxes, each of which must interact closely with the boxes surrounding it. The system software (which includes operating system software, language translators, and utility programs) surrounds and controls access to the hardware. Application software (such as software packages and services, apps, and the programming languages used to create them) must work via the system software to operate. End users work primarily with application software. Each type of software must be designed for a specific machine in order to ensure their compatibility. We examine each of these types of software in more detail in the following sections.

### OPERATING SYSTEM SOFTWARE

The system software that manages and controls a computer's activities is called the **operating system**. Other system software consists of computer language translation programs that convert programming languages into machine language that can be understood by the computer and utility programs that perform common processing tasks, such as copying, sorting, or computing a square root. The operating system allocates and assigns system resources, schedules the use of computer resources and computer jobs, and monitors computer system activities. The operating system also provides locations in primary memory for data and programs and controls the input and output devices, such as printers, displays, and telecommunication links. The operating system also coordinates the scheduling of work in various areas of the computer so that different parts of different jobs can be worked on simultaneously.

#### PC, Server, and Mobile Operating Systems

The operating system controls the way users interact with the computer. Contemporary PC operating systems and application software use a **graphical user interface**, which makes extensive use of icons, buttons, bars, and boxes to perform tasks.

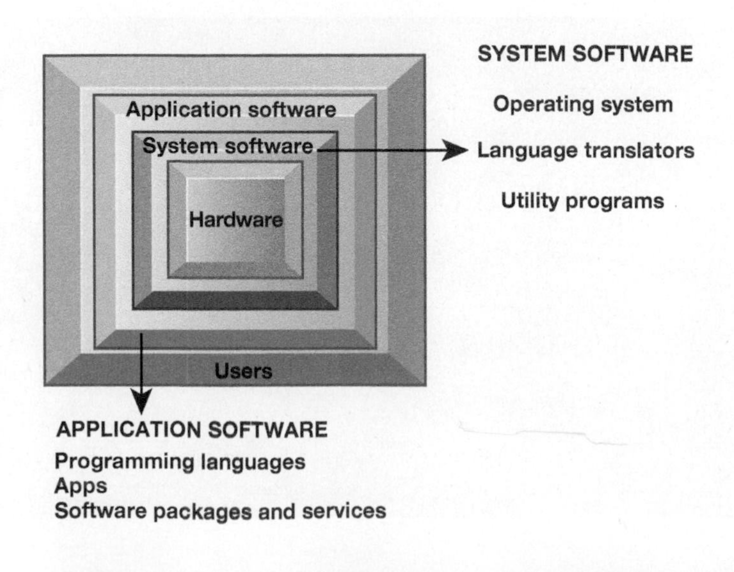


**Figure 5.6**
The Major Types of Software
*The relationship among the system software, application software, and users can be illustrated by a series of nested boxes. System software—consisting of operating systems, language translators, and utility programs—controls access to the hardware. Application software, including programming languages and software packages, must work via the system software to operate. The user interacts primarily with the application software.*

Conventional client operating system software was based on the use of a mouse and a keyboard, but this software is becoming more natural and intuitive by using **multitouch** technology. The multitouch interface on smartphones, tablet computers, and some PC models allows you to use one or more fingers to perform special gestures to manipulate lists or objects on a screen without using a mouse or a keyboard.

Table 5.3 compares leading PC and server operating systems. These include the Windows family of operating systems (Windows 10, Windows 11, and Windows Server), UNIX, Linux, and the operating system for the Macintosh computer.

The Microsoft Windows family of operating systems has both client and server versions and a streamlined graphical user interface that works with touch screens and mobile devices as well as with keyboards and traditional PCs. At the client level, most PCs use some form of the Microsoft Windows or Apple operating systems. The latest Windows client version is **Windows 11**.

Windows operating systems for servers provide network management functions, including support for virtualization and cloud computing. Windows Server has multiple versions for small, medium, and large businesses.

Today there is a much greater variety of operating systems than in the past, with new operating systems for cloud-connected computers and mobile devices. Google's **Chrome OS** provides a lightweight cloud-based operating system for Google's Chromebook laptop and tablet computers. Programs are not stored on the user's computer but are used over the Internet and accessed through the Chrome web browser. User data reside on cloud servers. **Android** is an open-source operating system for mobile devices such as smartphones and tablet computers developed by the Open Handset Alliance led by Google. It has become the most popular smartphone platform worldwide, competing with **iOS**, Apple's mobile operating system for the iPhone, iPad, and Apple Watch.

**UNIX** is a multiuser, multitasking operating system developed by Bell Laboratories in 1969 to connect various machines and is highly supportive of communications and networking. UNIX can run on many kinds of computers and can be easily customized. Application programs that run under UNIX can be ported from one computer to a different computer with little modification. In the past UNIX was frequently used on workstations, servers, and for running enterprise systems, but today, it has been largely supplanted by Linux.

**Linux** is a UNIX-like operating system that can be downloaded from the Internet free of charge or purchased for a small fee from companies that provide additional tools for the software. It is free, reliable, compactly designed, robust, and capable of running on many hardware platforms, including servers, mobile devices, medical

**TABLE 5.3**

Leading PC and Server Operating Systems

| Operating System | Features |
|---|---|
| Windows 11 | Most recent Windows client operating system, with a streamlined user interface, enhanced security, and the ability to integrate Android apps. |
| Windows Server | Windows operating system for servers. |
| UNIX | Used for PCs, workstations, and network servers. Supports multitasking, multiuser processing, and networking. Is portable to different models of computer hardware. |
| Linux | Open-source, reliable alternative to UNIX and Windows operating systems that runs on many types of computer hardware and can be modified by software developers. |
| macOS | Operating system for the Macintosh computer that is highly visual and user-friendly. Most recent version is macOS Monterey. |

equipment, and consumer electronics. Although Linux is not widely used for desktop systems, it is used extensively for servers, networks, and enterprise-level tasks, and, as previously noted, has largely supplanted UNIX. One hundred percent of the world's top 500 supercomputers run on Linux, as do more than 96 percent of the world's top one million servers and 23 of the world's top 25 websites. Ninety percent of the world's cloud computing infrastructure runs on Linux (Germain, 2021). The major hardware and software vendors offer versions of their products that can run on Linux.

## APPLICATION SOFTWARE AND DESKTOP PRODUCTIVITY TOOLS

Today, businesses have access to an array of tools for developing their application software. These include traditional programming languages, application software packages, and desktop productivity tools; software for developing Internet applications; and software for enterprise integration. It is important to know which software tools and programming languages are appropriate for the work your business wants to accomplish.

### Programming Languages for Business

Popular programming languages for business applications include C, C++, Visual Basic, and Java. **C** is a powerful and efficient language developed in the early 1970s that combines machine portability with rigid control and efficient use of computer resources. It is used primarily by professional programmers to create operating systems and application software, especially for PCs. **C++** is a newer version of C that has all the capabilities of C plus additional features for working with software objects. Unlike traditional programs, which separate data from the actions to be taken on the data, a software **object** combines data and procedures. **Visual Basic** is a widely used visual programming tool and environment for creating applications that run on Microsoft Windows operating systems. A **visual programming language** allows users to manipulate graphic or iconic elements to create programs. COBOL (COmmon Business Oriented Language) was developed in the early 1960s for business processing and can still be found in large legacy systems in banking, insurance, and retail.

**Java** has become a leading interactive programming environment for the web. The Java platform has migrated into mobile devices, automobiles, and game machines. Java software is designed to run on any computer or computing device, regardless of the specific microprocessor or operating system the device uses. For each of the computing environments in which Java is used, a Java Virtual Machine interprets Java programming code for that machine. In this manner, the code is written once and can be used on any machine for which there exists a Java Virtual Machine.

Other popular programming tools for web applications include Python and Ruby on Rails. Python is known for its flexibility in scaling web applications, its ability to run on most platforms, and its ease of use. Ruby on Rails provides a framework with tools for building feature-rich websites.

### Software Packages and Desktop Productivity Tools

Much of the software used in businesses today is not custom programmed but consists of application software packages and desktop productivity tools. A **software package** is a prewritten, precoded, commercially available set of programs that eliminates the need for individuals or organizations to write their own software programs for certain functions. Software packages that run on mainframes and larger computers usually require professional programmers for their installation and support, but desktop productivity software packages for consumer users can easily be installed and run by the users themselves. Table 5.4 describes the major desktop productivity software tools.

**TABLE 5.4**

Desktop Productivity Software

| Software Tool | Capabilities | Example |
|---|---|---|
| Word processing | Allows the user to make changes in a digital document, with various formatting options. | Microsoft Word<br>Apple Pages |
| Spreadsheet | Organizes data into a grid of columns and rows. When the user changes a value or values, all other related values on the spreadsheet are automatically recalculated. Used for modeling and what-if analysis and can also present numeric data graphically. | Microsoft Excel<br>Apple Numbers |
| Data management | Creates files and databases in which users can store, manipulate, and retrieve related data. Suitable for building small information systems. | Microsoft Access |
| Presentation graphics | Creates professional-quality digital graphics presentations and slide shows; can include multimedia displays of sound, animation, photos, and video clips. | Microsoft PowerPoint<br>Apple Keynote |
| Personal information management | Creates and maintains appointments, calendars, to-do lists, and business contact information; also used for email. | Microsoft Outlook |
| Collaboration and teamwork | Supports sharing documents and ideas, virtual meetings, messaging. | Microsoft Teams<br>Google Drive |

**Software Suites** The major desktop productivity tools are bundled together as a software suite. Microsoft Office is an example. Core Office tools include Microsoft Word **word processing software**, Excel **spreadsheet software**, Access database software, PowerPoint **presentation graphics** software, and Outlook, a set of tools for email, scheduling, and contact management. Microsoft now offers a hosted cloud version of its productivity and collaboration tools as a subscription service called **Microsoft 365**. Competing with Microsoft Office and Microsoft 365 are low-cost office productivity suites such as the open-source LibreOffice (downloadable free over the Internet) and cloud-based Google Docs (see Chapter 2).

**Web Browsers** A **web browser** is a type of easy-to-use software that enables users to access the web and other Internet resources. A web browser displays web pages that can include graphics, audio, and video as well as traditional text, and it allows users to click (or touch) on-screen buttons or highlighted words to link to related websites. Web browsers have become the primary interface for accessing the Internet and for using networked systems based on Internet technology. The leading web browsers today are Google Chrome, Microsoft Edge, Mozilla Firefox, and Apple Safari.

## HTML AND HTML5

**Hypertext Markup Language (HTML)** is a page description language for specifying how text, graphics, video, and sound should be placed on a web page and for creating dynamic links to other web pages and objects. Using these links, a user need only point at a highlighted keyword or graphic, click on it, and immediately be transported to another web page. Table 5.5 illustrates some sample HTML statements.

HTML programs can be custom written, but they also can be created using the HTML-authoring capabilities of web browsers or of popular word processing, spreadsheet, and data management software. HTML editors, such as Atom and web design software such as Adobe Dreamweaver, are more powerful HTML-authoring programs for creating web pages.

**TABLE 5.5**
Examples of HTML

| Plain English | HTML |
| --- | --- |
| Subcompact | <TITLE>Subcompact</TITLE> |
| 4 passenger | <LI>4 passenger</LI> |
| $16,800 | <LI>$16,800</LI> |

HTML was originally designed to create and link static documents composed largely of text. Today, however, the web is much more social and interactive, and many web pages have multimedia elements: images, audio, and video. In the past, third-party plug-in applications were required to integrate these rich media with web pages. However, these add-ons required additional programming. The most recent version of HTML, called **HTML5**, solves this problem by making it possible to embed images, audio, video, and other elements directly into a document without processor-intensive add-ons. HTML5 also makes it easier for web pages to function across different display devices, including mobile devices as well as desktops. Web pages execute more quickly, and web-based mobile apps work like web pages.

**JavaScript** is another core technology for making web pages more interactive. JavaScript can be used to instruct the computer on how to interact with the user upon receiving user input. The vast majority of websites use JavaScript for client-side page behavior, and all major web browsers have a built-in JavaScript engine to execute it very quickly.

## WEB SERVICES

**Web services** refer to a set of loosely coupled software components that exchange information with each other using universal web communication standards and languages. They can exchange data between two systems regardless of the operating systems or programming languages on which the systems are based. They can be used to build open-standard, web-based applications linking systems of different organizations, and they can also be used to create applications that link disparate systems within a single company. Different applications can use them to communicate with each other in a standard way without time-consuming custom coding.

The foundation technology for web services is **XML**, which stands for **Extensible Markup Language**. This language was developed in 1996 by the World Wide Web Consortium (W3C, the international body that oversees the development of the web) as a more powerful and flexible markup language than HTML for web pages. Whereas HTML is limited to describing how data should be presented in the form of web pages, XML can handle presentation, communication, and storage of data. In XML, a number is not simply a number: The XML tag specifies whether the number represents a price, a date, or a zip code. Table 5.6 illustrates some sample XML statements.

By tagging selected elements of the content of documents for their meanings, XML makes it possible for computers to manipulate and interpret their data automatically and perform operations on the data without human intervention. XML provides a standard format for data exchange, enabling web services to pass data from one process to another.

**TABLE 5.6**
Examples of XML

| Plain English | XML |
| --- | --- |
| Subcompact | <AUTOMOBILETYPE="Subcompact"> |
| 4 passenger | <PASSENGERUNIT="PASS">4</PASSENGER> |
| $16,800 | <PRICE CURRENCY="USD">$16,800</PRICE> |

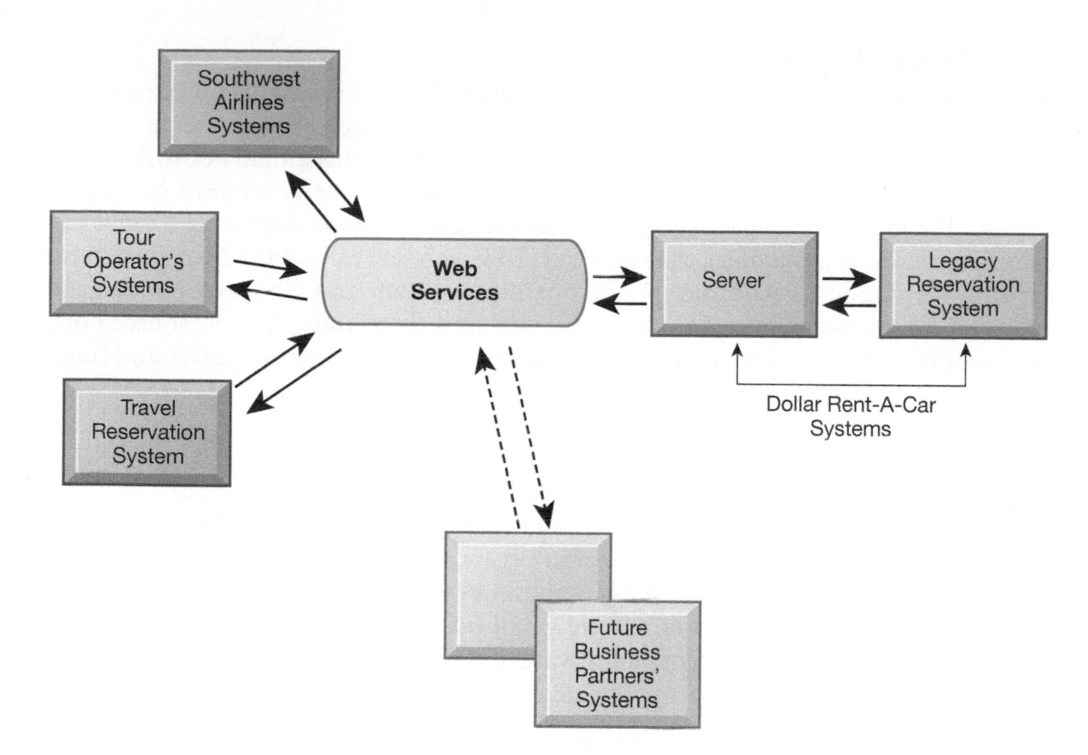


**Figure 5.7**
**How Dollar Rent-A-Car Uses Web Services**
*Dollar Rent-A-Car uses web services to provide a standard intermediate layer of software to talk to other companies' information systems. Dollar Rent-A-Car can use this set of web services to link to other companies' information systems without having to build a separate link to each firm's system.*

Web services communicate through XML messages over standard web protocols that allow a software application to connect freely to other applications without custom programming for each application with which it wants to communicate. Everyone shares the same standards.

The collection of web services that are used to build a firm's software systems constitutes what is known as a service-oriented architecture. A **service-oriented architecture (SOA)** is a set of self-contained services that communicate with each other to create a working software application. Software developers reuse these services in other combinations to assemble other applications as needed. Virtually all major software vendors, such as IBM, Microsoft, Oracle, and HP, provide tools and entire platforms for building and integrating software applications using web services.

Examine Figure 5.7, which illustrates how Dollar Rent-A-Car's systems use web services to link its online booking system with other companies' websites, such as the website of Southwest Airlines. Although both companies' systems are based on different technology platforms, a person booking a flight on Southwest.com can reserve a car from Dollar without leaving the airline's website. Instead of struggling to get Dollar's reservation system to share data with Southwest's information systems, Dollar used Microsoft.NET web services technology as an intermediary. Reservations from Southwest are translated into web services protocols, which are then translated into formats that Dollar's computers can understand.

Other car rental companies have linked their information systems to airline companies' websites before, but without web services, these connections had to be built one at a time. In contrast, web services provide a standard way for Dollar's computers to talk to other companies' information systems without having to build special links to each one. Dollar is expanding its use of web services to link directly to the systems of a small tour operator and a large travel reservation system. It does not have to write new software code for each new partner's information systems.

## SOFTWARE TRENDS

Today there are many more sources for obtaining software and many more capabilities for users to create their own customized software applications. Expanding use of open-source software and cloud-based software tools and services exemplify this trend.

## Open-Source Software

**Open-source software** provides all computer users with free access to its program code so that they can modify the code to fix errors or to make improvements. Open-source software is not owned by any company or individual. Instead, a global network of programmers and users manages and modifies the software, usually without being paid to do so. Popular open-source software tools include the Linux operating system (described earlier in this chapter), the Apache HTTP web server, and the Mozilla Firefox web browser. Google's Android mobile operating system and Chrome web browser are based on open-source code. Today you can find thousands of open-source computer programs to accomplish everything from e-commerce shopping carts and funds clearance to sales force management.

## Cloud-Based Software Services and Tools

In the past, software such as Microsoft Word or Adobe Acrobat came in a box and was designed to operate on a single computer. Today, you're more likely to download the software from the vendor's website or use the software as a cloud service delivered over the Internet. (Review our discussion of SaaS in Section 5-2.) In addition to the free or low-cost tools for individuals and small businesses, such as various software tools provided by Google, enterprise software and other software for larger businesses are available as cloud services from the major commercial software vendors.

**Apps and Mashups** **Apps** are software programs that can be accessed via the Internet. Google refers to its online services as apps, but when we talk about apps today, most of the attention goes to the apps that have been developed for the mobile platform. It is these apps that turn smartphones and other mobile devices into general-purpose computing tools. There are now millions of apps for the iOS and Android operating systems.

Some apps, once downloaded, do not access the web, but many do, providing a streamlined, nonbrowser pathway that enables faster access to online content and services than traditional web browsers. Today, the average amount of time US adults spend per day using mobile apps (more than three and a quarter hours) far outweighs the time they spend using mobile browsers (less than one hour). As a result, over the last several years, many companies have redesigned their websites to mimic the look, feel, and functionality of apps.

Many apps are free or available for a small fee, one that is much less than the cost of conventional software, which further adds to their appeal. There are currently almost 2 million apps for devices that use Apple's iOS operating system and a similar number that run on devices using the Android operating system. The success of these two competing mobile platforms depends in large part on the quantity and the quality of the apps available for each platform. However, apps tie a customer to a specific platform: Apps that use the iOS operating system are not compatible with Android devices, and apps that use the Android system cannot run on Apple devices. As users add more and more apps to their mobile devices, the cost of switching to a competing mobile platform rises.

The most frequently downloaded categories of consumer apps include games, news and weather, maps/navigation, social media, music, video/movies, as well as apps that enable the researching and buying of goods and services online. There are also apps for business users that make it possible to create and edit documents, connect to corporate systems, schedule and participate in meetings, track shipments, and dictate voice messages.

A **mashup** is a software application that integrates functionality and data from multiple applications to create a new, customized application. The idea behind mashups is the creation from different sources of a new work that is greater than the sum of its parts. You have created a mashup if you've ever personalized your Facebook profile by uploading photos stored on a cloud service or added videos or photos to a blog.

Web mashups combine the capabilities of two or more online applications to create a kind of hybrid that provides more customer value than the original sources provide alone. For instance, ZipRealty uses Google Maps and data provided by online real estate database Zillow.com to display a complete list of multiple listing service (MLS) real estate listings for any zip code the user specifies.

## 5-4 Identify the principal issues in managing hardware and software technology.

Selection and use of computer hardware and software technology have a profound impact on business performance. We now describe the most important issues you will face when managing hardware and software technology: engaging in capacity planning and scalability; determining the total cost of technology assets; determining whether to own and maintain your own hardware, software, and other infrastructure components or to lease or license them from an external technology service provider; and managing mobile platforms and software localization.

### CAPACITY PLANNING AND SCALABILITY

Compared to traditional commerce and business, e-commerce and e-business need much larger processing and storage resources to handle the volume of digital transactions flowing among different parts of the firm and among the firm and its customers and suppliers. Many people simultaneously using a website places strain on a computer system, as does hosting large numbers of interactive web pages with data-intensive graphics or video. Many such firms have turned to cloud services to deal with these demands.

Managers and information systems specialists now need to pay more attention to hardware capacity planning and scalability than before. From an IT perspective, **capacity planning** is the process of predicting when a computer hardware system will become saturated. It considers factors such as the maximum number of users that the system can accommodate at one time, the impact of existing and future software applications, and performance measures, such as minimum response time for processing business transactions. Capacity planning ensures that the firm has enough computing power for its current and future needs either on premises or in terms of the cloud resources that it employs. For example, the NASDAQ stock market performs ongoing capacity planning to identify peaks in the volume of stock trading transactions and to ensure that it has enough computing capacity to handle large surges in volume when trading is very heavy.

**Scalability** refers to the ability of a computer, product, or system to expand to serve a large number of users without breaking down. E-commerce and e-business both call for scalable IT infrastructures that have the capacity to grow with the business as the complexity of the website and the number of visitors increase. Organizations must make sure that they have sufficient computer processing, storage, and network resources to handle surges in digital transactions and to make such data immediately available online.

### TOTAL COST OF OWNERSHIP (TCO) OF TECHNOLOGY ASSETS

When you calculate how much your hardware and software cost, their purchase price is only the beginning. You must also consider ongoing administration costs for hardware and software upgrades, maintenance, technical support, training, and even utility and real estate costs for running and housing the technology. The **total cost of ownership (TCO)** model can be used to analyze these direct and indirect costs to help

**TABLE 5.7**

TCO Components

| | |
|---|---|
| Hardware acquisition | Purchase or licensing price of computer hardware equipment, including computers, displays, storage, and printers |
| Software acquisition | Purchase or license of software for each user |
| Installation | Cost to install computers and software or migrate to the cloud |
| Training | Cost to provide training to information systems specialists and end users |
| Support | Cost to provide ongoing technical support, help desks, and so forth |
| Maintenance | Cost to upgrade the hardware and software on premises or in the cloud |
| Infrastructure | Cost to acquire, maintain, and support related infrastructure, such as networks and specialized equipment (including storage backup units) |
| Downtime | Lost productivity if hardware or software failures cause the system to be unavailable for processing and user tasks |
| Space and energy | Real estate and utility costs for housing and providing power for the technology |

determine the actual cost of owning a specific technology. Table 5.7 describes the most important TCO components to consider in a TCO analysis, including TCO for cloud infrastructure.

When all these cost components are considered, the hidden costs for support staff, downtime, and additional network management could make distributed client/server architectures—especially those incorporating mobile devices—more expensive than centralized mainframe architectures.

Many large firms are saddled with redundant, incompatible hardware and software because of poor planning. These firms could reduce their TCO through greater centralization and standardization of their hardware and software resources. Companies could also reduce the size of the information systems staff required to support their infrastructure by minimizing the number of computer models and pieces of software that employees are allowed to use.

## USING TECHNOLOGY SERVICE PROVIDERS

Some of the most important questions facing managers are: "How should we acquire and maintain our technology assets?" "Should we build software applications ourselves or outsource them to an external contractor?" "Should we purchase and run them ourselves or rent them from external service providers?" In the past, most companies ran their own computer facilities and developed their own software. Today, however, more and more companies are obtaining their hardware and software technology from external service vendors.

### Outsourcing

A number of firms are **outsourcing** to external vendors the maintenance of their IT infrastructures and the development of new systems. They may contract with an external service provider to run their computer center and networks, to develop new software, or to manage all the components of their IT infrastructures. For example, FedEx outsourced 30 percent of its IT system operations and software development to external IT service providers.

Specialized web hosting services are available for companies that lack the financial or technical resources to operate their own websites. A **web hosting service** maintains a large web server, or a series of servers, and provides fee-paying subscribers with space to maintain their websites. The subscribing companies may create their own web pages or have the hosting service, or a web design firm, create them. Some services offer *colocation*, in which the firm owns the server computer housing its website but locates the server in the physical facility of the hosting service.

Firms often retain control over their hardware resources but outsource custom software development or maintenance to outside firms, frequently firms that operate offshore in low-wage areas of the world. When firms outsource software work outside their national borders, the practice is called **offshore software outsourcing**. Until recently, this type of software development involved lower-level maintenance, data entry, and call center operations, but with the growing sophistication and experience of offshore firms, particularly in India, more and more new program development is taking place offshore. Chapter 12 discusses offshore software outsourcing in more detail.

To manage their relationship with an outsourcer or technology service provider, firms will need a contract that includes a **service level agreement (SLA)**. The SLA is a formal contract between customers and their service providers that defines the specific responsibilities of the service provider and the level of service the customer expects. SLAs typically specify the nature and level of services provided, criteria for performance measurement, support options, provisions for security and disaster recovery, hardware and software ownership and upgrades, customer support, billing, and conditions for terminating the agreement.

### Choosing Cloud Services

Firms now have the option of maintaining their own IT infrastructures or using cloud-based hardware and software services and are steadily gravitating toward cloud-based services and tools. Companies considering the cloud computing model need to identify and assess the costs and benefits of external services carefully, weighing all management, organizational, and technology issues. These issues include the level of service and performance that is acceptable for the business and the expertise required to deploy and manage a new cloud-based IT infrastructure. New cloud users often overlook additional costs for networking, data transfer and storage, and the work required to upgrade or replace business application systems to run in a cloud environment. After moving workloads from on-premises to the cloud, a firm might find computing costs higher than originally estimated.

## MANAGING MOBILE PLATFORMS

Gains in productivity from equipping employees with mobile devices must be balanced against increased costs from integrating these devices into the firm's IT infrastructure and providing technical support. This is especially true when the organization allows employees to use their own personal devices for their jobs (BYOD).

For personal mobile devices to be able to access company information, the company's networks must be configured to receive connections from those devices. Firms need an efficient inventory management system that keeps track of which devices employees are using, where each device is, and what software is installed on it. They also need to know what pieces of corporate data are on those personal devices, and this is not always easy to determine. It is more difficult to protect the company's network and data when employees access them from their privately owned devices.

If a device is stolen or compromised, companies need to ensure that sensitive or confidential company information isn't exposed. Companies often use technologies that allow them to wipe data from devices remotely or encrypt data so that, if stolen, they cannot be used. You'll find out more about mobile security issues in Chapter 8.

Many companies allow employee mobile devices access only to a limited set of applications and noncritical corporate data. For more critical business systems, more company control is required, and firms often turn to **mobile device management (MDM)** software, which monitors, manages, and secures mobile devices that are deployed across the enterprise. It can be employed for multiple mobile service providers and across multiple mobile operating systems used in the organization. MDM tools also enable the IT department to monitor mobile usage, install or update mobile software, back up and restore mobile devices, and remove software and data from devices that are stolen or lost.

### MANAGING SOFTWARE LOCALIZATION FOR GLOBAL BUSINESS

If you are operating a global company, all the management issues we have just described will be affected by the need to create systems that can be realistically used by multiple business units in different countries. Although English has become a kind of standard business language, this is truer at higher levels of companies than at the middle and lower levels. Software may have to be built with local language interfaces before a new information system can be successfully implemented worldwide.

These interfaces can be costly and messy to build. Menu bars, commands, error messages, reports, queries, online data entry forms, and system documentation may need to be translated into all the languages of the countries where the system will be used. To be truly useful for enhancing the productivity of a global workforce, the software interfaces must be easily understood and mastered quickly. The entire process of converting software to operate in a second language is called **software localization**.

Global systems must also consider differences in local cultures and business processes. Cross-functional systems such as enterprise and supply chain management systems are not always compatible with differences in languages, cultural heritages, and business processes in other countries. In a global systems environment, all these factors add to the TCO and will influence decisions about whether to outsource or use technology service providers.

## 5-5 Understand how MIS can help your career.

Here is how this chapter and this book can help you find a job as a coordinating product manager at an online streaming service platform.

### THE COMPANY

Intertainment Direct is a subscription-based streaming entertainment service and production company based in Los Angeles. It offers a film and television series library through distribution deals as well as its own productions. A great deal of this content is maintained and delivered using cloud computing services.

### POSITION DESCRIPTION

One of the primary levers for growth is developing new subscription plans, with differentiated features and price points so that consumers can choose what is right for them. The product manager will help the company with designing, strategizing, executing, and prioritizing projects. The product manager will also be expected to build consensus and alignment with cross-functional experts to help push projects forward.

This job will consist of building a strategic roadmap; taking product concepts from idea through execution; writing strategy memos and presentations to represent your team's work to senior executives; and developing business cases to measure the impact of your product innovation with data science colleagues. Job responsibilities include:

- Developing, evaluating, launching, testing, and making iterations of potential new features for plan tiers
- Collaborating closely with the pricing team, integrating revenue objectives into plan development
- Interfacing daily with cross-functional experts
- Attaining consistent metrics and goals

### JOB REQUIREMENTS

- Bachelor's degree
- Strong interest in product, pricing, and behavioral economics
- Some product management experience

- Ability to quickly build credibility and trust with internal and external stakeholders
- Exceptional communication, interpersonal, and problem-solving skills
- Ability to use data to inform priorities and evaluate impact of initiatives
- Familiarity with cloud computing technology and services

### INTERVIEW QUESTIONS

1. What do you know about this company and the way it acquires and retains customers? What do you know about its technology for user personalization?
2. Do you have any work experience in product management? What exactly did you do?
3. Do you have any work experience using cloud computing services? What did you do with these services?
4. What experience do you have working with cross-functional teams? Can you describe the work you did with these teams?
5. Can you give an example of a business problem you solved and how you solved it?

### AUTHOR TIPS

1. Review this chapter and also Chapter 10 of this text, specifically the sections on cloud computing, software as a service (SaaS), and e-commerce marketing and business and revenue models.
2. Use the web to find out more about the company, its business strategy, and its subscription plans.
3. Review the company's social media profiles on LinkedIn, Twitter, and Facebook. What are the consistent themes across these social media channels focused on? Be prepared to discuss the kinds of business challenges and opportunities facing this company.
4. Learn what you can about the responsibilities outlined for this job. Indicate that you are very interested in learning more about how the company plans to capture new growth by developing new subscription plans with different features and price points.
5. Bring samples of your writing that demonstrate your analytical and problem-solving skills.

## Review Summary

5-1 **Identify the components of IT infrastructure.** IT infrastructure consists of the shared technology resources that provide the platform for the firm's specific information system applications. Major IT infrastructure components include computer hardware, computer software, data management technology, telecommunications and networking technology, and technology services.

5-2 **Describe the major computer hardware, data storage, input, and output technologies used in business and major hardware trends.** Computers come in an array of sizes with differing capabilities for processing information. PCs are desktop or laptop computers; workstations are desktop computers with powerful mathematical and graphic capabilities; server computers are specifically optimized to support a computer network or website; mainframes are large-capacity, high-performance computers; and supercomputers are sophisticated, powerful computers that can perform numerous and complex computations rapidly. Computing power can be further increased by creating a computational grid that combines the computing power of all the computers on a network. In the client/server model of computing, computer processing is split between clients and servers connected by a network. The exact division of tasks between client and server depends on the application.

Secondary storage technologies include magnetic disk, optical drives, solid state drives, USB flash drives, magnetic tape, and storage networks. Cloud-based storage services are increasingly being used for personal and corporate data. The principal input devices are keyboards, computer mice, touch screens (including those with multitouch), magnetic ink and optical character recognition devices, pen-based instruments, digital scanners, audio input devices, and sensors. The principal output devices are display screens, printers, and audio output devices.

Major hardware trends include the mobile platform, consumerization of IT and BYOD, virtualization, cloud computing, edge computing, green computing, high-performance computing (quantum computing and nanotechnology), and high-performance/power-saving processors. Cloud computing provides computer processing, storage, software, and other services as virtualized resources over a network, primarily the Internet, on an as-needed basis.

5-3 **Describe the major types of computer software used in business and major software trends.** The two major types of software are system software and application software. System software coordinates the various parts of the computer system and mediates between application software and computer hardware. Application software is used to develop specific business applications.

The system software that manages and controls the activities of the computer is called the operating system. Leading PC and server operating systems include Windows 10, Windows 11, Windows Server, UNIX, Linux, and the Macintosh operating system. Google's Chrome provides a lightweight, cloud-based operating system for its Chromebook laptop and tablet computers. Android and iOS are leading operating systems for mobile devices.

The principal programming languages used in business application software include C, C++, Visual Basic, and Java. Java is an operating system-independent and hardware-independent programming language that is a leading interactive programming environment for the web. Python and Ruby on Rails are used in web and cloud computing applications. PC and cloud-based productivity tools include word processing, spreadsheet, data management, presentation graphics, and web browser software. HTML is a page description language for creating web pages, and HTML5 and JavaScript are used to make web pages more interactive.

Web services are loosely coupled software components based on XML and open web standards that can work with any application software and operating system. They can be used as components of applications to link the systems of two organizations or to link disparate systems of a single company.

Software trends include the expanding use of open-source software and cloud-based software tools and services (including SaaS, apps, and mashups).

5-4 **Identify the principal issues in managing hardware and software technology.** Managers and information systems specialists need to pay special attention to hardware capacity planning and scalability to ensure that the firm has enough computing power for its current and future needs. Businesses also need to balance the costs and benefits of building and maintaining their own hardware and software versus outsourcing or using an on-demand cloud computing model. The total cost of ownership (TCO) of the organization's technology assets includes not only the original cost of computer hardware and software but also the costs for hardware and software upgrades, maintenance, technical support, and training, including the costs for managing and maintaining mobile devices. Companies with global operations need to manage software localization.

## Key Terms

Android, 175
Application server, 162
Application software, 159
Apps, 180
BYOD, 166
C, 176
C++, 176
Capacity planning, 181
Centralized processing, 162
Chrome OS, 175
Client, 162
Client/server computing, 162
Cloud computing, 167
Consumerization of IT, 166
Data center, 158
Data management software, 160
Distributed processing, 162
Edge computing, 170
Extensible Markup Language (XML), 178
Graphical user interface, 174
Green computing (green IT), 170
Grid computing, 161
HTML5, 178
Hybrid cloud, 170
Hypertext Markup Language (HTML), 177
Input devices, 163
iOS, 175
Java, 176
JavaScript, 178
Legacy systems, 160
Linux, 175
Mainframe, 161
Mashups, 180
Microsoft 365, 177
Mobile device management (MDM), 183
Multicore processor, 173
Multitouch, 175
Nanotechnology, 173
N-tier client/server architectures, 162
Object, 176
Offshore software outsourcing, 183
On-demand computing, 169
Open-source software, 180
Operating system, 174
Output devices, 163
Outsourcing, 182
Personal computer (PC), 161
Presentation graphics, 177
Private cloud, 169
Public cloud, 169
Quantum computing, 172
Scalability, 181
Server, 161
Service level agreement (SLA), 183
Service-oriented architecture (SOA), 179
Software as a service (SaaS), 168
Software-defined storage (SDS), 167
Software localization, 184
Software package, 176
Solid state drive (SSD), 163
Spreadsheet software, 177
Supercomputer, 161
System software, 159
Tablet computer, 164
Total cost of ownership (TCO), 181
UNIX, 175
Virtualization, 166
Visual Basic, 176
Visual programming language, 176
Web browsers, 177
Web hosting service, 182
Web server, 162
Web services, 178
Windows 11, 175
Word processing software, 177
Workstation, 161

## Review Questions

**5-1** Identify the components of IT infrastructure.

- Define information technology (IT) infrastructure, and describe each of its components.

**5-2** Describe the major computer hardware, data storage, input, and output technologies used in business and major hardware trends.

- List and describe the various types of computers available to businesses today.
- Define the client/server model of computing, and describe the difference between two-tiered and N-tier client/server architecture.
- Define and describe the mobile platform, BYOD, grid computing, cloud computing, edge computing, virtualization, green computing, nanotechnology, and multicore processing.

**5-3** Describe the major types of computer software used in business and major software trends.

- Distinguish between application software and system software, and explain the role the operating system of a computer plays.
- List and describe the major PC, server, mobile, and cloud operating systems.
- Name and describe the major desktop productivity software tools.
- Explain how Java, JavaScript, HTML, and HTML5 are used in building applications for the web.
- Define web services, describe the technologies they use, and explain how web services benefit businesses.

- Explain why open-source software is so important today and what its benefits for business are.
- Define and describe cloud computing software services, mashups, and apps, and explain how they benefit individuals and businesses.

**5-4** Identify the principal issues in managing hardware and software technology.
- Explain why managers need to pay attention to capacity planning and scalability of technology resources.
- Describe the cost components used to calculate the TCO of technology assets.
- Identify the benefits and challenges of using outsourcing, cloud computing services, and mobile platforms.
- Explain why software localization has become an important management issue for global companies.

**MyLab MIS**
To complete these problems, go to EOC Discussion Questions in **MyLab MIS**.

## Discussion Questions

**5-5** MyLab MIS Why is selecting computer hardware and software for the organization an important business decision? What people, organization, and technology issues should be considered when selecting computer hardware and software?

**5-6** MyLab MIS Should organizations use software service providers (including cloud services) for all their software needs? Why or why not? What people, organization, and technology factors should be considered when making this decision?

**5-7** MyLab MIS What are the advantages and disadvantages of cloud computing?

## Hands-On MIS Projects

The projects in this section give you hands-on experience in developing solutions for managing IT infrastructures, using spreadsheet software to evaluate the total cost of ownership (TCO) of alternative desktop systems, and using web research to budget for a sales conference. Visit MyLab MIS to access this chapter's Hands-On MIS Projects.

### MANAGEMENT DECISION PROBLEM

**5-8** EasyJet is the second-largest airline in the United Kingdom, operating more than 1,100 routes across more than 35 countries. A reliable and robust system for booking and managing reservations while keeping costs low is a key business requirement. EasyJet's legacy reservation system lacked the capability for customers to select their seats on a given flight when they made their reservations online. Adding this new service would require investing in an additional data center and software and modifying EasyJet's IT infrastructure. How could cloud computing services provide a solution? What people, organization, and technology factors and management decisions need to be considered when designing the solution?

### IMPROVING DECISION MAKING: USING A SPREADSHEET TO EVALUATE TOTAL COST OF OWNERSHIP (TCO) FOR PC DESKTOP SYSTEMS

Software skills: Spreadsheet formulas, SUM, VLOOKUP
Business skills: Technology pricing

**5-9** This project provides an opportunity to determine the total cost of ownership (TCO) of investing in new PC desktop systems over a three-year period. You'll use spreadsheet software to compare the total three-year cost of purchasing and maintaining new PC desktop systems with Windows 11 Pro and Microsoft 365 Business for 40 employees. You can find a list of specifications and cost components in the files EMIS15Chap5 Specifications and EMIS15Chap5 Question File in MyLab MIS.

Use your spreadsheet software to calculate the total cost of ownership (TCO) for investing in 40 new desktop productivity systems over a three-year period. In MyLab MIS you will also find on the spreadsheet file a price sheet with pricing alternatives for desktop all-in-one PC systems from five different vendors. It will also show the model and price of a selected printer that can be paired with each of the PC models. The cost of Windows 11 Pro is included in the desktop price. Position your price sheet in the lower-left portion of your worksheet, where it can serve as a lookup table. Use the same worksheet to calculate TCO costs. Management wants to compare the TCO of the HP and Dell desktop systems. Use VLOOKUP to select the pricing for each alternative.

Complete the worksheet by entering and calculating the values for the remaining one-time and recurring costs. Calculate the TCO over a three-year period and the TCO per user.

What other factors should the company consider besides cost in determining the PC system option to select? Explain your answer.

## IMPROVING DECISION MAKING: USING WEB RESEARCH TO BUDGET FOR A SALES CONFERENCE

Software skills: Internet-based software
Business skills: Researching transportation and lodging costs

**5-10** In this exercise, you'll use software at various online travel sites to obtain pricing for total travel and lodging costs for a sales conference.

The Foremost Composite Materials Company is planning a two-day sales conference on October 19–20, starting with a reception on the evening of October 18. The conference consists of all-day meetings that the entire sales force, numbering 120 sales representatives and their 16 managers, must attend. Each sales representative requires a separate room, and the company needs two common meeting rooms, one large enough to hold the entire sales force plus a few visitors (200 total) and the other able to hold half the sales force. Management has set a budget of $300,000 for the representatives' room rentals. The company would like to hold the conference in either Miami or Marco Island, Florida, at a Hilton- or Marriott-owned hotel.

Use the Hilton and Marriott websites to select a hotel in whichever of these cities would enable the company to hold its sales conference within its budget and to meet its other sales conference requirements. Then locate flights arriving the afternoon prior to the conference. Your attendees will be coming from Los Angeles (53), San Francisco (31), Seattle (21), Chicago (18), and Pittsburgh (13). Determine costs of each airline ticket from these cities. When you are finished, create a budget for the conference. The budget will include the cost of each airline ticket, the room cost, and $85 per attendee per day for food.

## COLLABORATION AND TEAMWORK PROJECT

Evaluating Server and Mobile Operating Systems

**5-11** Form a group with three or four of your classmates. Choose two server or mobile operating systems to evaluate. You might research and compare the capabilities and costs of Linux versus UNIX or the most recent version of the Windows operating system for servers. Alternatively, you could compare the capabilities of the Android mobile operating system with iOS for the iPhone. If possible, use Google Docs and Google Drive or Google Sites to brainstorm; organize and develop a presentation of your findings for the class.

# BUSINESS PROBLEM-SOLVING CASE

## HOW GREEN IS THE CLOUD?

Examples of cloud computing success stories abound. Shifting from an on-premises data center to the cloud in 2020 produced annual savings of $1 million for Emirates Group. a Dubai-based international aviation-holding company. Yedpay, a payment solution company offering a secure platform for e-commerce payments, reduced its computing costs by 40% after migrating to the cloud. Most manufacturers identify faulty parts after production and write off the cost. Instead, DataProphet uses AWS cloud computing services and AI machine learning to help customers define optimum production parameters that prevent faulty parts from ever being made in the first place. Manufacturers using DataProphet's services can reduce the cost of producing faulty parts by more than 40 percent. Cloud adoption can drive organizations toward sustainable growth while improving scalability and enhancing operational efficiency. Cloud computing has been praised for its positive impact on the environment.

Computers generate a great deal of heat, so data centers must be kept cool at all times. Cooling typically accounts for 40 percent of total data center energy consumption, and up to 80 percent if the natural climate where the data center is located is warm. A data center requires a 24/7 power supply to run servers and a cooling system to prevent overheating. It also has to dispose of outdated or malfunctioning equipment, which adds to e-waste. A public cloud uses resources more efficiently than an on-premises data center, where organizations may utilize only 10 percent of server computing capacity while keeping the equipment running all the time. Businesses sharing the resources of a cloud data center use 77 percent fewer servers, consume 84 percent less power, and reduce carbon emissions by 88 percent, compared to on-premises data centers. The move to cloud computing thus results in reduced infrastructure, physical space, and energy usage.

Cloud computing has the potential to improve organizations' abilities to support remote work. Having employees work at home reduces the need for large office spaces and commuting, which lowers carbon dioxide emissions. Research by Capgemini IT consultants suggests cloud computing could potentially remove 1 billion metric tons of CO2 between 2021 and 2024.

Cloud computing also contributes to sustainability through dematerialization, the replacement of high-carbon physical products with virtual equivalents. This helps reduce energy use and a firm's carbon footprint. Cloud services encourage people to use virtual services like videostreaming or server virtualization as opposed to resource-heavy physical products.

By reducing the number of physical products—hardware and other computing equipment—cloud computing reduces the amount of waste emanating from the disposal of these products. Approximately 50 million tons of e-waste are generated globally each year. Cloud computing helps organizations go paperless with the help of cloud storage options like Dropbox, iCloud, and Google Drive. Organizations no longer need to store paper documents because all documents are stored safely and securely in the cloud.

The benefits of cloud computing are further leveraged because it provides robust and affordable computing power to companies that offer services that help other companies pursue sustainability. For example, Ireland-based Crowley Carbon uses the AWS cloud to help businesses like Hilton Food Group, GE Healthcare, and Takeda Pharmaceutical Company reduce their carbon footprints while lowering their costs. Crowley Carbon provides energy optimization and sustainability solutions that aggregate data from a variety of sources—such as live production, weather, and factory systems—and use AI machine learning to provide insights into energy usage as well as opportunities for reductions.

Although cloud computing is a positive development for businesses and the environment, it's still not completely green: Increasing business efficiency does not necessarily mean reducing business activity. Instead, the increasing use of data centers has led to even more data centers, including more data centers in the cloud. Overall, the information technology sector is responsible for roughly 2 to 4 percent of global greenhouse gas emissions—about the same as the aviation industry. Data centers' global electricity use is expected to increase to between 3 percent and 13 percent of global electricity consumption by 2030. Without serious efforts to shift to clean energy sources, greenhouse gas emissions from data centers will increase at the same rate.

Data centers can't live without air conditioning (or another type of cooling system). Power-hungry computer room air conditioners or computer room air handlers are necessities in even the most advanced data centers. In North America, most data centers draw power from "dirty" electricity grids, especially in Virginia's "data center alley," the site through which passes 70 percent of the world's Internet traffic. To cool off, the cloud burns carbon.

In addition to cooling, there are other heavy energy requirements of cloud data centers. To ensure that customers' data and cloud services will be available anytime, anywhere, these data centers are designed to be hyper-redundant: If one system fails, another system must be on hand to take the first system's place at a moment's notice. There are redundant power systems like diesel generators and redundant servers ready to take over computational processes should others become unavailable, In some cases, only 6 to 12 percent of energy consumed by data centers is devoted to active computational processes. The remainder is allocated to cooling and maintaining redundant systems and tools to prevent costly downtime.

In many data centers today, chilled water is piped through the latticework of server racks for more efficient cooling because liquid is a better convective agent than air. This shift from cooling air to cooling water is an attempt to reduce the data centers' carbon footprint, but it is not without cost. Weathering historic drought and heat waves, water resources of communities in the Western United States are in increasingly short supply, and nearby water-cooled data centers add to the strain. Bluffdale, Utah, residents are struggling with water shortages and power outages as a result of the nearby Utah Data Center, a facility of the U.S. National Security Agency (NSA) that guzzles seven million gallons of water daily to operate. Data centers consume millions of gallons of precious Arizona water every day.

To help water-stressed communities, companies like Google are pledging to go "water-positive" by 2030 and to replenish 120 percent of the water they consume in their facilities and offices. Implementing costly "closed-loop" water cooling systems makes it possible for companies such as Google and Cyrus One to recycle some of the wastewater used in evaporative cooling. However, much of the water escapes into the atmosphere during the evaporation process. Google and others have promised to invest in water infrastructure and community resources to enhance "water stewardship" and "water security." AWS evaluates the climate patterns, along with local water management and availability for each AWS Region, to optimize their cooling systems so that they use minimal water.

Some of the most advanced, giant "hyperscale" data centers, like those maintained by Google, Facebook (Meta), and Amazon, have pledged to transition their sites to being carbon-neutral via carbon offsetting and investment in renewable energy sources like wind and solar. Amazon, Microsoft, and Google have begun implementing plans for their data centers to run on 100 percent carbon-free electricity.

Amazon claims to be the world's largest renewable energy purchaser, with the goal of powering its company with 100 percent renewables by 2025 and to become carbon net-zero by 2040. AWS has been setting up wind farms around the United States to serve as power sources for current and future data centers and has established its own Amazon Wind Farm business for this purpose. Microsoft has pledged to be carbon-negative by 2030 and to remove from the atmosphere all the carbon the company has ever emitted since it was founded in 1975. To achieve this, it plans on having all of its data centers running on 100 percent renewable energy by 2025.

Google had already met its 100 percent renewable energy target by 2018, although it did so in part by purchasing offsets to match those parts of its operations that still relied on fossil fuel electricity. By implementing load migration practices, Google has pledged that by 2030, all of the energy it uses will come from carbon-free sources.

Load migration involves shifting computer processing work among data centers to maximize energy efficiency and using renewable energy resources. Large data centers have begun using high-efficiency cooling systems or locating them underwater to keep servers cool or in places where renewable energy from wind or solar power is available, such as in a fjord above the Arctic Circle. These projects are highly capital-intensive, even if they are cost-beneficial in the long run. Relocating data centers to Nordic countries like Iceland or Sweden enables the centers to utilize ambient, cool air to minimize carbon footprint, a technique called "free cooling." However, delays in communication over long-distance networks make this green data center location unsuitable for meeting the computing and data storage demands of the wider world. Moving data centers to countries closer to the Arctic Circle also raises concerns about data security and privacy: Many countries, as well as the European Union, have passed laws that require citizen data to be stored on servers within their national borders.

Sources: Aws.amazon.com, accessed April 23, 2022; Steven Gonzalez Montserrate, "The Staggering Ecological Impacts of Computation and the Cloud," Thereader.mitpress.mit.edu, accessed April 19, 2022; Jamie Morgan, "5 Reasons Why the Cloud Is Environmentally Friendly," Missioncloud.com, accessed April 19, 2022; David Kuchta, "Does Cloud Computing Help or Harm the Environment?" *Treehugger*, January 13, 2022; Kaiyan Kumar, "Toward a Sustainable Transition to the Green Cloud," *CMS Wire*, October 19, 2021; Antoine Jeol, "The Environmental Impact of Cloud Computing," *Holori*, May 31, 2021; Sean Spicer," Environmental Benefits of Cloud Computing," *AgileIT*, April 21, 2019.

## CASE STUDY QUESTIONS

**5-12** How environment-friendly are data centers? Explain your answer.

**5-13** How green is cloud computing? How much does it help or hurt the environment? Explain your answer.

**5-14** What solutions have been proposed to make data centers greener? How effective are they?

## Chapter 5 References


Amazon Web Services. "Overview of Amazon Web Services." Docs.aws.amazon.com, accessed April 29, 2022.

Benitez, Jose, Gautam Ray, and Jörg Henseler. "Impact of Information Technology Infrastructure Flexibility on Mergers and Acquisitions." *MIS Quarterly* 42, No. 1 (March 2018).

Bova, Francesco, Avi Goldfarb, and Roger Melko." Quantum Computing Is Coming. What Can It Do?" *Harvard Business Review* (July 16, 2021).

Carr, Nicholas. *The Big Switch* (New York: Norton, 2008).

Gartner, Inc. "Gartner Forecasts Worldwide IT Spending to Grow 5.1 Percent in 2022." (January 18, 2022).

Germain, Jack M. "30 Years Later the Trajectory of Linux is Star Bound." *TechNewsWorld* (August 25, 2021).

Gregory, Robert Wayne, Evgeny Kaganer, Ola Henfridsson, and Thierry Jean Ruch. "IT Consumerization and the Transformation of IT Governance." *MIS Quarterly* 42, No. 4 (December 2018).

Guo, Zhiling, and Dan Ma. "A Model of Competition between Perpetual Software and Software as a Service." *MIS Quarterly* 42, No. 1 (March 2018).

Insider Intelligence/eMarketer. "Mobile Internet: Average Time Spent in US, App vs. Browser, 2019–2024." (April 28, 2022).

Kathuria, Abhishek, Arti Mann, Jiban Khuntia, Terence JV Saldanha, and Robert J. Kauffman. "A Strategic Value Appropriation Path for Cloud Computing." *Journal of Management Information Systems* 35, No. 3 (2018).

Krancher, Oliver, Pascal Luther, and Marc Jost. "Key Affordances of Platform-as-a-Service: Self-Organization and Continuous Feedback." *Journal of Management Information Systems* 35, No. 3 (2018).

Mecomber, Rebecca. "What Is the Role of Nanotechnology in Computers?" *EasyTechJunkie* (April 8, 2022).

Mell, Peter, and Tim Grance. "The NIST Definition of Cloud Computing, Version 15." NIST (October 17, 2009).

Monserrate, Steven Gonzalez. "The Staggering Ecological Impacts of Computation and the Cloud." *The MIT Press Reader* (April 19, 2022).

Moqri, Mahdi, Xiaowei Mei, Liangfei Qiu, and Subhajyoti Bandyopadhyay. "Effect of 'Following' on Contributions to Open Source Communities." *Journal of Management Information Systems* 35, No. 4 (2018).

Retana, German F., Chris Forman, Sridhar Narasimhan, Marius Florin Niculescu, and D. J. Wu. "Technology Support and Post-Adoption IT Service Use: Evidence from the Cloud." *MIS Quarterly* 42, No. 3 (September 2018).

Ruane, Jonathan, Andrew McAfee, and William D. Oliver. "Quantum Computing for Business Leaders." *Harvard Business Review* (January–Febuary 2022).

Schuff, David, and Robert St. Louis. "Centralization vs. Decentralization of Application Software." *Communications of the ACM* 44, No. 6 (June 2001).

Sherae, L. Daniel, Likoebe M. Maruping, Marcelo Cataldo, and Jin Herbsleb. "The Impact of Ideology Misfit on Open Source Software Communities and Companies." *MIS Quarterly* 42, No. 4 (December 2018).

Shipilov, Andrew, and Nathan Furr. "Making Quantum Computing a Reality." *Harvard Business Review* (April18, 2022).

Sumina, Vladimir. "26 Cloud Computing Statistics: Facts and Trends for 2022." Cloudwards.net (March 18, 2022).

Tucci, Linda. "The Shift to Edge Computing Is Happening Fast: Here's Why." SearchCIO.com (April 2019).

Weinschenk, Carl. "Mobile Device Management and the Enterprise." *IT Business Edge* (November 29, 2018).

CHAPTER

6

# Foundations of Business Intelligence: Databases and Information Management

## LEARNING OBJECTIVES

After completing this chapter, you will be able to:

**6-1** Define a database and explain how a relational database organizes data.

**6-2** Describe the principles of a database management system.

**6-3** Identify the principal tools and technologies for accessing information from databases to improve business performance and decision making.

**6-4** Explain why data governance and data quality assurance are essential for managing a firm's data resources.

**6-5** Understand how MIS can help your career.

## CHAPTER CASES

- Better Data, Better Decisions for the State of Maine
- New Cloud Database Tools Help Vodafone Fiji Make Better Decisions
- Higher Data Quality Helps Vyaire Save Lives
- Pursuing Sustainability with Blockchain

**MyLab MIS**

- **Video Case:** Stitch Fix CEO Sees Business of Personalization as Key to Success; Stitch Fix President on Booming Growth amid Pandemic
- Discussion Questions: 6-5, 6-6, 6-7
- Hands-On MIS Projects: 6-8, 6-9, 6-10, 6-11
- eText with Figure Videos

## BETTER DATA, BETTER DECISIONS FOR THE STATE OF MAINE

**Managers** and employees working in the public sector need easy access to accurate and complete data for decision making just like their private sector counterparts, and the state of Maine is no exception. Until recently, however, these data were not readily available.

The data required by state government decision makers had to be extracted from information systems that were more than 20 years old, and the legacy systems did not work well with more modern data analysis tools. Decision makers would have to move the data into their own tools for analysis, an inefficient, time-consuming process. If Maine's legislature wanted information, it could take days or weeks for the information systems to respond with the data, which often was retrieved by a single staffer who understood how the legacy system where the data resided actually worked.

Maine's Office of Information Technology was charged with modernizing the way the state government made real-time, data-based decisions across its operations. The Office implemented a solution based on Oracle Autonomous Data Warehouse running on Oracle Cloud Infrastructure (OCI). A data warehouse centralizes and consolidates large amounts of data from multiple sources for analysis. Oracle Autonomous Data Warehouse is an easy-to-use, fully autonomous data warehouse that scales elastically, quickly delivers answers to queries, and requires no database administration because it can perform many database management tasks on its own. OCI is a platform of cloud services that enables organizations to build and run a wide range of applications in a highly available, consistently high-performance environment.

The new data management system enabled Maine to consolidate data from multiple sources (including new and legacy systems for accounting, budgeting, payroll, and other human resources functions) into a high-performance data warehouse. The data warehouse made the data immediately available in the Oracle Analytics Cloud to decision makers across the state government, enabling them to build their own dashboards without having to rely on the IT department. The Oracle Analytics platform is a cloud-based service that provides the capabilities required by the entire analytics process. Maine now has a single source of consistent data to support operations and decision making.

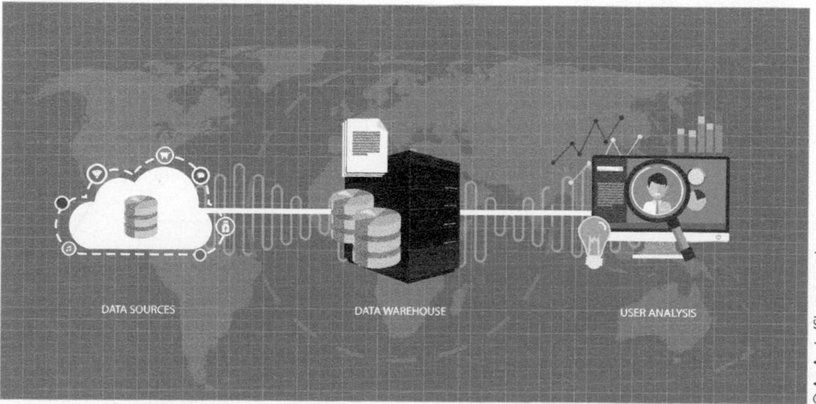

© Aa Amie/Shutterstock

In the state controller's office, monthly expenditure, revenue, and cash reports are now available daily for tens of thousands of individual state government programs. Generating the general ledger trial balance, which was previously a monthly process, is now completed in a matter of seconds. Because the data warehouse runs autonomously, the state saves the money it would have spent on third-party data management resources. With an autonomous database, there are fewer things that a human needs to do. These cost savings help the state of Maine use more of its tax dollars to provide services to its citizens, instead of spending the money on administrative issues.

Sources: Jeff Erickson, "How the Vision of One IT Team Brought Modern Analytics to the State of Maine" and "Oracle Cloud Helps Maine Make Data-Backed Decisions," Oracle.com, accessed May 1, 2022; Brian Roehm, "What Is Oracle Cloud and Why Should You Use It?" Cloudguru.com, June 11, 2021.

---

The experience of the state of Maine illustrates the importance of data management. Business performance depends on what a firm can or cannot do with its data. Both operational efficiency and management decision making were hampered by the state of Maine's fragmented data stored in outdated legacy systems that were difficult to access and analyze. How businesses store, organize, and manage their data has an enormous impact on organizational effectiveness.

The chapter-opening diagram calls attention to important points raised by this case and this chapter. Managers and employees working for the state were unable to easily obtain the data needed for decision making because the data were maintained in antiquated legacy systems that did not work with modern business intelligence tools. Decision makers had to extract the data from the legacy systems manually and put the data into their own tools for analysis. Decisions were not as effective as they could have been because it was so difficult to assemble and understand the data required to formulate a decision. The solution was to centralize and consolidate data from all sources in a data warehouse that provided a single source of data for reporting and analysis. Maine deployed new technologies such as Oracle Autonomous Data Warehouse and Oracle Analytics running on Oracle Cloud Infrastructure. Maine had to reorganize its data into a standard, enterprise-wide format; establish rules, responsibilities, and procedures for accessing and using the data; and provide tools for making the data accessible to users for querying and reporting. The solution improved operational efficiency and decision making and also lowered costs.

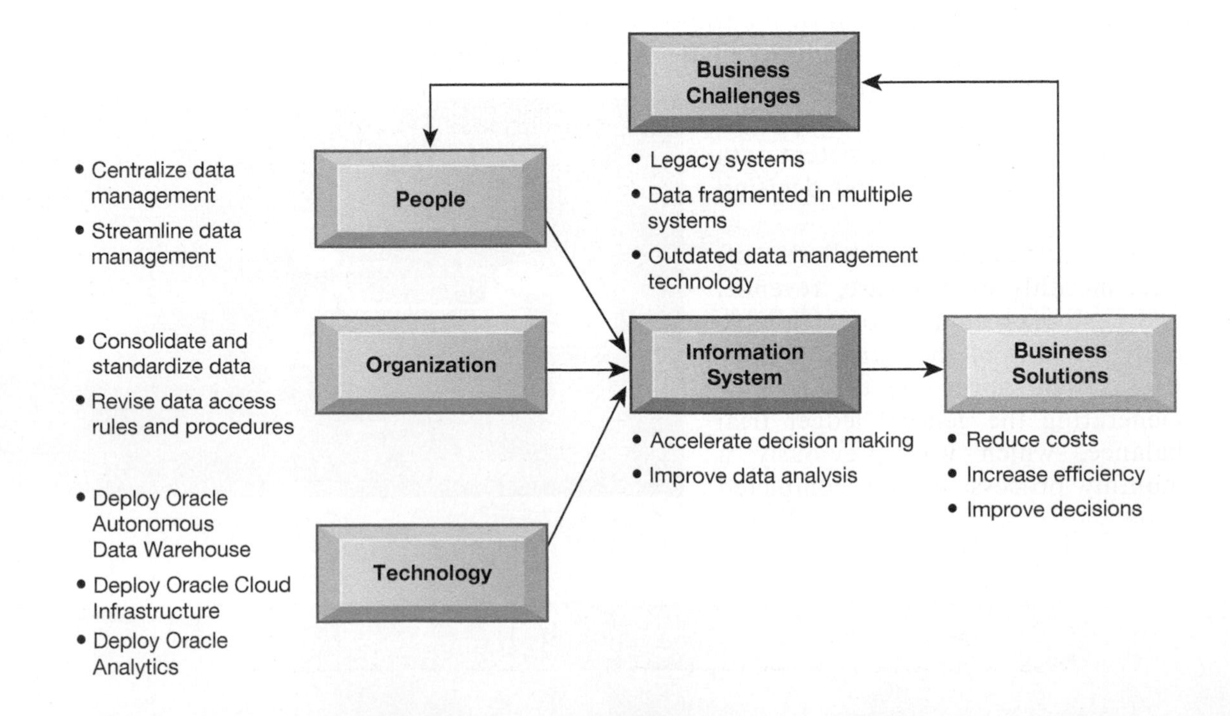

Here are some questions to think about: What was the business impact of Maine's data management problems? How did better data management technology improve operational efficiency and management decision making?

## 6-1 Define a database and explain how a relational database organizes data.

A computer system organizes data in a hierarchy that starts with bits and bytes and then progresses to fields, records, files, and databases. A **bit** (or binary digit) represents the smallest unit of data that a computer can handle: a value of 0 or 1. A group of bits, called a **byte**, represents a single character, which can be a letter, a number, or another symbol. A grouping of characters into a word, a group of words (such as person's name), or a complete number (such as a person's age) is called a **field**. A group of related fields comprises a **record**; a group of records of the same type is called a **file**. A group of related files makes up a **database**.

Examine Figure 6.1, which illustrates the data hierarchy used to create a student database. Individual bits form the base of the hierarchy and are grouped together to form a byte. A grouping of bytes forms a field, in this case a field that identifies a specific course. A group of related fields, here a specific student's ID, course, date, and grade fields, creates a record. A grouping of similar student records creates a student course file. The student course file is then grouped with files on students' personal histories and financial backgrounds to create a student database. Databases are at the heart of all information systems because they keep track of the people, places, and things that a business must deal with on a continuing, often instant basis. View the Figure 6.1 video in the eText for an animated and more detailed discussion of this figure.

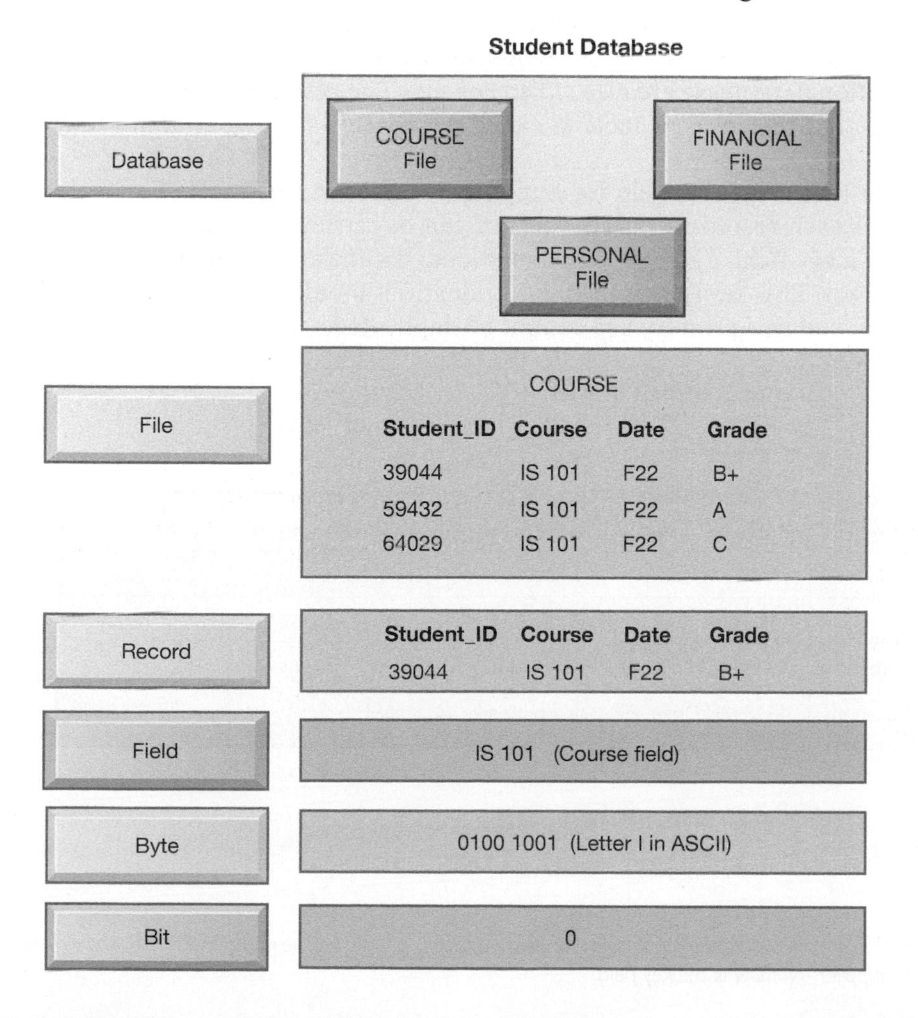


**Figure 6.1**
The Data Hierarchy
*A computer system organizes data in a hierarchy that starts with the bit, which represents either a 0 or a 1. Bits can be grouped to form a byte to represent one character, number, or symbol. Bytes can be grouped to form a field, and related fields can be grouped to form a record. Related records can be collected to form a file, and related files can be organized into a database.*

## ENTITIES AND ATTRIBUTES

To run a business, you most likely will be using data about categories of information such as customers, suppliers, employees, orders, products, shippers, and perhaps parts. Each of these generalized categories representing a person, a place, or a thing about which we store information is called an **entity**. Each entity has specific characteristics called **attributes**. Look again at the File section of Figure 6.1. Here, COURSE is an entity and Student_ID, Course, Date, and Grade are its attributes.

If you were a business keeping track of parts you used and their suppliers, the entity SUPPLIER would have attributes such as the supplier's name and address, which would most likely include the street, city, state, and zip code. The entity PART would typically have attributes such as part description, price of each part (unit price), and the supplier who produced the part.

## ORGANIZING DATA IN A RELATIONAL DATABASE

If you stored this information in paper files, you would probably have a file on each entity and its attributes. In an information system, a database organizes the data much the same way, grouping related pieces of data. The **relational database** is the most common type of database today. Relational databases organize data into two-dimensional tables (called *relations*) with columns and rows. Each table contains data about an entity and its attributes. For the most part, there is one table for each business entity, so, at the most basic level, you will have one table for customers and a table each for suppliers, parts in inventory, employees, and sales transactions.

Let's look at how a relational database would organize data about suppliers and parts. Examine Figure 6.2, which illustrates a SUPPLIER table. It consists of a grid of columns and rows of data. Each element of data about a supplier, such as the supplier name, street, city, state, and zip code, is stored as a separate field within the SUPPLIER table. Each field represents an attribute for the entity SUPPLIER. Fields in a relational database are also called *columns*.The actual information about a single supplier that resides in a table is called a *row*. Rows are commonly referred to as records.

Note that there is a field for Supplier_Number in this table. This field uniquely identifies each record so that the record can be retrieved, updated, or sorted, and it is called a **key field**. Each table in a relational database has one field designated as its **primary key**. This key field is the unique identifier for all the information in any row of the table, and this primary key cannot be duplicated.

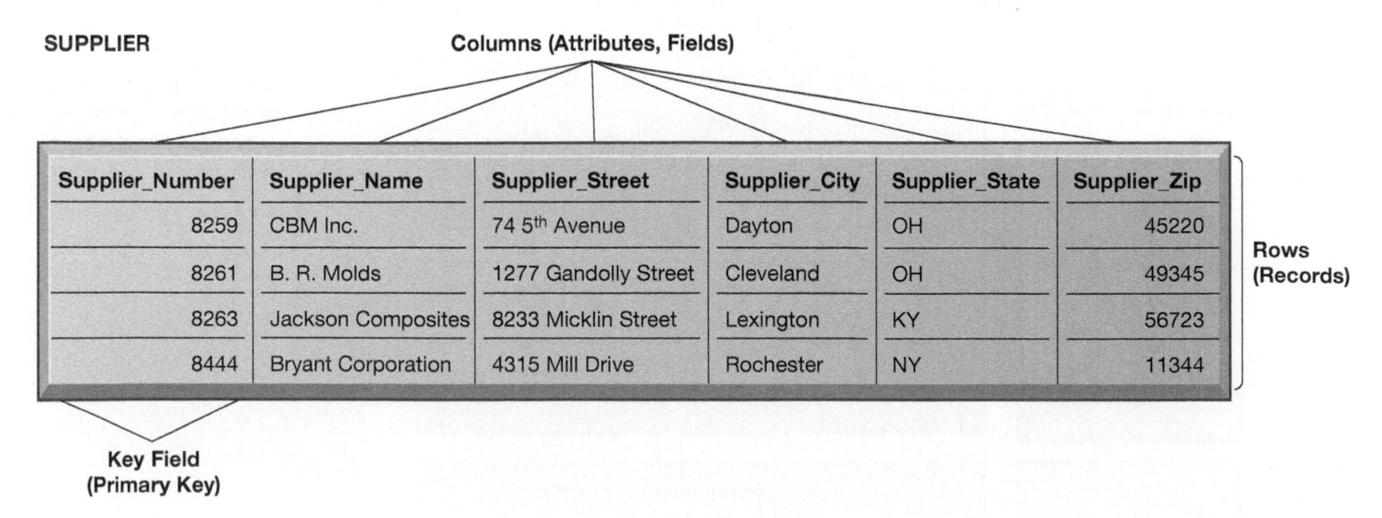


| Supplier_Number | Supplier_Name | Supplier_Street | Supplier_City | Supplier_State | Supplier_Zip |
|---|---|---|---|---|---|
| 8259 | CBM Inc. | 74 5th Avenue | Dayton | OH | 45220 |
| 8261 | B. R. Molds | 1277 Gandolly Street | Cleveland | OH | 49345 |
| 8263 | Jackson Composites | 8233 Micklin Street | Lexington | KY | 56723 |
| 8444 | Bryant Corporation | 4315 Mill Drive | Rochester | NY | 11344 |

**Figure 6.2**
A Relational Database Table
*A relational database organizes data in the form of two-dimensional tables. Illustrated here is a table for the entity SUPPLIER, showing how it represents the entity and its attributes. Supplier_Number is the key field.*

We could use the supplier's name as a key field. However, if two suppliers had the same name (which does happen from time to time), supplier name would not uniquely identify each, so it is necessary to assign a special identifier field for this purpose. For example, if you had two suppliers, both named "CBM," but one was based in Dayton and the other in St. Louis, it would be easy to confuse them. However, if each has a unique supplier number, such confusion is prevented.

We also see that the address information has been separated into four fields: Supplier_Street, Supplier_City, Supplier_State, and Supplier_Zip. Data are separated into the smallest elements that one would want to access separately to make it easy to select only the rows in the table that match the contents of one field, such as all the suppliers in Ohio (OH). The rows of data can also be sorted by the contents of the Supplier_State field to get a list of suppliers by state regardless of their cities.

So far, the SUPPLIER table does not have any information about the parts that a particular supplier provides for your company. PART is an entity separate from SUPPLIER, and fields with information about parts should be stored in a separate PART table.

Why not keep information on parts in the same table as information on suppliers? If we did that, each row of the table would contain the attributes of both PART and SUPPLIER. Because one supplier could supply more than one part, the table would need many extra rows for a single supplier to show all the parts that the supplier provided. We would be maintaining a great deal of redundant data about suppliers, and it would be difficult to search for the information on any individual part because we would not know whether this part is the first or the 50th part in this supplier's record. Examine Figure 6.3, which illustrates a separate table, PART, that should be created to store data about parts (in this case, part number, part name, and unit price) and solve this problem. View the Figures 6.2–6.3 video in the eText for an animated and more detailed discussion of these two figures.

The PART table would also have to contain another field, Supplier_Number, so that you would know the supplier for each part. It would not be necessary to keep repeating all the information about a supplier in each PART record because having a Supplier_ Number field in the PART table allows you to look up the data in the fields of the SUPPLIER table.

Notice that Supplier_Number appears in both the SUPPLIER and the PART tables. In the SUPPLIER table, Supplier_Number is the primary key. When the field Supplier_Number appears in the PART table, however, it is called a **foreign key** and is essentially a look-up field to find data about the supplier of a specific part. Note that the PART table would itself have its own primary key field, Part_Number, to identify each part uniquely. This key is not used to link PART with SUPPLIER but could be used to link PART with a different entity.

PART

| Part_Number | Part_Name | Unit Price | Supplier_Number |
|---|---|---|---|
| 137 | Door latch | 22.00 | 8259 |
| 145 | Side mirror | 12.00 | 8444 |
| 150 | Door molding | 6.00 | 8263 |
| 152 | Door lock | 31.00 | 8259 |
| 155 | Compressor | 54.00 | 8261 |
| 178 | Door handle | 10.00 | 8259 |

Primary Key — Foreign Key

**Figure 6.3**
The PART Table
*Data for the entity PART have their own separate table. Part_Number is the primary key, and Supplier_ Number is the foreign key, enabling users to find related information in the SUPPLIER table about the supplier for each part.*

As we organize data into tables, it is important to make sure that all the attributes for a particular entity apply only to that entity. If you were to keep the supplier's address with the PART record, that information would not really relate only to PART: It would relate to both PART and SUPPLIER. If the supplier's address were to change, it would be necessary to alter the data in every PART record rather than only once, in the SUPPLIER record.

## ESTABLISHING RELATIONSHIPS

Now that we've broken down our data into a SUPPLIER table and a PART table, we must make sure we understand the relationship between them. A schematic called an **entity-relationship diagram** clarifies table relationships in a relational data base. The most important piece of information that an entity-relationship diagram provides is the manner in which two tables are related to each other. Tables in a relational database may have one-to-one, one-to-many, and many-to-many relationships.

An example of a one-to-one relationship is a human resources system that stores confidential data about employees. The system stores data, such as the employee name, date of birth, address, and job position, in one table and confidential data about that employee, such as salary or pension benefits, in another table. These two tables pertaining to a single employee would have a one-to-one relationship because each record in the EMPLOYEE table with basic employee data has only one related record in the table storing confidential data.

The relationship between the SUPPLIER and the PART entities in our database is a one-to-many relationship. Each supplier can supply more than one part, but each part has only one supplier. For every record in the SUPPLIER table, many related records might be in the PART table.

Examine Figure 6.4, which illustrates how an entity-relationship diagram would depict this one-to-many relationship. The boxes represent entities. The lines connecting the boxes represent relationships. A line connecting two entities that ends in two short marks designates a one-to-one relationship. A line connecting two entities that ends with a crow's foot symbol preceded by a short mark indicates a one-to-many relationship. The text above the line is read left to right (i.e., SUPPLIER provides PART), while the text below the line is read right to left (i.e., PART is supplied by SUPPLIER). Figure 6.4 shows that each part has only one supplier, but the same supplier can provide many parts.

We would also see a one-to-many relationship if we wanted to add to our database a table about orders because one supplier services many orders. The ORDER table would contain only the Order_Number and Order_Date fields. Examine Figure 6.5, which illustrates a sample order report showing an order of parts from a supplier. If you look at the report, you can see that the information on the top-right portion of the report (the Order number and the Order date) comes from the ORDER table. The actual line items ordered are listed in the lower portion of the report.

Because one order can be for many parts from a supplier and because a single part can be ordered many times on different orders, this creates a many-to-many relationship between the PART and ORDER tables. Whenever a many-to-many relationship exists

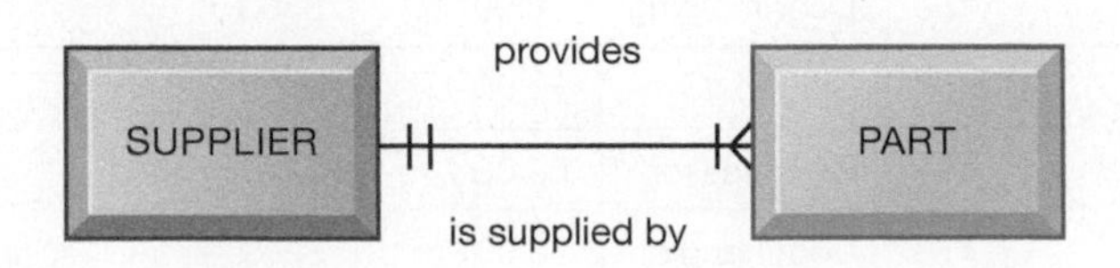


**Figure 6.4**
A Simple Entity-Relationship Diagram
*This diagram shows the relationship between the entities SUPPLIER and PART.*

Order Number: 3502
Order Date: 1/15/2023

Supplier Number: 8259
Supplier Name: CBM Inc.
Supplier Address: 74 5th Avenue, Dayton, OH 45220

| Order_Number | Part_Number | Part_Quantity | Part_Name | Unit_Price | Extended Price |
|---|---|---|---|---|---|
| 3502 | 137 | 10 | Door latch | 22.00 | $220.00 |
| 3502 | 152 | 20 | Door lock | 31.00 | 620.00 |
| 3502 | 178 | 5 | Door handle | 10.00 | 50.00 |
| | | | Order Total: | | $890.00 |

**Figure 6.5**
Sample Order Report
*The shaded areas show which data came from the ORDER, SUPPLIER, and LINE_ITEM tables. The database does not maintain data on extended price or order total because these data can be derived from other data in the tables.*

between two tables, it is necessary to link these two tables in a table that joins this information. Creating a separate table for a line item in the order would serve this purpose. This table is often called a *join table* or an *intersection relation*. This join table contains only three fields: Order_Number and Part_Number, which are used only to link the ORDER and PART tables, and Part_Quantity. If you look at the bottom-left part of the report, this is the information coming from the LINE_ITEM table.

We would thus wind up with a total of four tables in our database. Examine Figure 6.6, which illustrates the final set of tables. Note that the ORDER table does not contain data on extended price because that value can be calculated by

PART

| Part_Number | Part_Name | Unit_Price | Supplier_Number |
|---|---|---|---|
| 137 | Door latch | 22.00 | 8259 |
| 145 | Side mirror | 12.00 | 8444 |
| 150 | Door molding | 6.00 | 8263 |
| 152 | Door lock | 31.00 | 8259 |
| 155 | Compressor | 54.00 | 8261 |
| 178 | Door handle | 10.00 | 8259 |

LINE_ITEM

| Order_Number | Part_Number | Part_Quantity |
|---|---|---|
| 3502 | 137 | 10 |
| 3502 | 152 | 20 |
| 3502 | 178 | 5 |

ORDER

| Order_Number | Order_Date |
|---|---|
| 3502 | 1/15/2023 |
| 3503 | 1/16/2023 |
| 3504 | 1/17/2023 |

SUPPLIER

| Supplier_Number | Supplier_Name | Supplier_Street | Supplier_City | Supplier_State | Supplier_Zip |
|---|---|---|---|---|---|
| 8259 | CBM Inc. | 74 5th Avenue | Dayton | OH | 45220 |
| 8261 | B. R. Molds | 1277 Gandolly Street | Cleveland | OH | 49345 |
| 8263 | Jackson Components | 8233 Micklin Street | Lexington | KY | 56723 |
| 8444 | Bryant Corporation | 4315 Mill Drive | Rochester | NY | 11344 |

**Figure 6.6**
The Final Database Design with Sample Records
*The final design of the database for suppliers, parts, and orders has four tables. The LINE_ITEM table is a "join table" that eliminates the many-to-many relationship between ORDER and PART.*

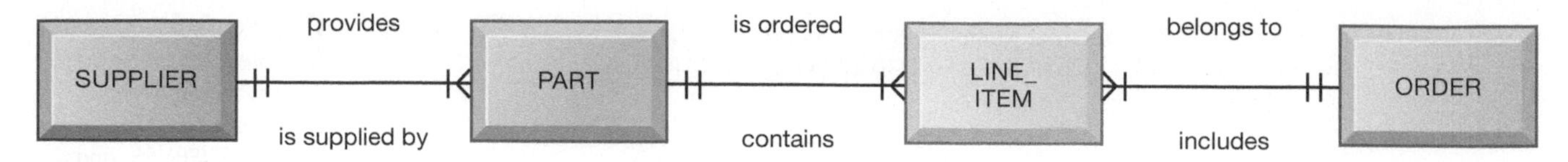


**Figure 6.7**
Entity-Relationship Diagram for the Database with Four Tables
*This diagram shows the relationships among the SUPPLIER, PART, LINE_ITEM, and ORDER entities.*

multiplying Unit_Price by Part_Quantity. This data element can be derived when needed, using information that already exists in the PART and LINE_ITEM tables. Order_Total is another derived field, calculated by totaling the extended prices for the items ordered. Finally, examine Figure 6.7, which shows what the entity-relationship diagram for this set of tables would look like. This diagram illustrates the relationships among the entities SUPPLIER, PART, LINE_ITEM, and ORDER. Figure 6.7 shows that one ORDER can contain many LINE_ITEMs. (A PART can be ordered many times and appear many times as a line item in a single order.) Each PART can have only one SUPPLIER, but many PARTs can be provided by the same SUPPLIER.

The process of streamlining complex groups of data to minimize redundant data elements and awkward many-to-many relationships and to increase stability and flexibility is called **normalization**. A properly designed and normalized database is easy to maintain and minimizes duplicate data.

Relational database systems enforce **referential integrity** rules to ensure that relationships between coupled tables remain consistent. When one table has a foreign key that points to another table, you may not add a record to the table with the foreign key unless there is a corresponding record in the linked table. In the database we have just created, the foreign key Supplier_Number links the PART table to the SUPPLIER table. We may not add a new record to the PART table for a part with supplier number 8266 unless there is a corresponding record in the SUPPLIER table for supplier number 8266. We must also delete the corresponding record in the PART table if we delete the record in the SUPPLIER table for supplier number 8266. In other words, we shouldn't have parts from nonexistent suppliers!

The example provided here for parts, orders, and suppliers is a simple one. Even in a small business, you will have tables for other important entities such as customers, shippers, and employees. A large corporation typically has databases with thousands of entities (tables) to maintain. What is important for any business, large or small, is to have a good data model that includes all its entities and the relationships among them, one that is organized to minimize redundancy, maximize accuracy, and make data easily accessible for reporting and analysis.

It cannot be emphasized enough: If the business does not get its data model right, the system will not be able to serve the business properly. The company's systems will not be as effective as they could be because they will have to work with data that may be inaccurate, incomplete, or difficult to retrieve. Understanding the organization's data and how they should be represented in a database is a very important lesson you can learn from this course.

For example, *The Globe and Mail*, Canada's largest newspaper, was unable until recently to execute ambitious marketing programs because it had trouble housing and managing the data on sales prospects. Much of the required data were stored in a mainframe system where the data were not easy to access and analyze. If users needed any information, they had to extract the data from the mainframe and bring it to one of a number of local databases—including those maintained in Microsoft Access and Excel—for analysis. This practice created numerous pockets of data maintained in isolated databases for specific purposes but no central repository where the most up-to-date data could be accessed from a single place. With data scattered in so many different systems throughout the company, it was very difficult to cross-reference subscribers with prospects when developing the mailing list for a marketing campaign.

## 6-2 Describe the principles of a database management system.

Now that you have started creating the files and identifying the data your business requires, you will need a database management system to help you manage and use the data. A **database management system (DBMS)** is a specific type of software for creating, storing, organizing, and accessing data from a database. Microsoft Access is a DBMS for desktop systems, whereas IBM Db2, Oracle Database, and Microsoft SQL Server are DBMS for large mainframes and midrange computers. MySQL is a popular open-source DBMS. All these products are relational DBMS that support a relational database.

By separating the logical view and the physical view of the data, the DBMS relieves the end user or programmer of the task of understanding where and how the data are actually stored. The *logical view* presents data as end users or business specialists would perceive them, whereas the *physical view* shows how data are actually organized and structured on physical storage media, such as a hard disk.

A DBMS makes the physical database available for different logical views required by users. Examine Figure 6.8, which illustrates a DBMS that enables different views of the data contained within a human resources database. Different members of the human resources team need access to different types of data. For instance, a benefits specialist typically will require a view consisting of an employee's name, social security number, and health insurance coverage. A payroll department member will need data such as an employee's name, social security number, gross pay, and net pay. The data for all of these views are stored in a single database, where the organization can manage it more easily.

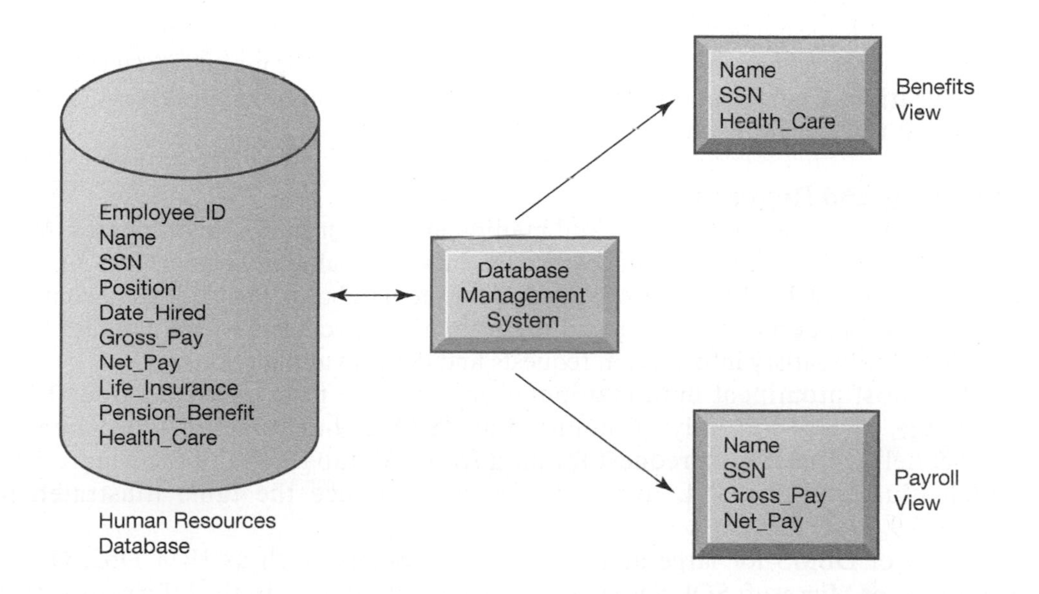


**Figure 6.8**
Human Resources Database with Multiple Views

*A single human resources database provides many views of data, depending on the information requirements of the user. Illustrated here are two possible views, one of interest to a benefits specialist and one of interest to a member of the company's payroll department.*

## OPERATIONS OF A RELATIONAL DBMS

In a relational database, tables can be easily combined to deliver data that users require, provided that any two tables share a common data element. Let's return to the database we set up earlier with SUPPLIER and PART tables illustrated, respectively, in Figures 6.2 and 6.3.

Suppose we wanted to find in this database the names of suppliers who could provide us with part number 137 or part number 150. We would need information from two tables: the SUPPLIER table and the PART table. Note that these two tables have a shared data element: Supplier_Number.

In a relational database, three basic operations are used to develop useful sets of data: select, project, and join. The *select* operation creates a subset consisting of all records in the file that meet stated criteria. Select creates, in other words, a subset of rows that meet certain criteria. The *project* operation creates a subset consisting of columns in a table, permitting the user to create new tables that contain only the information required. The *join* operation combines relational tables to provide the user with more information than is available in individual tables.

Examine Figure 6.9, which illustrates how these three basic operations can be used to find the names of the suppliers who provide part number 137 or part number 150. To do so, we need to select records (rows) from the PART table where the Part_Number equals 137 or 150. Next, we need to join the now-shortened PART table (only parts 137 or 150 are presented) and the SUPPLIER table into a single, new table. Finally, we extract from the new table only the following columns: Part_Number, Part_Name, Supplier_Number, and Supplier_Name.

## CAPABILITIES OF DATABASE MANAGEMENT SYSTEMS

A DBMS includes capabilities and tools for organizing, managing, and accessing the data in the database. The most important are its data definition capability, data dictionary, and data manipulation language.

DBMS have a **data definition** capability to specify the structure of the content of the database. This capability is used to create database tables and to define the characteristics of the fields in each table. This information about the database can be documented in a **data dictionary**. A data dictionary stores definitions of data elements and their characteristics. For instance, Microsoft Access has a rudimentary data dictionary capability that displays information about the name, description, size, type, format, and other properties of each field in a table. Examine Figure 6.10, which illustrates this capability for the SUPPLIER table. Data dictionaries for large corporate databases may capture additional information such as usage, ownership (who in the organization is responsible for maintaining the data), authorization, security, and the individuals, business functions, programs, and reports that use each data element.

### Querying and Reporting

DBMS include tools for accessing and manipulating information in databases. Most DBMS have a specialized language called a **data manipulation language** that is used to add, change, delete, and retrieve the data in the database. This language contains commands that permit end users and programming specialists to extract data from the database to satisfy information requests and develop applications.

The most prominent data manipulation language today is **Structured Query Language**, or **SQL** (usually pronounced as "S-Q-L" but sometimes pronounced as "Sequel"). A **query** is a request for data from a database. Examine Figure 6.11, which illustrates the SQL query that would produce the table illustrated in Figure 6.9.

Users of DBMS for large and midrange computers, such as IBM Db2, Oracle Database, or Microsoft SQL Server, can employ SQL to retrieve the information they

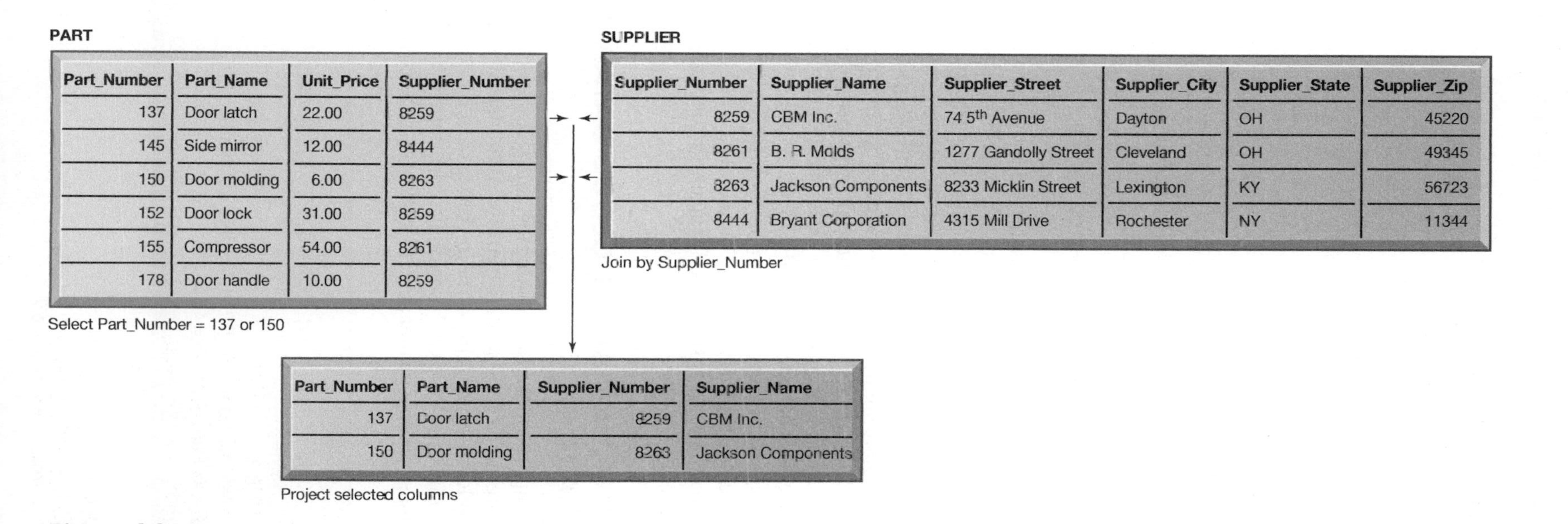

PART

| Part_Number | Part_Name | Unit_Price | Supplier_Number |
|---|---|---|---|
| 137 | Door latch | 22.00 | 8259 |
| 145 | Side mirror | 12.00 | 8444 |
| 150 | Door molding | 6.00 | 8263 |
| 152 | Door lock | 31.00 | 8259 |
| 155 | Compressor | 54.00 | 8261 |
| 178 | Door handle | 10.00 | 8259 |

Select Part_Number = 137 or 150

SUPPLIER

| Supplier_Number | Supplier_Name | Supplier_Street | Supplier_City | Supplier_State | Supplier_Zip |
|---|---|---|---|---|---|
| 8259 | CBM Inc. | 74 5th Avenue | Dayton | OH | 45220 |
| 8261 | B. R. Molds | 1277 Gandolly Street | Cleveland | OH | 49345 |
| 8263 | Jackson Components | 8233 Micklin Street | Lexington | KY | 56723 |
| 8444 | Bryant Corporation | 4315 Mill Drive | Rochester | NY | 11344 |

Join by Supplier_Number

| Part_Number | Part_Name | Supplier_Number | Supplier_Name |
|---|---|---|---|
| 137 | Door latch | 8259 | CBM Inc. |
| 150 | Door molding | 8263 | Jackson Components |

Project selected columns

**Figure 6.9**
The Three Basic Operations of a Relational DBMS
*The select, project, and join operations enable data from two tables to be combined and only selected attributes to be displayed.*

**Figure 6.10**
Microsoft Access Data Dictionary Features
*Microsoft Access has a rudimentary data dictionary capability that displays information about the size, format, and other characteristics of each field in a database. Displayed here is the information maintained in the SUPPLIER table. The small key icon to the left of Supplier_Number indicates that it is a key field.*

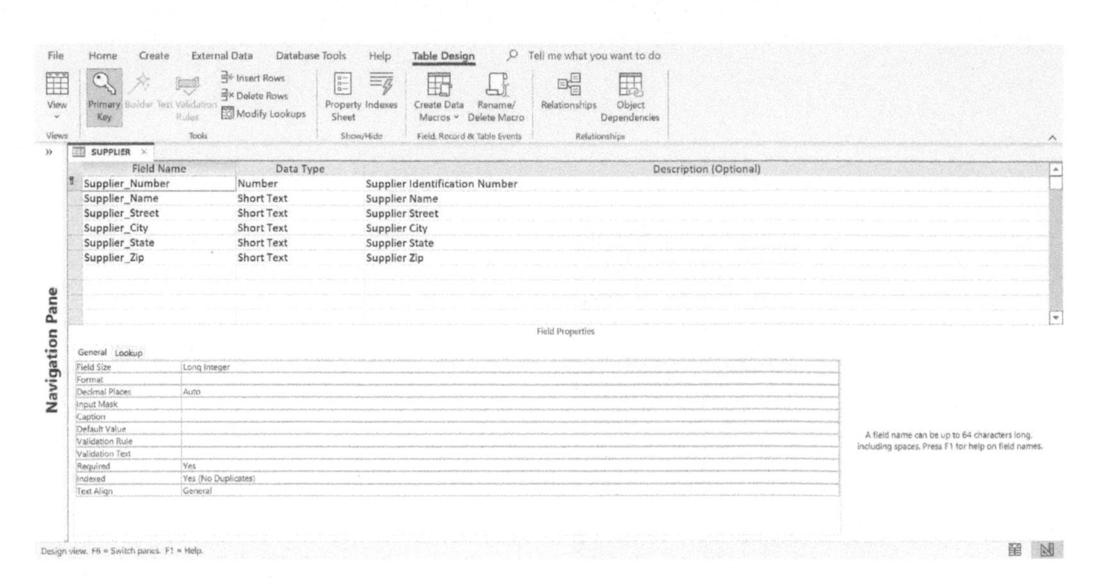


Used with permission from Microsoft.

need from the database. Microsoft Access also uses SQL, but it provides its own set of user-friendly tools for querying databases and for organizing data from databases into more polished reports.

Microsoft Access has capabilities to help users create queries by identifying the tables, fields, and results they want and then selecting the rows from the database that meet particular criteria. These actions in turn are translated into SQL commands. Examine Figure 6.12, which illustrates how the SQL query to select parts and suppliers in Figure 6.11 would be constructed using Microsoft Access.

DBMS typically include capabilities for report generation so that the data of interest can be displayed in a more structured and polished format than would be possible just by querying. SQL Server Reporting Services and Oracle Reports Builder are examples of **report generators** for corporate DBMS.

Microsoft Access also has capabilities for developing desktop system applications. These capabillities include tools for creating data entry screens and reports and developing the logic for processing transactions. Information systems specialists primarily use these capabilities.

## NONRELATIONAL DATABASES, CLOUD DATABASES, AND BLOCKCHAIN

Cloud computing, unprecedented data volumes, massive workloads for web services, and the need to store new types of data require database alternatives to the traditional relational model of organizing data in the form of tables, columns, and rows. Companies are turning to "NoSQL" nonrelational database technologies for this purpose. **Nonrelational database management systems** use a more flexible data model

```
SELECT PART.Part_Number, PART.Part_Name, SUPPLIER.Supplier_Number,
SUPPLIER.Supplier_Name
FROM PART, SUPPLIER
WHERE PART.Supplier_Number = SUPPLIER.Supplier_Number AND
Part_Number = 137 OR Part_Number = 150;
```

**Figure 6.11**
Example of a SQL Query
*Illustrated here are the SQL statements for a query to select suppliers for parts 137 or 150. They produce a list with the same results as those shown in Figure 6.9.*

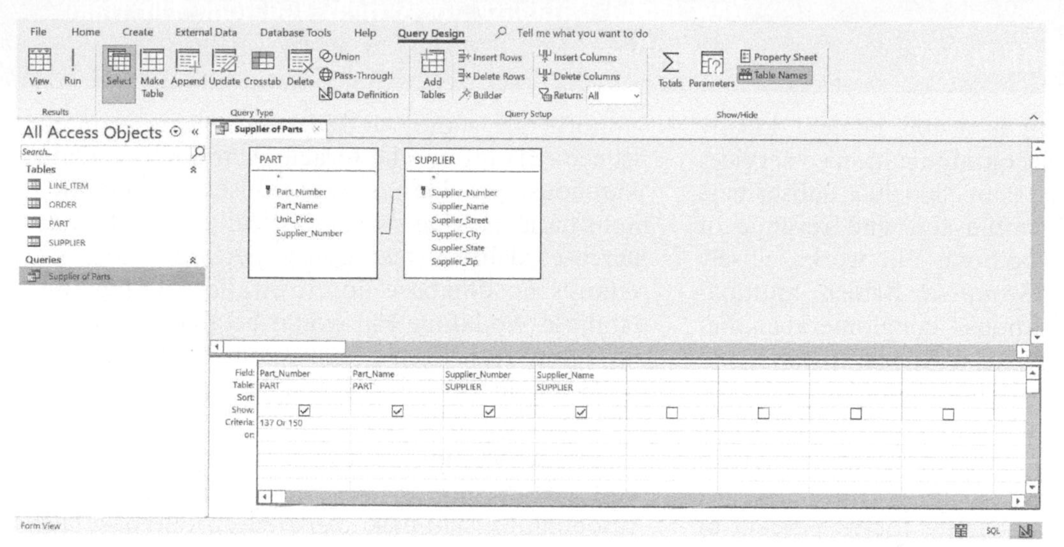


Used with permission from Microsoft.

**Figure 6.12**
**An Access Query**
*Illustrated here is how the query in Figure 6.11 would be constructed using Microsoft Access query-building tools. It shows the tables, fields, and selection criteria used for the query.*

and are designed for managing large data sets across many distributed machines and for easily scaling up or down. They are useful for accelerating simple queries against large volumes of structured and unstructured data, including web, social media, graphics, and other forms of data that are difficult to analyze with traditional SQL-based tools.

There are several different kinds of NoSQL databases, each with its own technical features and behavior. Oracle NoSQL Database is one example, as is Amazon Web Services' SimpleDB, a cloud-based NoSQL database that provides a simple web-based interface to create and store multiple data sets, query data easily, and return the results. There is no need to predefine a formal database structure or change that definition if new data are added later.

Leading Taiwanese mobile game developer Yile Technology Co. migrated to NoSQL MongoDB Atlas to improve the customer experience and increase the speed of playing. Yile's first prototype game, Online808, was initially based on a relational database management system (RDBMS). However, Yile's internal stress tests revealed performance issues related to using the RDBMS to handle the complex data for this type of application, which requires fast response times. Even when as few as 50 players connected to the game at the same time, the game response time could be as long as 3 seconds, far exceeding the 0.5-seond minimum the game designers had set. Yile was able to achieve a response time of less than 0.5 second when it switched to MongoDB's NoSQL database for the game's customer profile data.This responsiveness helped Online808 become the second-best-performing social casino game in Taiwan (MongoDB, 2022).

## Cloud Databases and Distributed Databases

Among the services that Amazon Web Services (AWS) and other cloud computing vendors provide are relational database engines. Amazon Relational Database Service (Amazon RDS) offers MySQL, Microsoft SQL Server, Oracle Database, PostgreSQL, MariaDB, and Amazon Aurora as database engines. Pricing is based on usage. Oracle has its own Database Service using its relational Oracle Database, and Microsoft Azure SQL Database is a cloud-based relational database service based on the Microsoft SQL Server DBMS. There is also a cloud version of IBM's Db2. Cloud-based data management services have special appeal for online startups or small to medium-sized businesses seeking database capabilities at a lower cost than in-house database products. (See the Spotlight on Technology case.)

Google now offers its Spanner distributed database technology as a cloud service. A **distributed database** is one that is stored in multiple physical locations. Parts or copies of the database are physically stored in one location, and other parts or copies are

## New Cloud Database Tools Help Vodafone Fiji Make Better Decisions

Vodafone Fiji Limited is a 100 percent locally owned wireless telecommunications services provider, with more than 800,000 subscribers on its network, 267 employees, and revenue of US $192.1 million. Vodafone Fiji works closely with the Vodafone Group—a British multinational telecommunications conglomerate that predominantly services Asia, Africa, Europe, and Oceania—and other operating companies to deliver cutting-edge technology to the people of Fiji. It has 85 percent of that country's market share in telecommunications services.

Prepaid customers account for 96 percent of Fiji's mobile communications market. (In comparison, only about 10 percent of US mobile customers use prepaid services.) Prepaid customers are not bound by mobile service contracts and are always looking for better ways to save. Consequently, Vodafone Fiji continually has to come up with better deals for its customers in order to persuade them to continue using its services. Unfortunately, Vodafone Fiji's systems were not able to deliver the information needed for the task.

The company lacked the computing power, storage, and data management tools to analyze data quickly to make informed decisions about which deals to offer to which customers and the right times to do so. The data were stored on premises in multiple databases. Making matters worse, Amalgamated Telecom Holdings (Vodafone Fiji's major shareholder) had recently acquired several other telecommunications companies serving nearby markets in Samoa, American Samoa, Vanuatu, Cook Islands, and Kiribati. The data to be managed and mined for insights increased threefold. All those businesses had different types of data in different systems and different formats.

Vodafone Fiji's chief commercial officer, Ronald Prasad, and his team calculated that it would cost the company about US $2.5 million to upgrade the company's on-premises systems. They chose Oracle Autonomous Data Warehouse and Oracle Analytics Cloud services as a much more suitable and cost-effective solution. Oracle Autonomous Data Warehouse Cloud (ADWC) is a fully managed, high-performance elastic cloud service providing analytical capability for data stored in the Oracle database. The environment is optimized for data warehouse workloads and supports all standard SQL and business intelligence (BI) tools. The Oracle Autonomous Data Warehouse provides an easy-to-use, fully autonomous database that scales elastically as workloads increase, delivers fast query performance, and requires no database administration. Using this database, Vodafone Fiji would be able to extract, move, and transform data from disparate sources in the cloud, where it could be analyzed much more rapidly.

By automating many of the routine tasks required to manage Oracle databases, Oracle Autonomous Database can free up database administrators (DBAs) to do higher-level and more strategic work because the warehouse system handles a great deal of tedious technical work on its own. Vodafone Fiji would not need to hire people with specialized database management skills, which would be challenging in a small job market such as Fiji's. In addition, an on-premises data warehouse would have taken the company two months to implement, whereas it set up Oracle Autonomous Data Warehouse Cloud within 30 minutes and reporting functionality within one week. Vodafone Fiji pays only for the computing resources consumed.

Oracle Analytics Cloud is a cloud-based platform that can handle data in almost any shape or size from almost any source (desktop, enterprise, data center) and has capabilities for collecting, consolidating, and transforming the data and for creating transactional and analytical reports and dashboards. Oracle Analytics Cloud provides self-service capabilities for users to perform what-if modeling and analysis. Users are empowered to visualize and discover data, including working with Big Data.

Vodafone Fiji can now easily obtain insights from the data collected and adjust its promotions to changing market dynamics. It is also able to target customers who are at risk of cancelling Vodafone services as well as those most likely to respond to a special offer. The data warehouse system can produce reports on call patterns to make sure customers are signed up for the optimal set of services or to target customers who have 4G plans and might be enticed by free trials to subscribe to a faster data plan. Promotional campaigns customized to specific cell sites can encourage customers to go online during low utilization periods.

More than 50 comprehensive reports are available to decision makers within minutes. Queries can now be completed five times faster than before. A data-mining procedure that used to take 125 minutes now takes only 25 minutes with Oracle Autonomous Data Warehouse; an aggregation query that used to take 294 seconds with the old, on-premises system now takes only 5 seconds. Examining customer service call data, data warehouse analytic tools have been able to identify trivial, repetitive inquiries where automated responses are possible, thus reducing inbound calls by 8 to 10 percent and freeing up customer service agents to focus on up-selling and cross-selling based on specific customer profiles. Prasad's same three-member team has been able to handle quadruple the preacquisition workload very effectively.

Sources: Vodafone.com.fj, accessed May 12, 2022; "Vodafone Fiji Speeds Decision Making with Oracle Cloud," Oracle.com, accessed May 8, 2022; Tara Swords, "Call to Action," *Profit Magazine*, Fall 2019.

## CASE STUDY QUESTIONS

1. Define the problem faced by Vodafone Fiji. What people, organization, and technology factors contributed to the problem?
2. Evaluate Oracle Autonomous Data Warehouse and Oracle Analytics Cloud as solutions for Vodafone Fiji.
3. How did the new Oracle tools change decision making at Vodafone Fiji?
4. Was using cloud services advantageous for Vodafone Fiji? Explain your answer.

maintained in other locations. Spanner makes it possible to store information across millions of machines in hundreds of data centers around the globe, with special time-keeping tools to synchronize the data precisely in all of its locations and ensure the data are always consistent. Google uses Spanner to support its various cloud services, including Google Photos, Google Ads (Google's online ad system), and Gmail, and is now making the technology available to other companies that might need such capabilities to run a global business.

### Blockchain

**Blockchain** is a distributed database technology that enables firms and organizations to create and verify transactions on a network nearly instantaneously without a central authority. The system stores transactions as a distributed ledger among a network of computers. The information held in the database is continually reconciled by the computers in the network.

The blockchain maintains a continuously growing list of records called blocks. Each block contains a timestamp and a link to a previous block. Once a block of data is recorded on the blockchain ledger, it cannot be altered retroactively. When someone wants to add a transaction, participants in the network (all of whom have copies of the existing blockchain) run algorithms to evaluate and verify the proposed transaction. Legitimate changes to the ledger are recorded across the blockchain in a matter of seconds or minutes, and records are protected through cryptography. What makes a blockchain system possible and attractive to business firms is encryption and authentication of the actors and participating firms, which ensure that only legitimate actors can enter information and that only validated transactions are accepted. Once recorded, the transaction cannot be changed.

Examine Figure 6.13, which illustrates how blockchain works when fulfilling an order. The first block in the chain is the order transaction record. The block is verified by the user and then broadcast to a peer-to-peer (P2P) network of computers that run an algorithm to evaluate and verify the block. Once the block is validated by the computers in the network, it is added to the chain of transactions for this user.

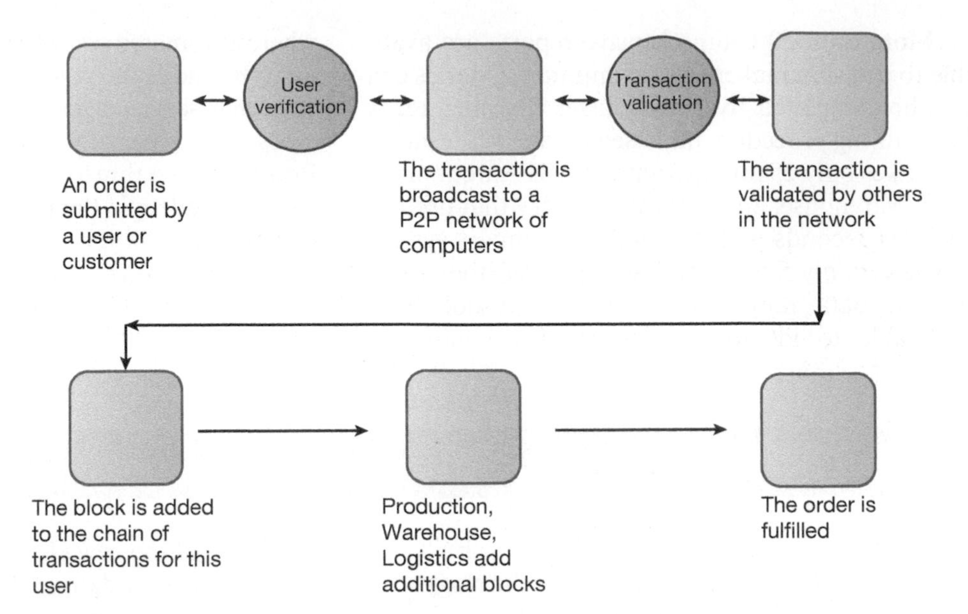


**Figure 6.13**
How Blockchain Works
*A blockchain system is a distributed database that records transactions in a peer-to-peer network of computers.*

Next, production, warehouse, and logistics add to the chain additional blocks related to the order, until the order is fulfilled.

There are many benefits to firms using blockchain databases. Blockchain networks reduce the costs of verifying users and validating transactions and also reduce the risks of storing and processing transaction information across thousands of firms. Instead of thousands of firms building their own private transaction systems and then integrating them with suppliers', shippers', and financial institutions' systems, blockchain can provide a single, simple, low-cost transaction system for participating firms. Standardization of recording transactions is aided by the use of *smart contracts.* Smart contracts are computer programs that implement the rules governing transactions among firms—e.g., what the price of products is, how they will be shipped, when the transaction will be completed, who will finance the transaction, what the financing terms are, and the like.

The simplicity, traceability, and security that blockchain offers has made it attractive for storing and securing financial transactions, supply chain transactions, medical records, and other types of data. Blockchain is becoming an increasingly popular way for businesses (particularly manufacturers and food producers) to meet consumer demand for transparency on how, where, and when their goods are sourced and produced. It also helps provide assurances about quality, ethical trade practices, and efforts to reduce carbon footprints (see the chapter-ending case study). Using blockchain's distributed ledger technology, businesses can collect all the relevant information about an end product as it makes its way through its supply chain in a trusted, traceable, and automated way, without the support or intervention of a centralized authority. Blockchain is a foundational technology for Bitcoin, Ethereum, and other cryptocurrencies. Chapter 8 provides more detail on securing transactions with blockchain.

## 6-3 Identify the principal tools and technologies for accessing information from databases to improve business performance and decision making.

Companies use their databases to keep track of basic transactions, such as paying suppliers, processing orders, serving customers, and paying employees, but companies also need databases to provide information that will help the company run the business more efficiently and help managers and employees make better decisions.

If a company wants to know which product is the most popular or who is its most profitable customer, the answer lies in the data.

## THE CHALLENGE OF BIG DATA

Most of the data that organizations once collected consisted of transaction data that could easily fit into rows and columns of relational database management systems. However, over the last two decades, there has been an explosion of data from many different sources, including web traffic, email messages, and social media content (tweets, status messages), as well as machine-generated data from sensors. These data may be unstructured or semistructured and not suitable for relational database products that organize data in the form of columns and rows. We now use the term **Big Data** to describe these data sets with volumes so huge that they are beyond the ability of typical DBMS to capture, store, and analyze.

Big Data is often characterized by the "3Vs": the extreme *volume* of the data, the wide *variety* of data types and sources, and the *velocity* at which the data must be processed. "Big Data" doesn't designate any specific quantity but usually refers to data in the petabyte and exabyte ranges—in other words, billions to trillions of records, respectively, from different sources. Big Data is produced in much larger quantities and much more rapidly than traditional data are. For example, a single jet engine is capable of generating 10 terabytes of data in just 30 minutes, and there are 45,000 airline flights each day. Twitter generates more than 12 terabytes of data daily. Digital information is growing exponentially, topping an expected 94 zettabytes worldwide in 2022. According to IDC, a leading technology research firm, the world's data are more than doubling every two years.

Businesses are interested in Big Data because these data can reveal more patterns and interesting relationships than can smaller data sets, with the potential to provide new insights into customer behavior, weather patterns, financial market activity, and other phenomena. For example, Shutterstock, the global online image marketplace, stores 200 million images and adds 55,000 more each day. To find ways to optimize the Shutterstock experience, it analyzes its Big Data to find out where its website visitors place their cursors and how long they hover over an image before making a purchase. Big Data is also being put to many uses in the public sector. For example, city governments are using Big Data to manage traffic flows and fight crime. (Review the Chapter 5 Spotlight on Technology case.)

However, to derive business value from Big Data, organizations need new technologies and tools that are capable of managing and analyzing nontraditional data along with their traditional enterprise data. They also need to know what questions to ask of the data and the limitations of Big Data. Capturing, storing, and analyzing Big Data can be expensive, and information from Big Data may not necessarily help decision makers. It's important to have a very clear understanding of the problems that Big Data will solve for the business. A number of companies have rushed to start Big Data projects without first establishing a business goal for this new information or key performance metrics to measure success. Swimming in numbers doesn't necessarily mean that the right information is being collected or that people will make smarter decisions. Organizations also won't benefit from Big Data that has not been properly cleansed, organized, and managed—think data quality.

## BUSINESS INTELLIGENCE TECHNOLOGY INFRASTRUCTURE

Suppose you wanted concise, reliable information about current operations, trends, and changes across the entire company. If you worked in a large company, the data you need might have to be pieced together from separate systems, such as sales, manufacturing, and accounting and even from external sources, such as demographic or competitor data. Increasingly, you might need to use Big Data. A contemporary technology infrastructure for business intelligence has an array of tools for obtaining useful information from all the different types of data used by businesses today, including semistructured and unstructured Big Data in vast quantities. These capabilities include data warehouses, data marts, and data lakes; Hadoop; in-memory computing; and analytical platforms. Some of these capabilities are available as cloud services.

### Data Warehouses, Data Marts, and Data Lakes

For the past three decades, the traditional tool for analyzing corporate data has been the data warehouse. A **data warehouse** is a database that stores current and historical data of potential interest to decision makers throughout the company. The data originate in many core operational transaction systems, such as systems for sales, customer accounts, and manufacturing, and may include data from website transactions. The data warehouse extracts current and historical data from multiple systems inside the organization. These data are combined with data from external sources and transformed by correcting inaccurate and incomplete data and structuring the data in a common repository for management reporting and analysis.

The data warehouse makes the data available for anyone to access as needed, but the data cannot be altered. A data warehouse system also provides a range of ad hoc and standardized query tools, analytical tools, and graphical reporting facilities. A data warehouse can be deployed on premises, in the cloud, or in a hybrid cloud environment (review Chapter 5).

Companies often build enterprise-wide data warehouses, where a central data warehouse serves the entire organization, or they create smaller, decentralized warehouses called data marts. A **data mart** is a subset of a data warehouse in which a summarized or highly focused portion of the organization's data is placed in a separate database for a specific population of users. For example, a company might develop marketing and sales data marts to deal with customer information. Bookseller Barnes & Noble used to maintain a series of data marts—one for point-of-sale data in retail stores, another for college bookstore sales, and a third for online sales.

A **data lake** is a repository for raw, unstructured data or structured data that for the most part have not yet been analyzed. The data lake stores these data in their native formats until they are needed and enables the data to be accessed in many ways.

### Hadoop

Relational DBMS and data warehouse products are not well suited for organizing and analyzing Big Data or data that do not easily fit into columns and rows used in their data models. For handling vast quantities of unstructured and semistructured data, as well as structured data, organizations are using **Hadoop**. Hadoop is an open-source software framework managed by the Apache Software Foundation that enables distributed, parallel processing of very large amounts of data across inexpensive computers. It breaks a Big Data problem down into subproblems, distributes them among up to thousands of inexpensive computer processing nodes, and then combines the results into a smaller data set that is easier to analyze. You've probably used Hadoop to find the best airfare on the Internet, do a search on Google, or connect with a friend on Facebook.

Hadoop consists of several key services, including the Hadoop Distributed File System (HDFS) for data storage and MapReduce for high-performance parallel data processing. HDFS links together the file systems on the numerous nodes in a Hadoop cluster to turn them into one big file system. HDFS is often used to store the data lake contents across a set of clustered computer nodes, and Hadoop clusters may be used to preprocess some of these data for use in a data warehouse, data mart, or analytic platform or for direct querying by power users. Hadoop's MapReduce was inspired by Google's MapReduce system for breaking down the processing of huge data sets and assigning work to the various nodes in a cluster. HBase, Hadoop's nonrelational database, provides rapid access to the data stored on HDFS and a transactional platform for running high-scale, real-time applications. Hadoop can process large quantities of any kind of data, including structured transactional data, loosely structured data such as Facebook and Twitter feeds, complex data such as web server log files, and unstructured audio and video data.

Hadoop runs on a cluster of inexpensive servers, and processors can be added or removed as needed. Companies use Hadoop for analyzing very large volumes of data as well as for a staging area for unstructured and semistructured data before they are loaded into a data warehouse. For instance, LinkedIn runs its Big Data analytics on Hadoop. It now stores more than 1 exabyte of data across all of its Hadoop clusters. Its largest cluster stores 500 petabytes of data and maintains one billion objects. eBay is another company that uses Hadoop. eBay recently transitioned from a vendor-based data warehouse used to store more than 20 petabytes of web analytics and transactional data, such as bids, checkouts, listings, users, and accounts, to an open-source-based Hadoop system. Companies offering Hadoop solutions include AWS, Cloudera, Microsoft, IBM, Oracle, and HPE. Other vendors offer tools for moving data into and out of Hadoop or for analyzing data within Hadoop.

Although Hadoop is still being used by many companies, it is not without its drawbacks. For instance, Hadoop is not very useful for smaller data sets. Apache Spark is a more recently developed data processing engine for Big Data. Like Hadoop, Spark splits up large tasks across different nodes. However, Spark uses random access memory (RAM) to cache and process data instead of a file system and can be much faster than Hadoop, particularly for smaller workloads. Spark also has a library of machine learning algorithms optimized for a variety of machine learning processes.

### In-Memory Computing

Another way of facilitating Big Data analysis is to use **in-memory computing**, which relies primarily on a computer's main memory (RAM) for data storage. (Conventional DBMS use disk storage systems.) Users access data stored in system's primary memory, thereby eliminating the bottlenecks that occur while retrieving and reading data in a traditional, disk-based database and dramatically shortening query response times. In-memory processing makes it possible for very large sets of data, amounting to the size of a data mart or a small data warehouse, to reside entirely in memory. Complex business calculations that used to take hours or days are able to be completed within seconds, and these calculations can be accomplished even on mobile devices.

The previous chapter describes some of the advances in contemporary computer hardware technology that make in-memory processing possible, such as powerful, high-speed processors, multicore processing, and falling computer memory prices. These technologies help companies optimize the use of memory and accelerate processing performance while lowering costs. Leading in-memory database products include SAP HANA, Oracle Database In-Memory, Azure SQL Database, and Teradata Intelligent Memory.

### Analytic Platforms

Commercial database vendors have developed specialized high-speed **analytic platforms** that use both relational and nonrelational technology that are optimized for analyzing large data sets. Analytic platforms feature preconfigured hardware–software systems that are specifically designed for query processing and analytics. For example, IBM PureData System for Analytics features tightly integrated database, server, and storage components that handle complex analytic queries 10 to 100 times more quickly than traditional systems can. Analytic platforms also include in-memory systems and NoSQL nonrelational database management systems. Analytic platforms are now available as cloud services.

Examine Figure 6.14, which illustrates a contemporary business intelligence technology infrastructure using the technologies we have just described. Current and historical data are extracted from multiple operational systems along with Internet of Things (IoT) machine-generated data, web and social media data, unstructured audio/visual data, and other data from external sources. All of these different types

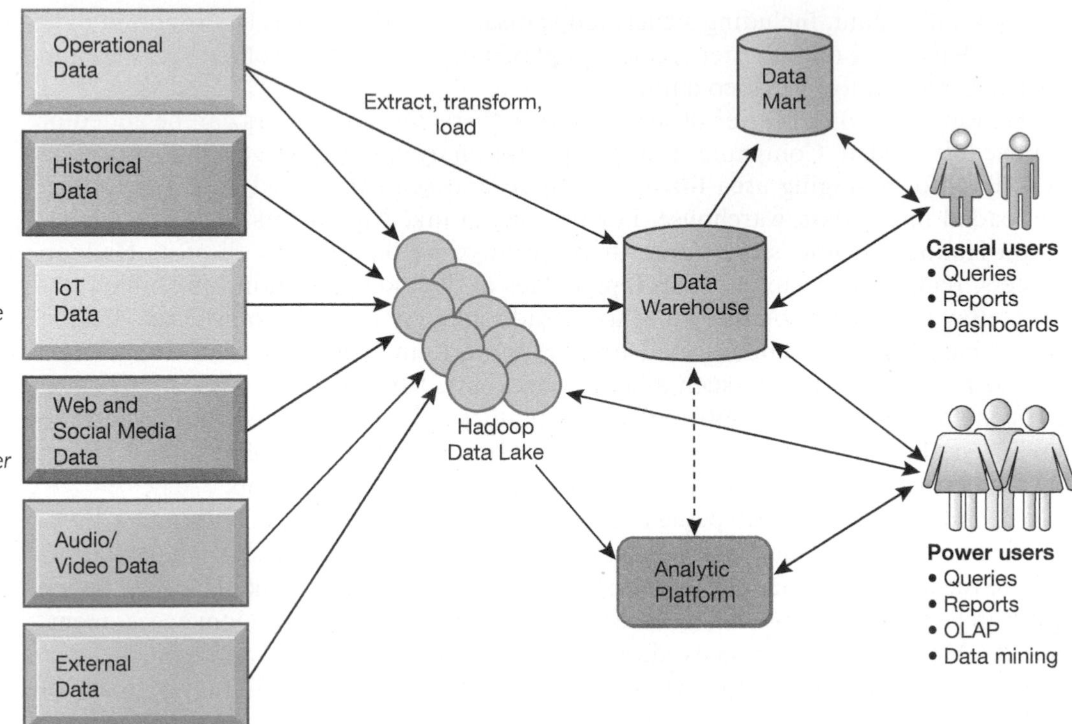


**Figure 6.14**
**Business Intelligence Technology Infrastructure**
*A contemporary business intelligence technology infrastructure features capabilities and tools to manage and analyze large quantities and different types of data from multiple sources. Easy-to-use query and reporting tools for casual business users and more sophisticated analytical toolsets for power users are included.*

of data are poured into a Hadoop data lake. The Hadoop Distributed File System (HDFS) is often used to store the data lake contents across a set of clustered computer nodes, and Hadoop clusters may be used to preprocess some of these data for use in a data warehouse, data marts, or an analytic platform or for direct querying by power users. Casual users can query the data for display on dashboards or to create reports, while power users typically have additional tools or can engage in direct querying. Outputs include reports and dashboards as well as query results. Chapter 11 discusses the various types of BI (business intelligence) users and BI reporting in greater detail.

## ANALYTICAL TOOLS: RELATIONSHIPS, PATTERNS, TRENDS

When data have been captured and organized using the business intelligence technologies we have just described, they are available for further analysis by using software for database querying and reporting, multidimensional data analysis (OLAP), and data mining. This section will introduce you to these tools, with more detail about business intelligence analytics and applications in Chapter 11.

### Online Analytical Processing (OLAP)

Suppose your company sells four products—nuts, bolts, washers, and screws—in the East, West, and Central regions of the United States. If you wanted to ask a straightforward question, such as how many washers were sold during the past quarter, you could easily find the answer by querying your sales database. However, what if you wanted to know how many washers were sold in each of your sales regions and to compare actual results with projected sales?

To obtain the answer, you could use **online analytical processing (OLAP)**. OLAP supports multidimensional data analysis, enabling users to view the same data in different ways using multiple dimensions. Each aspect of information—product, pricing, cost, region, or time period—represents a different dimension. A product manager could use a multidimensional data analysis tool to learn how many washers were sold in the East in June, how that compares with the previous month and the previous June, and how it compares with the sales forecast. OLAP enables users to obtain

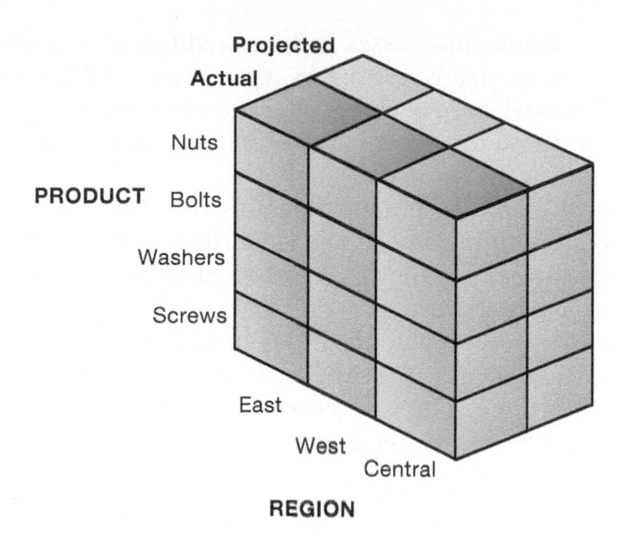


**Figure 6.15**
Multidimensional Data Model
*This view shows product versus region. If you rotate the cube 90 degrees, the face that will show is product versus actual and projected sales. If you rotate the cube 90 degrees again, you will see region versus actual and projected sales. Other views are possible.*

online answers to ad hoc questions such as these in rapid time, even when the data are stored in very large databases, such as those containing sales figures for multiple years.

Examine Figure 6.15, which shows a multidimensional data model that could be created to represent products, regions, actual sales, and projected sales. A matrix of actual sales can be stacked on top of a matrix of projected sales to form a cube with six faces. If you rotate the cube 90 degrees one way, the face showing will be product versus actual and projected sales. If you rotate the cube 90 degrees again, you will see region versus actual and projected sales. If you rotate the cube 180 degrees from the original view, you will see projected sales and product versus region. Cubes can be nested within cubes to build complex views of data. To create such models, a company can use either a specialized, multidimensional database or a tool that creates multidimensional views of data in relational databases.

## Data Mining

Traditional database queries answer such questions as "How many units of product number 403 were shipped in November 2022?" OLAP, or multidimensional analysis, supports much more complex requests for information, such as "Compare sales of product number 403 relative to the sales plan by quarter and sales region for the past two years." With OLAP and query-oriented data analysis, users need to have a good idea about the information for which they are looking.

**Data mining** is more discovery-driven. Data mining provides insights from data that cannot be obtained with OLAP by finding hidden patterns and relationships in large databases and inferring rules from those patterns and relationships to predict future behavior. The patterns and rules are used to guide decision making and forecast the effect of those decisions. The types of information obtainable from data mining include associations, sequences, classifications, clusters, and forecasts.

- *Associations* are occurrences linked to a single event. For instance, a study of supermarket purchasing patterns might reveal that when corn chips are purchased, a cola drink is purchased 65 percent of the time, but when there is a promotion, cola is purchased 85 percent of the time. This information helps managers make better decisions because they have learned the profitability of a promotion.
- In *sequences*, events are linked over time. We might find, for example, that if a house is purchased, a new refrigerator will be purchased within two weeks 65 percent of the time, and an oven will be bought within one month of the home purchase 45 percent of the time.
- *Classification* recognizes patterns that describe the group to which an item belongs by examining existing items that have been classified and by inferring a

set of rules. For example, businesses such as credit card or telephone companies worry about the loss of steady customers. Classification helps reveal the characteristics of customers who are likely to leave and can provide a model to help managers predict who those customers are so that the managers can devise special campaigns to retain such customers.

- *Clustering* works in a manner similar to classification but when no groups have yet been defined. A data-mining tool can reveal different groupings within data, such as affinity groups for bank cards or groups of customers based on demographics and types of personal investments.
- Although all of the preceding applications involve predictions, *forecasting* uses predictions in a different way. It uses a series of existing values to forecast what other values will be. For example, forecasting might find patterns in data to help managers estimate the future value of continuous variables, such as sales figures.

These systems perform high-level analyses of patterns or trends, and more sophisticated data-mining applications take advantage of artificial intelligence machine learning techniques (described in Chapter 11). They can also drill down to provide more detail when needed. There are data-mining applications for all the functional areas of business and for government and scientific work. One popular use of data mining is to provide detailed analyses of patterns in customer data for one-to-one marketing campaigns or for identifying profitable customers.

Forbes, the global media company that publishes *Forbes* magazine, relies on revenue from advertising and subscriptions. It mines the extensive data it has collected on individual readers. Forbes claims that it can understand each person who interacts with its brand. Whether that person is a subscriber or a registered website visitor, Forbes has information on that person's demographics, values, and lifestyle as well as how that person has interacted with Forbes over the years. These details help *Forbes* advertisers target their campaigns more precisely and also help Forbes' publications increase their circulation.

## Text Mining and Web Mining

Unstructured data, most often in the form of text files, are believed to account for more than 80 percent of useful organizational information and are some of the major sources of Big Data that firms want to analyze. Emails, memos, call center transcripts, survey responses, legal cases, patent descriptions, and service reports are all valuable for revealing patterns and trends that will help employees make better business decisions. **Text mining** tools that help businesses analyze these data are now available. These tools can extract key elements from unstructured Big Data sets, reveal patterns and relationships, and summarize the information.

Businesses might turn to text mining to analyze transcripts of calls to customer service centers to identify major service and repair issues or to measure customer sentiment about their company. **Sentiment analysis** software can mine text comments in an email message, blog, social media conversation, or survey form to determine favorable and unfavorable opinions about specific subjects. For example, Google Chrome's development team is constantly monitoring user feedback, viewing messages from users to determine positive or negative sentiment, wishes, and recommendations regarding the product in general and also specific elements such as security or the user interface. Google uses this information to help guide further research and development of the product (Vodovatova, 2022).

The web is another rich source of unstructured Big Data that can reveal patterns, trends, and insights into customer behavior. The discovery and analysis of useful patterns and information from the web is called **web mining**. Businesses might turn to web mining to help them understand customer behavior, evaluate the effectiveness of a particular website, or quantify the success of a marketing campaign. For instance, marketers use Google Trends, which tracks the popularity of various words and phrases used in Google search queries, to learn what people are interested in and what they are interested in buying.

Web mining looks for patterns in data through content mining, structure mining, and usage mining. Web content mining is the process of extracting knowledge from the content of web pages, which may include text, images, audio, and video data. Web structure mining examines data related to the structure of a particular website. For example, links pointing to a document indicate the popularity of the document; links coming out of a document indicate the richness or perhaps the variety of topics covered in the document. Web usage mining examines the user interaction data that a web server records whenever requests for a website's resources are received. The usage data records the user's behavior when the user browses or makes transactions on the website and collects the data in a server log. Analyzing such data can help companies determine the value of particular customers, cross-marketing strategies across products, and the effectiveness of promotional campaigns.

## DATABASES AND THE WEB

Many companies are using the web to make some of the information in their internal databases available to customers and business partners. Prospective customers might use a company's website to view the company's product catalog or to place an order. The company in turn might use the web to check inventory availability for that product from its supplier.

These actions involve accessing and (in the case of ordering) updating corporate databases via the web. Suppose, for example, a customer with a web browser wants to search an online retailer's database for pricing information. Examine Figure 6.16, which illustrates how that customer might access the retailer's internal database via the web. The customer accesses the retailer's website via the Internet using web browser software on a desktop computer or mobile device. The customer's web browser software then requests data from the organization's database using HTML commands to communicate with the web server.

Because many back-end databases cannot interpret commands written in HTML, the web server passes these requests for data to software that translates HTML commands into SQL so that the DBMS working with the database can process the commands. In a client/server environment, the DBMS often resides on a dedicated computer called a **database server**. The DBMS receives the SQL requests and provides the required data. The information is then transferred from the organization's internal database back to the web server for delivery in the form of a web page to the user.

Figure 6.16 also shows a dedicated computer, called an application server, between the web server and database server (see Chapter 5). The application server takes requests from the web server, runs the business logic to process transactions based on those requests, and provides connectivity to the organization's back-end systems or databases.

There are a number of advantages to using the web to access an organization's internal databases. First, almost everyone knows how to use web browser software, and employees require much less training than if they used proprietary query tools. Second, the web interface requires few or no changes to the internal database. Companies leverage their investments in older systems because it costs much less to add a web interface in front of a legacy system than to redesign and rebuild the system to improve user access.

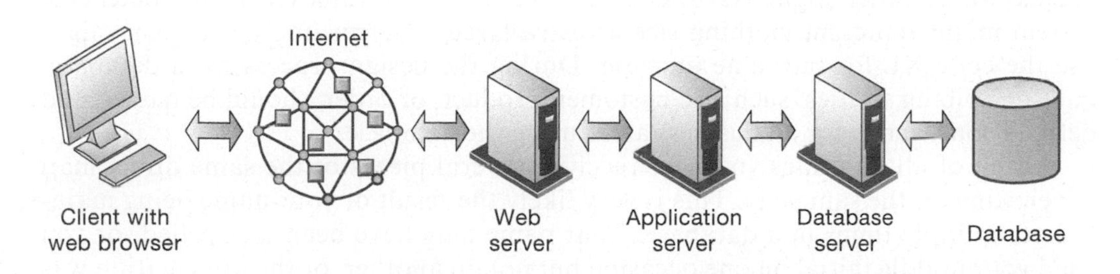


**Figure 6.16**
**Linking Internal Databases to the Web**
*Users access an organization's internal database via the web, using their desktop PCs or mobile devices and web browser software.*

Accessing corporate databases via the web is creating new efficiencies and opportunities, and, in some cases, it is even changing the way business is being done. Facebook (Meta) and Netflix are examples of entirely new businesses based on access to large databases via the web. Facebook maintains a massive database to house and manage all the data it collects about its 2.9 billion active users and their interests, friends, and photos. Netflix collects and stores data about the interests and video streaming subscription plans of its more than 220 million paid members.

## 6-4 Explain why data governance and data quality assurance are essential for managing a firm's data resources.

Setting up a database is only a start. To make sure that the data for your business remain accurate, reliable, and readily available to those who need them, your business will need special policies and procedures for data governance. **Data governance** encompasses policies and procedures through which data can be managed as an organizational resource. It establishes the organization's rules for sharing, disseminating, acquiring, standardizing, classifying, and inventorying information. These rules include identifying which users and organizational units can share information, where information can be distributed, who is responsible for updating and maintaining the information, and how data resources should be secured (see Chapter 8). A firm's information policy might specify, for example, that only selected members of the payroll and human resources department have the right to change or view sensitive employee data, such as an employee's salary or social security number, and that these departments are responsible for making sure that such employee data are accurate.

### ENSURING DATA QUALITY

With today's organizations relying so heavily on data to drive operations and decision making, data quality assurance is especially important. What would happen if a customer's telephone number or account balance were incorrect? What would be the impact if the database had the wrong price for the product being sold? Data that are inaccurate, untimely, or inconsistent with other sources of information create serious operational and financial problems for businesses, even those that have a well-designed database and information policy. When faulty data go unnoticed, they often lead to incorrect decisions, product recalls, and even financial losses.

Gartner Inc. has reported that more than 25 percent of the critical data in large *Fortune* 1000 companies' databases is inaccurate or incomplete, including bad product codes and product descriptions, faulty inventory descriptions, erroneous financial data, incorrect supplier information, and incorrect employee data. Some of these data quality problems are caused by redundant and inconsistent data produced by multiple systems. For example, the sales ordering system and the inventory management system might both maintain data on the organization's products. However, the sales ordering system might use the term *Item Number*, whereas the inventory system calls the same attribute *Product Number*. The sales, inventory, or manufacturing systems of a clothing retailer might use different codes to represent values of an attribute. One system might represent clothing size as extra large, whereas another system might use the code XL for the same purpose. During the design process for a database, data-describing entities, such as a customer, product, or order, should be named and defined consistently for all business areas using the database.

Think of all the times you have received several pieces of the same direct mail advertising on the same day. This is very likely the result of your name being maintained multiple times in a database. Your name may have been misspelled, or you used your middle initial on one occasion but not on another, or the information was initially entered on a paper form and not scanned properly into the system. Because of these inconsistencies, the database would treat you as three different people! We often receive redundant mail addressed to "Laudon," "Lavdon," "Lauden," or "Landon."

If a database is properly designed and enterprise-wide data standards are established, duplicate or inconsistent data elements should be minimal. Most data quality problems, however, such as misspelled names, transposed numbers, or incorrect or missing codes, stem from errors during data input. The incidence of such errors is rising as companies move their businesses to the web and allow customers and suppliers to enter data via their websites that directly update the company's internal systems.

Before a new database is in place, organizations need to identify and correct their faulty data and establish better routines for editing data once their database is in operation. Analysis of data quality often begins with a **data quality audit**, which is a structured survey of the accuracy and level of completeness of the data in an information system. Data quality audits can be performed by surveying entire data files, surveying samples from data files, or surveying end users for their perceptions of data quality.

**Data cleansing**, also known as *data scrubbing*, consists of processes for detecting and correcting data in a database that are incorrect, incomplete, improperly formatted, or redundant. Data cleansing not only corrects data but also enforces consistency among different sets of data that originated in separate information systems. Specialized data-cleansing software is available to survey data files automatically, correct errors in the data, and integrate the data in a consistent, companywide format.

Data quality problems are not just business problems. They also pose serious problems for individuals by affecting their financial condition and even their jobs. For example, inaccurate or outdated data about consumers' credit histories maintained by credit bureaus can prevent creditworthy individuals from obtaining loans or lower their chances of finding or keeping a job. The Spotlight on Organizations case shows how poor data quality can pose problems for manufacturing, health, and safety

## 6-5 Understand how MIS can help your career.

Here is how this chapter and this book can help you find a job as an entry-level data analyst.

### THE COMPANY

Mega Midwest Power, a large, diversified energy company headquartered in Cleveland, Ohio, has an open position for an entry-level data analyst. The company is involved in the distribution, transmission, and generation of electricity as well as energy management and other energy-related services for five million customers in the Midwest and mid-Atlantic regions.

### POSITION DESCRIPTION

Job responsibilities include:

- Maintaining the integrity of substation equipment and related data in multiple databases, including SAP.
- Querying databases in multiple systems.
- Modifying systems for proper data management and procedural controls.
- Recommending and implementing process changes based on data problems that are identified.

## SPOTLIGHT ON: ORGANIZATIONS Higher Data Quality Helps Vyaire Save Lives

Vyaire Medical, headquartered in Chicago, is one of the world's biggest medical device manufacturers, with operations in 40 countries. Vyaire was formed in 2016 as a standalone business that had previously been the respiratory products division of global medical device manufacturer Becton Dickinson. Vyaire's product line focuses exclusively on supporting breathing through every stage of life. Among the devices the company produces are the breathing ventilators that were in desperately high demand when the Covid-19 pandemic erupted in early 2020.

At that time, Vyaire was not producing nearly enough life-saving ventilators to meet medical needs—only 6 per day and 30 per week. One of the key metrics for scaling up ventilator production is a manufacturing benchmark called first pass yield, which measures the number of units produced in a production run that don't have defects. Vyaire made a breathing ventilator that had to pass through about 21 different manufacturing stations before it was completed. Each station had a test for product quality and regulatory controls. A device could move on to the next stage only after passing the previous stage's test. Only 20 percent of ventilator devices were getting through all the stages for a complete first pass yield. With such a low level of quality, Vyaire would never be able to scale up ventilator production for people affected by the pandemic.

Using good data enabled Vyaire engineering and manufacturing teams to pinpoint what was failing. More accurate, reliable data would help Vyaire make better decisions about how to improve production quality going forward. Vyaire also needed to address the remaining problems about the quality of the data itself.

As a relatively new standalone business, Vyaire in 2018 had a set of only basic capabilities for data management and data analytics. There were numerous data quality issues. Employees would choose data from systems and send Excel spreadsheets around. A key executive in the company received five reports from different employees every day about new orders. Each report said something different, causing the executive to have no idea which data were correct.

The company then created a Business Intelligence Center of Excellence to optimize the use of data across the organization using AWS cloud services such as Amazon Redshift and Amazon QuickSight. Amazon Redshift is AWS's data warehouse tool designed to handle large-scale data sets and analytic workloads and is the most widely used cloud data warehouse product in the world. Vyaire integrated data from a dozen ERP systems into a single AWS cloud-based Insights platform. Manufacturing and operations data funnel into a data lake, with Amazon Redshift used as a data warehouse.

Amazon QuickSight is a cloud-based business intelligence (BI) service that connects to an organization's data in the cloud and combines data from many different sources. QuickSight provides an interactive visual environment where decision makers can explore and interpret information with secure access to dashboards from any device, including mobile devices, on an organization's network. A single QuickSight data dashboard can include AWS data, third-party data, Big Data, spreadsheet data, SaaS data, B2B data, and more. Amazon QuickSight is a fully managed cloud-based service.

To achieve a high level of data quality throughout the enterprise, Vyaire deployed the Talend Data Fabric platform. Talend Data Fabric combines tools for data integration, integrity, and governance in a single, unified platform for maintaining high-quality data. It is able to cleanse and integrate data that are coming from different sources to ensure accuracy and supports Vyaire's master data management effort to identify duplicate data and consolidate information. Vyaire has to integrate data from many sources to obtain an enterprise-wide view, and Talend enables the company to integrate data as the data come into the cloud.

Vyaire's ability to generate consistent, clean data from multiple sources has improved sales, customer relationship management, and supply chain management as well as the company's ability to take advantage of new growth opportunities. Vyaire went from producing 6 ventilators per day to 600 per day for healthcare providers around the world.

Sources: Talend.com, accessed May 6, 2022; Vyaire.com, accessed May 6, 2022; Sean Michael Kerner, "Healthcare Device Maker Boosts Production with Data Quality," *TechTarget*, June 16, 2021.

## CASE STUDY QUESTIONS

1. Identify the problem in this case study. What people, organization, and technology factors contributed to this problem?
2. Describe the business impact of low data quality at Vyaire.
3. Describe the solution Vyaire selected. What was the role of data quality in the solution? How effective was it?

- Conducting business-specific research, gathering data, and compiling reports and summaries.
- Expanding knowledge of policies, practices, and procedures.

## JOB REQUIREMENTS

- BA/BS degree in business, finance, accounting, economics, engineering, or related discipline
- One to two years of professional work experience desirable
- Knowledge of Microsoft Office tools (Excel, PowerPoint, Access, and Word)
- Strong analytical capabilities, including attention to detail, problem solving, and decision making
- Strong oral and written communication and teamwork skills
- Familiarity with transmission-substation equipment desirable

## INTERVIEW QUESTIONS

1. What do you know about substation equipment? Have you ever worked with SAP for Utilities?
2. What do you know about data management and databases? Have you ever worked with data management software? If so, what exactly have you done with it?
3. Tell us what you can do with Access and Excel. What kinds of problems have you used these tools to solve? Did you take courses in Access or Excel?
4. What experience do you have with analyzing problems and developing specific solutions? Can you give an example of a problem you helped solve?

## AUTHOR TIPS

1. Do some research on the electric utility industry's equipment maintenance and software for electric utility asset management and predictive maintenance. Read blogs from IBM, Deloitte, and Intel about predictive maintenance, and watch YouTube videos from GE and IBM on this topic.
2. Review this chapter and this text on data management and databases, along with the Chapter 11 discussion of operational intelligence. Ask what you would be expected to do with databases in this job position.
3. Do some research on the capabilities of SAP for Utilities, and find out exactly how you would be using this software and what skills would be required. Watch SAP's YouTube video on SAP for Utilities.

## Review Summary

**6-1 Define a database and explain how a relational database organizes data.** A database is a group of related files that keeps track of people, places, and things (entities) about which organizations maintain information. The relational database is the primary method for organizing and maintaining data today in information systems. It organizes data in two-dimensional tables with rows and columns called relations. Each table contains data about an entity and its attributes. Each row represents a record, and each column represents an attribute or field. Each table also contains a key field to identify each record uniquely for retrieval or manipulation. An entity-relationship diagram graphically depicts the relationship among entities (tables) in a relational database. The process of breaking down complex groupings of data and streamlining them to minimize redundancy and awkward many-to-many relationships is called normalization.

**6-2 Describe the principles of a database management system.** A DBMS consists of software that permits centralization of data and data management so that businesses have a single, consistent source for all their data needs. A single database services multiple applications. The DBMS separates the logical and physical views of data so that the user does not have to be concerned about the data's physical location. The principal capabilities of a DBMS include a data definition capability, a data dictionary capability, and a data manipulation language. Nonrelational databases are becoming popular for managing types of data that can't be handled easily by the relational data model.

**6-3 Identify the principal tools and technologies for accessing information from databases to improve business performance and decision making.** Contemporary data management technology has an array of tools for obtaining useful information from all the types of data that businesses use today, including semistructured and unstructured Big Data in very large quantities from many different sources. These capabilities include data warehouses, data marts, and data lakes; Hadoop; in-memory computing; and analytical platforms. OLAP represents relationships among data as a multidimensional structure, which can be visualized as cubes of data and cubes within cubes of data. Data mining analyzes large pools of data, including the contents of data warehouses, to find patterns and rules that can be used to predict future behavior and guide decision making. Text mining tools help businesses analyze large, unstructured data sets consisting of text. Web mining tools focus on analyzing useful patterns and information from the web and examining the structure of websites, the activities of website users, and the contents of web pages. Conventional databases can be linked to the web or a web interface to facilitate user access to an organization's internal data.

**6-4 Explain why data governance and data quality assurance are essential for managing a firm's data resources.** Developing a database environment requires policies and procedures for managing organizational data as well as a good data model and database technology. Data governance encompasses organizational policies and procedures for the maintenance, distribution, and use of information in the organization. Data that are inaccurate, incomplete, or inconsistent create serious operational and financial problems for businesses if these data lead to bad decisions about the actions the firm should take. Ensuring data quality involves using enterprise-wide data standards, databases designed to minimize inconsistent and redundant data, data quality audits, and data cleansing software.

## Key Terms

Analytic platform, 213
Attributes, 198
Big Data, 211
Bit, 197
Blockchain, 209
Byte, 197
Database, 197
Database management system (DBMS), 203
Database server, 217
Data cleansing, 219
Data definition, 204
Data dictionary, 204
Data governance, 218
Data lake, 212
Data manipulation language, 204
Data mart, 212
Data mining, 215
Data quality audit, 219
Data warehouse, 212
Distributed database, 207
Entity, 198
Entity-relationship diagram, 200
Field, 197
File, 197
Foreign key, 199
Hadoop, 212
In-memory computing, 213
Key field, 198
Nonrelational database management systems, 206
Normalization, 202
Online analytical processing (OLAP), 214
Primary key, 198
Query, 204
Record, 197
Referential integrity, 202
Relational database, 198
Report generator, 206
Sentiment analysis, 216
Structured Query Language (SQL), 204
Text mining, 216
Web mining, 216

## Review Questions

**6-1** Define a database and explain how a relational database organizes data.
- Define a database.
- Define and explain the significance of entities, attributes, and key fields.
- Define a relational database, and explain how it organizes and stores information.
- Explain the role of entity-relationship diagrams and normalization in database design.

**6-2** Describe the principles of a database management system.
- Define a database management system (DBMS), describe how it works, and explain how it benefits organizations.
- Define and compare the logical and physical views of data.
- Define and describe the three operations of a relational database management system.
- Name and describe the three major capabilities of a DBMS.
- Define a nonrelational database management system, and explain how it differs from a relational DBMS.

**6-3** Identify the principal tools and technologies for accessing information from databases to improve business performance and decision making.
- Define Big Data, and describe the technologies for managing and analyzing Big Data.
- List and describe the components of a contemporary business intelligence technology infrastructure.
- Describe the capabilities of online analytical processing (OLAP).
- Define data mining, describe what types of information can be obtained from it, and explain how it differs from OLAP.
- Explain how text mining and web mining differ from conventional data mining.
- Explain how users can access information from a company's internal databases through the web.

**6-4** Explain why data governance and data quality assurance are essential for managing a firm's data resources.
- Define data governance, and explain how it helps organizations manage their data.
- List and describe the most common data quality problems.
- List and describe the most important tools and techniques for ensuring data quality.

## Discussion Questions

**6-5** It has been said that you do not need database management software to create a database environment. Discuss.
MyLab MIS

**6-6** To what extent should end users be involved in the selection of a database management system and database design?
MyLab MIS

**6-7** What are the consequences of an organization not having an information policy?
MyLab MIS

**MyLab MIS**
To complete these problems, go to EOC Discussion Questions in **MyLab MIS**.

## Hands-On MIS Projects

The projects in this section give you hands-on experience in establishing companywide data standards, creating a database for inventory management, and using the web to search online databases for overseas business resources. Visit MyLab MIS to access this chapter's Hands-On MIS Projects.

### MANAGEMENT DECISION PROBLEM

**6-8** Your industrial supply company wants to create a data warehouse from which management can obtain a single, corporate-wide view of critical sales information to identify best-selling products, key customers, and current sales trends. Your sales and product information are stored in two systems: a divisional sales system running on a Linux server and a corporate sales system running on an IBM mainframe. You would like to create a single, standard format that consolidates these data from both systems. In MyLab MIS, you can review the proposed format along with sample files from the two systems that would supply the data for the data warehouse. Then answer the following questions:

- What business problems are created by not having these data in a single, standard format?
- How easy would it be to create a database with a single, standard format that could store the data from both systems? Identify the problems that would have to be addressed.
- Should the problems be solved by database specialists or general business managers? Explain.
- Who should have the authority to finalize a single, corporate-wide format for this information in the data warehouse?

### ACHIEVING OPERATIONAL EXCELLENCE: BUILDING A RELATIONAL DATABASE FOR INVENTORY MANAGEMENT

Software skills: Database designing, querying, and reporting
Business skills: Inventory management

**6-9** In this exercise, you will use database software to design a database for managing inventory for a small business. Morgan's Skateboard Shop, located in Santa Cruz, California, sells skateboards for various skateboarding styles, including freestyle skateboarding, street skateboarding, park skateboarding, downhill skateboarding, and cruising. Using the information found in the tables in MyLab MIS, create queries and reports to manage information about Morgan's suppliers and products.

After you have reviewed the database, perform the following activities.

- Prepare a report that identifies the five-most-expensive skateboards. The report should list the skateboards in descending order from most expensive to least expensive, the quantity on hand for each, and the markup percentage for each.
- Prepare a report that lists each supplier, its products, the quantities on hand, and associated reorder levels. The report should be sorted alphabetically by supplier. Within each supplier category, the products should be sorted alphabetically.
- Prepare a report listing only the skateboards that are low in stock and need to be reordered. The report should provide supplier information for the identified items.
- Write a brief description of how the database could be enhanced to improve management of the business further. What tables or fields should be added? What additional reports would be useful?

## IMPROVING DECISION MAKING: SEARCHING ONLINE DATABASES FOR OVERSEAS BUSINESS RESOURCES

Software skills: Online databases
Business skills: Researching services for overseas operations

**6-10** This project develops skills in searching online web-enabled databases for information about products and services in faraway locations.

Your company is located in Greensboro, North Carolina, and manufactures office furniture of various types. You are considering opening a facility to manufacture and sell your products in Australia. You would like to contact organizations that offer a variety of services necessary for you to open your Australian office and manufacturing facility, including attorneys, accountants, import-export experts, and telecommunications equipment and support firms. Access the following online databases to locate companies that you would like to meet with during your upcoming trip: ShowMeLocal Australia, Australiatradenow, and the Nationwide Business Directory of Australia. If necessary, use search engines such as Google and Bing.

- List the companies you would contact on your trip to determine whether they can help you with these and any other functions you think are vital to establishing your office.
- Rate the databases you used for accuracy of name, completeness, ease of use, and general helpfulness.

## COLLABORATION AND TEAMWORK PROJECT

Identifying Entities and Attributes in an Online Database

**6-11** With your team of three or four students, select an online database to explore, such as Best Buy or IMDb (the Internet Movie Database). Explore one of these websites to see what information it provides. List the entities and attributes that the company running the website must keep track of in its databases. Diagram the relationships among the entities you have identified. If possible, use Google Docs and Google Drive or Google Sites to brainstorm, organize, and develop a presentation of your findings for the class.

## BUSINESS PROBLEM-SOLVING CASE

### PURSUING SUSTAINABILITY WITH BLOCKCHAIN

Unilever is one of the world's largest consumer goods companies, with more than 400 brands, such as Dove, Hellmann's Mayonnaise, and Seventh Generation, and operating in more than 190 countries. Among large corporations, Unilever is also a leader in sustainability and has made sustainability an important objective of its business strategy. In December 2020, the Unilever Board announced a Climate Transition Action Plan (CTAP) detailing the company's ambitious emissions reduction targets. The Climate Transition Action Plan sets out the steps the company is taking to reduce carbon emissions to zero within its own operations by 2030 and to reduce carbon emissions to net zero across its value chain (see Chapter 3) by 2039. Efforts to meet these ambitious goals focus on Unilever's operations, its brands and products, its value chain, and its wider influence on society. The company is transitioning to renewable energy for its operations; finding new, low-carbon ingredients; expanding its plant-based product range; and developing fossil fuel-free cleaning and laundry products.

One of Unilever's sustainability initiatives concerns the production of palm oil, which is derived from the fruit of the oil palm tree and is used in thousands of consumer products, including detergents, cosmetics, chocolate, and biofuel. Expanding palm oil production increases deforestation as rain forests are burned to clear the way for more palm plantations. Palm oil production has more than tripled over the past two decades because palm oil is one of the most common raw materials in so many household products. Expansion of palm oil production is a leading cause of carbon dioxide emissions, rain forest destruction, and human rights problems, Moreover, even if palm oil is sustainably sourced, it can be mixed with palm oil that isn't responsibly sourced, making it difficult for companies to create sustainable supply chains.

Most of Unilever's oil palm plantations are located in the Leuser rain forest in the Aceh province of Indonesia. The palm oil industry has poor production standards that have accelerated deforestation. A fall 2021 Unilever report on the environmental issues facing the palm oil industry stated that about 128,700 hectares of forests were destroyed by palm oil expansion between 2016 and 2020. (One hectare is equivalent to approximately 2.471 acres.)

Unilever has made a commitment to sustainably source all the palm oil it uses in its products and to use deforestation-free supply chains by 2023. The company has used technology such as satellite imaging and artificial intelligence (AI) to monitor deforestation and land conversion and is now turning to blockchain to increase transparency and traceability of its palm oil supply chain.

Supply chains for palm oil are long and complex, and the material can go through dozens of transformations as it moves from company to company from the time it's harvested until it ends up in soap or shampoo. Companies want to make sure that the palm oil they're sourcing comes from responsible sources, such as plantations that don't contribute to deforestation.

Unilever started using blockchain technology from GreenToken to track, verify, and report in near real time the origins and journey that palm oil takes through its long and complex supply chain. GreenToken is a blockchain-based supply chain traceability and chain-of-custody application that allows companies to record and track facts about the raw materials they use in their products. This can include information such as raw materials' origins, specific production attributes, and certifications. GreenToken is especially useful for tracking raw materials and commodities, such as palm, soy, coffee, sugar, and cocoa, that are linked to deforestation.

Other supply chain traceability solutions exist in the market, but they almost all rely on keeping tabs on batches. This approach doesn't work for tracking raw materials, which typically ship and are stored in bulk and are often commingled (mixed up) with materials from other sources. Even if batch identifiers for certain quantities are known, constant commingling generates an exponentially large number of records that need to be kept. No batch system is able to track this complexity. Companies that need to source raw materials and that use GreenToken are able to accurately report on sustainable or ethical procurement even if the materials are mixed or blended. GreenToken can also be used to measure the level of certain raw materials in finished goods. A consumer products producer would be able to demonstrate that it has a high volume of recycled material in the products themselves or in the packaging.

Companies can access GreenToken via a SaaS dashboard. Producers can create a digital token of physical goods that can include multiple attributes such as the original plantation and any certifications. Blockchain establishes a chain of custody for the tokens that cannot be altered as the materials move through the supply chain. Details about the commodity can also include whether it's coming from a mature plantation with no deforestation.

Responding to demands by governments and organizations such as the Rainforest Action Network (RAN) is one powerful reason for embracing sustainability. Germany's Supply Chain Due Diligence Act, which comes into effect in 2023, requires large public organizations to provide visibility of their supply chains in a digital and auditable format, and similar legislation is expected in other countries.

Another reason to embrace sustainability is to make money. A company that can demonstrate its commitment to sustainability will be rewarded by higher sales and brand loyalty. A recent IBM survey found that 57 percent of consumers were willing to change their shopping habits to reduce the environmental impact of the products they buy. This percentage will continue to rise. It is not enough for a company to manufacture green products. It must also prove to the consumer that it has done so. Using blockchain can demonstrate to environmentally conscious consumers that an item in question is what it claims to be.

Newlight Technologies, Inc. is the creator of AirCarbon, a biomaterial that is made out of greenhouse gases and that can substitute for fiber, leather, and synthetic plastic. AirCarbon is meltable into a variety of solid forms and is carbon-negative. That means its production captures or destroys more carbon dioxide than was emitted to make it. AirCarbon is currently being used in accessories such as sunglasses, handbags, and wallets sold by Covalent, a fashion brand launched by Newlight in September 2020.

When it came to marketing AirCarbon, Newlight needed a way to get the sustainability message out to potential customers because there was no way for consumers to verify Newlight's environmental claims. Blockchain appeared to provide a good solution for verifying the company's environmental impact. Blockchain creates a ledger of business transactions distributed among multiple computers, making the information easily visible yet impervious to tampering, and can be used to record virtually any type of business data. Its traceability makes it especially appealing to companies concerned about promoting their environmental responsibility.

Newlight worked with IBM, which was developing blockchains for a variety of industries. The blockchain was able to assign each Covalent product a blockchain-based number that could be used to display every step in the production process, along with third-party verification of the item's carbon impact. Each Covalent product has a unique, 12-digit number printed in it during production that shows the precise time when AirCarbon was used in production of the item. That number is then entered into Covalent's website, providing customers with the details about the product's movement through the supply chain, from initial molding to store shelf. IBM Blockchain provides visibility into not only the steps used to make each Covalent product but also the carbon impact that each specific product has on the environment. Newlight intends to roll out blockchain technology for items beyond the world of fashion, including food products.

Using blockchain to increase supply chain sustainability poses some challenges, such as how the insights will be used and how information will be shared among members of the supply chain. Companies like Unilever will need to learn how to work with suppliers to minimize deforestation while maintaining or improving the companies' economic well-being. There are also challenges concerning the use, management, and security of the data collected.

A copy of a blockchain is typically held on multiple computers, creating a scalability problem as the blockchain increases in size. Each computer on the network working to confirm transactions and keep accurate records of the blockchain must store data starting with the first block and moving to the most recent block. The redundancy creates a more secure system, but it also becomes increasingly inefficient as the network and blockchain grow.

When creating a new block on the blockchain, the computer that confirms the transactions must broadcast the new block to every other computer on the network. The other computers can then verify the transactions and add the block to the blockchain, which can consume substantial network resources as the network grows in size.

There are no universal blockchain standards to ensure interoperability, generate trust, and help ensure ease of use of the technology. Almost every implementation of blockchain technology is unique. That makes interoperability among blockchains difficult. If one company wants to share data with another company's blockchain, they'll likely need to develop additional tools to allow data to flow between the two blockchains. Nevertheless, companies that use blockchain applications correctly will be able to build stronger relationships with suppliers.

Sources: Green-token.io, accessed May 8, 2022; Unilever, "Climate Transition Action Plan," Assets.unilever.com, accessed May 7, 2022; Joe O'Donnell, "Unilever Pursues Supply Chain Sustainability with Blockchain," *SearchSAP*, March 30, 2022; "Unilever and SAP Partner for Blockchain Token Traceability Pilot," *Ledger Insights*, March 25, 2022; Adam Levy, "5 Problems with Blockchain Technology," *The Motley Fool*, February 28, 2022; Robert J. Bowman, "Tracing the Supply-Chain Journey of Green Products with Blockchain," *SupplyChainBrain*, March 3, 2021.

## CASE STUDY QUESTIONS

**6-12** Why is supply chain traceability so important for sustainability?

**6-13** Why is blockchain a useful technology for traceability?

**6-14** What people, organizational, and technology challenges should Unilever anticipate as it deploys blockchain for supply chain management?

**6-15** Should all businesses use blockchain? Explain your answer,

## Chapter 6 References

Bessens, Bart. "Improving Data Quality Using Data Governance." *Big Data Quarterly* (Spring 2018).

Buff, Anne. "The Conundrum of Data Governance." *Big Data Quarterly* (Fall 2019).

DalleMule, Leandro, and Thomas H. Davenport. "What's Your Data Strategy?" *Harvard Business Review* (May–June 2017).

Davenport, Thomas H. *Big Data at Work: Dispelling the Myths, Uncovering the Opportunities* (Boston, MA: Harvard Business School, 2014).

DeLanghe, Bart, and Stefano Puntoni. "Leading with Decision-Driven Analytics." *MIT Sloan Management Review* 62 No. 3 (Spring 2021).

Experian Information Solutions. "The 2019 Global Data Management Benchmark Report." (2019).

Hoffer, Jeffrey A., Ramesh Venkataraman, and Heikki Toppi. *Modern Database Management*, 13th ed. (Hoboken, NJ: Pearson Education, Inc., 2022).

Holwerda, Jacob A. "Big Data? Big Deal: Searching for Big Data's Performance Effects in HR." *Business Horizons* (July 14, 2021).

Kroenke, David M., David J. Auer., and Scott L. Vandenberg. *Database Processing: Fundamentals, Design, and Implementation*, 16th ed. (Hoboken, NJ: Pearson Education, Inc. 2022).

Lacity, Mary, and Remko Van Hoek. "What We've Learned So Far about Blockchain for Business." *MIT Sloan Management Review* 62 No. 3 (Spring 2021).

Lawton, George. "Data Quality for Big Data: Why It's a Must and How to Improve It." Searchdatamanagement.com (April 27, 2021).

Madnick, Stuart. "Blockchain Isn't as Unbreakable as You Think." *MIT Sloan Management Review* (Winter 2020).

Marcus, Gary, and Ernest Davis. "Eight (No, Nine!) Problems with Big Data." *New York Times* (April 6, 2014).

McKendrick, Joe. "Building a Data Lake for the Enterprise." *Big Data Quarterly* (Spring 2018).

MongoDB. "Yile Technology Relies on MongoDB Atlas to Win Global Online Gaming Market Share." Mongodb.com, accessed May 10, 2022.

Ross, Jeanne W., Cynthia M. Beath, and Anne Quaadgras. "You May Not Need Big Data after All." *Harvard Business Review* (December 2013).

Tapscott, Don, and Ricardo Viana Vargas. "Blockchain Is Changing How Companies Can Engage with Customers." *Harvard Business Review* (January 6, 2021).

Vial, Gregory. "Data Governance in the 21st Century."*MIT Sloan Management Review* (October 7, 2020).

Vodovatova, Elena. "Why Business Applies Sentiment Analysis? 5 Succeessful Examples." Theappsolutions.com, accessed May 11, 2022.

CHAPTER

7

# Telecommunications, the Internet, and Wireless Technology

## LEARNING OBJECTIVES

After completing this chapter, you will be able to:

7-1 Identify the principal components of telecommunications networks and key networking technologies.

7-2 Compare the different types of networks.

7-3 Explain how the Internet and Internet technology work, and how they support communication and e-business.

7-4 Identify the principal technologies and standards for wireless networking, communication, and Internet access.

7-5 Understand how MIS can help your career.

## CHAPTER CASES

- The National Hockey League Scores with Wireless Technology and the Internet of Things (IoT)
- Can Low Earth Orbit (LEO) Satellite Internet Systems Solve the Digital Divide?
- Monitoring Employees on Networks: Unethical or Good Business?
- Google, Apple, and Meta Battle for Your Internet Experience

**MyLab MIS**

- **Video Case:** Nokia CEO Suri Sees 5G Market Maturity in 2021
- Discussion Questions: 7-5, 7-6, 7-7
- Hands-On MIS Projects: 7-8, 7-9, 7-10, 7-11
- eText with Figure Videos

# THE NATIONAL HOCKEY LEAGUE SCORES WITH WIRELESS TECHNOLOGY AND THE INTERNET OF THINGS (IOT)

**Sports** such as baseball, basketball, and football have been intensive users of information technology to improve player and team performance and enhance the viewing experience. Today's sports fans don't want to just watch a sport: They want to engage with it, and they expect more information and interaction—data-enhanced viewing, live streaming, video on demand, mobile apps, and social media—than they used to in the past. Digital technology has become essential for attracting fans, sponsors, and broadcasters.

One could say, however, that hockey has been late to the game. The National Hockey League (NHL), comprising 32 teams (25 in the United States and 7 in Canada), was slow to use game-generated data and advanced metrics and lacked actual data to support many subjective opinions people had about how well players and teams performed during games. Now that's about to change. In 2020, the NHL began the process of deploying the NHL Edge Puck and Player Tracking system, based on technology developed by SMT (SportsMedia Technology). The system debuted at the 2020 NHL All-Star Game and was featured at the 2020 NHL conference finals. Now, all arenas in the league have been equipped with the technology. The system tracks every movement of the puck and each team's players during a game. The system is able to precisely and instantaneously detect puck and player positioning, speed, shot, skating distance, and time on ice. Exactly how quickly is a player skating? How long did he control the puck? How much time did he spend in the scoring zone? Now these questions can be answered.

The technology includes 14 to 16 antennae installed in arena rafters; four 4K cameras to support tracking functions; a wireless sensor sewn into the back of the jersey of each player on each team; and pucks embedded with a tiny battery, a tiny circuit board, and infrared and radio frequency sensors. After some initial difficulty with the first batch of pucks, which were recalled because they did not glide effectively, new pucks introduced in October 2021 have proved to be much better. The devices mounted in the rafters and on the upper tier of each arena record the *x*, *y*, and *z* coordinates of each sensor on the players and pucks hundreds of times per second. The system generates more than a million 3-D data points over the course of a single game ("Big Data"), which feed into SMT's OASIS software platform. OASIS collates and

© YES MARKET MEDIA/Shutterstock

analyzes the millions of coordinate data points to generate statistics, such as a player's top speed or total time of possession, and also triggers a separate system, SMT OPTICS, that automatically provides real-time graphics for use by broadcasters. The NHL has also partnered with Amazon's Amazon Web Services (AWS), naming it as its official cloud, artificial intelligence, and machine learning infrastructure provider in 2021. AWS is providing the NHL with a cloud-based video content delivery and storage system. AWS artificial intelligence (AI) technologies, coupled with the data from the Edge Puck and Player Tracking system, enabled the NHL to provide even more advanced shot and save analytics during the 2021–2022 hockey season.

The flood of data about player speed and execution is changing how coaches, broadcasters, and fans interact with the game. All the new data and metrics will of course be used by coaches and team managers to improve team performance, but these numbers will also enhance the fan experience inside the arena and among those watching on television or on mobile devices. In some cases, both types of viewing experiences may even include a digital stream broadcast dedicated to showcasing puck and player data.

According to NHL Commissioner Gary Bettman, player and puck tracking was initially designed to give people more insight into the game, show them how special the game is, and help them understand hockey a little better. But the business opportunities for using these data have expanded. For instance, according to NHL VP of Technology Keith Horstman, the new technology will enable in-play sports betting, allowing fans to wager on which player skates the fastest or shoots the hardest, or how far the puck travels on a particular shot. It also enables hockey fans to engage with the sport more deeply, whether that is through gaming, participating in fantasy leagues, or enriching their own education and insight into the game. The NHL hopes that this higher level of engagement will produce more revenue for the NHL and its clubs. The sport's Big Data, which are gathered via wireless technology, analyzed with AI techniques, and stored and distributed via cloud services, have indeed become a "game changer."

Sources: Amazon Web Services, "AWS Powers the NHL," Aws.amazon.com, accessed July 19, 2022; SMT, "Hockey Solutions," Smt.com, accessed July 19, 2022; Brandon Costa, "NHL Deploys Analytics-Powered Graphics for Puck/Player Tracking in Live Broadcast," Sportsvideo.org, October 25, 2021; Andrew Cohen, "NHL to Re-institute Puck Tracking after Last Season's Failed Sensor Debut," Sporttechie.com, October 11, 2021; Larry Lage, "NHL Tracking Pucks This Season, Opening up Gambling Options," Apnews.com, October 8, 2021; Zeeshan Hussain, "How the NHL Employs AWS to Maximize Its IoT Hardware and Engage Fans," Engineering.com, March 4, 2021; Stephen Whyno, "NHL Takes Big Stride on Data and Analytics with AWS Deal," Mbtmag.com, February 11, 2021.

The experience of the NHL illustrates some of the powerful capabilities and opportunities provided by contemporary networking technology. NHL games now use wireless networking and wireless sensor technology to closely track the speed and position of the players as well as of the hockey puck, along with other variables affecting game outcome, and to deliver information instantaneously to fans and broadcasters.

The chapter-opening diagram calls attention to important points raised by this case and this chapter. NHL hockey has many fans, but those fans are becoming more attuned to digital media and are calling for ways to use such media to enhance their experience with the sport. The NHL's management realized it could expand the NHL fan base and deepen fan engagement by taking advantage of opportunities presented by wireless networking technology and the Internet of Things (IoT). The NHL is now able to provide real-time game statistics and content for TV broadcasts and for streaming to mobile devices, thus increasing the popularity of the sport, fans' interest, and revenue.

Here are some questions to think about: Why has wireless technology played such a key role at the NHL's games? Describe how the technology has changed the way the NHL provides and uses data from its games.

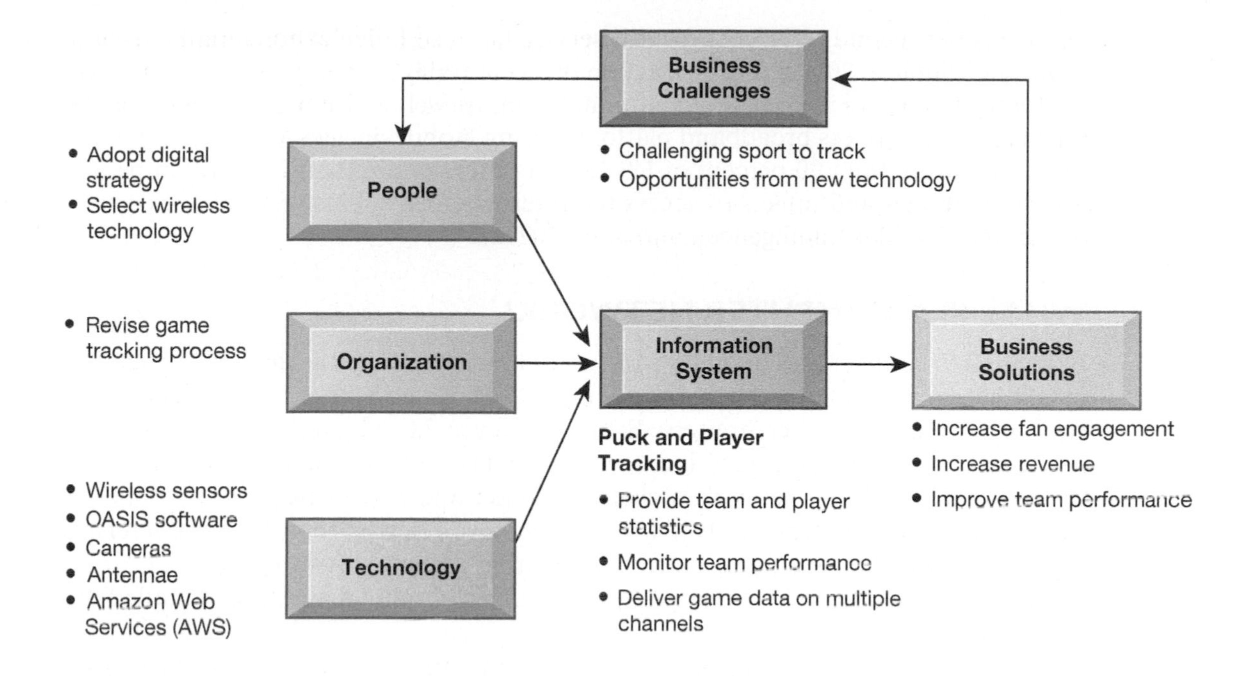


## 7-1 Identify the principal components of telecommunications networks and key networking technologies.

If you run a business, you can't do so without networks. You need to communicate rapidly with your customers, suppliers, and employees. Until about 1990, businesses used the postal system or telephone system with voice or fax for communication. Today, however, you and your employees use computers, email, text messaging, the Internet, and mobile devices connected to wireless networks to communicate. Networking and the Internet are now nearly synonymous with doing business.

### NETWORKING AND COMMUNICATION TRENDS

Firms in the past used two fundamentally different types of networks: telephone networks and computer networks. Telephone networks historically handled voice communication, and computer networks handled data traffic. During the twentieth century, telephone companies, which almost always operated as government-regulated monopolies, built telephone networks using voice transmission technologies (hardware and software). Computer companies originally built computer networks to transmit data among computers in different locations.

As a result of continuing telecommunications deregulation and information technology innovation, telephone and computer networks are converging into a single digital network using shared Internet-based standards and technology. Telecommunications providers that historically provided telephone services, such as AT&T and Verizon, now offer Internet access as well as voice services over both cellular networks and traditional "land lines." Cable companies that historically offered television services, such as Comcast, Charter Spectrum, and Altice, now offer Internet access and voice services over their cable networks. Computer networks no longer transmit just data. They now also provide Internet, telephone, and video services.

Both voice and data communication networks have also become more powerful (faster), more portable (smaller and more mobile), and less expensive. For instance, the typical Internet connection speed in 2000 was 56 Kbps (kilobits per second), but today the majority of US households have high-speed **broadband** connections (provided by telecommunications companies) running from 5 to more than 900 Mbps (megabits per second). The cost for this service has also fallen exponentially, from 50 cents per kilobit in 2000 to a tiny fraction of a cent today.

Increasingly, voice and data communication, as well as Internet access, are taking place over wireless broadband platforms using mobile devices and PCs in wireless networks. More than 90 percent of US Internet users (more than 275 million people) use smartphones and tablets to access the Internet either exclusively or at least some of the time (Insider Intelligence/eMarketer, 2022a).

## WHAT IS A COMPUTER NETWORK?

If you had to connect the computers for two or more employees in the same office, you would need a computer network. In its simplest form, a computer network consists of two or more connected computers. Examine Figure 7.1, which illustrates the major hardware, software, and transmission components of a simple network connected to the Internet or another network: a client computer and a dedicated server computer, network interface controllers (NICs, sometimes also referred to as network interface cards), a connection medium, network operating system software, switches, and a router. View the Figure 7.1 video in the eText for an animated and more detailed discussion of this figure.

Each computer on the network contains a NIC to link the computer to the network. The connection medium for linking network components can be a telephone wire, a coaxial cable, or a radio signal (in the case of cellphone and wireless local area networks [Wi-Fi networks]).

The **network operating system (NOS)** routes and manages communications on the network and coordinates network resources. It can reside on every computer in the network or primarily on a dedicated server computer for all the applications on the network. A server is a computer on a network that performs important network functions for client computers, such as displaying web pages, storing data, and storing the network operating system (hence controlling the network). Microsoft Windows Server and Linux are the most widely used network operating systems.

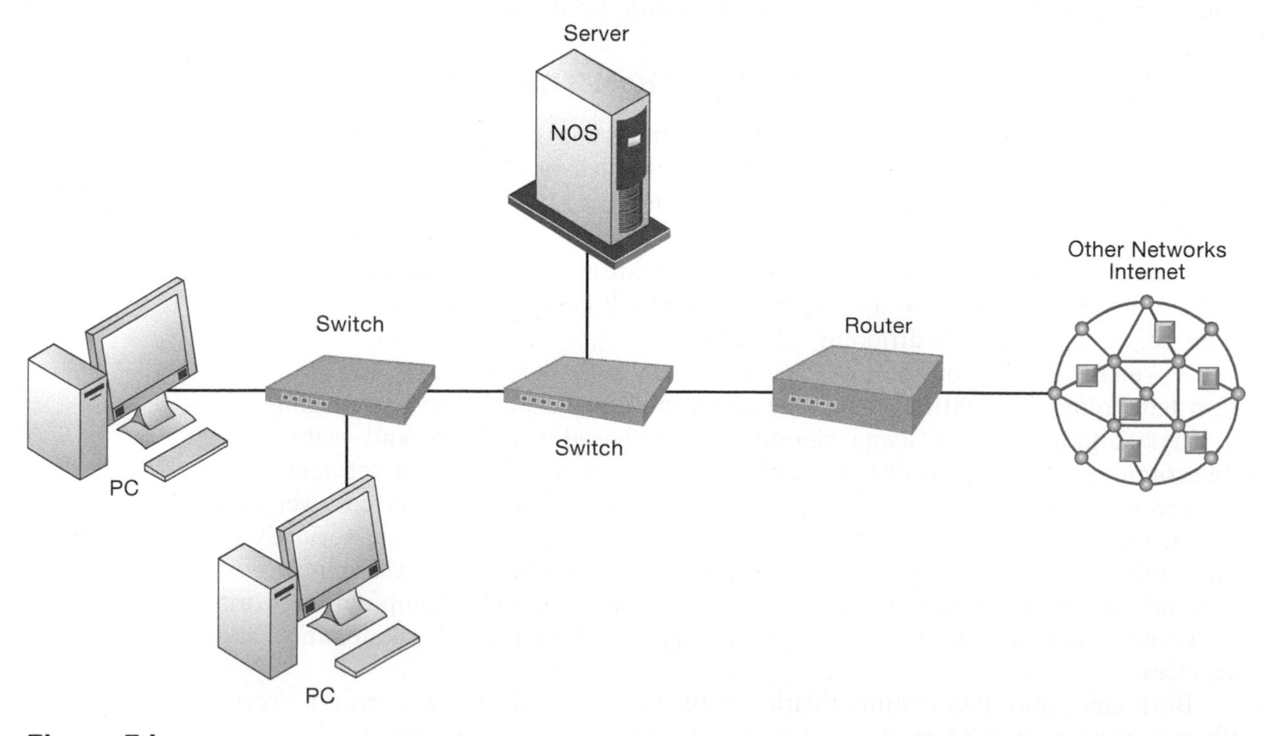


**Figure 7.1**
Components of a Simple Computer Network
*Illustrated here is a simple computer network consisting of computers, a network operating system (NOS) residing on a dedicated server computer, cable (wiring) connecting the devices, switches, and a router.*

Most networks also contain a switch or a hub acting as a connection point among the computers. **Hubs** are simple devices that connect network components, sending a packet of data to all other connected devices. A **switch** has more intelligence than a hub and can filter and forward data to a specified destination on the network.

What if you want to communicate with another network such as the Internet? You would need a router. A **router** is a communications processor that routes packets of data through different networks, ensuring that the packet of data sent gets to the correct address.

Network switches and routers have proprietary software built into their hardware for directing the movement of data on the network, which can create network bottlenecks and makes the process of configuring a network more complicated and time consuming. **Software-defined networking (SDN)** is a networking approach in which many of these control functions are managed by one central program, which can run on inexpensive servers that are separate from the network devices themselves. This is especially helpful in a cloud computing environment that has many pieces of hardware because SDN allows a network administrator to manage traffic loads in a flexible and more efficient manner.

## Networks in Large Companies

The network we've just described might be suitable for a small business, but what about large companies that have many locations and thousands of employees? As a firm grows, its small networks can be tied together into a corporate-wide networking infrastructure. The network infrastructure for a large company typically consists of a large number of these small local area networks linked to other local area networks and to firm-wide corporate networks. Powerful servers support a corporate website, a corporate intranet, and perhaps an extranet. Servers on the network may also link to other large computers supporting back-end systems.

Examine Figure 7.2, which provides an illustration of these more complex, larger-scale corporate-wide networks. Here the corporate network infrastructure supports a mobile sales force using mobile devices, mobile employees linking to the company website, and internal company Wi-Fi networks. In addition to these

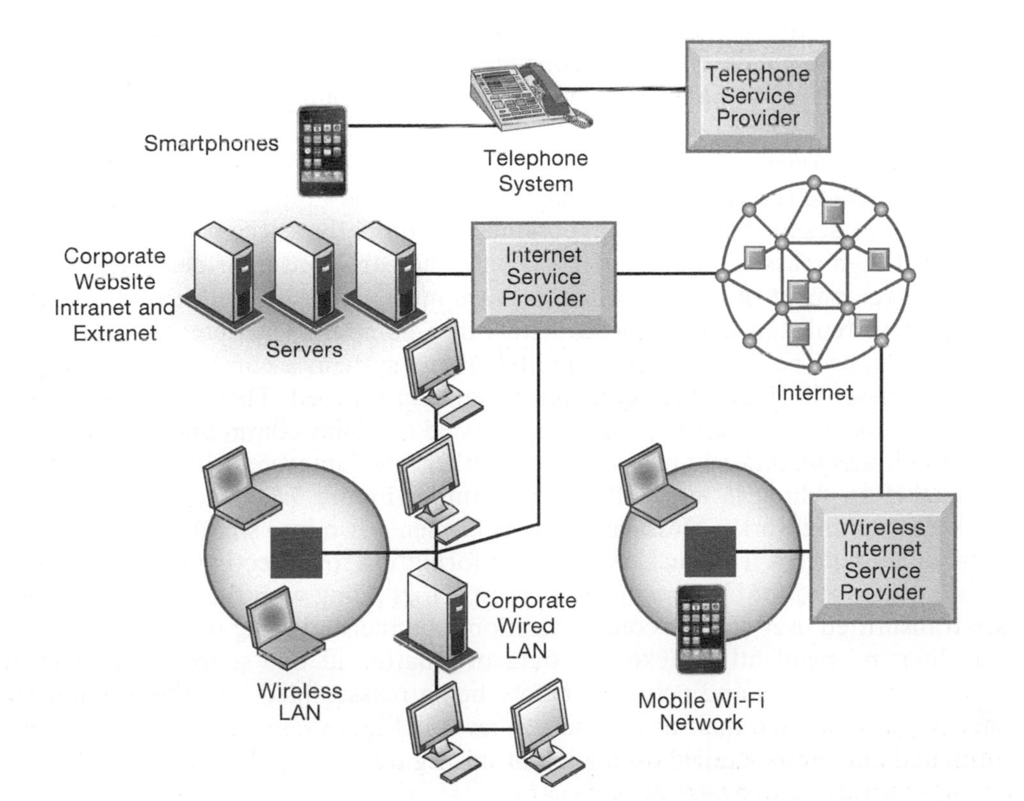


**Figure 7.2**
Corporate Network Infrastructure
*Today's corporate network infrastructure is a collection of many networks, from the public switched telephone network to the Internet to corporate local area networks linking workgroups, departments, or office floors.*

computer networks, the firm's infrastructure may include a separate telephone network that handles most voice data. Many firms are dispensing with their traditional telephone networks and using Internet telephones that run on their existing data networks (described later). View the Figure 7.2 video in the eText for an animated and more detailed discussion of this figure.

As you can see from this figure, a large corporate network infrastructure uses a wide variety of technologies—everything from ordinary telephone service and corporate data networks to Internet service, wireless Internet, and mobile devices. One of the major problems facing corporations today is how to integrate all these different communication networks and channels into a coherent system that enables information to flow from one part of the corporation to another and from one system to another.

## KEY DIGITAL NETWORKING TECHNOLOGIES

Contemporary digital networks and the Internet are based on three key technologies: client/server computing, the use of packet switching, and the development of widely used communications standards (the most important of which is Transmission Control Protocol/Internet Protocol, or TCP/IP) for linking disparate networks and computers.

### Client/Server Computing

Client/server computing, introduced in Chapter 5, is a distributed computing model in which some of the processing power is located within small, inexpensive client computers, such as desktop or laptop computers or mobile devices. These clients are linked to one another through a network that is controlled by a network server computer. This server sets the rules of communication for the network and provides every client with an address so that others can find it on the network.

Client/server computing has largely replaced centralized mainframe computing, where nearly all the processing takes place on a central, large mainframe computer. Client/server computing has extended computing to departments, workgroups, factory floors, and other parts of the business that could not be served by a centralized architecture. It also makes it possible for personal computing devices such as PCs, laptops, and mobile devices to be connected to networks such as the Internet. The Internet is the largest implementation of client/server computing.

### Packet Switching

**Packet switching** is a method of slicing digital messages into parcels called packets, sending the packets along different communication paths as the paths become available, and then reassembling the packets once they arrive at their destinations. Prior to the development of packet switching, computer networks used leased, dedicated telephone circuits to communicate with other computers in remote locations. In circuit-switched networks, such as the telephone system, a complete point-to-point circuit is assembled, and then communication can proceed. These dedicated circuit-switching techniques were expensive and wasted available communications capacity: the circuit was maintained regardless of whether any data were being sent.

Packet switching is more efficient. Examine Figure 7.3, which illustrates how packet switching works. Messages are first broken down into small, fixed bundles of data called packets. The packets include information for directing the packet to the right address and for checking transmission errors along with the data. The packets are transmitted over various communications channels by using routers, each packet traveling independently. Packets of data originating at one source will be routed through many paths and networks before being reassembled into the original message when they reach their destinations. View the Figure 7.3 video in the eText for an animated and more detailed discussion of this figure.

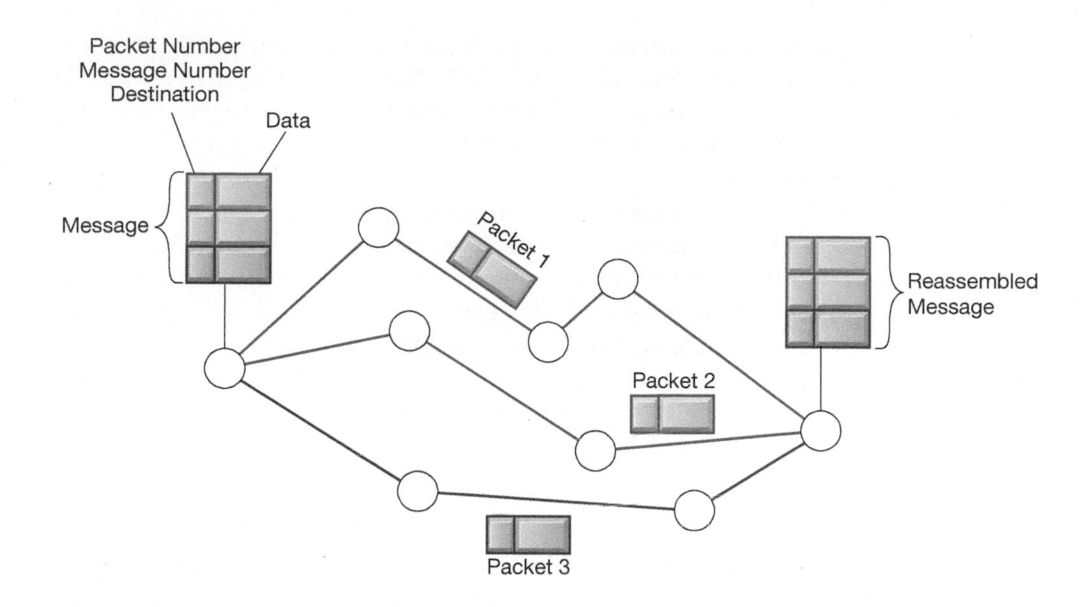


**Figure 7.3**
Packet-Switched Networks and Packet Communications
*Data are grouped into small packets, which are transmitted independently over various communications channels and then reassembled at their final destination.*

## TCP/IP and Connectivity

In a typical telecommunications network, diverse hardware and software components need to work together to transmit information. Different components in a network communicate with each other by adhering to a common set of rules called protocols. A **protocol** is a set of rules and procedures governing transmission of information between two points in a network.

In the past, diverse proprietary and incompatible protocols often forced business firms to purchase computing and communications equipment from a single vendor. However, today, corporate networks primarily use **Transmission Control Protocol/Internet Protocol (TCP/IP)**, a common worldwide standard. TCP/IP was developed during the early 1970s to support US Department of Defense Advanced Research Projects Agency (DARPA) efforts to help scientists transmit data among different types of computers over long distances.

TCP/IP uses a suite of protocols, the main ones being TCP (Transmission Control Protocol) and IP (Internet Protocol). TCP handles the movement of data among computers. TCP establishes a connection among the computers, sequences the transfer of packets, and acknowledges the packets sent. IP is responsible for the delivery of packets and includes the disassembling and reassembling of packets during transmission.

The Department of Defense's reference model conceptualizes the suite of protocols comprising TCP/IP as a series of four layers, described from top to bottom as follows:

1. **Application layer.** The Application layer enables client application programs to access the other layers and defines the protocols that the applications use to exchange data. One of these application protocols is the Hypertext Transfer Protocol (HTTP), which is used to transfer web page files.
2. **Transport layer.** The Transport layer is responsible for providing the Application layer with communication and packet services. This layer includes TCP and other protocols.
3. **Internet layer.** The Internet layer is responsible for addressing, routing, and packaging data packets called IP datagrams. The Internet Protocol is one of the protocols used in this layer.
4. **Network Interface layer.** At the bottom of the reference model, the Network Interface layer is responsible for placing packets on and receiving them from the network medium, which could be any networking technology.

Examine Figure 7.4, which illustrates how two computers using TCP/IP can communicate even if they are based on different hardware and software platforms.

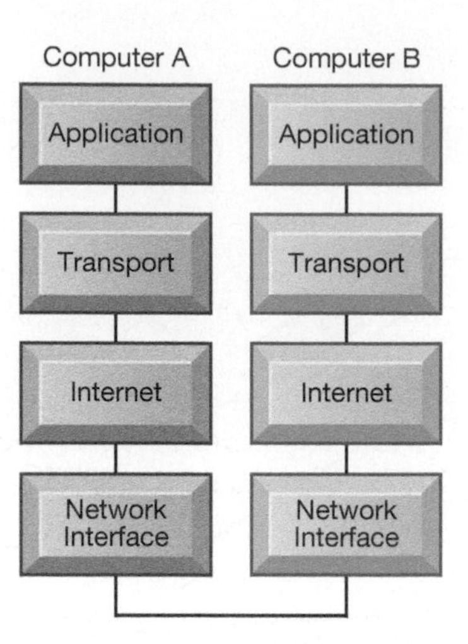


**Figure 7.4**
The Transmission Control Protocol/Internet Protocol (TCP/IP) Reference Model
*This figure illustrates the four layers of the TCP/IP reference model for communications.*

Data sent from one computer to the other pass downward through all four layers, starting with the sending computer's Application layer and passing through to the Network Interface layer. After the data reach the recipient host computer, they travel up the layers and are reassembled into a format the receiving computer can use. If the receiving computer finds a damaged packet, it asks the sending computer to retransmit it. This process is reversed when the receiving computer responds.

## 7-2 Compare the different types of networks.

Let's look more closely at the alternative networking technologies available to businesses.

### SIGNALS: DIGITAL VERSUS ANALOG

There are two ways to communicate a message in a network: via an analog signal or a digital signal. An *analog signal* is represented by a continuous waveform that passes through a communications medium and is used for audio communication. The most common analog devices are the traditional telephone handset, the speaker on your computer, and your iPhone earphone, all of which create analog waveforms that your ear can hear.

A *digital signal* is a discrete, binary waveform rather than a continuous waveform. Digital signals communicate information as strings of two discrete states: 1 bits and 0 bits, which are represented as on-off electrical pulses. Computers use digital signals and require a **modem** (which stands for "modulator-demodulator") to convert these digital signals into analog signals that can be sent over (or received from) telephone lines, cable lines, or wireless media that use analog signals. Examine Figure 7.5, which illustrates the function of a modem. Cable modems connect your computer to the Internet by using a cable network. DSL modems connect your computer to the Internet using a telephone company's landline network. Wireless modems perform the same function as traditional modems but connect your computer to a wireless network, which could be a cellular network or a Wi-Fi network.

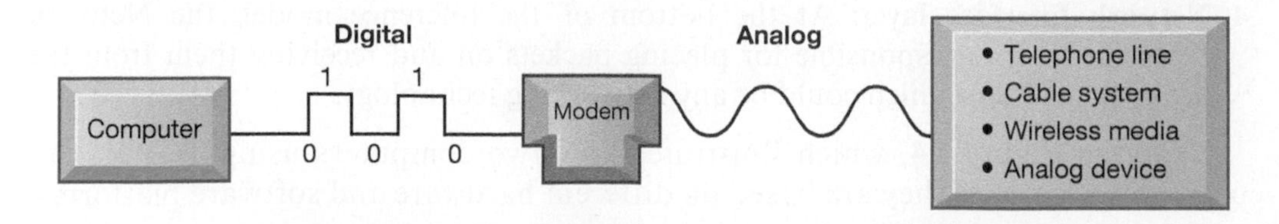


**Figure 7.5**
Functions of the Modem
*A modem is a device that translates digital signals into analog form (and vice versa) so that computers can transmit data over analog networks such as telephone and cable networks.*

## TYPES OF NETWORKS

There are many kinds of networks and ways of classifying them. One way of looking at networks is in terms of their geographic scope (see Table 7.1).

### Local Area Networks

If you work in a business that uses networking, you are probably connecting to other employees and groups via a local area network. A **local area network (LAN)** is designed to connect personal computers and other digital devices within a half-mile or 500-meter radius. LANs typically connect a few computers in a small office, all the computers in one building, or all the computers in several buildings in close proximity. LANs also are used to link to long-distance wide area networks (WANs, described later in this section) and other networks around the world, using the Internet.

Review Figure 7.1, which could serve as a model for a small LAN that might be used in an office. In this model, one computer is a dedicated network server, providing users with access to shared computing resources in the network, including software programs and data files.

The server determines who gets access to what and in which sequence. The router connects the LAN to other networks, which could be the Internet or another corporate network, so that the LAN can exchange information with networks external to it. The most common LAN operating systems are Windows and Linux.

Ethernet is the dominant LAN standard at the physical network level, specifying the physical medium to carry signals among computers, the access control rules, and a standardized set of bits that carry data over the system. Originally, Ethernet supported a data transfer rate of 10 Mbps (megabits per second). Newer versions, such as Gigabit Ethernet, support a data transfer rate of 1 Gbps (gigabits per second).

The LAN illustrated in Figure 7.1 uses a client/server architecture in which the network operating system resides primarily on a single server, and the server provides much of the control and resources for the network. Alternatively, LANs may use a **peer-to-peer** architecture. A peer-to-peer network treats all processors equally and is used primarily in small networks with 10 or fewer users. The various computers on the network can exchange data by direct access and can share peripheral devices without going through a separate server.

Larger LANs have many clients and multiple servers, with separate servers for specific services such as storing and managing files and databases (file servers or database servers), managing printers (print servers), storing and managing email (mail servers), or storing and managing web pages (web servers).

### Wide Area Networks and Metropolitan Area Networks

**Wide area networks (WANs)** span broad geographical distances—regions, states, continents, or the entire globe. The most universal and powerful WAN is the Internet. Computers connect to a WAN through public networks, such as the telephone system or private cable systems, or through leased lines or satellites. A **metropolitan area network (MAN)** is a network that spans a metropolitan area, usually a city and its major suburbs. Its geographic scope falls between that of a WAN and that of a LAN.

**TABLE 7.1**
Types of Networks

| Type | Area |
|---|---|
| Local area network (LAN) | Up to 500 meters (half a mile); an office or floor of a building |
| Campus area network (CAN) | Up to 1,000 meters (a mile); a college campus or corporate facility |
| Metropolitan area network (MAN) | A city or metropolitan area |
| Wide area network (WAN) | A regional, transcontinental, or global area |

**TABLE 7.2**

Physical Transmission Media

| Transmission Medium | Description | Speed |
|---|---|---|
| Twisted copper wire | Strands of copper wire twisted in pairs for voice and data communications. Includes CAT5, CAT6, CAT7, and CAT8 cabling, with CAT6–8 capable of gigabit speeds. | CAT5: 10 Mbps–100 Mbps<br>CAT8: Up to 40 Gbps |
| Coaxial cable | Thickly insulated copper wire, which is capable of high-speed data transmission and is less subject to interference than twisted wire. Currently used for cable TV and for networks with longer runs (more than 100 meters). | Up to 1 Gbps |
| Fiber-optic cable | Strands of clear glass fiber that transmit data as pulses of light generated by lasers. Useful for high-speed transmission of large quantities of data. More expensive than other physical transmission media; used for last-mile delivery to customers and the Internet backbone. | Up to 170+ Tbps |
| Wireless transmission media | Based on radio signals of various frequencies and includes both terrestrial and satellite microwave systems and cellular networks. Used for long-distance, wireless communication, and Internet access. | Up to 600+ Mbps |

## TRANSMISSION MEDIA AND TRANSMISSION SPEED

Networks use different kinds of physical transmission media, including twisted copper wire, coaxial cable, fiber-optic cable, and media for wireless transmission. Each has advantages and limitations. A wide range of transmission speeds (discussed in the next section) is possible for any given medium, depending on the software and hardware configuration.

### Bandwidth: Transmission Speed

The total amount of digital information that can be transmitted through any telecommunications medium is measured in bits per second (bps). One signal change, or cycle, is required to transmit one or several bits; therefore, the transmission capacity of each type of telecommunications medium is a function of its frequency. The number of cycles per second that can be sent through that medium is measured in **hertz**—one hertz is equal to one cycle of the medium.

The range of frequencies that can be accommodated on a particular telecommunications channel is called its **bandwidth**. The bandwidth is the difference between the highest and the lowest frequencies that can be accommodated on a single channel. The greater the range of frequencies, the greater the bandwidth and the greater the channel's transmission capacity.

Table 7.2 compares the different transmission media in terms of their various characteristics and speed capabilities.

# 7-3 Explain how the Internet and Internet technology work, and how they support communication and e-business.

The Internet has become an indispensable personal and business tool—but what exactly is the Internet? How does it work, and what does Internet technology have to offer for business? Let's look at the most important Internet features.

## WHAT IS THE INTERNET?

The Internet is the world's most extensive public communication system. It's also the world's largest implementation of client/server computing and internetworking, linking millions of individual networks all over the world. This global network of networks began in the early 1970s as a US Department of Defense project to link scientists and university professors around the world.

## INTERNET ADDRESSING AND ARCHITECTURE

The Internet is based on the TCP/IP networking protocol suite described earlier in this chapter. Every device connected to the Internet (or another TCP/IP network) is assigned a unique **Internet Protocol (IP) address** consisting of a string of numbers. There are two versions of IP addressing currently in use: IPv4 and IPv6. An IPv4 Internet address is a 32-bit number that appears as a series of four separate numbers marked off by periods, such as 64.49.254.91. However, the world is running out of available IP addresses using IPv4 because of the explosion in the number of devices now connected to the Internet. IPv4 is now being replaced by a new version of IP addressing called **IPv6** (Internet Protocol version 6), which contains 128-bit addresses (2 to the power of 128), or more than a quadrillion possible unique addresses. IPv6 is compatible with most modems and routers sold today and will fall back to the old addressing system if IPv6 is not available on local networks. The transition to IPv6 is expected to take several years.

When a user sends a message to another user on the Internet or another TCP/IP network, the message is first decomposed into packets. Each packet contains its destination address. The packets are then sent from the client to the network server and from there to as many other servers as necessary to arrive at a specific computer with a known address. At the destination address, the packets are reassembled into the original message.

### The Domain Name System

Because it would be incredibly difficult for Internet users to remember long strings of numbers, an IP address can be represented by a natural language convention called a **domain name**. The **Domain Name System (DNS)** converts domain names to IP addresses. DNS servers maintain a database containing IP addresses mapped to their corresponding domain names. To access a computer on the Internet, users need only specify its domain name, such as Expedia.com.

DNS has a hierarchical structure. Examine Figure 7.6, which illustrates this hierarchical structure. At the top of the DNS hierarchy is the root domain, represented

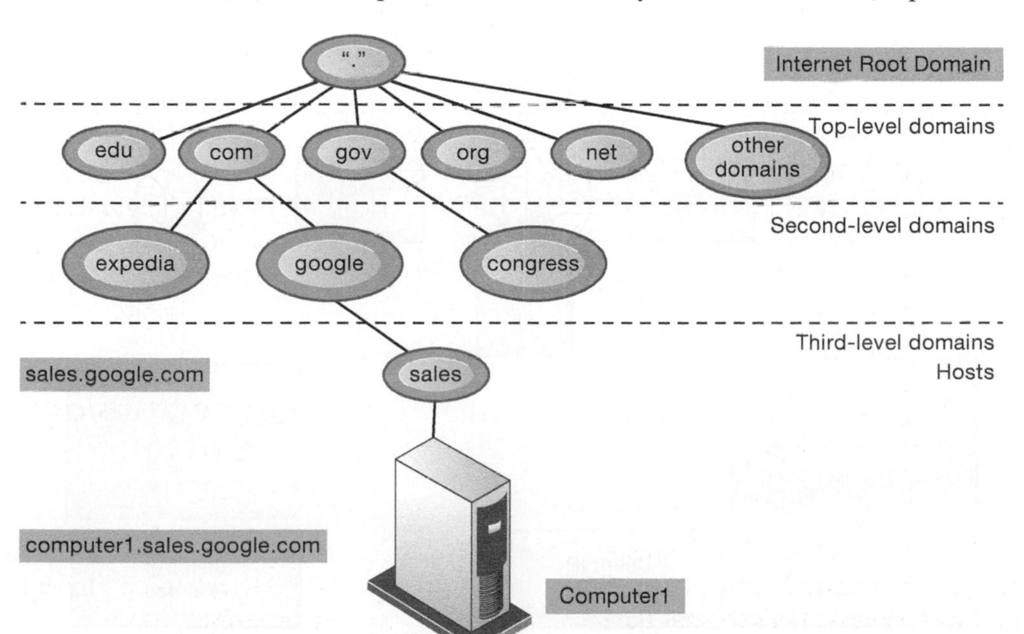


**Figure 7.6**
**The Domain Name System**
*The Domain Name System is a hierarchical system with a root domain, top-level domains, second-level domains, and host computers at the third level.*

by a "dot." The child domain of the root is called a top-level domain. Top-level domains are the two- and three-character names you are familiar with from surfing the web—for example, .com, .edu, .gov, and the various country codes such as .ca for Canada, .it for Italy, and .br for Brazil. The child domain of a top-level domain is called a second-level domain. Second-level domains have two parts, which designate a top-level name and a second-level name—such as buy.com, nyu.edu, or amazon.ca. Third-level domains, which are at the bottom of the hierarchy, are used to designate the host name of a specific computer on either the Internet or a private network.

The following list shows the most common domain extensions currently available and officially approved:

.com Commercial organizations/businesses
.edu Educational institutions
.gov U.S. government agencies
.mil U.S. military
.net Network computers
.org Any type of organization
.biz Business firms
.info Information providers

Countries also have domain names such as .uk, .au, and .fr (United Kingdom, Australia, and France, respectively). ICANN, the international body that governs the issuance of domains, has also approved a number of new top-level domains, such as .xyz, .online, .shop, .kids, and ones that use non-English characters. As many as 1,300 new top-level domains are expected to become available in the next few years.

## Internet Architecture

Examine Figure 7.7, which illustrates some of the main physical elements of today's Internet. Internet data traffic is carried over transcontinental, high-speed backbone networks that generally operate in the gigabit range. These trunk lines are typically owned by long-distance telecommunications companies (called *Tier 1 Internet service providers [Tier 1 ISPs]*) or by national governments. Tier 1 ISPs have "peering"

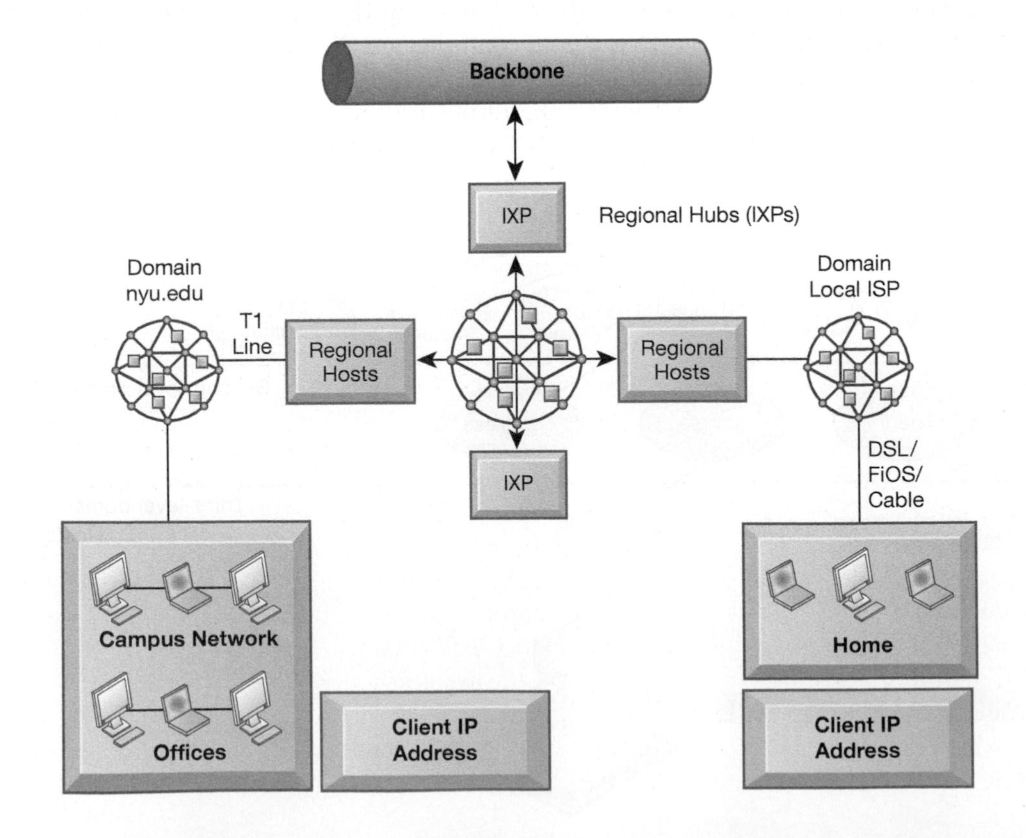


**Figure 7.7**
Internet Network Architecture
*The Internet backbone connects to regional networks, which in turn provide access to Internet service providers, large firms, and government institutions. Internet Exchange Points (IXPs) are hubs where the backbone intersects with regional and local networks and where backbone owners connect with one another.*

arrangements with other Tier 1 ISPs to allow Internet traffic to flow through each other's cables and equipment without charge. Tier 1 ISPs physically connect with one another and to regional (Tier 2) ISPs at regional hubs, commonly referred to as Internet Exchange Points (IXPs). Tier 2 ISPs connect Tier 1 ISPs with Tier 3 ISPs, which deal with "the last mile of service" and provide Internet access to consumers and business. Tier 3 ISPs are popularly referred to as just **Internet service providers**. Major Tier 3 ISPs in the United States include Comcast, Charter Spectrum, AT&T, Verizon, Altice (Optimum Online), CenturyLink, and Cox. (It's important to note that many Tier 3 ISPs are also Tier I ISPs; the two roles are not mutually exclusive.) There are also thousands of much smaller, local access ISPs.

Each organization pays for its own networks and its own local Internet connection services, a part of which is paid to the long-distance trunk line owners. Individual Internet users pay ISPs for using their service, and they generally pay a flat subscription fee no matter how much or how little they use the Internet. There are a variety of services for ISP Internet connections. Connecting via a traditional telephone line and modem, at a speed of 56.6 kilobits per second (Kbps), used to be the most common form of connection worldwide, but high-speed broadband connections have largely replaced it. Today, digital subscriber line (DSL) and fiber-optic service (FiOS), cable Internet, satellite Internet, and T lines provide these broadband services. A variety of new methods are also being explored to provide Internet access to areas that are not well served by wired or cellular networks, including low earth orbit (LEO) satellites and drones. The Spotlight on Technology case focuses on the race to create a LEO satellite network.

**Digital subscriber line (DSL)** technologies operate over existing telephone lines to carry voice, data, and video at transmission rates ranging from 385 Kbps to more than 100 Mbps, depending on usage patterns and distance. FiOS (Verizon's fiber-optic service) can deliver more than 900 Mbps, although most home service delivers 100 Mbps. **Cable Internet connections** provided by cable providers use digital cable coaxial lines to deliver high-speed Internet access to homes and businesses. These lines can provide high-speed access to the Internet at between 20 and 600 Mbps, with high-speed cable Internet download speeds approaching 1 Gbps. Where DSL and cable services are unavailable, it is possible to access the Internet via satellite, although most satellite Internet connections have slower upload speeds than other broadband services have.

T1 and T3 are international telephone standards for digital communication. They are leased, dedicated lines suitable for businesses or government agencies requiring high-speed, guaranteed service levels. **T1 lines** offer guaranteed delivery at 1.54 Mbps, and **T3 lines** offer delivery at 45 Mbps. The Internet does not provide similar guaranteed service levels but, simply, the best effort.

### Internet Governance

No one owns the Internet, and it has no formal management. However, worldwide Internet policies are established by a number of professional organizations and government bodies, including the Internet Architecture Board (IAB), which helps define the overall structure of the Internet; the Internet Corporation for Assigned Names and Numbers (ICANN), which manages the domain name system; and the World Wide Web Consortium (W3C), which sets Hypertext Markup Language (HTML) and other programming standards for the web.

These organizations influence government agencies, network owners, ISPs, and software developers with the goal of keeping the Internet operating as efficiently as possible. The Internet must also conform to the laws of the sovereign nation-states in which it operates as well as to the technical infrastructures that exist within the nation-states. Although in the early years of the Internet and the web there was very little governmental interference, this situation is changing as the Internet plays a growing role in the distribution of information and knowledge, including content that some find objectionable.

## Can Low Earth Orbit (LEO) Satellite Internet Systems Solve the Digital Divide?

The Covid-19 pandemic has highlighted the importance of broadband Internet access for businesses as well as everyday life. However, many people and businesses in the United States and throughout the world are located in areas that conventional fiber-optic cable and wireless networks cannot cost-effectively reach. In the United States, an estimated 42 million people lack any broadband Internet access whatsoever. In addition, in many cases where broadband access is supposedly available, it does not actually operate at broadband speed, defined by the US Federal Communications Commission (FCC) as being at least 25 Mbps. According to Microsoft, this is the case for more than 120 million people (more than one-third of the US population). The term "digital divide" refers to this gap. In the United States, the digital divide disproportionately impacts those who live in rural areas or who have low incomes.

Over the last several years, there has been a rush of companies seeking to address this problem using a very high-tech alternative: creating a network of low earth orbit (LEO) satellites that can beam broadband Internet access to places where it was not previously available. Traditional satellite Internet access relies on geostationary (GEO) satellites that orbit over the Earth's equator at a very high altitude (22,236 miles above the Earth) and move at the speed of the Earth's rotation, thereby remaining at a fixed location relative to the Earth. Although traditional satellite Internet can deliver broadband service, it has limitations. For instance, because of the distance that data must travel to the satellite and back, traditional satellite Internet service has greater latency—characterized by the lag between when an action is taken and when the result is shown—than other types of broadband Internet service. In addition, weather conditions, such as rain and snow, can cause interruptions in service. Because GEO satellites are located over the equator, they do not provide good service to very northerly or southerly locations. Traditional satellite Internet service is also more costly than other types of broadband service, in terms of both equipment fees and monthly costs.

Unlike GEO satellites, LEO satellite systems orbit at a much lower height (anywhere from about 300 to 1,200 miles above the Earth), move more quickly than the Earth's rotation, and are not restricted to orbiting over the equator. Because each LEO satellite is accessible for only a short period of time from any one point on Earth, LEO satellite Internet systems rely instead on a network of satellites, which together can provide continuous, uninterrupted coverage, similar to how a cellular network deals with a moving person. Users connect their devices via specially designed antennas directly to the satellites, which communicate with each other and are linked to the Internet on Earth via ground stations. LEO satellite Internet systems have several advantages compared to GEO satellites. Because they orbit at a much lower height, they offer both faster broadband speed and much less latency. They are less likely to be impacted by the weather and provide expanded coverage to higher- and lower-latitude areas. They are also eventually expected to be less expensive for consumers. LEO satellite systems are additionally expected to play a significant role in the development of 5G networks by providing redundancy in critical segments and alternatives for connectivity for IoT devices and data traffic.

Elon Musk's SpaceX has thus far taken the lead in deploying a LEO satellite system, which it calls Starlink. Starlink already has almost 1,900 satellites in orbit and more than 100,000 customers in 14 different countries. Starlink has a license from the FCC for 12,000 satellites and has filed for permission for an additional 30,000. Starlink expects that it will initially be able to provide broadband speeds of 50 to 150 Mbps and ultimately to reach 300 Mbps, a significant increase over traditional satellite Internet speeds, although not as fast as land-based fiber. Amazon's Project Kuiper, which is expected to launch its first two prototype satellites toward the end of 2022, is positioning itself as a primary competitor to Starlink. Amazon has received approval from the FCC for more than 3,200 satellites and has applied for permission for more than 7,700 additional satellites. Amazon has teamed up with Verizon to combine Project Kuiper with Verizon's 5G network. Other major companies seeking to establish their LEO footprint in space include British-owned OneWeb, which has a joint venture with Airbus and has deployed more than 300 satellites thus far; Boeing, which has received FCC approval for a constellation of 147 satellites and has applied for permission to have an additional 5,790; and Canadian-owned Telesat, which launched its first LEO satellite in 2018 and hopes to be able to commence comprehensive service in 2022. Other companies, such as Immarsat,

Intelsat, and Hughes have also filed FCC applications for permission to have satellites. In addition, several Chinese companies have announced plans for their own LEO satellite projects. Low earth orbit may become a very crowded place!

There are a number of issues with respect to the race to create LEO satellite systems. For instance, the sheer number of satellites that potentially may be launched has raised concerns about safety. In 2021, a Chinese spacecraft had to maneuver on two separate occasions to avoid potential collisions with Starlink satellites. LEO systems will add a significant number of objects to the 28,100 objects (as well as millions of pieces of "space junk") already orbiting Earth, which raises the risk for collision. Astronomers have also pointed to light pollution of the night sky, changes to the chemistry of the Earth's upper atmosphere, and increased dangers to the Earth's surface from falling space debris as concerns.

A related concern is the issue of governance. Although the International Telecommunications Union (ITU) has some overall regulatory authority, individual countries are the ones granting permission for specific systems and do not necessarily assess global impact. The current regulatory regime, which was created to govern single satellites, is not well-equipped to handle mega-constellations such as LEO systems.

In addition, the implementation of LEO Internet systems will add a whole new dimension to global Internet infrastructure. Who should govern the delivery of Internet services from international space? LEO satellites will be using ground-based infrastructure in different countries, and customers may be anywhere around the world. Which country's laws will apply, and who can enforce them? For instance, LEO satellite systems will provide new levels of location accuracy. What privacy laws will apply to Internet traffic in space?

Finally, the question remains whether LEO satellite Internet systems can actually make a real dent in the digital divide. Some critics question whether these systems will be able to consistently deliver the claimed performance enhancements in Internet service. Even if they are able to do so, they are meant primarily for sparsely populated regions and will be able to serve only a limited number of customers in high-density areas. And although the costs are lower than those of traditional satellite Internet, costs of LEO satellite systems will still be quite high. Starlink will initially cost $99 a month, plus a $499 equipment fee, leading many to be concerned that rural and low-income customers will still be priced out.

Sources: Eric Mack, "China's Space Station Has Had to Dodge SpaceX Starlink Satellites Twice," Cnet.com, December 28, 2021; Daniel Voelson, "Internet from Space: How New Satellite Connections Could Affect Global Internet Governance," SWP Research Paper 2021/RP 03, Swp-berlin.org, December 4, 2021; Ry Crist, "Starlink Explained: Everything You Should Know about Elon Musk's Satellite Internet Venture," Cnet.com, December 3, 2021; Shira Ovide, "Satellite Hopes Meet Internet Reality," *New York Times*, November 9, 2021; Michael Sheetz, "In Race to Provide Internet from Space, Companies Ask FCC for about 38,000 New Broadband Satellites," Cnbc.com, November 5, 2021; Mike Dano, "Verizon, Amazon to Integrate LEO Satellites with 5G," Lightreading.com, October 26, 2021; Melissa Griffith and Christopher Hocking, "The Role of Satellites in 5G Networks," Wilsoncenter.org, October 1, 2021; Neel Patel, "Who Is Starlink Really For?" *MIT Technology Review*, September 6, 2021; Aaron Boley and Michael Byers, "Satellite Mega-Constellations Create Risks in Low Earth Orbit, the Atmosphere and on Earth," *Nature*, May 20, 2021; Marco Hogewoning, "The Challenges Satellite Internet Must Address for Take Off," Telecoms.com, April 19, 2021; Neil Lappage, "Will Low-Earth Orbit Satellites Fly Under the Privacy Radar?" Isaca.org, January 4, 2021.

## CASE STUDY QUESTIONS

**1.** How does Internet access relate to the concept of a digital divide? In what ways may LEO satellite Internet systems help bridge that divide?

**2.** How do LEO satellite Internet systems differ from traditional satellite Internet systems?

**3.** What ethical, legal, and political issues are raised by LEO satellite Internet systems?

## INTERNET SERVICES AND COMMUNICATION TOOLS

The Internet is based on client/server technology. Individuals using the Internet control what they do through client applications, such as web browser software, on their computers. The data, including email messages and web pages, are stored on servers. A client uses the Internet to request information from a particular web server on a distant computer, and the server sends the requested information back to the client over the Internet. Client platforms today include not only PCs and other computers but also smartphones and tablets.

**TABLE 7.3**

Major Internet Services

| Capability | Functions Supported |
| --- | --- |
| Email | Person-to-person messaging; document sharing |
| Instant messaging | Interactive conversations |
| Telnet | Enables remote login on another computer. Rarely used today because of security concerns. |
| File Transfer Protocol (FTP) | Transferring files from computer to computer. Has been mostly supplanted by more secure protocols such as FTP over SSL (FTPS) and SSH File Transfer Protocol (SFTP). |
| World Wide Web (the web) | Retrieving, formatting, and displaying information (including text, audio, graphics, and video) by using hypertext links |

## Internet Services

A client computer connecting to the Internet has access to a variety of services. These services include email, instant messaging, **Telnet**, **File Transfer Protocol (FTP)**, and the web. Table 7.3 provides a brief description of these services.

Each Internet service is implemented by one or more software programs. Examine Figure 7.8, which illustrates one way these services can operate within a multitiered client/server architecture. All the services may run on a single server computer (as shown in Figure 7.8) or different services may be allocated to different servers.

**Email** enables messages to be exchanged from computer to computer, with capabilities for routing messages to multiple recipients, forwarding messages, and attaching text documents or multimedia files to messages. Most email today is sent through the Internet. The cost of email is far lower than the equivalent voice, postal, or overnight delivery costs, and email messages can arrive anywhere in the world in a matter of seconds. Internet protocols used for email include Simple Mail Transfer Protocol (SMTP), Post Office Protocol 3 (POP3), and Internet Message Access Protocol (IMAP).

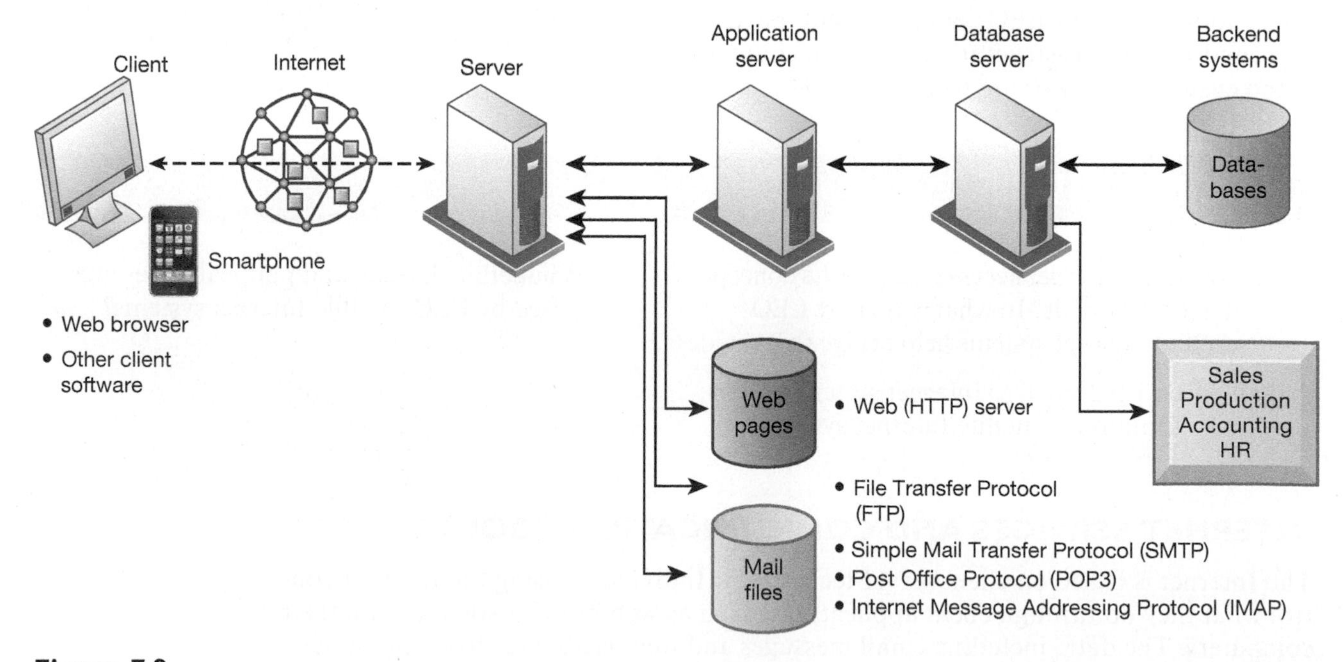


**Figure 7.8**
Client/Server Computing on the Internet
*Client computers running web browsers and other software can access an array of services on servers over the Internet. These services may all run on a single server or on multiple specialized servers.*

**Instant messaging** is a type of chat service that enables participants to create their own private chat channels. The instant messaging system alerts users whenever someone on their private list is online so that the user can initiate a chat session with other individuals. Instant messaging systems for consumers include Google Chat, Facebook Messenger, and Skype. Companies concerned with security use proprietary communications and messaging systems such as HCL Sametime, Microsoft Teams, and Slack.

Employee use of email, instant messaging, and the Internet is supposed to increase worker productivity, but the Spotlight on People case shows that this may not always be the case. Many companies believe they need to monitor and even regulate their employees' online activity, but is this ethical? Although there are some strong business reasons that companies may need to monitor their employees' email and web activities, what does this monitoring mean for employee privacy? The issue has become even more pressing because of the number of people now working remotely and because of businesses' desires to ensure that their employees and contractors are, in fact, working as required when they are not physically present.

## Voice over IP (VoIP)

The Internet has also become a popular platform for voice transmission and corporate networking. **Voice over IP (VoIP)** technology delivers voice information in digital form using packet switching, thus avoiding charges from local and long-distance telephone networks. Examine Figure 7.9, which illustrates how VoIP works. Voice calls can be made and received with a computer equipped with a microphone and speakers, a mobile device with a VoIP app, or with a VoIP-enabled telephone. Analog voice calls that would ordinarily be transmitted over public telephone networks are instead digitized into packets and routed via a VoIP gateway over a corporate network based on the Internet protocol or over the public Internet to the IP address of the receiving device or telephone.

Many telecommunications companies such as AT&T, Verizon, Charter Spectrum, Comcast, and Altice provide VoIP service bundled with their high-speed Internet offerings. Skype offers free VoIP worldwide using a peer-to-peer network, and Google has its own free VoIP service.

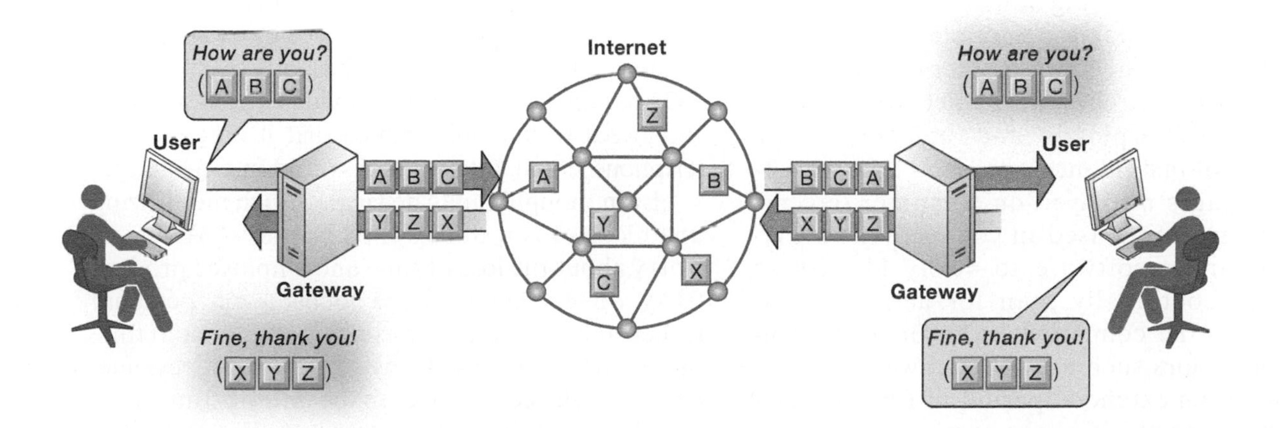


**Figure 7.9**
How Voice over IP Works
*A VoIP phone call digitizes and breaks up a voice message into data packets that may travel along different routes before being reassembled at the final destination. A processor nearest the call's destination, called a gateway, arranges the packets in the proper order and directs them to the telephone number of the receiver or the IP address of the receiving computer.*

## SPOTLIGHT ON: PEOPLE

## Monitoring Employees on Networks: Unethical or Good Business?

The Internet has become an extremely valuable business tool and never more so than during the pandemic, when it provided a lifeline for many businesses whose employees were forced to shift to working from home. But the Internet can also be a huge distraction, enabling employees to waste valuable company time by surfing inappropriate websites (social networks, shopping, sports, etc.), sending and receiving personal email, instant messaging with friends, and streaming videos and music.

Many businesses fear that because work-from-home employees do not have to worry about coworkers or managers possibly peering over their shoulders, such behavior will become even more prevalent. To guard against this, companies have turned to increased employee monitoring. According to Gartner, since the beginning of the pandemic, the number of large employers monitoring workers has doubled to 60 percent, and Gartner expects that percentage to continue rising over the next three years to 70 percent. As a result, the market for remote employee-monitoring software is expected to skyrocket, from about $265 million in 2019 to almost $1.4 billion by 2027.

You may not be aware of all the ways an employer can monitor you. For instance, if your employer has an enterprise email account with Google or Microsoft, authorized administrators can access all the emails you send and receive via that account. Add-on monitoring tools such as Teramind, Hubstaff, TimeCamp, InterGuard, ActivTrak, Mobistealth, and Work Examiner, which extend monitoring abilities, are also available. Such tools enable companies to track the movements of your mouse; record your keystrokes, online searches, and video conferences; monitor file downloads and uploads; create transcripts of instant messages; and take screenshots of images displayed on computer screens. Webcams are being used in connection with facial recognition software to verify identity as well as to continually scan for activities that might break the company's rules or for "suspicious" behaviors such as looking away from the monitor for an extended period of time, using a smartphone, or blocking the camera's view.

Even more invasive tracking may be in store in the future. For instance, Fujitsu has created an AI algorithm that reportedly can detect concentration levels by tracking facial muscle movements, which it is touting as applicable for online meetings and sales activities. Microsoft has patented a biometric wellness insights service that uses biometric data, such as heart rate and blood pressure, and haptic feedback, such as how hard keys on a keyboard are pressed, to detect a worker's stress levels and generate an employee anxiety score.

Collaboration tools such as Slack, Microsoft Teams, Microsoft 365, Google Workspace, and others provide another way for companies to monitor employees. For instance, Microsoft collects and analyzes data on the frequency of chats, emails, and meetings between its staff and clients using its own Office 365 services to measure employee productivity, management effectiveness, and work–life balance. Tracking emails, chats, and calendar appointments can show how employees spend an average of 20 hours of their work time each week. Microsoft also has enhanced machine-learning capabilities for its Microsoft 365 compliance center tool in the works, which will use artificial intelligence to better detect employee actions.

Microsoft also sells workplace analytics software to other companies, such as Macy's which analyzed data on staff work–life balance by measuring how many hours employees spent sending emails and logging in online outside of business hours. Mortgage giant Freddie Mac used Microsoft's tools to gauge how much time workers spent in meetings and to try to determine whether some of those gatherings were redundant.

In general, US companies have the legal right to monitor employee Internet and email activity while they are at work, although many states are passing new legislation regulating employee monitoring. For instance, in New York, a law that took effect in May 2022 requires employers to provide current and new employees with written notice that it may monitor telephone, email, Internet access, and Internet usage.

Even though it may be legal, is such monitoring unethical, or is it simply good business? Managers worry about the loss of time and employee productivity when employees are focusing on personal rather than company business. Too much time on personal business translates into lost revenue. Some employees may even be billing time they spend pursuing personal interests online to clients, thus overcharging them.

If personal traffic on company networks is too high, it can also clog the company's network so that legitimate business work cannot be performed. GMI Insurance Services, which serves the US transportation industry, found that employees were downloading a great deal of music and streaming video and storing the files on company servers. GMI's server backup space was being eaten up.

When employees use email or the web (including social networks) at employer facilities or with employer equipment, anything employees do, including anything illegal, carries the company's name. Therefore, the employer can be traced and held liable. Management in many firms fear that racist, sexually explicit, or other potentially offensive material accessed or distributed by their employees could result in adverse publicity and even lawsuits for the firm. Even if the company is found not to be liable, responding to lawsuits can run up huge legal bills. Companies also fear leakage of confidential information and trade secrets through email or social networks. The question is whether electronic surveillance is an appropriate tool for maintaining an efficient and positive workplace atmosphere. Some companies try to ban all personal activities on corporate networks—zero tolerance. Others block employee access to specific websites or social networks, closely monitor email messages, or limit personal time on the web.

Should all employees be monitored while working? Not necessarily. Not every workforce, workplace, or work culture and environment is a candidate for electronic surveillance. The decision depends on the company and the work environment an employer wants to create. A major concern of some employers is the potential damage to a work culture that fosters trust, employee commitment, and motivation. Electronic surveillance of employees could prove highly counterproductive to such an environment.

No solution is problem-free, but many consultants believe companies should write corporate policies on employee email, social media, and Internet use. The policies should include explicit ground rules that state, by position or by level, under what circumstances employees can use company facilities for email, social media posts, or web surfing. The policies should also inform employees whether these activities are monitored and explain why.

The rules should be tailored to specific business needs and organizational cultures. For example, an investment firm might need to allow many of its employees access to other stock-trading sites. A company dependent on widespread information sharing, innovation, and independence could very well find that monitoring creates more problems than it solves.

Sources: ReportsandData, "Employee Remote Monitoring Software Market Is Expected to Reach USD $1,396.2 Million by 2027," Einnews.com, December 21, 2021; Matthew Finnegan, "Rise in Employee Monitoring Prompts Call for New Rules to Protect Workers," Computerworld.com, November 30, 2021; "New York Imposes New Requirements for Employee Monitoring," *National Law Review* November 23, 2021; Chris Matyszczyk, "Microsoft Will Now Snitch on You at Work Like Never Before," Zdnet.com, November 7, 2021; Barclay Ballard, "Microsoft Teams May Be Tracking More of Your Info than You Think," Techradar.com, October 6, 2021; Danielle Abril and Drew Harwell, "Keystroke Tracking, Screenshots, and Facial Recognition: The Boss May Be Watching Long after the Pandemic Ends," *Washington Post*, September 24, 2021; Tatum Hunter, "Here Are All the Ways Your Boss Can Legally Monitor You," *Washington Post*, September 24, 2021; Matthew Finnegan, "Microsoft Patents Biometric 'Wellness Insights' Tool for Workers," Computerworld.com, April 28, 2021; Matthew Finnegan, "Collaboration Analytics: Yes, You Can Track Employees. Should You?," Computerworld.com, April 12, 2021.

## CASE STUDY QUESTIONS

1. Should managers monitor employee email and Internet usage? Why or why not?
2. Describe an effective email and web use policy for a company.
3. Should managers inform employees that their web behavior is being monitored? Or should managers monitor secretly? Why or why not?

Although up-front investments are required for an IP phone system, VoIP can reduce communication and network management costs while providing more communication capabilities. For example, VoIP provider Nextiva helps Titan Solar Power service its quickly growing customer base. Prior to adopting Nextiva, the Arizona-based company had no way of tracking why people were calling or the volume of calls to and from customers. Using Nextiva's tools to track these data, Titan can see the number of inbound and outbound calls and the number of calls in queue. The company can then take steps to reduce customer wait times. Additionally, Nextiva provides capabilities for recording customer calls and for directly routing customers to the right departments.

### Unified Communications

In the past, each of a firm's networks for wired and wireless data, voice communications, and videoconferencing operated independently of each other and had to be managed separately by the information systems department. Now, however, firms can merge disparate communications modes into a single, universally accessible service using unified communications technology. A **unified communications** system integrates disparate channels for voice communications, data communications, instant messaging, email, and teleconferencing into a single experience by which users can seamlessly switch back and forth between different communication modes. Presence technology shows whether a person is available to receive a call. Unified communications as a service (UCaaS) provides cloud-delivered unified applications.

InspereX, a fixed-income market fnancial services firm with 200 employees and three different office locations, selected Avaya's all-in-one UCaaS solution to replace its legacy on-premises phone system, which required remote access points to be set up for employees to be able use their office phones when out of the office. The Avaya system provides easier and better connectivity across InspereX's office locations and enhances the ability to make and receive calls anywhere, anytime, and using any device, which has added flexibility for users and boosted users' work productivity and satisfaction. The solution also provides built-in analytics that enable InspereX to gain a deeper understanding of the organization's communications regardless of where meetings or calls are taking place. Because the solution is provided as a cloud-based service, it also reduces the burden on InspereX's IT infrastructure and reduces costs (Cardoza, 2021).

### Virtual Private Networks

What if you had a marketing group, with members spread across the United States, charged with developing new products and services for your firm? You would want them to be able to email each other and communicate with the home office without any chance that outsiders could intercept the communications. Large private networking firms offer secure, private, dedicated networks to customers, but this type of network is expensive. A lower-cost solution is to create a virtual private network within the public Internet.

A **virtual private network (VPN)** is a secure, encrypted, private network that has been configured within a public network to take advantage of the economies of scale and the management facilities of large networks, such as the Internet. Examine Figure 7.10, which illustrates how a VPN works. Computers in a private network can use VPN protocols to encrypt the data being transferred and to establish a secure "tunnel" that shield the data as they travel over the Internet. A VPN provides your firm with secure, encrypted communications at a much lower cost than the same capabilities offered by traditional non-Internet providers that use their private networks to secure communications. VPNs also provide a network infrastructure for combining voice and data networks.

**Figure 7.10**
A Virtual Private Network Using the Internet
*This VPN is a private network of computers linked using a secure tunnel connection through the Internet. It protects data transmitted over the public Internet by encoding the data and wrapping them within the Internet protocol. By adding a wrapper around a network message to hide its content, organizations can create a private connection that travels through the public Internet.*

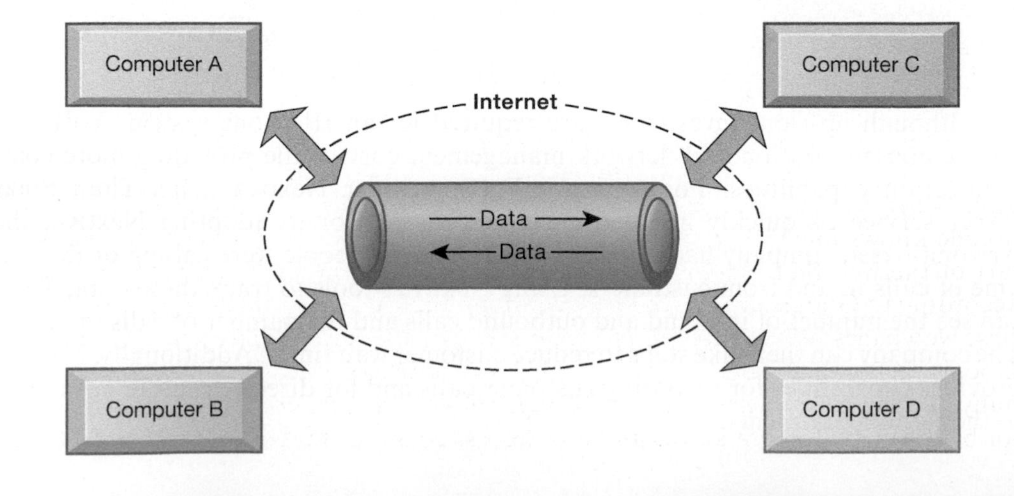

A variety of protocols are used to protect data transmitted over the public Internet, including Point-to-Point Tunneling Protocol (PPTP). In a process called *tunneling*, packets of data are encrypted and wrapped inside IP packets. By adding this wrapper around a network message to hide its content, business firms create a private connection that travels through the public Internet.

## THE WEB

The web is the most popular Internet service. It's a system with universally accepted standards for storing, retrieving, formatting, and displaying information by using a client/server architecture. Web pages are formatted using hypertext, which consists of embedded links that connect documents to one another and that also link pages to other objects, such as sound, video, or animation files. When you click a graphic and a video clip plays, you have clicked a hyperlink. A typical **website** is a collection of web pages linked to a home page.

### Hypertext

Web pages are based on Hypertext Markup Language (HTML), which formats documents and incorporates dynamic links to other documents and other objects stored in the same or remote computers (see Chapter 5). Web pages are accessible through the Internet because the web browser software operating on your computer can request web pages stored on an Internet host server by using the **Hypertext Transfer Protocol (HTTP)**. HTTP is the communications standard that transfers pages on the web. For example, when you type a web address, such as http://www.sec.gov, in your browser, your browser sends an HTTP request to the sec.gov server requesting the home page of sec.gov.

HTTP (or HTTPS, a more secure version of HTTP that we will discuss further in Chapter 8) is the first set of letters at the start of every web address, followed by the domain name, which specifies the organization's server computer that is storing the web page. Most companies have a domain name that is the same as, or closely related to, their official corporate name. The directory path and web page name are two more pieces of information within the web address that help the browser track down the requested page. Together, the address is called a **uniform resource locator (URL)**. When typed into a browser, a URL tells the browser software exactly where to look for the information. For example, in the fictitious URL http://www.megacorp.com/content/features/082610.html, *http* names the protocol that displays web pages, www.megacorp.com is the domain name, content/features is the directory path that identifies where on the domain web server the page is stored, and 082610.html is the web page name and the name of the format it is in. (It is an HTML page.)

### Web Servers

A web server is software for locating and managing stored web pages. It locates the web page a user requests on the computer where it is stored and delivers the web page to the user's computer. Server applications usually run on dedicated computers, although these applications can all reside on a single computer in small organizations.

The leading web servers in use today are Microsoft Internet Information Services (IIS) and Apache HTTP Server. Apache is an open-source product that is free of charge and can be downloaded from the web.

### Searching for Information on the Web

No one knows for sure how many web pages there really are. The surface web is the part of the web that search engines visit and about which information is recorded. According to Google, its search index contains hundreds of billions of web pages, and these web pages reflect a large portion of the publicly accessible web page population (Google, 2022). But there is a "deep web" that contains trillions of additional pages, many of them proprietary (such as the pages of *Wall Street Journal Online*, which cannot be visited without a subscription or access code), or that are stored in protected corporate databases. Facebook users can block Google and other search engines from accessing the content they post. A portion of the deep web called the **dark web** has been intentionally hidden from search engines, uses masked IP addresses, and is accessible only with a special web browser to preserve anonymity. The dark web has become a haven for criminals because it allows the buying and selling of illicit goods, including credit card and social security numbers, with complete anonymity.

**Search Engines** Obviously, with so many web pages, finding, nearly instantly, the specific ones that can help you or your business is an important problem. The question is, how can you find the one or two pages you really want and need out of billions of indexed web pages? **Search engines** attempt to solve the problem of finding useful information on the web nearly instantly. Today's search engines can sift through HTML files; files of Microsoft Office applications; PDF files; and audio, video, and image files. There are hundreds of search engines in the world, but the vast majority of search results come from Google (more than 85 percent), Microsoft's Bing (7 percent), Yahoo (3 percent), and China's Baidu (1 percent). Also, while we typically think of Amazon as an online store, it is also a powerful product search engine (Johnson, 2021a).

Web search engines started out in the early 1990s as relatively simple software programs that roamed the nascent web, visiting pages and gathering information about the content of each page. The first search engines were simple keyword indexes of all the pages they visited, leaving users with lists of pages that may or may not have been truly relevant to their search.

In 1994, Stanford University computer science students David Filo and Jerry Yang created a hand-selected list of their favorite web pages and called it "Yet Another Hierarchical Officious Oracle," or Yahoo. Yahoo was not initially a search engine but rather an edited selection of websites organized by categories. Currently, Yahoo relies on Microsoft's Bing for search results.

In 1998, Larry Page and Sergey Brin, two other Stanford computer science students, released their first version of Google. This search engine was different: Not only did it index each web page's words, but it also ranked search results based on the relevance of each page. Page patented the idea of a page-ranking system (called *PageRank System*), which essentially measures the popularity of a web page by calculating the number of sites that link to that page as well as the number of pages to which it links. The premise is that popular web pages are more relevant to users. Brin contributed a unique web crawler program that indexed not only keywords on a page but also combinations of words (such as the names of authors and the titles of their articles). These two ideas became the foundation of the Google search engine. Examine Figure 7.11, which illustrates what happens when you enter a search query using Google.

**Mobile Search** Mobile search via smartphones and tablets makes up more than 60 percent of all searches and is expected to continue expanding rapidly in the next few years. Google, Amazon, and Yahoo have developed new search interfaces to make searching and shopping from smartphones more convenient. Google revised its search algorithm to favor sites that look good on smartphone screens. Mobile e-commerce (m-commerce) now comprises an ever-increasing percentage of e-commerce, accounting for more than 40 percent of retail e-commerce revenues in 2021, with more than 80 percent of those revenues derived from purchases made via smartphone (Johnson, 2021b; Insider Intelligence/eMarketer, 2022b).

**Semantic Search** Another way for search engines to become more discriminating and helpful is to make them capable of understanding what we are really looking for. Called **semantic search**, its goal is to build a search engine that can really understand human language and behavior. Google and other search engine firms are attempting to refine their search engine algorithms to capture more of what the user intended and the meaning of a search. Rather than evaluate each word in a search separately, Google's Hummingbird search algorithm tries to evaluate an entire sentence, focusing

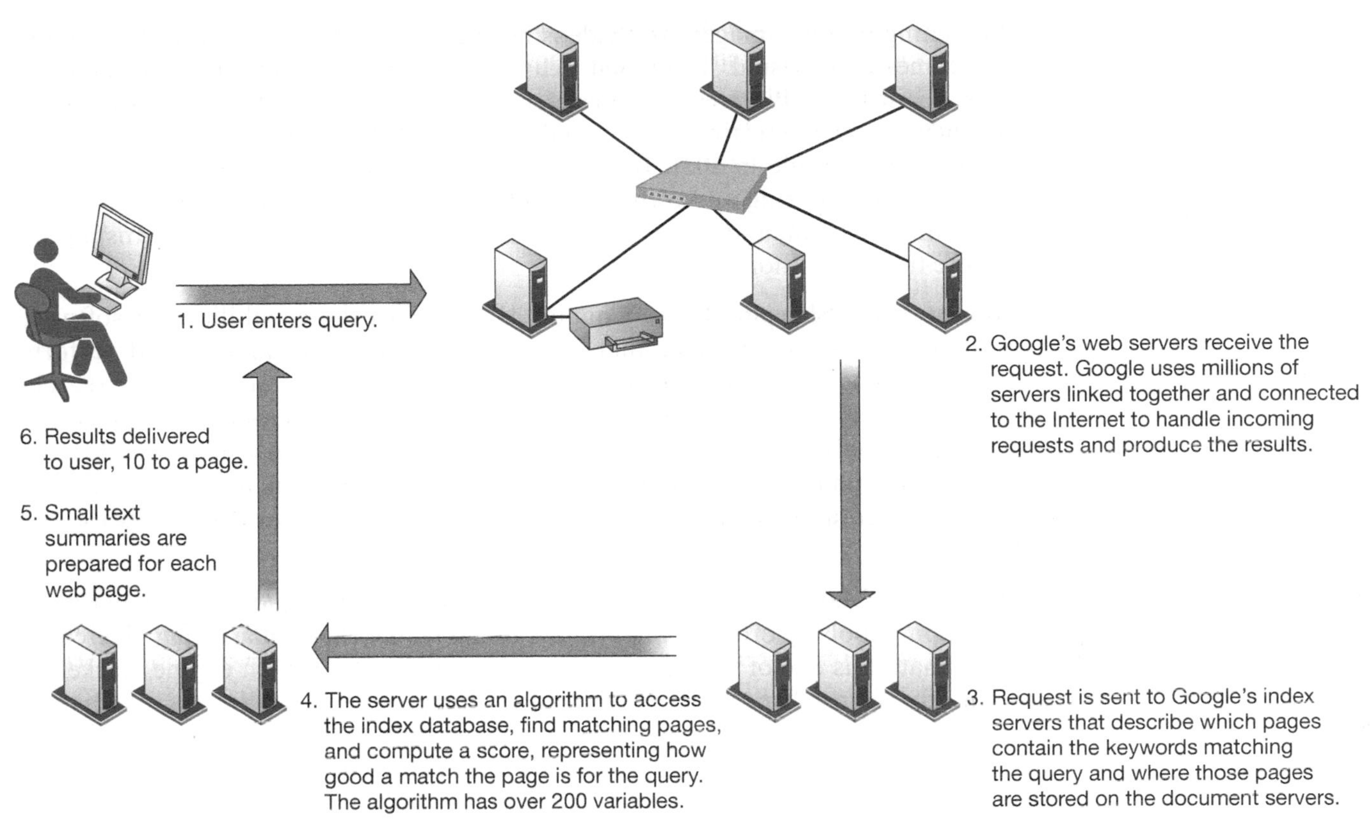
1. User enters query.
2. Google's web servers receive the request. Google uses millions of servers linked together and connected to the Internet to handle incoming requests and produce the results.
3. Request is sent to Google's index servers that describe which pages contain the keywords matching the query and where those pages are stored on the document servers.
4. The server uses an algorithm to access the index database, find matching pages, and compute a score, representing how good a match the page is for the query. The algorithm has over 200 variables.
5. Small text summaries are prepared for each web page.
6. Results delivered to user, 10 to a page.

**Figure 7.11**
How Google Works
*The Google search engine is continuously crawling the web, indexing the content of each page it locates, calculating each page's popularity, and storing the pages so that it can respond quickly to user requests to see a page. The entire process takes about half a second.*

on the meaning behind the words. For instance, if your search is a long sentence like "Google annual report selected financial data 2021," Hummingbird should be able to figure out that you really want Google parent company Alphabet's Form 10-K report filed with the Securities and Exchange Commission in February 2022.

Google searches also take advantage of Knowledge Graph, an effort of the search algorithm to anticipate what you might want to know more about as you search on a topic. Results of the Knowledge Graph appear on the right of the screen on many search result pages and contain more information about the topic or person you are searching for. For example, if you search "Lake Tahoe," the search engine will return basic facts about Lake Tahoe (its altitude, average temperature, and local fish, for example), a map, and hotel accommodations. Google has made **predictive search** part of most search results. This part of the search algorithm guesses what you are looking for and suggests search terms as you enter your search words. Google has continued to refine its search algorithms using mathematical modeling and artificial intelligence machine learning techniques (see Chapter 11) to make its responses to queries even more accurate and intelligent (Moz.com, 2022; Copeland, 2019).

**Visual Search and the Visual Web** Although search engines were originally designed to search text documents, the explosion of photos and videos on the Internet created a demand for searching and classifying these visual objects. Facial recognition software can create a digital version of a human face. Although you previously could search for people on Facebook by using their digital image to find and identify them, in November 2021, Meta, the parent company of Facebook, shut down Facebook's facial recognition system as a result of increasing criticism. However, Meta continues to develop the system for other purposes.

**Social networks** enable users to build communities of friends and professional colleagues. Members typically create a profile—a page for posting photos, videos, audio files, and text—and then share these profiles with others on the service who are identified as their friends or contacts. Social networks are highly interactive, offer real-time user control, rely on user-generated content, and are broadly based on social participation and the sharing of content and opinions. Leading social networks include Facebook, Instagram, Twitter, Pinterest, and LinkedIn (for professional contacts).

Social networks have radically changed how people spend their time online; how people communicate and with whom; how businesses stay in touch with their customers, suppliers, and employees; how providers of goods and services learn about their customers; and how advertisers reach potential customers. The large social networks are also application development platforms. Facebook alone has millions of apps and websites integrated into it, including applications for gaming, video sharing, and communicating with friends and family. We talk more about business applications of social networks in Chapters 2 and 10, and you can find discussions about social networks in many other chapters of this book.

### The Future Internet/Web Ecosystem

The Internet and web are not static. They are constantly evolving as new technologies are developed. For instance, **Internet2** is an advanced networking consortium focused on the development of next-generation Internet technologies. More than 320 US universities, 60 government agencies, 44 regional and state education networks, 50 leading corporations, and 70 national research and education network partners from more than 100 countries participate in Internet2. To connect these communities, Internet2 developed a high-capacity, 100-Gbps network that serves as a test bed for leading-edge technologies that may eventually migrate to the public Internet, including large-scale network performance measurement and management tools, secure identity and access management tools, and capabilities such as scheduling high-bandwidth, high-performance circuits. Internet2 is currently in the process of testing a 400-Gbps network on a coast-to-coast basis and has also deployed an 800-Gbps network on a pilot basis as a part of its Next Generation Infrastructure program.

The future web will be more visible. Its key features are more tools for individuals to make sense of the billions to trillions of pages on the web and the millions of apps available for smartphones and tablets and a visual, three-dimensional (3-D) virtual reality that has been dubbed the **metaverse**. Facebook is so convinced that the metaverse will be at the center of the future web that it has rebranded itself as Meta, and other companies are also jumping on the bandwagon. At the same time, ideas behind blockchain (see Chapter 6) are driving a different movement that some are calling Web3. **Web3** would be even more decentralized than the current Internet and would be controlled by its users rather than dominated by big corporations (White, 2021; Allyn, 2021).

Closer in time is a pervasive Internet/web that controls everything from a city's traffic lights and water usage to the lights in your living room and your car's rearview mirror. This is referred to as the Internet of Things (IoT), which we introduced in Chapter 1, and is based on billions of Internet-connected sensors throughout our physical world. Objects, animals, or people are provided with unique identifiers and the ability to transfer data over a network without requiring human-to-human or human-to-computer interaction.

General Electric, IBM, HP, Oracle, and hundreds of smaller companies are building smart machines, factories, and cities through the extensive use of remote sensors and fast cloud computing. While estimates of the number of IoT devices vary widely, some experts believe that by 2025, more than 30 billion IoT devices will be connected to the web. Over time, more and more everyday physical objects will be connected to the Internet and will be able to identify themselves to other devices, creating networks that can sense and respond as data change (Vailshery, 2021).

Apps are another element in the future Internet/web ecosystem. The growth of apps within the mobile platform is astounding. More than 75 percent of the time spent on mobile devices in the United States is spent using apps as opposed to using a mobile browser. Apps give users direct access to content and are much faster than loading a browser and searching for content (Insider Intelligence/eMarketer, 2022d).

Other complementary trends leading toward the future Internet/web include more widespread use of cloud computing and software as a service (SaaS) business models, ubiquitous connectivity among mobile platforms and Internet-access devices, and the transformation of the web from a network of separate, siloed applications and content into a more seamless and interoperable whole.

## 7-4 Identify the principal technologies and standards for wireless networking, communication, and Internet access.

Welcome to the wireless revolution! Smartphones, tablets, and wireless-enabled laptop computers now allow you to perform many of the computing tasks you used to do at your desk—and a whole lot more. We introduced smartphones in Chapters 1 and 5 as part of our discussion of the mobile platform. **Smartphones** such as the iPhone and Android phones combine the functionality of a cell phone with that of a laptop computer that has Wi-Fi capability. This makes it possible to combine music, video, Internet access, and telephone service in one device. The Internet has become a mobile, access-anywhere, broadband service for the delivery of video, audio, and web search.

### CELLULAR SYSTEMS

Today, almost 85 percent of the US population owns a mobile phone, and 75 percent owns a smartphone. Mobile is now the leading digital platform, with total activity on smartphones and tablets accounting for more than 50 percent of digital media time spent (Insider Intelligence/eMarketer, 2022e).

Earlier generations of cellular systems were designed primarily for voice and limited data transmission in the form of short text messages. **3G networks**, which have transmission speeds ranging from 144 Kbps for mobile users in, say, a car, to more than 2 Mbps for stationary users, offer transmission speeds appropriate for email and web browsing but are too slow for videos. 3G networks are currently being phased out by the major telecommunications carriers in the United States. **4G networks** have much higher speeds, up to 100 Mbps for downloading and 50 Mbps for uploading, with more than enough capacity for watching high-definition videos on your smartphone.

The newest generation of wireless technology, called **5G**, is designed to support transmission of much larger amounts of data in the gigabit range—with fewer transmission delays and the ability to connect many more devices (such as sensors and smart devices) at once—than existing cellular systems can support. 5G technology will aid the continued development of self-driving vehicles, smart cities, and the extensive use of the Internet of Things (IoT), as well as improve the speed and intensive data-handling ability of smartphones. T-Mobile, AT&T, and Verizon have all launched 5G networks with varying degrees of functionality and they are expected to enhance them further in 2022.

## WIRELESS COMPUTER NETWORKS AND INTERNET ACCESS

An array of technologies provides high-speed wireless access to the Internet for PCs and mobile devices. These high-speed services have extended Internet access to numerous locations that could not be covered by traditional, wired Internet services and have made anywhere, anytime computing a reality.

### Bluetooth

**Bluetooth** is the popular name for the 802.15 wireless networking standard that is useful for creating small **personal area networks (PANs)**. A Bluetooth-enabled device can link up to seven other Bluetooth devices within a 10-meter area using low-power, radio-based communication and can transmit up to 722 Kbps in the 2.4-GHz band. The Bluetooth Low Energy (BLE) specification reduces the energy usage of Bluetooth peripherals. Bluetooth 5 extends the ability to use BLE to a variety of devices, such as wireless headphones, and increases the range over which these devices can communicate to up to 240 meters, as well as data transfer speeds up to 2 Mbps.

Wireless phones, pagers, computers, printers, and computing devices using Bluetooth communicate with each other and even operate each other without direct user intervention. Examine Figure 7.12, which illustrates how a person can use a smartphone or a computer's wireless keyboard and mouse to send a document file wirelessly to a printer.

Although Bluetooth lends itself to personal networking, it also has uses in large corporations. For example, FedEx drivers use Bluetooth to transmit the delivery data captured by their handheld computers to cellular transmitters, which forward the data to corporate computers. Drivers no longer need to spend time physically docking their handheld units into the transmitters, and Bluetooth has saved FedEx $20 million per year.

### Wi-Fi and Wireless Internet Access

The 802.11 set of standards for wireless LANs and wireless Internet access is also known as **Wi-Fi**. The first of these standards to be widely adopted was 802.11b, which can transmit up to 11 Mbps in the 2.4-GHz band and has an effective distance of 30

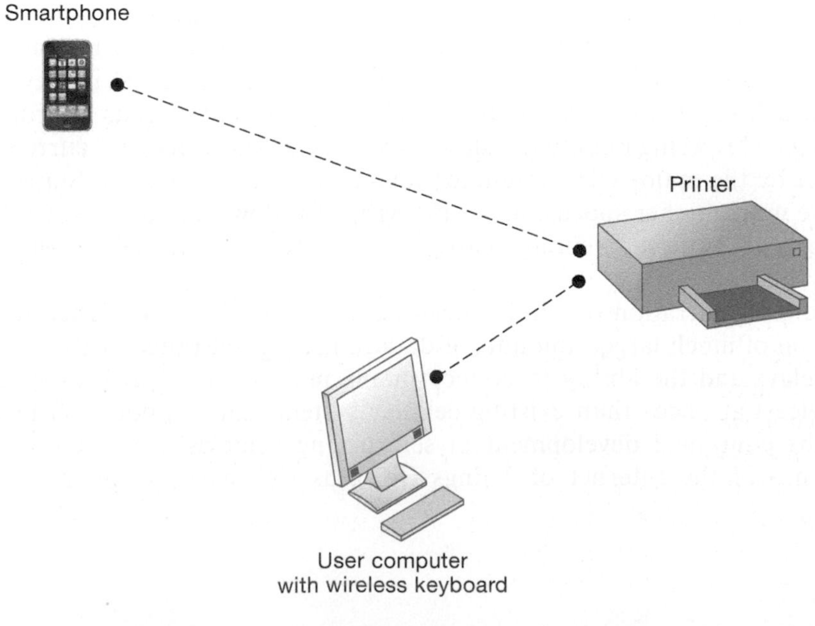

Smartphone

Printer

User computer
with wireless keyboard

**Figure 7.12**
A Bluetooth Network (PAN)
*Bluetooth enables a variety of devices, including smartphones, wireless keyboards and mice, PCs, and printers, to interact wirelessly with each other within a small, 30-foot (10-meter) area. In addition to the links shown, Bluetooth can be used to network similar devices to, for example, send data from one PC to another.*

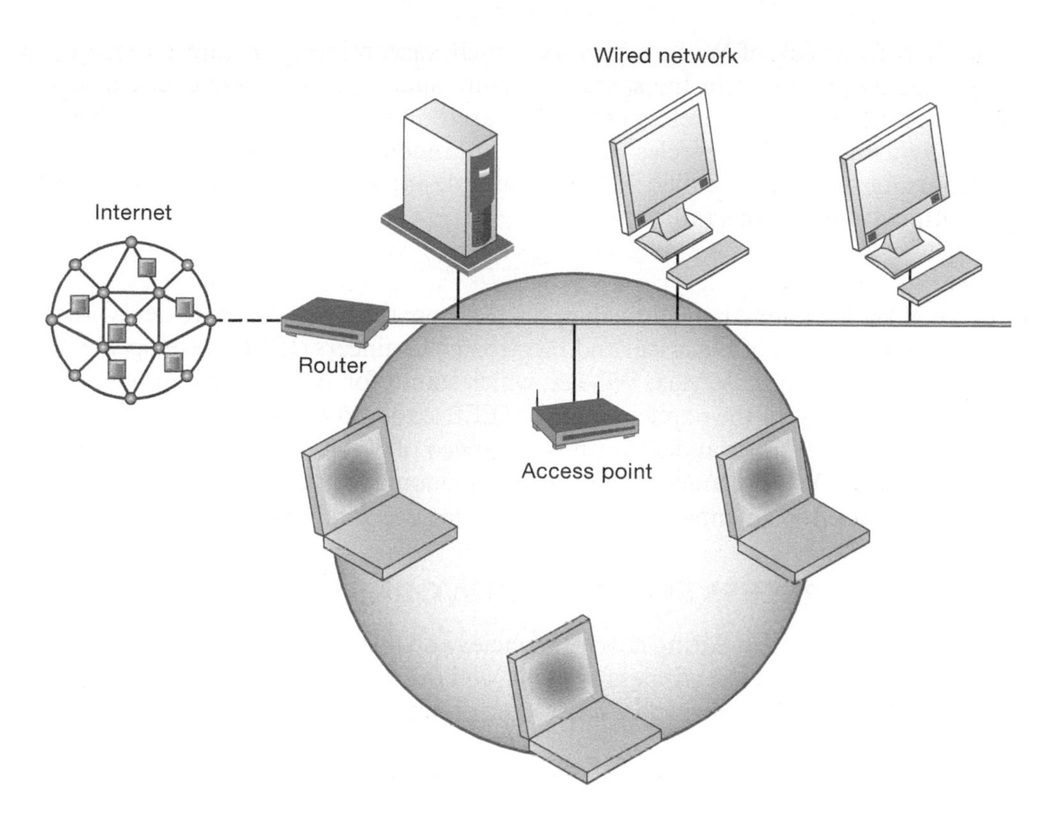


**Figure 7.13**
An 802.11 Wireless Lan
*In a Wi-Fi network, mobile wireless devices such as laptop computers communicate with a wired LAN and the Internet via an access point. The access point uses radio waves to communicate with both the mobile devices and the wired network.*

to 50 meters. The most current 802.11 standards are 802.11ac (Wi-Fi 5) and 802.11ax (Wi-Fi 6). Wi-Fi 5 provides for throughputs of 500 Mbps to 1 Gbps, while Wi-Fi 6 can operate over broader ranges of frequencies and achieve theoretical throughputs of up to 10 Gbps. Today's PCs and tablets have built-in support for Wi-Fi, as do the iPhone, and other smartphones.

In most Wi-Fi communication, wireless devices communicate with a wired LAN using access points. An access point is a box consisting of a radio receiver/transmitter and antennas that link to a wired network, router, or hub.

Examine Figure 7.13, which illustrates an 802.11 (Wi-Fi) wireless LAN that connects laptop computers equipped with network interface adapters to a wired LAN and to the Internet. Most wireless devices in a Wi-Fi LAN are clients. The servers that the mobile client stations need to use are on the wired LAN. The access point uses radio waves to transmit network signals from the wired network to the client adapters, which convert the signals to data that the mobile device can understand. The client adapter then transmits the data from the mobile device back to the access point, which forwards the data to the wired network.

The most popular use for Wi-Fi today is high-speed wireless Internet service. In this instance, the access point plugs into an Internet connection, which could come from a cable service, DSL or FiOs, or satellite service. Computers within range of the access point use it to link wirelessly to the Internet.

**Hotspots** are locations with one or more access points providing wireless Internet access and are often in public places. Some hotspots are free or do not require any additional software to use; others may require activation and the establishment of a user account by providing a credit card number over the web.

Businesses of all sizes are using Wi-Fi networks to provide low-cost wireless LANs and Internet access. Wi-Fi hotspots can be found in hotels, airport lounges, libraries, cafés, and college campuses. Dartmouth College is one of many campuses where students now use Wi-Fi for research, course work, and entertainment.

Wi-Fi technology poses several challenges, however. One is Wi-Fi's weak security, which makes these wireless networks vulnerable to intruders. We provide more detail about Wi-Fi security issues in Chapter 8.

Another drawback of Wi-Fi networks is their susceptibility to interference from nearby systems operating in the same spectrum, such as microwave ovens or other wireless LANs. However, wireless networks based on the 802.11n standard solve this problem by using multiple wireless antennas in tandem to transmit and receive data and also using technology called MIMO (multiple-input multiple-output) to coordinate multiple simultaneous radio signals.

### WiMax

The range of Wi-Fi systems is no more than 300 feet from the base station. To deal with this constraint, the Institute of Electrical and Electronics Engineers (IEEE) developed a family of standards known as WiMax. **WiMax**, which stands for Worldwide Interoperability for Microwave Access, is the popular term for IEEE standard 802.16. It has a wireless access range of up to 31 miles and a transmission speed of 30–40 Mbps (and up to 1 Gbps for fixed stations). WiMax antennas are powerful enough to beam high-speed Internet connections to rooftop antennas of homes and businesses that are miles away.

## RFID AND WIRELESS SENSOR NETWORKS

Mobile technologies are creating new efficiencies and ways of working throughout enterprises. In addition to the wireless systems we have just described, radio frequency identification (RFID) systems and wireless sensor networks are having major impacts.

### Radio Frequency Identification (RFID) and Near Field Communication (NFC)

**Radio frequency identification (RFID)** systems provide technology for tracking the movement of goods throughout the supply chain. To transmit radio signals over a short distance to RFID readers, RFID systems use tiny tags with embedded microchips containing data about an item and its location. The RFID readers then pass the data over a network to a computer for processing. Unlike bar codes, RFID tags do not need line-of-sight contact to be read.

Examine Figure 7.14, which illustrates how RFID works. The RFID tag is programmed with information that can uniquely identify an item, plus other information about the item such as its location, where and when it was made, or its status during production. An RFID reader also includes an antenna that constantly transmits radio waves in ranges anywhere from 1 inch to 100 feet. When an RFID tag comes

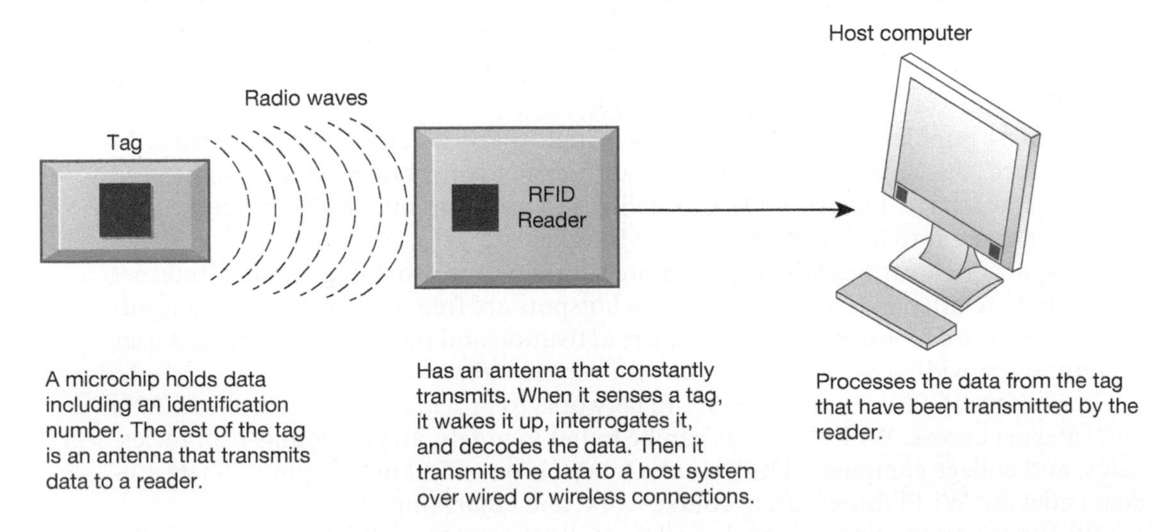

Host computer
Radio waves
Tag
RFID Reader
A microchip holds data including an identification number. The rest of the tag is an antenna that transmits data to a reader.
Has an antenna that constantly transmits. When it senses a tag, it wakes it up, interrogates it, and decodes the data. Then it transmits the data to a host system over wired or wireless connections.
Processes the data from the tag that have been transmitted by the reader.


**Figure 7.14**
How RFID Works
*RFID uses low-powered radio transmitters to read data stored in a tag at distances ranging from 1 inch to 100 feet. The reader captures the data from the tag and sends them over a network to a host computer for processing.*

within the range of the reader, the reader activates the tag, which starts sending data. The reader captures these data, decodes them, and sends them back over a wired or wireless network to a host computer for further processing. Both RFID tags and antennas come in a variety of shapes and sizes.

In inventory control and supply chain management, RFID systems capture and manage more detailed information about items in warehouses or in production than bar coding systems. If a large number of items are shipped together, RFID systems can track each pallet, lot, or even unit in the shipment. This technology helps companies improve receiving and storage operations by improving their ability to see exactly what stock is stored in warehouses and on retail store shelves.

For instance, Walmart has installed RFID readers at receiving docks at its stores to record the arrival of pallets and cases of goods shipped with RFID tags. The RFID reader reads the tags a second time just as the cases are brought onto the sales floor from backroom storage areas. Software then combines sales data from Walmart's point-of-sale systems and the RFID data regarding the number of cases brought out to the sales floor. The software program determines which items will soon be depleted and automatically generates a list of items to obtain from the warehouse in order to replenish store shelves before they are empty. This information helps Walmart reduce out-of-stock items, increase sales, and further shrink its costs.

The cost of RFID tags used to be too high for widespread use, but now the cost starts at around 7 cents per tag in the United States. As the price decreases, RFID is starting to become cost-effective for many applications.

In addition to installing RFID readers and tagging systems, companies may need to upgrade their hardware and software to process the massive amounts of data produced by RFID systems—data that could add up to tens or hundreds of terabytes.

Software is used to filter, aggregate, and prevent RFID data from overloading business networks and system applications. Applications often need to be redesigned to accept large volumes of frequently generated RFID data and to share those data with other applications. Major enterprise software vendors now offer RFID-ready versions of their supply chain management applications.

Tap-and-go services like Apple Pay and Google Wallet use an RFID-related technology called **near field communication (NFC)**. NFC is a short-range wireless connectivity standard that uses electromagnetic radio fields to enable two compatible devices to exchange data when brought within a few centimeters of each other. A smartphone or other NFC-compatible device sends out radio frequency signals that interact with an NFC tag found in compatible card readers or smart posters. The signals then create a current that flows through the NFC tag, allowing the device and the tag to communicate with one another. In most cases the tag is passive and can only send out information whereas the other device (such as a smartphone) is active and can both send and receive information. (There are NFC systems in which both components are active.)

NFC is used for wireless payment services, for retrieving information, and even for on-the-go exchanging of videos or information with friends. You could share a website link by passing your phone over a friend's phone and then waving your phone in front of a poster or display containing an NFC tag. You could also share information about what you're viewing at a museum or exhibit.

### Wireless Sensor Networks

If your company wanted state-of-the art technology to monitor building security or detect hazardous substances in the air, it might deploy a wireless sensor network. **Wireless sensor networks (WSNs)** are networks of interconnected wireless devices that are embedded in the physical environment to provide measurements of many points over large spaces. These devices have built-in processing, storage, and radio frequency sensors and antennas. They are linked into an interconnected network that routes the data that the devices capture to a computer for analysis. These networks range from

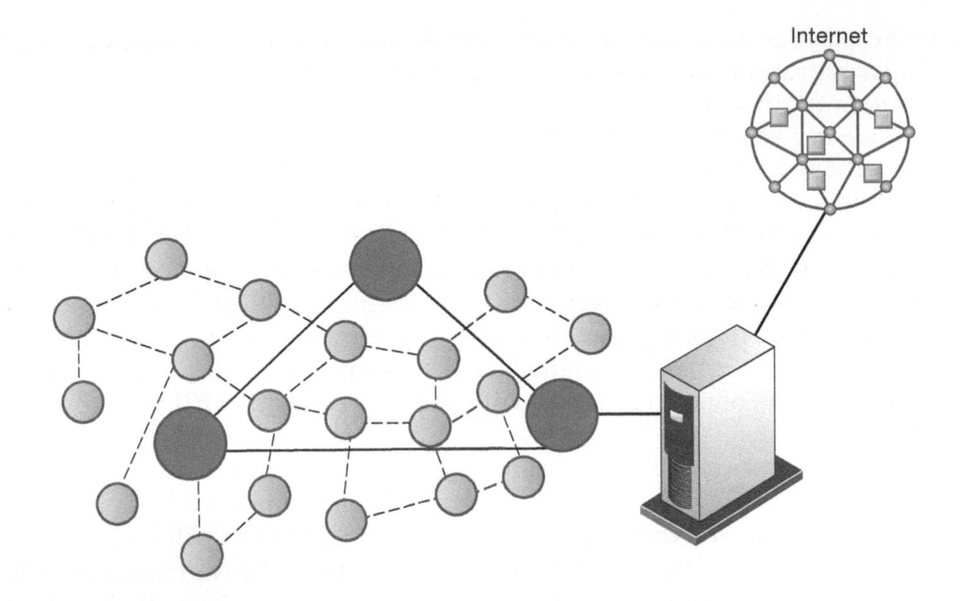


**Figure 7.15**
A Wireless Sensor Network
*Wireless sensor networks can contain hundreds to thousands of interconnected wireless sensors (nodes) that route the data they capture to a computer for analysis.*

hundreds to thousands of nodes. Examine Figure 7.15, which illustrates one type of wireless sensor network. The small circles in Figure 7.15 represent lower-level nodes, and the larger circles represent higher-level nodes. Lower-level nodes forward data to each other or to higher-level nodes, which transmit data to a server with greater processing power and which can also act as a gateway to a network based on Internet technology. Using only the higher-level nodes to transmit data to the server enables the data to be transmitted more rapidly and speeds up network performance.

Wireless sensor networks are valuable for uses such as monitoring environmental changes; monitoring traffic or military activity; protecting property; efficiently operating and managing machinery and vehicles; establishing security perimeters; monitoring supply chain management; and detecting chemical, biological, or radiological material.

Output from RFID systems and wireless networks is fueling the Internet of Things (IoT) (introduced in Chapter 1), in which machines such as jet engines, power plant turbines, and agricultural sensors constantly gather data and send the data over the Internet for analysis. The NHL Puck and Player Tracking system, described in the chapter-opening case, is an example of an IoT application. You'll find more examples of the Internet of Things in Chapters 2 and 11.

## 7-5 Understand how MIS can help your career.

Here is how this chapter and this book can help you find a job as a web developer.

### THE COMPANY

Best Body Fitness (BBF), a local fitness company that previously focused on in-person exercise classes but has expanded its virtual offerings because of the pandemic, is looking for a web developer and search engine marketing/search engine optimization (SEM/SEO) specialist to help build its business. BBF previously relied primarily on social media marketing via its Facebook page and Instagram feed, and it currently has only a rudimentary website.

### POSITION DESCRIPTION

BBF is looking for someone with knowledge of WordPress and website development tools to develop its website and help it evolve its platform with SEM/SEO best practices. The job involves optimizing, updating, and creating web pages and similar elements, as well as keeping a close eye on web analytics and maintenance. Specific job responsibilities include the following:

- Creating website layout/user interfaces by using standard WordPress/HTML practices
- Working with SEM/SEO elements such as metadata and site speed optimization and performing ongoing keyword discovery, expansion, and optimization
- Working with a variety of hosting environments and domain registars to set up, migrate, and/or launch websites
- Publishing blogs, landing pages, ads, and related content for high SEO scores
- Executing tests; collecting and analyzing data and results; and identifying trends and insights to achieve maximum return on investment for paid search campaigns
- Tracking, reporting, and analyzing website analytics
- Following emerging technologies/industry trends and applying them to operations and activities

## JOB REQUIREMENTS

- College graduate with a degree in MIS, marketing, computer science, or a related field
- Experience building or maintaining a successful WordPress website integrated with Google and social media trackers; proficiency with other webmaster tools
- Familiarity with current, past, and projected trends in the SEO industry
- Familiarity with how web applications, including security and best development practices, work
- Adequate knowledge of relational database systems and web application development
- Familiarity with performance marketing, conversion, and online customer acquisition
- Ability to work independently in a fast-paced, remote environment and master diverse web technologies and techniques
- Aggressive problem diagnosis and creative problem-solving skills

## INTERVIEW QUESTIONS

1. What kinds of web development and digital marketing courses have you taken?
2. What is your level of proficiency with WordPress? What other website development tools are you familiar with?
3. What experience do you have using SEM and SEO?
4. What emerging technologies do you think may have an impact in the next year?
5. Have you ever maintained a blog? If so, what did you do with the blog?
6. What mobile strategies would you suggest to grow our business?

## AUTHOR TIPS

1. Review the discussions of the web, search, search engine marketing, and blogs in this chapter and also the discussions of e-commerce marketing and building an e-commerce presence in Chapter 10.
2. Use the web to learn more about WordPress, website development tools, and SEM and SEO. Look into how to generate metrics reports using standardized tools and how to put together analyses and recommendations based on website data.

3. Look at how different fitness firms in the region are using websites and other online marketing tools. Which channels appear to be generating the highest levels of audience engagement?
4. Find out exactly what you would have to do for website management and the required software skills.
5. Inquire about the relational database skills you would need for this job.

## Review Summary

7-1 **Identify the principal components of telecommunications networks and key networking technologies.** A simple network consists of two or more connected computers. Basic network components include computers, network interface controllers (NICs), a connection medium, network operating system software, either a hub or a switch, and a router. The networking infrastructure for a large company typically includes an array of local and wide area networks, including the Internet, mobile cellular communication, wireless local area networks, and videoconferencing systems. The networking infrastructure supports applications such as a corporate website, a corporate intranet, and perhaps a corporate extranet. The firm's infrastructure may also include a separate telephone network that handles voice data.

Contemporary networks have been shaped by the rise of client/server computing, the use of packet switching, and the adoption of Transmission Control Protocol/Internet Protocol (TCP/IP) as a universal communications standard for linking disparate computers and networks, including the Internet. Protocols provide a common set of rules that enable communication among diverse components in a telecommunications network.

7-2 **Compare the different types of networks.** Local area networks (LANs) connect PCs and other digital devices within a 500-meter radius and are used today for many corporate computing tasks. Wide area networks (WANs) span broad geographical distances, ranging from several miles to entire continents, and are often private networks that are independently managed. Metropolitan area networks (MANs) span a single urban area.

The principal physical transmission media are twisted copper wire, coaxial cable, fiber-optic cable, and wireless transmission media. A wide range of transmission speeds is possible for any given medium, depending on the software and hardware configuration.

7-3 **Explain how the Internet and Internet technology work, and how they support communication and e-business.** The Internet is a worldwide network of networks that uses the client/server model of computing and the TCP/IP network reference model. Every computer on the Internet is assigned a unique numeric IP address. The Domain Name System (DNS) converts IP addresses to more user-friendly domain names. Internet data traffic is carried over transcontinental, high-speed backbone networks. The Internet backbone connects to regional networks, which in turn provide access to Internet service providers, large firms, and government institutions. Digital subscriber line (DSL) technologies, cable Internet connections, satellite Internet, and T1 and T3 lines are often used for high-capacity Internet connections. Worldwide Internet policies are established by organizations such as the Internet Architecture Board (IAB), the Internet Corporation for Assigned Names and Numbers (ICANN), and the World Wide Web Consortium (W3C).

Major Internet services include email, instant messaging, Telnet, FTP, and the web. Firms are also starting to use VoIP technology for voice transmission, unified communications systems, and virtual private networks (VPNs) as low-cost alternatives to private WANs.

The web is the most popular Internet service. Web pages are based on Hypertext Markup Language (HTML) and can display text, graphics, video, and audio. A website is a collection of web pages. Search engines help users locate on the web the information they need. Blogs, RSS, wikis, and social networks are examples of the information-sharing capabilities of the web. The future Internet/web ecosystem will feature higher-speed networks; visual, 3-D virtual reality, the prevalence of apps, and the interconnectedness of many different devices (Internet of Things).

7-4 **Identify the principal technologies and standards for wireless networking, communication, and Internet access.** Cellular networks are evolving toward high-speed, high-bandwidth, digital packet-switched transmission. 4G networks are capable of transmission speeds of 100 Mbps, and 5G networks, which are capable of transmitting data in the gigabit range among many devices, are starting to be rolled out.

Standards for wireless computer networks include Bluetooth (802.15) for small personal area networks (PANs), Wi-Fi (802.11) for local area networks (LANs), and WiMax (802.16) for metropolitan area networks (MANs).

Radio frequency identification (RFID) systems provide technology for tracking the movements of items by using tiny tags with embedded data about an item and its location. RFID readers read the radio signals transmitted by these tags and pass the data over a network to a computer for processing. Wireless sensor networks (WSNs) are networks of interconnected, wireless sensing and transmitting devices that are embedded in the physical environment to provide measurements of many points over large areas.

## Key Terms

3G networks, 257
4G networks, 257
5G, 257
Bandwidth, 240
Blog, 255
Bluetooth, 258
Broadband, 233
Cable Internet connections, 243
Dark web, 252
Digital subscriber line (DSL), 243
Domain name, 241
Domain Name System (DNS), 241
Email, 246
File Transfer Protocol (FTP), 246
Hertz, 240
Hotspots, 259
Hubs, 235
Hypertext Transfer Protocol (HTTP), 251
Instant messaging, 247
Internet Protocol (IP) address, 241
Internet service provider (ISP), 243
Internet2, 256
IPv6, 241
Local area network (LAN), 239
Metaverse, 256
Metropolitan area network (MAN), 239
Modem, 238
Near field communication (NFC), 261
Network operating system (NOS), 234
Packet switching, 236
Peer-to-peer, 239
Personal area networks (PANs), 258
Predictive search, 253
Protocol, 237
Radio frequency identification (RFID), 260
Router, 235
RSS, 255
Search engine, 252
Search engine marketing, 254
Search engine optimization (SEO), 254
Semantic search, 252
Shopping bots, 254
Smartphones, 257
Social network, 256
Software-defined networking (SDN), 235
Switch, 235
T1 lines, 243
T3 lines, 243
Telnet, 246
Transmission Control Protocol/Internet Protocol (TCP/IP), 237
Unified communications, 250
Uniform resource locator (URL), 251
Virtual private network (VPN), 250
Visual web, 254
Voice over IP (VoIP), 247
Website, 251
Web3, 256
Wide area networks (WANs), 239
Wi-Fi, 258
Wiki, 255
WiMax, 260
Wireless sensor networks (WSNs), 261

## Review Questions

**7-1** Identify the principal components of telecommunications networks and key networking technologies.
- Describe the features of a simple network and the network infrastructure for a large company.
- Name and describe the principal technologies and trends that have shaped contemporary telecommunications systems.

**7-2** Compare the different types of networks.
- Define an analog signal and a digital signal.
- Distinguish among a LAN, a MAN, and a WAN.

**7-3** Explain how the Internet and Internet technology work, and how they support communication and e-business.
- Define the Internet, describe how it works, and explain how it provides business value.
- Explain how the Domain Name System (DNS) and the IP addressing system work.
- List and describe the principal Internet services.
- Define and describe VoIP and virtual private networks, and explain how they provide value to businesses.
- List and describe alternative ways of locating information on the web.
- Describe how online search technologies are used for marketing.

**7-4** Identify the principal technologies and standards for wireless networking, communication, and Internet access.
- Define Bluetooth, Wi-Fi, WiMax, and 3G, 4G, and 5G networks.
- Describe the capabilities of each technology and for which types of applications each is best suited.
- Define RFID, explain how it works, and describe how it provides value to businesses.
- Define near field communication (NFC), and explain how it works.
- Define WSNs, explain how they work, and describe the kinds of applications that use them.

**MyLab MIS**
To complete these problems, go to EOC Discussion Questions in **MyLab MIS**.

## Discussion Questions

**7-5** (MyLab MIS) It has been said that smartphones have become the single-most-important digital device we own. Discuss the implications of this statement.

**7-6** (MyLab MIS) Should all major retailing and manufacturing companies switch to RFID? Why or why not?

**7-7** (MyLab MIS) What are some of the issues to consider in determining how the Internet can provide your business with a competitive advantage?

## Hands-On MIS Projects

The projects in this section give you hands-on experience evaluating and selecting communications technology, using spreadsheet software to improve selection of telecommunications services, and using web search engines for business research. Visit MyLab MIS to access this chapter's Hands-On MIS Projects.

### MANAGEMENT DECISION PROBLEM

**7-8** San Juan, the capital of Puerto Rico, includes the Old San Juan historic district, parts of which date back 500 years. The city was unable to provide broadband connectivity to tourists and to residents of this special district. Digging restrictions limited the amount of fiber-optic cable that could be laid down, and Old San Juan's network infrastructure was very outdated. However, the city was able to solve this problem by implementing a broadband Wi-Fi network that included technology supplied by Facebook Terragraph. Visit Terragraph.com to learn more about Facebook Terragraph, and explain why it provided a good solution for Old San Juan. Additionally, describe the people, organization, and technology issues that the city had to address when selecting a solution.

## IMPROVING DECISION MAKING: USING SPREADSHEET SOFTWARE TO EVALUATE WIRELESS SERVICES

Software skills: Spreadsheet formulas, formatting
Business skills: Analyzing wireless services and costs

**7-9** In this project, you'll use the web to research alternative wireless services and use spreadsheet software to calculate wireless service plan and smartphone costs.

You would like to provide your sales force of 20, based in Birmingham, Alabama, with smartphones that have wireless service plans that include unlimited talk, text, and data. You would like to equip each sales force member with a new iPhone 14 (128 gigabytes). Use the web to research the smartphone and service plans offered by Verizon, AT&T, and T-Mobile. Use your spreadsheet software to determine the wireless service plan and smartphone plan (purchase outright or pay monthly installments) that will offer the best pricing per user over a two-year period. For the purposes of this exercise, you do not need to consider corporate discounts.

## ACHIEVING OPERATIONAL EXCELLENCE: USING SEARCH ENGINES FOR BUSINESS RESEARCH

Software skills: Web search tools
Business skills: Researching new technologies

**7-10** This project will help develop your Internet skills in using search engines for business research.

Use Google and Bing to obtain information about biofuels as alternative fuels for motor vehicles. If you wish, try some other search engines as well. Compare the volume and quality of information you find using each search tool. Which tool is the easiest to use? Which produced the best results for your research? Why?

## COLLABORATION AND TEAMWORK PROJECT

Evaluating Smartphones

**7-11** Form a group with three or four of your classmates. Choose an Apple iPhone model that is currently available, and compare its capabilities to the capabilities of a similar model smartphone from another vendor. Your analysis should consider the purchase cost of each device, the wireless networks where each device can operate, the plan costs, and the services available for each device. You should also consider other capabilities of each device, including available software, security features, and the ability to integrate the device with existing corporate or PC applications. Which device would you select? On what criteria would you base your selection? If possible, use Google Docs and Google Drive or Google Sites to brainstorm, organize, and develop a presentation of your findings for the class.

## BUSINESS PROBLEM-SOLVING CASE

### GOOGLE, APPLE, AND META BATTLE FOR YOUR INTERNET EXPERIENCE

Three titans—Google, Apple, and Meta (the new company brand for the various platforms owned by Facebook)—are in an epic struggle to dominate your online experience. Caught in the crossfire are search, music, video, and other media along with the devices you use for all of these things. Mobile devices with advanced functionality have overtaken traditional desktop computers as the most popular form of computing and method of Internet access. Today, people spend more than two-thirds of their time online using mobile devices. The online experience is inextricably intertwined with, and driven by, online advertising, and the three titans are battling for control on that front as well.

Apple, which started as a personal computer company, quickly expanded into software and consumer electronics. After upending the music industry with its iPod MP3 player and the iTunes digital music service, Apple took mobile computing by storm with the iPhone and iPad. Apple's goal is to be the computing platform of choice for the Internet.

Apple's competitive strength is based not only on its hardware platform but also on its superior user interface, integrated services, and mobile software applications, in which it is a revenue leader. Apple's App Store offers more than two million apps for mobile and tablet devices. Apps greatly enrich the experience of using a mobile device, and whichever company offers the most appealing set of devices and applications will derive a significant competitive advantage over rival companies.

Apple thrives on its legacy of innovation. In 2011, it unveiled Siri (Speech Interpretation and Recognition Interface), a combination search/navigation tool and personal assistant. Siri provides personalized recommendations that improve as Siri gains user familiarity—all from a verbal command. Google countered by quickly releasing its own intelligent assistant tools, Google Now and then Google Assistant.

Apple faces strong competition for its phones and tablets from Android-based phones, such as Google's Pixel and Samsung's, that have just as sophisticated capabilities. However, Apple is not counting solely on hardware devices for future growth. Services have always played a large part in the Apple ecosystem, and they have emerged as a major revenue source. Apple has more than 1 billion smartphones, and a total of 1.65 billion active iOS-based devices, in circulation worldwide, creating a huge installed base of users willing to purchase services and, thus, a source of new revenue streams. Apple's services business, which includes Apple's music (both downloads and subscriptions), video sales and rentals, books, apps (including in-app purchases, subscriptions, and advertising), iCloud storage, and payments, brought in $58 billion in revenue for fiscal year 2021, nearly triple the amount in fiscal year 2016. As Apple rolls out more devices, such as the Apple Watch, Apple TV, HomePod, and AirTags, its services revenue will continue to expand and diversify, deepening ties with Apple users.

Although Apple does not derive substantial revenues directly from online advertising, Apple plays a major role in the online advertising ecosystem, as its various devices provide an important platform for that advertising. Apple has recently made a number of changes to prevent advertisers from tracking and collecting user data on its devices and via its Safari web browser without user consent, which some are predicting will have a significant impact on the online experience.

Google got its start as a software company, developing what has become the world's leading search engine. Today, more than 85 percent of web searches on laptop and desktop computers and more than 90 percent of mobile searches take place on Google Search. Although Google is by far the most dominant in search, it does face some competition in the search engine marketplace. Many consumers now begin their search for online products on Amazon rather than on Google. Pinterest hopes to compete with Google as a visual search engine, and Meta's Facebook, which has its own internal search functionality, has also become an important gateway to the web.

In 2005, Google purchased the Android open-source mobile operating system. Google provides Android at no cost to smartphone manufacturers, generating revenue indirectly through app purchases and advertising. Many different manufacturers have adopted Android as a standard. In contrast, Apple allows only its own devices to use its proprietary operating system, and all the apps it sells can run only on Apple products. Android is the most common operating system for smartphones and tablets and is deployed on more than three billion different devices around the world, including watches, smart speakers, car dashboards, and TVs. Google wants to extend Android to as many devices as possible. Android is expected to gain even more market share in the coming years, which could be problematic for Apple as it tries to maintain customer loyalty and keep software developers focused on the iOS platform. Whoever has the dominant smartphone operating system will have control over the apps where smartphone users spend most of their time and the built-in channels those apps provide for serving ads to mobile devices.

Google and Meta dominate online advertising. Together, the two firms account for more than 50 percent of all digital advertising revenue. Online advertising is the most important revenue stream of Alphabet (Google's parent company). More than 80 percent of Alphabet's revenue comes from online advertising, most of it from Google's search engine. Apple's changes to eliminate tracking on its mobile devices without user consent has advantaged Google in the short term, as many brands have shifted more ad spending to Google because its search ad business does not rely on data collected from app and web tracking.

Seven Google products and services, including Search, YouTube, and Maps, have more than a billion users each. Google's ultimate goal is to knit all of its services and devices together so that Google users will seemlessly interact with the company all day long and so that everyone will want to use Google. Much of Google's efforts to make its search and related services more powerful and user-friendly are based on the company's investments in artificial intelligence and machine learning (see Chapter 11). The goal is to evolve search into more of a smart assistance capability in which computers can understand what people are saying and respond conversationally with the right information at the right moment. Google Assistant is meant to provide a continuous, conversational dialogue between users and the search engine.

Meta's Facebook is the world's largest social network, with 2.9 billion monthly active users worldwide as of the first quarter of 2022. People use Facebook to stay connected with their friends and family and to express their opinions on things that matter to them. Facebook Platform enables developers to build applications and websites that integrate with Facebook to reach its global network of users and to build personalized and social products. Facebook is so pervasive and appealing that it has become many users' primary gateway to the Internet. For a lot of people, Facebook *is* the Internet: Whatever they do on the Internet is through Facebook.

Meta has persistently worked on ways to convert Facebook's popularity and trove of user data into advertising dollars. Facebook ads allow companies to target Facebook users based on their real identities and expressed interests rather than on educated guesses derived from web-browsing habits and other online behavior. Facebook continually tweaks the Feed (which acts as a user's "home page" on Facebook) to give advertisers more opportunities and more information with which to target markets. In 2021, almost 98 percent of Meta's global revenue came from advertising, with more than 95 percent of that ad revenue from mobile advertising. Facebook, together with Instagram, which Meta also owns, now leads Google in the mobile ad market and is just narrowly behind Google in terms of overall digital ad revenue share.

Meta is continually expanding advertising in its various products such Instagram, WhatsApp, Facebook Watch video on demand, and Messenger, although the majority of ad revenue still comes from its ads displayed in the Feed. Facebook CEO Mark Zuckerberg is convinced that social networks are the ideal way to use the web and to consume all of the other content people might desire, including news and video. That makes it an ideal marketing platform for companies. But he also knows that Meta can't continue to grow at the rate that it has grown in the past based on social networking alone. Rebranding Facebook as Meta signals its increased focus on the "metaverse," which Zuckerberg believes will entail moving the Internet experience beyond 2-D screens and toward immersive experiences like augmented and virtual reality.

Meta is challenging Google's YouTube as the premier destination for both personal and online advertising videos. Meta is also focused on making its Messenger service, which allows you to chat with friends or a business, send money securely, and share pictures and videos, "smarter" by deploying chatbots. Chatbots are stripped-down software agents that understand what you type or say, respond by answering questions or executing tasks, and run in Messenger's background (see Chapter 11). Zuckerberg also has said that he intends to help bring the next billion people online by attracting users in developing countries with affordable web connectivity. Meta has launched several Internet access initiatives designed to get more people online so that they can explore web applications, including Meta's social networks.

Monetization of personal data drives both Meta's and Google's online advertising–dominated business models. However, this practice also threatens individual privacy. The consumer surveillance underlying Meta's and Google's free services has come under siege from users and governments on both sides of the Atlantic and, as previously noted, from competitors such as Apple. Calls for restricting Meta's and Google's collection and use of personal data have gathered steam, especially after revelations about Russian agents trying to use Facebook to sway American voters and about Facebook's uncontrolled sharing of user data with third-party companies (see the Chapter 4 ending case study). Both companies are also still grappling with the impact of the European Union's General Data Protection Regulation (GDPR), which went into full effect in 2018. The GDPR requires companies to obtain consent from users before processing their data and has already inspired more stringent privacy legislation in the United States, such as the California Consumer Privacy Act and the California Privacy Rights Act. Efforts to curb the use of consumer data put the business model of the ad-supported Internet—and possibly Meta and Google—at risk.

These tech giants are also being scrutinized for monopolistic behavior. As already noted, together, Meta and Google dominate the digital ad industry and have been responsible for almost all of its growth. In the United States, Google and Apple provide 99 percent of mobile phone operating systems and the two dominant app store platforms. Google drives almost 90 percent of Internet search, while 95 percent of young adults on the Internet use a Meta platform.

Critics have called for new laws regulating anti-competitive behavior by Big Tech companies and even to break them up. Apple, Google, and Meta also face a plethora of lawsuits. For instance, Apple and Epic Games are embroiled in a lawsuit based on Apple's requirement that all payments for apps be made through its App Store, thus enabling Apple to claim a 30 percent commission. Epic has filed a similar lawsuit against Google. The attorneys-general of 36 states have filed an antitrust lawsuit against Google asserting that its app store abuses its market power and forces unfair terms on software developers. The US Department of Justice is also reportedly preparing a lawsuit targeting Google's digital advertising practices. In October 2021, the US Federal Trade Commission refiled a lawsuit against Meta, claiming that its acquisitions of Instagram and WhatsApp were part of an attempt to create a social network monopoly.

In Europe, Apple faces many antitrust investigations and lawsuits from the European Union and in the UK, Germany, and Japan. EU regulators have fined Google billions of dollars over the years, including $5 billion for forcing cellphone makers that use the company's Android operating system to install Google search and browser apps and an additional $1.7 billion for restrictive advertising practices in its AdSense business unit. Meta also faces antitrust investigation in the EU and UK.

Have these companies become so large that they are squeezing out consumers and innovations? How governments answer this question affects how Apple, Google, and Meta fare in the future and what kind of Internet experience they will be able to provide.

Sources: Jon Swartz, "Google Enters 2022 Battling Antitrust Actions on Multiple Fronts—with More Likely to Come," Marketwatch.com, January 1, 2022; Joe Rossignol, "Apple's Services Achieve All-Time Quarterly Revenue Record," Macrumors.com, October 28, 2021; Patience Haggin, "Why Apple's Privacy Changes Hurt Snap and Facebook but Benefited Google," *Wall Street Journal*, October 27, 2021; Joseph Johnson, "Global Market Share of Search Engines 2012–2021," Statista.com, October 14, 2921; Joseph Johnson, "U.S. Market Share of Mobile Search Engines 2010–2021," Statista.com, October 8, 2021; Lauren Goode, "Facebook Renews Its Ambitions to Connect the World," Wired.com, October 7, 2021; Cecelia Kang, "U.S. Revives Facebook Suit, Adding Details to Back Claim of Monopoly", *New York Times*, August 19, 2021; Daniel Newman, Opinion: Who's More Anti-competitive—Alphabet or Apple?" Marketwatch.com, July 21, 2021; Shira Ovide, "Apple's Strategy Bends the World," *New York Times*, June 29, 2021; Patience Haggin and Suzanne Vranica, "Apple's Moves to Tighten Flow of User Data Leave Advertisers Anxious," *Wall Street Journal*, June 9, 2021; Adam Satariano, "Facebook Faces Two Antitrust Inquiries in Europe," *New York Times*, June 4, 2021; "From California to Brazil, Europe's Privacy Laws Have Created a Recipe for the World," Cnbc.com, April 8, 2021.

## CASE STUDY QUESTIONS

**7-12** Compare the business models and core competencies of Google, Apple, and Meta.

**7-13** Why is mobile computing so important to these three firms? Evaluate the mobile strategies of each firm.

**7-14** Which company and business model do you think is most likely to dominate the Internet, and why?

**7-15** What difference would it make to a business or to an individual consumer if Apple, Google, or Meta dominated the Internet experience? Explain your answer.

# Chapter 7 References


Allyn, Bobby. "People Are Talking about Web3. Is It the Internet of the Future or Just a Buzzword?" Npr.org (November 21, 2021).

Cardoza, Edlyn. "Avaya Cloud Office Enables FinTech Firms to Improve Communication, Flexibility, and Reduce Risk." Ibisintelligence.com (October 6, 2021).

Chiang, I. Robert, and Jhih-Hua Jhang-Li. "Delivery Consolidation and Service Competition among Internet Service Providers." *Journal of Management Information Systems* 34, No. 3 (Winter 2014).

Copeland, Rob. "Google Lifts Veil, a Little, into Secretive Search Algorithm Changes." *Wall Street Journal* (October 25, 2019).

Insider Intelligence/eMarketer. "US Dual Mobile Device & Desktop/Laptop Users and Penetration" (February 2022); "US Mobile-Only Internet Users and Penetration" (February 2022) (2022a).

Insider Intelligence/eMarketer. "US Retail Mcommerce Sales" (February 2022); "US Smartphone Retail Mcommerce Sales" (February 2022) (2022b).

Insider Intelligence/eMarketer. "US Search Ad Spending" (October 2022c).

Insider Intelligence/eMarketer. "Mobile Internet: Average Time Spent in US, App vs. Browser, 2019–2024" (April 28, 2022d).

Insider Intelligence/eMarketer. "US Mobile Phone Users and Penetration" (February 2022); "US Smartphone Users and Penetration" (February 2022); "US Share of Average Time Spent per Day with Media (April 2022) (2022e).

Frick, Walter. "Fixing the Internet." *Harvard Business Review* (July–August 2019).

Gong, Jing, Vibhanshu Abhisek, and Beibei Li. "Examining the Impact of Keyword Ambiguity on Search Advertising Performance: A Topic Model Approach." *MIS Quarterly* 42, No. 3 (September 2018).

Google. "How Search Works," Google.com (accessed February 21, 2022).

Huang, Keman, Michael Siegel, Keri Pearlson, and Stuart Madnick. "Casting the Dark Web in a New Light." *MIT Sloan Management Review* 61 No.1 (Fall 2019).

Johnson, Joseph. "Global Market Share of Search Engines 2010–2021." Statista.com (October 8, 2021a).

______________. "Mobile Share of U.S. Organic Search Engine Visits 2013–2021." Statista.com (July 27, 2021b).

McKinsey & Company. "The Impact of Internet Technologies: Search" (July 2011).

Moz. "Google Algorithm Update History," Moz.com (accessed February 21, 2022).

National Telecommunications and Information Agency. "NTIA Announces Intent to Transition Key Internet Domain Name Functions." (March 14, 2014).

Nextiva. "How Titan Solar Power Used Nextiva to Grow Headcount by 450% Over 2 Years." Nextiva.com (accessed February 21, 2022).

Panko, Raymond R., and Julia L. Panko. *Business Data Networks and Security*, 11th ed. (Upper Saddle River, NJ: Prentice-Hall, 2019).

Pearl, Robert, and Brian Wayling. "The Telehealth Era Is Just Beginning." *Harvard Business Review* (May–June 2022).

Segan, Sascha. "5G Year in Review: 2021's Winners and Losers." Pcmag.com (December 22, 2021).

Stackpole, Thomas. "Exploring the Metaverse." *Harvard Business Review* (July-August 2022).

Vailshery, Lionel Sujay. "IoT and Non-IoT Connections Worldwide 2010–2025." Statista.com (March 8, 2021).

White, Monica. "What Is the Metaverse? A Deep Dive into the 'Future of the Internet.'" Digitaltrends.com (November 23, 2021).

CHAPTER

# Securing Information Systems

## LEARNING OBJECTIVES

After completing this chapter, you will be able to:

**8-1** Explain why information systems are vulnerable to destruction, error, and abuse.

**8-2** Describe the business value of security and control.

**8-3** Identify the components of an organizational framework for security and control.

**8-4** Identify the most important tools and technologies for safeguarding information resources.

**8-5** Understand how MIS can help your career.

## CHAPTER CASES

- Ransomware Everywhere
- Race Against Time: The Scramble to Fix the Log4Shell Zero-Day Vulnerability
- PayPal Ups Its Digital Resiliency
- SolarWinds Hack Shines a Light on Software Supply Chain Attacks

**MyLab MIS**

- **Video Case:** Fastly Internet Outage Exposes Vulnerability of Major Websites; Ransomware Is a Worldwide Problem: Palo Alto Networks
- Discussion Questions: 8-5, 8-6, 8-7
- Hands-On MIS Projects: 8-8, 8-9, 8-10, 8-11
- eText with Figure Videos

## RANSOMWARE EVERYWHERE

In 2013, a "new" form of malware began to attract public attention. Ransomware blocks or limits access to a computer or network by encrypting files and then demands a ransom payment, typically in a cryptocurrency such as Bitcoin, in exchange for the decryption key. Ransomware has become a significant threat, and in 2021, two high-profile attacks thrust it to the forefront of public consciousness.

Colonial Pipeline operates a 5,500-mile pipeline system that runs from Texas to New Jersey and transports nearly half of all fuel for the East Coast. On May 6, 2021, hackers launched an attack on Colonial Pipeline, first stealing 100 gigabytes of data and then locking its computers with ransomware and demanding payment. In response, the company proactively shut down some of its computer systems, which resulted in the complete shutdown of all its pipeline operations. The impact of the shutdown was immediate. It set off a cascading crisis that led to long lines and panicked buying at gas pumps, a jump in gas prices, shortages of airline fuel, and emergency meetings at the White House. On May 7, Colonial Pipeline agreed to pay the hackers the ransom fee–75 Bitcoin, worth nearly $5 million–and began the process of bringing its various systems back online. On May 12, Colonial Pipeline restarted its pipeline operations, although it took several days thereafter for the fuel delivery supply chain to return to normal.

JBS is the world's largest meat-processing company. JBS processes roughly 25 percent of the US beef supply and 20 percent of its pork supply, and its subsidiary Pilgrim's Pride is the second-largest US poultry processor. On May 30, 2021, hackers hit JBS with ransomware. Its IT team immediately began to shut down various systems to slow the attack's advance and halted operations at most of its plants in North America as well as in Australia. The attack put pressure on meat and poultry supplies at a time when the food supply chain was already under strain because of the Covid-19 pandemic. Although JBS maintained a secondary backup of all its data and was able to restart operations using those backup systems, it decided to pay $11 million in ransom to avert further risk.

Subsequent investigations showed that the Colonial Pipeline hackers, identified as part of a hacking group known as DarkSide (which was linked to the notorious Russian-based REvil/ Sodinokibi ransomware gang), were able to enter Colonial Pipeline's network through a single compromised password. The account did not require multifactor authentication, which made gaining access easier. Similarly, researchers believe the JBS ransomware attack began with leaked credentials in March 2021. Hackers then began to extract data from JBS's networks over a period of three months, finally launching the attack at the end of May. According to the US government, the REvil/Sodinokibi group was responsible for this attack as well.

© Andrey_Popov/Shutterstock

The attacks highlight the risks posted by ransomware to critical infrastructure, with US government officials characterizing it as an urgent national security threat. In the energy and food supply sectors, mergers and acquisitions have created industry giants, making them prime targets for hackers. Historically, these industries have not considered themselves as targets, nor have they been subject to as stringent cybersecurity regulations as have some other industry sectors, such as the financial services industry. Local and regional governments, hospitals, school systems, and managed service providers have also increasingly been attacked. Many of these organizations do not have advanced security safeguards in place.

Thus far, there has been only limited progress in combating ransomware. But there have been some successes. In June 2021, the FBI was able to retrieve about half of the ransom fee paid by Colonial Pipeline. In November 2021, the US Justice Department indicted an alleged REvil member and seized $6.1 million in digital currency. In January 2022, the Russian government arrested 14 members of the REvil/Sodinokibi ransomware gang and seized cash, cryptocurrency wallets, and high-end cars purchased by members of the group. But whether these efforts will have any long-lasting impact remains unknown, as new ransomware groups and threats seem to continually emerge. Experts suggest that effective ransomware prevention will require continuing international cooperation among all nations; more advanced oversight over cryptocurrencies; and much better security practices by organizations.

Sources: Dustin Volz and Robert McMillan, "Russia Arrests Hackers Tied to Major U.S. Ransomware Attacks, Including Colonial Pipeline Disruption," *Wall Street Journal*, January 14, 2022; Andrada Fiscutean, "How to Control Ransomware? International Cooperation, Disrupting Payments Are Key, Experts Say," Csoonline.com, July 5, 2021; Ryan Sherstobitoff, "JBS Ransomware Attack Started in March and Much Larger in Scope than Previously Identified," Securityscorecard.com, June 8, 2021; Dev Kundaliya, "Colonial Pipeline Hackers Entered Network through a Single Compromised Password," Computing.co.uk, June 7, 2021; Daniel Volz et al, "U.S. Retrieves Millions in Ransom Paid to Colonial Pipeline Hackers," *Wall Street Journal*, June 7, 2021; Lydia Mulvany and David Wethe, "How Mass Consolidations Turned Food, Energy Firms Into Hacking Targets," Bloomberg.com, June 3, 2021; Catherine Thorbecke and Luke Barr, "What We Know about the Colonial Pipeline Ransomware Attack," Abcnews.com, May 10, 2021.

Foreign hacker efforts to penetrate and disrupt critical industries such as the US energy system and meat processing food supply chain illustrate some of the reasons that organizations need to pay special attention to information systems security. The IT security breaches that enabled hackers to break into the Colonial Pipeline and JBS information systems have the potential to disrupt critical infrastructure throughout the country, paralyzing business, government, and daily life. Weak IT security has been responsible for many billions of dollars of corporate and consumer financial losses in other areas as well.

The chapter-opening diagram calls attention to important points raised by this case and this chapter. The US energy and food supply chain is complex, with many unguarded points of entry for malicious intruders. Organizations that lack or fail to effectively implement security awareness, resources, and tools create significant risk not only for their own organizations but also for their supply chain partners, for their ultimate customers, and for society at large.

Here are some questions to think about: What security vulnerabilities were exploited by the hackers? What people, organizational, and technological factors contributed to these security weaknesses? What was the business impact of these problems? Could the ransomware attacks against Colonial Pipeline and JBS have been prevented?

## 8-1 Explain why information systems are vulnerable to destruction, error, and abuse.

Can you imagine what would happen if you linked to the Internet without a firewall or antivirus software? It is likely that your computer would be disabled within a few seconds, and it might take you many days to recover your computer systems and files.

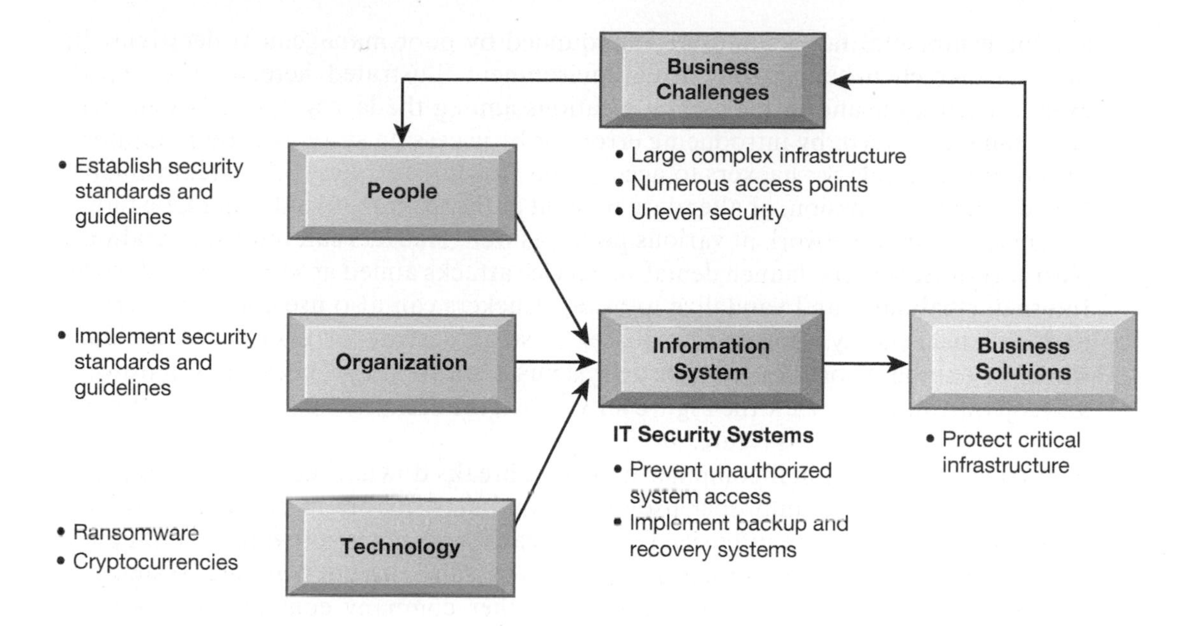


If you used the computer to run your business, you might not be able to sell to your customers or place orders with your suppliers while your computer was down. And you might find that your computer system had been penetrated by outsiders, who perhaps stole or destroyed valuable data, including confidential payment data from your customers. If too much data were destroyed or divulged, your business might never be able to recover!

In short, if you operate a business today, you need to make security and control top priorities. **Security** refers to the policies, procedures, and technical measures used to prevent unauthorized access, alteration, theft, or physical damage to information systems. **Controls** are the methods, policies, and organizational procedures that ensure the safety of the organization's assets, the accuracy and reliability of its records, and operational adherence to management standards.

## WHY SYSTEMS ARE VULNERABLE

Information systems and the data stored on them are vulnerable to many kinds of threats. For instance, information systems in different locations are interconnected via communications networks. The potential for unauthorized access or damage is not limited to a single location but can occur at many access points in the network.

Examine Figure 8.1, which illustrates the most common threats against contemporary information systems. These threats can stem from technical, organizational,

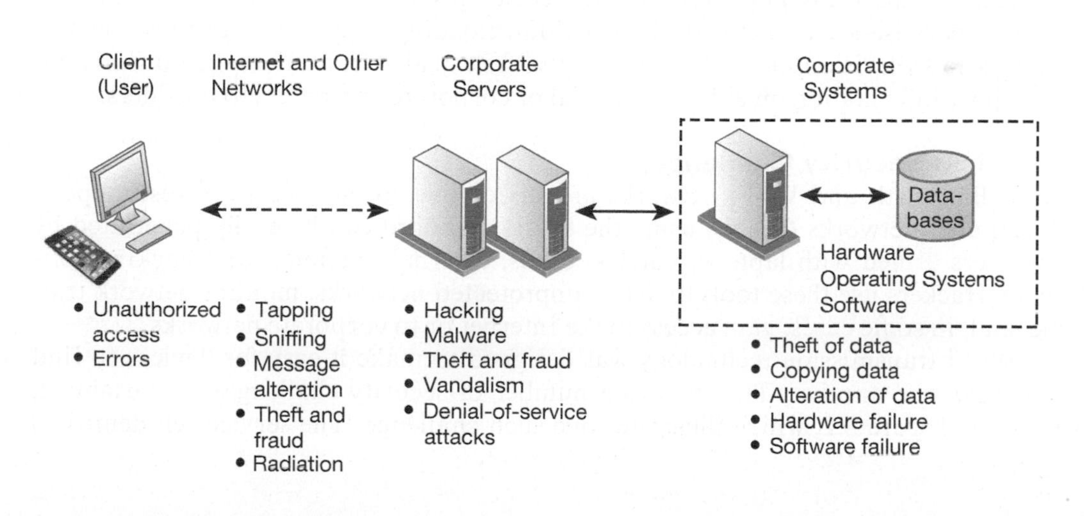


**Figure 8.1**
**Contemporary Security Challenges and Vulnerabilities**
*The architecture of a web-based application typically includes a web client, a server, and corporate information systems linked to databases. Each of these components presents security challenges and vulnerabilities. Floods, fires, power failures, and other electrical problems can also cause disruptions at any point in the network.*

and environmental factors and be compounded by poor management decisions. In the multitier client/server computing environment illustrated here, vulnerabilities exist at each layer and in the communications among the layers. Users at the client layer can cause harm by introducing errors or by accessing systems without authorization. It is possible for hackers to access data flowing over networks, steal valuable data during transmission, or alter data without authorization. Electromagnetic radiation may disrupt a network at various points as well. Hackers can introduce malware onto corporate servers, launch denial-of-service attacks aimed at servers linked to the Internet, steal data, and vandalize websites. Hackers can also use corporate servers to access backend systems and databases to steal, destroy, or alter corporate data stored in databases or files. Corporate systems are also at risk because of hardware and software failures. View the Figure 8.1 video in the eText for an animated and more detailed discussion of this figure.

Systems malfunction if computer hardware breaks down, is not configured properly, or is damaged by improper use or criminal acts. Errors in programming, improper installation, or unauthorized changes cause computer software to fail. Power failures, floods, fires, or other natural disasters can also disrupt computer systems.

Domestic or offshore partnering with another company contributes to system vulnerability if valuable information resides on networks and computers outside the organization's control. Without strong safeguards, valuable data could be lost or destroyed or fall into the wrong hands, revealing important trade secrets or information that violates personal privacy.

Portability makes mobile devices easy to lose or steal. Smartphones share many of the same security weaknesses as other Internet-connected devices and are vulnerable to malicious software and penetration from outsiders. Smartphones that corporate employees use often contain sensitive data such as sales figures, customer names, phone numbers, and email addresses. Intruders may also be able to access internal corporate systems via these devices.

### Internet Vulnerabilities

Large public networks, such as the Internet, are more vulnerable than internal networks because they are virtually open to anyone. The Internet is so widespread that when abuses do occur, they can have a huge impact. When the Internet links to the corporate network, the organization's information systems are even more vulnerable to actions from outsiders.

Vulnerability has also increased from widespread use of email, instant messaging (IM), and peer-to-peer (P2P) file-sharing programs. Email may contain attachments or links that serve as springboards for malicious software or unauthorized access to internal corporate systems. Employees may use email messages to transmit valuable trade secrets, financial data, or confidential customer information to unauthorized recipients. Instant messaging over the Internet can in some cases be used as a back door to an otherwise-secure network. Illegally downloading copyrighted content, such as movies and television shows, from P2P networks can also transmit malicious software or expose information on either individual or corporate computers to outsiders.

### Wireless Security Challenges

Both Bluetooth and Wi-Fi networks are susceptible to hacking by eavesdroppers. Local area networks (LANs) using the 802.11 standard can be easily penetrated by outsiders armed with laptops, wireless cards, external antennas, and hacking software. Hackers use these tools to detect unprotected networks, monitor network traffic, and, in some cases, gain access to the Internet or to corporate networks.

Wi-Fi transmission technology was designed to make it easy for devices to find and hear one another. This creates a number of security challenges. For example, examine Figure 8.2, which illustrates one such challenge. The service set identifiers

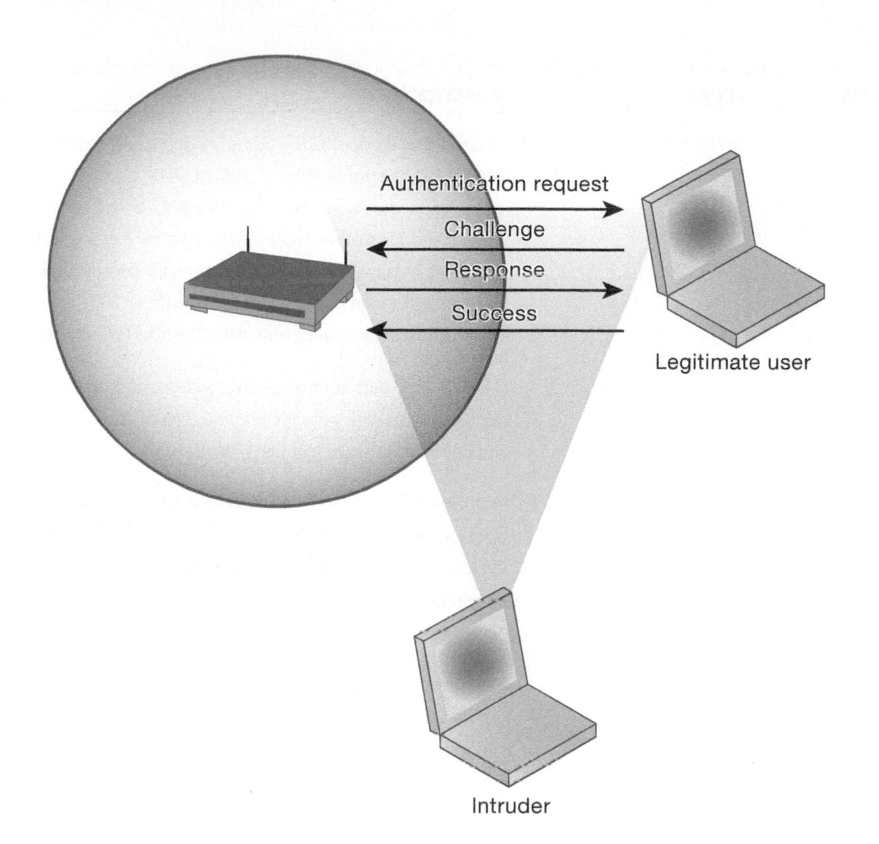


**Figure 8.2**
Wi-Fi Security Challenges
*Many Wi-Fi networks can be penetrated easily by intruders using sniffer programs to obtain an address to access the resources of a network without authorization.*

(SSIDs) that identify the access points in a Wi-Fi network are broadcast multiple times and can be picked up fairly easily by intruders' sniffer programs. Many wireless networks do not have basic protections against **war driving**, in which eavesdroppers drive by buildings or park outside them and try to intercept wireless network traffic.

An intruder who has associated with an access point by using the correct SSID can access other resources on the network. For example, the intruder could use the system's network operating system to determine which other users are connected to the network, access their computers, and open or copy their files.

Intruders also use the information they have gleaned to set up rogue access points on a different radio channel in physical locations near users to force a user's radio network interface controller (NIC) to associate with the rogue access point. Once this association occurs, hackers using the rogue access point can capture the names and passwords of unsuspecting users.

## MALICIOUS SOFTWARE (MALWARE)

Malicious software programs are referred to as **malware** and include a variety of threats such as computer viruses, worms, and Trojan horses. (See Table 8.1.) A **computer virus** is a rogue software program that attaches itself to other software programs or data files to be executed, usually without user knowledge or permission. Most computer viruses deliver a payload. The payload may be relatively benign, such as instructions to display a message or image, or it may be highly destructive—destroying programs or data, clogging computer memory, reformatting a computer's hard drive, or causing programs to run improperly. Viruses typically spread from computer to computer when humans take an action, such as sending an email attachment or copying an infected file.

**Worms** are independent computer programs that copy themselves from one computer to other computers over a network. Unlike viruses, worms can operate on their own without attaching to other computer program files and rely less on human behavior to spread rapidly from computer to computer. Worms destroy data and programs as well as disrupt or even halt the operation of computer networks.

**TABLE 8.1**

Notable Examples of Malicious Code

| Name | Type | Description |
|---|---|---|
| Emotet | Trojan/Botnet | Described by Europol as the world's most dangerous malware. Initially used to steal bank login credentials by surreptitiously capturing people's keystrokes. Later versions added malware delivery services, including other Trojans and ransomware, and were spread via botnet. In January 2021, an internationally coordinated effort took down many Emotet botnets, but by November 2021, it had once again resurfaced, indicating how difficult it is to fully eradicate. |
| Cryptolocker | Ransomware/Trojan | Hijacks users' data such as photos, videos, and text documents; encrypts them with virtually unbreakable asymmetric encryption; and demands ransom payment for them. |
| Conficker | Worm | First detected in November 2008 and still a problem for users of older, unpatched Windows operating systems. Uses flaws in Windows software to take over machines and link them into a virtual computer that can be commanded remotely. Had nearly 10 million computers worldwide under its control. Difficult to eradicate. |
| Sasser.ftp | Worm | First appeared in 2004. Spread over the Internet by attacking random IP addresses. Caused computers to continually crash and reboot and infected computers to search for more victims. Affected millions of computers worldwide and caused an estimated $14.8 billion to $18.6 billion in damages. |
| ILOVEYOU | Virus | First detected on May 3, 2000. Script virus written in Visual Basic script and transmitted as an email attachment with the subject line ILOVEYOU. Overwrote music, image, and other files with a copy of itself and did an estimated $10 billion to $15 billion in damage. |

Worms and viruses are often spread over the Internet from files of downloaded software; from files attached to email transmissions; or from compromised email messages, online ads, or instant messaging. Viruses have also invaded information systems from infected external storage devices or infected machines. Especially prevalent today are **drive-by downloads**, consisting of malware that comes with a downloaded file that a user intentionally or unintentionally requests.

Hackers can do to a smartphone just about anything they can do to any other Internet-connected device: request malicious files without user intervention, delete files, transmit files, install programs running in the background to monitor user actions, and potentially convert the smartphone to a robot in a botnet to send email and text messages to anyone. According to IT security experts, mobile devices now pose the greatest security risks, outpacing those from computers. More than 90 percent of global organizations surveyed suffered at least one mobile malware attack, according to security firm Check Point, while McAfee reported that it detected 43 million different variants of mobile malware, a record number (Check Point, 2021; McAfee, 2022).

Android, which is the world's leading mobile operating system, is the mobile platform targeted by most hackers. Mobile device viruses pose serious threats to enterprise computing because so many wireless devices are now linked to corporate information systems.

Social networks such as Facebook, Twitter, and LinkedIn have emerged as new conduits for malware. Members are more likely to trust messages they receive from friends even if this communication is not legitimate. For example, the Facebook Messenger virus is spread by phishing messages sent from a hijacked Facebook account. Malicious spam messages are sent to everybody in the victim's Facebook contact list (2-Spyware, 2021).

Many malware infections are Trojan horses. A **Trojan horse** is a software program that appears to be benign but then does something unexpected. The Trojan horse is not itself a virus because it does not replicate, but it is often a way for viruses or other malicious code to be introduced into a computer system. The term is based on the Trojan Horse, a huge wooden horse that the Greeks used to trick the Trojans into opening the gates to their fortified city during the Trojan War.

An example of a modern-day Trojan horse is the ZeuS (Zbot) Trojan, which infected more than 3.6 million computers in 2009 and still poses a threat. It has been used to steal login credentials for banking by surreptitiously capturing people's keystrokes as they use their computers. ZeuS is spread mainly through drive-by downloads and phishing, and recent variants have been difficult to eradicate.

**SQL injection attacks** exploit vulnerabilities in poorly coded web application software to introduce malicious program code into a company's systems and networks. These vulnerabilities occur when a web application fails to validate properly or filter data a user enters on a web page, which might occur when ordering something online. An attacker uses this input validation error to send a rogue SQL query to the underlying database to access the database, plant malicious code, or access other systems on the network.

As described in the opening case, malware known as **ransomware** is proliferating on both desktop and mobile devices. Ransomware tries to extort money from users by taking control of their computers, blocking access to files, or displaying annoying pop-up messages.

Some types of **spyware** also act as malicious software. These programs install themselves surreptitiously on computers to monitor user web-surfing activity and display advertising. Thousands of forms of spyware have been documented. Many users find such spyware annoying and an infringement on their privacy. Some forms of spyware are especially nefarious. **Keyloggers** record every keystroke made on a computer to steal serial numbers for software, to launch Internet attacks, to gain access to email accounts, to obtain passwords to protected computer systems, or to pick up personal information such as credit card or bank account numbers. The ZeuS trojan described earlier uses keylogging. Other spyware programs reset web browser home pages, redirect search requests, or slow performance by taking up too many computer resources.

Malware is also likely to be targeted at the 3-D virtual reality environment known as the metaverse as it develops further (see Chapter 7). The hardware needed for virtual reality and augmented reality platforms will create new endpoints that hackers will seek to exploit. Attackers could potentially manipulate platforms to create physical dangers. Participants in the metaverse may also be subject to various forms of harassment by malicious actors. The privacy of participants and the security of their personal information are also concerns.

## HACKERS AND COMPUTER CRIME

A **hacker** is an individual who intends to gain unauthorized access to a computer system. Within the hacking community, the term *cracker* is typically used to denote a hacker with criminal intent, although in the public press, the terms *hacker* and *cracker* are used interchangeably. Hackers gain unauthorized access by finding weaknesses in the security protections that websites and computer systems employ. Hacker activities have broadened beyond mere system intrusion to include theft of goods and information as well as system damage and **cybervandalism**, the intentional disruption, defacement, or even destruction of a website or corporate information system. Organizations sometime hire *ethical hackers* to test their own security measures. These types of hackers work under an agreement with the target firms that they will not be prosecuted for their efforts to break in. Firms such as Apple, Microsoft, Intel, HP, and many others are also often willing to pay monetary bounties ("bug bounties") to hackers who discover bugs in their software and hardware.

### Spoofing and Sniffing

Hackers attempting to hide their true identities often spoof, or misrepresent, themselves by using fake email addresses or masquerading as someone else. **Spoofing** may also involve redirecting a web link to an address different from the intended one, with the site masquerading as the intended destination. For example, if hackers redirect customers to a fake website that looks almost exactly like the true site, the hackers can then collect and process orders, effectively stealing business as well as sensitive customer information from the true site. We will provide more details about other forms of spoofing in our discussion of computer crime.

A **sniffer** is a type of eavesdropping program that monitors the information traveling over a network. When used legitimately, sniffers help identify potential network trouble spots or criminal activity on networks, but when used for criminal purposes, sniffers can be damaging and very difficult to detect. Sniffers enable hackers to steal proprietary information, including email messages, company files, and confidential reports, from anywhere on a network.

### Denial-of-Service Attacks

In a **denial-of-service (DoS) attack**, hackers flood a network server or web server with many thousands of false communications or requests for services in order to crash the network. The network receives so many queries that it cannot keep up with them and is thus unavailable to service legitimate requests. A **distributed denial-of-service (DDoS) attack** uses numerous computers to inundate and overwhelm a network from many launch points.

Although DoS/DDoS attacks do not destroy information or access restricted areas of a company's information systems, they often cause a website to shut down, making it impossible for legitimate users to access the site. For busy e-commerce sites, these attacks are costly. While the site is shut down, customers cannot make purchases. Especially vulnerable are small and midsize businesses whose networks tend to be less protected than those of large corporations.

Perpetrators of DDoS attacks often use thousands of PCs infected with malicious software without their owners' knowledge and organize these PCs into a **botnet**. Hackers create these botnets by infecting other people's computers with bot malware that opens a back door through which an attacker can give instructions to the infected computer. When hackers infect enough computers, they can use the amassed resources of the botnet to launch DDoS attacks, phishing campaigns, or unsolicited, spam email. One example is the Mirai botnet, which infected numerous Internet of Things (IoT) devices (such as Internet-connected surveillance cameras) and then used these devices to launch a DDoS attack against Dyn, whose servers monitor and reroute Internet traffic. The Mirai botnet overwhelmed the Dyn servers, taking down Etsy, GitHub, Netflix, Shopify, SoundCloud, Spotify, Twitter, and a number of other major websites. In 2021, Internet infrastructure firm Cloudflare had to fight off another massive DDoS attack launched by a Mirai botnet comprised of more than 20,000 bots in 125 countries.

### Computer Crime

Most hacker activities are criminal offenses, and the system vulnerabilities we have just described make systems targets for other types of computer crime as well. The US Department of Justice defines computer crime as "any violations of criminal law that involve a knowledge of computer technology for their perpetration, investigation, or prosecution." The US Congress addressed the threat of computer crime in 1986 with the Computer Fraud and Abuse Act, which makes it illegal to access a computer system without authorization. Most states have similar laws, and nations in Europe have comparable legislation. Table 8.2 provides examples of the computer as both a target of, and an instrument of, crime.

**TABLE 8.2**

Examples of Computer Crime

| Computers as Targets of Crime |
| --- |
| Breaching the confidentiality of protected data |
| Accessing a computer system without authority to do so |
| Knowingly accessing a protected computer to commit fraud |
| Intentionally accessing a protected computer and causing damage negligently or deliberately |
| Knowingly transmitting a program, program code, or command that intentionally causes damage to a protected computer |
| Threatening to cause damage to a protected computer |
| **Computers as Instruments of Crime** |
| Stealing trade secrets |
| Unauthorized copying of software or copyrighted intellectual property, such as articles, books, music, and video |
| Scheming to defraud |
| Using email or messaging for threats or harassment |
| Intentionally attempting to intercept electronic communication |
| Illegally accessing stored electronic communications, including email and voice mail |
| Transmitting or possessing child pornography by using a computer |

US legislation, such as the Wiretap Act, Wire Fraud Act, Economic Espionage Act, Electronic Communications Privacy Act, CAN-SPAM Act, and Protect Act, covers computer crimes involving intercepting electronic communication, using electronic communication to defraud, stealing trade secrets, illegally accessing stored electronic communications, using email for threats or harassment, and transmitting or possessing child pornography. The National Information Infrastructure Protection Act makes malware distribution and hacker attacks to disable websites federal crimes. All 50 states, the District of Columbia, Guam, Puerto Rico, and the Virgin Islands have enacted legislation requiring private or governmental entities to notify individuals of security breaches involving personally identifiable information.

No one knows the magnitude of computer crime—how many systems are invaded, how many people engage in the practice, or the total economic damage that results from computer crime. According to a Center for Strategic and International Studies (CSIS)/McAfee report, the global cost of cybercrime in 2020 was more than $1 trillion. Other researchers believe the cost is even higher: for instance, Cybersecurity Ventures estimates the global cost in 2021 to be more than $6 trillion and that the cost will grow to almost $11 trillion by 2025 (CSIS/McAfee, 2020; CyberSecurity Ventures, 2021).

Many companies are reluctant to report computer crimes because the crimes may involve employees or because publicizing vulnerability will hurt their reputations. The most economically damaging kinds of computer crime are DDoS attacks, activities of malicious insiders, and web-based attacks.

## Data Breaches and Identity Theft

With the growth of the Internet and e-commerce, data breaches and identity theft has become especially troubling. A **data breach** occurs whenever an organization loses control over corporate information to outsiders. According to the IBM Security's 2021 Cost of a Data Breach Report, the total average cost of a data breach among the

537 companies surveyed globally was $4.24 million (IBM Security, 2021). Moreover, brand damage can be significant although hard to quantify.

Data breaches also enable credential stuffing attacks. Credential stuffing is a brute force attack that hackers launch via botnets and automated tools using known username and password combinations obtained from data breaches. Credential stuffing attacks are becoming increasingly common, especially in the financial services industries. Table 8.3 describes some major data breaches.

Data breaches are typically linked to identity theft. **Identity theft** is a crime in which an imposter obtains key pieces of personal information, such as social security numbers, driver's license numbers, or credit card numbers, to impersonate someone else. The information may be used to obtain credit, merchandise, or services in the name of the victim or to provide the thief with false credentials. Identity theft has flourished on the Internet, with credit card files a major target of website hackers. According to the 2022 Identity Fraud Study by Javelin Strategy & Research, identity fraud losses reached $52 billion in 2021 (Javelin Strategy & Research, 2022).

One popular tactic is a form of spoofing called **phishing**. Phishing involves setting up fake websites or sending email messages that look like those of legitimate businesses to ask users for confidential personal data. The email message instructs recipients to update or confirm records by providing social security numbers, bank and credit card information, and other confidential data by responding to the email message, entering the information at a bogus website, or calling a telephone number. eBay, PayPal, Amazon, Walmart, and a variety of banks have been among the top spoofed companies. In 2021, Meta, Facebook's parent company, filed a lawsuit seeking to disrupt phishing schemes designed to deceive users into sharing their user credentials on fake login pages for Facebook, Messenger, Instagram, and WhatsApp.

**TABLE 8.3**

Major Data Breaches

| Data Breach | Description |
|---|---|
| LinkedIn | In 2021, personal data, including email addresses, phone numbers, geolocation data, and social media details that were associated with more than 700 million LinkedIn users (more than 90 percent of LinkedIn's user accounts), were discovered on a Dark Web forum posted by a hacker who exploited a vulnerability in the site's application programming interface using data-scraping techniques. |
| T-Mobile | In 2021, T-Mobile said hackers stole the personal data of 54 million people, including 7.8 million current customers and 40 million former or prospective customers. The attackers broke into T-Mobile's servers through an open access point. Data stolen included social security numbers, phone numbers, and driver's license/ID information as well as account PINs of certain customers. |
| Marriott | In 2018, Marriott, the world's largest hotel company, revealed that a hack in the reservation database for its Starwood properties may have exposed the personal information of up to 500 million guests. Exposed data included names, phone numbers, email addresses, passport numbers, date of birth, and credit card numbers. State-sponsored Chinese hackers copied and encrypted the data and took steps toward removing them. In 2020, Marriott was hacked again after the login credentials of two employees were used. |
| Equifax | In 2017, Equifax, one of the largest major credit reporting and scoring firms, reported that hackers had gained access to some of its systems and, potentially, the personal information of more than 140 million US consumers (about 45 percent of the US population), including social security numbers and driver's license numbers. Credit card numbers for more than 200,000 consumers and personal information used in disputes for more than 180,000 people were also compromised. |

The scheme involved the creation of more than 39,000 different websites. In a more targeted form of phishing called *spear phishing*, messages appear to come from a trusted source, such as an individual within the recipient's own company or a friend.

Phishing techniques called evil twins and pharming are harder to detect. **Evil twins** are wireless networks that pretend to offer trustworthy Wi-Fi connections to the Internet, such as those in airport lounges, hotels, or coffee shops. The bogus network looks identical to a legitimate public network. Fraudsters try to capture passwords or credit card numbers of unwitting users who log on to the network.

**Pharming** redirects users to a bogus web page even when the users type the correct web page address into their browser. This is possible when pharming perpetrators gain access to the IP address information that Internet service providers (ISPs) store to speed up web browsing and when flawed software on ISP servers allows the fraudsters to hack in and change those addresses.

## Global Threats: Cyberterrorism and Cyberwarfare

The cyber criminal activities we have described—launching malware, DoS/DDoS attacks, and phishing probes—are borderless. Attack servers for malware are now hosted in more than 200 countries and territories. The countries that are the leading sources of malware attacks include China, Brazil, Russia, Poland, Iran, India, Nigeria, Vietnam, the United States, and Germany (Plis, 2021). The global nature of the Internet makes it possible for cybercriminals to operate—and do harm—anywhere in the world.

Internet vulnerabilities have also turned individuals and even entire nation-states into easy targets for politically motivated hackers to conduct sabotage and espionage. **Cyberwarfare** is a state-sponsored activity designed to cripple and defeat another state or nation by penetrating its computers or networks to cause damage and disruption. Examples include the efforts of Russian hackers to penetrate the US energy industry and food supply chain, as described in the chapter-opening case, and to infiltrate companies via software supply chain attacks described in the chapter-ending case study. In January 2022, during a period of heightened tensions between Russia and Ukraine, the Microsoft Threat Intelligence Center announced that it had identified evidence of a destructive malware operation targeting multiple government, nonprofit, and information technology organizations based in Ukraine that was poised for launch and advised those organizations to take proactive steps to protect their information systems. Cyberwarfare also includes defending against these types of attacks.

Cyberwarfare is more complex than conventional warfare. Although many potential targets are military, a country's power grids, dams, financial systems, communications networks, and voting systems can also be crippled. Nonstate actors such as terrorists or criminal groups can mount attacks, and it is often difficult to identify who is responsible. Therefore, nations must constantly be on the alert for new malware and other technologies that could be used against them, especially because some of these technologies developed by skilled hacker groups are openly for sale to interested governments.

Cyberwarfare attacks have become much more widespread, sophisticated, and potentially devastating. Foreign hackers have stolen source code and blueprints to the oil and water pipelines and power grid of the United States and infiltrated the Department of Energy's networks hundreds of times. Over the years, hackers have stolen plans for missile tracking systems, satellite navigation devices, surveillance drones, and leading-edge jet fighters.

According to US intelligence, more than 30 countries, including Russia, China, Iran, and North Korea, are developing offensive cyberattack capabilities. Their cyberarsenals include collections of malware for penetrating industrial, military, and critical civilian infrastructure controllers; email lists and text for phishing attacks on important targets; and algorithms for DoS/DDoS attacks. US cyberwarfare efforts are concentrated in the United States Cyber Command, which coordinates and directs the operations and defense of Department of Defense information networks and prepares for military cyberspace operations. Cyberwarfare poses a serious threat to the infrastructure of modern societies because their major financial, health, government, and industrial institutions rely on the Internet for daily operations.

## INTERNAL THREATS: EMPLOYEES

We tend to think that the security threats to a business originate outside the organization. In fact, company insiders pose serious security problems. Studies have found that users' lack of knowledge is the single-greatest cause of network security breaches. Many employees forget the passwords they need to access computer systems or allow coworkers to use them, which compromises the system. Malicious intruders seeking system access sometimes trick employees into revealing their passwords by pretending to be legitimate members of the company in need of information. This practice is called **social engineering**.

Insiders bent on harm have also exploited their knowledge of the company to break into corporate systems, including those running in the cloud. For instance, a former employee at Amazon Web Services (AWS) used her knowledge of Amazon cloud security to steal many millions of Capital One Financial customer records stored by AWS. In 2022, the US Department of Justice announced the arrest of a publishing industry employee who had, over the course of five years, exploited knowledge gleaned from various employers to create phishing emails, fake identities, and websites to steal the intellectual property from their clients.

## SOFTWARE VULNERABILITY

Software errors pose a constant threat to information systems, causing untold losses in productivity and sometimes endangering people who use or depend on these systems. The growing complexity and size of software programs, coupled with demands for rapid delivery to markets, have contributed to an increase in software flaws or vulnerabilities.

A major problem with software is the presence of hidden **bugs**, or program code defects. Studies have shown that it is virtually impossible to eliminate all bugs from large programs. The main source of bugs is the complexity of decision-making code. Even a relatively small program of several hundred lines will contain tens of decisions leading to hundreds or even thousands of paths. Important programs within most corporations are usually much larger, containing tens of thousands or even millions of lines of code, each with many times the choices and paths of smaller programs.

Zero defects cannot be achieved in larger programs. Complete testing simply is not possible. Fully testing programs that contain thousands of choices and millions of paths would require thousands of years. Even with rigorous testing, you would not know for sure that a piece of software was dependable until the product proved itself after much operational use.

Flaws in commercial software not only impede performance but also create security vulnerabilities that open networks to intruders. Each year security firms identify thousands of software vulnerabilities. Especially troublesome are **zero-day vulnerabilities**, which are holes in the software unknown to its creator. Hackers then exploit this security hole before the vendor becomes aware of the problem and hurries to fix it. This type of vulnerability is called *zero-day* because the author of the software has zero days after learning about it to patch the code before it can be exploited in an attack. Sometimes security researchers spot the software holes, but more often, the holes remain undetected until an attack has occurred. The Spotlight on Technology case discusses a critical zero-day vulnerability, named Log4Shell, in a Java logging tool,

SPOTLIGHT ON: TECHNOLOGY

## Race Against Time: The Scramble to Fix the Log4Shell Zero-Day Vulnerability

One Friday afternoon in December 2021, Jordan LaFontaine, the endpoint operations team leader at Southern New Hampshire University, got an urgent notice about a recently discovered software vulnerability, Log4Shell, that could potentially impact more than 7,500 computing devices under his team's supervision. LaFontaine was just one of many security professionals across the globe scrambling that weekend and in the days that followed to deal with a major zero-day vulnerability, one that was given the highest-possible severity score by the US National Institute of Standards and Technology and characterized by the US Cybersecurity and Infrastructure Security Agency as one of the most serious vulnerabilities seen in decades.

The vulnerability was discovered in Apache Software Foundation's (ASF's) Log4j, an open-source utility program created for the Java software development environment. ASF is an all-volunteer community that acts as a "steward" for open-source code and makes it available to programmers and end users free of charge. Log4j allows users to create a built-in "log," or record, of activity to troubleshoot issues and track data within their programs.

On November 24, 2021, a user had reported to ASF the discovery of a zero-day vulnerability in Log4j's software code. The Log4Shell vulnerability, as it came to be known, was very easy to exploit and required very little technical expertise. On December 9, the existence of the vulnerability was publicly disclosed, with Microsoft's Minecraft identified as the first big-name victim. The same day, ASF released a patch.

However, despite the existence of the patch, combating the Log4Shell vulnerability faced a number of difficulties. For starters, the Log4j program is used in a broad range of Java-based software products and web services, from security software to networking tools and to videogame servers. The exact number of users of Log4j is unknown, but the software has reportedly been downloaded millions of times. A public catalog of products known to have the flaw has received more than 2,800 submissions. Apple's iCloud, Microsoft's Minecraft, and Google, Amazon, Twitter, LinkedIn, as well as many others, were among the firms impacted. Microsoft warned that many organizations might not even be aware that Log4j was part of the applications they were using, meaning that they could be vulnerable without even knowing it. Even applications not written in Java often are hosted in web containers that use Java, meaning that a project could have no apparent dependence on Log4j but still be vulnerable. Security experts cautioned that it might take months, or even years, to fully eradicate the vulnerability.

As soon as the announcement was made, security firms began reporting on hacker attempts to exploit the flaw. Hackers know that organizations are often slow to patch even critical security flaws. By December 20, 2021, Check Point posted that it had stopped more than 4.3 million breach attempts, with known hacking groups accounting for more than 45 percent of those attempts. Some attacks involved exploiting the flaw to install crypto-mining malware. Other attacks included delivering Cobalt Strike, a penetration-testing tool that hackers often use to steal usernames and passwords to gain further access to networks. Other security firms reported similar activity. Cloudflare said that its researchers were seeing around 1,000 attempts per second to actively exploit the flaw. Bitdefender said that it had detected multiple attempts by attackers to use the Log4Shell vulnerability to deploy a new family of ransomware known as Khonsairi on vulnerable systems. Microsoft reported that it had observed hacking groups linked to China, Iran, North Korea, and Turkey launching attacks. The Belgian Ministry of Defense confirmed an attack on its computer network. Akamai researchers said that they had found evidence suggesting that attackers were using the Log4Shell vulnerability in Zyxel networking devices to spread malware used by the Mirai botnet. To make matters worse, between December 9 and December 28, three additional Log4j vulnerabilities were discovered, necessitating three separate, additional patches.

In addition to the patches issued by ASF, the larger open-source community sprang into action with various resources. For instance, open-source security provider WhiteSource released a free developer tool that organizations could use to detect and resolve Log4j vulnerabilities. Third-party vendors were also quick to release tools. CrowdStrike released a Log4j scanner just before Christmas, and shortly thereafter, Microsoft rolled out a Log4j dashboard for threat and vulnerability management. Cisco, Oracle, and VMware also issued patches and fixes to secure their own products.

The Federal Trade Commission (FTC) threatened legal action against companies that did not take steps to fix the vulnerability, and the Securities and Exchange Commission indicated that it might do the same.

The Log4Shell vulnerability has highlighted security issues surrounding the use of open-source software. The issues are regarded as so serious that the White House national security advisor has described it as a key national security concern. The FTC has also expressed alarm, noting that although open-source software and services are critically important parts of the world's digital infrastructure (with the average application using 528 different open-source components), the software is created and maintained by volunteers, who don't always have adequate resources and personnel for incident response and proactive maintenance.

But some have a contrary view. They believe that open source is, in fact, the best way to create secure software. They argue that even the most fully funded software has bugs and vulnerabilities and that focusing on the volunteer nature of open-source code takes away one of its greatest strengths: developer passion. Others point to the speed with which the open-source community mobilized to deal with the crisis. The effort helped organizations to remain ahead of the curve and proactively mitigate problems.

Google has recently pledged $100 million to support open-source development and fix vulnerabilities, while another Google project is focusing on how to audit and improve critical open-source projects. It may be that the Log4Shell vulnerability could ultimately have a silver lining: that of focusing attention on both the critical and the ubiquitous nature of open-source code in today's digital infrastructure and encouraging both governments and technology companies to take steps to try to ensure that a similar crisis does not happen in the future.

Sources: Corin Faife, "DHS Creates Cyber Safety Review Board, Targets Log4j Exploit for Its First Report, Theverge.com, February 4, 2022; Ariel Assaraf, "The Log4Shell Vulnerability: A PostMortem," Venturebeat.com, January 22. 2022; John Leonard, "Fixing Log4Shell: How a University Patched All Its Endpoints over a Weekend," Computing.co.uk, January 13, 2022; Dustin Volz, "Cyber Officials Warn of Long-Term Fallout from Log4j Cyber Flaw," *Wall Street Journal*, January 10, 2022; "FTC Warns Companies to Remediate Log4j Vulnerability," Ftc.gov, January 4, 2022; Liam Tung, "Log4j Flaw Attack Levels Remain High, Microsoft Warns," Zdnet.com, January 4, 2022; Patrick Howell O'Neill, "The Internet Runs on Free Open-Source Software. Who Pays to Fix It?" Technologyreview.com, December 17, 2021; Matt Asay, "Log4j Vulnerability: Why Your Hot Take on It Is Wrong," Techrepublic.com, December 15, 2021; Ojasiv Nath, "Log4j Flaw: Critical Zero-Day Leaves Millions of Systems at Risk," Toolbox.com, December 15, 2021; Lance Whitney, "Critical Log4Shell Security Flaw Lets Hackers Compromise Vulnerable Servers," Techrepublic.com, December 13, 2021.

## CASE STUDY QUESTIONS

**1.** What people, organization, and technology factors were responsible for the Log4j software vulnerability?

**2.** Why is open-source software a potential security risk?

**3.** What steps could be taken in the future to prevent future open-source software vulnerabilities?

Apache Log4j, that allows attackers to take control of susceptible servers. Because of its prevalence, experts think it may take months—if not years—to fix all the systems that use the tool.

To correct software flaws once they are identified, the software vendor creates software called **patches** to repair the flaws without disturbing the proper operation of the software. It is up to users of the software to track these vulnerabilities, test the system for them, and then apply all patches. This process is called *patch management.*

Because a company's IT infrastructure is typically laden with multiple business applications, operating system installations, and other system services, maintaining patches on all devices and services that a company uses is often time consuming and costly. Malware is being created so rapidly that companies have very little time—between the time a vulnerability and a patch are announced and the time malicious software appears to exploit the vulnerability—to respond to the vulnerability.

## 8-2 Describe the business value of security and control.

Companies have very valuable information assets to protect. Systems often house confidential information about individuals' taxes, financial assets, medical records, and job performance reviews. The systems also can contain information on corporate operations including trade secrets, new product development plans, and marketing strategies. Government systems may store information on weapons systems, intelligence operations, and military targets. These information assets have tremendous value, and the repercussions can be devastating if they are lost, destroyed, or placed into the wrong hands. Systems that are unable to function because of security breaches, disasters, or malfunctioning technology can have permanent impacts on a company's financial health. Some experts believe that 40 percent of all businesses will not recover from application or data losses that are not repaired within three days.

Inadequate security and control may result in serious legal liability. Businesses must protect not only their own information assets but also those of customers, employees, and business partners. Failure to do so may open the firm to costly litigation for data exposure or theft. An organization can be held liable for needless risk and harm created if the organization fails to take appropriate protective action to prevent loss of confidential information, data corruption, or breach of privacy. For example, Target had to pay more than $150 million to settle various legal claims filed by banks, credit card companies, and consumers against Target related to a massive hack of Target's payment systems. Developing a sound security and control framework that protects business information assets is of critical importance to the entire enterprise, including senior management. It can no longer be limited to the IT department (Rothrock et al., 2018).

### LEGAL AND REGULATORY REQUIREMENTS FOR ELECTRONIC RECORDS MANAGEMENT

Government regulations are forcing companies to take security and control more seriously by mandating the protection of data from abuse, exposure, and unauthorized access. Firms face legal obligations to retain and store electronic records, known as electronic records management (ERM).

If you work in the healthcare industry, your firm will need to comply with the Health Insurance Portability and Accountability Act (HIPAA). **HIPAA** outlines medical security and privacy rules and procedures for simplifying the administration of healthcare billing and automating the transfer of healthcare data among healthcare providers, payers, and plans. It requires members of the healthcare industry to retain patient information for six years and ensure the confidentiality of that information. It also specifies privacy, security, and electronic transaction standards for healthcare providers that handle patient information and provides penalties for breaches of medical privacy, disclosure of patient records by email, or unauthorized network access.

If you work in a firm providing financial services, your firm will need to comply with the Financial Services Modernization Act, better known as the **Gramm-Leach-Bliley Act**. This act requires financial institutions to ensure the security and confidentiality of customer data. Data must be stored on a secure medium, and special security measures must be enforced to protect such data on storage media and during transmittal.

If you work in a publicly traded company, your company will need to comply with the Public Company Accounting Reform and Investor Protection Act, better known as the **Sarbanes-Oxley Act**. This act is designed to protect investors after the financial scandals at Enron, WorldCom, and other public companies. It imposes responsibility on companies and their management to safeguard the accuracy and integrity of the financial information that is used internally and released externally. The Sarbanes-Oxley Act is fundamentally about ensuring that internal controls are in place to govern the creation and documentation of information in financial statements. Because information systems are used to generate, store, and transport such data, the legislation requires firms to consider information systems security and other controls needed to ensure the integrity, confidentiality, and accuracy of their data. Each system application that deals with critical financial reporting data requires controls to make sure the data are accurate. Controls to secure the corporate network, prevent unauthorized access to systems and data, and ensure data integrity and availability in the event of disaster or other disruption of service are essential as well.

Firms in all industries must also be aware of, and comply with, data storage and retention requirements imposed by various privacy laws and regulations, such as the European Union's General Data Protection Regulation (GDPR), which applies to all organizations that operate within the European Union, as well as various federal and state laws enacted in the United States (see Chapter 4). For instance, California's Privacy Rights Act, which goes into full effect in January 2023, requires companies that operate in California to have a detailed and transparent data retention schedule and data destruction program. Courts have ruled that simply holding onto data longer than a specified retention period violates a person's privacy, even if no other breach or use has occurred.

## ELECTRONIC EVIDENCE AND COMPUTER FORENSICS

Security, control, and ERM have become essential in connection with legal actions. Most of the evidence today for stock fraud, embezzlement, theft of company trade secrets, computer crime, and many civil cases is in digital form. Legal cases increasingly rely on evidence represented as digital data (including emails, text messages, instant messages, and social media) stored on mobile devices, portable storage devices, computers and/or in the cloud (referred to as electronically stored information, or ESI).

In a legal action, a firm is obligated to respond to a discovery request for ESI that may be used as evidence, and the company is required by law to produce that data. The cost of responding to a discovery request can be enormous if the company has trouble assembling the required data or the data have been corrupted or destroyed. Courts now impose severe financial and even criminal penalties for improper destruction of ESI.

An effective ESI retention policy ensures that ESI is well organized, accessible, and neither retained too long nor discarded too soon. It also reflects an awareness of how to preserve potential evidence for computer forensics. **Computer forensics** is the scientific collection, examination, authentication, preservation, and analysis of ESI in such a way that the information can be used as evidence in a court of law. It deals with the following issues:

- Recovering data from computers while preserving evidential integrity
- Securely storing and handling recovered electronic data
- Finding significant information in a large volume of electronic data
- Presenting the information to a court of law

ESI may reside on computer storage media in the form of computer files and as *ambient data*, which are not visible to the average user. An example might be a file that has been deleted from a hard drive. Data that a computer user may have deleted on computer storage media can often be recovered through various techniques. Computer forensics experts try to recover such hidden data for presentation as evidence.

An awareness of computer forensics should be incorporated into a firm's contingency planning process. The CIO, security specialists, information systems staff, and corporate legal counsel should all work together to have a plan in place that can be executed if a legal need arises.

## 8-3 Identify the components of an organizational framework for security and control.

Even with the best security tools, your information systems won't be reliable and secure unless you know how and where to deploy those tools. You'll need to know where your company is at risk and what controls you must have in place to protect your information systems. You'll also need to develop a security policy and plans for keeping your business running if your information systems aren't operational.

### INFORMATION SYSTEMS CONTROLS

Information systems controls are both manual and automated and consist of general and application controls. **General controls** govern the design, security, and use of computer programs and the security of data files in general throughout the organization's information technology infrastructure. On the whole, general controls apply to all applications and consist of a combination of hardware, software, and manual procedures that create the overall control environment.

General controls include software controls, physical hardware controls, computer operations controls, data security controls, controls over the systems development process, and administrative controls. Table 8.4 describes the functions of each of these controls. View the Table 8.4 video in the eText for an animated and more detailed discussion of this table.

**Application controls** are specific controls unique to each application, such as payroll or order processing. They include both automated and manual procedures that ensure that only authorized data are completely and accurately processed by that application. Application controls can be classified as (1) input controls, (2) processing controls, and (3) output controls.

*Input controls* check data for accuracy and completeness when the data enter the system. There are specific input controls for input authorization, data conversion,

**TABLE 8.4**
General Controls

| Type of General Control | Description |
|---|---|
| Software controls | Monitor the use of system software and prevent the unauthorized access and use of software programs, system software, and computer programs. |
| Hardware controls | Ensure that computer hardware is physically secure and check for equipment malfunction. Organizations that are critically dependent on their computers must also make provisions for backup or continued operation to maintain constant service. |
| Computer operations controls | Oversee the work of the computer department to ensure that programmed procedures are consistently and correctly applied to the storage and processing of data. They include controls over the setup of computer-processing jobs and backup and recovery procedures for processing that ends abnormally. |
| Data security controls | Ensure that valuable business data files maintained internally or by an external hosting service are not subject to unauthorized access, change, or destruction while they are in use or in storage. |
| Implementation controls | Audit the systems development process at various points to ensure that the process is properly controlled and managed. |
| Administrative controls | Formalize standards, rules, procedures, and control disciplines to ensure that the organization's general and application controls are properly executed and enforced. |

data editing, and error handling. *Processing controls* establish that data are complete and accurate during updating. *Output controls* ensure that the results of computer processing are accurate, complete, and properly distributed.

Information systems controls should not be an afterthought. They need to be incorporated into the design of a system and should consider not only the system's performance under all possible conditions but also the behavior of the organizations and people using the system.

## RISK ASSESSMENT

Before your company commits resources to security and information systems controls, it must know which assets require protection and the extent to which these assets are vulnerable. A risk assessment helps answer these questions and determine the most cost-effective set of controls for protecting assets.

A **risk assessment** determines the level of risk to a firm if a specific activity or process is not properly controlled. Not all risks can be anticipated and measured, but most businesses will be able to acquire some understanding of the risks they face. Business managers working with information systems specialists should try to determine the value of information assets, points of vulnerability, the likely frequency of a problem, and the potential for damage. For example, if an event is likely to occur no more than once a year, with a maximum of a $1,000 loss to the organization, it would not be wise to spend $20,000 on the design and maintenance of a control to protect against that event. However, if that same event could occur at least once a day, with a potential loss of more than $300,000 a year, $100,000 spent on a control might be entirely appropriate.

Table 8.5 illustrates sample results of a risk assessment for an online order-processing system that processes 30,000 orders per day. The likelihood of each exposure occurring over a one-year period is expressed as a percentage. The next column shows the highest and lowest possible losses that could be expected each time the exposure occurred as well as an average loss calculated by adding the highest and lowest figures and dividing by two. The expected annual loss for each exposure can be determined by multiplying the average loss by its probability of occurrence.

This risk assessment shows that the probability of a power failure occurring in a one-year period is 30 percent. Loss of order transactions while power is down could range from $5,000 to $200,000 (averaging $102,500) for each occurrence, depending on how long processing is halted. The probability of embezzlement occurring over a yearly period is about 5 percent, with potential losses ranging from $1,000 to $50,000 (and averaging $25,500) for each occurrence. User errors have a 98 percent chance of occurring over a yearly period, with losses ranging from $200 to $40,000 (and averaging $20,100) for each occurrence.

After the risks have been assessed, system builders will concentrate on the control points with the greatest vulnerability and potential for loss. In this case, controls should focus on ways to minimize the risk of power failures and user errors because anticipated annual losses are highest for these areas.

**TABLE 8.5**

Online Order-Processing Risk Assessment

| Exposure | Probability of Occurrence (%) | Loss Range/Average ($) | Expected Annual Loss ($) |
|---|---|---|---|
| Power failure | 30% | $5,000 – $200,000 ($102,500) | $30,750 |
| Embezzlement | 5% | $1,000 – $50,000 ($25,500) | $1,275 |
| User error | 98% | $200 – $40,000 ($20,100) | $19,698 |

## SECURITY POLICY

After you've identified the main risks to your systems, your company will need to develop a security policy for protecting the company's assets. A **security policy** consists of statements ranking information risks, identifying acceptable security goals, and identifying the mechanisms for achieving these goals. What are the firm's most important information assets? Who in the firm generates and controls this information? What existing security policies are in place to protect the information? What level of risk is management willing to accept for each of these assets? Is it willing, for instance, to lose customer credit card data once every 10 years? Or will it build a security system for credit card data that can withstand the once-in-a-hundred-years disaster? Management must estimate how much it will cost to achieve this level of acceptable risk.

The security policy drives other policies determining the acceptable use of the firm's information resources and which members of the firm will have access to its information assets. An **acceptable use policy (AUP)** defines acceptable uses of the firm's information resources and computing equipment, including desktop and laptop computers, mobile devices, telephones, and the Internet. A good AUP defines unacceptable and acceptable actions for every user and specifies consequences for noncompliance.

Examine Figure 8.3, which shows one example of how an organization might specify the access rules for different levels of users in the human resources function. It specifies what portions of the human resources database each user is permitted to access, based on the information required to perform that person's job. The database contains sensitive personal information such as employees' salaries, benefits, and medical histories.

The access rules illustrated here are for two sets of users. One set of users consists of all employees who perform clerical functions, such as inputting employee data into the system. All individuals with this type of profile can update the system but can neither read nor update sensitive fields, such as salary, medical history, or earnings data. Another set of users consists of divisional managers, who cannot update the system but can read all employee data fields, including medical history and salary, for their division. We provide more detail about the technologies for user authentication later in this chapter.

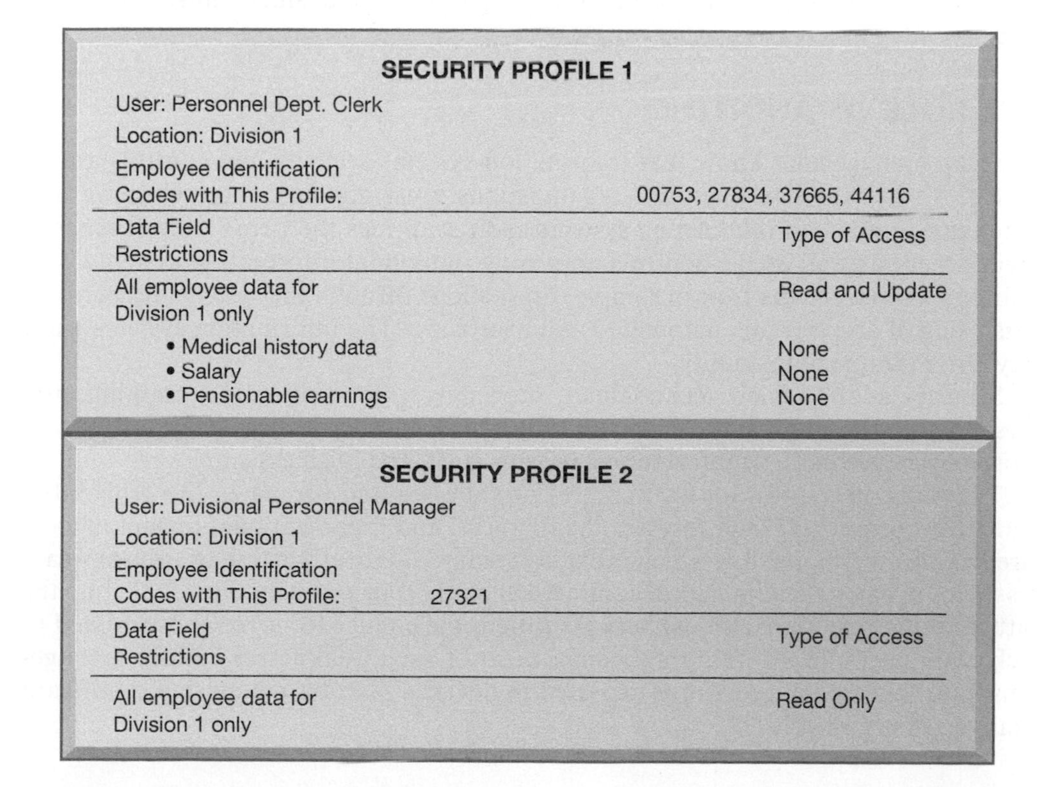

**SECURITY PROFILE 1**

User: Personnel Dept. Clerk
Location: Division 1
Employee Identification Codes with This Profile: 00753, 27834, 37665, 44116

| Data Field Restrictions | Type of Access |
|---|---|
| All employee data for Division 1 only | Read and Update |
| • Medical history data | None |
| • Salary | None |
| • Pensionable earnings | None |

**SECURITY PROFILE 2**

User: Divisional Personnel Manager
Location: Division 1
Employee Identification Codes with This Profile: 27321

| Data Field Restrictions | Type of Access |
|---|---|
| All employee data for Division 1 only | Read Only |

**Figure 8.3**
**Access Rules for a Personnel System**
*These two examples represent two security profiles or data security patterns that might be found in a personnel system. Depending on the security profile, a user would have certain restrictions on access to various systems, locations, or data in an organization.*

## DISASTER RECOVERY PLANNING AND BUSINESS CONTINUITY PLANNING

If you run a business, you need to plan for events—such as power outages; natural disasters such as tornados, floods, earthquakes, and wildfires; and terrorist attacks—that may prevent your information systems and your business from operating. **Disaster recovery planning** devises plans for the restoration of disrupted computing and communications services. Disaster recovery plans focus primarily on the technical issues involved in keeping systems up and running, such as which files to back up and the maintenance of backup computer systems or disaster recovery services.

For example, MasterCard maintains a duplicate data center in Kansas City, Missouri, to serve as an emergency backup to its global technology and operations headquarters near St. Louis. Rather than build their own backup facilities, many firms contract with cloud-based disaster recovery services or firms such as SunGard Availability Services that provide sites with spare computers around the country where subscribing firms can run their critical applications in an emergency.

**Business continuity planning** focuses on how the company can restore business operations after a disaster strikes. The business continuity plan identifies critical business processes and determines action plans for handling mission-critical functions if systems go down. For example, Southern Glazer's Wine & Spirits, a major wine and spirits distributor that has more than 200 different locations in 44 states and Canada, wanted to ensure that it could withstand any business disruptions. It undertook a critical analysis of every process across the organization, and then, using software tools from Fusion Risk Management, built a business continuity plan with real-time reporting capabilities, a mass notification system to alert all of its sites to potential risks, and a methodology to break down potential silos across the organization to improve overall efficiency.

Business managers and information technology specialists need to work together on both types of plans to determine which systems and business processes are most critical to the company. They must conduct a business impact analysis to identify the firm's most critical systems and the impact that a systems outage would have on the business. Management must determine the maximum amount of time the business can survive with its systems down and which parts of the business must be restored first.

## THE ROLE OF AUDITING

How can management know that information systems security and controls are effective? To answer this question, organizations must conduct comprehensive and systematic audits. An **information systems audit** examines the firm's overall security environment as well as the controls governing individual information systems. The auditor should trace the flow of sample transactions through the system and perform tests using, if appropriate, automated audit software. The information systems audit may also examine data quality.

Security audits review technologies, procedures, documentation, training, and personnel. A thorough audit will even simulate an attack or disaster to test the response of the technology, information systems staff, and business employees.

The audit lists and ranks all control weaknesses and estimates the probability of their occurrence. It then assesses the financial and organizational impact of each threat. Examine Figure 8.4, which provides a sample auditor's listing of control weaknesses for a loan system in a local commercial bank. It includes sections for listing the nature and impact of each weakness identified, the chance for error/abuse related to such weakness, and for notifying management of such weaknesses and for management's response. Management is expected to devise a plan for countering significant weaknesses in controls.

| Function: Loans<br>Location: Peoria, IL | Prepared by: J. Callejas<br>Date: May 15, 2022 | | Received by: T. Matoshi<br>Review date: June 1, 2022 | |
|---|---|---|---|---|
| Nature of Weakness and Impact | Chance for Error/Abuse | | Notification to Management | |
| | Yes/No | Justification | Report date | Management response |
| User accounts with missing passwords | Yes | Leaves system open to unauthorized outsiders or attackers | 5/10/22 | Eliminate accounts without passwords |
| Network configured to allow some sharing of system files | Yes | Exposes critical system files to hostile parties connected to the network | 5/10/22 | Ensure only required directories are shared and that they are protected with strong passwords |
| Software patches can update production programs without final approval from Standards and Controls group | No | All production programs require management approval; Standards and Controls group assigns such cases to a temporary production status | | |

**Figure 8.4**
Sample Auditor's List of Control Weaknesses
*This chart is a sample page from a list of control weaknesses that an auditor might find in a loan system in a local commercial bank. This form helps auditors record and evaluate control weaknesses and shows the results of discussing those weaknesses with management as well as any corrective actions management takes.*

## 8-4 Identify the most important tools and technologies for safeguarding information resources.

Businesses have an array of technologies available for protecting their information resources. These technologies include tools for managing user identities, preventing unauthorized access to systems and data, ensuring system availability, and ensuring software quality.

### IDENTITY AND ACCESS MANAGEMENT AND AUTHENTICATION

Midsize and large companies have complex IT infrastructures and many systems, each with its own set of users. **Identity and access management (IAM)** software automates the process of keeping track of all these users and their system privileges, assigning each user a unique digital identity for accessing each system. It also includes tools for authenticating users, protecting user identities, and controlling access to system resources. **Zero trust** is a popular cybersecurity framework based on the principle of maintaining strict access controls and not trusting anyone or anything by default, even those behind a corporate firewall. The concept of *least privilege access* is a core part of the zero-trust framework. Least privilege access means that no user should have access to system resources beyond what is absolutely necessary to fulfill that user's specified tasks.

To gain access to a system, a user must be authorized and authenticated. **Authentication** refers to the ability to know that people are who they claim to be. Authentication is often established by using **passwords** known only to authorized users. An end user uses a password to log on to a computer system and may also use passwords for accessing specific systems and files. However, users often forget passwords, share them, or choose poor passwords that are easy to guess, which compromises security. Password systems that are too rigorous hinder employee productivity. When employees must change complex passwords frequently, they often take shortcuts, such as choosing passwords that are easy to guess or keeping their passwords at their workstations in plain view. Passwords can also be sniffed out if they are transmitted over a network or stolen through social engineering.

Authentication technologies, such as security tokens, smart cards, and biometric authentication, overcome some of these problems. A **security token** is a physical

*This smartphone has a biometric fingerprint reader for fast yet secure access to files and networks.*

© Pro-stock Studio/Shutterstock

device, similar to an identification card, that is designed to prove the identity of a single user. Security tokens are small gadgets that typically fit on key rings and display passcodes that change frequently. A **smart card** is a device about the size of a credit card that contains a chip formatted with access permission and other data. (Smart cards are also used in digital payment systems.) A reader device interprets the data on the smart card and allows or denies access.

**Biometric authentication** uses systems that read and interpret individual human traits, such as facial features, fingerprints, irises, and voices, to grant or deny access. Biometric authentication is based on the measurement of a physical or behavioral trait that makes each individual unique. It compares a person's unique characteristics, such as facial structure, fingerprints, voice, or retinal image, against a stored profile of those characteristics to determine any differences between the characteristics and the stored profile. If the two profiles match, access is granted. Fingerprint and facial recognition technologies are now being used for many applications, with many mobile devices equipped with fingerprint identification and facial recognition software. Financial services firms such as Vanguard and Fidelity have implemented voice authentication systems for their clients.

The steady stream of incidents in which hackers have been able to access traditional passwords highlights the need for more secure means of authentication. **Multifactor authentication (MFA)** tools increase security by validating users via a multistep process. Authentication credentials might include something the user knows, such as a password or personal identification number (PIN); something the user possesses, such as a smartphone or security token; and something the user "has," such as a physical characteristic embodied in biometric data like fingerprints, iris prints, or voice prints. **Two-factor authentication (2FA)** is a subset of MFA that requires two credentials. A common example of two-factor authentication is a bank card. The card itself is the physical item that the user possesses, and the PIN is the credential that goes with the physical item.

## FIREWALLS, INTRUSION DETECTION AND PREVENTION SYSTEMS, AND ANTI-MALWARE SOFTWARE

Without protection against malware and intruders, connecting to the Internet would be very dangerous. Firewalls, intrusion detection and prevention systems, and anti-malware software have become essential business tools.

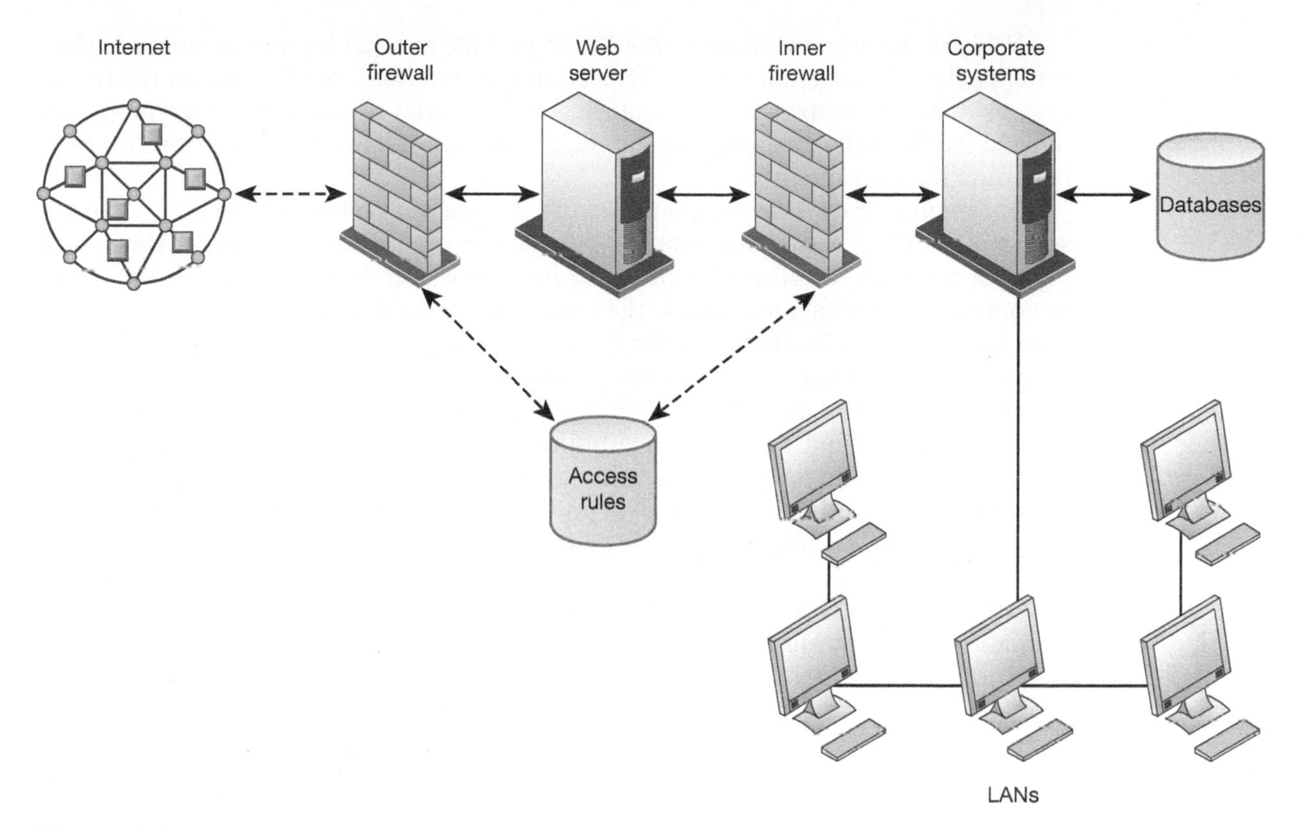


**Figure 8.5**
A Corporate Firewall
*The firewall is placed between the firm's private network and the public Internet or another distrusted network to protect against unauthorized traffic*

## Firewalls

**Firewalls** prevent unauthorized users from accessing private networks. A firewall is a combination of hardware and software that controls the flow of incoming and outgoing network traffic. Examine Figure 8.5, which illustrates how a firewall works. An outer firewall is generally placed between the organization's private internal networks and distrusted external networks such as the Internet, although inner firewalls can also be used to protect one part of a company's network from the rest of the network.

The firewall acts like a gatekeeper that examines each user's credentials before it grants access to a network. The firewall identifies names, IP addresses, applications, and other characteristics of incoming traffic. It checks this information against the access rules that the network administrator has programmed into the system. The firewall prevents unauthorized communication going into and out of the network.

In large organizations, the firewall often resides on a specially designated computer that is separate from the rest of the network so that no incoming request directly accesses private network resources. There are a number of firewall screening technologies, including static packet filtering, stateful inspection, Network Address Translation, and application proxy filtering. They are frequently used in combination to provide firewall protection.

*Static packet filtering* examines selected fields in the headers of data packets flowing back and forth between the trusted network and the Internet, examining individual packets in isolation. This filtering technology can miss many types of attacks.

*Stateful inspection* provides additional security by determining whether packets are part of an ongoing dialogue between a sender and a receiver. It sets up state tables to track information over multiple packets. Packets are accepted or rejected based on whether they are part of an approved conversation or attempting to establish a legitimate connection.

*Network Address Translation (NAT)* can provide another layer of protection when static packet filtering and stateful inspection are employed. NAT conceals the IP addresses of the organization's internal host computer(s) to prevent sniffer programs outside the firewall from ascertaining the IP addresses and using that information to penetrate internal systems.

*Application proxy filtering* examines the application content of packets. A proxy server stops data packets originating outside the organization, inspects them, and passes a proxy to the other side of the firewall. If a user outside the company wants to communicate with a user inside the company, the outside user first communicates with the proxy application, and the proxy application then communicates with the firm's internal computer. Likewise, a computer user inside the organization goes through the proxy to connect to computers on the outside of the organization.

To create a good firewall, an administrator must maintain detailed internal rules identifying the people, applications, or addresses that are allowed or rejected. Firewalls can deter, but not completely prevent, network penetration by outsiders and should be viewed as only one element of an overall security plan.

### Intrusion Detection and Prevention Systems

In addition to firewalls, commercial security vendors now provide intrusion detection tools and services to protect against suspicious network traffic and attempts to access files and databases. **Intrusion detection systems (IDS)** feature full-time monitoring tools placed at the most vulnerable points or hot spots of corporate networks to detect and deter intruders continuously. The system generates an alarm if it finds a suspicious or anomalous event. Scanning software looks for patterns indicative of known methods of computer attacks such as bad passwords, checks to see whether important files have been removed or modified, and sends warnings of vandalism or system administration errors. The intrusion detection tool can also be customized to shut down a particularly sensitive part of a network if it receives unauthorized traffic. An **intrusion prevention system (IPS)** has all the functionalities of an IDS, with the additional ability to take steps to prevent and block suspicious activities. For instance, an IPS can terminate a session and reset a connection, block traffic from a suspicious IP address, or reconfigure firewall or router security controls.

### Anti-Malware Software

Defensive technology plans for both individuals and businesses must include anti-malware protection for every computer. **Anti-malware software** prevents, detects, and removes malware, including computer viruses, computer worms, Trojans, spyware, and adware. However, most anti-malware software is effective only against malware already known when the software was written. To remain effective, the software must be continually updated. Even then it is not always effective because some malware can evade detection. Organizations need to use additional malware detection tools for better protection.

### Unified Threat Management Systems

To help businesses reduce costs and improve manageability, security vendors have combined various security tools, including firewalls, virtual private networks, intrusion detection and prevention systems, and web content filtering and anti-spam software, into a single product. These comprehensive security management products are called **unified threat management (UTM)** systems. UTM products are available for all sizes of networks. Leading UTM vendors include Fortinent, Sophos, and Check Point, and networking vendors such as Cisco Systems and Juniper Networks provide some UTM capabilities in their products.

## SECURING WIRELESS NETWORKS

The initial security standard developed for Wi-Fi, called Wired Equivalent Privacy (WEP), was not very effective because its encryption keys were relatively easy to crack. WEP has been replaced by Wi-Fi Protected Access 2 (WPA2), which provides stronger security standards.

Instead of the static encryption keys used in WEP, WPA2 uses much longer keys that continually change, making them harder to crack. The most recent specification is WPA3, introduced in 2018. However, even the updated WPA3 standard has vulnerabilities that could allow attackers to recover passwords. Companies can further improve Wi-Fi security by using Wi-Fi in conjunction with virtual private network (VPN) technology when accessing internal corporate data.

## ENCRYPTION AND PUBLIC KEY INFRASTRUCTURE

Many businesses use encryption to protect digital information that they store, physically transfer, or send over the Internet. **Encryption** is the process of transforming plain text or data into cipher text that cannot be read by anyone other than the sender and the intended receiver. Data are encrypted by using a secret numerical code, called an encryption key, that transforms plain data into cipher text. The message must be decrypted by the receiver.

Two methods for encrypting network traffic on the web are TLS and HTTPS. **Transport Layer Security (TLS)**, the successor to Secure Sockets Layer (SSL), enables client and server computers to manage encryption and decryption activities as they communicate with each other during a secure web session. TLS is used in conjunction with **HTTPS**, a secure version of the HTTP protocol that uses TLS for encryption and authentication. It is implemented by a server adopting the HTTP Strict Transport Security (HSTS) feature, which forces browsers to access the server using only HTTPS. Today, almost 80 percent of websites use HTTPS as their default protocol (W3techs.com, 2022).

The capability to generate secure sessions is built into Internet client browser software and servers. The client and the server negotiate what key and what level of security to use. Once a secure session is established between the client and the server, all messages in that session are encrypted.

Two methods of encryption are symmetric key encryption and public key encryption. In symmetric key encryption, the sender and the receiver establish a secure Internet session by creating a single encryption key and sending it to the receiver so that both the sender and the receiver share the encryption key. The strength of the encryption key is measured by its bit length. Today, a typical private key will be 56 to 256 bits long (a string of from 56 to 256 binary digits) depending on the level of security desired. The longer the key, the more difficult it is to break the key. The downside is that the longer the key, the more computing power it takes for legitimate users to process the information.

The problem with all symmetric encryption systems is that the key itself must be shared somehow among the senders and receivers, which exposes the key to outsiders who just might be able to intercept and decrypt the key. A more secure form of encryption called **public key encryption** uses two keys: one shared (or public) and one totally private. The keys are mathematically related so that data encrypted with one key can be decrypted using only the other key. RSA is a public key encryption algorithm that is widely used. The typical size of RSA public keys is 2,048 bits, although RSA public keys can range from 1,024 bits to 4,096 bits. Examine Figure 8.6, which illustrates how public key encryption works. To send and receive messages, communicators first create separate pairs of private and public keys. The public key is kept in a directory, and the private key must be kept secret. The sender encrypts a message with the recipient's public key. Upon receiving the message, the recipient uses their private key to decrypt it. View the Figure 8.6 video in the eText for an animated and more detailed discussion of this figure.

**Digital certificates** are data files used to establish the identity of users and digital assets to protect online transactions. A digital certificate system uses a trusted third party, known as a certificate authority (CA), to validate a user's identity. There are many CAs in the United States and around the world, including DigiCert, GoDaddy, and Comodo.

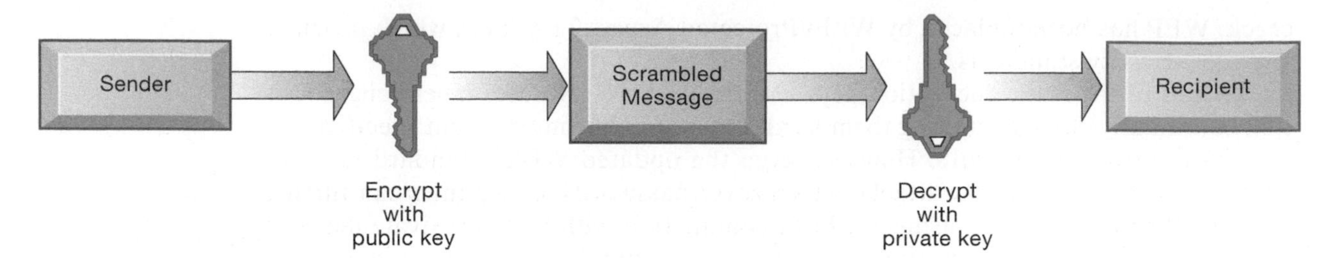


**Figure 8.6**
Public Key Encryption
*A public key encryption system can be viewed as a series of public and private keys that lock data when they are transmitted and unlock data when they are received. The sender locates the recipient's public key in a directory and uses the key to encrypt a message. The message is sent in encrypted form over the Internet or a private network. When the encrypted message arrives, the recipient uses their private key to decrypt the data and read the message.*

Examine Figure 8.7, which illustrates how digital certificates work. An institution or an individual requests a digital certificate. The CA verifies the user's identity offline. This information is put into a CA server, which generates an encrypted digital certificate containing owner identification information and a copy of the owner's public key. The certificate authenticates that the public key belongs to the designated owner. The CA makes its own public key available either in print or perhaps on the Internet. The recipient of an encrypted message uses the CA's public key to decode the digital certificate attached to the message, verifies it was issued by the CA, and then obtains the sender's public key and identification information contained in the certificate. By using this information, the recipient can send an encrypted reply. The digital certificate system enables, for example, a credit card user and a merchant to validate that their digital certificates were issued by an authorized and trusted third party before they exchange data. **Public key infrastructure (PKI)**, the use of public key encryption working with a CA, is now widely used in e-commerce.

## SECURING TRANSACTIONS WITH BLOCKCHAIN

Blockchain, which we introduced in Chapter 6, is gaining attention as an alternative approach for securing transactions and establishing trust among multiple parties. A blockchain is a chain of digital "blocks" that contain records of transactions. Each block is connected to all the blocks before and after it, and the blockchains are continually updated and kept in sync. This makes it difficult to tamper with a single

**Figure 8.7**
Digital Certificates
*Digital certificates help establish the identity of people or digital assets. They protect online transactions by providing secure, encrypted, online communication.*

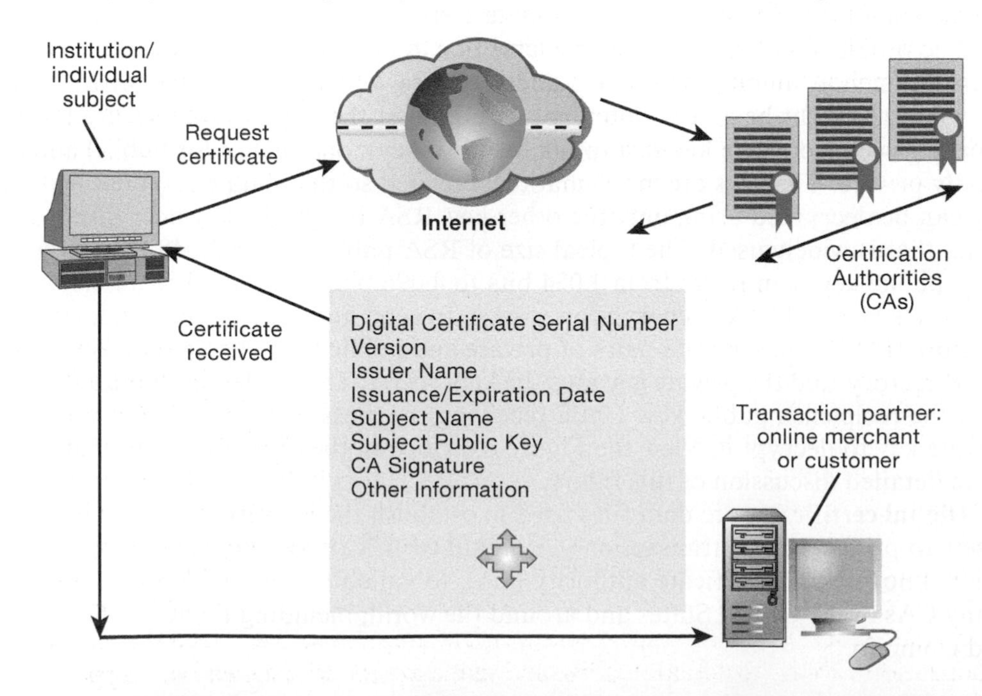

record because one would have to change the block containing that record as well as those linked to it to avoid detection.

Once recorded, a blockchain transaction cannot be changed. The records in a blockchain are secured through cryptography, and all transactions are encrypted. Blockchain network participants have their own private keys that are assigned to the transactions they create and act as a personal digital signature. If a record is altered, the signature will become invalid, and the blockchain network will know immediately that something is amiss. Because blockchains aren't contained in a central location, they don't have a single point of failure and cannot be changed from a single computer. Researchers point out, however, that blockchain is vulnerable in some of the same ways that conventional, centralized record-keeping systems are vulnerable. Blockchain systems also require the same careful attention to security and control that is given to other systems with high security requirements (Madnick, 2020).

**Cryptocurrencies** are digital assets that use blockchain technology and cryptography to create a medium of exchange (currency). Bitcoin is the most prominent example of cryptocurrency in use today, but many other, alternative cryptocurrencies have emerged in recent years.

## ENSURING SYSTEM AVAILABILITY

As companies increasingly rely on digital networks for revenue and operations, they need to take additional steps to ensure that their systems and applications are always available. Firms such as those in the airline and financial services industries, which have critical applications requiring online transaction processing, have traditionally used fault-tolerant computer systems to ensure 100 percent availability. In **online transaction processing**, transactions entered online are immediately processed by the computer. Multitudinous changes to databases, reporting, and requests for information occur each instant.

**Fault-tolerant computer systems** contain redundant hardware, software, and power supply components that create an environment that provides continuous, uninterrupted service. Fault-tolerant computers use special software routines or self-checking logic built into their circuitry to detect hardware failures and automatically switch to a backup device. In addition, parts from these computers can be removed and repaired without disruption to the computer or downtime. **Downtime** refers to periods of time in which a system is not operational.

### Security Outsourcing

Many companies, especially small businesses, lack the resources or expertise to create a secure, high-availability computing environment on their own. However, they can outsource many security functions to **managed security service providers (MSSPs)**, which monitor network activity and perform vulnerability testing and intrusion detection. SecureWorks, AT&T Cybersecurity, Verizon Managed Security, IBM Security, and Accenture are leading providers of MSSP services.

## ACHIEVING DIGITAL RESILIENCY

Today's organizations are much more hypernetworked and interconnected than organizations were in the past, with important parts of their IT infrastructures maintained remotely in the cloud, managed by outsiders, and accessible by mobile devices. Firms are embracing the concept of **digital resiliency** to deal with the realities of this new digital environment. Digital resiliency deals with maintaining and increasing the resilience of an organization and its business processes in an all-pervasive digital environment, not just the resilience of the IT function. In addition to computing, storing, and networking technologies, digital resiliency calls attention to managerial and organizational issues such as corporate policies and goals, business processes, organizational culture, business requirements, accountability, and business risk management. These factors can affect an organization's ability to actually utilize and manage network connectivity, applications, databases, and data centers; its ability to provide 24/7 availability to its customers; and its ability to respond to changing business conditions. If resiliency has not been explicitly designed in, measured, and tested, a single weak link in this chain can cause an outage or prevent the firm from responding to new challenges and opportunities. For instance, Facebook endured a six-hour outage in October 2021 because of an automated configuration change on the backbone routers that coordinated network traffic among its data centers, which had a cascading effect on the way the data centers communicate.

In another example, many companies whose employees were required to work from home to avoid exposure to the coronavirus were unsure of whether they had enough server capacity to support thousands of people working remotely. If the companies had used a digital resiliency approach, this contingency would have been better anticipated and planned for.

The Spotlight on Organizations case illustrates how PayPal, a heavily technology-driven business, was able to increase its digital resiliency by paying more attention to measuring the operational effectiveness of its data center teams and the reduction of employee errors.

## INTERNET OF THINGS, CLOUD COMPUTING, AND MOBILE PLATFORM SECURITY ISSUES

Although the Internet of Things (IoT), cloud computing, and the mobile platform deliver powerful benefits, they also pose new challenges to system security and reliability. We now describe some of these challenges and how they should be addressed.

### Internet of Things (IoT) Security Issues

The Internet of Things (IoT) introduces a number of security challenges: from the Internet-linked devices themselves, their platforms and operating systems, and their communications with the systems to which they're connected. New security tools will be required to protect IoT devices and platforms from both information attacks and physical tampering, to encrypt their communications, and to address new challenges such as attacks that drain batteries. Many IoT devices such as sensors have simple processors and operating systems that may not support sophisticated security approaches. Table 8.6 describes some of the unique security challenges posed by IoT.

**TABLE 8.6**

IoT Security Challenges

| Challenge | Implications |
|---|---|
| Many IoT devices, such as sensors, are deployed on a much wider scale than traditional Internet-connected devices are, thus creating a vast quantity of interconnected links that can be exploited. | New tools, methods, and strategies need to be developed to deal with the unprecedented scale presented by IoT devices. |
| Many IoT installations consist of collections of identical devices that share the same characteristics. | The potential impact of a security vulnerability is magnified. |
| Many IoT devices have a much longer service life than typical equipment does. | Devices may outlive their manufacturers, leaving them without long-term support, which could create persistent vulnerabilities. |
| Many IoT devices cannot be upgraded, or the upgrade process is difficult. | Raises the possibility that vulnerable devices cannot or will not be fixed, leaving them perpetually vulnerable. |
| Many IoT devices don't provide visibility into how the device works or the data being produced; nor do they alert the user when a security problem arises. | Users may believe an IoT device is functioning as intended when, in fact, it may be performing in a malicious manner. |
| Some IoT devices are so unobtrusive that a user many not even be aware of the device. | A security breach might persist for a long time before being noticed. |

## PayPal Ups Its Digital Resiliency

PayPal is a US company operating a worldwide online payments system that supports online money transfers and serves as a digital alternative to traditional payment methods like checks and money orders. The company operates as a payment processor for online vendors, auction sites, and many other commercial users, as well as enables person-to-person payments. You've probably used PayPal if you've bought something from eBay or from an e-commerce company. It is a very well-established and widely accepted payment system. As of the end of 2022, PayPal had 426 million active accounts (392 million consumer and 34 million merchant). In 2021, PayPal processed 19.3 billion payment transactions, with a transaction value of $1.25 trillion.

Obviously, this is a company that has to work vigilantly to make its services ultrasecure and available 24/7 throughout the world, and PayPal does maintain very high standards of security and reliability. But management wanted to make sure that the company was doing the best job possible, so it turned to Uptime Institute to evaluate the way PayPal ran its data centers and its level of digital resiliency. Uptime Institute is a consulting group focusing on improving the performance, efficiency, and reliability of business-critical infrastructure through innovation, collaboration, and independent performance certifications.

Although data centers try to operate sites that are full of cost-saving technologies and innovative, new approaches, they still struggle with the ongoing performance and reliability of their sites because of the operational plans in place. Various levels of staffing and experience, along with limited or inaccurate written documentation of operational processes, create inconsistent behaviors and service outages.

Uptime Institute has found that the majority of reported data center outages are directly related to human error. This could be operator error or management error in its decisions regarding staffing, maintenance, training, or overall rigor of operation. With human error responsible for so many data center incidents, organizations need to take a more holistic approach to staffing, organizational practices, maintenance and operations activities, management, and planning.

Sean Tugwell, PayPal's Director of Data Center Architecture and Engineering, wanted to make sure that his company had achieved a high level of digital resiliency and 99.999 percent availability across all data centers. He also wanted to make sure that the colos working with PayPal were sufficiently resilient as well. (A colocation data center, often referred to as a "colo," is a large data center facility that rents out rack space to other businesses for their servers or other computing equipment.)

Uptime Institute's M&O Assessment measures the operational effectiveness of the teams within a data center, focusing on five behaviors that should be proactive, practiced, and informed. These behaviors apply to staffing and organization, maintenance, training, planning, coordination and management, and operating conditions. After an extensive review, PayPal received Uptime Institute's M&O Stamp of Approval for its PHX01 data center, earning a very high first-time score of 96.2 percent. Several other PayPal data centers received high first-time scores of 100 percent, further indicating that PayPal had a high level of data center sophistication and maturity.

PayPal also scored well in its implementation of the Service Now platform, which is used for improving procedural approval workflows, maintenance management for critical data center infrastructure, incident management, and space and power planning. One area where PayPal shines is its approach to staffing and organization. Its Facilities Operations group has at least three facilities technicians on site at all times, and these technicians undergo rigorous training to ensure they have comprehensive and in-depth knowledge of a variety of systems and equipment.

PayPal's Facilities Operations group is also responsible for preventive and corrective maintenance at a data center. The group creates and implements on-site maintenance standards and procedures to make sure data center maintenance work is successfully completed and documented. In this area, too, PayPal scored very well: all its data centers had no items for deferred maintenance. PayPal's Preventive Maintenance Program helps the company keep equipment in like-new condition.

New hires at PayPal's Facility Operations Team are required to complete an initial training program before they are allowed to work on shift. PayPal also has a training program for vendors who will be performing on-site maintenance.

To promote effective planning, coordination, and management that will result in greater uptime,

PayPal has developed various procedures and standards, which are available in its ServiceNow Knowledge Base. This helps promote consistency across all of PayPal's data centers and also reduces the chance of human error at each site. PayPal's financial team creates, reviews, and tracks budgets to ensure that the budget for each data center is appropriate for supporting the firm's business objectives.

Consistency has become even more important for PayPal as it uses more colocation companies to add to its computing capacity. PayPal was able to use the M&O Stamp of Approval program to evaluate its colocation data center vendors. New vendor contracts now require a data center vendor to achieve Uptime Institute's M&O Stamp of Approval and maintain that rating as long as PayPal is a customer.

A high priority of PayPal's Data Center Services Team is reducing business costs. Any data center downtime, whether from a company data center or a colo, creates a loss to the business in terms of number of transactions processed, customer service, and the time and resources required to solve the downtime problem. Uptime Institute's M&O Stamp of Approval promotes this goal and also helps teams make sure that there are no surprises when new teams and operational practices are introduced to the company.

Given the critical importance of digital resiliency, in addition to its M&O Assessment, Uptime Institute also offers a Digital Resiliency Assessment program. This program specifically focuses on the resiliency of a firm's internal and cloud-based digital infrastructure, spanning its end users, networks, applications, databases, and data centers, to identify any weak links in the chain and validate the resiliency of the entire system.

Sources: Uptime Institute, "Digital Resiliency Assessment" and "Client Showcase: PayPal M&O Stamp of Approval," Uptimeinstitute.com, accessed July 18, 2022; PayPal Holdings, Inc. Form 10-K for the fiscal year ending December 31, 2021, filed with the Securities and Exchange Commission, February 3, 2022.

## CASE STUDY QUESTIONS

1. Why is digital resiliency so important for a company such as PayPal?
2. How did PayPal benefit from measuring its digital resiliency? What issues did PayPal address?
3. What are the roles of management and organizational issues in making an organization's IT infrastructure more resilient?

### Security in the Cloud

Using the public cloud disrupts the traditional cybersecurity models that companies may have created over the years. Although many organizations know how to manage the security for their own data center, they're often unsure of exactly what they need to do when they shift computing work to the cloud. They need new tool sets and skill sets to manage cloud security from their end to configure and launch cloud instances (a virtual server that runs workloads in the cloud), manage identity and access controls, update security controls to match configuration changes, and protect workloads and data. Companies that use a hybrid cloud model, which includes using both the company's own data centers as well as a public cloud service, face particular challenges because of the cloud's added complexity and the need to blend various components and frameworks.

In addition, there's still a misconception held by some IT departments that whatever happens in the cloud is not their responsibility. This is not the case. The category of cloud service (IaaS, PaaS, or SaaS) affects exactly how these responsibilities are shared. For IaaS, for example, the cloud service provider (CSP) typically supplies and is responsible for securing basic IT resources such as machines, storage systems, and networks. The cloud services customer is typically responsible for its operating system, applications, and corporate data placed into the cloud computing environment. This means that most of the responsibility for securing the applications and the corporate data falls on the customer. Therefore, cloud service customers should carefully review their cloud services agreement with their CSP to make sure their applications and data hosted in cloud services are secured in accordance with their own security and compliance policies.

It is essential for companies to revise their cybersecurity practices so that they can both protect critical data and fully exploit the speed and agility that cloud services provide. Basic hybrid cloud security best practices include running continuous audits to have real-time visibility into threats and any irregularities in the company's cloud environment, adopting zero-trust principles including least privilege access (which, in the context of hybrid cloud security, requires that all interactions between the company's private data center and the public cloud be as limited as possible while still being able to achieve operational goals), establishing unified security management, and embracing artificial intelligence and automation to help detect and resolve security risks.

Cloud computing also raises issues about regulatory compliance with respect to data security and protection. Public cloud applications reside in remote data centers that supply business services and data management to multiple corporate clients. To save money and keep costs low, CSPs often distribute work to data centers around the globe where the work can be accomplished most efficiently. Cloud users need to confirm that regardless of where their data are stored, they are protected at a level that meets their corporate requirements. Users should stipulate that the CSP store and process data in specific jurisdictions according to the privacy rules of those jurisdictions. Cloud clients should also find out how the CSP segregates their corporate data from those of other companies and ask for proof that encryption mechanisms are sound.

In addition, it's important to know how the CSP will respond if a disaster strikes, whether the provider will be able to restore the client's data completely, and how long this should take. Cloud users should also ask whether CSPs will submit to external audits and security certifications. These kinds of controls can be written into the service level agreement (SLA) before signing with a CSP. The Cloud Security Alliance (CSA) has created industry-wide standards for cloud security, specifying best practices to secure cloud computing.

### Securing Mobile Platforms

Just like desktops and laptops, mobile devices need to be secured against malware, theft, accidental loss, unauthorized access, and hacking attempts. Mobile devices accessing corporate systems and data require special protection. Companies should make sure that their corporate security policy includes mobile devices, with additional details on how mobile devices should be supported, protected, and used. Companies will need mobile device management (MDM) tools to authorize all devices in use; to maintain accurate inventory records on all mobile devices, users, and applications; to control updates to applications; and to lock down or erase lost or stolen devices so that they can't be compromised. Data loss prevention technology can identify where critical data are saved, who is accessing the data, how data are leaving the company, and where the data are going. Firms should develop guidelines stipulating approved mobile platforms and software applications as well as the required software and procedures for remote access of corporate systems. The organization's mobile security policy should also forbid employees from using unsecured, consumer-based applications for transferring and storing corporate documents and files or sending such documents and files to oneself by email without encryption. Companies should encrypt communication whenever possible. All mobile device users should be required to use the most advanced security features available for their particular smartphones. Mobile security has become even more important in the wake of the Covid-19 pandemic, as more of the workforce continues to work remotely, in many cases using mobile devices.

## ENSURING SOFTWARE QUALITY

In addition to implementing effective security and controls, organizations can improve system quality and reliability by employing software metrics and rigorous software testing. Software metrics are objective assessments of the system in the form of quantified measurements. Ongoing use of metrics allows the information systems department and end users to measure the performance of the system jointly and to identify problems as they occur. Examples of software metrics include the number of transactions that can be processed in a specified unit of time, the online response time, the number of payroll checks printed per hour, and the number of known bugs per hundred lines of program code. For metrics to be successful, they must be carefully designed, formal, objective, and used consistently.

Early, regular, and thorough testing will contribute significantly to system quality. Many view testing as a way to prove the correctness of the work they have done. In fact, all sizable software is riddled with errors and must be tested to uncover these errors.

Good testing begins before a software program is even written, by using a *walkthrough*—a review of a specification or design document by a small group of people carefully selected based on the skills needed for the particular objectives being tested. When developers start writing software programs, coding walkthroughs can also be used to review program code. However, code must be tested by computer runs. When errors are discovered, the source is found and eliminated through a process called *debugging*. You can find out more about the various stages of testing required to put an information system into operation in Chapter 12.

# 8-5 Understand how MIS can help your career.

Here is how this chapter and this book can help you find an entry-level job as an identity and access management (IAM) support specialist.

## THE COMPANY

OptiCare Medical Group, a regional healthcare provider serving more than 1 million patients in 30 different locations, is looking to fill an entry-level position for an identity and access management (IAM) support specialist. OptiCare uses Microsoft Active Directory as its IAM service for employees and a separate cloud-based, SaaS service for IAM for its patients, who use a patient portal to access their records.

## POSITION DESCRIPTION

The IAM support specialist will be responsible for monitoring the company's IAM system to ensure that the company is meeting its audit and compliance requirements. This position reports to the company's security operations manager. Job responsibilities include:

- Using the user support ticket system to track, classify, and document user support requests with respect to IAM
- Performing data integrity testing of IAM system integrations with other applications
- Assisting in processes associated with identity management using Microsoft Active Directory
- Maintaining information on system user roles and privileges
- Assisting in the assessment of existing IAM processes and technologies and the drafting of policies and standards to enhance those processes

### JOB REQUIREMENTS

- Bachelor's degree in MIS, CIS, or a related field
- Familiarity with Microsoft Windows Server and Microsoft Active Directory
- Ability to multitask and work independently in a remote environment if required
- Attention to detail
- Strong time management skills
- Ability to communicate with both technical and nontechnical staff and with clients

### INTERVIEW QUESTIONS

1. What do you know about authentication and IAM? Have you ever worked with IAM tools or other IT security systems? What did you do with this software?
2. Have you ever worked with Microsoft Active Directory? What exactly did you do with this software?
3. What knowledge and experience do you have with ensuring data integrity?
4. Can you give an example of a situation in which you had to multitask and manage your time? Please describe how you handled it.
5. What software tools have you worked with?

### AUTHOR TIPS

1. Review the last two sections of this chapter, especially the discussions of IAM and authentication. Also review the Chapter 6 discussions of data integrity and data quality.
2. Use the web to find out more about IAM, data integrity testing, leading IAM software tools, and Microsoft Active Directory.
3. Use the web to find out more about the company, the kinds of systems it uses, and who might be using those systems.

## Review Summary

8-1 **Explain why information systems are vulnerable to destruction, error, and abuse.** Information systems and the data stored on them are vulnerable to destruction, misuse, error, fraud, and hardware or software failures. The Internet is designed to be an open system and makes internal corporate systems more vulnerable to actions from outsiders. Wi-Fi networks can easily be penetrated by intruders using sniffer programs to obtain an address to access the resources of the network. Malware such as viruses, worms, Trojan horses, ransomware, and spyware can disable systems and websites, with mobile devices a major target. Hackers can unleash denial-of-service (DoS) and distributed denial-of-service (DDoS) attacks or penetrate corporate networks, causing serious system disruptions. Most hacker activities are criminal offenses, and the vulnerabilities of systems make them targets for other types of computer crime as well, such as identity theft and data breaches. Cyberterrorism and cyberwarfare have become global threats. Software presents problems because software bugs may be impossible to eliminate and because software vulnerabilities can be exploited by hackers and malicious software. In addition, end users often introduce errors.

8-2 **Describe the business value of security and control.** Lack of sound security and control can cause firms relying on computer systems for their core business functions to lose sales and productivity. Information assets, such as confidential employee records, trade secrets, or business plans, lose much of their value if they are revealed to outsiders or if they expose the firm to legal liability. Laws such as HIPAA, the Sarbanes-Oxley Act, and the Gramm-Leach-Bliley Act require companies to practice stringent electronic records management and adhere to strict standards for security, privacy, and control. Legal actions requiring electronic evidence and computer forensics also require firms to pay more attention to security and electronic records management.

8-3 **Identify the components of an organizational framework for security and control.** Firms need to establish a good set of both general and application controls for their information systems. A risk assessment evaluates information assets, identifies control points and control weaknesses, and determines the most cost-effective set of controls. Firms must also develop a coherent corporate security policy and plans for continuing business operations in the event of disaster or disruption. The security policy includes policies for acceptable use and identity management. Comprehensive and systematic information systems auditing helps organizations determine the effectiveness of security and controls for their information systems.

8-4 **Identify the most important tools and technologies for safeguarding information resources.** Technologies for protecting businesses' information resources include using tools for managing user identities and access, preventing unauthorized access to systems and data, ensuring system availability, and ensuring software quality. Passwords, security tokens, smart cards, and biometric authentication are used to authenticate system users. Firewalls prevent unauthorized users from accessing a private network when it is linked to the Internet. Intrusion detection and prevention systems monitor private networks for suspicious network traffic and attempts to access corporate systems. Anti-malware software checks computer systems for infections by viruses and worms and often eliminates the malicious software. Encryption, the coding and scrambling of messages, is a widely used technology for securing digital transmissions over unprotected networks. Digital certificates combined with public key encryption provide further protection of digital transactions by authenticating a user's identity. Blockchain technology enables companies to create and verify tamperproof transactions on a network without a central authority. To ensure system availability, companies can use fault-tolerant computer systems, engage managed security service providers, and embrace the concept of digital resiliency. Companies must also address the new challenges posed by the Internet of Things, cloud computing, and the moble platform. Use of software metrics and rigorous software testing help improve software quality and reliability.

## Key Terms

Acceptable use policy (AUP), 291
Anti-malware software, 296
Application controls, 289
Authentication, 293
Biometric authentication, 294
Botnet, 280
Bugs, 284
Business continuity planning, 292
Computer forensics, 288
Computer virus, 277
Controls, 275
Cryptocurrencies, 299
Cybervandalism, 279
Cyberwarfare, 283
Data breach, 281
Denial-of-service (DoS) attack, 280
Digital certificates, 297
Digital resiliency, 299
Disaster recovery planning, 292
Distributed denial-of-service (DDoS) attack, 280
Downtime, 299
Drive-by download, 278
Encryption, 297
Evil twins, 283
Fault-tolerant computer systems, 299
Firewalls, 295
General controls, 289
Gramm-Leach-Bliley Act, 287
Hacker, 279
HIPAA, 287
HTTPS, 297
Identity and access management (IAM), 293
Identity theft, 282
Information systems audit, 292
Intrusion detection systems (IDS), 296

Intrusion prevention systems (IPS), 296
Keyloggers, 279
Malware, 277
Managed security service providers (MSSPs), 299
Multi-factor authentication (MFA), 294
Online transaction processing, 299
Password, 293
Patches, 286
Pharming, 283
Phishing, 282
Public key encryption, 297
Public key infrastructure (PKI), 298
Ransomware, 279
Risk assessment, 290
Sarbanes-Oxley Act, 287
Security, 275
Security policy, 291
Security token, 293
Smart card, 294
Sniffer, 280
Social engineering, 284
Spoofing, 280
Spyware, 279
SQL injection attack, 279
Transport Layer Security (TLS), 297
Trojan horse, 279
Two-factor authentication (2FA), 294
Unified threat management (UTM), 296
War driving, 277
Worm, 277
Zero-day vulnerabilities, 284
Zero trust, 293

## Review Questions

**8-1** Explain why information systems are vulnerable to destruction, error, and abuse.

- List and describe the most common threats against contemporary information systems.
- Define malware and distinguish among a virus, a worm, and a Trojan horse.
- Define hacker, and explain how hackers create security problems and damage systems.
- Define computer crime. Provide two examples of crime in which computers are targets and two examples in which computers are used as instruments of crime.
- Define identity theft and phishing, and explain why identity theft is such a big problem today.
- Describe the security and system reliability problems that employees create.
- Explain how software defects affect system reliability and security.

**8-2** Describe the business value of security and control.

- Explain how security and control provide value for businesses.
- Describe the relationships among security and control, recent US government regulatory requirements, and computer forensics.

**8-3** Identify the components of an organizational framework for security and control.

- Define general controls, and describe each type of general control.
- Define application controls, and describe each type of application control.
- Describe the function of risk assessment, and explain how it is conducted for information systems.
- Define and describe the following: security policy, acceptable use policy, and identity and access management.
- Explain how information systems auditing promotes security and control.

**8-4** Identify the most important tools and technologies for safeguarding information resources.

- Name and describe three authentication methods.
- Describe the roles of firewalls, intrusion detection and prevention systems, and anti-malware software in promoting security.
- Explain how encryption protects information.
- Describe the roles of encryption and digital certificates in a public key infrastructure.
- Distinguish between disaster recovery planning and business continuity planning.
- Define digital resiliency, and describe its benefits for an organization.
- Identify and describe the security challenges posed by Internet of Things (IoT) devices.
- Describe measures for improving software quality and reliability.

## Discussion Questions

**8-5** Security isn't simply a technology issue; it's a business issue. Discuss.
**MyLab MIS**

**8-6** If you were developing a business continuity plan for your company, where would you start? What aspects of the business would the plan address?
**MyLab MIS**

**8-7** Your business is about to launch an e-commerce website where it will sell goods and accept credit card payments. Discuss the major security threats to this website and their potential impacts. What can be done to minimize these threats?
**MyLab MIS**

**MyLab MIS**
To complete these problems, go to the EOC Discussion Questions in **MyLab MIS**.

## Hands-On MIS Projects

The projects in this section give you hands-on experience analyzing security vulnerabilities, using spreadsheet software for risk analysis, and using web tools to research security outsourcing services. Visit MyLab MIS to access this chapter's Hands-On MIS Projects.

### MANAGEMENT DECISION PROBLEM

**8-8** Scopely is a leading online gaming company offering a variety of popular games such as Scrabble Go, Star Trek Fleet Command, and Looney Tunes World of Mayhem. Although Scopely primarily focuses on the mobile platform, with Apple iOS and Android versions of its games, some games are also available for play on desktop/laptop computers. Scopely also has a corporate website and a social media presence on Facebook, Twitter, and LinkedIn. Prepare a security analysis for Scopely. What kinds of threats should it anticipate? What would be their impact on the business? What steps can it take to prevent damage to its operations?

### IMPROVING DECISION MAKING: USING SPREADSHEET SOFTWARE TO ANALYZE SECURITY EVENTS

Software skills: Spreadsheet sorting and data filtering
Business skills: Monitoring and analyzing security events

**8-9** Events that occur in computer systems are commonly recorded in log files. Each operating system, including Windows, uses its own log files, and applications and hardware devices also generate logs. Security teams can use security logs to track users on the corporate network, identify suspicious activity, and detect vulnerabilities. The security log lists events related to security, including login attempts or file deletions. System administrators use their organization's audit policy to determine which events to enter into their security log.

Your company's Information Systems Department has compiled from the company's Windows security event log a list of security events that merit special attention of the organization. In MyLabMIS you will find a spreadsheet showing the Security Event, Event ID, and Potential Criticality of each event on that list.

The "Event ID" is the code Windows assigns to each particular security event. The "Potential Criticality" column identifies whether the event should be considered to have Low, Medium, or High criticality in detecting attacks. A potential criticality of High means that even one occurrence of the event should be investigated. Potential criticality of Medium or Low means that these events should be investigated only if they occur unexpectedly or in numbers that significantly exceed the expected baseline in a measured period of time. All organizations should test these recommendations in their environments before creating alerts that require mandatory investigative responses. Every environment is different, and some of the events ranked with a potential criticality of High may occur because of other harmless events.

- Sort the data by Potential Criticality from High to Low.
- Filter the data to display all events of High Potential Criticality.
- Select one event of High Potential Criticality, and use the web to find out more about the event's importance and significance. Write a paragraph to explain why businesses should pay close attention to this type of security event.

## IMPROVING DECISION MAKING: EVALUATING SECURITY OUTSOURCING SERVICES

Software skills: Web browser and presentation software
Business skills: Evaluating business outsourcing services

**8-10** This project will help develop your skills in using the web to research and evaluate security outsourcing services.

- You have been asked to help your company's management decide whether to outsource security or keep the security function within the firm. Search the web to find information to help you decide whether to outsource security and to locate security outsourcing services.
- Present a brief summary of the arguments for and against outsourcing computer security for your company.
- Select two firms that offer computer security outsourcing services, and compare them and their services.
- Prepare a PowerPoint or other form of presentation for management, summarizing your findings. Your presentation should address whether your company should outsource computer security. If you believe your company should outsource, the presentation should identify which security outsourcing service you selected and justify your decision. If you believe the company should not outsource, you should identify the security tools it should use to ensure appropriate security.

## COLLABORATION AND TEAMWORK PROJECT

Evaluating Security Software Tools

**8-11** With a group of three or four students, use the web to research and evaluate two competing vendors' security products, such as anti-malware software, firewalls, or intrustion detection and prevention software. For each product, describe its capabilities, for what types of businesses it is best suited, and its cost to purchase and install. Which is the best product? Why? If possible, use Google Docs and Google Drive or Google Sites to brainstorm, organize, and develop a presentation of your findings for the class.

## BUSINESS PROBLEM-SOLVING CASE

### SOLARWINDS SHINES A LIGHT ON SOFTWARE SUPPLY CHAIN ATTACKS

In December 2020, cybersecurity firm FireEye made a disturbing discovery. Its "Red Team" toolkit, containing sophisticated hacking tools used to conduct penetration testing for its clients, had been stolen. As it investigated further, it discovered something even more ominous. The attackers had gained entry into FireEye's network via a "back door" in the network and applications monitoring software platform it was using, from a major software company named SolarWinds.

SolarWinds develops software that helps businesses manage their networks, systems, and IT infrastructure. A US-based public company, SolarWinds has more than 300,000 customers worldwide. Its Orion network and applications monitoring platform is used by 425 of the US *Fortune* 500, the top-10 telecommunications companies, the top-5 US accounting firms, all branches of the US military, and many US federal agencies, as well as hundreds of universities and colleges worldwide.

After being alerted by FireEye, SolarWinds discovered that in October 2019, hackers had infiltrated SolarWinds' Orion platform "build" server (the server used by the firm's developers to create updated versions of the software) via a malware implant. SolarWinds does not know precisely how the hacker gained initial entry to the server, but its investigations uncovered evidence that the hackers had, for at least nine months prior to October 2019, used compromised credentials to access its software development environment and its internal systems in order to conduct research and surveillance on SolarWinds' systems.

Sometime between March and June 2020, the malware implant replaced one of the source files in the Orion software with a backdoor (so named because it provides hackers a covert way into a system). As customers using the Orion platform downloaded updates to the platform running on their networks, the backdoor spread, providing the hackers with access to those customers' networks and data. Ultimately, the hackers were able to access about 100 networks, including those of the US Departments of Justice, Defense, State, Homeland Security, Treasury, Energy, and Commerce; the National Nuclear Security Administration; the National Institutes of Health; NASA; and Microsoft, as well as private sector networks throughout North America, Europe, Asia, and the Middle East. The SolarWinds hackers are believed to be a group known as APT29 (also sometimes called CozyBear, UNC2452, or Nobelium) and affiliated with the hacking arm of Russia's foreign intelligence service. As such, the attack is considered to be an act of nation-state cyberwarfare. In April 2021, the White House instituted a range of sanctions on Russian officials and assets in response to the hack. Russia has denied the allegations, saying it had no involvement.

Although SolarWinds immediately issued two hot fixes to address the vulnerability, a long time had already elapsed between the time the attack began and the time it was discovered. According to FireEye, the attack was the most sophisticated of its kind to that date and was very hard to detect because the hackers focused on detection evasion and leveraging existing trust relationships. Although SolarWinds took immediate steps once it learned of the attack, security experts believe that some systems might continue to be vulnerable, even if they were patched, because the attackers could have pivoted and maintained a presence in other areas of a company's network without the company knowing it.

The SolarWinds hack highlights a relatively new and menacing threat: software supply chain attacks. Most people are familiar with supply chains in the context of physical goods, but few people are aware that the concept also applies to the development of software. A software supply chain involves many different parties who are involved in six basic software phases: design, development and production, distribution, acquisition and deployment, maintenance, and disposal. A software supply chain attack occurs when a hacker infiltrates the software supply chain at the development and production stage and compromises the software before it is acquired and deployed by the customer. This may occur in either newly acquired software, as part of an update, or even as part of a patch. What makes the attack so insidious is that the malicious code is inserted into a trusted piece of software via trusted mechanisms, such as the update or patch process. In addition, the hack of a single software vendor can provide hackers with a springboard into the networks of hundreds or even thousands of the vendor's customers.

In addition to compromising update servers, other techniques have been used in software supply chain attacks. Some attacks begin with the corruption of tools used to create software in the first instance. For example, in 2015, hackers distributed a fake version of a tool used to build iOS applications that then was used to implant malicious code into dozens of Chinese iPhone apps. In another example, hackers corrupted a version of the Microsoft Visual Studio compiler, enabling malware to be hidden in video games.

Hackers have also targeted open-source code, hoping that unsuspecting developers will mistakenly add the malicious blocks of code into their own projects. For example, in 2018, researchers discovered 12 malicious Django Python code libraries with slightly misspelled names that had the same code and functionality as those they were impersonating but also had additional, malicious functionality.

Another technique involves undermining the code signing process, which involves applying a cryptographic signature to software components. Code signing is used to validate the identify of the software developer and the integrity of the code. Users and security tools typically trust signed code. However, hackers have been able to undermine the process by stealing code signing keys and certificates or by breaking into signing systems, enabling hackers to create compromised code signing certificates impersonating those of a trusted vendor. According to security researchers, a hacking group based in China, variously known as APT 41, Winnti Group, or Barium, has spent more than a decade refining code signing attack methods and has been involved in a number of software supply chain attacks.

Although the SolarWinds attack was not the first software supply chain attack, it may have been a turning point in raising awareness of such attacks. Governments and organizations are now focusing intently on what steps they can take to prevent future attacks. For instance, prior to the SolarWinds attack, few organizations, if any, incorporated the possibility of such an attack into their risk assessment procedures. In May 2021, the Biden administration issued an executive order addressing numerous aspects of government cybersecurity, with a section specifically on software supply chains, and setting new security standards for any company that wants to sell software to the federal government. In August 2021, Google announced that it would invest $10 billion in security initiatives over a five-year period and singled out software supply chains as a high-priority area. GitHub, which is owned by Microsoft, has launched an initiative aimed at automatically spotting security vulnerabilities in open-source projects.

Some security experts note that organizational efforts are equally as important as technological efforts in the fight against software supply chain attacks. From a software development standpoint, the increasing use of software development frameworks that value speed and agility, such as DevOps and agile development, adds to software supply chain risk. Organizations that have been wary of putting in place security measures that slow down development may need to rethink their processes. Experts say that software developers need to start thinking more about how to protect the integrity of software code and how to minimize risks to customers. The Cybersecurity and Infrastructure Security Agency (CISA) recommends that software developers incorporate secure practices throughout all phases of the software development process; take steps to identify and disclose vulnerabilities, including having a product vulnerability response program; participate in the CVE (Common Vulnerabilities and Exposures) database; submit products for third-party assessments; and use proactive exploit-mitigation technologies.

Organizations who acquire software also must take steps to protect themselves. Prior to acquiring software, organizations should evaluate software vendors and hold them to specified standards, similar to how manufacturing companies seek to control and limit their supply chains to ensure reliability. CISA recommends some specific actions for organizations to take, including establishing a formal Cybersecurity Supply Chain Risk Management program that involves executives and managers across the organization, establishing a set of security requirements or controls for all software suppliers, and using supplier certifications to ensure that vendors adhere to best practices.

Software supply chain attacks exploit two major aspects of third-party software products. First, to operate effectively, most common third-party software products require privileged access to various parts of the customer's network. Customers often accept third-party software defaults without questioning them. Second, most third-party software products require connectivity between the vendor and the customer for updates and patches. Security experts say that companies need to start thinking about applying zero-trust principles and role-based access controls not just to users but also to applications and servers. Just as not every user or device should be able to access any application or server on a network, not every server or application should be able to access other servers and applications on the network. Where feasible, an organization should segment its network to isolate certain parts of the network. If hackers still find a way in, the organization should have business resiliency plans in place that identify alternative suppliers for critical software and failover processes and workarounds in the event specific software becomes unavailable.

Despite the spotlight on software supply chain attacks and more focused efforts to prevent them, they are likely to continue to occur. For instance, in July 2021, a hack of Kaseya, which provides software solutions to more than 40,000 organizations (including thousands of managed service providers [MSPs]), resulted in a number of damaging ransomware attacks. As a provider of technology to MSPs, which typically serve hundreds of other companies, Kaseya is an important link in those companies' software supply chain. The ransomware was inserted via an automated, fake, and malicious software update using VSA, Kayesa's unifed, remote monitoring and management tool for networks and endpoints. The fake update then propagated to many of the MSP clients' systems, with more than 1,000 companies having servers and workstations encrypted by the ransomware.

A 2021 global survey of more than 2,200 IT executives by security firm Crowdstrike highlights the dangers ahead. Of the organizations surveyed, 45 percent had experienced at least one software supply chain attack in the previous year, and almost 60 percent of the organizations suffering such an attack for the first time did not have a response strategy in place. Almost 85 percent believe that software supply chain attacks will be one of the biggest cyber threats to their organizations within the next three years.

Sources: Nathan Eddy, "Software Supply Chain Security Becomes Prime Concern in 2022," Insights.dice.com, January 10, 2022; Jeff Burt, "SolarWinds-Like Supply Chain Attacks Will Peak in 2022, Apiiro Security Chief Predicts," Esecurityplanet.com, December 23, 2021; Vishal Salvi, "Ensuring Cybersecurity Defenses Permeate an Organization," Techradar.com, December 22, 2021; Crowdstrike, "What Is a Supply Chain Attack," Crowdstrike.com, December 8, 2021; Lily Hay Newman, "A Year after the SolarWinds Hack, Supply Chain Threats Still Loom", Wired.com, December 8, 2021; Charlie Osborne, "Updated Kaseya Ransomware Attack FAQ: What We Know Now," Zdnet.com July 23, 2021; Pam Baker, "The SolarWinds Hack Timeline: Who Knew What, and When?" Csoonline.com, June 4, 2021; Andy Greenberg, "Hacker Lexicon: What Is a Supply Chain Attack," Wired.com, May 31, 2021; Sudhakar Ramakrishna, "An Investigative Update on the Cyberattack," Orangematter.solarwinds.com, May 7, 2021; Russell Brandom, "US Institutes New Russia Sanctions in Response to SolarWinds Hack," Theverge.com, April 15, 2021; Cybersecurity and Infrastructure Security Agency (CISA), "Defending against Software Supply Chain Attacks," Cisa.gov, April 2021; Lucian Constantin, "SolarWinds Attack Explained: And Why It Was So Hard to Detect," Csoonline.com, December 15, 2020; Andy Greenberg, "A Mysterious Hacker Group Is on a Supply Chain Hijacking Spree," Wired.com, May 3, 2019.

## CASE STUDY QUESTIONS

**8-12** Identify and describe the security and control weaknesses discussed in this case.

**8-13** What people, organization, and technology factors contributed to these weaknesses?

**8-14** Discuss the impact of the SolarWinds hack.

**8-15** How can future software supply chain attacks like those discussed in the case be prevented? Explain your answer.

## Chapter 8 References


2-Spyware. "Messenger Virus: 2021 Update. A New Threat for Facebook Users." 2-spyware.com (August 5, 2021).

Bose, Idranil, and Alvin Chung Man Leung. "Adoption of Identity Theft Countermeasures and Its Short- and Long-Term Impact on Firm Value." *MIS Quarterly* 43, No. 1 (March 2019).

Burgess, Christopher. "Facebook Outage a Prime Example of Insider Threat by Machine." Csoonline.com (November 4, 2021).

Burgess, Christopher. "FBI Arrests Social Engineer Who Allegedly Stole Unpublished Manuscripts from Authors." Csoonline.com (January 13, 2022).

Center for Strategic and International Studies/McAfee (Zhanna Malekos Smith and Eugenia Lostri). "The Hidden Cost of Cybercrime." (2020).

Check Point. "Mobile Security Report 2021." Checkpoint.com (April 2021).

Cichy, Patrick, Torsten Oliver Salge, and Rajiv Kohli. "Privacy Concerns and Data Sharing in the Internet of Things: Mixed Methods Evidence from Connected Cars." *MIS Quarterly* 45 No. 4 (Dcember 2021).

CyberSecurity Ventures (Steve Morgan). "2021 Report: Cyberwarfare in the C-Suite." Cybersecurityventures.com (January 21, 2021).

IBM Security. "Cost of a Data Breach Report 2021." Ibm.com (2021).

Javelin Strategy & Research. "Identity Fraud Losses Total $52 Billion in 2021, Impacting 43 Million U.S. Adults." Javelinstrategy.com (March 29, 2022).

Madnick, Stuart. "Blockchain Isn't as Unbreakable as You Think." *MIT Sloan Management Review* 61 No. 2 (Winter 2020).

McAfee. "The McAfee Consumer Mobile Threat Report." (February 2022).

Microsoft Security. "Destructive Malware Targeting Ukrainian Organizations." Microsoft.com (January 15, 2022).

Nichols, Shaun. "Metaverse Rollout Brings New Security Risks, Challenges." Techtarget.com (February 7, 2022).

Oracle and KPMG. "Oracle and KPMG Cloud Threat Report 2020." Oracle.com (2020).

Palmer, Danny. "Emotet, Once the World's Most Dangerous Malware, Is Back." Zdnet.com (November 16, 2021).

Panko, Raymond R., and Julie L. Panko. *Business Data Networks and Security*, 11th ed. (Hoboken, NJ: Pearson Education, Inc., 2019).

Pearlson, Keri, and Keman Huang. "Design for Cybersecurity from the Start." *MIT Sloan Management Review* 63, No. 2 (Winter 2022).

Pearlson, Dr. Keri, Sean Sposito, Masha Arbisman, and Josh A. Schwartz. "How Yahoo Built a Culture of Cyberecurity." *Harvard Business Review* (September 30, 2021).

Plis. "Top 10 Countries Where Security Hackers Come from & Their Types." Cyberkite.com.au (July 22, 2021).

Romero, Jessica. "Taking Legal Action against Phishing Attacks." About.fb.com (December 20, 2021).

Rothrock, Ray A., James Kaplan, and Friso Van der Oord. "The Board's Role in Managing Cybersecurity Risks." *MIT Sloan Management Review* (Winter 2018).

Sanyal, Pallab, Nirup Menan, and Mikko Siponin. "An Empirical Examination of the Economics of Mobile Application Security." *MIS Quarterly* 54, No. 4 (December 2021).

Tozzi, Chris. "Popular Security Best Practices for Hybrid Cloud." Techtarget.com (December 14, 2021).

Uptime Institute. "Digital Infrastructure Resiliency." (accessed July 18, 2022).

Vial, Gregory. "Managing the Risks of Software Reuse." *MIT Sloan Management Review* 63, No. 4 (Summer 2022).

W3Techs. "Usage Statistics of Default Protocol HTTPS for Websites." W3techs.com (accessed July 18, 2022).

# PART III



# Key System Applications for the Digital Age

Part III examines the core information system applications businesses are using today to improve operational excellence and decision making. These applications include enterprise systems; systems for supply chain management, customer relationship management, and artificial intelligence; e-commerce applications; and business intelligence systems to enhance decision making. This part answers questions such as these: How can enterprise applications improve business performance? How do firms use e-commerce to extend the reach of their businesses? How can systems improve decision making and help companies benefit from artificial intelligence?

CHAPTER

# 9

# Achieving Operational Excellence and Customer Intimacy: Enterprise Applications

## LEARNING OBJECTIVES

After completing this chapter, you will be able to:

9-1 Explain how enterprise systems help businesses achieve operational excellence.

9-2 Describe how supply chain management systems coordinate planning, production, and logistics with suppliers.

9-3 Explain how customer relationship management systems help firms achieve customer intimacy.

9-4 Describe the challenges that enterprise applications pose and how enterprise applications are taking advantage of new technologies.

9-5 Understand how MIS can help your career.

## CHAPTER CASES

- Lenzing Sustainably Balances Supply and Demand
- MillerKnoll Uses Salesforce.com to Transform Its Business Strategy
- Versum's ERP Transformation
- The Coronavirus Pandemic Disrupts Supply Chains Around the World

**MyLab MIS**

- **Video Case:** Software Startup Freshworks Not in a Rush to Raise Capital
- Discussion Questions: 9-5, 9-6, 9-7
- Hands-On MIS Projects: 9-8, 9-9, 9-10, 9-11
- eText with Figure Videos

## LENZING SUSTAINABLY BALANCES SUPPLY AND DEMAND

The Austria-based Lenzing Group supplies high-quality specialty fibers for the global fashion industry, sports and outdoor wear, and protection wear. Its botanic cellulose fibers, used in the production of textile and nonwoven products, have much less damaging impact on the environment than traditional materials do. The company has production sites in major global markets along with a worldwide network of sales and marketing offices, 7,400 employees, and 2021 revenue surpassing $1.8 billion. Lenzing's supply chain involves many different members, including spinners, weavers, millers, dye workers and converters, as well as fashion brands and retailers.

According to Robert van de Kerkhof, Lenzing's chief commercial officer, the fashion industry is the second-largest polluter worldwide. Its production processes generate waste, and the end products themselves are generally nonbiodegradable and rarely recycled. Lenzing adheres to very high environmental standards and has numerous international sustainability certifications for its business processes. It is considered the most sustainable company in its sector.

Lenzing wants to do more and is focusing on innovations in both its products and its processes to minimize its environmental impact while still meeting the needs of its end consumers. The company needed to create an end-to-end supply chain planning process that precisely matches supply and demand, minimizing inefficiencies while maximizing profitability. Lenzing had been using Excel spreadsheets, but manually intensive processes could not do the job, given the complexity of the company's worldwide business model and supply chain. More powerful software tools were required to digitally link demand forecasting, sales planning, and operations planning in order to create a highly accurate and efficient end-to-end supply chain.

Lenzing needed to overhaul its Sales & Operations Planning (S&OP) processes to fully integrate business planning processes and completely eliminate manual work. The S&OP process enables companies to realize revenue, margin, and operating performance gains through improved decision support and cross-functional alignment. Lenzing selected Blue Yonder Sales & Operations Planning for this purpose.

© Petrmalinak/Shutterstock

Blue Yonder Sales & Operations Planning supports six distinct enterprise processes and associated scenario planning: demand review, supply review, demand-supply balancing, financial review, continuous plan refinement, and business performance management. Companies using this software are able to take a cross-functional approach to integrated business planning that unites all the moving parts across their supply chain to meet demand across markets, serving both immediate and long-term strategic goals. When risks, opportunities, or threats interfere with strategic execution, the entire organization can act swiftly and decisively to get back on track. The system provides visibility across departments and can identify performance gaps and provide "what-if" resolution. There is improved sales forecast accuracy, transparency, business alignment, and decision making.

Lenzing was able to quickly launch its new Sales & Operations Planning system using the Blue Yonder Cloud Software as a Service (SaaS). Returns on investment were immediate, with a 50 percent reduction in planning and decision-making time. The company had better visibility into its supply chain, leading to greater forecast accuracy and better decisions. Profitability and resource utilization improved, with better capabilities for allocating critical resources to the most profitable markets and product applications. The system helped Lenzing minimize waste, and its entire supply chain became leaner.

Sources: "Lenzing: Fashioning a Transformation" and "Lenzing: Boundaryless Planning Supporting Sustainabiliry: Challenges in the Supply Chain," Blueyonder.com, accessed January 8, 2022; Lenzing.com, accessed January 8, 2022; Bloomberg News, "Greener Fashion Industry Could Unlock $100 Billion in Value," Supplychainbrain.com, January 22, 2020.

Lenzing's problems with balancing supply and demand in a global marketplace illustrate the critical role of supply chain management systems in business. Lenzing's business performance was impeded because it could not accurately predict exactly the quantity of fiber materials that other members of its supply chain needed in many different locations around the world. Lenzing's existing systems were highly manual and lacked the flexibility and power to predict this, which sometimes left the company holding too much inventory it couldn't sell or not having enough inventory at the right time or place to fulfill customer orders. This not only added to costs but was also environmentally wasteful.

The chapter-opening diagram calls attention to important points raised by this case and this chapter. Lenzing's supply chain is far-reaching, servicing customers ordering fibers in many different locations around the globe. Lenzing's legacy systems were unable to coordinate demand, inventory, and supply planning across its entire global enterprise. The way it managed its supply chain was highly wasteful and ran counter to its corporate mission to promote sustainability. Implementing Blue Yonder software tools for Sales & Operations Planning has made it much easier for Lenzing's managers to access and analyze data for forecasting, inventory planning, and fulfillment, greatly improving both decision making and operational efficiency across the global enterprise.

Here are some questions to think about: How were Lenzing's business model and objectives affected by having an inefficient supply chain? How did Blue Yonder software tools improve operations and decision making at Lenzing?

## 9-1 Explain how enterprise systems help businesses achieve operational excellence.

Around the globe, companies are increasingly becoming more connected, both internally and with other companies. If you run a business, you'll want to be able to react instantaneously when a customer places a large order or when a shipment from a supplier is delayed. You may also want to know the impact of these events on every part of your business and how your business is performing at any point in time, especially if you're running a large company. Enterprise systems provide the integration to make this possible. Let's look at how they work and what they can do for a firm.

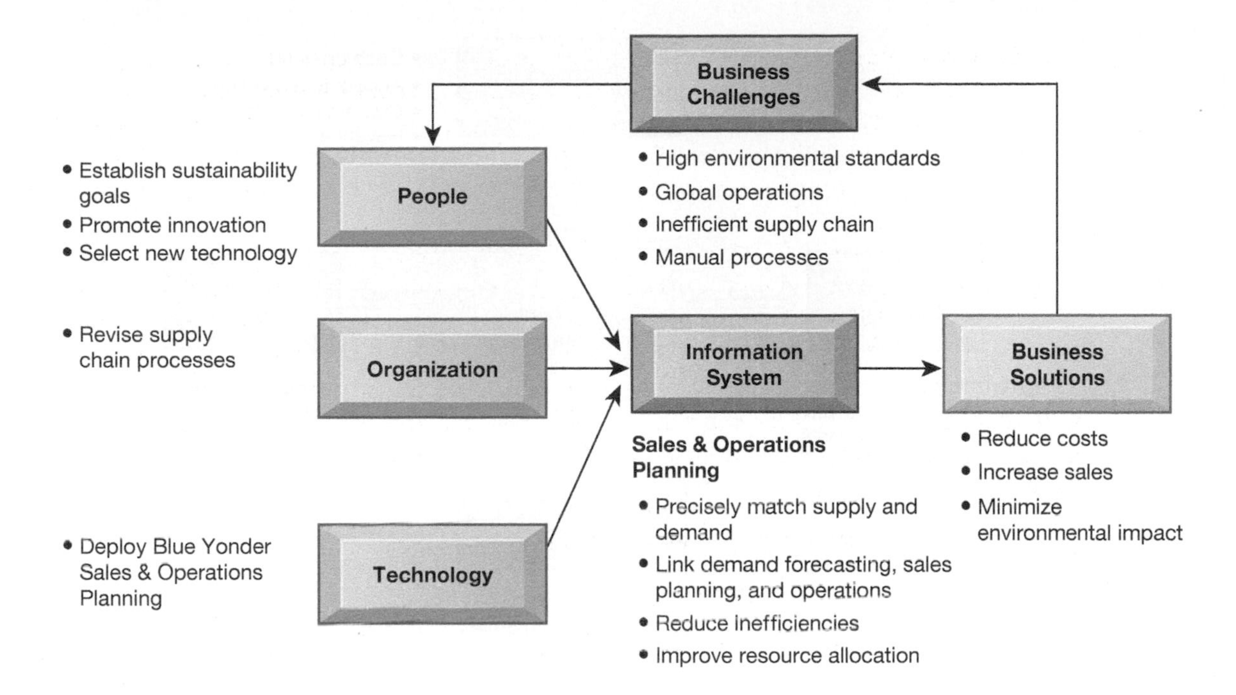


## WHAT ARE ENTERPRISE SYSTEMS?

Imagine that you had to run a business based on information from tens or even hundreds of databases and systems, none of which could communicate with one another. Imagine your company had 10 major product lines, each produced in separate factories and each with separate and incompatible sets of systems controlling production, warehousing, and distribution.

Your decision making would likely have to be based on manual, hard-copy reports that might be out of date, and it would be difficult to understand what is happening in the business as a whole. Sales personnel might not be able to tell at the time they place an order whether the ordered items are in inventory, and manufacturing could not easily use sales data to plan for new production. You now have a good idea of why firms need a special enterprise system to integrate information.

Chapter 2 introduced enterprise systems, also known as enterprise resource planning (ERP) systems, which are based on a suite of integrated software modules and a common, central database. Examine Figure 9.1, which illustrates how enterprise systems work. The database collects data from many divisions and departments in a firm and from a large number of key business processes in manufacturing and production, finance and accounting, sales and marketing, and human resources, making the data available for applications that support nearly all of an organization's internal business activities. When new information is entered by one process, the information is made immediately available to other business processes View the Figure 9.1 video in the eText for an animated and more detailed discussion of this figure.

If a sales representative places an order for tire rims, for example, the system verifies the customer's credit limit, schedules the shipment, identifies the best shipping route, and reserves the necessary items from inventory. If inventory stock is insufficient to fill the order, the system schedules the manufacture of more rims, ordering the needed materials and components from suppliers. Sales and production forecasts are immediately adjusted. General ledger and corporate cash levels are automatically updated with the revenue and cost information from the order. Users can tap into the system and find out where that particular order is at any minute, and management can obtain information at any point in time about how the business is operating. The system can also generate enterprise-wide data for management analyses of product cost and profitability.

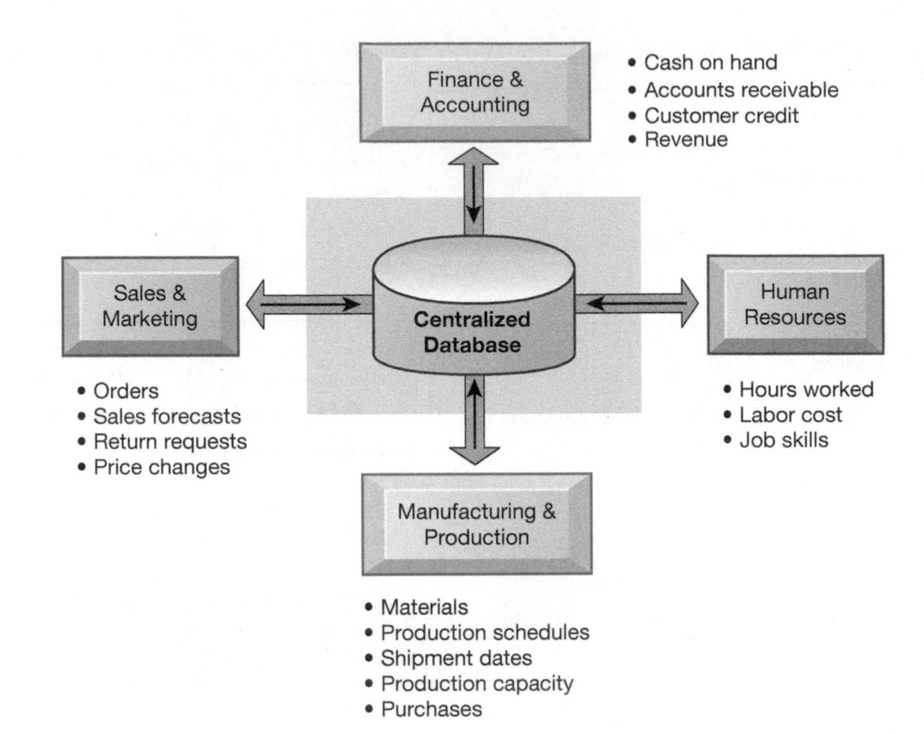


**Figure 9.1**
How Enterprise Systems Work
*Enterprise systems feature a set of integrated software modules and a central database through which business processes and functional areas throughout the enterprise can share data.*

## ENTERPRISE SOFTWARE

**Enterprise software** is built around thousands of predefined business processes that reflect best practices. Table 9.1 describes some of the major business processes that enterprise software supports.

Companies implementing this software first have to select the functions of the system they wish to use and then map their business processes to the predefined business processes in the software. Configuration tables provided by the software vendor enable the firm to tailor a particular aspect of the system to the way the firm does business. For example, the firm could use these tables to select whether it wants to track revenue by product line, geographical unit, or distribution channel.

If the enterprise software does not support the way the organization does business, the software can sometimes be rewritten to support the way its business processes work. However, enterprise software is unusually complex, and extensive customization may degrade system performance, compromising the information and process integration that are the main benefits of the system. If companies want to reap the

**TABLE 9.1**
Business Processes Supported by Enterprise Systems

**Financial and accounting processes,** including general ledger, accounts payable, accounts receivable, fixed assets, cash management and forecasting, product-cost accounting, cost-center accounting, asset accounting, tax accounting, credit management, and financial reporting

**Human resources processes,** including personnel administration, time accounting, payroll, personnel planning and development, benefits accounting, applicant tracking, time management, compensation, workforce planning, performance management, and travel expense reporting

**Manufacturing and production processes,** including procurement, inventory management, purchasing, shipping, production planning, production scheduling, material requirements planning, quality control, distribution, transportation execution, and plant and equipment maintenance

**Sales and marketing processes,** including order processing, quotes, contracts, product configuration, pricing, billing, credit checking, incentive and commission management, and sales planning

maximum benefits from enterprise software, they must change the way they work to conform to the business processes defined by the software.

To implement a new enterprise system, Tasty Baking Company identified its existing business processes and then translated them into the business processes built into the SAP ERP software it had selected. To ensure that it obtained the maximum benefits from the enterprise software, Tasty Baking Company deliberately planned for customizing less than 5 percent of the system and made very few changes to the SAP software itself. It used as many tools and features that were already built into the SAP software as it could. SAP has more than 3,000 configuration tables for its enterprise software.

Leading enterprise software vendors include SAP, Oracle, IBM, Infor Global Solutions, and Microsoft. There are also versions of enterprise software packages designed for small and medium-sized businesses and on-demand software services running in the cloud (see Section 9-4).

## BUSINESS VALUE OF ENTERPRISE SYSTEMS

Enterprise systems provide value by both increasing operational efficiency and providing firmwide information to help managers make better decisions. Large companies with many operating units in different locations use enterprise systems to enforce standard practices and data so that everyone does business the same way worldwide.

Coca-Cola, for instance, implemented an SAP enterprise system to standardize and coordinate important business processes in 200 countries. Lack of standard, companywide business processes had prevented the company from using its worldwide buying power to obtain lower prices for raw materials and from reacting rapidly to market changes.

Enterprise systems help firms respond rapidly to customer requests for information or products. Because the system integrates order, manufacturing, and delivery data, manufacturing is better able to produce only what customers have ordered, procure exactly the right number of components or raw materials to fill actual orders, stage production, and minimize the time that components or finished products are in inventory.

AbbVie, a large global biopharmaceutical company, with operations spanning 70 countries, had been working with more than 50 legacy systems for mission-critical processes in more than 100 worldwide locations. The company decided to standardize its business processes for all its affiliates and manufacturing facilities globally and to support those processes with a single, global SAP ERP system. The new ERP system provides a set of key metrics, such as the length of time to create new customers, vendor payments, payment terms, and order fulfillments. There are dashboards available that enable managers to look at every country, measure results, find the root cause of problems, and take corrective action more easily. Reporting from the system is also more accurate (AbbVie, 2019).

Enterprise systems provide much valuable information for improving management decision making. Corporate headquarters can access to up-to-the-minute data on sales, inventory, and production and use this information to create more accurate sales and production forecasts. Enterprise software includes analytical tools to use the data the system captures to evaluate overall organizational performance. Enterprise system data have common, standardized definitions and formats that are accepted by the entire organization. Performance statistics mean the same thing across the company. Enterprise systems also allow senior management to find out easily and at any moment how a particular organizational unit is performing, to determine which products are most or least profitable, and to calculate costs for the company as a whole.

## 9-2 Describe how supply chain management systems coordinate planning, production, and logistics with suppliers.

If you manage a small firm that makes a few products or sells a few services, chances are you will have a small number of suppliers. You could coordinate your supplier orders and deliveries by using just a telephone and a fax machine. But if you manage a firm that produces more complex products and services, you will have hundreds of suppliers, and each of your suppliers will have its own set of suppliers. Suddenly, you will need to coordinate the activities of hundreds or even thousands of other firms to produce your products and services. Supply chain management (SCM) systems, which we introduced in Chapter 2, are an answer to the problems of supply chain complexity and scale.

### THE SUPPLY CHAIN

A firm's **supply chain** is a network of organizations and business processes that procure raw materials, transform these materials into intermediate and finished products, and distribute the finished products to customers. It links suppliers, manufacturing plants, distribution centers, retail outlets, and customers to supply goods and services from source through consumption. Materials, information, and payments flow through the supply chain in both directions.

Goods start out as raw materials and, as they move through the supply chain, are transformed into intermediate products (also referred to as components or parts), and, finally, are transformed into finished products. The finished products are shipped to distribution centers and from there to retailers and customers. Returned items flow in the reverse direction, from the buyer back to the seller.

Let's look at the supply chain for Nike sneakers as an example. Nike designs, markets, and sells sneakers, socks, athletic clothing, and accessories throughout the world. Its primary suppliers are contract manufacturers with factories in China, Thailand, Indonesia, Brazil, and other countries. These companies fashion Nike's finished products.

Nike's contract suppliers do not manufacture sneakers from scratch. Rather, they obtain components for the sneakers—the laces, eyelets, uppers, and soles—from other suppliers and then assemble the components into finished sneakers. These suppliers in turn have their own suppliers. For example, the suppliers of soles have suppliers of synthetic rubber, suppliers of the chemicals used to melt the rubber for molding, and suppliers of the molds into which to pour the rubber. Suppliers of laces have suppliers of their thread, dyes, and plastic lace tips.

Examine Figure 9.2, which provides a simplified illustration of Nike's supply chain for sneakers. It shows the flow of information and materials among suppliers, Nike, Nike's distributors, retailers, and customers. Nike's contract manufacturers are its primary suppliers. The suppliers of soles, eyelets, uppers, and laces are the secondary (Tier 2) suppliers. Suppliers to these suppliers are the tertiary (Tier 3) suppliers.

The *upstream* portion of the supply chain includes the company's primary, Tier 2, and Tier 3 suppliers and the processes for managing relationships with them. The *downstream* portion consists of the organizations and processes for distributing and delivering the products to the final customers. Companies that manufacture, such as Nike's contract suppliers of sneakers, also manage their own *internal supply chain processes* for transforming the materials, components, and services that their suppliers furnish into finished products or intermediate products (components or parts) for their customers and for managing materials and inventory.

The supply chain illustrated in Figure 9.2 has been simplified. It shows only two contract manufacturers for sneakers and only the upstream supply chain for sneaker soles. In reality, Nike has hundreds of contract manufacturers turning out finished

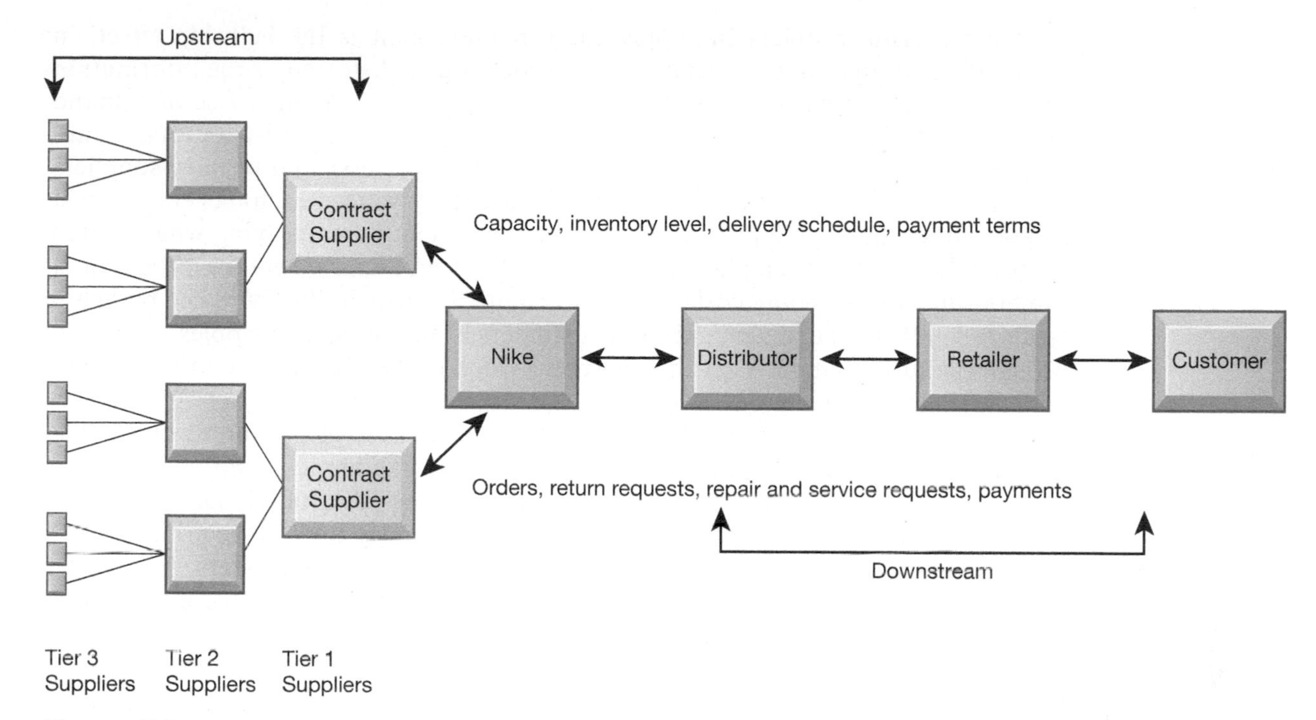


**Figure 9.2**
Nike's Supply Chain
*This figure illustrates the major entities in Nike's supply chain and the flow of information upstream and downstream to coordinate the activities involved in buying, making, and moving a product. Shown here is a simplified supply chain, with the upstream portion focusing only on the suppliers for sneakers and sneaker soles.*

sneakers, socks, and athletic clothing, each with its own set of suppliers. The upstream portion of Nike's supply chain actually comprises thousands of entities. Nike also has numerous distributors and many thousands of retail stores where its products are sold, so the downstream portion of its supply chain is also large and complex. View the Figure 9.2 video in the eText for an animated and more detailed discussion of this figure.

## INFORMATION SYSTEMS AND SUPPLY CHAIN MANAGEMENT

Inefficiencies in the supply chain, such as parts shortages, underused plant capacity, excessive finished goods inventory, or high transportation costs, are caused by inaccurate or untimely information. For example, manufacturers may keep too many parts in inventory because they do not know exactly when they will receive their next shipments from their suppliers. Suppliers may order too few raw materials because they do not have precise information about demand. These supply chain inefficiencies waste as much as 25 percent of a company's operating costs.

If a manufacturer had perfect information about exactly how many units of product customers wanted, when they wanted the products, and when the products could be produced, it would be possible to implement a highly efficient, **just-in-time strategy**. Components would arrive exactly at the moment they were needed, and finished goods would be shipped as they left the assembly line.

In a supply chain, however, uncertainties arise because many events cannot be foreseen—uncertain product demand, late shipments from suppliers, defective parts or raw materials, or production process breakdowns. To satisfy customers, manufacturers often deal with such uncertainties and unforeseen events by keeping more material or products in inventory than they think they may actually need. The *safety stock* acts as a buffer for the lack of flexibility in the supply chain. Although excess inventory is expensive, low fill rates are also costly because business may be lost from canceled orders.

One recurring problem in supply chain management is the **bullwhip effect**, in which information about the demand for a product gets distorted as the information passes from one entity to the next across the supply chain. A slight rise in demand for an item might cause different members in the supply chain—distributors, manufacturers, suppliers, secondary suppliers (suppliers' suppliers), and tertiary suppliers (suppliers' suppliers' suppliers)—to stockpile inventory so that each has enough just in case. These changes ripple throughout the supply chain, magnifying what started out as a small change from planned orders and creating excess inventory, production, warehousing, and shipping costs. Examine Figure 9.3, which illustrates the bullwhip effect. A small change in demand from customers is amplified as it ripples backward through the supply chain. View the Figure 9.3 video in the eText for an animated and more detailed discussion of this figure.

Some experts believe the bullwhip effect was one cause of the supply chain disruptions that emerged during and after the Covid-19 pandemic. In mid-2022, some large retailers such as Walmart, Target, and Gap over-estimated the amount of inventory they would need, leading them to significantly over-order from wholesalers. The wholesalers in turn over-ordered from their own suppliers, leading to a major mismatch between consumers' actual demand and inventories on hand (McCormick, 2022).

The bullwhip effect is tamed by reducing uncertainties about demand and supply because all members of the supply chain have accurate and up-to-date information. If all supply chain members share dynamic information about inventory levels,

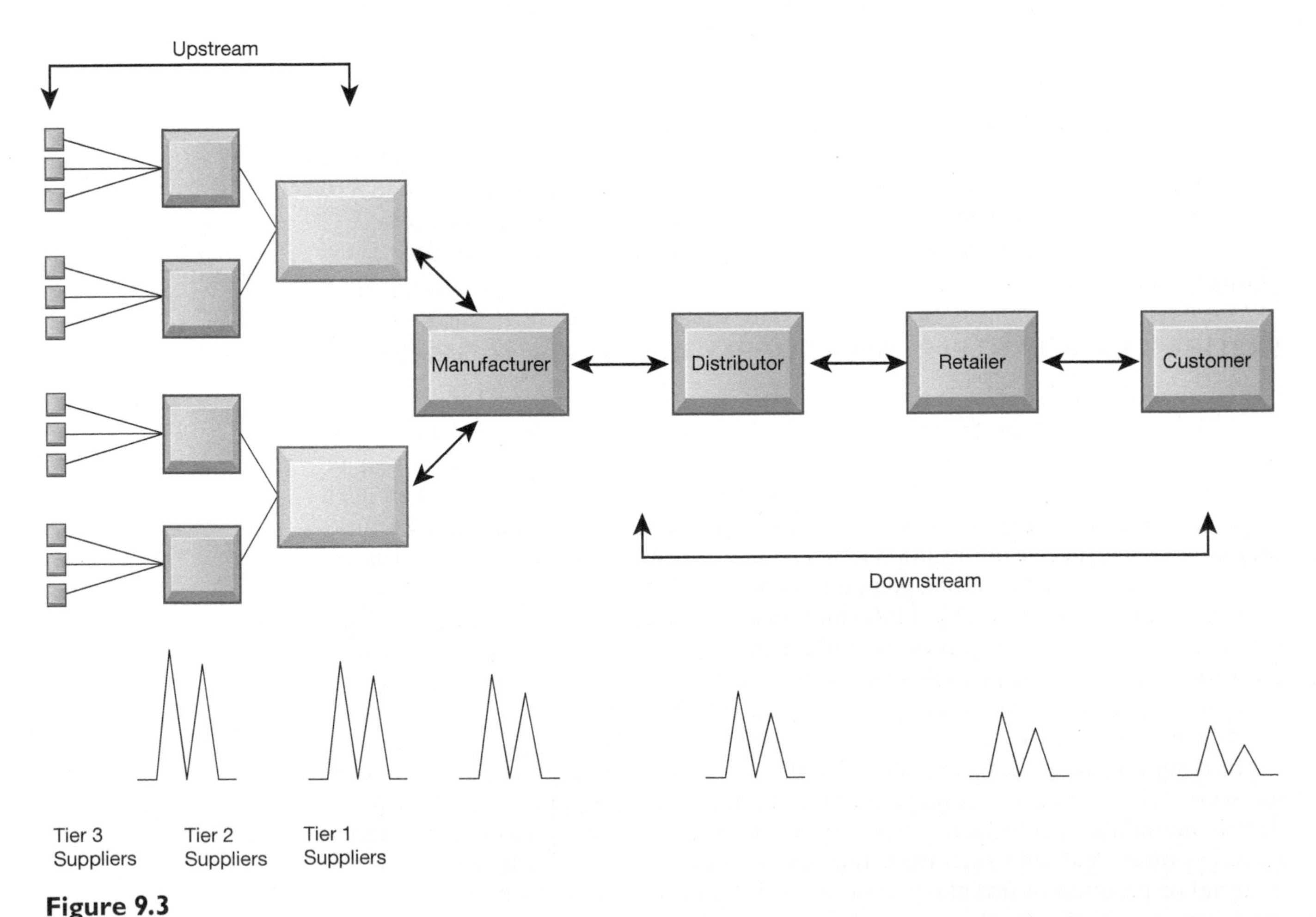


**Figure 9.3**
The Bullwhip Effect
*Inaccurate information can cause minor fluctuations in demand for a product to be amplified as one moves further back in the supply chain. For example, minor fluctuations in retail sales of a product can create excess inventory for distributors, manufacturers, and suppliers.*

schedules, forecasts, and shipments, they have more precise knowledge about how to adjust their sourcing, manufacturing, and distribution plans. Supply chain management systems provide the kind of information that helps members of the supply chain make better purchasing and scheduling decisions.

## SUPPLY CHAIN MANAGEMENT SOFTWARE

Supply chain software is classified as either software to help businesses plan their supply chains (supply chain planning) or software to help businesses execute the supply chain steps (supply chain execution). **Supply chain planning systems** enable the firm to model its existing supply chain, generate demand forecasts for products, and develop optimal sourcing and manufacturing plans. Such systems help companies make better decisions, such as determining how much of a specific product to manufacture in a given time period; establishing needed inventory levels of raw materials, intermediate products, and finished goods; determining where to store finished goods; and identifying the transportation mode to use for product delivery (see the chapter-opening case study).

For example, if a large customer places a larger order than usual or changes that order on short notice, this change can have a widespread impact throughout the supply chain. Additional raw materials or a different mix of raw materials may need to be ordered from suppliers. Manufacturing may have to change job scheduling. A transportation carrier may have to reschedule deliveries. Supply chain planning software makes these necessary adjustments to production and distribution plans. In addition, information about changes is shared among the relevant supply chain members so that their work can be coordinated. One of the most important—and complex—supply chain planning functions is **demand planning**, which determines how much product a business needs to make in order to satisfy all its customers' demands. Blue Yonder Software, SAP, and Oracle all offer supply chain management solutions.

**Supply chain execution systems** manage the flow of products through distribution centers and warehouses to ensure that products are delivered to the right locations in the most efficient manner. These systems track the physical status of goods, the management of materials, warehouse and transportation operations, and financial information involving all parties. An example is the Warehouse Management System (WMS) that Haworth Incorporated uses. Haworth is a world-leading manufacturer and designer of office furniture, with distribution centers in four states. The WMS tracks and controls the flow of finished goods from Haworth's distribution centers to its customers. Acting on shipping plans for customer orders, the WMS directs the movement of goods based on immediate conditions of space, equipment, inventory, and personnel.

## GLOBAL SUPPLY CHAINS AND THE INTERNET

Before the Internet, supply chain coordination was hampered by the difficulties of making information flow smoothly among disparate internal supply chain systems for purchasing, materials management, manufacturing, and distribution. It was also difficult to share information with external supply chain partners because the systems of suppliers, distributors, or logistics providers were based on incompatible technology platforms and standards. Enterprise and supply chain management systems enhanced with Internet technology supply some of this integration.

A manager uses a web interface to tap into suppliers' systems to determine whether inventory and production capabilities match demand for the firm's products. Business partners use web-based supply chain management tools to collaborate online on forecasts. Sales representatives access suppliers' production schedules and logistics information to monitor customers' order status.

### Global Supply Chain Issues

More and more companies are entering international markets, outsourcing manufacturing operations, and obtaining supplies from other countries as well as selling abroad. Their supply chains extend across multiple countries and regions, which leads to additional complexities and challenges to managing a global supply chain.

Global supply chains typically span greater geographic distances and time differences than domestic supply chains do and have participants from a number of countries. Performance standards may vary from region to region or from nation to nation. Supply chain management may need to reflect foreign government regulations and cultural differences.

The Internet helps companies manage many aspects of their global supply chains, including sourcing, transportation, communications, and international finance. Today's apparel industry, for example, relies heavily on outsourcing to contract manufacturers in China and other low-wage countries. Apparel companies are using the web to manage their global supply chain and production issues. (Review the discussion of Li & Fung in Chapter 3.)

In addition to contract manufacturing, globalization has encouraged outsourcing warehouse management, transportation management, and related operations to third-party logistics providers such as UPS Supply Chain Solutions and Schneider National. These logistics services offer web-based software to give their customers a better view of their global supply chains. Customers can check a secure website to monitor inventory and shipments, helping them run their global supply chains more efficiently.

### Demand-Driven Supply Chains: From Push to Pull Manufacturing and Efficient Customer Response

In addition to reducing costs, supply chain management systems facilitate efficient customer response, enabling the workings of the business to be driven more by customer demand. (We introduced efficient customer response systems in Chapter 3.)

Examine Figure 9.4, which illustrates the difference between a push-based model (also known as build-to-stock), which drove earlier supply chain management systems, and a pull-based model. In a **push-based model**, supply, production, inventory, and stock are all driven by forecasts or best guesses of demand for products, rather than what the customer orders. Instead, products are pushed to customers. With new flows of information made possible by web-based tools, supply chain management can now more easily follow a pull-based model. In a **pull-based model**, also known as a demand-driven or build-to-order model, actual customer orders or purchases trigger events in the supply chain. Transactions to produce and deliver only what customers have ordered move up the supply chain from retailers to distributors to manufacturers and eventually to suppliers. Only products that fulfill these orders move back down the supply chain to the retailer. Manufacturers use actual order demand information to drive their production schedules and the procurement of components or raw materials. Walmart's continuous replenishment system described in Chapter 3 is an example of the pull-based model.

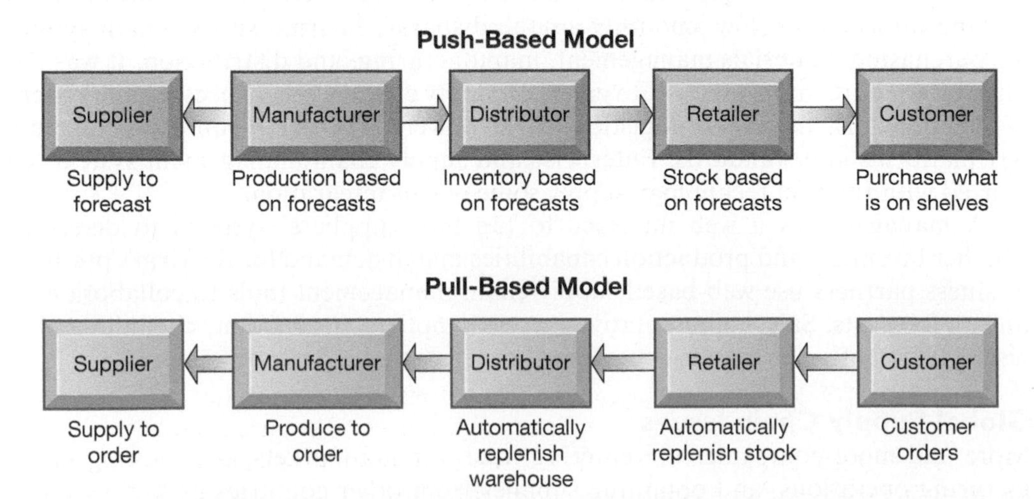


**Figure 9.4**
Push- Versus Pull-Based Supply Chain Models
*The difference between push- and pull-based models is summarized by the slogan "Make what we sell, not sell what we make."*

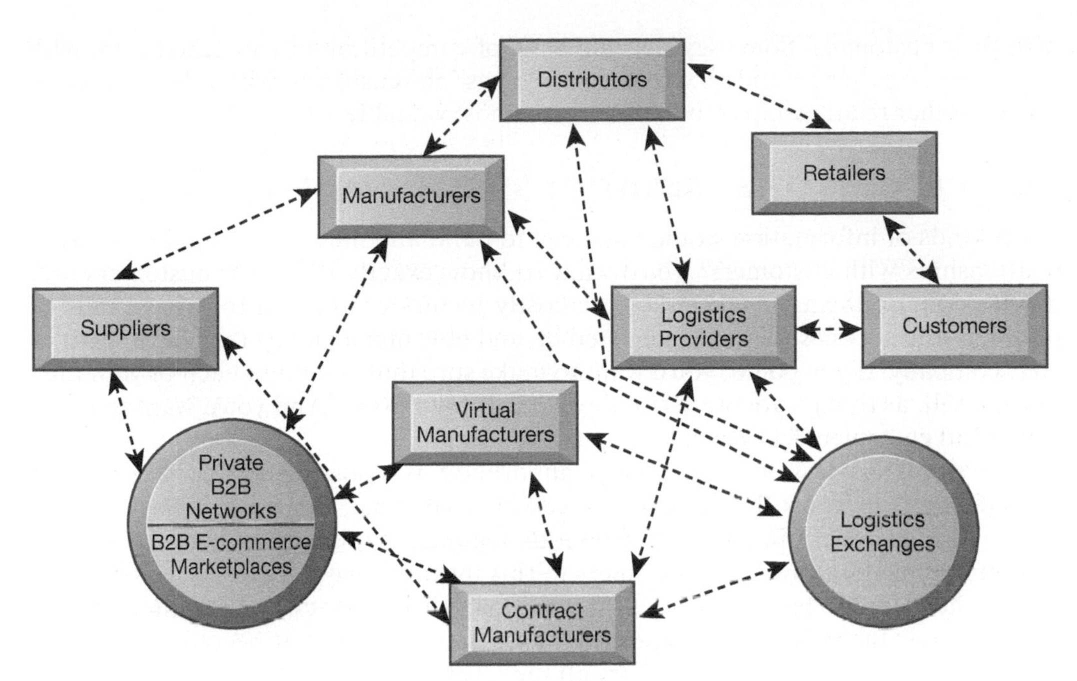


**Figure 9.5**
The Internet-Driven Supply Chain

*The Internet-driven supply chain operates like a digital logistics nervous system. It provides multidirectional communication among firms, networks of firms, and e-marketplaces so that entire networks of supply chain partners can immediately adjust inventories, orders, and capacities.*

The Internet and Internet technology make it possible to move from sequential supply chains, where information and materials flow sequentially from company to company, to concurrent supply chains, where information flows in many directions simultaneously among members of a supply chain network. Complex supply networks of manufacturers, logistics suppliers, outsourced manufacturers, retailers, and distributors can adjust immediately to changes in schedules or orders. Examine Figure 9.5, which illustrates an Internet-driven supply chain that operates like a digital logistics nervous system. The system provides for multidimensional communication among supply chain partners and enables them to make real-time adjustments to production, inventory, and orders.

## BUSINESS VALUE OF SUPPLY CHAIN MANAGEMENT SYSTEMS

You have just seen how supply chain management systems enable firms to streamline both their internal and their external supply chain processes and provide management with more accurate information about what to produce, store, and move. By implementing a networked and integrated supply chain management system, companies match supply to demand, reduce inventory levels, improve delivery service, speed product time to market, and use assets more effectively.

Total supply chain costs represent the majority of operating expenses for many businesses and in some industries approach 75 percent of the total operating budget. Reducing supply chain costs has a major impact on firm profitability.

In addition to reducing costs, supply chain management systems help increase sales. If a product is not available when a customer wants it, the customer will often try to purchase it from someone else. More precise control of the supply chain enhances a firm's ability to have the right product available for customer purchases at the right time.

# 9-3 Explain how customer relationship management systems help firms achieve customer intimacy.

You've probably heard phrases such as "the customer is always right" and "the customer comes first." Today these words ring truer than ever. Because competitive advantage based on an innovative new product or service is often very short lived, companies are realizing that their most enduring competitive strength may be their relationships with their customers. Some say that the basis of competition has switched from who sells the most products and services to who "owns" the customer, with others stressing that customer relationships represent a firm's most valuable assets.

## WHAT IS CUSTOMER RELATIONSHIP MANAGEMENT?

What kinds of information would you need to build and nurture strong, long-lasting relationships with customers? You'd want to know exactly who your customers are, how to contact them, whether they are costly to service and sell to, what kinds of products and services they are interested in, and how much money they have spent at your company. If you could, you'd want to make sure that you knew each of your customers well, as though you were running a small-town store. And you'd want to make your good customers feel special.

In a small business operating in a neighborhood, it is possible for business owners and managers to know their customers well on a personal, face-to-face basis, but in a large business operating on a metropolitan, regional, national, or even global basis, it is impossible to know your customers in this intimate way. In these kinds of businesses, there are too many customers and too many ways that customers interact with the firm (over the web, the phone, email, blogs, and in person). It becomes especially difficult to integrate information from all these sources and deal with the large number of customers.

A large business's processes for sales, service, and marketing tend to be highly compartmentalized, and these departments may not share much essential customer information. Some information on a specific customer might be stored and organized in terms of that person's account with the company. Other pieces of information about the same customer might be organized by products that the customer purchased. In this traditional business environment, there is no convenient way to consolidate all this information to provide a unified view of a customer across the company.

This is where customer relationship management (CRM) systems help. CRM systems, which we introduced in Chapter 2, capture and integrate customer data from all over the organization, consolidate the data, analyze the data, and then distribute the results to various systems and customer touch points across the enterprise. A **touch point** (also known as a contact point) is a method of interaction with the customer, such as through telephone, email, customer service desk, conventional mail, Facebook, Twitter, website, mobile device, or retail store. Well-designed CRM systems provide a single enterprise view of customers that is useful for improving both sales and customer service. Examine Figure 9.6, which illustrates the various types of customer data that CRM systems collect: sales data such as telephone, web, retail store, and field sales; service data such as call center, web self-service, mobile, field service, and social network data; and marketing data such as data generated from campaigns, content marketing, and data analysis.

Good CRM systems provide data and analytical tools for answering questions such as these: What is the value of a particular customer to the firm over their lifetime? Who are our most loyal customers? Who are our most profitable customers? What do these profitable customers want to buy? Firms use the answers to these questions to acquire new customers, provide better service and support to existing customers, customize their offerings more precisely to customer preferences, and provide ongoing value to retain profitable customers.

## CUSTOMER RELATIONSHIP MANAGEMENT SOFTWARE

Commercial CRM software packages range from niche tools that perform limited functions, such as personalizing websites for specific customers, to large-scale enterprise applications that capture myriad interactions with customers, analyze the interactions using sophisticated reporting tools, and link to other major enterprise applications, such as supply chain management and enterprise systems. Major CRM

**Figure 9.6**
Customer Relationship Management (CRM)
*CRM systems examine customers from a multifaceted perspective. These systems use a set of integrated applications to address all aspects of the customer relationship, including customer service, sales, and marketing.*

software vendors include Oracle, SAP, Salesforce.com, and Microsoft Dynamics CRM. CRM systems typically provide software and online tools for sales, customer service, and marketing. The more comprehensive CRM packages contain modules for partner relationship management and employee relationship management. We briefly describe some of these capabilities.

## Sales Force Automation

**Sales force automation (SFA)** modules in CRM systems help sales staff increase productivity by focusing their sales efforts on the most profitable customers, those who are good candidates for sales and services. SFA modules provide sales prospect and contact information, product information, product configuration capabilities, and sales quote generation capabilities. Such software can assemble information about a particular customer's past purchases to help the salesperson make personalized recommendations. SFA modules enable sales, marketing, and shipping departments to share customer and prospect information easily. SFA increases each salesperson's efficiency by reducing the cost per sale as well as the cost of acquiring new customers and retaining old ones. SFA modules also provide capabilities for sales forecasting, territory management, and team selling.

## Customer Service

Customer service modules in CRM systems provide information and tools to increase the efficiency of call centers, help desks, and customer support staff. They also have capabilities for assigning and managing customer service requests.

One such capability is an appointment or advice telephone line. When a customer calls a standard phone number, the system routes the call to the correct serviceperson, who inputs information about that customer into the system only once. When the customer's data are in the system, any service representative can handle the customer relationship. Improved access to consistent and accurate customer information helps call centers handle more calls per day and decrease the duration of each call. Thus, call centers and customer service groups achieve greater productivity, reduced transaction time, and higher quality of service at lower cost. Customers are happier because they spend less time on the phone restating their problem to customer service representatives.

CRM systems may also include web-based self-service capabilities. The company website can be set up to provide inquiring customers personalized support information as well as the option to contact customer service staff by phone or online chat for additional assistance.

## Marketing

CRM systems support marketing campaigns by providing capabilities that enable companies to capture prospect and customer data, provide product and service information, qualify leads for targeted marketing, and schedule and track direct-marketing mailings or email. Examine Figure 9.7, which illustrates a sample produced by a CRM system analyzing the results of a promotional campaign. The analysis shows that the most productive channel was telephone, followed by direct mail, email, and the web. Social media was the least effective channel. Marketing modules also include tools for analyzing marketing and customer data, identifying profitable and unprofitable customers, designing products and services to satisfy specific customer needs and interests, and identifying opportunities for cross-selling.

**Cross-selling** is the marketing of complementary products to customers. (For example, in financial services, a customer with a checking account might be sold a money market account or a home improvement loan.) CRM tools also help firms manage and execute marketing campaigns at all stages, from planning to determining the rate of success for each campaign.

Examine Figure 9.8, which illustrates the most important capabilities of sales, marketing, and service processes found in major CRM software products. Like enterprise software, CRM software is business-process driven, incorporating hundreds of business processes thought to represent best practices in each of these areas.

To achieve maximum benefit, companies need to revise and model their business processes to conform to the best-practice business processes in the CRM software. Examine Figure 9.9, which illustrates how a best practice for increasing customer loyalty through customer service might be modeled by CRM software. Directly servicing customers provides firms with opportunities to increase customer retention by singling out profitable, long-term customers for preferential treatment. CRM software can assign each customer a score based on that person's value and loyalty to the company and provide that information to help call centers route customer service requests to agents who can best handle the customers' needs. The system can automatically provide the service agent with detailed profiles of customers that include their scores for value and loyalty. The service agent can then use this information to present special offers or additional services to the customers to encourage the customers to keep transacting business with the company.

## Partner Relationship Management and Employee Relationship Management

**Partner relationship management (PRM)** modules use many of the same data, tools, and systems that a CRM system uses to enhance collaboration between a company and its selling partners. If a company does not sell directly to customers but rather works through distributors or retailers, PRM helps these channels sell to customers

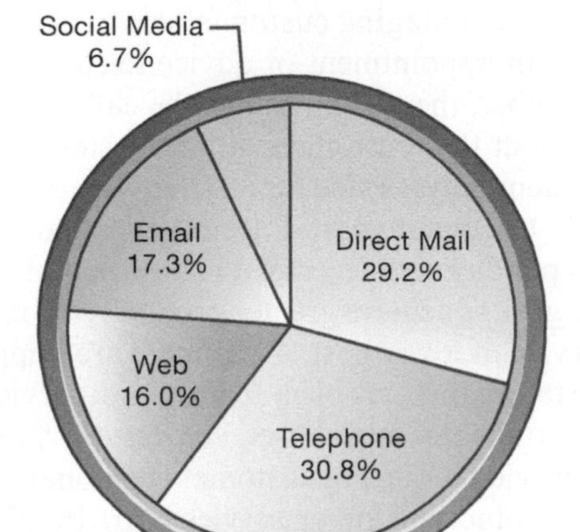


**Figure 9.7**
How CRM Systems Support Marketing
*Customer relationship management software provides a single point for users to manage and evaluate marketing campaigns across multiple channels, including email, direct mail, telephone, the web, and social media.*

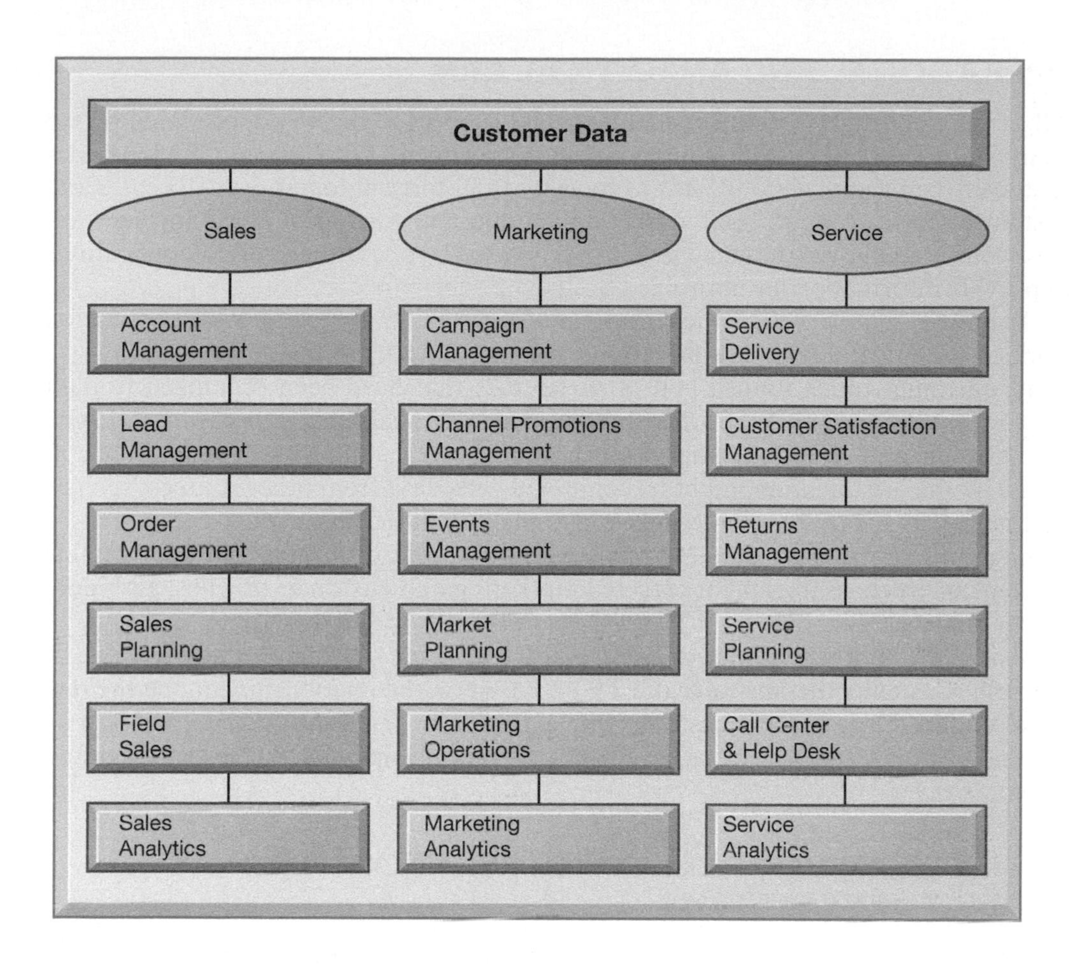


**Figure 9.8**
CRM Software Capabilities
*The major CRM software products support business processes in sales, service, and marketing by integrating customer information from many sources. Included is support for both the operational and the analytical aspects of CRM.*

directly. It provides a company and its selling partners with the ability to trade information and distribute leads and data about customers, integrating lead generation, pricing, promotions, order configurations, and availability. It also provides a firm with tools to assess its partners' performances so that it can make sure its best partners receive the support they need to close more business deals.

**Employee relationship management (ERM)** modules deal with employee issues that are closely related to CRM, such as setting objectives, employee performance management, performance-based compensation, and employee training.

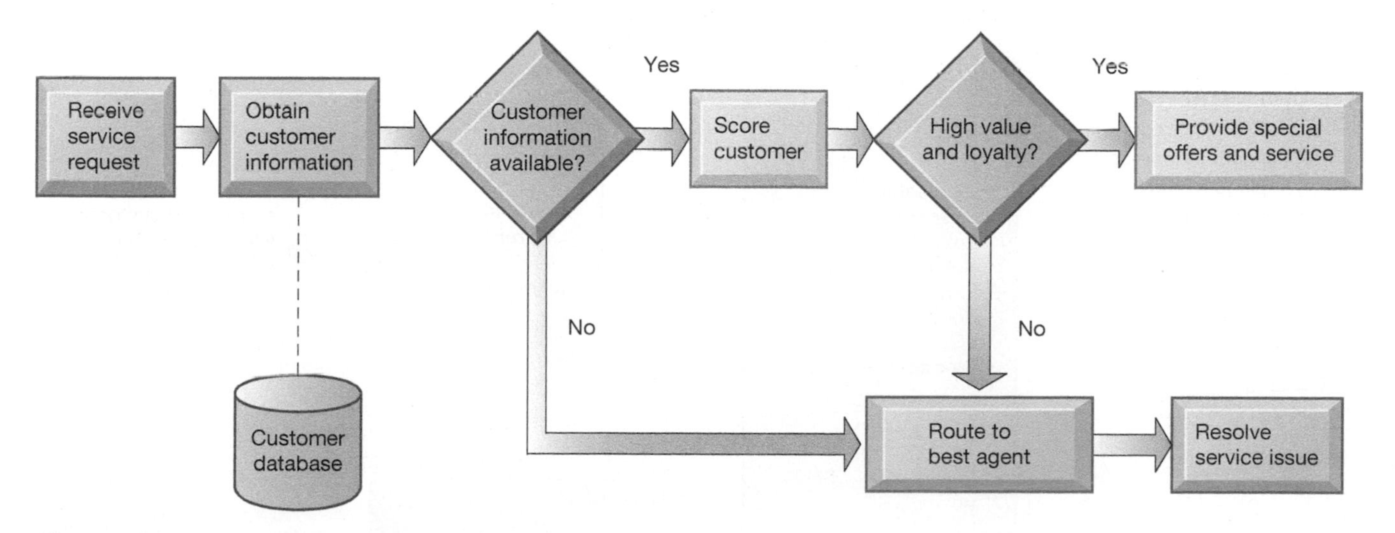


**Figure 9.9**
Customer Loyalty Management Process Map
*This process map shows how a best practice for promoting customer loyalty through customer service would be modeled by customer relationship management software. The CRM software helps firms identify high-value customers for preferential treatment.*

## OPERATIONAL AND ANALYTICAL CRM

All of the applications we have just described support either the operational or the analytical aspects of customer relationship management. **Operational CRM** includes customer-facing applications, such as tools for sales force automation, call center and customer service support, and marketing automation. **Analytical CRM** includes applications that analyze the customer data generated by operational CRM applications to provide information for improving business performance.

Examine Figure 9.10, which illustrates how an analytical CRM application works. Analytical CRMs are based on customer data from operational CRM systems, customer touch points, and other sources that have been organized in data warehouses or analytic platforms for use in online analytical processing (OLAP), data mining, and other data analysis techniques (see Chapter 6). Customer data collected by the organization might be combined with data from other sources, such as customer lists for direct-marketing campaigns purchased from other companies or demographic data. Such data are analyzed to pinpoint profitable and unprofitable customers, create segments for targeted marketing and customer profiles, and identify buying patterns.

Another important output of analytical CRM is the customer's lifetime value to the firm. **Customer lifetime value (CLTV)** is based on the relationship among the revenue produced by a specific customer, the expenses incurred in acquiring and servicing that customer, and the expected life of the relationship between the customer and the company.

## BUSINESS VALUE OF CUSTOMER RELATIONSHIP MANAGEMENT SYSTEMS

Companies with effective CRM systems realize many benefits, including increased customer satisfaction, reduced direct-marketing costs, more effective marketing, and lower costs for customer acquisition and retention. Information from CRM systems increases sales revenue by identifying the most profitable customers and segments for focused marketing and cross-selling (see the Spotlight on Organizations case).

Customer churn is reduced as sales, service, and marketing respond better to customer needs. The **churn rate** measures the number of customers who stop using or purchasing products or services from a company. It is an important indicator of the growth or decline of a firm's customer base.

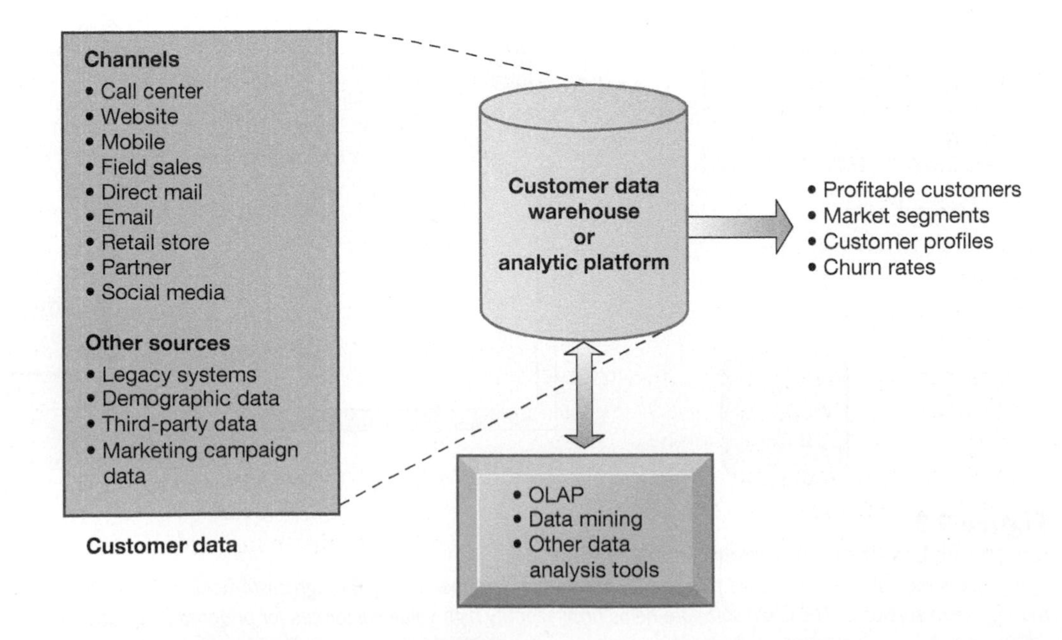


**Figure 9.10**
Analytical CRM
*Analytical CRM uses a customer data warehouse or analytic platform and tools to analyze customer data collected from the firm's customer touch points and from other sources.*

## SPOTLIGHT ON: ORGANIZATIONS MillerKnoll Uses Salesforce.com to Transform Its Business Strategy

MillerKnoll, formerly known as Herman Miller, Inc., is a Michigan-headquartered company that produces innovative office furniture, equipment, and home furnishings. MillerKnoll has subsidiaries and corporate offices, independent dealers, retailers, and licensees in more than 40 countries in North America, Asia Pacific, Europe, the Middle East, and Latin America. Its products and services are sold through retail stores, e-commerce websites, catalogs, and a global dealer network of independent businesses. In fiscal 2021, the company generated $2.5 billion in revenue and employed 7,600 people worldwide.

When the Covid-19 pandemic forced people all over the world to work at home, MillerKnoll's furniture was in high demand because of its ability to blend design, functionality, and comfort. E-commerce sales skyrocketed, inspiring the company to transform its business strategy and online store. The strategy called for providing a frictionless omni-channel experience that was more convenient and personalized for customers. MillerKnoll wanted to be able to engage with customers from the first moment of design inspiration; operate a user-friendly, leading-edge digital storefront; and provide exceptional customer service during and after the purchase process.

To serve customers better and manage all of its relationships with them, MillerKnoll turned to Salesforce.com, which features cloud-based tools for customer relationship management (CRM) and application development. Salesforce.com helps MillerKnoll identify key customer segments, develop closer ties to those customers, and design differentiated experiences tailored to each customer's needs. Salesforce CRM tools make it possible for MillerKnoll to have a single, consistent view of each customer across all the various channels through which that person interacts with the company.

MillerKnoll uses Salesforce Marketing Cloud to develop complex customer journeys (customer experiences with a brand across all touch points) and sophisticated email campaigns. Salesforce Marketing Cloud is a CRM platform that allows marketers to create and manage marketing relationships with customers and campaigns. The Marketing Cloud incorporates integrated solutions for customer journey management, email, mobile, social media, web personalization, advertising, content creation, content management, and data analysis. Every imaginable customer interaction and engagement is covered. The software includes predictive analytics to help make decisions such as, for example, what channel would be preferable for a given message. A component called Journey Builder helps marketers tailor campaigns to customers' behaviors and needs, demographics, and communication channel preferences.

Each of MillerKnoll's brands has a different customer base, and these bases largely do not overlap. Marketing Cloud helps the company make sure that it is treating each of these brands and customers differently, understands preferences across different channels, and uses that data to personalize customer interactions. Previously, MillerKnoll merchandising operations were manual and required too much time to put products in the right place and build interactions that had to change frequently.

The Marketing Cloud is connected to Salesforce.com's Sales Cloud and Service Cloud to provide a unified experience and prevent customers from being contacted separately by representatives from sales, marketing, and service groups. Service Cloud is a platform for customer service and support. Companies using Service Cloud can automate service processes, streamline workflows, and locate key articles, topics, and experts with information to help the company's 1,100 customer service agents. Service Cloud can "listen" and respond to customers across a variety of social platforms and automatically route cases to the appropriate agent. Service for Apps makes it possible to embed customer support software into mobile applications, including features for live agent video chat, screen sharing, and on-screen guided assistance. Service Cloud makes it possible to deliver service that is more personalized and convenient in whatever form each customer prefers—email, web, social media, or telephone—all from a single application. Service Cloud helped MillerKnoll handle the spike in customer service requests during the Covid-19 pandemic.

Sales Cloud is a cloud-based application designed to help salespeople sell more quickly by centralizing customer information, logging their interactions with your company, and automating many of the tasks salespeople do every day. This means they'll spend less time on administration and more time closing deals. For sales managers, Sales Cloud gives real-time visibility into their team's activities, so forecasting sales with confidence is not as difficult. Sales Cloud is easy to use and customizable to the way salespeople work. MillerKnoll uses Sales Cloud to manage sales leads and opportunities across brands, monitor team performance, track sales insights, and automate tasks through a single dashboard.

MillerKnoll relaunched its websites on Salesforce Commerce Cloud, which has resulted in increased site speed and a richer shopping process. Shoppers can browse and purchase products more easily, with more tailored product recommendations, a streamlined checkout process, and tools for live video chat with sales associates.

Having unified customer data with a powerful platform to manage customer relationships helped MillerKnoll retain and use vital customer information as the company continued to grow globally. With sales, marketing, and service teams united around the customer, MillerKnoll was able to meet customer expectations for personalized service as it expanded. According to Ben Groom, MillerKnoll's chief digital officer, Salesforce helped the company become more integrated across all of its brands, and its online shopping experience mirrors the same quality as its products

Sources: "Herman Miller Designs Experiences to Help People Do Their Best Work," Salesforce.com, accessed January 2, 2022; Millerknoll.com, accessed January 2, 2022; Mark Bowen, "Michigan-Based MillerKnoll Increased E-commerce Sales by over 300% after Deploying Solutions from Salesforce," *Intelligent CIO*, December 13, 2021.

## CASE STUDY QUESTIONS

**1.** What is MillerKnoll's business strategy? What is the role of customer relationship management in that strategy?

**2.** How did using Salesforce.com change the way MillerKnoll ran its business?

**3.** Give examples of two business decisions at MillerKnoll that were improved by using Salesforce.com.

# 9-4 Describe the challenges that enterprise applications pose and how enterprise applications are taking advantage of new technologies.

Many firms have implemented enterprise systems and systems for supply chain and customer relationship management because these systems are such powerful instruments for achieving operational excellence and enhancing decision making. But precisely because they are so powerful in changing the way the organization works, they are challenging to implement. Let's briefly examine some of these challenges as well as new ways of obtaining value from these systems.

## ENTERPRISE APPLICATION CHALLENGES

Promises of dramatic reductions in inventory costs, order-to-delivery time, more efficient customer response, and higher product and customer profitability make enterprise systems and systems for SCM and CRM very alluring. But to obtain this value, you must clearly understand how your business has to change to use these systems effectively.

Enterprise applications involve complex pieces of software that are expensive to purchase and implement. According to a 2022 survey of 140 ERP users conducted by Panorama Consulting Group, 41 percent of ERP projects that completed implementation experienced cost overruns, and 9.3 percent of the projects had significant overruns. Thirty-six percent of the projects were completed later than anticipated. Of those that were over budget, the most common reason was organizational issues (Panorama Consulting Group, 2022). Changes in project scope and additional customization work add to implementation delays and costs.

Enterprise applications require not only deep-seated technological changes but also fundamental changes in the way the business operates. Companies must make sweeping changes to their business processes to work with the software. Employees must accept new job functions and responsibilities. They must learn how to perform a new set of work activities and understand how the information they enter into the system can affect other parts of the company. This requires new organizational learning and should also be factored into ERP implementation costs.

SCM systems require multiple organizations to share information and business processes. Each participant in the system may have to change some of its processes and the way it uses information to create a system that best serves the supply chain as a whole.

Some firms experienced enormous operating problems and losses when they first implemented enterprise applications because they didn't understand how much organizational change was required. For example, Lidl, a major German grocery chain, started working with SAP in 2011 to replace an antiquated in-house inventory system. In 2018, after spending nearly €500 million (approximately US $500 million), Lidl canceled the project. The problem was traced to Lidl's record-keeping process. Lidl had based its inventory systems on the price it pays for goods, whereas most companies base their inventory systems on the retail price for which they sell the goods. Lidl didn't want to change its way of doing business, which meant that the SAP implementation had to be customized, thus triggering a cascade of implementation problems. Woolworth's Australia also encountered data-related problems when it transitioned from an antiquated home-grown ERP system to SAP. Weekly profit-and-loss reports tailored for individual stores couldn't be generated for nearly 18 months. The company then had to change its data collection procedures but failed to understand its own processes or to document these business processes properly (Fruhlinger et al., 2020).

Enterprise applications also introduce switching costs. When you adopt an enterprise application from a single vendor, such as SAP, Oracle, or others, it is very costly to switch vendors, and your firm becomes dependent on the vendor to upgrade its product and maintain your installation.

Enterprise applications are based on organization-wide definitions of data. You'll need to understand exactly how your business uses its data and how the data would be organized in a CRM, SCM, or ERP system. CRM systems typically require some data-cleansing work (see Chapter 6).

Enterprise software vendors are addressing these problems by offering pared-down versions of their software, fast-start programs for small and medium-sized businesses, and best-practice guidelines for larger companies. Companies are also achieving more flexibility by using cloud applications for functions not addressed by the basic enterprise software so that companies are not constrained by a single, do-it-all type of system.

Companies adopting enterprise applications can also save time and money by keeping customizations to a minimum. For example, Kennametal, a $2 billion metal-cutting tools company in Pennsylvania, had spent $10 million over 13 years maintaining an ERP system with more than 6,400 customizations. The company replaced this ERP system with a plain-vanilla, uncustomized version of SAP enterprise software and changed its business processes to conform to the software. Office Depot avoided customization when it moved from in-house systems to the Oracle ERP Cloud. The retailer is using best practices embedded in Oracle's Supply Chain Management Cloud and in its cloud-based Human Capital Management (HCM) and Enterprise Performance Management (EPM) systems. By not customizing its Oracle ERP applications, Office Depot simplified its information systems and reduced the cost of maintaining and managing those systems (Thibodeau, 2018).

## NEXT-GENERATION ENTERPRISE APPLICATIONS

Today, enterprise application vendors are delivering more value by becoming more flexible, user-friendly, web-enabled, mobile, and capable of integration with other systems. Stand-alone enterprise systems, CRM systems, and SCM systems are becoming a thing of the past. The major enterprise software vendors have created what they call *enterprise solutions, enterprise suites,* or *e-business suites* to make their CRM, SCM, and ERP systems work closely with each other and link to systems of customers and suppliers.

Next-generation enterprise applications also include cloud solutions as well as more functionality that is available on mobile platforms. Large enterprise software vendors such as SAP, Oracle, and Microsoft now feature cloud versions of their flagship enterprise applications and also cloud-based products for small and medium-sized businesses. SAP, for example, offers SAP S/4HANA Cloud for large companies and SAP Business ByDesign and SAP Business One enterprise software for medium-sized and small businesses. Microsoft offers the Dynamics 365 cloud version of its enterprise software. Cloud-based enterprise systems are also offered by smaller vendors such as NetSuite. The Spotlight on Technology case describes how Versum Materials was able to transform its business using a new, cloud-based ERP system.

The undisputed global market leader in cloud-based CRM systems is Salesforce.com, which we described in Chapter 5. Salesforce.com is widely used by small, medium-sized, and large enterprises. As cloud-based products mature, more companies, including very large *Fortune* 500 firms, are choosing to run all or parts of their enterprise applications in the cloud.

## Social CRM

CRM software vendors are enhancing their products to take advantage of social network technologies. These social enhancements help firms identify new ideas more rapidly, improve team productivity, and deepen interactions with customers (see Chapter 10). Using **social CRM** tools, businesses can better engage with their customers by, for example, analyzing their sentiments about the businesses' products and services.

Social CRM tools enable a business to connect customer conversations and relationships expressed in social networks to CRM processes. The leading CRM vendors now offer such tools to link data from social networks to their CRM software. SAP, Salesforce.com, and Oracle CRM products feature technology to monitor, track, and analyze social media activity on Facebook, LinkedIn, Twitter, YouTube, and other social networks. Business intelligence and analytics software vendors such as SAS also have capabilities for social media analytics (with several measures of customer engagement across a variety of social networks) along with campaign management tools for testing and optimizing both social and traditional web-based campaigns.

Salesforce.com connected its system for tracking leads in the sales process with social-listening and social-media marketing tools, enabling users to tailor their social-marketing dollars to core customers and observe the resulting comments. If an ad agency wants to run a targeted Facebook or Twitter ad, these capabilities make it possible to aim the ad specifically at people in the client's lead pipeline who are already being tracked in the CRM system. Users are able to view tweets as they take place in real time and perhaps uncover new leads. Clients can also manage multiple campaigns and compare them all to figure out which ones generate the highest click-through rates and cost per click.

## Business Intelligence in Enterprise Applications

Enterprise application vendors have added business intelligence features to help managers obtain more meaningful information from the massive amounts of data these systems generate, including data from the Internet of Things (IoT). SAP now makes it possible for its enterprise applications to use HANA in-memory computing technology so that they are capable of much more rapid and complex data analysis. Included are tools for flexible reporting; ad hoc analysis; interactive dashboards; what-if scenario analysis; data visualization; and AI machine learning to analyze very large sets of data, make connections, make predictions, and provide more accurate recommendations for customer purchases and operations optimization (see Chapter 11). AI-enhanced supply chain management software is helping companies

## Versum's ERP Transformation

Versum Materials is an electronic materials company supplying premier specialty gases and products to the semiconductor industry's chip manufacturing process. It merged with Merck in late 2019, is headquartered in Billerica, Massachusetts, and operates a network of offices, manufacturing sites, and research and development (R&D) centers in Asia, Europe, and North America. In October 2016 Versum was divested from its parent company, Air Products, and transitioned from a $10 billion industrial gas and chemical company that predominantly sold large volumes of inexpensive products, such as cylinders of pure oxygen, nitrogen, hydrogen, and helium, to an independent $1 billion company selling highly regulated gases and chemicals for semiconductor manufacturing.

After separating from Air Products, Versum continued to rely on its former parent company's information systems under a transition service agreement (TSA). This was only an interim solution, however, since Versum really needed its own IT infrastructure and enterprise system to support its new business model and plans for growth. The company was given a very ambitious deadline of 18–24 months to get off the TSA and have its own systems in place.

When determining what ERP system to put in place, Versum Materials had a few options. The first was to copy Air Products' existing, 20+-year-old SAP ERP system. The second was to bring the same customizations and functionality to a new SAP ERP system, and the third was to start completely from scratch with all-new servers, software, and processes. Versum selected the third option because that option provided an opportunity to wipe the slate clean and implement processes that worked for its specific business.

The legacy ERP system had been heavily customized with various upgrades over the years, and most of those customizations were for the parent company's industrial gas division, which had very different information requirements than the new Versum. Versum Materials management didn't want to continue using that system when it was not really suited for the business.

To find the new, replacement system, Versum Materials looked at different ERP providers and products. Management ultimately elected to stick with SAP software because it had all the capabilities for satisfying the industry's legal regulatory requirements and because it appeared to be the quickest and easiest to implement in a short timeframe. If Versum had tried to put in a totally new, non-SAP ERP system, employees and IT staff would have had to learn an entirely new, unfamiliar system, which could not be easily accomplished in short order. The training and the issues with people's understanding and knowledge for a new software package would have dramatically impacted the business.

Versum Materials instead decided to make SAP S/4HANA the foundation of its new system. S/4HANA is an integrated ERP system that runs on SAP's in-memory database, SAP HANA (see Chapter 6). Instead of running the new SAP ERP system in a data center of its own, Versum Materials opted to implement the system in a private cloud, with a hosting provider that would manage the cloud SAP S/4HANA platform running in its own physical data center. This saved the company from making a large up-front capital investment in the new system and the responsibility for managing the system.

Versum implemented its SAP S/4HANA system by taking advantage of SAP and industry best practices embedded in the software and creating consistent processes across its business units. The typical SAP implementation is rolled out in phases, often country by country for global enterprises. Companies with global supply chains where materials or products are manufactured or shipped from one country to another must convert their supply chain data from the old system to the new system quickly and cleanly to keep products flowing. Versum Materials has a global supply chain spanning more than 10 countries, so a phased rollout would have required putting temporary interfaces in place to feed data from the previous system to the new system to enable the new system to go live country by country. Versum's transition services agreement (TSA) did not allow enough time for that option, so the new ERP system had to go live everywhere all at once.

A "big-bang" system rollout also was advantageous because it forced Versum to keep the system simple, thus having fewer problems to address. Dragging the program out and rolling out Taiwan one month and Korea the next, for example, would have taken much longer and been much more complicated, notes Sally Giamalis, SAP Program Director of Versum Materials. Versum had anticipated that the new system would initially have problems taking orders, creating purchase orders, or shipping products to customers. However, by working with Accenture consultants, Versum found that the implementation went smoothly and with minimal disruption to its business.

When Versum went live with SAP S/4HANA, it also rolled out other SAP solutions, including modules for SAP Business Warehouse optimized for SAP HANA, SAP Business Planning and Consolidation, SAP Global Trade Services, the SAP Advanced Planning and Optimization component of SAP Supply Chain Management, SAP Solution Manager, SAP Process Integration, SAP Data Services, and SAP governance, risk, and compliance (GRC), as well as some non-SAP applications. The entire implementation took just 15 months. During the course of implementing the new system, Versum did, however, run up against some challenges, especially with converting the data from the old system. For example, Versum had to go through three cycles of data cleansing to obtain sufficiently accurate invoicing, address, and contact information. In addition, the data structure in the old SAP ERP system was not the same as that for the new SAP S/4HANA system, which had a new hierarchical Business Partner structure. In the old SAP ERP system, the Customer structure (ship to, bill to, contact, and vendor) consisted of many different data objects. In the new SAP S/4HAHA structure, all those different data objects are part of a new object called Business Partner. In the old structure, a customer and the vendor could be the same thing even though they were two different entities. In the new structure, if the company sells to a customer but also buys from that customer, that customer has a single Business Partner record.

The new SAP S/4HANA system makes it easier for Versum to optimize operations, manage costs, and take advantage of real-time analytics. The company now has a solid foundation for operating as a standalone company providing specialty materials.

Sources: Accenture.com, accessed January 3, 2022; Versummaterials.com, accessed January 3, 2022; "Business Transformation through SAP S/4HANA," Accenture.com, accessed January 2, 2022; Lauren Bonneau, "How the Guiding Principles of Standardize, Harmonize, Simplify, and Scalable Led to a 'Ghostly Quiet' Go-Live," *SAP Insider*, February 26, 2019.

## CASE STUDY QUESTIONS

1. Define the problem in this case study. What people, organization, and technology factors contributed to the problem?
2. Was the SAP S/4HANA SaaS solution a good one for Versum? Explain your answer.
3. What challenges did Versum encounter when implementing the new system?
4. How did the new system change the way Versum ran its business?

see patterns and determine inefficiencies. Salesforce.com now offers an Einstein AI tool in its core cloud products for tasks such as lead and opportunity scoring and optimizing product recommendations. For example, with Einstein Recommendations, furniture seller Room & Board's automated customer recommendations boosted web sales by 40 percent and in-store sales by 60 percent (Salesforce, 2022).

## 9-5 Understand how MIS can help your career.

Here is how this chapter and this book can help you find a job as a supply chain analyst.

### THE COMPANY

LightningFast Supply Chain Solutions, headquartered in Toledo, Ohio, has an open supply chain analyst position for a recent college graduate. With operations in North America, South America, Europe, and Asia, LightningFast provides supply chain management and logistics services to leading companies around the world. The company delivers value through its design, planning, and execution in transportation, warehousing, and freight management.

## POSITION DESCRIPTION

The company is looking for someone to coordinate supply chain processes along with tracking, auditing, researching, and troubleshooting operational issues. This job requires understanding supply chain distribution, customer analytics, customer relations, process improvement, and supplier/carrier relations. Job responsibilities include:

- Following daily parts control, including track and trace, production schedule analysis, and exception management
- Providing parts follow-up to ensure optimal material flow
- Assisting in the development and presentation of customer analytics, key trends, and solution proposals during internal and external meetings
- Assisting in continuous improvement and cost-saving initiatives
- Working with customers to resolve stock shortages, meet deadlines, and answer questions
- Working with carriers to resolve issues in transit or at a customer's facility
- Analyzing current processes and recommending improvements to reduce inefficiencies

## JOB REQUIREMENTS

- Bachelor's degree preferred
- Fundamental understanding of transportation, supply chains, and logistics
- One year of inventory/transportation logistics experience preferred
- Ability to learn and operate various inventory and management systems
- Proficiency in Excel and Outlook required
- Experience analyzing large sets of data preferred
- High-energy, self-motivated team player with the ability to multitask
- Strong problem-solving and critical-thinking skills and attention to detail
- Willingness and ability to work remotely

## INTERVIEW QUESTIONS

1. Describe a project you worked on that involved analyzing large sets of data. What role did you play in this project? What did you do to help your team achieve its goal?
2. What do you know about process improvement and improving customer inventory and material flow processes?
3. What are you able to do with Excel? Have you ever used it on the job? If so, how did you use it? What kind of experience do you have with Outlook?
4. Have you ever pulled together data on customer analytics, key trends, and solution proposals for internal and external meetings? If so, how did you use the data to assist the company in continuous improvement and cost savings initiatives?

## AUTHOR TIPS

1. Do some research on the company, its industry, and the kinds of challenges it faces. Who are its main competitors? Visit the company's website, and also review this chapter's discussion of supply chain management.
2. Look on the company's social media to see how it is using app-based technology to improve customer relationship management.
3. Look at the company's social media for research reports it prepared that forecast supply chain trends for next year.
4. What are the key trends on the company's LinkedIn posts over the past six months?
5. View YouTube videos created by major IT consulting firms that discuss the latest trends in global supply chain technology.

## Review Summary

**9-1 Explain how enterprise systems help businesses achieve operational excellence.** Enterprise software is based on a suite of integrated software modules and a common central database. The database collects data from and feeds the data into numerous applications that can support nearly all of an organization's internal business processes. When one process enters new information, the information is made available immediately to the other business processes.

Enterprise systems support organizational centralization by enforcing uniform data standards and business processes throughout the company and a single, unified technology platform. The firmwide data that enterprise systems generate help managers evaluate organizational performance.

**9-2 Describe how supply chain management systems coordinate planning, production, and logistics with suppliers.** Supply chain management (SCM) systems automate the flow of information among members of the supply chain so that they can use the information to make better decisions about when and how much to purchase, produce, or ship. More accurate information from SCM systems reduces uncertainty and the impact of the bullwhip effect.

SCM software includes software for supply chain planning and for supply chain execution. Internet technology facilitates the management of global supply chains by providing the connectivity for organizations in different countries to share supply chain information. Improved communication among supply chain members also facilitates efficient customer response and movement toward a demand-driven model.

**9-3 Explain how customer relationship management systems help firms achieve customer intimacy.** Customer relationship management (CRM) systems integrate and automate customer-facing processes in sales, customer service, and marketing, providing an enterprise-wide view of customers. Companies can use this customer knowledge when they interact with customers to provide them with better service or to sell new products and services. These systems also identify profitable or unprofitable customers or opportunities to reduce the churn rate.

The major customer relationship management software packages provide capabilities for both operational CRM and analytical CRM. They often include modules for managing relationships with selling partners (partner relationship management) and for employee relationship management.

**9-4 Describe the challenges that enterprise applications pose and how enterprise applications are taking advantage of new technologies.** Enterprise applications are difficult to implement. They require extensive organizational change; large, new software investments; and careful assessment of how these systems will enhance organizational performance. Enterprise applications cannot provide value if they are implemented atop flawed processes or if firms do not know how to use these systems to measure performance improvements. Employees require training to prepare for new procedures and roles. Attention to data management is essential.

Enterprise applications are now more flexible, web-enabled and capable of integration with other systems.. They can run in cloud infrastructures or on mobile platforms. CRM software has added social networking capabilities to enhance internal collaboration, deepen interactions with customers, and use data from social networks. Enterprise applications are incorporating business intelligence capabilities to analyze the large quantities of data that enterprise applications generate.

## Key Terms

Analytical CRM, 332
Bullwhip effect, 324
Churn rate, 332
Cross-selling, 330
Customer lifetime value (CLTV), 332
Demand planning, 325
Employee relationship management (ERM), 331
Enterprise software, 320
Just-in-time strategy, 323
Operational CRM, 332
Partner relationship management (PRM), 330
Pull-based model, 326
Push-based model, 326
Sales force automation (SFA), 329
Social CRM, 336
Supply chain, 322
Supply chain execution systems, 325
Supply chain planning systems, 325
Touch point, 328

## Review Questions

**9-1 Explain how enterprise systems help businesses achieve operational excellence.**
- Define an enterprise system, and explain how enterprise software works.
- Describe how enterprise systems provide value for a business.

**9-2 Describe how supply chain management systems coordinate planning, production, and logistics with suppliers.**
- Define a supply chain, and identify each of its components.
- Explain how supply chain management systems help reduce the bullwhip effect and how they provide value for a business.
- Define and compare supply chain planning systems and supply chain execution systems.
- Describe the challenges of global supply chains and how Internet technology can help companies manage them better.
- Distinguish between a push-based and a pull-based model of supply chain management, and explain how contemporary supply chain management systems facilitate a pull-based model.

**9-3 Explain how customer relationship management systems help firms achieve customer intimacy.**
- Define customer relationship management, and explain why customer relationships are so important today.
- Describe the tools and capabilities of customer relationship management software for sales, marketing, and customer service.
- Describe how partner relationship management (PRM) and employee relationship management (ERM) are related to customer relationship management (CRM).
- Distinguish between operational and analytical CRM.

**9-4 Describe the challenges that enterprise applications pose and how enterprise applications are taking advantage of new technologies.**
- List and describe the challenges that enterprise applications pose.
- Explain how these challenges can be addressed.
- Describe how enterprise applications are taking advantage of cloud computing and business intelligence.
- Define social CRM, and explain how customer relationship management systems are using social networking.

## Discussion Questions

**9-5** MyLab MIS Supply chain management is less about managing the physical movement of goods and more about managing information. Discuss the implications of this statement.

**9-6** MyLab MIS If a company wants to implement an enterprise application, it had better do its homework. Discuss the implications of this statement.

**9-7** MyLab MIS Which enterprise application should a business install first: ERP, SCM, or CRM? Explain your answer.

**MyLab MIS**
To complete these problems, go to EOC Discussion Questions in **MyLab MIS**.

## Hands-On MIS Projects

The projects in this section give you hands-on experience analyzing business process problems, using spreadsheet software to select suppliers, and evaluating supply chain management business services. Visit MyLab MIS to access this chapter's Hands-On MIS Projects.

### MANAGEMENT DECISION PROBLEM

**9-8** Management at your agricultural chemicals corporation has been dissatisfied with production planning. Production plans are created using best guesses of demand for each product, which are based on how much of each product has been ordered in the past. If a customer places an unexpected order or requests a change to an existing order after it has been placed, there is no way to adjust production plans. The company may have to tell customers it can't fill their orders, or it may run up extra costs maintaining additional inventory to prevent stock-outs.

At the end of each month, orders are totaled and manually keyed into the company's production planning system. Data from the past month's production and inventory systems are manually entered into the firm's order management system. Analysts from the sales department and from the production department analyze the data from their respective systems to determine what the sales targets and production targets should be for the next month. These estimates are usually different. The analysts then get together at a high-level planning meeting to revise the production and sales targets to take into account senior management's goals for market share, revenues, and profits. The outcome of the meeting is a finalized production master schedule.

The entire production planning process takes 17 business days to complete. Nine of these days are required to enter and validate the data. The remaining days are spent developing and reconciling the production and sales targets and finalizing the production master schedule.

- Draw a diagram of the existing production planning process.
- Analyze the problems this process creates for the company.
- Describe how an enterprise system could solve these problems. In what ways could it lower costs? Diagram what the production planning process might look like if the company implemented enterprise software.

### IMPROVING DECISION MAKING: USING SPREADHEET SOFTWARE TO SELECT SUPPLIERS

Software skills: Spreadsheet date functions, data filtering, AVERAGE function
Business skills: Analyzing supplier performance and pricing

**9-9** In this exercise, you'll use spreadsheet software to improve management decisions about selecting suppliers. You will filter transactional data about suppliers based on several criteria to select the best suppliers for your company.

You run a company that manufactures aircraft components. You have many competitors who are trying to offer lower prices and better service to customers, and you are trying to determine whether you can benefit from better supply chain management. In MyLab MIS, you will find a spreadsheet file for this chapter that contains a list of all the items that your firm has ordered from its suppliers during the past five months. The fields in the spreadsheet file include vendor name, vendor identification number, purchaser's order number, item identification number and item description (for each item ordered from the vendor), cost per item, number of units of the item ordered (quantity), total cost of each order, vendor's accounts payable terms, order date, and actual arrival date for each order.

Prepare a recommendation of how you can use the data in this spreadsheet file to improve your decisions about selecting suppliers. Calculate the delivery days (days required to deliver each order). Some criteria to consider for identifying preferred suppliers include the suppliers taking the least amount of time to deliver a product, the suppliers offering the best accounts payable terms, all suppliers producing a specific component such as gaskets and their pricing, and the average number of days for all suppliers to deliver. How can these reports help your company manage its supply chain?

### ACHIEVING OPERATIONAL EXCELLENCE: EVALUATING SUPPLY CHAIN MANAGEMENT SERVICES

Software skills: Web browser and presentation software
Business skills: Evaluating supply chain management services

**9-10** In addition to carrying goods from one place to another, some trucking companies provide supply chain management services and help their customers manage their information. In this project, you'll use the web to research and evaluate two of these business services. Investigate the websites of two companies, UPS and Schneider National, to see how these companies' services can be used for supply chain management. Then prepare a presentation that responds to the following questions:

- What supply chain processes can each of these companies provide for its clients?
- How can customers use the websites of each company to help with supply chain management?
- Compare the supply chain management services these companies provide. Which company would you select to help your firm manage its supply chain? Why?

### COLLABORATION AND TEAMWORK PROJECT

Analyzing Enterprise Application Vendors

**9-11** With a group of three or four other students, use the web to research and evaluate the products of two vendors of enterprise application software. You could compare, for example, the SAP and Oracle enterprise systems, the supply chain management systems from Blue Yonder and SAP, or the customer relationship management systems of Oracle and Salesforce.com. Use what you have learned from these companies' websites to compare the software products you have selected in terms of business functions supported, technology platforms, cost, and ease of use. Which vendor would you select? Why? Would you select the same vendor for a small business (50–300 employees) as for a large one? If possible, use Google Docs and Google Drive or Google Sites to brainstorm, organize, and develop a presentation of your findings for the class.

## BUSINESS PROBLEM-SOLVING CASE

### THE CORONAVIRUS PANDEMIC DISRUPTS SUPPLY CHAINS AROUND THE WORLD

The Covid-19 pandemic has tested supply chains like no other event in recent history. Entire populations were isolating and quarantining, creating spikes in demand for certain products (such as hand sanitizer) and large drops in demand for others. Many businesses were shuttered for months, with small businesses, retail stores, and restaurants especially hard hit. Large drops in demand, shortfalls in cash flow, worldwide port congestion, factory shutdowns, and disruptions to air cargo, trucking, and rail services paralyzed companies all over the world.

Even when most businesses reopened, supply chain shortages continued, especially for computer chips, automobiles. exercise equipment, and breakfast cereal. From the Port of Los Angeles to the Suez Canal, bottlenecks that gummed up the world economy in 2020 worsened, and a supply chain crunch that many expected to be temporary lasted well into 2022. Prices—and inflation—skyrocketed.

Because of the pandemic, customers changed their purchasing habits. Many started spending more on essentials, creating shortages across both e-commerce and brick-and-mortar retail stores. Changes in consumer spending behavior in turn upended predictive models as customers shifted their spending to new stores, channels, and product lines. At the same time, companies that didn't deal in essentials faced weakened demand, and millions of people found themselves out of work.

According to a March 2020, analysis by trading platform Forex.com, nearly 75 percent of all companies had already reported supply chain disruptions, and that number was expected to rise to 80 percent. Manufacturing firms worldwide were especially affected by the shutdown of industrial activity in Wuhan, China, where the pandemic started. These firms had depended on components, materials, and finished goods made in China.

Because of the multitude of supply chain disruptions caused by the coronavirus pandemic, most companies were unable to respond quickly and flexibly, something that can be done only when the entire supply chain is visible. Most companies don't have supply chain visibility. (Supply chain visibility is the ability of parts, components, or products in transit to be tracked from the manufacturer to their final destination.) The majority of enterprises have only 20 percent visibility into their supply chains. Experts believe 70 to 80 percent visibility is required to deal with major supply chain disruptions.

The modern supply chain is incredibly fragile. Companies have created global supply chains based on outsourcing to external suppliers and incredibly thin margins of safety stock. (Safety stock is an additional quantity of an item held by a company in inventory in order to reduce the risk that the item will be out of stock when customers place orders for it.) The prevailing wisdom in supply chain management has embraced "lean" principles that try to optimize costs by minimizing safety stock, using "just-in-time" delivery to keep only 15–30 days of products on hand, and concentrating sourcing in a few countries. For example, more than 80 percent of the manufacturing facilities that produce components for drugs used in the United States are located abroad, mainly in China. Many companies have found it cheaper to manufacture goods in China, and elsewhere in Asia, rather than do so closer to home. Auto parts, fashion, technology, medical gear, and drug components are especially vulnerable to supply chain disruptions in Asia.

To make supply chains more resilient, businesses need to eliminate their dependence on sourcing from a single supplier, region, or country. Large companies can build regional supply chains and diversify the location of their manufacturing plants and their suppliers. They should also consider pulling back from inventory-optimization and safety stock calculations that optimize costs by keeping stock to a minimum and building some level of reserves to absorb shocks, even if doing so increases costs.

Dealmed, the largest independent medical supply distributor in New York, New Jersey, Connecticut, and Pennsylvania, suffered less supply chain disruptions than other medical supply companies by not relying on just-in-time supply. Dealmed instead opted for a slightly bulkier inventory, with more safety stock, and did not rely exclusively on a single region such as China for products. Dealmed sources products in the United States where it can, as well as in Europe and other Asian countries. Sometimes the product is a little more expensive, which is unfortunate because Dealmed is a low-cost leader. But this approach paid off handsomely during the pandemic.

The cost of manufacturing has been one of the key justifications for moving manufacturing offshore. However, the labor cost component of manufacturing has been steadily growing smaller as new automation tools have been developed. Thirty years ago, when labor costs represented 30 to 40 percent of the cost to manufacture goods, U.S. manufacturers were tempted to move production offshore to Chinese factories replete with low-cost laborers assembling products by hand. Today, the trend is toward more automated factories, which lower the labor component and reduce profit-and-loss pressures. US leadership in factory automation will undoubtedly help bring some offshore manufacturing back home.

Switching to more digital tools for supply chain management can also be helpful. A contemporary supply chain management system increases transparency and responsiveness because all the activities in the supply chain are able to interact with one another in near real time. There are new digital applications and platforms that help companies establish interconnected networks of what had been discrete, siloed supply chain processes and manage their supply chains more flexibly. Gartner Inc. predicts that by 2023, at least 50 percent of global companies will be using artificial intelligence (see Chapter 11), advanced analytics, and the Internet of Things (IoT) in supply chain operations. Firms such as Procter & Gamble (P&G) are using artificial intelligence machine-learning algorithms to perform demand planning for products such as Tide detergent multiple times per day. Other companies are implementing IoT technologies such as GPS and radio-frequency identification (RFID—see Chapter 7) devices to identify and track items in stores and warehouses as well as real-time data on variables such as speed of delivery.

A word of caution: Even if a company uses digital supply chain management tools, the tools may need updating and fine-tuning in order to deal with major global shutdowns. The algorithms used by the supply chain management systems of large companies didn't work during the coronavirus pandemic. For example, Walmart, noted for its efficient, state-of-the-art supply chain management systems (see Chapter 3), found that disruptions during the pandemic made these systems unable to accurately predict how many diapers and garden hoses, for example, it needed to keep on store shelves.

Normally Walmart's system is able to accurately analyze inventory levels, historical purchasing trends, and discounts to recommend how much of a product to order. But the worldwide disruption caused by the Covid-19 pandemic caused the software's recommendations to change more frequently.

Most retail companies base their predictions of what customers will want and how much to order on some type of model or algorithm. Their models incorporate some understanding of how shocks like natural disasters disrupt supply chains and impact demand, using historical data to predict future trends. Under normal conditions these algorithms work fairly well. But global pandemics are something new that the models don't know how to take into account. Disasters like floods or hurricanes tend to be regional, but the Covid-19 pandemic disrupted the entire world. Production, transportation, and people's behavior changed dramatically during the pandemic. Because of these massive, worldwide disruptions, the normal data feeding the models, including historical buying patterns, aren't as relevant.

The models in supply chain management software can still be used, but the types of data collected need to be changed. The people who manage supply chains will need to be more active in interpreting the projections rather than assuming the models will be able to capture everything that is going on. For example, Alloy, a consumer goods analytics company, has worked with a company that saw sales for its product rise 40 percent at a major retailer in March 2020 as the pandemic started to surge in the United States. The retailer placed a very large order for April to handle the spike in sales, but Alloy's analysts knew that demand for the product had plummeted and that the retailer wouldn't be able to sell everything it had ordered. Alloy thus told the retailer not to purchase so much of the product.

Technology for strengthening supply chains, in the form of innovations such as analytics, artificial intelligence, and machine learning alone, won't shore up vulnerabilities and inefficiencies. Companies must rethink their strategies and redesign supply chains so that they're able to source products from multiple locations, depending on where a disruption occurs. One key lies in supply chain mapping, without which companies can't devise workable recovery plans.

The small number of companies that had mapped their supply networks prior to the pandemic were better prepared to deal with the resulting disruptions. These companies were able to determine exactly which suppliers, sites, products, and parts were at risk, which helped them arrive at a solution more quickly. A company might assume that its biggest vulnerability lies with a primary supplier. But a detailed breakdown of its supply chain could show instead that the highest risk comes from a small, lower-tier supplier of a critical component that costs 10 cents.

However, supply network mapping is time consuming and expensive, and most companies have not used it. (After the 2011 earthquake and tsunami, it took a team of 100 people at a Japanese semiconductor manufacturer more than a year to map the company's supply networks into sub-tiers.) Instead, firms rely on human-supplied (and often anecdotal) information from their top-tier and a few lower-tier suppliers.

Sources: Helen Atkinson, "When Globalization Meets Just-in-Time in the Medical Supply Chain," *Supply Chain Brain*, February 16, 2022; Patrick Cason, "How the Internet of Things Boosts Supply Chain Visibility," *Supply Chain Brain*, January 10, 2022; Adam Compain, "The Short- and Long-Term Solutions to Supply Chain Bottlenecks," *Supply Chain Brain*, January 6, 2022; Danny Shields, "The Four Technologies Shaping Next-Gen Supply Chains," *Supply Chain Brain*, September 5, 2021; Ashish Rastogi, "How Digital Solutions Are Creating More Resilient Supply Chains," *Supply Chain Brain*, April 30, 2020; Nicole Wetsman, "The Algorithms Big Companies Use to Manage Their Supply Chains Don't Work during Pandemics," Theverge.com, April 27, 2020; David Parker, "In 2020, Supply-Chain Disruption Is No Longer Optional," *Supply Chain Brain*, April 2, 2020; Thomas Y. Choi, Dale Rogers, and Bindiya Vakil," Coronavirus Is a Wake-Up Call for Supply Chain Management," *Harvard Business Review*, March 27, 2020; Fred Schmalz, "The Coronavirus Outbreak Is Disrupting Supply Chains around the World—Here's How Companies Can Adjust and Prosper," Kellogg Insights, March 26, 2020; and Lizzie O'Leary, "The Modern Supply Chain Is Snapping," *The Atlantic*, March 19, 2020.

## CASE STUDY QUESTIONS

**9-13** Define the problem described in this case study. What people, organization, and technology factors contributed to this problem?

**9-14** To what extent can information technology solve this problem? Explain your answer.

**9-15** What people, organization, and technology issues should be addressed to redesign supply chains to deal with major disruptions such as the coronavirus pandemic?

## Chapter 9 References

"AbbVie Builds a Global Pharmaceuticals Company on New Foundations with SAP and IBM." IBM.com, accessed January 6, 2019.

Bowers, Melissa R., Adam G. Petrie, and Mary C. Holcomb. "Unleashing the Potential of Supply Chain Analytics." *MIT Sloan Management Review* 59 No. 1 (Fall 2017).

Bozarth, Cecil, and Robert B. Handfield. *Introduction to Operations and Supply Chain Management*, 5th ed. (Hoboken, NJ: Pearson Education, Inc., 2019).

Cao, Chengxin, Gautam Ray, Mani Subramani, and Alok Gupta. "Enterprise Systems and the Likelihood of Horizontal, Vertical, and Conglomerate Mergers and Acquisitions." *MIS Quarterly* 46 No. 2 (June 2022).

Chen, Liwei, J. Po-An Hsieh, Arun Rai, and Sean Xin Xu. "How Does Employee Infusion Use of CRM Systems Drive Customer Satisfaction? Mechanism Differences between Face-to-Face and Virtual Channels." *MIS Quarterly* 45 No. 2 (June 2021).

Fruhlinger, Josh, Peter Sayer, and Thomas Wailgum. "16 Famous ERP Disasters, Dustups and Disappointments." *CIO* (March 20, 2020).

Gaur, Vishal, and Abhinay Gaiha. "Building a Transparent Supply Chain." *Harvard Business Review* (May–June 2020).

Hitt, Lorin, D. J. Wu, and Xiaoge Zhou. "Investment in Enterprise Resource Planning: Business Impact and Productivity Measures." *Journal of Management Information Systems* 19, No. 1 (Summer 2002).

Isik, Tuncay. "The Role of AI, Analytics in Advancing Supply Chains." *Supply Chain Brain* (February 17, 2022).

Kitchens, Brent, David Dobolyi, Jingjing Li, and Ahmed Abbasi. "Advanced Customer Analytics: Strategic Value through Integration of Relationship-Oriented Big Data." *Journal of Management Information Systems* 35, No. 2 (2018).

Laudon, Kenneth C. "The Promise and Potential of Enterprise Systems and Industrial Networks." Working paper, The Concours Group. Copyright Kenneth C. Laudon (1999).

Lee, Hau L., V. Padmanabhan, and Seugin Whang. "The Bullwhip Effect in Supply Chains." *MIT Sloan Management Review* 38, No. 3 (Spring 1997).

McCormick, Emily. "Inflation Will Probably Fall, but It Won't Be the Fed's Doing." Finance.yahoo.com, (June 28, 2022).

Panorama Consulting Group. "The 2022 ERP Report." (2022).

Ramasubbu, Narayan, and Chris F. Kemerer. "Controlling Technical Debt Remediation in Outsourced Enterprise Systems Maintenance: An Empirical Analysis." *Journal of Management Information Systems* 38, No. 1 (2021).

Ranganathan, C., and Carol V. Brown. "ERP Investments and the Market Value of Firms: Toward an Understanding of Influential ERP Project Variables." *Information Systems Research* 17, No. 2 (June 2006).

Rohm, Ted. "How IoT Will Dramatically Impact Enteprise Resource Planning (ERP) Systems." Technology Research Centers (August 14, 2019).

"Room & Board Is a Trailblazer." Salesforce.com, accessed January 4, 2022.

Shih,Willy. "Global Supply Chains in a Post-Pandemic World." *Harvard Business Review* (September–October 2020).

Sykes, Tracy Ann. "Enterprise System Implementation and Employee Job Outcomes: Understanding the Role of Formal and Informal Support Structures Using the Job Strain Model." *MIS Quarterly* 44, No. 4 (December 2020).

Tan, Barney, Shan L. Pan, Wenbo Chen, and Lihua Huang. "Organizational Sensemaking in ERP Implementation: The Influence of Sensemaking Structure." *MIS Quarterly* 44, No. 4 (December 2020).

Thibodeau, Patrick. "Office Depot Says 'No' to Oracle ERP Cloud Customizations." Techtarget.com, (February 1, 2018).

Tian, Feng, and Sean Xin Xu. "How Do Enterprise Resource Planning Systems Affect Firm Risk? Post-Implementation Impact." *MIS Quarterly* 39, No. 1 (March 2015).

Wailgum, Thomas. "What Is ERP? A Guide to Enterprise Resource Planning Systems." *CIO* (July 27, 2017).

Zhang, Jonathan Z., George F. Watson IV, and Robert W. Palmatier. "Customer Relationships Evolve—So Must Your CRM Strategy." *MIT Sloan Management Review* 59, No. 4 (May 1, 2018).

CHAPTER

# 10 E-commerce: Digital Markets, Digital Goods

## LEARNING OBJECTIVES

After completing this chapter, you will be able to:

**10-1** Identify the unique features of e-commerce, digital markets, and digital goods.

**10-2** Compare the principal e-commerce business and revenue models.

**10-3** Explain how e-commerce has transformed marketing.

**10-4** Describe how e-commerce has affected business-to-business transactions.

**10-5** Describe the role m-commerce has in business and the most important m-commerce applications.

**10-6** Identify the issues that must be addressed when building an e-commerce presence.

**10-7** Understand how MIS can help your career.

## CHAPTER CASES

- Chewy Blends B2C and B2B E-commerce
- Lemonade Disrupts the Insurance Industry with "Insurtech"
- Engaging "Socially" with Customers on TikTok
- Uber Discovers that Becoming the Uber of Everything Is Not So Easy

**MyLab MIS**

- **Video Case:** Shopify Earnings Soar Even as Economies Reopen; Shopify Is Writing the Future of Commerce, says President
- Discussion Questions: 10-7, 10-8, 10-9
- Hands-On MIS Projects: 10-10, 10-11, 10-12, 10-13
- eText with Figure and Table Videos

## CHEWY BLENDS B2C AND B2B E-COMMERCE

**Businesses** are always on the lookout for ways to expand their revenues. Chewy is best known as an online retailer of pet food and supplies. When it began business, its primary focus was on business-to-consumer (B2C) e-commerce. But competition in the pet products business is very strong. Chewy faces competition from other online retailers, omni-channel retailers, suppliers that sell directly to consumers, traditional bricks-and-mortar retailers such as local pet stores, and even veterinarians. In the face of all this competition, which Chewy expects to continue to increase, Chewy decided that it needed to diversify its business model.

When Chewy's founders started the company in 2011, they faced the specter of competing with Amazon. They planned to differentiate Chewy by providing a more personalized experience than Amazon could provide. They poured resources into call centers, live chat representatives, and prompt responses to customer email. Today, that ethos remains, with Chewy's hallmarks including 24/7 customer service provided by agents who are familiar with the company's products, handwritten notes when a customer makes a first purchase, and condolence cards and flowers when a pet passes away. Chewy grew rapidly and in 2017 was purchased by Petsmart, a bricks-and-mortar retailer, for $3.35 billion. It was, at the time, the highest amount ever paid for an online retailer. Part of the deal was that Chewy would continue to be operated independently. Chewy's revenues continued to grow, and in 2019 it was spun off into its own publicly traded business. In 2020, Chewy and Petsmart officially split into two unrelated businesses.

Chewy began the process of diversifying its business model in 2018. In 2017, although its revenues had more than doubled compared to the previous year's, it had lost almost $340 billion. Its first step was to establish Chewy Health, a new healthcare business that includes a variety of services, including an online pet pharmacy. In 2020, Chewy launched Connect with a Vet, a proprietary telehealth platform that enables customers to connect with veterinarians. Expanding further into pet healthcare is an important part of Chewy's strategy. In 2022, it also intends to offer pet health insurance.

In 2021, it took another major step, this time in the business-to-business (B2B) arena. It began the rollout of Practice Hub, a complete e-commerce solution for veterinarians that can be integrated into a vet's existing practice management solution. Chewy characterizes Practice Hub as one of the most innovative projects it has launched within Chewy Health. Using Chewy's proprietary app, veterinarians can easily create

© 11.studio/Shutterstock

and manage preapproved prescriptions all in one place. They can earn revenue as a seller when one of their customers places an order with Chewy for those prescriptions. Chewy also benefits by being able to engage with those customers, many of whom may be new to Chewy, on its platform. Chewy handles all inventory, fulfillment, shipping, and customer service, leveraging its existing infrastructure. Chewy worked closely with a team of veterinarians in developing Practice Hub, with the vets helping to test features and provide feedback on enhancements. Several nationwide veterinary practices have already started to use the new platform.

Although Chewy's revenues have consistently risen over the years, it has not yet shown a profit, although the amount of money that it lost in 2021 decreased by about 20 percent compared to the amount it lost in 2020. As with many other sectors of the economy, it has been hampered by supply chain woes and increasing inflation. As a result, although as of May 2022 it still has a market value of more than $10 billion, its stock price had fallen to half the price it was in May 2021. Nevertheless, many analysts believe that Chewy is undervalued, pointing in part to Chewy Healthcare and its new Practice Hub and noting that these parts of the company provide new opportunities for Chewy that can both accelerate its growth and increase its profit margins.

Sources: Chewy, Inc. "Form 10-K for the Fiscal Year Ended January 30, 2022," Sec.gov, March 29, 2022; Amy Lamare, "If You Own a Pet You've Heard of Chewy—Here's Why Ryan Cohen and Michael Day Succeeded Where Pets.com Failed," Businessofbusiness.com, November 17, 2021; Brendan Meyer, "How Collaborating with Veterinarians Led to Chewy's Latest Innovative Product," Builtinboston.com, October 26, 2021; "Chewy Launches Innovative Marketplace Service for Veterinarians to Grow Clinic Revenues, Streamline Shopping Experience," Businesswire.com, September 15, 2021; Mary Hansbury, "The Cofounder of Chewy Explains How He Created a $10.2 Billion Empire Selling Pet Food to Millennials Who Treat Their Animals Like Their Firstborn Child," Businessinsider.com, November 30, 2019.

---

Chewy is an example of a company that is using a variety of Internet-enabled e-commerce business models. When Chewy began operations, it focused on B2C e-commerce, and its primary business model was online retail. As it grew, it diversified into providing services and most recently has established a B2B offering. In addition, Chewy exemplifies the use of unique e-commerce technologies such as personalization and customization, which play a primary role in its business strategy and provide it with a competitive advantage.

The chapter-opening diagram calls attention to important points raised by this case and this chapter. The challenge facing Chewy is how to deal with the reality of ever-increasing competition in its primary market. One step Chewy has chosen to take is to diversify its business model, using its already-existing digital infrastructure, and expand into new areas: healthcare services and B2B e-commerce.

Here are some questions to think about: What people, organization, and technology issues did Chewy need to address in expanding into, first, services and then into B2B e-commerce? What are the potential advantages and disadvantages of Chewy's decision to do so?

## 10-1 Identify the unique features of e-commerce, digital markets, and digital goods.

Today, purchasing goods and services online by using smartphones, tablets, and desktop computers has become ubiquitous. In 2022, an estimated 240 million people in the United States (about 85 percent of the US population) will shop online, and 215 million will purchase something online, as will more than 2.5 billion people worldwide. Although most purchases still take place through traditional channels, e-commerce continues to grow rapidly and to transform the way many companies do business. E-commerce is composed of three major segments: retail goods, travel

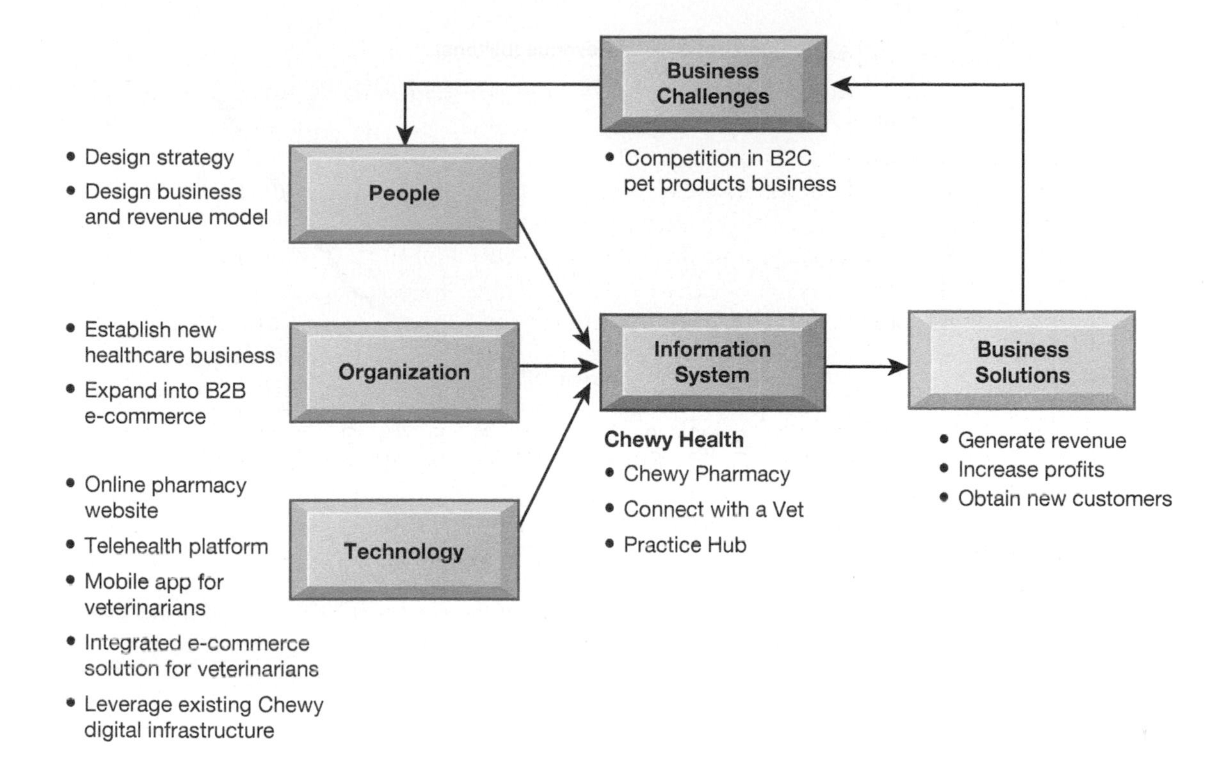


and other services, and online content. In 2022, e-commerce retail sales of goods and services ($1.05 trillion) and travel (about $225 billion) will total about $1.3 trillion.

Online sales of retail goods will be about 15 percent of total US retail sales of $7 trillion and are expected to grow by around 12 percent annually between 2022 and 2026 (compared with 3 percent for traditional retailers). E-commerce is still a small part of the much larger retail goods market that takes place in physical stores. However, e-commerce has expanded from the desktop and home computer to mobile devices, from an isolated activity into a new more social form of commerce, and from commerce focused on a national audience to commerce focused on local merchants and consumers whose location is known to mobile devices. More than 40 percent of all retail e-commerce sales are now mobile, with about 85 percent taking place on a smartphone. The key words for understanding e-commerce today are "social, mobile, local" (Insider Intelligence/eMarketer, 2022a, 2022b, 2022c, 2022d).

## E-COMMERCE TODAY: SOCIAL, LOCAL, MOBILE

E-commerce refers to the use of the Internet, the web, and mobile apps and browsers to transact business. More formally, e-commerce is about digitally enabled commercial transactions between and among organizations and individuals. For the most part, e-commerce refers to transactions that occur over the Internet, the web, and/or mobile devices. Commercial transactions involve the exchange of value (e.g., money) across organizational or individual boundaries in return for products and services.

E-commerce began in 1995 when one of the first Internet portals, Netscape.com, accepted the first ads from major corporations and popularized the idea that the web could be used as a new medium for advertising and sales. No one envisioned at the time what would turn out to be an exponential growth curve for e-commerce retail sales. Examine Figure 10.1, which illustrates the growth of e-commerce from 1995 through 2025. E-commerce grew at double-digit rates from 1995 until the recession of 2008–2009, when growth slowed, but since then has been growing every year in the 12 to 15 percent range. In comparison, offline retail sales have been increasing in the range of just 2 to 4 percent.

One of the biggest changes is the extent to which e-commerce has become more social, mobile, and local. Online marketing once consisted largely of creating

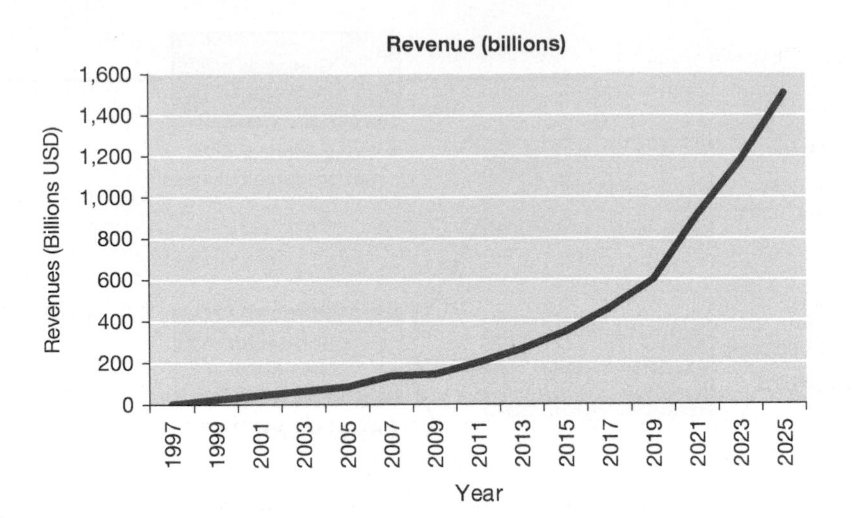


**Figure 10.1**
The Growth of Retail E-commerce
*B2C retail e-commerce revenues grew 15–25 percent per year until the recession of 2008–2009, when they slowed measurably. From 2022 through 2025, e-commerce revenues are expected to grow an estimated 12 percent.*
*Sources:* Based on data from Insider Intelligence/eMarketer, "US Retail Ecommerce Sales," June 2022; authors' estimates.

a corporate website, buying display ads on portal sites such as Yahoo, purchasing search-related ads on Google, and sending email messages. The workhorse of online marketing was the display ad. Display ads were based on television ads, where brand messages were flashed before millions of users who were not expected to interact with them. The primary measure of success was how many unique visitors a website produced and how many impressions a marketing campaign generated. (An impression was one ad shown to one person.) Both of these measures were carryovers from the world of television, which measures marketing in terms of audience size and ad views.

After 2007, all this changed with the rapid growth of Facebook and other social networks; the explosive growth of smartphones, beginning with the Apple iPhone; and the growing interest in local marketing. What's different about the world of social-mobile-local e-commerce is the dual and related concepts of conversations and engagement. In the popular literature, this is often referred to as *conversational commerce.* In the past, firms could tightly control their brand messaging and lead consumers down a funnel of cues that ended in a purchase. This level of control is not possible with social marketing, which instead is based on firms engaging in multiple online conversations with their customers, potential customers, and even critics. Your brand is being talked about on the web and in social media (that's the conversation part); and marketing your firm and building and maintaining your brand require you to locate, identify, and participate in these conversations. Social marketing means all things social—listening, discussing, interacting, empathizing, and engaging—and is all about firms participating in and shaping this social process. Consumer purchase decisions are increasingly driven by the conversations, choices, tastes, and opinions of their social networks.

The emphasis in online marketing has shifted from a focus on unique visitors to a focus on participating in customer-oriented conversations. In this sense, social marketing is not simply an ad channel but a collection of technology-based tools for communicating with shoppers. The leading social e-commerce platforms are Facebook, Instagram, Pinterest, TikTok, and Twitter.

Social, mobile, and local e-commerce are connected. As mobile devices become more powerful, they are more useful for accessing Facebook and other social sites. As mobile devices become more widely adopted, customers can use them to find local merchants, and merchants can use them to alert customers in their neighborhood of special offers. Mobile advertising and marketing now constitute more than two-thirds of online ad spending (Insider Intelligence/eMarketer, 2022e).

**TABLE 10.1**

The Growth of E-commerce

| **Business Transformation** |
| --- |
| E-commerce remains the fastest-growing form of commerce, with. mobile, social, and local e-commerce becoming major channels. |
| The breadth of e-commerce offerings is growing, especially in the services economy. On-demand services like Uber, Lyft, and Airbnb; meal delivery services;, and pet care services have further expanded online service offerings. |
| E-commerce business models are refined further to achieve higher levels of profitability. Traditional retail firms such as Walmart, Target, and Home Depot, have developed omni-channel business models to strengthen their dominant physical retail assets. |
| Small businesses and entrepreneurs continue to flood the e-commerce marketplace, often riding on infrastructures created by industry giants such as Amazon, Apple, Google, and eBay and are increasingly employing cloud-based computing resources. |
| Mobile e-commerce (m-commerce) has taken off in the United States with location-based services and entertainment downloads, including those of e-books, movies, music, and television shows. Mobile retail e-commerce is expected to generate more than $415 billion in 2022. |
| The traditional advertising industry is disrupted because online advertising now comprises more than 70 percent of all ad spending; Google, Yahoo, and Facebook display trillions of ads a year. |
| Two-thirds of the US population has joined an online social network, created blogs, and shared photos and music. In 2022, social networks will account for an estimated 15 percent of time spent with digital media. Social networks have become the primary gateway to news, music, and, increasingly, products and services. |
| The growth of online entertainment business models is enabled by the cooperation of the major copyright owners in Hollywood and New York, as well as the movement of companies such as Netflix, Amazon, Apple, and YouTube into movie and TV production. Cable television is in decline, as some viewers cut or reduce their cable subscriptions and rely on Internet-based streaming alternatives such as Hulu or YouTube TV. |
| Newspapers and other traditional media adopt online, interactive models but continue to lose advertising revenues despite gaining online readers. Book publishing continues to grow slowly because of the growth in e-books and the continuing appeal of traditional trade books. |
| **Technology Foundations** |
| Wireless Internet connections (Wi-Fi, WiMax, 4G, and 5G smartphones) continue to expand. |
| Smartphones and tablet computers provide access to music, web surfing, and entertainment, as well as voice communication. Streaming and podcasting take off as platforms for distribution of video, radio, and user-generated content. |
| Mobile devices expand to include wearable computers such as Apple Watch and Fitbit trackers along with in home devices such as Amazon Alexa and Google Assistant. |
| The Internet broadband foundation becomes stronger in households and businesses as communication prices fall. |
| Social networks such as Facebook, Twitter, LinkedIn, Instagram, TikTok, and others are becoming major new platforms for e-commerce, marketing, and advertising. |
| Internet-based models of computing, such as smartphone apps, cloud computing, software as a service (SaaS), and platform as a service (PaaS), reduce the costs of building and maintaining e-commerce websites. |

## WHY E-COMMERCE IS DIFFERENT

Why has e-commerce grown so rapidly? The answer lies in the unique nature of the Internet and the web. Simply put, the Internet and web are much richer and more powerful than previous technology revolutions such as radio, television, and the telephone. Table 10.2 describes the unique features of the Internet and the web that enable e-commerce. View the Table 10.2 video in the eText for an animated and more detailed discussion of this table. Now, let's explore each of these unique features.

**TABLE 10.2**
Eight Unique Features of E-commerce Technology

| E-commerce Technology Dimension | Business Significance |
|---|---|
| *Ubiquity.* Internet/web technology is available everywhere: at work, at home, and, via mobile devices, elsewhere. | The marketplace is extended beyond traditional boundaries and is removed from a temporal, geographic location. Marketspace is created. Shopping can take place anytime, anywhere. Customer convenience is enhanced, and shopping costs are reduced. |
| *Global Reach.* The technology reaches across national boundaries and around the earth. | Commerce is enabled across cultural and national boundaries seamlessly and without modification. The marketspace includes, potentially, billions of consumers and millions of businesses worldwide. |
| *Universal Standards.* There is one set of technology standards, namely, Internet standards. | With one set of technical standards across the globe, disparate computer systems can easily communicate with each other. |
| *Richness.* Video, audio, and text messages are possible. | Video, audio, and text marketing messages are integrated into a single marketing message and consumer experience. |
| *Interactivity.* The technology works through interaction with the user. | Consumers are engaged in a dialogue that dynamically adjusts the experience to the individual and makes the consumer a participant in the process of delivering goods to the market. |
| *Information Density.* The technology reduces information costs and raises quality. | Information-processing, storage, and communication costs drop dramatically, whereas currency, accuracy, and timeliness improve greatly. Information becomes plentiful, cheap, and more accurate. |
| *Personalization/Customization.* The technology allows personalized messages to be delivered to individuals as well as to groups. | Personalization of marketing messages and customization of products and services are based on individual characteristics. |
| *Social Technology.* The technology supports content generation and social networking. | Internet social business models enable user content creation and distribution and support social networks. |

### Ubiquity

In traditional commerce, a marketplace is a physical place, such as a retail store, that you visit to transact business. E-commerce is ubiquitous, meaning that it is available just about everywhere all the time. It makes it possible to shop from your desktop, at home, at work, or even in your car, using smartphones. The result is called a **marketspace**—a marketplace extended beyond traditional boundaries and removed from a temporal, geographic location.

From a consumer point of view, ubiquity reduces **transaction costs**—the costs of participating in a market. To transact business, it is no longer necessary to spend time or money traveling to a market, and much less mental effort is required to make a purchase in a marketspace.

### Global Reach

E-commerce technology permits commercial transactions to cross cultural and national boundaries far more conveniently and cost-effectively than is possible in traditional commerce. As a result, the potential market size for e-commerce merchants is roughly equal to the size of the world's online population (estimated to be more than 4.5 billion).

In contrast, most traditional commerce is local or regional—it involves local merchants or national merchants with local outlets. Television stations, radio stations, and newspapers, for instance, are primarily local and regional institutions with limited, but powerful, national networks that can attract a national audience but not easily cross national boundaries to a global audience.

### Universal Standards

One strikingly unusual feature of e-commerce technologies is that the technical standards of the Internet and, therefore, the technical standards for conducting e-commerce are universal standards. All nations around the world share them and enable any computer to link with any other computer regardless of the technology platform each computer is using. In contrast, most traditional commerce technologies differ from one nation to the next. For instance, television and radio standards differ around the world, as does cellular telephone technology.

The universal technical standards of the Internet and e-commerce greatly lower **market entry costs**—the costs merchants must pay simply to bring their goods to market. At the same time, for consumers, universal standards reduce **search costs**—the effort required to find suitable products.

### Richness

Information **richness** refers to the complexity and content of a message. Traditional markets, national sales forces, and small retail stores have great richness: They can provide personal, face-to-face service, using aural and visual cues when making a sale. The richness of traditional markets makes them powerful selling or commercial environments. Prior to the development of the web, there was a trade-off between richness and reach—the larger the audience reached, the less rich the message. But the web makes it possible to deliver rich messages with text, audio, and video simultaneously to large numbers of people.

### Interactivity

Unlike any of the commercial technologies of the twentieth century, with the possible exception of the telephone, e-commerce technologies are interactive, meaning they allow for two-way communication between merchant and consumer and peer-to-peer communication among friends. Television, for instance, cannot ask viewers any questions or enter conversations with them, and it cannot request customer information to be entered on a form. In contrast, all these activities are possible on an e-commerce website or mobile app. Interactivity allows an online merchant to engage a consumer in ways similar to a face-to-face experience but on a massive, global scale.

### Information Density

The Internet and the web vastly increase **information density**—the total amount and quality of information available to all market participants, consumers and merchants alike. E-commerce technologies reduce information collection, storage, processing, and communication costs while greatly increasing the currency, accuracy, and timeliness of information.

Information density in e-commerce markets make prices and costs more transparent. **Price transparency** refers to the ease with which consumers can find out the variety of prices in a market; **cost transparency** refers to the ability of consumers to discover the actual costs that merchants pay for products.

There are advantages for merchants as well. Online merchants can discover much more about consumers than was possible in the past. This allows merchants to segment the market into groups that are willing to pay different prices and permits the merchants to engage in **price discrimination**—selling the same goods, or nearly the same goods, to different targeted groups at different prices. For instance, an online merchant can discover a consumer's avid interest in expensive, exotic vacations and then pitch high-end vacation plans to that consumer at a premium price, knowing this person is willing to pay extra for such a vacation. At the same time, the online merchant can pitch the same vacation plan at a lower price to a more price-sensitive consumer. Information density also helps merchants differentiate their products in terms of cost, brand, and quality.

### Personalization/Customization

E-commerce technologies permit **personalization**. Merchants can target their marketing messages to specific individuals by adjusting the message to a person's clickstream behavior, name, interests, and past purchases. The technology also permits **customization**—changing the delivered product or service based on a user's preferences or prior behavior. Given the interactive nature of e-commerce technology, much information about the consumer can be gathered in the marketplace at the moment of purchase. With the increase in information density, a great deal of information about the consumer's past purchases and behavior can be stored and used by online merchants.

The result is a level of personalization and customization unthinkable with traditional commerce technologies. For instance, you may be able to shape what you see on television by selecting a channel, but you cannot change the content of the channel you have chosen. In contrast, online news outlets such as the *Wall Street Journal Online* allow you to select the type of news stories you want to see first and also give you the opportunity to be alerted when certain events happen.

### Social Technology: User Content Generation and Social Networking

In contrast to previous technologies, the Internet and e-commerce technologies have evolved to be much more social by allowing users to create and share with their friends (and a larger worldwide community) content in the form of text, videos, music, and photos. By using these forms of communication, users can create new social networks and strengthen existing ones.

All previous mass media, including the printing press, use a broadcast (one-to-many) model in which content is created in a central location by experts (professional writers, editors, directors, and producers), with audiences concentrated in huge numbers to consume a standardized product. Internet and e-commerce technologies empower users to create and distribute content on a large scale and permit users to program their own content consumption. The Internet provides a unique many-to-many model of mass communications.

## KEY CONCEPTS IN E-COMMERCE: DIGITAL MARKETS AND DIGITAL GOODS IN A GLOBAL MARKETPLACE

The location, timing, and revenue models of business are based in some part on the cost and distribution of information. The Internet has created a digital marketplace where millions of people all over the world can exchange massive amounts of information directly, instantly, and free of charge. As a result, the Internet has changed the way companies conduct business and increased their global reach.

The Internet reduces information asymmetry. An **information asymmetry** exists when one party in a transaction has more information that is important for the transaction than the other party. That information helps determine each party's relative bargaining power. In digital markets, consumers and suppliers can see the prices being charged for goods, and in that sense, digital markets are said to be more transparent than traditional markets.

For example, before companies began advertising and selling cars online, there was significant information asymmetry between auto dealers and customers. Only the auto dealers knew the manufacturers' prices, and it was difficult for consumers to shop around for the best price. Auto dealers' profit margins depended on this asymmetry of information. Today's consumers have access to a legion of websites providing competitive pricing information, and more than 90 percent of US auto buyers use the Internet to research prospective car purchases. Thus, the web has reduced the information asymmetry surrounding an auto purchase. The Internet has also helped businesses seeking to purchase from other businesses reduce information asymmetries and locate better prices and terms.

## DIGITAL MARKETS

Digital markets are flexible and efficient because they operate with reduced search and transaction costs, lower **menu costs** (merchants' costs of changing prices), greater price discrimination, and the ability to change prices dynamically based on market conditions. In **dynamic pricing**, the price of a product varies depending on the demand characteristics of the customer or the supply situation of the seller. For instance, online retailers from Amazon to Walmart change prices on thousands of products based on time of day, demand for the product, and users' prior visits to their sites. Using Big Data analytics, some online firms can adjust prices at the individual level based on behavioral targeting parameters such as whether the consumer is a price haggler (who will receive a lower price offer) versus a person who accepts offered prices and does not search for lower prices. Prices can also vary by zip code. Uber, along with other ride services, uses surge pricing to adjust prices of a ride based on demand (which always rises during storms and major conventions).

Digital markets such as these can either reduce or increase switching costs, depending on the nature of the product or service being sold, and they might cause some extra delay in gratification due to shipping times. Unlike a physical market, you can't immediately consume a product such as clothing that is purchased over the web (although immediate consumption is possible with digital products).

Digital markets provide many opportunities to sell directly to the consumer, bypassing intermediaries such as distributors or retail outlets. Eliminating intermediaries in the distribution channel can significantly lower purchase transaction costs. To pay for all the steps in a traditional distribution channel, a product may have to be priced as high as 135 percent of its original cost to manufacture.

Examine Figure 10.2, which illustrates how much savings result from eliminating various layers, such as distributors and retailers, in the distribution process. By selling directly to consumers or reducing the number of intermediaries, companies can raise profits while charging lower prices. The removal of organizations or business process layers responsible for intermediary steps in a value chain is called

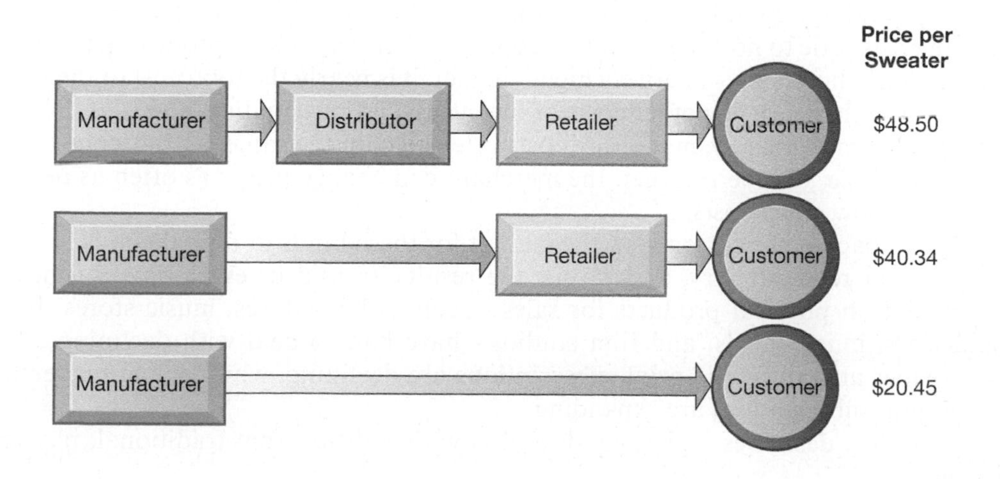


**Figure 10.2** The Benefits of Disintermediation to the Consumer

*The typical distribution channel has several intermediary layers, each of which adds to the final cost of a product, such as a sweater. Removing layers lowers the final cost to the customer.*

**TABLE 10.3**

Digital Markets Compared with Traditional Markets

| | Digital Markets | Traditional Markets |
|---|---|---|
| Information asymmetry | Asymmetry reduced | Asymmetry high |
| Search costs | Low | High |
| Transaction costs | Low (sometimes virtually nothing) | High (time, travel) |
| Delayed gratification | High (or lower in the case of a digital good) | Lower: purchase now |
| Menu costs | Low | High |
| Dynamic pricing | Low cost, instant | High cost, delayed |
| Price discrimination | Low cost, instant | High cost, delayed |
| Market segmentation | Low cost, moderate precision | High cost, less precision |
| Switching costs | Higher/lower (depending on product characteristics) | High |
| Network effects | Strong | Weaker |
| Disintermediation | More possible/likely | Less possible/unlikely |

**disintermediation**. E-commerce has also given rise to a new set of new intermediaries in the retail industry such as Amazon and eBay.

Disintermediation is affecting the market for services. Airlines and hotels operating their own reservation sites online earn more per ticket because they have eliminated travel agents as intermediaries. However, as in the retail industry, e-commerce has given rise to a new set of intermediaries such as Expedia, Orbitz, and Hotels.com. Table 10.3 summarizes the differences between digital markets and traditional markets.

## DIGITAL GOODS

The Internet's digital marketplace has greatly expanded sales of **digital goods**—goods that can be delivered over a digital network. Music, video, movies, software, newspapers, magazines, and books can all be expressed, stored, delivered, and sold as purely digital products. Digital goods are intellectual property, which is protected from misappropriation by copyright, patent, trademark, and trade secret laws (see Chapter 4). Today, all these products are delivered as digital streams or downloads while their physical counterparts decline in sales.

In general, for digital goods, the marginal cost of producing another unit is about zero (it costs little to nothing to make a copy of a file). The cost of producing the original first unit, however, is relatively high—in fact, it is nearly the total cost of the product because there are few other costs of inventory and distribution. Costs of delivery over the Internet are low, marketing costs often remain the same, and pricing can be highly variable. On the Internet, the merchant can change prices as often as desired because of low menu costs.

The impact of the Internet on the market for these kinds of digital goods is nothing short of revolutionary, and we see the results around us every day. Businesses dependent on physical products for sales—such as bookstores, music stores, book publishers, music labels, and film studios—have had to deal with declining sales. Newspaper and magazine print subscriptions are declining, while online readership and digital subscriptions are expanding.

Table 10.4 describes digital goods and how they differ from traditional, physical goods.

**TABLE 10.4**

How the Internet Changes the Markets for Digital Goods

| | Digital Goods | Traditional Goods |
|---|---|---|
| Marginal cost/unit | Zero | Greater than zero, high |
| Cost of production | High (most of the cost) | Variable |
| Copying cost | Approximately zero | Greater than zero, high |
| Distributed delivery cost | Low | High |
| Inventory cost | Low | High |
| Marketing cost | Variable | Variable |
| Pricing | More variable (bundling, random pricing) | Fixed, based on unit costs |

# 10-2 Compare the principal e-commerce business and revenue models.

E-commerce is a fascinating combination of business models and information technologies. Let's start with a basic understanding of the types of e-commerce and then describe e-commerce business and revenue models.

## TYPES OF E-COMMERCE

There are many ways to classify e-commerce transactions. One way involves looking at the nature of the participants. The three major e-commerce categories are business-to-consumer (B2C) e-commerce, business-to-business (B2B) e-commerce, and consumer-to-consumer (C2C) e-commerce.

- **Business-to-consumer (B2C) e-commerce** involves selling products and services to individual shoppers. Amazon, Walmart, and Apple Music are examples of B2C e-commerce.
- **Business-to-business (B2B) e-commerce** involves sales of goods and services among businesses. Grainger, which focuses on the purchase and sale of maintenance, repair, and operating (MRO) products to businesses, is an example of one type of B2B e-commerce.
- **Consumer-to-consumer (C2C) e-commerce** involves consumers selling directly to other consumers. For example, eBay enables people to sell their goods to other consumers by auctioning their merchandise off to the highest bidder or by selling it for a fixed price. eBay acts as an intermediary by creating a digital platform for peer-to-peer commerce. Craigslist is another platform that consumers use to buy from and sell directly to others.

Another way of classifying e-commerce transactions is in terms of the platforms that participants use in a transaction. Prior to 2007 and the introduction of the smartphone, e-commerce transactions took place using a desktop computer connected to the Internet over a wired network. Since then, smartphones and tablet computers have become a popular alternative, today accounting for more than 40 percent of all retail e-commerce sales. The use of mobile devices to purchase goods and services from any location is referred to as **mobile e-commerce** or **m-commerce**, which we discuss in detail in Section 10-5.

## E-COMMERCE BUSINESS MODELS

Changes in the economics of information described earlier have created the conditions for entirely new business models to appear while disrupting, in some cases, older business models. Table 10.5 describes some of the primary business models that have

**TABLE 10.5**

Internet Business Models

| Category | Description | Examples |
|---|---|---|
| E-tailer | Sells physical products directly to consumers or to individual businesses. | Amazon<br>Wayfair |
| Transaction broker | Saves users money and time by processing online sales transactions and generating a fee each time a transaction occurs. | E*Trade<br>Expedia |
| Market creator | Provides a digital environment where buyers and sellers can meet, search for products, display products, and establish prices for those products, generating revenue from transaction fees. | eBay<br>eBid<br>Etsy |
| Content provider | Creates revenue by providing digital content, such as news, music, photos, or video, over the web. The customer may pay to access the content, or revenue may be generated by selling advertising space. | WSJ.com<br>Shutterfly<br>Apple Music<br>Scopely |
| Community provider/ social network | Provides an online meeting place where people with similar interests can communicate and find useful information. | Facebook<br>Twitter |
| Portal | Provides initial point of entry to the web along with specialized content and other services. | Yahoo<br>MSN<br>AOL |
| Service provider | Provides online services such as financial services, career services, on-demand services, and online data storage and backup. | Uber<br>RocketMortgage<br>Dropbox |

emerged. Note that in some cases, the business models overlap. For instance, transaction brokers can also be considered service providers, as can some market creators. All, in one way or another, use the Internet (including apps on mobile devices) to add extra value to existing products and services or to provide the foundation for new products and services. View the Table 10.5 video in the eText for an animated and more detailed discussion of Internet business models.

### E-tailer

**Online retailers**, often called **e-tailers**, come in all sizes, from giant Amazon (with 2021 retail sales revenues of more than $365 billion—about 40 percent of all retail e-commerce) to tiny local stores that have websites. There are four main types of online retail business models: virtual merchants such as Wayfair that do not have a physical presence, omni-channel merchants such as Walmart or Target (sometimes referred to as bricks-and-clicks) that have both physical stores as well as online offerings, catalog merchants such as LandsEnd or L.L. Bean, and manufacturer-direct (sometimes referred to as direct-to-consumer, or D2C/DTC), where the manufacturer of a product, such as Nike, sells directly to the consumer. Altogether, online retail (the sale of physical goods online) will generate more than $1 trillion in revenues in 2022. The value proposition of online retailers is to provide convenient, low-cost shopping 24/7; large selections; and consumer choice.

### Transaction Broker

Transaction brokers provide online processing for transactions that previously were handled in person, by phone, or by mail. The most common industries using this model are financial services and travel services. The online transaction broker's primary value propositions are savings of money and time and providing an extraordinary inventory of financial products or travel packages in a single location. Online stockbrokers and travel booking services charge fees that are considerably less than traditional versions of these services. E*Trade and Expedia are two examples of online financial and travel service firms based on a transaction broker model.

### Market Creator

**Market creators** build a digital environment in which buyers and sellers can meet, display products, search for products, and establish prices. The value proposition of online market creators is that they provide a platform where sellers can easily display their wares and purchasers can buy directly from sellers. Online auction markets such as eBay and eBid are good examples of the market creator business model. Another example is Amazon's Merchants platform (and similar programs at eBay), where merchants set up stores on Amazon Marketplace and sell goods at fixed prices to consumers. Companies that enable on-demand services, exemplified by Uber (described in the chapter-ending case study) and Airbnb, can be viewed as being based on the idea of a market creator building a digital platform where supply meets demand; for instance, spare auto or room rental capacity finds individuals who want transportation or lodging. Crowdfunding markets such as Kickstarter bring together private investors and entrepreneurs in a funding marketplace.

### Content Provider

E-commerce has increasingly become a global content channel. *Content* is defined broadly to include all forms of intellectual property. Intellectual property refers to products of the mind in which the creator claims a property right. Content providers distribute information content—such as digital video, music, audio files in the form of podcasts, photos, text, and artwork—via the Internet to computers and mobile devices. The value proposition of online content providers is that consumers can conveniently find and purchase a wide range of content online that can be played or viewed on multiple devices.

Providers do not have to be the creators of the content (although sometimes they are, such as Disney). Instead, many content providers act merely as a platform for the distribution of content created by others. For example, Apple Music distributes music, but it does not create or commission new music. Netflix, on the other hand, both distributes original content that it produces, as well as content produced by other.

Content providers typically make content available online either via download or streaming. The **download** method involves transferring a file from a web server to the user's device, where it resides and can be accessed by the user at any time. **Streaming** is a publishing method for music and video files that flows as a continuous stream of content to a user's device without being stored locally on the device.

### Service Provider

Whereas e-tailers sell products online, service providers offer services online. The major service industry groups include finance, insurance, real estate, travel, professional services such as legal and accounting, business services, healthcare services, and educational services. While online banking and banking services are dominated by large banks with millions of customers, startup financial technology service firms, also known as **fintech** firms, have used IT innovatively to compete with banks for peer-to-peer (P2P) bill payment, money transfer, lending, crowdfunding, financial advice, and account aggregation services. According to a recent survey, almost 90 percent of US consumers are now using fintech solutions, with digital payment services the most popular, followed by digital savings tools, and online investing (Plaid and the Harris Poll, 2021). Notable fintech companies include P2P payment services providers Venmo and Zelle (see the discussion of mobile payment app systems in section 10-5). Fintech firms are often purchased by larger financial service firms who want to acquire the fintech's technology and customer base. Insurtech is a subset of fintech that focuses on the insurance services industry. The Spotlight on Technology case focuses on one such insurtech company, Lemonade, that is using cutting-edge technology to disrupt the insurance industry.

## SPOTLIGHT ON: TECHNOLOGY

## Lemonade Disrupts the Insurance Industry with "Insurtech."

Insurance is one of the largest industries in the world. Property, casualty, and life insurance premiums amount to approximately $5 trillion globally. In the United States, 14 of the *Fortune* 100 companies are insurance companies. Most of the leading insurance companies have been in business for more than a century.

Insurance products can be very complex, and in the past, the insurance business has been driven by person-to-person interactions between consumers and insurance agents or brokers. In addition, the leading insurance companies have years of investments in legacy IT systems as well as corporate cultures developed over many decades. The process of grafting new technologies and integrating new ways of doing business in such an environment is typically a slow one, leaving the industry ripe for a disruptive player.

Enter Lemonade. Founded in 2015, Lemonade is attempting to take the insurance industry by storm by leveraging a variety of cutting-edge technologies, such as Big Data collection and analysis, artificial intelligence, chatbots, and a sophisticated mobile app. Its digital platform spans marketing to customer care to claims processing, collecting and deploying data at each step along the way. Lemonade refers to itself as a digitally native and customer-centric insurance company.

Lemonade originally focused on the homeowners and renters insurance market but has since expanded into pet, term life, and car insurance on its way to becoming a one-stop shop for consumer insurance needs. In 2020 Lemonade went public, and in May 2022, it had a market valuation of more than $1.3 billion. Lemonade is also a Certified B Corp. Certified B Corps are companies certified by B Lab (an independent, nonprofit organization) as meeting rigorous standards of social and environmental performance, accountability, and transparency.

The typical Lemonade customer is younger than 35. For more than 90 percent of these customers, Lemonade is their first experience obtaining insurance. Compared to traditional insurers, Lemonade has pioneered a fundamentally different approach to the process. For instance, the application process for a standard homeowners policy is typically based on a form with between 20 and 50 fields (such as name, address, birthdate). In contrast, Lemonade has dispensed with forms altogether. Instead, prospective customers use their mobile app to interact with a chatbot named AI Maya. Maya uses a natural language interface and typically asks only about 13 questions before being able to give customers a personalized quote for home insurance. Lemonade is able to streamline the process by using artificial intelligence. Although Maya may only ask a limited number of questions, the interaction generates almost 1,700 data points, all of which are logged, aggregated, and mined for correlation to claims. Maya provides the interface for all types of insurance that Lemonade offers.

The claims process is handled by a different chatbot, AI Jim. Jim takes the first notice of loss from a customer making a claim in almost all cases and is typically able to handle the entire claim through resolution for about one-third of the cases. In some cases, Jim can even authorize and pay a claim in as little as three seconds. Jim assigns the rest to human claims experts based on Jim's analysis of the expert's specialty, qualifications, workloads, and schedule.

Lemonade also has a chatbot platform, CX.AI, that is designed to understand and instantly resolve customer requests for assistance without human intervention. As of the end of 2021, about 30 percent of all customer inquiries were handled in this manner.

Lemonade's digital infrastructure is driven by three key proprietary applications: Forensic Graph, Blender, and Cooper. Forensic Graph uses Big Data, artificial intelligence, and behavioral economics to predict, detect, and block fraud throughout the customer application process. Blender is a built-from-scratch insurance management system designed to be a cohesive and streamlined management tool for Lemonade's customer experience, including underwriting, claims, growth, marketing, finance, and risk teams. Cooper is an internal AI bot that handles complex as well as repetitive tasks, from helping Lemonade's customer experience teams handle manual processes such as processing paper checks, to automatically running tens of thousands of tests on each release of its software. The latter is a particularly important task because Lemonade averaged more than 40 code releases per day in 2021. The frequent refreshing of its code means that Lemonade can promptly and efficiently modify customer onboarding questions, underwriting guidelines, claims handling, and other elements of its platform, enabling it to respond to changes in customer needs and market conditions at a faster rate than its traditional insurance competitors can.

But Lemonade's approach is not without controversy. In May 2021, a tweet from its Twitter account created outrage when it appeared to boast about how Lemonade's AI system boosted its profits by automatically denying claims based on analyzing videos submitted by customers. Lemonade quickly deleted the tweet and apologized, stating that Lemonade's AI does not use physical or personal features to deny claims, and that its systems do not evaluate claims based on background, gender, appearance, skin tone, disability, or any other physical characteristic. However, in the same statement, Lemonade confirmed that it was using facial recognition technology to flag certain claims. Not long thereafter, Lemonade was sued in class action lawsuits in both Illinois and New York, alleging that it had violated biometric privacy laws in both states by collecting such data without users' knowledge and consent. In May 2022, Lemonade agreed to pay $4 million to settle some of those suits.

Despite these missteps, and even though thus far it has not shown a profit, Lemonade continues to grow. It has increased its annual revenues by more than 500 percent since 2018, and projects that those revenues will continue to grow by more than 66 percent in 2022. It has also garnered many accolades, including recognition by various publications as "best-in-class" renters, homeowners, and pet insurance.

Sources: Josh Liberatore, "Insurer Lemonade Reaches $4M Deal in Ill. Biometric Data Suit," Law360.com, May 18, 2022; "Lemonade Stock Looks Like a Bargain at Current Pricing," Schaeffersresearch.com, May 18, 2022; Lemonade, Inc., "Form 10-K for the Fiscal Year Ended December 31, 2021", Sec.gov, March 1, 2022; Carlton Fields and Michael Yaeger, "AI Insurance Company Faces Class Action for Use of Biometric Data," Jdsupra.com, February 1, 2022; Jonathan Greig, "Lemonade Insurance Faces Backlash for Claiming AI System Could Automatically Deny Claims," Zdnet.com, May 27, 2021.

## CASE STUDY QUESTIONS

1. What distinguishes the "insurtech" services described in this case from traditional insurers? Explain your answer.
2. How has Lemonade used information technology to innovate?
3. What issues are raised by Lemonade's use of artificial intelligence?
4. What factors would you consider in deciding whether to use an "insurtech" service such as Lemonade's?

On-demand services, in addition to being viewed as an example of the market creator business model, can also be seen as a variation on the service provider business model, as are companies that offer online, cloud-based software, storage, and data backup. Google has led the way in developing online software service applications such as G Suite, Google Sites, Gmail, and online data storage services. Salesforce.com is a major provider of cloud-based software for customer management (see Chapters 5 and 9).

### Community Providers (Social Networks)

**Community providers** create a digital online environment where people with similar interests can transact (buy and sell goods); share interests, photos, and videos; communicate with like-minded people; receive interest-related information; and even play out fantasies by adopting online personalities called *avatars* (see Chapter 2). Social networks Facebook, Instagram, LinkedIn, Pinterest, TikTok, Twitter, and hundreds of other smaller, niche networks all offer users community-building tools and services.

### Portal

Portals are gateways to the web and are often defined as those sites that users set as their home page. Portals such as Google, Yahoo, MSN, and AOL offer search tools as well as an integrated package of content and services such as news, email, instant messaging, maps, calendars, shopping, music downloads, videostreaming, and more, all in one place. Today's portals hope to be a destination where users start their web searching and linger to read news, find entertainment, meet other people, and, of course, be exposed to advertising. Facebook is a different kind of portal, one that is based on social networking. Millions of users have set Facebook as their home page, and about 70 percent of Facebook users say they visit it daily, including around half who do so several times a day (Gramlich, 2021). Portals generate revenue primarily by attracting very large audiences, charging advertisers for display ad placement (similar to traditional newspapers), collecting referral fees for steering customers to other sites, and charging for premium services. Although there are hundreds of portal/search engine sites, the top portals (Google, MSN, Yahoo, and AOL) gather more than 90 percent of the Internet portal traffic because of their superior brand recognition.

## E-COMMERCE REVENUE MODELS

A firm's **revenue model** describes how the firm will earn revenue, generate profits, and produce a superior return on investment. Although many e-commerce revenue models have been developed, most companies rely on one, or some combination, of the following six revenue models: advertising, sales, subscription, free/freemium, transaction fee, and affiliate.

### Advertising Revenue Model

In the **advertising revenue model**, an online business generates revenue by attracting a large audience of visitors to its website and/or apps, who can then be exposed to advertisements. The advertising model is the most widely used revenue model in e-commerce, and arguably, without advertising revenues, the web would be a vastly different experience from what it is now because people would be asked to pay for access to content. Much of the content available online today—everything from news to videos and opinions—is free because advertisers pay the production and distribution costs in return for the right to expose visitors to ads. Companies are expected to spend almost $250 billion on online advertising in 2022, up 18 percent from 2021. More than $190 billion of this will be for mobile ads, which will account for about 68 percent of all digital advertising. In the past five years, advertisers have increased their online spending while maintaining outlays on traditional channels such as radio and television (cutting their print ads). In 2022, online advertising is expected to constitute more than 70 percent of all advertising in the United States (Insider Intelligence/eMarketer, 2022e).

Websites with the largest number of unique visitors or that are able to retain user attention (stickiness) can charge higher advertising rates. Yahoo, for instance, derives nearly all its revenue from display ads (banner ads), video ads, and, to a lesser extent, search engine text ads. Google and Facebook each derive more than 90 percent of their revenue from advertising, including selling keywords (AdWords), ad spaces (AdSense), and display ads to advertisers.

### Sales Revenue Model

In the **sales revenue model**, companies derive revenue by selling goods, information, or services to customers. Companies such as Amazon, Net-A-Porter (luxury fashion), and Wayfair (furniture) all have sales revenue models. Content providers make money by charging for downloading content such as music (Apple Music), books (Amazon Kindle), or movies (Apple TV).

### Subscription Revenue Model

In the **subscription revenue model**, a company offering content or services charges a subscription fee for access to some or all of its offerings on an ongoing basis. Content providers often use this revenue model. For instance, the online version of *Consumer*

*Reports* provides access to premium content, such as detailed ratings, reviews, and recommendations, only to subscribers for a $39 annual fee. Netflix is one of the most successful users of the subscription revenue model, with more than 220 million customers worldwide in 2022. To be successful, the subscription model requires the content to be perceived as differentiated, having high added value, and not readily available elsewhere or easily replicated. Other companies offering content or services online on a subscription basis include Match (dating services), Ancestry (genealogy research), and Microsoft Xbox Live.

### Free/Freemium Revenue Model

In the **free/freemium revenue model**, firms offer basic services or content for free but charge a premium for advanced or special features. For example, Google offers free applications but charges for premium services. Pandora, the subscription radio service, offers a free service with limited play time and advertising and a premium service with unlimited play. The idea is to attract large audiences with free services and then convince some of this audience to pay a subscription for premium services. One problem with this model is convincing people to become paying customers.

### Transaction Fee Revenue Model

In the **transaction fee revenue model**, a company receives a fee for enabling or executing a transaction. For example, eBay provides an online auction marketplace and receives a small transaction fee from a seller when the seller is successful in selling an item. The transaction revenue model enjoys wide acceptance in part because the true cost of using the platform is not immediately apparent to the user. Online financial services, from banking to payment systems, typically rely on a transaction fee model.

### Affiliate Revenue Model

In the **affiliate revenue model**, websites (called *affiliate websites*) send visitors to other websites in return for a referral fee or percentage of the revenue from any resulting sales. Referral fees are also referred to as lead generation fees. For example, MyPoints makes money by connecting companies to potential customers by offering special deals to MyPoints members. When MyPoint members take advantage of an offer and make a purchase, they earn points they can redeem for free products and services, and MyPoints receives a referral fee. Companies such as Yelp, which publishes crowd-sourced reviews of businesses, receive much of their revenue from steering potential customers to websites where the customers make a purchase. Amazon uses affiliates that steer business to Amazon by placing the Amazon logo on their blogs. Personal blogs often contain display ads as part of affiliate programs. Some bloggers are paid directly by manufacturers, or receive free products, for speaking highly of the manufacturers' products and providing links to sales channels.

## 10-3 Explain how e-commerce has transformed marketing.

Although e-commerce and the Internet have changed entire industries and enabled new business models, no industry has been more affected than marketing and marketing communications.

The Internet provides marketers with ways of identifying and communicating with millions of potential customers at costs far lower than traditional media, including search engine marketing, data mining, recommender systems, and targeted email. For instance, the Internet enables **long tail marketing**. Before the Internet, reaching a large audience was expensive, and marketers had to focus on attracting the largest number of consumers with popular, hit products, whether music, Hollywood movies, books, or cars. In contrast, the Internet allows marketers to find potential customers

**TABLE 10.6**

Online Ad Spending by Formats

*Source:* Based on Insider Intelligence/eMarketer, "U.S. Digital Ad Spending by Format," March 2022.

| Marketing Format | 2022 Estimated Revenue (billions) | Description |
|---|---|---|
| Search engine | $99 | Text ads targeted at precisely what the customer is looking for at the moment of shopping and purchasing. Sales oriented. |
| Display ads | $52.3 | Banner ads (pop-ups and leave-behinds) with interactive features; increasingly behaviorally targeted to individual web activity. Brand development and sales. Includes social media and blog display ads. |
| Video | $76.2 | Fastest-growing format, engaging and entertaining; behaviorally targeted, interactive. Branding and sales. |
| Classified | $2.4 | Job, real estate, and services ads; interactive, rich media, and personalized to user searches. Sales and branding. |
| Rich media | $11.1 | Animations, games, and puzzles. Interactive, targeted, and entertaining. Branding orientation. |
| Lead generation | $2.98 | Marketing firms that gather sales and marketing leads online and then sell them to online marketers for a variety of campaign types. Sales or branding orientation. |
| Sponsorships | $3.96 | Online articles, games, puzzles, and contests sponsored by firms to promote products. Sales orientation. |
| Email | $0.59 | Effective, targeted marketing tool with interactive and rich media potential. Sales oriented. |

inexpensively for products where demand is low. For instance, the Internet makes it possible to sell independent music profitably to small audiences. There's always some demand for almost any product. Put a string of such long tail sales together, and you have a profitable business.

The Internet also provides ways—often instantaneous and spontaneous—to gather information from customers, adjust product offerings, and increase customer value. Table 10.6 describes the leading marketing and advertising formats used in e-commerce.

## BEHAVIORAL TARGETING

Many e-commerce marketing firms use **behavioral targeting** techniques to increase the effectiveness of banners, rich media, and video ads. Behavioral targeting refers to tracking the clickstreams (history of clicking behavior) of individuals on thousands of websites to understand their interests and intentions and to expose them to advertisements that are uniquely suited to their online behavior. Marketers and most researchers believe this more precise understanding of the customer leads to more efficient marketing (the firm pays for ads that are directed only to those shoppers who are most interested in their products) and increased sales and revenues. Unfortunately, behavioral targeting also leads to the invasion of personal privacy without user consent. When consumers lose trust in their online experience, they tend not to purchase anything. Backlash is growing against the aggressive uses of personal information as consumers seek out safer havens for purchasing and messaging.

Behavioral targeting takes place at two levels: at individual websites or from within apps (first-party tracking) and on various advertising networks that track users across thousands of websites (third-party tracking). All websites collect data on visitor browser activity and store it in a database. They have tools to record the site that users visited prior to coming to the website, where these users go when they leave that site, the type of operating system they use, browser information, and even some location data. They also record the specific pages visited on the particular site, the time spent on each page of the site, the types of pages visited, and what the visitors purchased. Firms analyze this information about customer interests and behavior to develop precise profiles of existing and potential customers. In addition, most major websites have hundreds of tracking programs on their home pages, which track your clickstream behavior across the web by following you from site to site and retarget ads to you by showing you the same ads on different sites. The leading online advertising network is Google's Marketing Platform. Examine Figure 10.3, which illustrates the steps that occur to track website visitors, both on the site they visit and then across the web.

This information enables firms to understand how well their website is working; create unique, personalized web pages that display content or ads for products or services of special interest to each user; improve the customer's experience; and create additional value through a better understanding of the shopper. Examine Figure 10.4, which provides several examples of personalized content that a company might display based on previous data they have collected about a user. By using personalization technology to modify the web pages presented to each customer, marketers achieve some of the benefits of using individual salespeople but at dramatically lower costs. For instance, General Motors may show a Chevrolet display ad emphasizing safety and utility to people over the age of 50, whereas young adults in their twenties may receive ads emphasizing sportiness or performance.

It's a short step from ad networks to programmatic ad buying. Ad networks create real-time bidding (RTB) platforms, where marketers bid in an automated environment for highly targeted slots available from web publishers. Here, ad platforms can predict how many targeted individuals will view the ads, and ad buyers can estimate how much this exposure is worth to them.

What if you are a large national advertising company or global manufacturer trying to reach millions of consumers? With millions of websites, working with each one would be impractical. Advertising networks solve this problem by creating a network of several thousand of the most popular websites (which millions of people visit),

Click 1
The shopper clicks on the home page. The store can tell that the shopper arrived from Yahoo.com at 2:30 PM (which might help determine staffing for customer service centers) and how long she lingered on the home page (which might indicate trouble navigating the site). Tracking beacons load cookies on the shopper's browser to follow her across the Web.
Click 2
Click 3
Click 4
Click 5
The shopper clicks on blouses, then clicks to view a woman's pink blouse. The shopper clicks to select this item in a size 10 in pink and clicks to place it in her shopping cart. This information can help the store determine which sizes and colors are most popular. If the visitor moves to a different site, ads for pink blouses will appear from the same or a different vendor.
Click 6
From the shopping cart page, the shopper clicks to close the browser to leave the website without purchasing the blouse. This action could indicate the shopper changed her mind or that she had a problem with the website's checkout and payment process. Such behavior might signal that the website was not well designed.

**Figure 10.3**
Website Visitor Tracking
*Tools are available to track a shopper's every step through an online store and then across the web as shoppers move from site to site. Close examination of customer behavior at a website selling women's clothing, for example, shows what the store might learn at each step and what actions it could take to increase sales.*

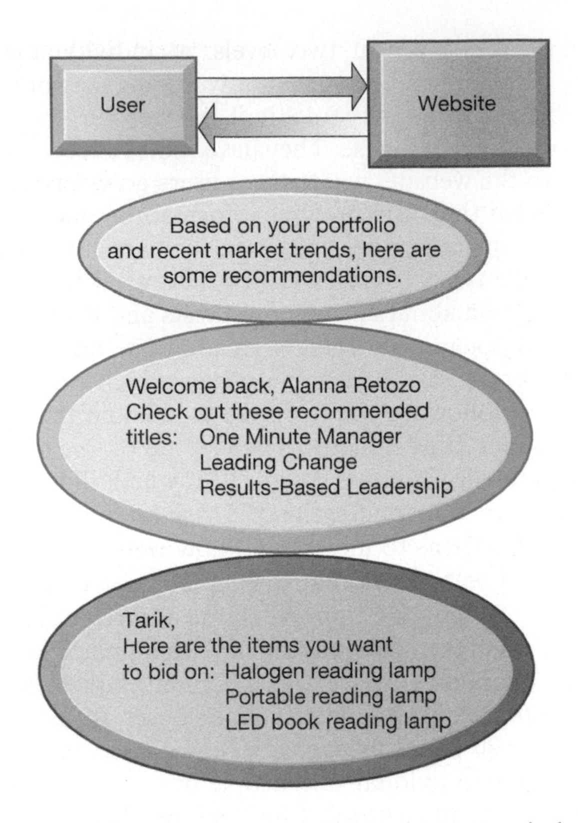


**Figure 10.4** Website Personalization
*Firms can create unique, personalized web pages that display content or ads for products or services of special interest to individual users, improving the customer experience and creating additional value.*

tracking the behavior of these users across the entire network, building profiles of each user, and then selling these profiles to advertisers in a real-time bidding environment. Popular websites download dozens of web-tracking cookies, bugs, and beacons, which report users' online behavior to remote servers without the users' knowledge. Looking for young, single consumers with college degrees, living in the Northeast, in the 18–34 age range who are interested in purchasing a European car? Advertising networks can identify and deliver thousands of people who fit this profile and expose them to ads for European cars as they move from one website to another. Estimates vary, but behaviorally targeted ads are generally 10 times more likely to produce a consumer response than a randomly chosen banner or video ad. Examine Figure 10.5, which illustrates how an advertising network works. When a consumer requests a web page from a website that is a member of an advertising network such as Google Marketing Platform, the website's server connects to the ad network's ad server. That ad server reads the cookie on the originating website and checks its database for a profile. It then selects and serves an appropriate display ad based on the profile. All of this happens within milliseconds. The ad network follows the consumer from site to site through the use of tracking files. View the Figure 10.5 video in the eText for an animated and more detailed discussion of this figure. In 2021, about 90 percent of spending on online display advertising was for targeted ads served by programmatic ad networks, while the rest depended on the context of the pages shoppers visited—the estimated demographics of visitors, or on so-called blast-and-scatter advertising—which is placed randomly on any available page with minimal targeting, such as time of day or season.

It's another short step to **native advertising**. Native advertising involves placing ads in social network newsfeeds or within traditional editorial content, such as a newspaper article. This is also referred to as organic advertising, where content and advertising are in close proximity or are integrated together.

Around 90 percent of US adults are concerned about online privacy. More than half of US adults have decided not to use a product or service because they are worried about how much personal information would be collected about them. Another survey found that 54 percent felt that ads that follow them around are "creepy." (Commisso, 2021; Perrin, 2020; Insider Intelligence/eMarketer, 2021). See Chapter 4 for a more detailed discussion of Internet challenges to privacy, as well as technological responses

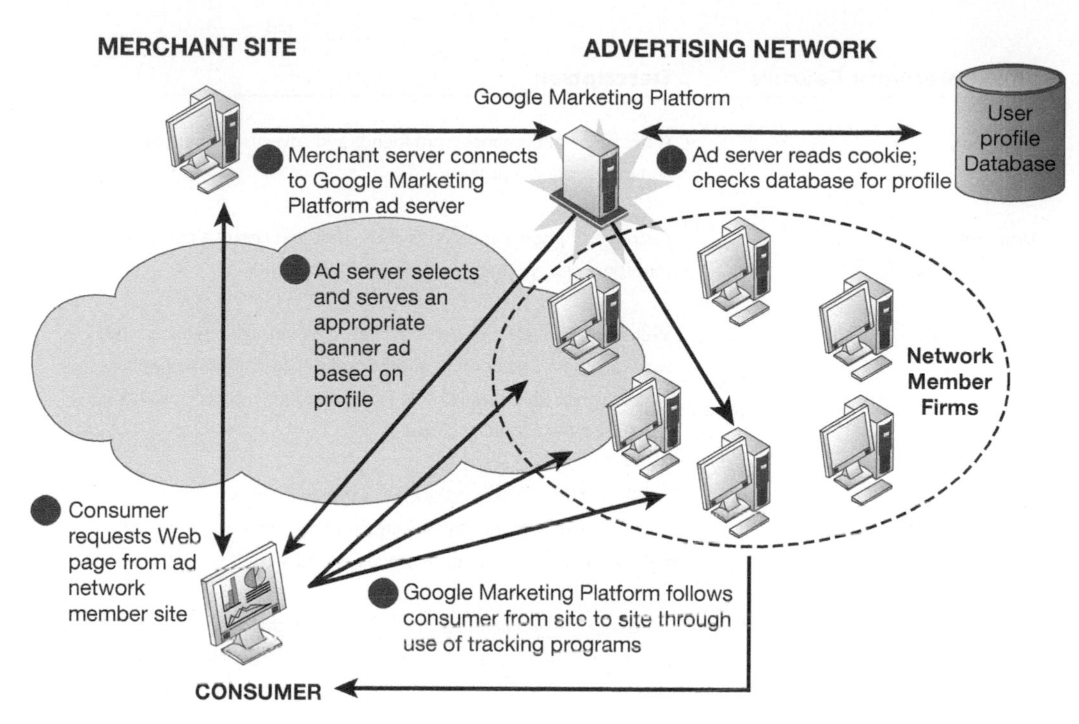
MERCHANT SITE
ADVERTISING NETWORK
Google Marketing Platform
User profile Database
Merchant server connects to Google Marketing Platform ad server
Ad server reads cookie; checks database for profile
Ad server selects and serves an appropriate banner ad based on profile
Network Member Firms
Consumer requests Web page from ad network member site
Google Marketing Platform follows consumer from site to site through use of tracking programs
CONSUMER

**Figure 10.5**
How an Advertising Network Works
*Advertising networks and their use of tracking programs have become controversial among privacy advocates because of these programs' ability to track individual consumers across the Internet.*

to issues related to behavioral tracking, such as Apple's App Tracking Transparency initiative, Google's Privacy Sandbox, and the phasing out of third-party cookies.

## SOCIAL NETWORK MARKETING AND SOCIAL E-COMMERCE

Social media is one of the fastest-growing media for branding and marketing. Companies will spend an estimated $75 billion in 2022 using social networks such as Facebook, Instagram, Pinterest, Twitter, TikTok, and LinkedIn to reach millions of consumers who spend hours a day on social network sites and apps. Social networks in the offline world are collections of people who voluntarily communicate with one another over an extended period of time. Online social networks, along with other sites with social components, use websites and apps that enable users to communicate with one another, form group and individual relationships, and share interests, values, and ideas. Table 10.7 lists and describes some of the primary features of social networks.

**Social e-commerce** is commerce based on the idea of the digital **social graph**, a mapping of all significant online social relationships. The social graph is synonymous with the idea of a social network used to describe offline relationships. You can map your own social graph (network) by drawing lines from yourself to the 10 closest people you know. If they know one another, draw lines between these people. If you are ambitious, ask these 10 friends to list and draw in the names of the 10 people closest to them. What emerges from this exercise is a preliminary map of your social network. Now imagine if everyone on the Internet did the same and posted the results to a large database with a website. Ultimately, you would end up with Facebook or a site like it.

According to small world theory, you are only six links away from any other person on Earth. If you entered your personal address book, which has, say, 100 names in it, into a list and sent it to your friends, and they in turn entered 50 new names of their friends, and so on, five times, the social network created would encompass 31 billion people! The social graph is therefore a collection of millions of personal social graphs (and all the people in them).

If you understand the interconnectedness of people, you will see just how important this concept is to e-commerce. The products and services you buy will influence the decisions of your friends, and their decisions will in turn influence you. If you are a marketer trying to build and strengthen a brand, you can take advantage of the fact

**TABLE 10.7**

Features of Social Networks

| Social Network Feature | Description |
|---|---|
| Newsfeed | A stream of notifications from friends and advertisers that social users find on their home pages. |
| Timelines | A stream of photos and events in the past that create a personal history for users, one that can be shared with friends. |
| Social sign-on | Websites allow users to sign into their sites through their social network pages on Facebook or another social site. This allows websites to receive valuable social profile information from Facebook and use it in their own marketing efforts. |
| Collaborative shopping | An environment where consumers can share their shopping experiences with one another by viewing products, chatting, or texting. Friends can chat online about brands, products, and services. |
| Network notification | An environment where consumers can share their approval (or disapproval) of products, services, or content or share their geolocation, perhaps a restaurant or club, with friends. Facebook's ubiquitous "like" button is an example, as are Twitter's tweets and followers. |
| Social search (recommendations) | An environment where consumers can ask their friends for advice on purchases of products, services, and content. Although Google can help you find things, social search can help you evaluate the quality of things by listening to the evaluations of your friends or their friends. For instance, Amazon's social recommender system can use your Facebook social profile to recommend products. |

that people are enmeshed in social networks, share interests and values, and communicate and influence one another. As a marketer, your target audience is not a million isolated people watching a TV show but the social network of people who watch the show and the viewers' personal networks. Moreover, online social networks are where the largest Internet audiences are located.

Facebook, with more than 175 million monthly US users in 2022, receives most of the public attention given to social networks. Other top social networks include Instagram, with more than 125 million US monthly active users; TikTok, with around 95 million; Snapchat, with around 90 million; Pinterest, with around 85 million; LinkedIn, with around 68 million; and Twitter, with about 60 million. According to analysts, 14.5 percent of the total time spent online in the United States is spent on social network sites, and social networking is one of the most common online activities.

Social e-commerce has taken off within the last few years and is growing at more than 25 percent a year. In 2022, more than 95 million consumers are expected to make a purchase via a social network, and social retail e-commerce revenues are expected to top $54 billion. All the major social networks now include social e-commerce functionality. For instance, in 2020 Facebook launched both Facebook Shops and Instagram Shops, which enable retailers to post their catalogs directly on the social network. Retailers can choose to route customers to their own website or sell directly on the platform. (Insider Intelligence/eMarketer, 2022f, 2022g, 2022h, 2022i).

TikTok is rapidly becoming a major venture for both social marketing and social e-commerce. The Spotlight on Organizations case provides some examples of how companies are using TikTok to connect with their customers. Doing so, however, presents new challenges for companies, which also want to protect their brands.

## SPOTLIGHT ON: ORGANIZATIONS Engaging "Socially" with Customers on TikTok

Social networks such as TikTok offer myriad opportunities for companies to engage consumers, amplify product messages, discover trends and influencers, build brand awareness, respond to customer requests and recommendations, and, increasingly, to sell their products directly to consumers. Social media monitoring helps companies and marketers understand more about buyers' likes, dislikes, and complaints concerning products, additional products or product modifications customers want, and what people are saying about a brand (positive or negative sentiment).

TikTok, the third-most-popular social network in the United States behind Facebook and Instagram, is also one of the fastest growing, with 95 million US users and more than 750 million users worldwide. Launched in 2017, TikTok is a short-form video sharing app owned by Chinese company Bytedance. TikTok videos were initially limited to 15 seconds but can now be up to 10 minutes long. Many TikTok videos feature music, with users lip-syncing, singing, and dancing; others focus on comedy and creativity. Users can "remix" posts from other users and put their own spin on them, using the app's array of editing tools, filters, and other effects. Algorithms analyze the viewing habits of each user and provide content that is customized based on their activity. TikTok skews much younger than other social networks and is the most popular network in the United States among children, teens, and young adults. Almost 45 percent of TikTok's US users are under the age of 25.

Like other social networks, TikTok is full of communities, or subgroups, made up of people who share a common interest, organized by hashtags, such as #BeautyTok, #FashionTok, and #PlantTok. These communities help brands understand what type of content resonates with their potential audience.

Books are not something that is typically associated with members of the Gen Z and Millennial generations. However, the #BookTok community on TikTok, comprised mostly of women in their teens and twenties, had more than 53 billion views as of May 2022. #BookTok members often record time lapses of themselves reading or their emotional reactions to something they have just read. #BookTok has had a significant impact on the book industry, with some books featured in videos seeing a resurgence in sales even though the books may have been published years previously. For instance, *The Song of Achilles*, by Marilyn Miller, was first published in 2012. After being featured in a TikTok video called "books that will make you sob" in August 2021, its sales grew to nine times as much as they were when it was first released. The creator of the "books that will make you sob" video, Selene Velez, known as @moongirlreads on TikTok, now has more than 220,000 followers and has started making videos that publishers pay her to create.

Initially, the vast majority of #BookTok videos happened organically, but as the book publishing industry has become more aware of the trend, it has jumped on board as well. For instance, Barnes and Noble now has #BookTok tables in their stores. Book of the Month began working with #BookTok influencers in late 2020 and in 2021 created a formal #BookTok influencer program. At first, Book of the Month gave its influencers specific guidelines, but it soon discovered that this limited creativity. Once they allowed the creators more freedom, their videos generated many more views. According to Samantha Boures, Book of the Month's manager of media and influencer marketing, the impact of its #BookTok influencers on book sales and subscriptions has been very positive.

Using social networks such as TikTok for brand marketing is not without its drawbacks, however. Keeping a brand's reputation safe in the process is a major concern. For instance, viral "TikTok Challenge" videos are one of TikTok's hallmarks. Some of the challenges have led to serious injury or death. Even if a brand has not sponsored the specific challenge, merely being associated with it can be an issue. The "Sleepy Chicken/NyQuil Chicken" challenge is one such example. The challenge involves cooking chicken in a sedative cough syrup such as NyQuil, a very dangerous action. The manufacturer of NyQuil was forced to issue a statement advising consumers not to use its product in this manner. TikTok has also been plagued with concerns about its content moderation policies and recently agreed to pay $92 million to settle a federal class action suit based on its failure to protect its users' biometric and personal data. TikTok has also been fined by various regulatory agencies for lax data protections for its underage users.

TikTok has attempted to address concerns about brand safety by establishing a Brand Safety Center. It now offers a variety of solutions to help ensure that branded content on the platform is displayed adjacent to suitable videos. It also has introduced an array of initiatives aimed at keeping the TikTok community safe, such as age-related privacy and safety settings and tools to promote kindness, combat bullying, and curb the spread of misinformation. Whether these solutions and tools will truly be effective remains to be seen. However, many brands, lured by the opportunity to gain the exposure to the lucrative Gen Z and Millennial demographics that TikTok provides, appear willing to take the risk.

Sources: Arielle Feger, "Behind BookTok's Popularity," Insider Intelligence/eMarketer, July 27, 2022; Jasmine Enberg, Phoebe Bain, "How BookTok Changed Book of the Month's Influencer Marketing Strategy," Marketingbrew.com, March 7, 2022; Andrew Hutchinson, "TikTok Launches New Brand Safety Center to Provide a Central Hub for Its Various Resources," Socialmediatoday.com, February 15, 2022; Griffin Davis, "TikTok 'Sleepy Chicken' Challenge Warning! Experts Warn Consumers to Avoid 'Nyquil Chicken' Trend", Techtimes.com, January 15, 2022; "BookTokers Are Completely Changing Publishing," Mic.com, November 25, 2021; "Federal Court Gives Preliminary Approval of $92 Million TikTok MDL Settlement Over Objections," *National Law Review*, October 5, 2021; Elizabeth Harris, "How Crying on TikTok Sells Books," *New York Times*, March 20, 2021.

## CASE STUDY QUESTIONS

1. Assess the people, organization, and technology issues for using social media technology to engage with customers.
2. What are the advantages and disadvantages of using social media for advertising, brand building, market research, and customer service?
3. What kinds of companies are best suited to use social network platforms for marketing and customer srevice?

### The Wisdom of Crowds

Creating sites and apps where thousands, even millions, of people can interact offers business firms new ways to market and advertise and to discover who likes (or hates) their products. Some argue that, in a phenomenon called the **wisdom of crowds**, large numbers of people can make better decisions about a wide range of topics or products than a single person or even a small committee of experts can.

Obviously, this is not always the case, but it can happen in interesting ways. In marketing, the wisdom of crowds concept suggests that firms should consult with thousands of their customers first as a way of establishing a relationship with them and, second, to understand better how their products and services are used and appreciated (or rejected). One of most significant benefits of social media is the ability for marketers to understand what people are saying about a brand as well as its competitors, sometimes referred to as **social listening**. According to a recent survey, more than 60 percent of businesses have a social listening system in place. Actively soliciting the comments of your customers builds trust and sends the message to your customers that you care about what they are thinking and that you need their advice (Hutchinson, 2022).

Beyond merely soliciting advice, firms can be actively helped in solving some business problems by using **crowdsourcing**. For instance, BMW launched a crowdsourcing project to enlist the aid of customers in designing an urban vehicle for 2025. In 2021, it created a Crowd Innovation platform to tap into the wisdom of both internal and external innovation crowds. Kickstarter is arguably one of the most famous crowdfunding sites where visitors can help fund creative projects such as films, music, art, books, and the creation of new, innovative products. Other crowdsourcing examples include Caterpillar working with customers to design better machinery, Lego's Lego Idea platform for developing new toys and games, and Unilever's use of a digital crowdsourcing platform to create a community of local people in Indonesia to track palm oil throughout the production process (See the Chapter 6 ending case study).

## 10-4 Describe how e-commerce has affected business-to-business transactions.

Trade among business firms (business-to-business commerce, or B2B) represents a huge marketplace. The total amount of B2B trade in the United States in 2022 is estimated to be about $16 trillion, with B2B e-commerce (online B2B) contributing about $8.5 trillion. B2B e-commerce is expected to grow, reaching about $10 trillion in the United States by 2026 (Insider Intelligence/eMarketer, 2022j; U.S. Bureau of the Census, 2021; authors' estimates).

The process of conducting trade among business firms is complex and requires considerable human intervention; therefore, it consumes significant resources. Some firms estimate that each corporate purchase order for support products costs them, on average, at least $100 in administrative overhead, including processing paper, approving purchase decisions, using the telephone and fax machines to search for products and arrange for purchases, arranging for shipping, and receiving the goods. Across the economy, this adds up to trillions of dollars spent annually for procurement processes that could be automated. If even just a portion of interfirm trade were automated and parts of the entire procurement process were assisted by the Internet, literally trillions of dollars might be released for more productive uses; consumer prices potentially would fall; productivity would increase; and the economic wealth of the nation would expand. This is the promise of B2B e-commerce. The challenge of B2B e-commerce is changing existing patterns and systems of procurement and designing and implementing new Internet and cloud-based B2B solutions.

### ELECTRONIC DATA INTERCHANGE (EDI)

B2B e-commerce refers to the commercial transactions that occur among business firms. Increasingly, these transactions are flowing through a variety of Internet-enabled mechanisms. About 75 percent of online B2B e-commerce is still based on proprietary systems for **Electronic Data Interchange (EDI)**. EDI enables the computer-to-computer exchange of standard transactions such as invoices, bills of lading, shipment schedules, or purchase orders, between two organizations. Transactions are automatically transmitted from one information system to another through a network, eliminating the printing and handling of paper at one end and the inputting of data at the other. Each major industry in the United States and much of the rest of the world has EDI standards that define the structure and information fields of digital transactions for that industry.

EDI originally automated the exchange of documents such as purchase orders, invoices, and shipping notices. Although many companies still use EDI for document automation, firms engaged in just-in-time inventory replenishment and continuous production use EDI as a system for continuous replenishment. Examine Figure 10.6, which illustrates the interchange of shipping data, payment data, and production/inventory requirements between supplier and purchasing firms. Suppliers have online access to selected parts of the purchasing firm's production and delivery schedules and automatically ship materials and goods to meet prespecified targets without intervention by firm purchasing agents.

Although many organizations still use private networks for EDI, these networks are increasingly web-enabled because Internet technology provides a much more flexible and low-cost platform for linking to other firms. Businesses can extend digital technology to a wider range of activities and broaden their circle of trading partners.

Procurement, for example, involves not only purchasing goods and materials but also sourcing, negotiating with suppliers, paying for goods, and making delivery arrangements. Businesses can now use the Internet to locate the lowest-cost supplier,

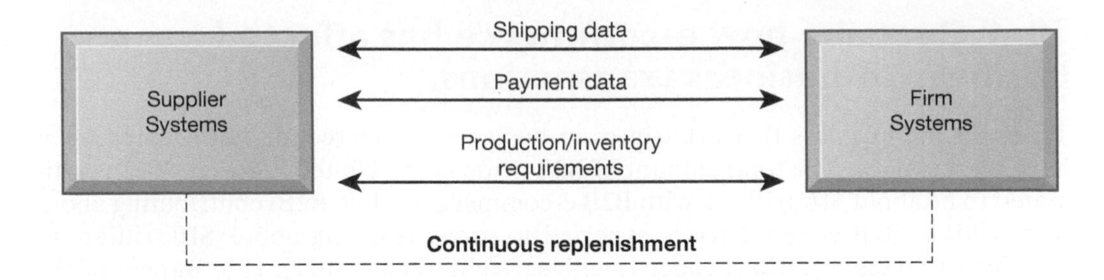


**Figure 10.6** Electronic Data Interchange (EDI)
*Companies use EDI to automate transactions for B2B e-commerce and continuous inventory replenishment. Suppliers can automatically send data about shipments to purchasing firms. The purchasing firms can use EDI to provide production and inventory requirements and payment data to suppliers.*

search online catalogs of supplier products, negotiate with suppliers, place orders, make payments, and arrange transportation. They are no longer limited to partners linked by traditional EDI networks.

## NEW WAYS OF B2B BUYING AND SELLING

The Internet and web technology enable businesses to create online storefronts for selling to other businesses using the same techniques as used for B2C commerce. Alternatively, businesses can use Internet technology to create extranets or digital marketplaces for linking to other businesses for purchase and sale transactions.

**Private B2B networks** typically consist of a large firm using a secure website to link to its suppliers and other key business partners. Examine Figure 10.7, which illustrates a private B2B network. The buyer owns the network, which permits the firm and its designated suppliers, distributors, and other business partners to share product design and development, marketing, production scheduling, inventory management, and unstructured communication, including graphics and email.

An example is VW Group Supply, which links the Volkswagen Group and its suppliers. VW Group Supply handles 90 percent of all global purchasing for Volkswagen, including all automotive and parts components.

**B2B e-commerce marketplaces** provide a single, digital marketplace based on Internet technology for many buyers and sellers. They are industry-owned or operate as independent intermediaries between buyers and sellers. B2B e-commerce marketplaces generate revenue from purchase and sale transactions and other

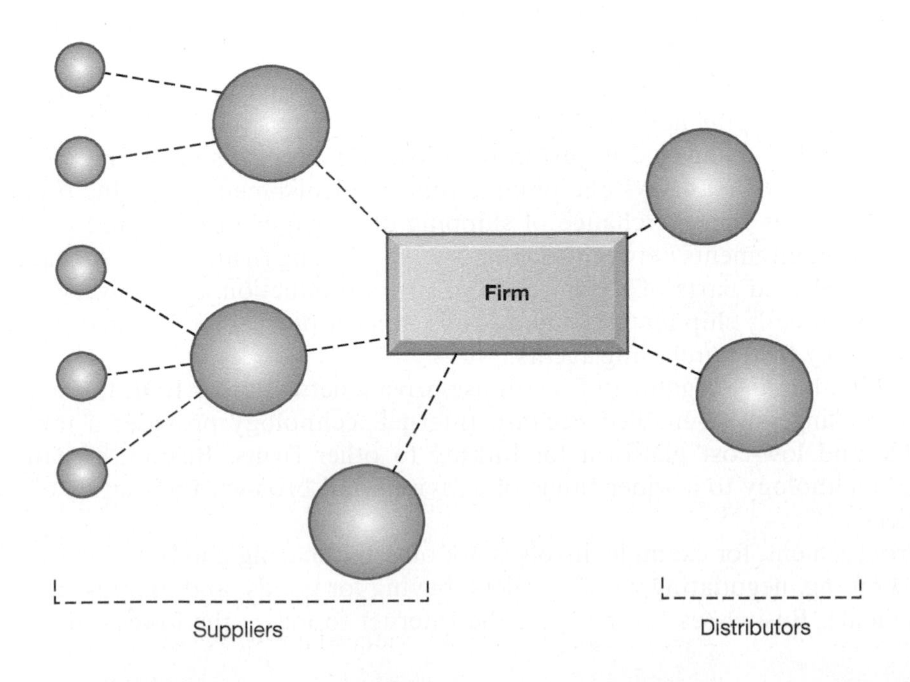


**Figure 10.7** A Private B2B Network
*A private B2B network links a firm to its suppliers, distributors, and other key business partners for efficient supply chain management and other collaborative commerce activities.*

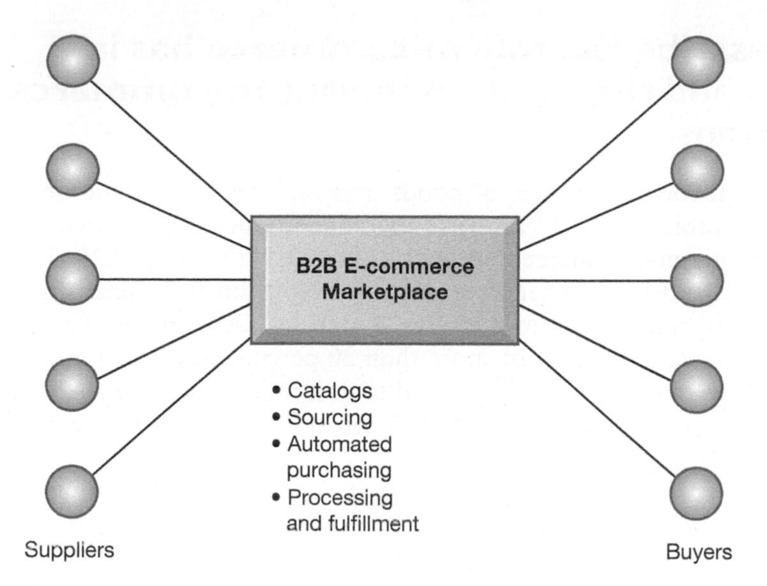


**Figure 10.8**
**A B2B E-commerce Marketplace**
*B2B e-commerce marketplaces are online marketplaces where multiple buyers can purchase from multiple sellers.*

services provided to clients. Participants in B2B e-commerce marketplaces can establish prices through online negotiations, auctions, or requests for price quotes, or they can use fixed prices. Examine Figure 10.8, which illustrates how a B2B e-commerce marketplace connects multiple suppliers and multiple buyers, providing online catalogs of goods as well as tools for sourcing, automated processing, and fulfillment.

There are many types of B2B e-commerce marketplaces and ways of classifying them. Some sell direct goods and some sell indirect goods. **Direct goods** are goods used in a production process, such as sheet steel for auto body production. **Indirect goods** are all other goods not directly involved in the production process, such as office supplies or products for maintenance and repair. Some B2B e-commerce marketplaces support contractual purchasing based on long-term relationships with designated suppliers, and others support short-term spot purchasing, where goods are purchased based on immediate needs, often from many suppliers.

Some B2B e-commerce marketplaces serve vertical markets for specific industries, such as automobiles, telecommunications, or machine tools, whereas others serve horizontal markets for goods and services, such as office equipment or transportation, that can be found in many industries.

SupplyOn is an example of an industry-owned B2B e-commerce marketplace, focusing on long-term contract purchasing relationships and on providing common networks and computing platforms for reducing supply chain inefficiencies. SupplyOn was founded by automative suppliers Bosch, Continental, Schaeffler, and ZF and originally focused on the automotive industry, SupplyOn has since expanded to serve firms in the aerospace, railway, and engineering industries. More than 140,000 businesses worldwide uses SupplyOn's supply chain network and solutions, and it counts among its customers such well-known companies as Airbus, BMW Group, BorgWarner, Siemens, and Thales

**Exchanges** are independently owned third-party B2B e-commerce marketplaces that connect thousands of suppliers and buyers for spot purchasing. Many exchanges provide vertical markets for a single industry, such as food, electronics, or industrial equipment, and they primarily deal with direct inputs. For example, Go2Paper enables a spot market for paper, board, and craft among buyers and sellers in the paper industries from more than 75 countries.

Exchanges proliferated during the early years of e-commerce, but many failed. However, in recent years, exchanges have experienced a resurgence.

## 10-5 Describe the role m-commerce has in business and the most important m-commerce applications.

M-commerce involves the sales of goods and services via mobile devices, such as smartphones, tablets, and wearables. Examine Figure 10.9, which illustrates the growth of retail m-commerce revenues from 2019 through 2025. In 2022, retail m-commerce is expected to generate about $420 billion in revenues, which is about 40 percent of all retail e-commerce sales. Retail m-commerce is the fastest-growing form of e-commerce, growing by more than 30 percent from 2019 to 2021, primarily because of the pandemic, and is expected to continue increasing at the rate of about 17 percent per year between 2022 and 2025, when it is estimated to reach $666 billion (Insider Intelligence/eMarketer, 2022c).

All the top retailers now have mobile websites that enable shoppers to use their mobile devices to shop and place orders. Virtually all large traditional and online retailers also have apps that enable m-commerce.

Banks and credit card companies have developed services that let customers manage their accounts from their mobile devices. JPMorgan Chase and Bank of America customers can use their smartphones to check account balances, transfer funds, and pay bills. Apple Pay for the iPhone and Apple Watch, along with Android smartphone apps, allow users to charge items to their credit card accounts with a swipe of their phone. (For more informaton on these types of apps, see the following section on mobile app payment systems.)

The main areas of growth in m-commerce are mass market retailing such as Amazon's; sales of digital content such as music, TV shows, movies, and e-books; and in-app sales via mobile devices. On-demand firms such as Uber (described in the chapter-ending case) and Airbnb are location-based services and examples of m-commerce as well. Larger mobile screens and more-convenient payment procedures also play roles in the expansion of m-commerce.

The mobile advertising market is the fastest-growing online ad platform, racking up an estimated $170 billion in ad revenue in 2022. Meta's Facebook/Instagram is the largest mobile advertising market, expected to post about $55 billion in mobile ad revenue (more than 95 percent of its total ad revenue) in 2022, with Google number two at about $45 billion (65 percent of its total digital ad business). Google is displaying ads linked to mobile searches. Ads are embedded in games, videos, and other mobile applications (Insider Intelligence/eMarketer, 2022k).

**Figure 10.9**
Retail M-commerce Revenues
*Retail m-commerce is the fastest-growing type of B2C e-commerce. By 2025, retail m-commerce sales are expected to account for almost 45 percent of total e-commerce sales.*

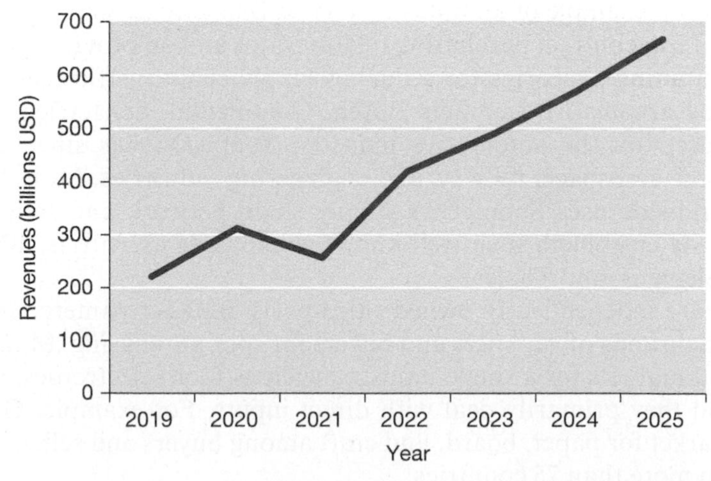


*Source:* Data from Insider Intelligence/eMarketer chart "US Retail Mcommerce Sales," 2022.

## LOCATION-BASED SERVICES AND APPLICATIONS

**Location-based services** include geosocial, geoadvertising, and geoinformation services. Seventy-four percent of smartphone owners use location-based services. What ties these activities together and is the foundation for mobile commerce is the global positioning system (GPS)-enabled map services available on smartphones. A **geosocial service** can tell you where your friends are meeting. **Geoadvertising services** can tell you where to find the nearest Italian restaurant, and **geoinformation services** can tell you the price of a house you are looking at or about special exhibits at a museum you are passing. Some of the most well-known location-based services are on-demand firms such as Uber, Lyft, Airbnb, and hundreds more that provide services to users in local areas and are based on the user's location (or, in the case of Airbnb, the user's intended travel location).

Waze is an example of a popular, social geoinformation service. Waze is a GPS-based map and navigational app for smartphones and is now owned by Google. Waze locates the user's car on a digital map using GPS and, like other navigation programs, collects information on the user's speed and direction continuously. What makes Waze different is that it collects traffic information from users who submit accident reports, speed traps, landmarks, street fairs, protests, and even addresses. Waze uses this information to suggest alternative routes, give travel times, and issue warnings and can even make recommendations for gas stations along the way. The Waze app is used extensively by Uber and Lyft drivers and by more than 140 million other drivers in the United States.

Foursquare is an example of geosocial services. Geosocial services help you find friends, or help your friends find you, by checking in to the service and announcing your presence in a restaurant or other place. Your friends are instantly notified. Foursquare's Swarm app provides a location-based social network service that enables its users to connect with friends, update their location, and provide reviews and tips for enjoying a location. Points are awarded for checking in at designated venues. Users choose to post their check-ins on their accounts on Twitter, Facebook, or both. Users also earn badges by checking in at locations with certain tags, for check-in frequency, or for the time of check-in.

Connecting people to local merchants in the form of geoadvertising is the economic foundation for m-commerce. Geoadvertising sends ads to users based on their GPS locations. Smartphones report their locations back to Google and Apple. Merchants buy access to these consumers when they come within range of a merchant. For instance, Kiehl Stores, a cosmetics retailer, sent special offers and announcements to customers who came within 100 yards of their store. Shopkick is a mobile app that enables retailers such as Best Buy and Macy's to offer coupons to people when they walk into their stores. The Shopkick app automatically recognizes when the user has entered a partner retail store and offers a new virtual currency called kickbucks, which can be redeemed for store gift cards.

## MOBILE APP PAYMENT SYSTEMS

Mobile app payment systems use mobile apps to replace credit cards and traditional banking services. There are three main types of mobile payment apps. Near-field communication (NFC)-driven systems enable NFC-enabled smartphones and other mobile devices to make contactless payments by communicating with an NFC-enabled reader at a merchant's point-of-sale (POS) terminal in close physical proximity. (Review the discussion of radio-frequency identification [RFID] and near field communication [NFC] in Chapter 7.) Apple Pay and

**TABLE 10.8**

Types of Mobile App Payment Systems

| Type of Mobile App Payment System | Description | Examples |
|---|---|---|
| Near Field Communication (NFC) | System using technology driven by near field communication (NFC) chips in both payers' mobile devices and merchants' point-of-sale (POS) reader devices. When close together and activated, these NFC chips exchange encrypted data to complete a payment. Can be used with many different merchants if they use NFC readers and software for accepting payments. | Apple Pay, Google Pay, Samsung Pay |
| QR Code | Quick Response (OR) code technology uses a two-dimensional barcode in which information is encoded to perform contactless transactions using a code-scanning and generation app on a smartphone and a compatible merchant device. After a merchant enters the payment amount, the customer opens an app that displays a QR code generated for the transaction. The merchant scans the QR code, and the payment amount is deducted from the customer's mobile wallet. Alternatively, the customer opens the app and scans a QR code displayed by the merchant, enabling the app to identify the merchant. The customer then supplies the amount and completes the payment. | Starbucks, Walmart, Target, Dunkin' Donuts |
| Peer-to-peer (P2P) payment system | Technology allowing individuals to transfer funds from their bank accounts to other accounts on the same platform via the Internet. P2P users establish a secure account with a trusted third-party vendor, designating their bank account or credit card account to send and accept funds. Using the third-party app, users can send money to another person or to a merchant's account. Users are generally identified by their email address or mobile phone number. | Venmo, Zelle |

Google Pay are examples. QR Code payment systems, such as Walmart Pay, use a contactless payment method. A payment is initiated scanning a two-dimensional barcode called a QR (Quick Response) code using a mobile app on the payer's smartphone. Peer-to-peer (P2P) payment systems such as Venmo or Zelle QuickPay are used for transferring money among individuals who have installed a proprietary app. Table 10.8 compares these three types of mobile app payment systems.

## 10-6 Identify the issues that must be addressed when building an e-commerce presence.

Building a successful e-commerce presence requires a keen understanding of business, technology, and social issues as well as a systematic approach. Today, an e-commerce presence is not just a corporate website but also includes a social network account on Facebook/Instagram, a Twitter feed, and smartphone apps where customers can access your services. Developing and coordinating all these customer venues can be difficult. A complete treatment of the topic is beyond the

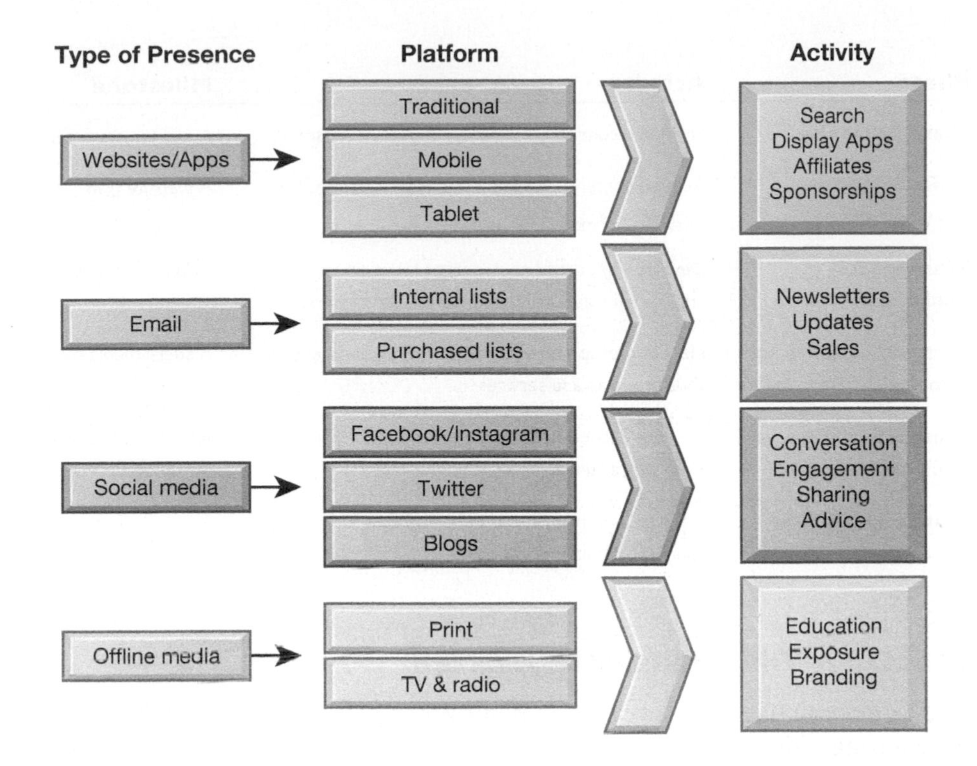


**Figure 10.10**
E-commerce Presence Map
*An e-commerce presence requires firms to consider four types of presence, with specific platforms and activities associated with each.*

scope of this text, and students should consult books devoted to just this topic (Laudon and Traver, 2024). The two most important management challenges in building a successful e-commerce presence are (1) developing a clear understanding of your business objectives and (2) knowing how to choose the right technology to achieve those objectives.

## DEVELOP AN E-COMMERCE PRESENCE MAP

E-commerce has moved from being a PC-centric activity on the web to a mobile and tablet-based activity. Currently, more than 90 percent of Internet users in the United States use a mobile device to shop for goods and services, look up prices, enjoy entertainment, and access social networks. Your potential customers use these various devices at different times during the day and involve themselves in different conversations, depending on what they are doing—touching base with friends, tweeting, or reading a blog. Each of these is a touch point where you can meet the customer, and you have to think about how you develop a presence in these different virtual places. Figure 10.10 provides a roadmap to the platforms and related activities you will need to think about when developing your e-commerce presence.

Examine Figure 10.10, which illustrates four kinds of e-commerce presence: websites/apps, email, social media, and offline media. You must address different platforms for each of these types. For instance, in the case of website presence, there are three platforms: traditional desktop, tablets, and smartphones, each with different capabilities. Mobile apps can be in conjunction with a website or as standalone presence. Moreover, for each type of e-commerce presence, there are related activities you will need to consider. For instance, in the case of websites, you will want to engage in search engine marketing, display ads, affiliate programs, and sponsorships. Offline media, the fourth type of e-commerce presence, is included here because many firms use multiplatform or integrated marketing by which print ads refer customers to websites.

**TABLE 10.9**

E-commerce Presence Timeline

| Phase | Activity | Milestone |
|---|---|---|
| **Phase 1: Planning** | Envision e-commerce presence; determine personnel. | A mission statement |
| **Phase 2: Website development** | Acquire content; develop a site design; arrange for hosting the site. | A website plan |
| **Phase 3: Web implementation** | Develop keywords and metatags; focus on search engine optimization; identify potential sponsors. | A functional website |
| **Phase 4: Social media plan** | Identify appropriate social platforms and content for your products and services. | A social media plan |
| **Phase 5: Social media implementation** | Develop Facebook/Instagram, Twitter, TikTok, and Pinterest presence. | A functioning social media presence |
| **Phase 6: Mobile plan** | Develop a mobile plan; consider options for porting your website to smartphones. | A mobile media plan |

### DEVELOP A TIMELINE: MILESTONES

Where would you like to be a year from now? When you begin, it's very helpful for you to have a rough idea of the timeframe for developing your e-commerce presence. You should break your project down into a small number of phases that could be completed within a specified time. Table 10.9 illustrates a one-year timeline for the development of an e-commerce presence for a startup company devoted to fashions for teenagers.

## 10-7 Understand how MIS can help your career.

Here is how this chapter and this text can help you find a job as a junior e-commerce associate.

### THE COMPANY

Leila X Designs is a small, rapidly growing company that offers an upscale but affordable collection of area rugs, lighting fixtures, and accent furniture. Leila X sells the products it offers via a variety of methods: through physical retailers such as Pier 1, online marketplaces such as Wayfair, and directly to consumers via its website, which uses the Shopify platform. Leila X is looking for a recent college graduate to fill a junior e-commerce associate position to help it continue growing.

### POSITION DESCRIPTION

The junior e-commerce associate will work with Leila X's e-commerce team to optimize the day-to-day operations of its website and enhance its customers' digital experience. Job responsibilities include:

- Collaborating with the buying, marketing, and e-commerce operations team.
- Uploading contents such as product descriptions, details, and images to the Shopify platform.
- Refreshing product pages to maintain newness on the website.
- Creating compelling landing pages for product and category launches.
- Identifying opportunities to increase website conversion.
- Maintaining a calendar to track product launches, site updates, home page refreshes, and sales/promotions.
- Collaborating with search optimization agency to plan and help execute site optimazations and implement best website practices.

### JOB REQUIREMENTS

- Bachelor's degree in MIS, e-commerce, or digital marketing
- Experience with website management and operations
- Familiarity with Shopify
- Knowledge of Google Analytics
- Proficiency with Microsoft Excel, Microsoft Word, and Adobe Photoshop
- Ability to work remotely
- Strong communication and organizational skills

### INTERVIEW QUESTIONS

1. Why do you think you would be a good fit for this position? What courses have you taken that have helped prepare you for this position?
2. What experience do you have developing website content and/or managing and operating a website?
3. Have you ever analyzed data about website performance or online customer behavior? What tools did you use to do so?
4. What experience do you have using Shopify? Are there any new features or functionalities of the platform that you think we should adopt?
5. How would you propose working with our nontechnical teams in telling a story about customer data insights so that those teams are able to drive customer engagement and loyalty and execute more effectively?
6. What do you know about search engine optimization? Have you ever used SEO tools on the job? What did you do with them?

### AUTHOR TIPS

1. Review this chapter and also the discussion of search and search engine marketing in Chapter 7.
2. Use the web to do more research on this company. Try to find out more about its strategy, competitors, and business challenges.
3. Visit Shopify to learn about the features and functionality that it offers.
4. Research other online marketplaces that might be appropriate for the company to use to sell its products.

## Review Summary

**10-1 Identify the unique features of e-commerce, digital markets, and digital goods.** E-commerce involves digitally enabled commercial transactions between and among organizations and individuals. Unique features of e-commerce technology include ubiquity, global reach, universal technology standards, richness, interactivity, information density, capabilities for personalization and customization, and social technology. E-commerce is becoming increasingly social, mobile, and local.

Digital markets are said to be more transparent than traditional markets, with reduced information asymmetry, search costs, transaction costs, and menu costs along with the ability to change prices dynamically based on market conditions. Digital goods, such as music, video, software, and books, can be delivered over a digital network. Once a digital product has been produced, the cost of delivering that product digitally is extremely low.

**10-2 Compare the principal e-commerce business and revenue models.** E-commerce business models include online retailers, transaction brokers, market creators, content providers, community providers, service providers, and portals. The principal e-commerce revenue models are advertising, sales, subscription, free/freemium, transaction fee, and affiliate.

**10-3 Explain how e-commerce has transformed marketing.** The Internet provides marketers with new ways of identifying and communicating with millions of potential customers at costs far lower than traditional media costs. Crowdsourcing using the wisdom of crowds helps companies learn from customers how to improve product offerings and increase customer value. Behavioral targeting techniques increase the effectiveness of display and video ads. Social network advertising uses social networks to improve targeting of products and services and to drive social e-commerce.

**10-4 Describe how e-commerce has affected business-to-business transactions.** B2B e-commerce generates efficiencies by enabling companies to use the Internet and other digital tools to locate suppliers, solicit bids, place orders, and track shipments in transit. Private B2B networks link a firm with its suppliers and other strategic business partners to develop highly efficient and responsive supply chains. B2B e-commerce marketplaces provide a single, digital marketplace for many buyers and sellers. Private B2B networks link a firm with its suppliers and other strategic business partners to develop highly efficient and responsive supply chains.

**10-5 Describe the role m-commerce has in business and the most important m-commerce applications.** M-commerce involves the sales of goods and services via mobile devices, such as smartphones, tablets, and wearables. M-commerce is especially well suited for location-based services and apps. Mobile devices are being used for mobile bill payment, banking, securities trading, transportation schedule updates, and downloads of digital content such as music, games, and video clips. Mobile payment systems include technologies for near field communication, QR codes, and peer-to-peer payments. The GPS capabilities of smartphones make geoadvertising, geosocial, and geoinformation services possible.

**10-6 Identify the issues that must be addressed when building an e-commerce presence.** Building a successful e-commerce presence requires a clear understanding of the business objectives to be achieved and selecting the right platforms, activities, and timeline to achieve those objectives. An e-commerce presence includes not only a corporate website but also a presence on Facebook, Instagram, Twitter, and other social networks.

## Key Terms

Advertising revenue model, 364
Affiliate revenue model, 365
B2B e-commerce marketplaces, 374
Behavioral targeting, 366
Business-to-business (B2B) e-commerce, 359
Business-to-consumer (B2C) e-commerce, 359
Community providers, 363
Consumer-to-consumer (C2C) e-commerce, 359
Cost transparency, 355
Crowdsourcing, 372
Customization, 356
Digital goods, 358
Direct goods, 375
Disintermediation, 358
Download, 361
Dynamic pricing, 357
Electronic Data Interchange (EDI), 373
E-tailers, 360
Exchanges, 375
Fintech, 361
Free/freemium revenue model, 365
Geoadvertising services, 377
Geoinformation services, 377
Geosocial services, 377
Indirect goods, 375
Information asymmetry, 356
Information density, 355
Location-based services, 377
Long tail marketing, 365
Market creator, 361
Market entry costs, 355
Marketspace, 354
Menu costs, 357
Mobile commerce (m-commerce), 359

Native advertising, 368
Online retailer, 360
Personalization, 356
Price discrimination, 355
Price transparency, 355
Private B2B networks, 374
Revenue model, 364
Richness, 355
Sales revenue model, 364
Search costs, 355
Social e-commerce, 369
Social graph, 369
Social listening, 372
Streaming, 361
Subscription revenue model, 364
Transaction costs, 354
Transaction fee revenue model, 365
Wisdom of crowds, 372

## Review Questions

**10-1 Identify the unique features of e-commerce, digital markets, and digital goods.**
- Name and describe four business trends and three technology trends shaping e-commerce today.
- List and describe the eight unique features of e-commerce.
- Define a digital market and digital goods and describe their distinguishing features.

**10-2 Compare the principal e-commerce business and revenue models.**
- Name and describe the principal e-commerce business models.
- Name and describe the e-commerce revenue models.

**10-3 Explain how e-commerce has transformed marketing.**
- Explain how social networks and the wisdom of crowds help companies improve their marketing.
- Define behavioral targeting and explain how it works at individual websites and on advertising networks.
- Define the social graph and explain how it is used in e-commerce marketing.

**10-4 Describe how e-commerce has affected business-to-business transactions.**
- Explain how Internet technology supports business-to-business e-commerce.
- Define and describe B2B e-commerce marketplaces, and explain how they differ from private industrial networks (private exchanges).

**10-5 Describe the role m-commerce has in business and the most important m-commerce applications.**
- List and describe important types of m-commerce services and applications.
- List and describe three types of mobile app payment systems.

**10-6 Identify the issues that must be addressed when building an e-commerce presence.**
- List and describe four types of e-commerce presence and the platforms and activities associated with each.

## Discussion Questions

**10-7** MyLab MIS How does the Internet change consumer and supplier relationships?

**10-8** MyLab MIS Marketers defend the use of behavioral targeting as a method to expose people to advertising that is uniquely suited to them. Do you agree? Why or why not?

**10-9** MyLab MIS How have social technologies changed e-commerce?

**MyLab MIS**
To complete these problems, go to EOC Discussion Questions in **MyLab MIS.**

# Hands-On MIS Projects

The projects in this section give you hands-on experience developing e-commerce strategies for businesses, using spreadsheet software to research the profitability of an e-commerce company, and using web tools to research and evaluate e-commerce hosting services. Visit MyLab MIS to access this chapter's Hands-On MIS Projects.

## MANAGEMENT DECISION PROBLEM

**10-10** T.J. Maxx, the giant off-price clothing retailer with nearly 1,300 stores around the world, has experienced problems implementing an e-commerce site for online sales. An off-price store like T.J. Maxx buys excess inventory and last year's fashions from a vast network of department stores and manufacturers. However, it buys these items in much smaller lots than traditional retailers such as Nordstrom or Walmart do, with much of its inventory consisting of one-time items in small quantities. T.J. Maxx has new brand-name and designer fashions arriving every week. The inventory varies a great deal from one store to the next, and you never know what you'll find when you visit a T.J. Maxx store. Shoppers are lured into the stores in the hope that they might find a really hot bargain that might be available for only a few days, and that is part of TJ Maxx's special appeal. An online storefront can also cannibalize in-store sales. On the other hand, ignoring e-commerce can mean losing market share to competitors. Define the problem for TJ Max using the problem-solving steps outlined in Chapter 1 and suggest a solution. What people, organization, and technology factors should be considered??

## IMPROVING DECISION MAKING: USING SPREADSHEET SOFTWARE TO ANALYZE AN E-COMMERCE BUSINESS

Software skills: Spreadsheet downloading, formatting, and formulas
Business skills: Financial statement analysis

**10-11** Pick an e-commerce company that is publicly traded in the United States—for example, Warby Parker, Carvana, or Etsy. Visit the company's website and review the web pages that describe the company and explain its purpose and structure. Use the web to find articles that comment on the company. Then visit the Securities and Exchange Commission's website at Sec.gov to access the company's 10-K (annual report) form showing income statements and balance sheets. Select only the sections of the 10-K form containing the desired portions of financial statements that you need to examine and download them into your spreadsheet. (MyLab MIS provides more detailed instructions on how to download this 10-K data into a spreadsheet.) Create simplified spreadsheets of the company's balance sheets and income statements for the past three years.

- Is the company a success, borderline business, or failure? What information provides the basis of your decision? Why? When answering these questions, pay special attention to the company's three-year trends in revenues, costs of sales, gross margins, operating expenses, and net margins.
- Prepare a presentation (with a minimum of five slides), including appropriate spreadsheets or charts, and present your work to your professor and classmates.

## ACHIEVING OPERATIONAL EXCELLENCE: EVALUATING E-COMMERCE HOSTING SERVICES

Software skills: Web browser software
Business skills: Evaluating e-commerce hosting services

**10-12** This project will help develop your Internet skills in evaluating commercial services for hosting an e-commerce site for a small startup company.

You would like to set up a website to sell towels, linens, pottery, and tableware from Portugal and are examining services for hosting small-business website storefronts. Your website should be able to take secure credit card payments and calculate shipping costs and taxes. Initially, you would like to display photos and descriptions of 40 products. Visit Wix, GoDaddy, and Square, and compare the range of e-commerce hosting services they offer to small businesses, their capabilities, and their costs. Examine the tools they provide for creating an e-commerce site. Compare these services and decide which you would use if you were actually establishing a web store. Write a brief report indicating your choice and explaining the strengths and weaknesses of each service.

## COLLABORATION AND TEAMWORK PROJECT

Performing a Competitive Analysis of E-commerce Sites

**10-13** Form a group with three or four of your classmates. Select two businesses that are competitors in the same industry and that use their websites for e-commerce. Visit these websites. You might compare, for example, the websites for Pandora and Spotify, Petsmart and Chewy, or E*Trade and TD Ameritrade. Prepare an evaluation of each business's website in terms of its functions, user-friendliness, and ability to support the company's business strategy. Which website does a better job? Why? Can you make some recommendations to improve these websites? If possible, use Google Docs and Google Drive or Google Sites to brainstorm, organize, and develop a presentation of your findings for the class.

# BUSINESS PROBLEM-SOLVING CASE

## UBER DISCOVERS THAT BECOMING THE UBER OF EVERYTHING IS NOT SO EASY

You're in New York, Paris, Chicago, or another major city and need a ride. Instead of trying to hail a cab, you pull out your smartphone and tap the Uber app. A Google map pops up displaying your nearby surroundings. You select a spot on the screen designating an available driver, and the app secures the ride, showing how long it will take for the ride to arrive and how much it will cost. Once you reach your destination, the fare is automatically charged to your credit card. No fumbling for money. Uber is an example of an on-demand service provider. It offers a platform that connects consumers who require a service with suppliers who can quickly fulfill that demand.

Many decisions required for running the business do not require humans, relying instead on finely tuned computer algorithms. Uber's systems decide how much to price a ride in periods of peak and slow demand and where drivers should relocate to find more ride-hailing passengers. Uber drivers receive in-app notifications, heat maps, and emails with real-time and predictive information. Uber systems also use the accelerometer in drivers' phones along with GPS and gyroscope to track drivers' performance and send them safe driving reports. Uber's driver rating system also is automated. In certain Uber services, if drivers fall below 4.6 stars on a 5-star rating system, they may be "deactivated."

Uber runs much leaner than a traditional taxi company does. Uber does not own taxis. It categorizes its drivers as independent contractors (often referred to as "gig workers"); rather than paying them a salary, it instead gives them a cut of each fare. Uber is not encumbered with employee costs for drivers such as workers' compensation, minimum wage requirements, training, health insurance, or commercial licensing costs. Uber has shifted the costs of running a taxi service almost entirely to the drivers. Drivers pay for their own cars, fuel, and insurance, something that has become increasingly problematic in 2022 as fuel prices dramatically escalated. Although Uber began adding a fuel surcharge in March 2022 of between 35 to 55 cents per ride or delivery, drivers say that it is far from adequate.

Uber is headquartered in San Francisco and was founded in 2009 by Travis Kalanick and Garrett Camp. In 2022, it has an estimated 4 million drivers working in around 10,000 cities in 72 countries worldwide. Uber's business strategy has been to expand as quickly as possible, forgoing short-term profits in favor of laying the groundwork for long-term returns. In 2021, it recorded an operating loss of $3.8 billion, and it has an accumulated deficit of an astounding $23.6 billion.

Although Uber began business solely as an alternative to traditional taxis, it has expanded its horizons to envisioning itself as a platform for a variety of different services associated with the movement of people and things from one point to another. While its flagship offering is still mobility services that provide rides for consumers in a variety of vehicles, it now is almost as equally focused on restaurant food delivery services and freight services. It sees itself as the "Amazon" of transportation, with the potential to become the dominant force in all forms of transportation services. But Uber faces significant challenges in each of these areas and has made some major missteps along the way.

By digitally disrupting a traditional and highly regulated industry, Uber ignited a firestorm of opposition from existing taxi services in the United States and around the world. Uber also has been accused of denying its drivers the benefits of employee status by classifying them as contractors, violating public transportation laws and regulations throughout the world, abusing the personal information it has collected on ordinary people, increasing traffic congestion, undermining public transportation, and failing to protect public safety by refusing to perform sufficient criminal, medical, and financial background checks on its drivers. Uber's brand image had been further tarnished by negative publicity about its aggressive, unrestrained workplace culture.

Critics fear that Uber and other on-demand firms have the potential for creating a society of part-time, low-paid temp work, displacing traditionally full-time, secure jobs—the so-called Uberization of work. According to one study, half of Uber drivers earn less than the minimum wage in their state. Uber contends that it is lowering the cost of transportation, expanding the demand for ride services, and expanding opportunities for car drivers, whose pay is about the same as that of other taxi drivers. Uber has also taken some remediating steps. It enhanced its app to make it easier for drivers to take breaks while they are on the job. Drivers can now also be paid instantly for each ride they complete rather than weekly and see on the app's dashboard how much they have earned. Uber also added an option to its app that allows passengers to tip its US drivers.

Uber has a number of competitors in the ride-hailing business, including Lyft in the United States and local firms in Asia and Europe. Established taxi firms in New York and other cities are launching their own hailing apps and trumpeting their fixed-rate prices. The Covid-19 pandemic has had an adverse impact on this segment of Uber's business, drastically reducing the demand for ride services. As of mid-2022, demand had not yet recovered to pre-pandemic levels, although its financial results for this segment improved compared to 2020's. It has also experienced a shortage of drivers as a result of the pandemic. The classification of its drivers as independent contractors, which significantly reduces Uber's costs, is currently being challenged in courts, by legislators, and by government agencies in the United States and around the world. Uber is apparently so desperate to maintain this classification that it is backing bills that would classify its drivers as such in exchange for agreeing not to try to block their efforts to unionize.

Prior to the pandemic, Uber had invested more than $1 billion in an effort to develop its own autonomous, self-driving car. It had scaled back that effort somewhat after a self-driving Uber car struck and killed a woman in 2018, and in 2020, decided to abandon it altogether to refocus on its core businesses. It sold its self-driving car unit to a startup company, a major shift for a company that had just a few years previously championed the development of self-driving technology as key to its long-term profitability. Around the same time it also sold its flying taxi business, Uber Elevate, which was developing an all-electric, vertical take-off and landing passenger aircraft, to a different startup company. It had also previously divested itself of its e-bikes and e-scooters business.

In the wake of the pandemic, Uber turned instead to UberEats, its online food ordering and delivery service, which became much more in demand. In December 2020, Uber acquired competitor Postmates for $2.65 billion. It had previously acquired Careem, Uber's rival in the Middle East, for $3.1 billion. In October 2021, it diversified even further, purchasing The Drizly Group, which operates an on-demand alcohol marketplace in North America. Uber's revenues from its food delivery services now outpace its mobility segment, accounting for $8.3 billion in revenue, although it too continues to operate at a loss. Uber faces stiff competition in the delivery business, including from DoorDash, Deliveroo, Instacart, Grubhub, and many others. In a sign that Uber may be finding it hard to let go of its dreams of dominating all sorts of transportation-related services, it has recently announced a variety of new features for Uber Eats that expand its core value proposition. For instance, it has partnered with a direct-to-consumer telehealth company to deliver health and wellness products in 12 markets across the United States via the Uber Eats app. It has also teamed up with digital pharmacy startups to deliver prescription medications. In May 2022, it announced expansion of its partnerships with Albertsons to include grocery delivery for more than 2,000 Albertsons.

One of Uber's newer lines of business is its Uber Freight segment, which it launched in 2017. Uber is aiming to revolutionize the logistics industry much in the way it did to the ride-hailing business: by providing an on-demand platform to automate logistics transactions. The platform connects shippers with carriers and gives shippers upfront, transparent pricing; the ability to book a shipment with just a few clicks; and the ability to track shipments in real time from pickup to delivery. To date, Uber has invested heavily in its Freight segment, and in 2021, it acquired Transplace, a managed transportation and logistics network, for $2.25 billion. Like Uber's other lines of business, Uber Freight's revenue is growing, particularly as a result of its acquisition of Transplace, but it is still operating at a loss. Uber Freight also faces significant competition from a number of already entrenched global and North American freight brokers.

In May 2022, Uber released its financial results for the first quarter of 2022. It was in many senses a promising report, as revenue for all of its segments grew, to $6.9 billion. But it yet again recorded a loss from operations. Will Uber ever be able to consistently turn a profit?

Sources: Uber Technologies, Inc. "Form 10-Q for the Quarterly Period Ended March 31, 2022; Sec.gov, May 5, 2022; Kellen Browning, "Uber Continues Its Recovery from the Pandemic Lull but Loses $5.6 Billion from Investments," *New York Times*, May 4, 2022; Bridget Goldschmidt, "Albertsons, Uber Expand Partnership to 2,000+ Stores," Progressivegrocer.com, May 3, 2022; Gabrielle Bienasz, "Uber Drivers Are Slamming the Company's Fuel Surcharge as 'Woefully Inadequate'", Businessinsider.com, March 16, 2022; Zak Stambor, "Gig Worker-focused Business Models Face a Host of Challenges," Insider Intelligence/eMarketer, March 9, 2022; Uber Technologies, Inc. "Form 10-K for the Fiscal Year Ended December 31, 2021," Sec.gov, February 24, 2022; Rhea Patel, "Uber Is Driving the Rise of On-Demand Healthcare," Insider Intelligence/eMarketer, November 30, 2021; Kirsten Korosec, "Uber Sells Air Taxi Business Elevate to Joby Aviation, Shedding Its Last Moonshot," Techcrunch.com, December 8, 2020; Andrew Hawking, "Uber's Fraught and Deadly Pursuit of Self-Driving Cars Is Over," Theverge.com, December 7, 2020; Eliot Brown, "Uber Wants to Be the Uber of Everything—But Can It Make a Profit?" *Wall Street Journal*, May 4, 2019; Alex Rosenblat, "When Your Boss Is an Algorithm," *New York Times*, October 12, 2018.

## CASE STUDY QUESTIONS

**10-14** Analyze Uber using the competitive forces and value chain models. What is its competitive advantage?

**10-15** What is the relationship between information technology and Uber's business model? Explain your answer.

**10-16** How disruptive is Uber?

**10-17** Are any ethical and social issues raised by Uber and its business model? Does Uber's business model create an ethical dilemma? Explain your answers.

**10-18** Is Uber a viable business? Explain your answer.

## Chapter 10 References


Adomavicius, Gediminas, Jesse C. Bockstedt, Shawn P. Curley, and Jingjing Zhang. "Effects of Personalized Recommendations versus Aggregate Ratings on Post-Consumption Preference Responses." *MIS Quarterly* 46 No. 1 (March 2022).

Almquist, Eric, Jamie Cleghorn, and Lori Sherer. "The B2B Elements of Value." *Harvard Business Review* (March–April 2018).

Bapna, Ravi, Jui Ramaprasad, and Akmed Umyarov. "Monetizing Freemium Communities: Does Paying for Premium Increase Social Engagement?" *MIS Quarterly* 42, No. 3 (September 2018).

Commisso, Danielle. "Concerns Grow Over Consumer Privacy and Facial Recognition." Civicscience.com (April 20, 2021).

Dennis, Alan R., Lingyao (Ivy), Xuan Feng, Eric Webb, and Christine J. Hsieh. "Digital Nudging: Numeric and Semantic Priming in E-Commerce." *Journal of Management Information Systems* 37 No. 1 (2020).

Fay, Brad, Ed Keller, Rick Larkin, and Koen Pauwels. "Deriving Value from Conversations about Your Brand." *MIT Sloan Management Review* (Winter 2019).

Gomber, Peter, Robert J. Kauffman, Chris Parker, and Bruce W. Weber. "On the FinTech Revolution: Interpreting the Forces of Innovation, Disruption, and Transformation in Financial Services." *Journal of Management Information Systems* 35 No. 1 (2018).

Gramlich, John. "10 Facts About Americans and Facebook." Pew Research Center (June 1, 2021).

Hutchinson, Andrew. "The State of Social Listening in 2022—Report." (May 17, 2022).

Insider Intelligence/eMarketer. "Digital Shoppers & Buyers, US." (February 2022a).

________. "Digital Shoppers & Buyers, Worldwide." (February 2022b).

________. "Retail & Ecommerce Sales, US."(June 2022c).

________. "Digital Travel Sales, US." (May 2022d).

________. "Digital Ad Spending, US." (March 2022e).

________. "Social Network Users, US." (April 2022f).

________. "Time Spent with Social Networks, US." (January 2022g).

________. "Social Commerce Sales, US." (July 2022h).

________. "Social Network Ad Spending, US." (March 2022i).

________. "U.S. B2B Electronic Sales." (July 2022j).

________. "Digital Ad Revenues, by Company." (March 2022k).

________. "Consumer Attitudes toward Digital Advertising 2021." (July 2021).

Keränen, Joona, Harri Terho, and Antti Saurama."Three Ways to Sell Value in B2B Markets." *MIT Sloan Management Review* 63 No. 1 (Fall 2021).

Koh, Byungwan, Il-Horn Hann, and Srinivasan Raghunathan. "Digitization of Music: Consumer Adoption amidst Piracy, Unbundling, and Rebundling." *MIS Quarterly* 43 No. 1 (March 2019).

Laudon, Kenneth C., and Carol Guercio Traver. *E-commerce 2023: business. technology. society.* 17th ed. (Hoboken, NJ: Pearson Education, Inc., 2024).

Li, Huifang, Yulin Fang, Kai H. Lim, and Youwei Wang. "Platform-Based Function Repertoire, Reputation, and Sales Performance of E-Marketplace Sellers." *MIS Quarterly* 43 No. 1 (March 2019).

Perrin, Andrew. "Half of Americans Have Decided Not to Use a Product or Service Because of Privacy Concerns." Pewresearch.org (April 14, 2020)

Plaid and the Harris Poll. "The Fintech Effect." (October 12, 2021).

Rangan, V. Kasturi, Daniel Corsten, Matt Higgins, and Leonard A. Schlesinger. "How Direct-to-Consumer Brands Can Continue to Grow." *Harvard Business Review* (November–December 2021).

Rhue, Lauren, and Arun Sundararajan. "Playing to the Crowd: Digital Visibility and the Social Dynamics of Purchase Disclosure." *MIS Quarterly* 43 No. 4 (December 2019).

US Bureau of the Census. "E-Stats." Census.gov, accessed September 23, 2021.

CHAPTER

# 11

# Improving Decision Making and Managing Artificial Intelligence

## LEARNING OBJECTIVES

After completing this chapter, you will be able to:

11-1 Identify the different types of decisions and explain how the decision-making process works.

11-2 Describe how business intelligence and business analytics support decision making.

11-3 Define artificial intelligence (AI) and explain how it differs from human intelligence.

11-4 Identify the major types of AI techniques and show how they benefit organizations.

11-5 Understand how MIS can help your career.

## CHAPTER CASES

- Big Data Analytics: A New Way to Fight Wildfires
- What Happened to Watson Health?
- Do You Know Who Is Using Your Face?
- Should an Algorithm Make Our Decisions?

**MyLab MIS**

- **Video Case:** Predictive Tech Can Save $3B–$4B A Year: Tom Siebel
- Discussion Questions: 11-5, 11-6, 11-7
- Hands-On MIS Projects: 11-8, 11-9, 11-10, 11-11
- eText with Figure Videos

## BIG DATA ANALYTICS: A NEW WAY TO FIGHT WILDFIRES

**According** to a new analysis based on a model built by the First Street Foundation, a broad swath of the United States not typically associated with wildfires is now under threat. Nearly one in six US residents live in areas with significant wildfire risk, and that risk is disproportionately higher for communities of color—about 44 percent of all Native Americans and nearly 25 percent of Hispanic people by 2052.

Wildfires are becoming more severe and frequent because of climate change, drought conditions, past forest mismanagement, and large population movements into wildfire-prone areas. They're also becoming more unpredictable. When a wildfire swept through drought-stricken towns near Boulder, Colorado, in late December 2021, it didn't matter that it was winter or that many of the 1,000 homes and other structures lost were situated in suburban areas rather than in forested enclaves. The old rules no longer apply. Wildfires are now spreading more quickly than normal, changing direction suddenly, and breaching natural barriers that once stopped them. The Caldor Fire that burned through the Eldorado National Forest and other areas of the Sierra Nevada in late summer 2021 destroyed more than 600 homes in Grizzly Flats. It spread more than 40 miles, cresting the Sierra Nevada for only the second time on record for a fire. This fire burned so hot and quickly that traditional fire breaks, such as a two-mile ridgeline, were no longer effective barriers. (A ridgeline is a line formed along the highest points of a mountain ridge.)

Fire crews are increasingly relying on data-driven models to help them predict and control wildfires. Until recently, wildland firefighters relied on their past experiences, known among firefighters as "slide decks"—recollections of what they did for a particular fire—to determine what strategy to use when fighting a particular blaze. As fires became more erratic, this process became less reliable. Earlier computer models used fire, weather, and topographic information to predict the path of a wildfire, but the information they produced was insufficiently detailed.

© Robert Wilder Jr/Shutterstock

Thanks to Big Data analytics, the wildfire computer models have become more sophisticated. Matthew Thompson, Dave Calkin, and other Forest Service researchers found that if they added more data to their computer programs, including topographical maps of past fires and wind conditions, fire crews would be able to make more accurate decisions in real time. Fire analysts added data on fuel moisture, weather, vegetation, and where a fire was ignited, along with data on topography, what the fire has been doing in recent days, and data on previous fire history in the area. (Weather data is especially critical.) The programs can then generate various models to predict what the fire will do in the short, near, and long terms. Analysts can also use the models to simulate thousands of artificial scenarios during an even longer period to predict where a fire might spread under various conditions. Instead of taking four to five hours to draw a fire map, a computer program can spit out the data in about five seconds.

A remaining challenge for the firefighting models is that to date, they have relied on historical data from past fire seasons, which does not fully account for the extreme dryness of the fuels amid drought conditions that have rarely been so severe. And models can't always prevent destruction. According to Forest Service fire behavior analyst Robert Scott, "the models are good, but you have to give the model the correct information." Additional information from fire crews on how well the models worked will help make the models more accurate.

Sources: John Muyskens, Andrew Ba Tran, Naema Ahmed, and Anna Phillips. "1 in 6 Americans Live in Areas with Significant Wildfire Risk," *Washington Post*, May 16, 2022; Jim Carlton and Dan Frosch, "Moneyball Analytics Help Fight Wildfires: This Year's Blazes Are Testing Their Limits," *Wall Street Journal*, September 9, 2021; Rachel Leven, "Computing and Data Science Improve What We Know about Wildfires and How to Fight Them," Data.berkeley.edu, July 13, 2021.

---

Predicting wildfire behavior and developing mitigation strategies with Big Data are powerful illustrations of how information systems can dramatically improve decision making. More sophisticated computer models driven by Big Data have created systems that can generate firefighting decision-making scenarios to help make many of these decisions much more rapidly and accurately.

The chapter-opening diagram calls attention to important points raised by this case and this chapter. Wildfires are increasing in severity and unpredictability because of climate change, population increase, and bad forest management practices. The information for developing a firefighting strategy derived from wildland firefighters' experience and outdated computer models was insufficient for dealing with this problem. Using updated data-driven computer models helped the US Forest Service and wildland firefighters make better decisions about how to contain and extinguish the fires.

Here are some questions to think about: How did using Big Data analytics change the way wildfires are fought and managed? What are the social, environmental, and business benefits of using Big Data analytics for this purpose?

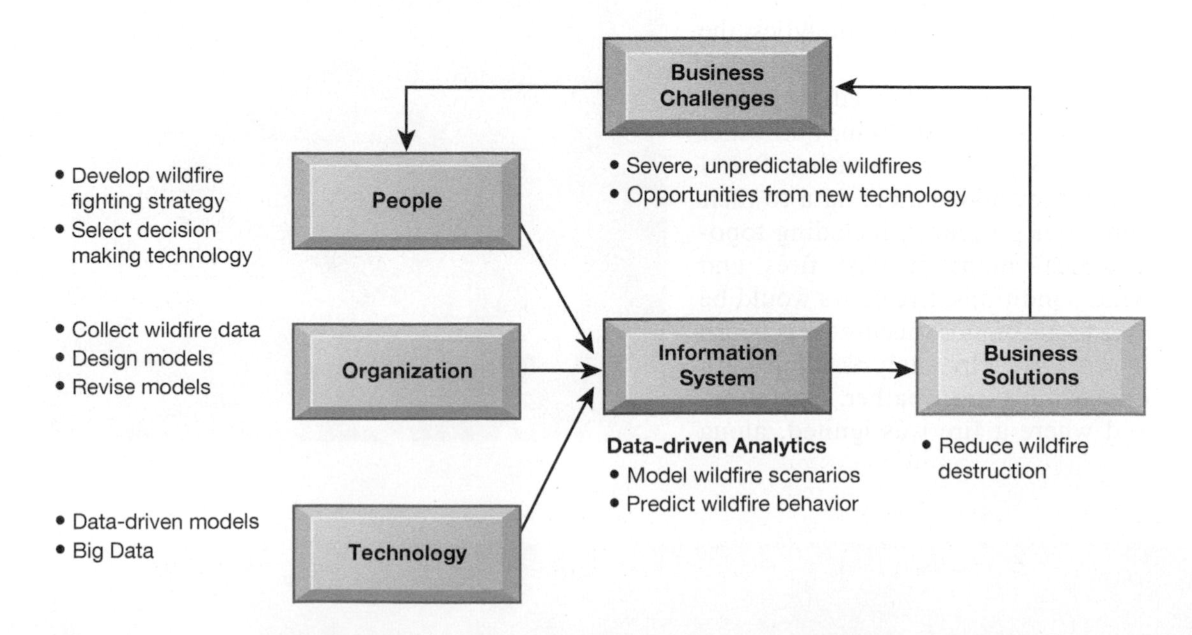

## 11-1 Identify the different types of decisions and explain how the decision-making process works.

One of the main contributions of information systems has been to improve decision making for both individuals and groups. Decision making in businesses used to be limited to management. Today, lower-level employees are responsible for some of these decisions as information systems make information available to lower levels of the organization. But what do we mean by better decision making? How does decision making take place in businesses and other organizations? Let's take a closer look.

### BUSINESS VALUE OF IMPROVED DECISION MAKING

What does it mean to a business to be able to make a better decision? What is the monetary value to a business of improved decision making? Table 11.1 measures the monetary value of improved decision making for a small US manufacturing firm with $280 million in annual revenue and 140 employees. The firm has identified a number of key decisions where new system investments might improve the quality of decision making. The table provides selected estimates of annual value (in the form of cost savings or increased revenue) from improved decision making in selected areas of the business.

We can see from Table 11.1 that those decisions are made at all levels of the firm and that some of these decisions are common, routine, and numerous. Although the value of improving any single decision may be small, improving hundreds of thousands of small decisions adds up to a large annual value for the business.

### TYPES OF DECISIONS

Chapter 2 described the different levels in an organization, such as senior management, middle management, and operational management. Each of these levels has different information requirements for decision support and responsibility for different types of decisions. Decisions are classified as structured, semistructured, and unstructured. Examine Figure 11.1, which illustrates the types of decisions each level of management is typically faced with. View the Figure 11.1 video in the eText for an animated and more detailed discussion of this figure.

**TABLE 11.1**

Business Value of Enhanced Decision Making

| Example Decision Value | Decision Maker | Number of Annual Decisions | Estimated Value to Firm of a Single Improved Decision | Annual |
|---|---|---|---|---|
| Allocate support to most valuable customers. | Accounts manager | 12 | $100,000 | $1,200,000 |
| Predict call center daily demand. | Call center management | 4 | $150,000 | $600,000 |
| Decide parts inventory levels daily. | Inventory manager | 365 | $5,000 | $1,825,000 |
| Identify competitive bids from major suppliers. | Senior management | 1 | $2,000,000 | $2,000,000 |
| Schedule production to fill orders. | Manufacturing manager | 150 | $10,000 | $1,500,000 |
| Allocate labor to complete a job. | Production floor manager | 100 | $4,000 | $400,000 |

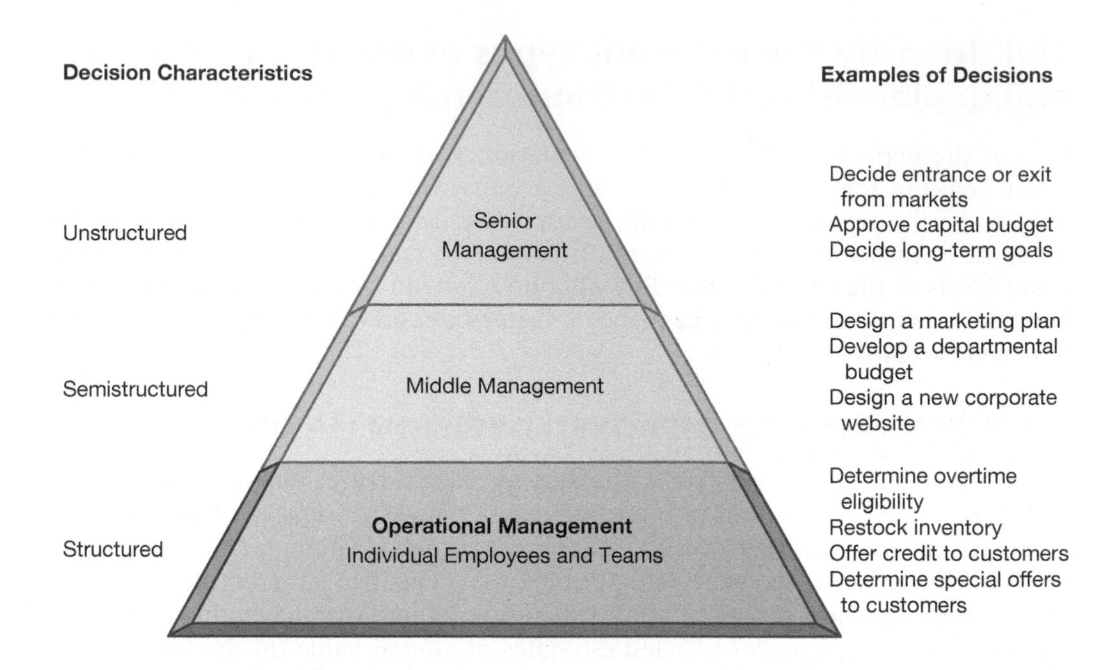


**Figure 11.1** Information Requirements of Key Decision-Making Groups in a Firm
*Senior managers, middle managers, operational managers, and employees have different types of decisions and information requirements.*

**Unstructured decisions** are those in which the decision maker must provide judgment, evaluation, and insight to solve the problem. Each of these decisions is novel, important, and not routine, and there is no well-understood or agreed-on procedure for making them.

**Structured decisions**, by contrast, are repetitive and routine, and they involve a definite procedure for handling them so that they do not have to be treated each time as if they were new. **Semistructured decisions** have elements of both types: Only part of the problem has a clear-cut answer provided by an accepted procedure. In general, structured decisions are more prevalent at lower organizational levels, whereas unstructured problems are more common at higher levels of the firm.

Senior executives face many unstructured decision situations, such as establishing the firm's five-year or 10-year goals or deciding new markets to enter. Answering the question, "Should we enter a new market?" requires access to news, government reports, and industry views as well as high-level summaries of firm performance. But the answer also requires senior managers to use their own best judgment and poll other managers for their opinions.

Middle management faces more structured decision scenarios, but their decisions may include unstructured components. A typical middle-level management decision might be "Why is the reported order fulfillment showing a decline over the past six months at a distribution center in Minneapolis?" This middle manager could obtain a report from the firm's enterprise system or distribution management system on order activity and operational efficiency at the Minneapolis distribution center. This is the structured part of the decision, but before arriving at an answer, this middle manager will have to interview employees and gather more unstructured information from external sources about local economic conditions or sales trends.

Operational management and rank-and-file employees tend to make more structured decisions. For example, a supervisor on an assembly line has to decide whether an hourly paid worker is entitled to overtime pay. If the employee worked more than eight hours on a particular day, the supervisor would routinely grant overtime pay for any time beyond eight hours that was clocked on that day.

A sales account representative often has to make decisions about extending credit to customers by consulting the firm's customer database that contains credit information. If the customer met the firm's specific criteria for granting credit, the account representative would grant that customer credit to make a purchase. In both instances, the decisions are highly structured and routinely made thousands of times each day in most large firms. The answer has been programmed into the firm's payroll and accounts receivable systems.

## THE DECISION-MAKING PROCESS

Making a decision is a multistep process. Simon (1960) described four stages in decision making: intelligence, design, choice, and implementation. Examine Figure 11.2, which illustrates these stages and their relationship to the four steps in the problem-solving process used throughout this book.

**Intelligence** consists of discovering, identifying, and understanding the problems occurring in the organization—why the problem exists, where, and what effects it is having on the firm. **Design** involves identifying and exploring various solutions to the problem. **Choice** consists of choosing among alternative solutions. **Implementation** involves making the chosen alternative work and continuing to monitor how well the solution is working.

What happens if the solution you have chosen does not work? Figure 11.2 shows that you can return to an earlier stage in the decision-making process and repeat it if necessary. For instance, in the face of declining sales, a sales management team may decide to pay the sales force a higher commission for making more sales to spur on the sales effort. If this does not increase sales, managers would need to investigate whether the problem stems from poor product design, inadequate customer support, or a host of other causes that call for a different solution.

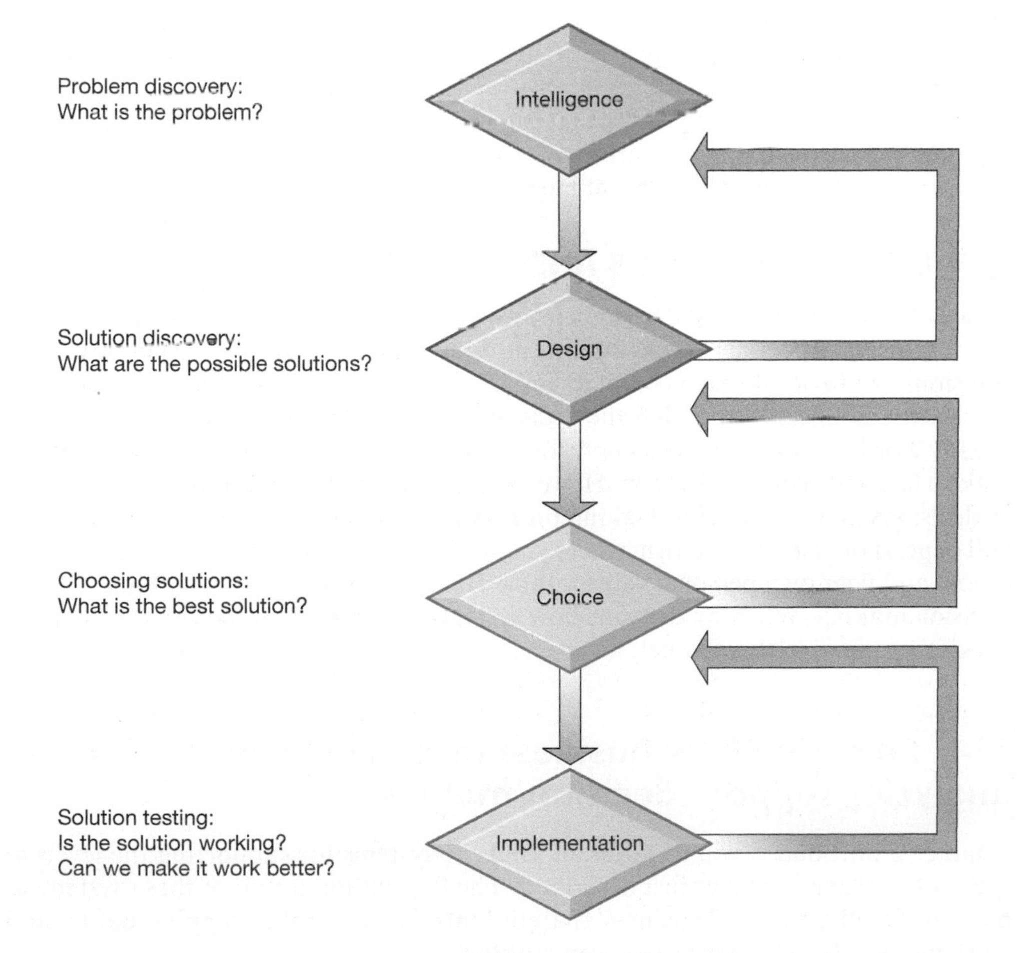

Problem discovery:
What is the problem?
Intelligence
Solution discovery:
What are the possible solutions?
Design
Choosing solutions:
What is the best solution?
Choice
Solution testing:
Is the solution working?
Can we make it work better?
Implementation

**Figure 11.2**
Stages in Decision Making
*The decision-making process can be broken down into four stages.*

**TABLE 11.2**

Qualities of Decisions and the Decision-Making Process

| Quality Dimension | Description |
|---|---|
| Accuracy | Decision reflects reality |
| Comprehensiveness | Decision reflects a full consideration of the facts and circumstances |
| Fairness | Decision faithfully reflects the concerns and interests of affected parties |
| Speed (efficiency) | Decision making is efficient with respect to time and other resources, including the time and resources of affected parties, such as customers |
| Coherence | Decision reflects a rational process that can be explained to others and made understandable |
| Due process | Decision is the result of a known process and can be appealed to a higher authority |

## HIGH-VELOCITY AUTOMATED DECISION MAKING

Today, many of the decisions made in an organization are not made by managers or any humans. For instance, when you enter a query into Google's search engine, Google's computer system has to decide which URLs to display in about half a second on average (500 milliseconds). High-frequency trading programs at stock exchanges in the United States execute trades within nanoseconds. Humans are eliminated from the decision chain because they are too slow.

In these high-speed automated decisions, the intelligence, design, choice, and implementation parts of the decision-making process are captured by computer algorithms that precisely define the steps to be followed to produce a decision. The people who wrote the software identified the problem, designed a method for finding a solution, defined a range of acceptable solutions, and implemented the solution. In these situations, organizations are making decisions more quickly than managers can monitor or control, and great care needs to be taken to ensure the proper operation of these systems to prevent significant harm.

## QUALITY OF DECISIONS AND DECISION MAKING

How can you tell whether a decision has become better or the decision-making process has improved? Accuracy is one important dimension of quality; in general, we think decisions are better if they accurately reflect the real-world data. Speed is another dimension: We tend to think that the decision-making process should be efficient, even speedy. For instance, when you apply for car insurance, you want the insurance firm to make a fast and accurate decision. However, there are many other dimensions of quality in decisions and the decision-making process to consider. Which is important for you will depend on the business firm where you work, the various parties involved in the decision, and your own personal values. Table 11.2 describes some quality dimensions for decision making. When we describe how systems "improve decisions and the decision-making process" in this chapter, we are referencing the dimensions in this table.

# 11-2 Describe how business intelligence and business analytics support decision making.

Chapter 2 introduced you to different kinds of systems for supporting the levels and types of decisions we have just described. The foundation for all of these systems is a business intelligence and business analytics infrastructure that supplies data and the analytic tools for supporting decision making.

## WHAT IS BUSINESS INTELLIGENCE?

"Business intelligence" (BI) is a term that hardware and software vendors and information technology consultants use to describe the infrastructure for warehousing, integrating, reporting, and analyzing data that come from the business environment. The foundation infrastructure collects, stores, cleans, and makes available relevant data to managers. Think databases, data warehouses, data marts, data lakes, Hadoop, and analytic platforms, which we described in Chapter 6. "Business analytics" (BA) is also a vendor-defined term. BA focuses more on tools and techniques for analyzing and understanding data. Think OLAP (online analytical processing), statistics, models, and data mining, which we also introduced in Chapter 6.

BI and analytics are essentially about integrating all the information streams a firm produces into a single, coherent, enterprise-wide set of data and then using modeling, statistical analysis, and data-mining tools to make sense out of all these data so that managers can make better decisions and better plans.

It is important to remember that BI and analytics are products defined by technology vendors and consulting firms. The largest five providers of these products are SAP, Oracle, IBM, SAS, and Microsoft. A number of BI and BA products now have cloud and mobile versions.

## THE BUSINESS INTELLIGENCE ENVIRONMENT

Examine Figure 11.3, which gives an overview of a BI environment, highlighting the kinds of hardware, software, and management capabilities that the major vendors offer and that firms develop over time. There are six elements in this BI environment:

- **Data from the business environment:** Businesses must deal with both structured and unstructured data from many sources, such as call centers, websites, mobile devices, social media data, stores, suppliers, and governmental and economic data, including Big Data. The data need to be integrated and organized so that they can be analyzed and used by human decision makers.
- **Business intelligence infrastructure:** The underlying foundation of BI is a powerful database system that captures all the relevant data to operate the business. The data may be stored in transactional databases or combined and integrated into an enterprise data warehouse, series of interrelated data marts, or analytic platforms.

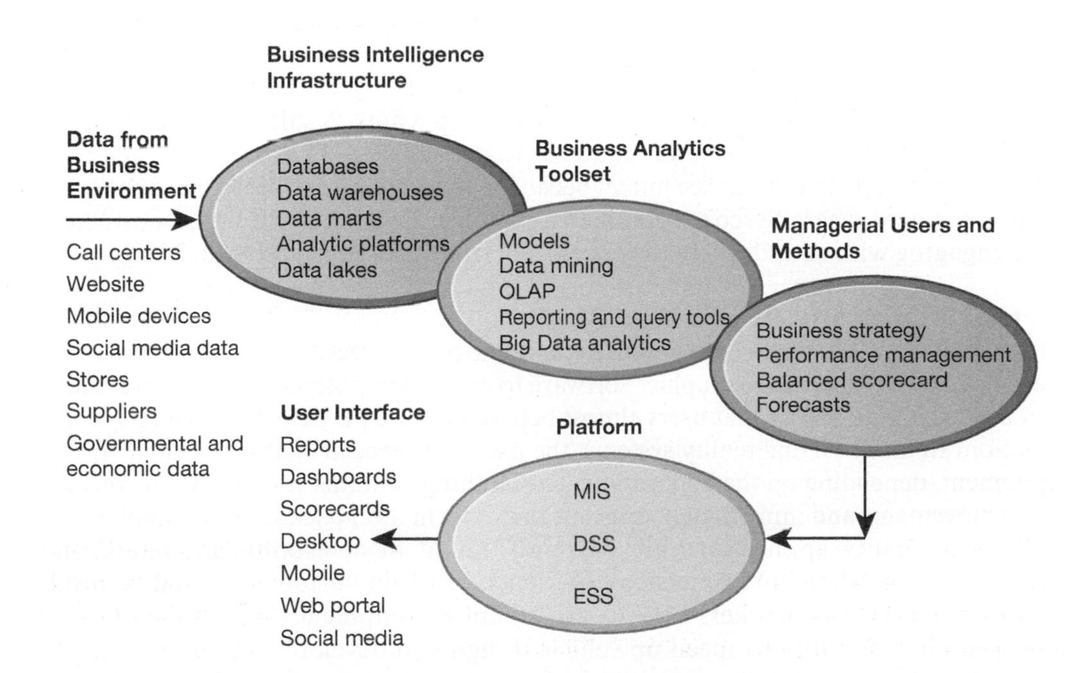


**Figure 11.3**
Business Intelligence and Analytics for Decision Support
*Business intelligence and analytics require a strong database foundation, a set of analytic tools, and an involved management team that can ask intelligent questions and analyze data.*

- **Business analytics toolset:** A set of software tools, such as models, data mining, OLAP, reporting and query tools, and Big Data analytics, is used to analyze data and produce reports, respond to questions managers pose, and track the progress of the business by using key indicators of performance.
- **Managerial users and methods:** BI hardware and software are only as intelligent as the human beings who use them. Managers impose order on the analysis of data by using a variety of managerial methods that define strategic business goals and specify how progress will be measured. These include enterprise performance management and balanced scorecard approaches that focus on key performance indicators, with special attention given to competitors.
- **Delivery platform—MIS, DSS, ESS:** The results from BI and analytics are delivered to managers and employees in a variety of ways, depending on what they need to know to perform their jobs. MIS, decision-support systems (DSS), and executive support systems (ESS), which we introduced in Chapter 2, deliver information and knowledge to different people and levels in the firm—operational employees, middle managers, and senior executives. In the past, these systems could not easily share data and operated as independent systems. Today, business intelligence and analytics tools can integrate all this information and bring it to managers' desktops or mobile platforms.
- **User interface:** Businesspeople often learn more quickly from a visual representation of data than from a dry report with columns and rows of information. Today's business analytics software suites feature **data visualization** tools, such as rich graphs, charts, dashboards, and maps. They can deliver reports on mobile devices as well as on the firm's website.

A leading visualization tool is from Tableau Software, which was acquired by Salesforce in 2019. Tableau is a data visualization and business intelligence tool that can be used for reporting and analyzing very large volumes of data. Tableau helps users create different charts, graphs, maps, dashboards, and stories for visualizing and analyzing data to help make business decisions. Nontechnical users are able to quickly and easily create and share customized, interactive dashboards to provide business insights from a broad spectrum of data, including data from spreadsheets, corporate databases, and the web. Tableau can query relational databases, online analytical processing cubes, cloud databases, and spreadsheets to generate graph-type data visualizations, and it can join and blend different datasets.

Global restaurant and food service supplier Sysco uses Tableau to enable customers to self-serve their inventory needs instead of relying on assistance from a sales representative. For customer orders placed across multiple tracking systems for sales accounting, Tableau can visualize order activity more clearly and in greater detail, providing more insights for the sales and sales leadership teams. A sales-facing dashboard shows all orders placed by customers and internal associates. The dashboard helps the company promote its self-service model because the company can see by account or by geographic area where Sysco's e-commerce platform is being used, which customers were engaging with it, and where they need help from sales reps (Tableau, 2022).

## Virtual Reality and Augmented Reality

**Virtual reality (VR) systems** have powerful capabilities for three-dimensional data visualization. They use interactive graphics software to create computer-generated simulations that are so close to reality that users almost believe they are participating in a real-world situation. In many virtual reality systems, the user dons special clothing, headgear, and equipment, depending on the application. The clothing contains sensors that record the user's movements and immediately transmit that information back to the computer.

Virtual reality applications are currently found in entertainment, retail, and manufacturing, where an immersive experience can help customers visualize products or teach factory workers how to use complex equipment. Volkswagen Group has used virtual reality to speed up vehicle design and development and to identify

*Data visualization tools facilitate the creation of graphs, charts, dashboards, and maps to make it easier for users to obtain insights from data.*



potentially costly design problems earlier in the development cycle. Volkswagen has been able to cut out costly physical prototypes and replace them with immersive, 360-degree views of digitally constructed interior and exterior components of a vehicle using virtual reality HTC Vive headsets. Virtual components of a car, including interior and exterior parts such as buttons, lights, or consoles, can be switched out and replaced easily with a few lines of software code during the design process.

**Augmented reality (AR)** is a related technology for enhancing visualization by overlaying digital data and images onto a physical real-world environment. The digital technology provides additional information to enhance the perception of reality, making the surrounding real world of the user more interactive and meaningful. The yellow first-down markers shown on televised football games are examples of augmented reality, as are medical procedures like image-guided surgery, where data acquired from computerized tomography (CT) and magnetic resonance imaging (MRI) scans or from ultrasound imaging are superimposed on the patient in the operating room.

Other industries where AR has caught on include military training, engineering design, robotics, and consumer design. Leading retail brands like IKEA, Amazon, and Sephora provide e-commerce shopping experiences over augmented reality mobile apps or wearable glasses to help shoppers make buying decisions in real time. For example, the IKEA Place mobile app takes pictures of a living room, measures the space, and recommends the furniture that fits the space. Users can thus virtually "place" furnishings into their space.

The virtual and augmented reality applications attracting much attention today are those for the metaverse, which we described in Chapters 2, 4, and 7. The metaverse is conceived as a graphically rich, virtual, 3-D space, with some degree of verisimilitude, where people can work, play, shop, and socialize—basically, doing the things humans like to do together in real life but using digital avatars. Metaverse proponents often focus on the concept of "presence" as a defining factor: feeling like you're really there and feeling like other people are really there with you, too. The infrastructure underpinning the metaverse, including virtual reality glasses and augmented reality software, will rely on data showing how users interact with their surroundings in fictional worlds, digital workplaces, virtual doctors' appointments, and elsewhere.

Facebook's corporate name change to Meta Platforms Inc. is a sign that businesses behind games, office tools, and other services will increasingly invest in this new interactive technology. Google, Microsoft, and Apple have all been working on metaverse-related products.

**TABLE 11.3**

Examples of Predefined Business Intelligence Production Reports

| Business Functional Area | Production Reports |
|---|---|
| Sales | Sales forecasts, sales team performance, cross-selling, sales cycle times |
| Service/Call Center | Customer satisfaction, service cost, resolution rates, churn rates |
| Marketing | Campaign effectiveness, loyalty and attrition, market basket analysis |
| Procurement and Support | Direct and indirect spending, off-contract purchases, supplier performance |
| Supply Chain | Backlog, fulfillment status, order cycle time, bill of materials analysis |
| Financials | General ledger, accounts receivable and payable, cash flow, profitability |
| Human Resources | Employee productivity, compensation, workforce demographics, retention |

## BUSINESS INTELLIGENCE AND ANALYTICS CAPABILITIES

BI and analytics promise to deliver correct, nearly real-time information to decision makers, and the analytic tools help them quickly understand the information and take action. There are six analytic functionalities that BI systems deliver to achieve these ends:

**Production reports:** These are predefined reports based on industry-specific requirements (see Table 11.3).

**Parameterized reports:** Users enter several parameters to filter data and isolate impacts of parameters. For instance, you might want to enter region and time of day to understand how sales of a product vary by region and time. If you were Starbucks, you might find that customers in the Eastern United States buy most of their coffee in the morning, whereas in the Northwest, customers buy coffee throughout the day. This finding might lead to different marketing and ad campaigns in each region. (See the discussion of pivot tables later in this section.)

**Dashboards/scorecards:** These are visual tools for presenting performance data that users define.

**Ad hoc query/search/report creation:** This allows users to create their own reports based on queries and searches.

**Drill down:** This is the ability to move from a high-level summary to a more detailed view.

**Forecasts, scenarios, models:** These include capabilities for linear forecasting, what-if scenario analysis, and data analysis, using standard statistical tools.

### Predictive Analytics

An important capability of BI analytics is the ability to model future events and behaviors, such as the probability that a customer will respond to an offer to purchase a product. **Predictive analytics** use statistical analysis, data-mining techniques, historical data, and assumptions about future conditions to predict future trends and behavior patterns. Variables that can be measured to predict future behavior are identified. For example, an insurance company might use variables such as age, gender, and driving record as predictors of driving safety when issuing auto insurance policies. A collection of such predictors is combined into a predictive model for forecasting future probabilities with an acceptable level of reliability. Netflix uses predictive analytics models in its recommendation engine to predict user preferences for TV shows and movies based on the user's past viewing habits. Royal Dutch Shell PLC is using the Microsoft Azure cloud platform and the C3 IoT platform-as-a-service

(PaaS) application development platform to monitor and predict where and when maintenance is needed for compressors, valves, and other equipment.

Predictive analytics are being incorporated into numerous BI applications for sales, marketing, finance, fraud detection, and healthcare. One of the best-known applications is credit scoring, which is used throughout the financial services industry. When you apply for a new credit card, scoring models process your credit history, loan application, and purchase data to determine your likelihood of making future credit payments on time. Healthcare insurers have been analyzing data for years to identify which patients are most likely to generate high costs.

Many companies employ predictive analytics to predict responses to marketing campaigns and other efforts to cultivate customers. By identifying customers more likely to respond, companies can lower their marketing and sales costs by focusing their resources on customers who have been identified as more promising. For instance, FedEx has been using predictive analytics to develop models that predict how customers will respond to price changes and new services, which customers are most at risk of switching to competitors, and how much revenue will be generated by new storefront or drop-box locations. The accuracy rate of FedEx's predictive analytics system ranges from 65 to 90 percent.

## Big Data Analytics

Predictive analytics are starting to use Big Data from both private and public sectors, including data from social media, customer transactions, and output from sensors and machines. In e-commerce, many online retailers have capabilities for making personalized online product recommendations to their website visitors to help stimulate purchases and guide the retailers' decisions about what merchandise to stock. Most of these product recommendations, however, are based on the behaviors of similar groups of customers, such as those with incomes under $50,000 or those whose ages are between 18 and 25. Now some firms are starting to analyze the tremendous quantities of online and in-store customer data they collect along with social media data to make these recommendations more individualized. These efforts are translating into higher customer spending and retention rates. Table 11.4 provides examples of companies using Big Data analytics.

**TABLE 11.4**

What Big Data Analytics Can Do

| Organization | Big Data Capabilities |
|---|---|
| Rolls Royce | Uses Big Data to reduce the amount of carbon its aircraft engines produce while optimizing maintenance to prolong the life of an engine. Monitors how each engine flies, the conditions under which it is flying, and how pilots use it. Customizes maintenance regime for individual engines. |
| Hertz | Uses data from web surveys, email, text messages, website traffic patterns, and data gathered at all of its 8,500 rental car locations to adjust staffing levels at individual locations in order to accommodate times of the day when demand peaks or falls. |
| Purdue University College of Agriculture | Captures terabytes of data about seed growth, water levels, fertilizer quantities, and soil types daily from sensors, cameras, and human inputs for analysis by a HPE supercomputer. Helps farmers make data-driven decisions about how much fertilizer to apply, how deep to plant, and how much water to use for small sections of fields and individual plants. Automated equipment can apply the ideal treatment to eliminate various weeds. |
| EHarmony | Online dating website analyzes personal and behavioral data provided by 10 million active users to match couples based on features of compatibility found in thousands of successful relationships. Processes more than 15 million matches daily. |

In the public sector, Big Data analytics are driving the movement toward smart cities, which make extensive use of digital technology and public record data stores to make better decisions about running cities and serving their residents. (See the Chapter 5 Spotlight on Technology case.) Municipalities are capturing more data through sensors, location data from mobile phones, and targeted smartphone apps. Predictive modeling programs now inform utility management, transportation operation, healthcare delivery, and public safety. For example, irrigation systems built into the city of Barcelona's parks monitor soil moisture and turn on sprinklers when water is needed. The city expects to reduce its water bill by 25 percent per year after installing sensors in local parks.

### Operational Intelligence and Analytics

Many decisions deal with how to run cities and businesses on a day-to-day basis. These are largely operational decisions, and this type of business activity monitoring is called **operational intelligence**. The Internet of Things is creating huge streams of data from web activities, smartphones, sensors, gauges, and monitoring devices that can be used for operational intelligence about activities inside and outside the organization. Software for operational intelligence and analytics enables organizations to analyze these streams of Big Data as they are generated in real time. Companies can set trigger alerts for particular events or have them fed into live dashboards to help managers with their decisions.

An example of operational intelligence is the use of data generated by sensors on trucks, trailers, and intermodal containers owned by Schneider National, one of North America's largest truckload, logistics, and intermodal services providers. The sensors monitor location, driving behaviors, fuel levels, and whether a trailer or container is loaded or empty. Data from fuel tank sensors help Schneider identify the optimal location at which a driver should stop for fuel based on how much is left in the tank, the truck's destination, and fuel prices en route. Schneider's sensors also capture hard braking in a moving truck and relay the data to corporate headquarters, where the data are tracked in dashboards monitoring safety metrics. The event initiates a conversation between the driver and that person's supervisor.

### Location Analytics and Geographic Information Systems

Big Data analytics include **location analytics**, the ability to gain business insight from the location (geographic) component of data, including location data from smartphones, output from sensors or scanning devices, and data from maps. For example, location analytics might help a marketer determine which people to target with mobile ads about nearby restaurants and stores or quantify the impact of mobile ads on in-store visits. Location analytics would help a utility company identify, view, and measure outages and their associated costs as related to customer location to help prioritize marketing, system upgrades, and customer service efforts. UPS's package-tracking and delivery-routing systems, described in Chapter 1, use location analytics, as does an application that Starbucks uses to determine where to open new stores. (The system identifies geographic locations that will produce a high sales-to-investment ratio and per-store sales volume.)

The Starbucks and utility company applications are examples of **geographic information systems (GIS)**. GIS provide tools to help decision makers visualize problems that benefit from mapping. GIS software ties location data about the distribution of people or other resources to points, lines, and areas on a map. Some GIS have modeling capabilities for changing the data and automatically revising business scenarios. GIS might be used to help state and local governments calculate response times to natural disasters and other emergencies, to help banks identify the best locations for new branches or ATM terminals, or to help police forces pinpoint locations with the highest incidence of crime.

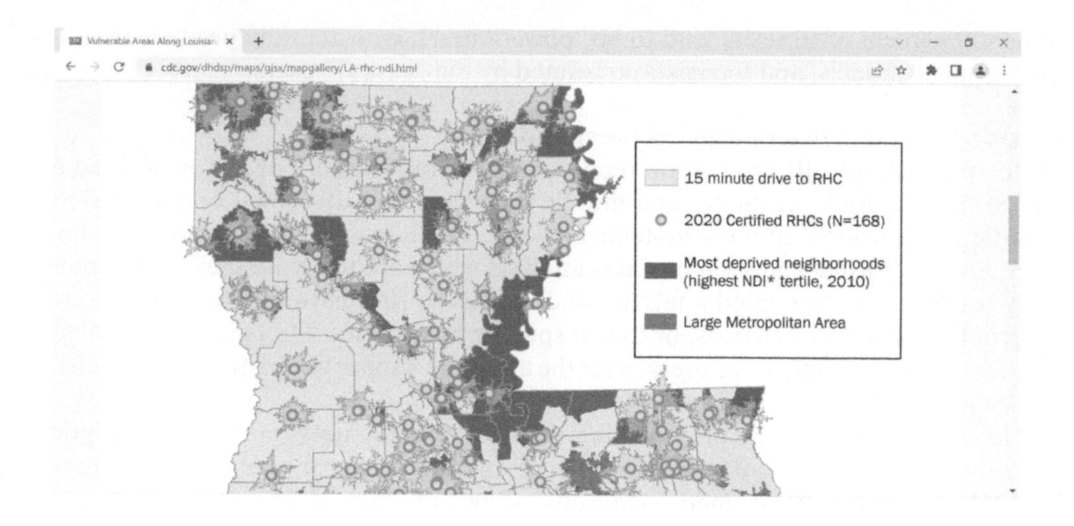


*This map shows the locations of vulnerable areas along Louisiana's Mississippi Delta that lack geographic access to Rural Health Clinics (RHCs). RHCs provide primary care access points for rural area residents. Although many rural areas have access to nearby RHCs, there are still rural neighborhoods along the Mississippi Delta with no nearby clinics.*

*Source: Anna L. Hamilton, Well-Ahead Louisiana. Accessed from the Centers for Disease Control and Prevention's Chronic Disease GIS Exchange https://www.cdc.gov/dhdsp/maps/gisx/mapgallery/LA-rhc-ndi.html*

Land O'Lakes, an agricultural cooperative serving the dairy industry, uses GIS for strategic asset management. In the agricultural industry, efficiencies and profits are deeply affected by the distance between a farm and various assets, such as feed lots, grain elevators, or crop nutrient facilities. Existing storage facilities may be located in suboptimal areas for serving farms' current customers, and new facilities are not always built in the most strategic locations. GIS can show co-op members mapping and locational information about trade areas—where the trade area is located, where members' facilities are situated within the trade area, and where competitors are based. GIS will be used for analyzing drive-time routes within specific trade territories, route planning, and identifying the optimal locations for new facilities.

## BUSINESS INTELLIGENCE USERS

Examine Figure 11.4, which illustrates the most common users of BI and the purposes for which they use it. For instance, 80 percent of the audience for BI consists of "casual" users. Senior executives tend to use BI to monitor firm activities by using visual interfaces such as dashboards and scorecards. Middle managers and analysts are much more likely to be immersed in the data and software, entering queries and slicing and dicing the data along different dimensions. Operational employees will, along with customers and suppliers, be looking mostly at prepackaged reports. The

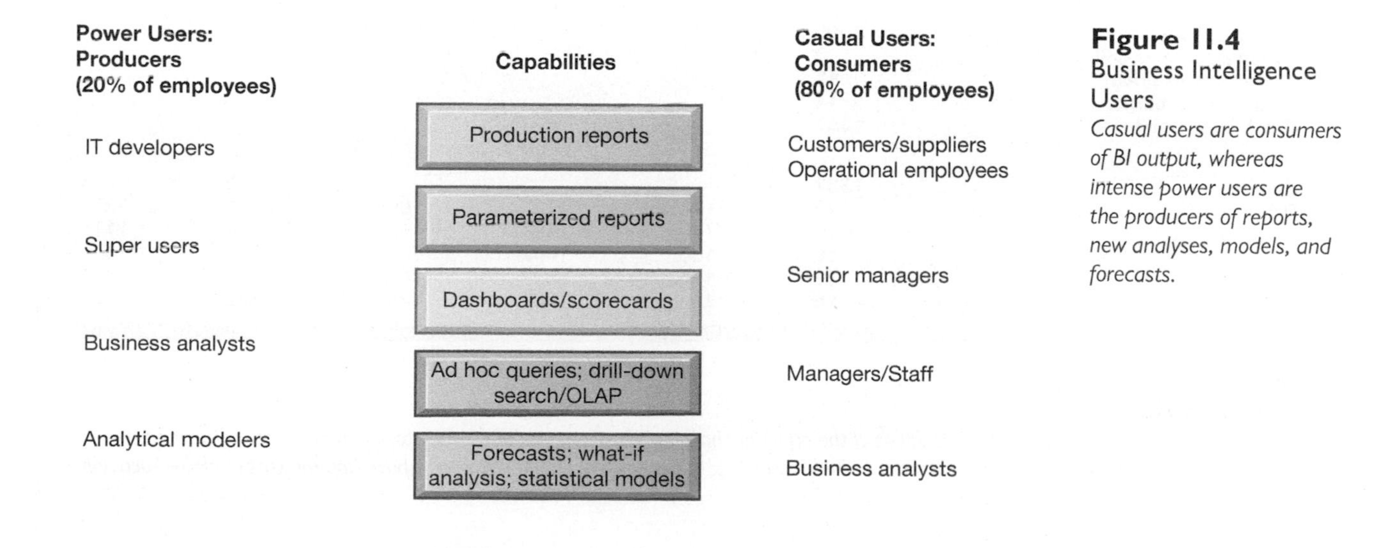


**Figure 11.4**
Business Intelligence Users
*Casual users are consumers of BI output, whereas intense power users are the producers of reports, new analyses, models, and forecasts.*

other 20 percent of BI users tend to be "power" users, who actually produce the various reports, models, and forecasts consumed by the "casual" users.

### Support for Semistructured Decisions

Many prepackaged BI production reports are MIS reports supporting structured decision making for operational and middle managers. We described operational and middle management, and the systems they use, in Chapter 2. Some managers, however, are super users and keen business analysts who want to create their own reports; they use more sophisticated analytics and models to find patterns in data, to model alternative business scenarios, or to test specific hypotheses. DSS are the BI delivery platform for this category of users, with the ability to support semistructured decision making.

DSS rely more heavily on modeling than MIS, using mathematical or analytical models to perform what-if or other kinds of analysis. What-if analysis, working forward from known or assumed conditions, allows the user to vary certain values in order to test results and predict outcomes if changes occur in those values. What happens if we raise product prices by 5 percent or increase the advertising budget by $1 million?

**Sensitivity analysis** models ask what-if questions repeatedly to predict a range of outcomes when one or more variables are changed multiple times. Examine Figure 11.5, which illustrates a sample sensitivity analysis that examines the effect of changing the sales price and cost per unit of a specific product on the product's break-even point. Backward sensitivity analysis helps decision makers with goal seeking: If I want to sell 1 million product units next year, how much must I reduce the price of the product?

Chapter 6 identified OLAP as a key business intelligence technology. Spreadsheets have a similar feature for multidimensional analysis, called a **pivot table**, which super-user managers and analysts employ to identify and understand patterns in business information that may be useful for semistructured decision making.

Examine Figure 11.6, which depicts a Microsoft Excel pivot table that examines a large list of order transactions for a company selling online management training videos and books. It shows the relationship between two dimensions: the sales region and the source of contact (web display ad or email) for each customer order. It answers the question of whether the source of the customer contact—in addition to region—make a difference. The pivot table in this figure shows that most customers

| | | | | | | |
|---|---|---|---|---|---|---|
| **Total fixed costs** | **19000** | | | | | |
| **Variable cost per unit** | **3** | | | | | |
| **Average sales price** | **17** | | | | | |
| **Contribution margin** | **14** | | | | | |
| **Break-even point** | **1357** | | | | | |
| | | | **Variable Cost per Unit** | | | |
| **Sales** | **1357** | **2** | **3** | **4** | **5** | **6** |
| **Price** | **14** | 1583 | 1727 | 1900 | 2111 | 2375 |
| | **15** | 1462 | 1583 | 1727 | 1900 | 2111 |
| | **16** | 1357 | 1462 | 1583 | 1727 | 1900 |
| | **17** | 1267 | 1357 | 1462 | 1583 | 1727 |
| | **18** | 1188 | 1267 | 1357 | 1462 | 1583 |

**Figure 11.5**
Sensitivity Analysis
*This table displays the results of a sensitivity analysis of the effect of changing the sales price of a beach towel and the cost per unit on the product's break-even point. It answers the question, "What happens to the break-even point if the sales price and the cost to make each unit increase or decrease?"*

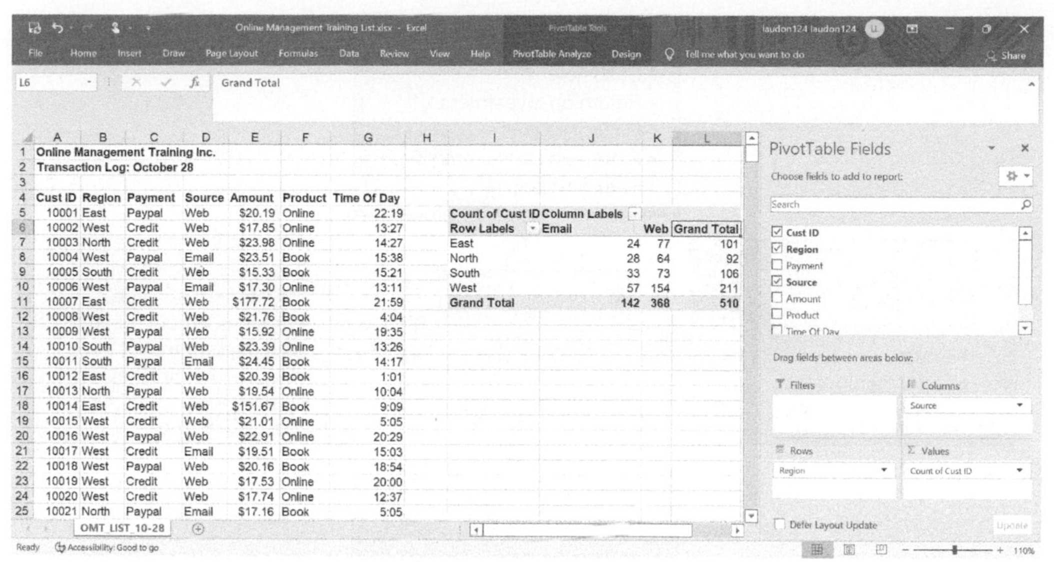

Online Management Training Inc.
Transaction Log: October 28

| Cust ID | Region | Payment | Source | Amount | Product | Time Of Day |
|---|---|---|---|---|---|---|
| 10001 | East | Paypal | Web | $20.19 | Online | 22:19 |
| 10002 | West | Credit | Web | $17.85 | Online | 13:27 |
| 10003 | North | Credit | Web | $23.98 | Online | 14:27 |
| 10004 | West | Paypal | Email | $23.51 | Book | 15:38 |
| 10005 | South | Credit | Web | $15.33 | Book | 15:21 |
| 10006 | West | Paypal | Email | $17.30 | Online | 13:11 |
| 10007 | East | Credit | Web | $177.72 | Book | 21:59 |
| 10008 | West | Credit | Web | $21.76 | Book | 4:04 |
| 10009 | West | Paypal | Web | $15.92 | Online | 19:35 |
| 10010 | South | Paypal | Web | $23.39 | Online | 13:26 |
| 10011 | South | Paypal | Email | $24.45 | Book | 14:17 |
| 10012 | East | Credit | Web | $20.39 | Book | 1:01 |
| 10013 | North | Paypal | Web | $19.54 | Online | 10:04 |
| 10014 | East | Credit | Web | $151.67 | Book | 9:09 |
| 10015 | West | Credit | Web | $21.01 | Online | 5:05 |
| 10016 | West | Paypal | Web | $22.91 | Online | 20:29 |
| 10017 | West | Credit | Email | $19.51 | Book | 15:03 |
| 10018 | West | Paypal | Web | $20.16 | Book | 18:54 |
| 10019 | West | Credit | Web | $17.53 | Online | 20:00 |
| 10020 | West | Credit | Web | $17.74 | Online | 12:37 |
| 10021 | North | Paypal | Email | $17.16 | Book | 5:05 |

| Count of Cust ID | Column Labels | | |
|---|---|---|---|
| Row Labels | Email | Web | Grand Total |
| East | 24 | 77 | 101 |
| North | 28 | 64 | 92 |
| South | 33 | 73 | 106 |
| West | 57 | 154 | 211 |
| Grand Total | 142 | 368 | 510 |



Used with permission from Microsoft

**Figure 11.6**
A Pivot Table that Examines Customer Regional Distribution and Advertising Source

*In this pivot table, we can examine where an online training company's customers come from in terms of region and advertising source.*

come from the West and that web display advertising produces most of the customers in all the regions.

One of the Hands-On MIS projects for this chapter asks you to use a pivot table to find answers to a number of other questions by using the same list of transactions for the online training company that we used in this discussion. The complete Excel file for these transactions is available in MyLab MIS.

In the past, much of this modeling was done with spreadsheets and small, stand-alone databases. Today these capabilities are incorporated into large enterprise BI systems, and they can analyze data from large corporate databases. Such capabilities helped Monster.com, one of the world's largest job listing sites, develop a more targeted, multichannel approach to advertising. Most of Monster's revenue comes from employers who pay to post job listings and to search its résumé database. Monster had been sending generic messages about its services to large groups of companies, but Monster now tries to find new employers and candidates using much more personalized email, direct mail, social engagement, and prioritized telemarketing. SAS statistical modeling software identifies existing and prospective customers who are most likely to purchase job listings and other services in a specific quarter. An IBM Unica marketing database maintains data on when the email campaigns ran, which email recipients responded to the email messages, and who clicked through. By analzying these data, Unica is able to generate mailing lists based on criteria such as past response behavior and an "opportunity score."

## Decision Support for Senior Management: The Balanced Scorecard and Enterprise Performance Management

BI delivered in the form of ESS helps senior executives focus on the most important performance information that affects the overall profitability and success of the firm. A leading methodology for understanding the important information that a firm's executives need is called the **balanced scorecard method**. Examine Figure 11.7, which illustrates the balanced scorecard framework. In this figure, a firm's strategic plan is operationalized by focusing on measurable outcomes of four dimensions of firm performance: financial, business process, customer, and learning and growth.

Performance of each dimension is measured using **key performance indicators (KPIs)**, which are the measures proposed by senior management for understanding how well the firm is performing along any given dimension. Look at each of the

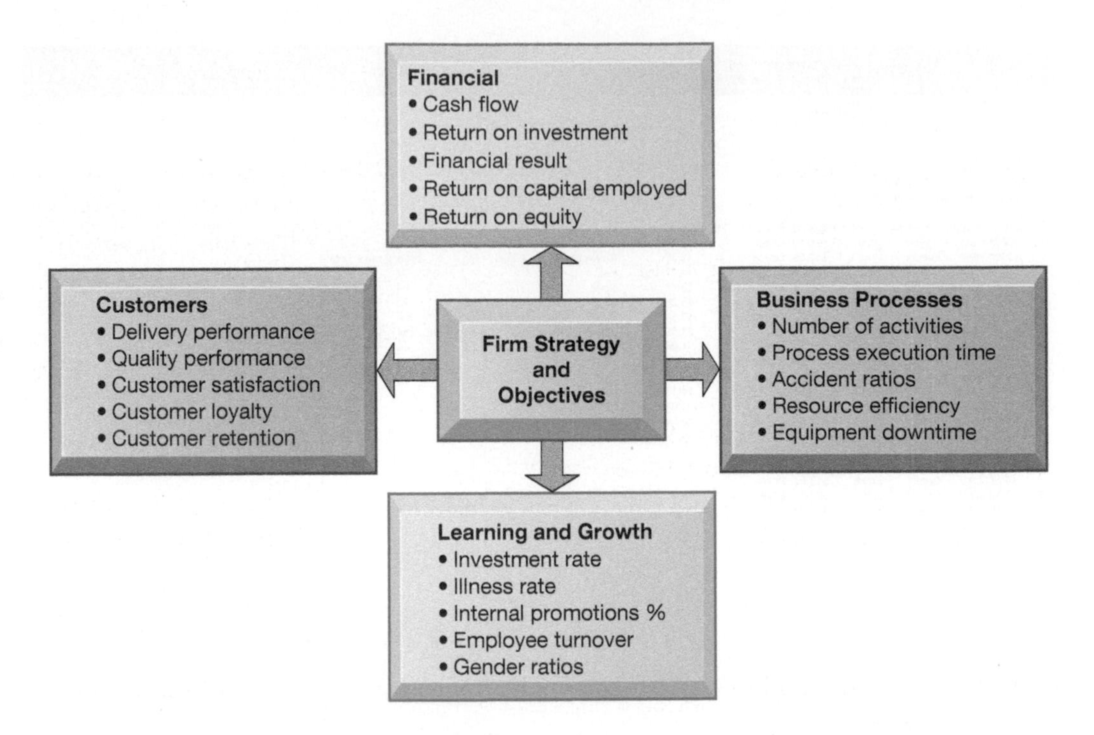


**Figure 11.7**
The Balanced Scorecard Framework
*In the balanced scorecard framework, the firm's strategic objectives are operationalized along four dimensions: financial, business process, customers, and learning and growth. Each dimension is measured using several KPIs.*

dimensions shown in Figure 11.7 for a list of various KPIs related to that dimension. For instance, one key indicator of how well an online retail firm is meeting its customer performance objectives is the average length of time required to deliver a package to a customer. If your firm is a bank, one KPI of business process performance is the length of time required to perform a basic function such as creating a new customer account.

The balanced scorecard framework is thought to be balanced because it causes managers to focus on more than just financial performance. In this view, financial performance is past history—the result of past actions—and managers should focus on the things they can influence today, such as business process efficiency, customer satisfaction, and employee training. Once consultants and senior executives develop a scorecard, the next step is automating a flow of information to executives and other managers for each of the key performance indicators.

Apple used the balanced scorecard to provide managers with a framework that they could apply to develop a holistic strategy rather than optimizing just one part of the business. Apple used five performance indicators: customer satisfaction, core competencies, employee commitment and alignment, market share, and shareholder value (Tudorache, 2021).

Corporate data for contemporary ESS are supplied by the firm's existing enterprise applications (enterprise resource planning, supply chain management, and customer relationship management). ESS also provide access to news services, financial market databases, economic information, and whatever other external data senior executives require. ESS have significant **drill-down** capabilities if managers need more detailed views of data.

Well-designed ESS help senior executives monitor organizational performance, track activities of competitors, recognize changing market conditions, and identify problems and opportunities. Employees lower down in the corporate hierarchy also use these systems to monitor and measure business performance in their areas of responsibility. For these and other business intelligence systems to be truly useful, the information must be actionable—readily available and easy to use when making decisions. If users have difficulty identifying critical metrics within the reports they receive, employee productivity and business performance will suffer.

## 11-3 Define artificial intelligence (AI) and explain how it differs from human intelligence.

**"Intelligent" techniques** are often described as **artificial intelligence (AI)**. There are many definitions of artificial intelligence. In the most ambitious vision, AI involves the attempt to build computer systems that think and act like humans. Humans see, hear, and communicate with natural languages, make decisions, plan for the future, achieve goals, perceive patterns in their environments, and learn, among many other capabilities. Humans also love, hate, and choose what objectives they want to pursue. These are the foundations of what is called "human intelligence" and what is called "common sense" or generalized intelligence.

So far the "Grand Vision" of AI remains a distant dream: There are no computer programs that have demonstrated generalized human intelligence or common sense. Human intelligence is vastly more complex than the most sophisticated computer programs and covers a broader range of activities than is currently possible with "intelligent" computer systems and devices.

A narrow definition of artificial intelligence is far more realistic and useful. Stripped of all the hyperbole, artificial intelligence programs are like all computer programs: They take data input from the environment, process that data, and produce outputs. AI programs differ from traditional software programs in the techniques they use to input and process data. AI systems today can perform many tasks that would be impossible for humans to accomplish and can equal or come close to humans in tasks such as interpreting CT scans, recognizing faces and voices, playing games like chess or Go, or besting human experts in certain well-defined tasks. In many industries they are transforming how business is done, where people are employed, and how people do their jobs.

### EVOLUTION OF AI

In the last decade, significant progress has been made within this limited vision of AI. The major forces driving the rapid evolution of AI are the development of Big Data databases generated by the Internet, e-commerce, the Internet of Things (IoT), and social media. Secondary drivers include the drastic reduction in the cost of computer processing and the growth in the power of processors. And finally, the growth of AI has relied on the refinement of algorithms by tens of thousands of AI software engineers and university AI research centers, along with significant investment from businesses and governments. There have been few fundamental conceptual breakthroughs in AI in this period, or in understanding how humans think. Many of the algorithms and statistical techniques were developed decades earlier but could not be implemented and refined on such a large scale as is currently possible.

Large-scale image recognition programs have gone from 25 percent error rates down to less than 2 percent and even lower for some facial recognition systems (see the Spotlight on People case). Natural language speech recognition errors have dropped from 15 percent to 5 percent; and in translation among common languages, Google's Translate program achieves about 85 percent accuracy compared to humans. These advances have made possible personal assistants like Siri (Apple), Alexa (Amazon), Cortana (Microsoft), and Google Assistant, as well as speech-activated systems in automobiles.

## 11-4 Identify the major types of AI techniques and show how they benefit organizations.

Artificial intelligence is a family of programming techniques and technologies, each of which has advantages in select applications. Table 11.5 describes the major types of AI: expert systems, machine learning, neural networks and deep learning, genetic

**TABLE 11.5**

Major Types of AI Techniques

| | |
|---|---|
| Expert systems | Represent the knowledge of experts as a set of rules that can be programmed so that a computer can assist human decision makers. |
| Machine learning | Software that can identify patterns in very large databases without explicit programming although with significant human training. |
| Neural networks and deep learning | Loosely based on human neurons, algorithms that can be trained to classify objects into known categories based on data inputs. Deep learning uses multiple layers of neural networks to reveal the underlying patterns in data and, in some limited cases, identify patterns without human training. |
| Genetic algorithms | Algorithms based loosely on evolutionary natural selection and mutation, commonly used to generate high-quality solutions to optimization and search problems. |
| Natural language processing | Algorithms that make it possible for a computer to understand and analyze natural human language. |
| Computer vision systems | Systems that can view and extract information from real-world images. |
| Robotics | Use of machines that can substitute for human movements as well as computer systems for their control and information processing. |
| Intelligent agents | Software agents that use built-in or learned knowledge to perform specific tasks or services for an individual. |

algorithms, natural language processing, computer vision systems, robotics, and intelligent agents. Let's take a look at each type of AI and understand how it is used by businesses and other organizations.

## EXPERT SYSTEMS

**Expert systems** were developed in the 1970s and were the first large-scale applications of AI in businesses and other organizations. They account for an estimated 20 percent of all AI systems today. Expert systems capture the knowledge of individual experts in an organization through in-depth interviews and represent that knowledge as sets of rules. These rules are then converted into computer code in the form of IF-THEN rules. Such programs are often used to develop apps that walk users through a process of decision making.

Expert systems provide benefits such as improved decisions, reduced errors, reduced costs, reduced training time, and better quality and service. They have been used in applications for making decisions about granting credit and for diagnosing equipment problems as well as in medical diagnostics, legal research, civil engineering, building maintenance, drawing up building plans, and educational technology (personalized learning and responsive testing) (Maor, 2003). For instance, if you were the project manager of a 14-story office building and were given the task of configuring the building's air conditioning system, which has hundreds of parts and subassemblies, an expert system could walk you through the process by asking a series of questions, producing an order to suppliers, and providing an overall cost estimate for the project, all in a matter of hours rather than weeks.

Examine Figure 11.8, which illustrates an expert system used to decide whether a person should be granted a credit line and the amount that should be granted, based on the answers to a series of related questions. The rules are interconnected, the number of outcomes is known in advance and is limited, there are multiple paths to the same outcome, and the system can consider multiple rules at a single time. View

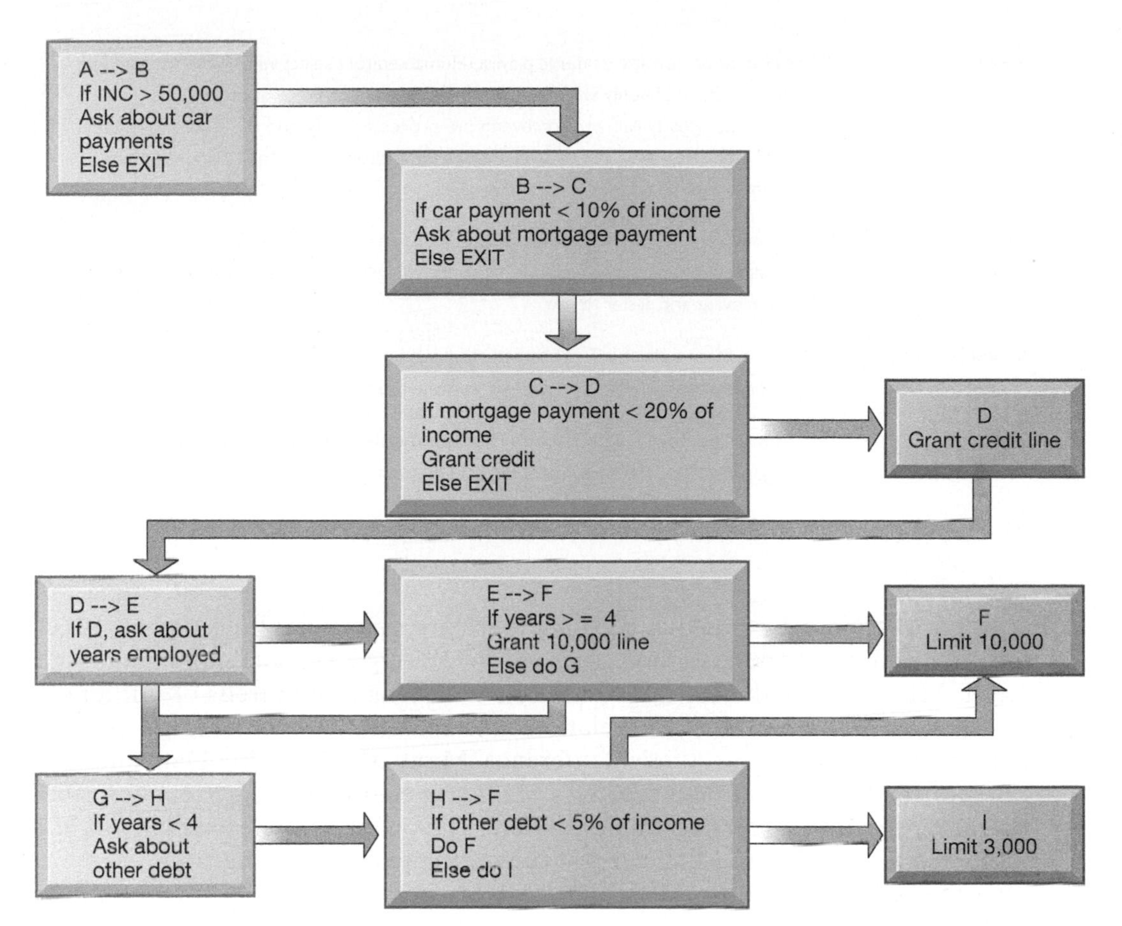


**Figure 11.8**
Rules in an Expert System
*An expert system contains a number of rules to be followed. The rules are interconnected, the number of outcomes is known in advance and is limited, there are multiple paths to the same outcome, and the system can consider multiple rules at a single time. The rules illustrated are for simple credit-granting expert systems.*

the Figure 11.8 video in the eText for an animated and more detailed discussion of this figure.

### How Expert Systems Work

Expert systems model human knowledge as a set of rules that collectively are called the **knowledge base**. Expert systems can have from a handful to many thousands of rules, depending on the complexity of the decision-making problem. The strategy used to search through the collection of rules and formulate conclusions is called the **inference engine**. The inference engine works by searching through the rules and firing those rules that are triggered by facts the user gathers and enters.

Expert systems have a number of limitations, the most important of which is that even experts can't explain how they make decisions: They know more than they can say. People drive cars, for instance, but are challenged to say how they do it. The knowledge base can become chaotic as the number of rules can reach into the thousands. In rapidly changing environments, say, medical diagnosis, the rules change and need to be continually updated. Expert systems are not useful for dealing with unstructured problems that managers and employees typically encounter, and these systems do not use real-time data to guide their decisions. Expert systems do not scale well to the kinds of very large data sets produced by the Internet and the Internet of Things (IoT), and they are expensive to build. For these reasons, expert system development has slowed to small domains of expert knowledge such as automobile diagnosis.

## MACHINE LEARNING

Most AI development today involves some kind of **machine learning (ML)**, with the main focus on finding patterns in data and classifying data inputs into known (and unknown) outputs. Machine learning is based on an entirely different AI paradigm

**TABLE 11.6**

Examples of Machine Learning

| | |
|---|---|
| Spotify | Uses ML to personalize its music playlist. Human editors select which songs are perfect for which list, and Spotify will automatically adjust the playlist to each listener's preferences. Nearly half a trillion events are processed daily, and the more information Spotify's models gather, the smarter they become about making associations among different artists, songs, podcasts, and playlists. |
| Zendesk | Customer service software company uses ML to analyze the data from customer support tickets, generating actionable insights for agents to make customer help articles more relevant and easier to use. |
| Schindler Group | Monitors thousands of IoT-connected elevators and walkways using GE's Predix operating system and machine learning to make predictions about needed maintenance. |
| PayPal | Uses machine learning algorithms to identify patterns of fraud for 426 million customers who generate 19.3 billion payment transactions annually. |

than expert systems. In machine learning there are no experts, and there is no effort to write computer code to represent rules reflecting an expert's understanding. Instead, ML begins with very large data sets with tens to hundreds of millions of data points and finds patterns and relationships in the data by analyzing a large set of examples and making a statistical inference. Many of today's Big Data analytics applications such as Royal Dutch Shell's predictive maintenance system described earlier use machine learning technology. Table 11.6 provides some examples of how leading business firms are using various types of machine learning.

Facebook has more than 2.9 billion monthly active users who spend an average of 35 minutes on the site daily. The firm displays an estimated 1 billion ads monthly to this audience, and it decides which ads to show each person in less than one second. For each person, Facebook bases this decision on the prior behavior of the user, including information shared (posts, comments, Likes), the activity of their social network friends, background information supplied to Facebook (age, gender, location, devices used), information supplied by advertisers (email address, prior purchases), and user activity on apps and other websites that Facebook can track. Facebook uses ML to identify patterns in the data set and to estimate the probability that any specific user will click on a particular ad based on the patterns of behavior Facebook has identified. At the end of this process is a simple show ad/no show ad result.

All of the very large e-commerce firms, including Amazon, Alphabet's Google, Microsoft, Alibaba, Tencent, Netflix, and Baidu, use similar ML algorithms. Obviously, no human or group of humans could achieve these results given the enormous database size, the speed of transactions, or the complexity of working in real time. The benefits of ML illustrated by this brief example come down to its extraordinary ability to recognize patterns at the scale of millions of people in a matter of seconds and to classify objects into discrete categories.

### Supervised and Unsupervised Learning

Nearly all machine learning today involves **supervised learning**, in which the system is "trained" by receiving specific examples of desired inputs and outputs identified by humans in advance. A very large database is developed with, say, 10 million photos posted on the Internet and then split into two sections, one a development database and the other a test database. Humans select a target, let's say to identify all photos that contain a car image. Humans feed a large collection of verified pictures, some of which contain a car image, into a neural network (described next) that proceeds iteratively through the development database in millions of cycles, until eventually the system can identify photos with a car at an acceptable rate. The machine learning system is then tested using the test database to ensure the algorithms can achieve the same results with a different set of photos. In many cases, but not all, machine learning can come close to or equal human efforts but on a very much larger scale and much more quickly. Over time, with tweaking by programmers, by making the database even bigger, and by using ever-larger computing systems, the system will improve its performance and, in that sense, will learn. Supervised learning is one technique used to develop autonomous vehicles that need to be able to recognize objects around them, such as people, other cars, buildings, and lines on the pavement to guide them.

In **unsupervised learning**, the same procedures are followed, except that humans do not feed the system examples. Instead, the system is asked to process the development database and report whatever patterns it finds. For example, suppose an unsupervised learning algorithm is given an input data set containing images of different types of cats and dogs. The algorithm is never trained upon the given data set, so it has no idea about the features of the data set. The task of the unsupervised learning algorithm is to identify the image features on its own. The unsupervised learning algorithm will perform this task by clustering the image data set into groups according to similarities among images.

One of the earliest unsupervised learning efforts (often referred to as "The Cat Paper") involved researchers collecting 10 million YouTube photos from videos and building a ML system that could detect human faces without labeling or "teaching" the machine with verified human face photos. The result was a system that could detect human faces, as well as cat faces and human bodies, in photos. The system was then tested on 22,000 object images on ImageNet (a large online visual database) and achieved a 16 percent accuracy rate. Since then, the facial recognition accuracy rate has improved considerably. (See the Spotlight on People case.) In principle, then, it is possible to create machine learning systems that can "teach themselves." Unsupervised learning enables businesses to identify patterns in large volumes of data more quickly when compared to manual observation. Some of the most common real-world applications of unsupervised learning are listed in Table 11.7.

To put this in perspective, one-year old human babies can recognize faces, cats, tables, doors, windows, and hundreds of other objects they have been exposed to and continuously catalog new experiences that they seek out on their own for recognition in the future. Babies have a huge computational advantage over the most advanced ML systems, as do adults. The human adult brain consumes about 7 watts of energy when operating and has an estimated 86 billion neurons, each with more than 10,000 connections to other neurons (synapses), and more than one trillion

**TABLE 11.7**

Examples of Unsupervised Learning Applications

| Unsupervised Learning Application | Description |
|---|---|
| Computer vision | Visual perception tasks such as object recognition |
| Medical imaging | Capabilities for medical imaging devices, such as image detection, classification, and segmentation. Used in radiology and pathology to diagnose patients quickly and accurately. |
| News aggregation | Google News uses unsupervised learning to categorize articles on the same topic from various online news outlets. For example, the results of a presidential election could be categorized under their label for "US" news. |
| Anomaly detection | Unsupervised learning models can sift through large amounts of data and discover atypical data points within a dataset. These anomalies can call attention to faulty equipment, human error, or breaches in security. |

total connections in its network (brain). Modern *homo sapiens* have been programmed (by nature) for an estimated 300,000 years, and their mammalian predecessors for 2.5 million years. ML systems are still much less capable than the human brain in the ability to learn and apply knowledge and experience in a generalized manner, and, in contrast to the human brain, require very large databases, teams of software and system engineers to create and support ML algorithms, and large computing facilities that consume up to several hundred thousand watts of energy.

## NEURAL NETWORKS

A neural network is composed of interconnected units called neurons. Each neuron can take data from other neurons and transfer data to other neurons in the system. The artificial neurons are not biological, physical entities as in the human brain but instead are software programs and mathematical models that perform the input and output function of neurons. The strength of the connections (the weight) can be controlled by researchers using a Learning Rule, an algorithm that systematically alters the strength of the connections among the neurons to produce the final, desired output, which could be identifying a picture of a cancerous tumor, fraudulent credit card transactions, or suspicious telephone calling patterns.

By using machine learning algorithms and computational models that are loosely based on how the biological human brain is thought to operate, **neural networks** find patterns and relationships in very large amounts of data that would be too complicated and difficult for a human being to analyze. Neural networks are pattern detection programs and identify patterns in large quantities of data by sifting through the data and ultimately finding pathways through the network of thousands of neurons. Some pathways are more successful than others in their ability to identify objects like cars, animals, faces, and voices. There may be millions of pathways through the data. The Learning Rule identifies these successful paths and strengthens the connection among neurons in these pathways. This process is repeated thousands or millions of times until only the most successful pathways are identified. The Learning Rule identifies the best or optimal pathways through the data. At some point, after millions of pathways are analyzed, the process stops when an acceptable level of pattern recognition is reached: for instance, successfully identifying cancerous tumors about as well as humans, or even better than humans, can.

Examine Figure 11.9, which represents one type of neural network comprising an input layer, a "hidden" processing layer, and an output layer. Humans train the network by inputting a set of outcomes they want the machine to learn. For instance, if the objective is to build a system that can identify patterns in fraudulent credit card purchases, the system is trained using actual examples of fraudulent transactions. The data set may be composed of a million examples of fraudulent transactions. The

**Figure 11.9**
How a Neural Network Works
*A neural network uses rules it "learns" from patterns in data to construct a hidden layer of logic. The hidden layer then processes inputs, classifying them based on the experience of the model. In this example, the neural network has been trained to distinguish between valid and fraudulent credit card purchases.*

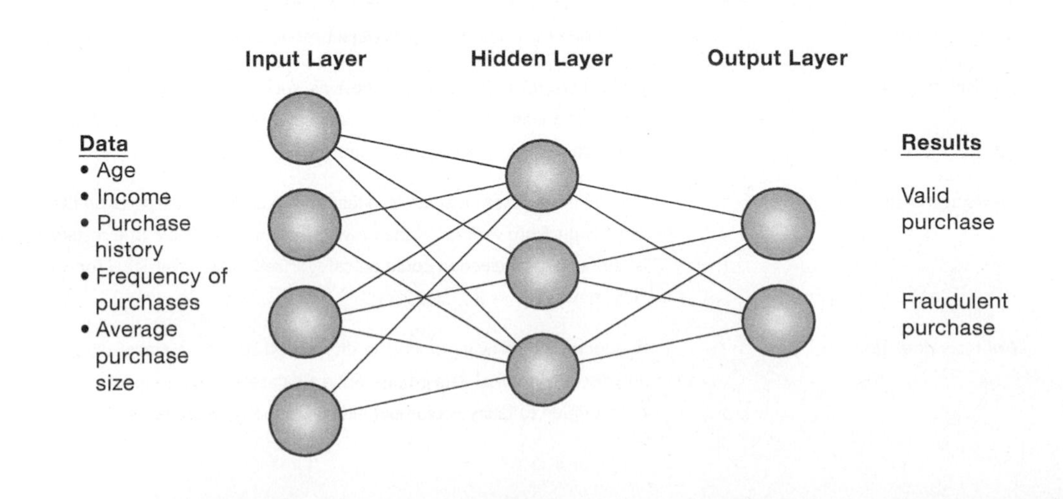

data set is then divided into two segments: a training data set and a test data set. The training data set is used to train the system. After millions of test runs, the program hopefully will identify the best path through the data. To verify the accuracy of the system, it is then used on the test data set, which the system has not analyzed before. If successful, the system will be tested on new data sets. The neural network in Figure 11.9 has learned how to identify a likely fraudulent credit card purchase.

Neural network applications in medicine, science, and business address problems in pattern classification, prediction probabilities, and control and optimization. In medicine, neural network applications are used to screen patients for coronary artery disease, to diagnose epilepsy and Alzheimer's disease, and to perform pattern recognition of pathology images, including those of certain cancers. The financial industry uses neural networks to discern patterns in vast pools of data that might help investment firms predict the performance of equities, corporate bond ratings, or corporate bankruptcies. Visa International uses a neural network to help detect credit card fraud by monitoring all Visa transactions for sudden changes in the buying patterns of cardholders. Table 11.8 provides examples of neural networks.

### "Deep Learning" Neural Networks

**"Deep learning"** neural networks are more complex, with many layers of transformation of the input data to produce a target output. Collections of neurons are called nodes or layers. Deep learning networks are in their infancy. They are used primarily at this point for pattern detection on unlabeled data where the system is not told what to look for specifically but to simply discover patterns in the data. The system is expected to be self-taught.

For instance, in our earlier example of unsupervised learning involving a machine learning system that could identify cats and other objects without training, the system used was a deep learning network. Examine Figure 11.10, which illustrates this system. It consists of three layers of neural networks (layers 1, 2, and 3). Each of these layers has two levels of pattern detection (levels 1 and 2). Each level was developed to identify a low-level feature of the photos: Layer 1 identified lines in the photos, and

**TABLE 11.8**

Examples of Neural Networks

| Functionality | Inputs | Process | Outputs/ Application |
|---|---|---|---|
| Computer vision | Millions of digital images, videos, or sensors | Recognize patterns in images, and objects | Photo tagging, facial recognition, autonomous vehicles |
| Speech recognition | Digital soundtracks, voices | Recognize patterns and meaning in soundtracks and speech | Digital assistants, chatbots, help centers |
| Machine controls, diagnostics | Internet of Things, thousands of sensors | Identify operational status, patterns of failure | Preventive maintenance, quality control |
| Language translation | Millions of sentences in various languages | Identify patterns in multiple languages | Translate sentences from one language to another |
| Transaction analysis | Millions of loan applications, stock trades, phone calls | Identify patterns in financial and other transactions | Fraud control, theft of services, stock market predictions |
| Targeted online ads | Millions of browser histories | Identify clusters of consumers; preferences | Programmatic advertising |

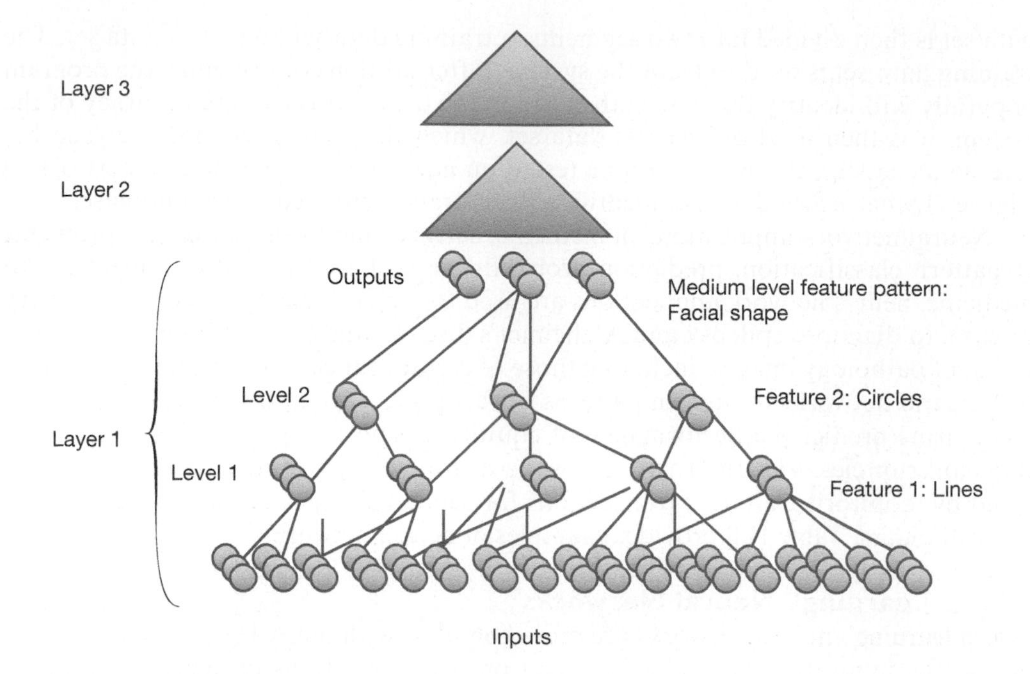


**Figure 11.10**
A Deep Learning Network
*Deep learning networks consist of many layers of neural networks working in a hierarchical fashion to detect patterns. Shown here is an expanded look at layer 1. Other layers have the same structure.*

layer 2 identified circles. The result of the first layer may be blobs and fuzzy edges. Second and third layers refine the images emerging from the first layer, until at the end of the process the system can distinguish cats, dogs, and humans, although not very well in this early effort, with a 16 percent accuracy rate.

Some experts believe deep learning networks will be able to come close to the "Grand Vision" of AI, where ML systems would be capable of learning like a human being. Others who work in ML and deep learning are more critical (Larson, 2021; Marcus, 2018; Pearl 2016).

### Limitations of Neural Networks and Machine Learning

Neural networks have a number of limitations currently. They require very large data sets to identify patterns. There are often many patterns in large data sets that are nonsensical, and it takes humans to choose which patterns "make sense." Many patterns in large data sets are ephemeral: There may be a pattern in the stock market or in the performance of professional sports teams, but they do not last long. In many important decision situations, there are no large data sets. For instance, should you apply to College A or College B? Should we merge with another company? Answers to many important questions are difficult to specify, or describe, and in that sense are only semistructured at best and depend greatly on human assessments, judgments, and sentiments.

Systems engineers do not always know why neural networks and machine learning arrive at a particular conclusion. For instance, in the case of the IBM Watson computer playing Jeopardy, researchers could not say exactly why Watson chose the answers it did, only that they were either right or wrong. Most real-world ML applications in business involve classifying digital objects into simple binary categories (yes or no; 0 or 1). But many of the significant problems facing managers, firms, and organizations do not have binary solutions. Neural networks may not perform well if their training covers too little or too much data. AI systems have no sense of ethics: They may recommend actions that are illegal or immoral. In most current applications, AI systems are best used as tools for relatively low-level decisions in systems where errors do not have catastrophic consequences like death or injury, aiding but not substituting for managers.

## GENETIC ALGORITHMS

**Genetic algorithms** are another form of machine learning. Genetic algorithms are useful for finding the optimal solution for a specific problem by examining a very large number of alternative solutions to that problem. Their method of solving problems is based on ideas inspired by evolutionary biology such as inheritance, mutation, selection, and crossover (recombination).

A genetic algorithm works by searching a population of randomly generated strings of binary digits to identify the string representing the best possible solution for the problem. As solutions alter and combine, the worst ones are discarded, and the better ones survive to go on to produce even better solutions.

Examine Figure 11.11, which illustrates the components of a genetic algorithm. Each string corresponds to one of the variables in the problem. One applies a test for fitness, ranking the strings in the population according to their level of desirability as possible solutions. After the initial population is evaluated for fitness, the algorithm then produces the next generation of strings, consisting of strings that survived the fitness test plus offspring strings produced from mating pairs of strings, and tests their fitness. The process continues until a solution is reached.

Genetic algorithms are used to solve problems that are dynamic and complex and involve hundreds or thousands of variables or formulas. The problem must be one whose range of possible solutions can be represented genetically and for which criteria can be established for evaluating fitness. Genetic algorithms expedite the solution because they can evaluate many solution alternatives quickly to find the best one. For example, General Electric engineers used genetic algorithms to help optimize the design for jet turbine aircraft engines in which each design change required changes in up to 100 variables. The supply chain management software from Blue Yonder uses genetic algorithms to optimize production-scheduling models, incorporating hundreds of thousands of details about customer orders, material and resource availability, manufacturing and distribution capability, and delivery dates.

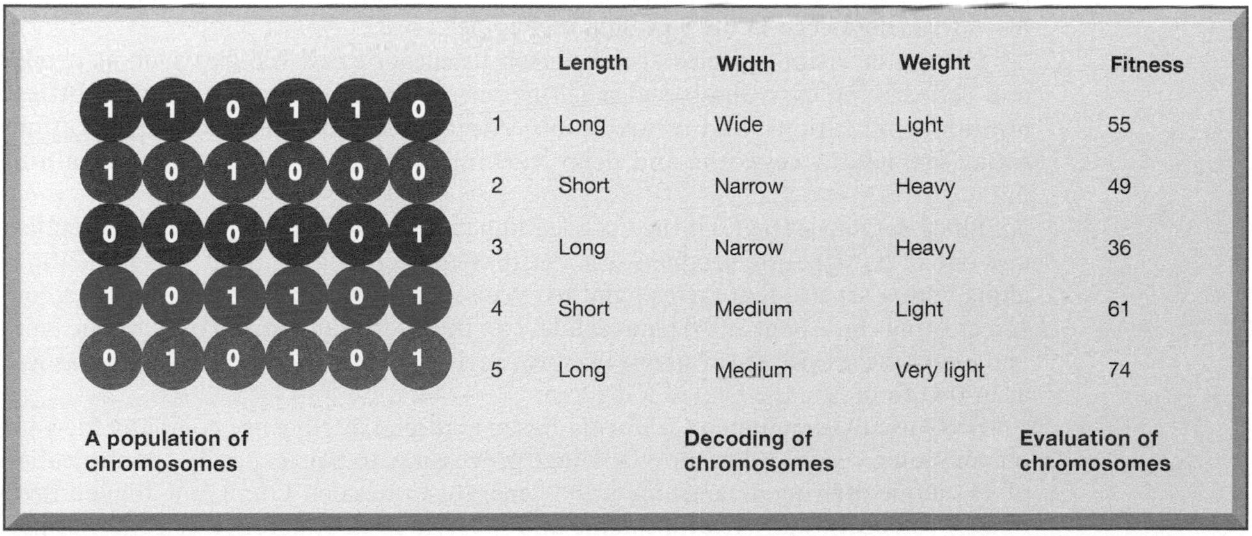

| A population of chromosomes | | Length | Width | Weight | Fitness |
|---|---|---|---|---|---|
| 1 1 0 1 1 0 | 1 | Long | Wide | Light | 55 |
| 1 0 1 0 0 0 | 2 | Short | Narrow | Heavy | 49 |
| 0 0 0 1 0 1 | 3 | Long | Narrow | Heavy | 36 |
| 1 0 1 1 0 1 | 4 | Short | Medium | Light | 61 |
| 0 1 0 1 0 1 | 5 | Long | Medium | Very light | 74 |

A population of chromosomes — Decoding of chromosomes — Evaluation of chromosomes

**Figure 11.11**
The Components of a Genetic Algorithm
*This example illustrates an initial population of "chromosomes," each representing a different solution. The genetic algorithm uses an iterative process to refine the initial solutions so that the better ones, those with the higher fitness, are more likely to emerge as the best solution.*

## NATURAL LANGUAGE PROCESSING, COMPUTER VISION SYSTEMS, AND ROBOTICS

Other important AI techniques include natural language processing, computer vision systems, and robotics.

### Natural Language Processing

Human language is not always precise. It is often ambiguous, and meanings of words can depend on complex variables such as slang, regional dialects, and social context. **Natural language processing (NLP)** makes it possible for a computer to analyze natural language—language that human beings instinctively use, not language specially formatted to be understood by computers. NLP algorithms are typically based on machine learning, including deep learning, which can learn how to identify a speaker's intent from many examples. Akershus University Hospital in Norway used NLP and IBM Watson Explorer to sift through thousands of medical records with unstructured textual data expressed in everyday language like natural speech. The algorithms could read text on a medical record and classify its meaning. Akershus staff used this information to verify whether they had made correct treatment decisions. You can also see natural language processing at work in leading search engines such as Google, spam filtering systems, and text-mining sentiment analysis (discussed in Chapter 6).

IBM's Watson became one of the most famous examples of natural language processing when the Watson computer defeated the two top players of the game Jeopardy in February 2011. The Spotlight on Technology case provides more detail about the technology used in Watson and the challenges IBM encountered when developing Watson applications for healthcare.

### Computer Vision Systems

**Computer vision systems** deal with how computers can emulate the human visual system to view and extract information from real-world images. Such systems incorporate image processing, pattern recognition, and image understanding. An example is Facebook's facial recognition tool called DeepFace, which is nearly as accurate as the human brain in recognizing a face. However, facial recognition technology is controversial because it can be misused in mass surveillance systems that pose threats to privacy, as described in the Spotlight on People case.

Computer vision systems are also used in autonomous vehicles such as drones and self-driving cars, industrial machine vision systems (e.g., inspecting bottles), military applications, and robotic tools. Almost all computer vision applications today use neural networks and deep learning with massive amounts of training data.

Block & Quayle (B&Q) is the UK's leading home improvement and garden retailer, and one of its signature products is its Valspar emulsion paint. B&Q Valspar features along with a service that mixes paint to precisely match the color of any item a consumer brings in. About 2,000 paint shades can be ordered and mixed online. The paint and matching service are available in stores such as Lowe's in the United States as well as in B&Q stores in the United Kingdom.

B&Q used GumGum, a California-based artificial intelligence company focusing on computer vision and natural language processing, to help expand communication of its unique offering to consumers and generate more sales. GumGum aligned B&Q images to contextually relevant home and lifestyle content across a selection of premium publishers. (Contextual advertising refers to the practice of placing ads on web pages based on the content of those pages.) GumGum's image recognition technology analyzed the exact color of the items in the image and dynamically changed the color of ads to match it. The campaign helped increase purchase intent from 64.9 percent to 74.7 percent. B&Q's online ad awareness increased from 26.1 percent to 35.4 percent,

## What Happened to Watson Health?

Watson was an IBM computer that made history in February 2011 by handily defeating the two-most-decorated champions of the game show *Jeopardy!*, Ken Jennings and Brad Rutter. Watson's achievement marked a milestone in the ability of computers to process and interpret human language. The Watson version used in Jeopardy took 20 IBM engineers three years to build at an $18 million labor cost, with an estimated $1 million in equipment. Watson had to be able to register the intent of a question, search through millions of lines of text and data, pick up nuances of meaning and context, and rank potential responses for a user to select, all in less than three seconds.

Watson was able to learn from its mistakes as well as its successes. Watson analyzed both Jeopardy questions and answers to determine patterns or similarities among clues. Using these patterns, it assigned varying degrees of confidence to the answers it gave. Although Watson was initially able only to correctly answer a small fraction of the questions it was given, machine learning allowed the system to continue to improve until it reached Jeopardy champion level.

IBM viewed its investment in Watson as a stepping stone to broader commercial uses of its AI technology, including applications for healthcare, financial services, or any industry where sifting through large amounts of data (including unstructured data) to answer questions is important. Watson was expected to become more useful and powerful by learning from new sets of experts in new fields of knowledge. IBM allocated one-third of its overall research efforts to Watson and made a huge bet that Watson could transform IBM's business from an aging hardware company into an AI leader. Former IBM CEO Virginia Rometty referred to Watson as "our moonshot." At one point Watson Health had 7,000 employees.

Healthcare appeared to provide the biggest, best opportunity for demonstrating the "new" IBM. The US healthcare industry collects a tremendous amount of information every day on the care of hundreds of millions of people. However, that information is very fragmented and unable to link individuals across all the domains in which they get care to develop a holistic picture of who they are, their diseases, the best treatments, and how to ensure the best care at the lowest possible cost. There is no connectivity right now that can do that at scale. The US healthcare market is a $3 trillion business that has legacy technology infrastructure and insufficient connectivity. Tech companies such as Microsoft, Google, and IBM all want a piece of the healthcare action.

To develop Watson for the healthcare field, IBM needed massive amounts of data on which to train Watson. It obtained that data through acquisitions, eventually spending some $5 billion buying a series of health data companies such as Truven, Phytel, Explorys, and Merge. Truven had the biggest insurance database in the United States, with 300 million covered individuals. Explorys provided a clinical data set of actual electronic health records kept by health systems that represented about 50 million patients. Phytel added more and Merge had a huge imaging database. The idea was to expose Watson to all these data to find patterns that physicians and anyone else couldn't possibly find when looking at that data, given all the variables.

Armed with Watson and Big Data, Watson Health was touted as a revolutionary healthcare solution that could help diagnose patients, recommend treatment options, improve drug development, and match patients with clinical trials. But it didn't work out: After billions invested over the course of a decade and a series of setbacks, Watson Health was sold for its data and assets for only around $1 billion to private equity firm Francisco Partners in January 2022—much, much less than what IBM had invested over the years.

IBM's partnership with MD Anderson Cancer Center in Texas is instructive. Participating physicians said that there weren't enough data for the program to make sound recommendations and that Watson had trouble with the complexity of patient files. The participants complained that Watson's recommendations were just not relevant. Watson might, for example, suggest a particular kind of treatment that wasn't available in the locality for which it was making the recommendation, or the recommendation did not square with the treatment protocols used at that local institution. The physicians also said that Watson was not telling them anything they didn't already know. An artificial intelligence tool exposed to data on patients who were cared for on the upper east side of Manhattan may not be able to derive any meaningful insights to treat patients in India. You need to have representative data, and the data limited to New York are not necessarily going to apply to different kinds of patients all the way across the world. The Anderson partnership was eventually audited and terminated. Other high-profile hospital partners also ended their collaborations with Watson.

IBM continued to pour money into marketing Watson Health without having proven that it could live up to the hype. The Watson Health investment was too big to fail, but it eventually did. Does this mean that using AI is too difficult for healthcare? Experts don't think so: Microsoft and a large group of hospitals have formed a coalition to promote the use of AI in healthcare by providing recommendations, tools, and best practices. The generation of AI technology used in Watson Health was nowhere near ready to accomplish what IBM had promised. Nevertheless, AI can make significant improvements in healthcare if companies learn from IBM's mistakes.

Sources: Shravani Durbhakula, "IBM Dumping Watson Is an Opportunity to Re-Evaluate Artificial Intelligence." *MedCity News*, March 27, 2022; Sandeep Konam, "Where Did IBM Go Wrong with Watson Health?" *Quartz*, March 2, 2022; Lizzie O'Leary, "How IBM's Watson Went from the Future of Health Care to Sold Off for Parts," Slate.com, January 31, 2022; Elly Yates-Roberts, "Microsoft Forms New Coalition for AI in Healthcare," *Technology Record*, January 17, 2022.

## CASE STUDY QUESTIONS

1. One critic has described Watson Health as "a hammer looking for a nail" and said that it is more effective to define and understand a problem before building an AI application. Discuss.
2. How could IBM Watson Health have benefited from using the four-step problem-solving method introduced in Chapter 1?
3. To what extent was Watson Health a technology problem? A people problem? An organizational problem? Explain your answer.
4. How can organizations using AI in healthcare avoid the mistakes IBM made?

and its brand association as the "Place to Buy Paint" increased by 8.5 percent. The engagement rate was nearly three times higher than the industry benchmark of 2.8 percent (GumGum, 2022)

### Robotics

**Robotics** deals with the design, construction, operation, and use of movable machines that can substitute for humans, along with computer systems for their control, sensory feedback, and information processing. Robots cannot substitute entirely for people but are programmed to perform a specific series of actions automatically. They are often used in dangerous environments (such as bomb detection and deactivation), manufacturing processes, military operations (drones), and medical procedures (surgical robots). Many employees now wonder whether robots will replace people entirely and take away their jobs (see the Chapter 1 Spotlight on People case).

The most widespread use of robotic technology has been in manufacturing and logistics. For example, automobile assembly lines employ robots to do heavy lifting, welding, applying glue, and painting. People still do most of the final assembly of cars, especially when installing small parts or wiring that needs to be guided into place. Amazon, FedEx, and Walmart use robots to speed up package deliveries. Amazon deploys more than 200,000 robots to help with deliveries. The company has been setting up Amazon Robotic Fulfillment Centers strategically located across the United States and Europe. Although the Amazon Robotic Fulfillment Centers have human employees, they are in large part warehouses operated by different kinds of robots. The robots are assigned tasks such as heavy lifting, pulling carts of plastic totes from one side of a warehouse to another, and carrying inventory pods to picking stations. Human workers do the picking and perform other tasks that are too complex for the robots to handle (Knight, 2021).

Facial recognition is an artificial intelligence application that can uniquely identify a person by analyzing patterns based on the person's facial textures and shape. Facial recognition systems can be used to identify people in photos, video, or in real time. A facial recognition system uses biometrics to map facial features from a photograph or video. It compares the information with a database of known faces to find a match. The system uses computer algorithms to highlight specific, distinctive details about a person's face, such as the distance between the eyes or the shape of the chin. (Some algorithms explicitly map the face, measuring the distances between the eyes, the nose and mouth, and so on. Others map the face using more abstract features.) The system converts these details to a mathematical representation and compares these details to data on other faces stored in a face recognition database. The data about a particular face are called a face template, and it can be compared to other templates on file. Facial recognition technology learns how to identify people by analyzing as many digital pictures as possible using neural networks, which are complex mathematical systems that require vast amounts of data to build pattern recognition.

Face recognition tools are now frequently used in routine policing. Police compare mugshots of arrestees to local, state, and federal face recognition databases. Law enforcement can query these mugshot databases to identify people in photos taken from social media, traffic cameras, and closed-circuit television surveillance cameras in stores, parks, and other places. There are systems to compare faces in real time with "hot lists" of people suspected of illegal activity. Face recognition has been used in airports, border crossings, and events such as the Olympic Games. The FBI spent more than a decade using such systems to compare driver's license and visa photos against the faces of suspected criminals.

Facial recognition systems can make products safer and more secure. For example, face authentication can ensure that only the right person gets access to sensitive information meant just for them. It can also be used for social good: There are nonprofits using facial recognition to combat trafficking of minors. However, these systems also have limitations that can do harm as well.

Dozens of databases of people's faces are being compiled by companies and researchers, with many of the images then being shared around the world. The databases are populated with images from social networks, photo websites, dating services like OkCupid, and cameras placed in restaurants and on college quads. While there is no precise count of the data sets, privacy activists have pinpointed repositories that were built by Microsoft, Stanford University, and others, with one repository holding more than 10 million images and another with more than two million. Georgetown University has estimated that photos of nearly half of all US adults have been entered into at least one face recognition database.

Tech giants like Facebook and Google are reputed to have amassed the largest facial data sets, which they do not distribute, according to research papers. But other companies and universities have widely shared their image troves with researchers, governments, and private enterprises in Australia, China, India, Singapore, and Switzerland for training artificial intelligence, according to academics, activists, and public papers.

Startup Clearview AI created a powerful facial recognition app that enables the user to take a picture of a person, upload it, and be able to view public photos of that person along with links to where those photos appeared. The system uses a database of more than three billion images that Clearview claims to have scraped from Facebook, YouTube, Venmo, and millions of other websites. Federal and state law enforcement officers have used the Clearview app to help solve shoplifting, identity theft, credit card fraud, murder, and child exploitation cases.

A website called PimEyes provides similar functionality that's available to anyone in the general public willing to pay $29.99 per month. PimEyes finds photos of a person from across the Internet, along with links to where the photos appear on the Internet. The results are highly accurate, and the photos may include some the person does not want exposed.

Companies and labs have gathered facial images for more than a decade, and image databases are an essential component of facial recognition technology. But people often have no idea that their faces are in these databases. And although names are typically not attached to the photos, individuals can be recognized because each face is unique to a person. There is no oversight of these facial recognition data repositories.

Privacy advocates worry that facial recognition systems are being misused. A database called Brainwash was created by Stanford University researchers in 2014. The researchers captured more than 10,000 images using a camera located in San Francisco's Brainwash Café (now closed). It is unclear whether the patrons knew their images were being captured and used for research. The Stanford researchers shared Brainwash with Chinese academics associated with the National University of Defense Technology and Megvii, an AI company that provided surveillance technology for racial profiling of China's Uighur Muslim population. Stanford removed Brainwash from its public online archives in mid-2019.

Using eight cameras on campus to collect images, Duke University researchers gathered more than two million video frames with images of more than 2,700 people. The database, called Duke MTMC, was reported to have been used to train AI systems in the United States, Japan, China, and elsewhere. The cameras were identified with signs, which gave a phone number or email that people could use to opt out.

Moreover, facial recognition systems are not entirely accurate. These systems have varying abilities to identify people under challenging conditions such as poor lighting, low-quality image resolution, and suboptimal angle of view, which might occur if a photograph was taken from above looking down on an unknown person. Facial recognition software is also not as accurate when identifying people of color as well as women and young people. Facial recognition becomes less accurate as the number of people in the database increases. Many people around the world look alike. As the likelihood of similar faces goes up, matching accuracy goes down.

Nevertheless, facial recognition technology continues to improve. As of April 2020, the best face identification algorithm had an error rate of just 0.08 percent, compared to 4.1 percent for the leading algorithm in 2014, according to tests by the National Institute of Standards and Technology (NIST). These improvements need to be taken into account when considering the best way to regulate the technology.

Sources: Kashmir Hill, "A Face Search Engine Anyone Can Use Is Alarmingly Accurate," *New York Times*, May 26, 2022; Tulsee Doshi, "Improving" Skin Tone Representation across Google," Ai.google.com, accessed May 15, 2022; "About Face" Eff.org, accessed May 11, 2022; William Crumpler,"How Accurate Are Facial Recognition Systems and Why Does It Matter?," Center for Strategic and International Studies, April 14, 2020; Kashmir Hill, "The Secretive Company that Might End Privacy as We Know It," *New York Times*, January 18, 2020; Cate Metz, "Facial Recognition Tech Is Growing Stronger, Thanks to Your Face," *New York Times*, July 13, 2019.

## CASE STUDY QUESTIONS

1. Explain the key technologies used in facial recognition systems.
2. What are the benefits of using facial recognition systems? How do they help organizations improve operations and decision making? What problems can they help solve?
3. Identify and describe the disadvantages of using facial recognition systems and facial databases.

## INTELLIGENT AGENTS

**Intelligent agents** are software programs that work in the background without direct human intervention to carry out specific tasks for an individual user, business process, or software application. The agent uses a limited, built-in or learned knowledge base to accomplish tasks or make decisions on the user's behalf, such as deleting junk email, scheduling appointments, or finding the cheapest airfare to California.

There are many intelligent agent applications today in operating systems, application software, email systems, mobile computing software, and network tools. Of special interest to businesses are intelligent agent bots that search for information on the Internet. Chapter 7 describes how shopping bots help consumers find products they want and assist them in comparing prices and other features.

Although some software agents are programmed to follow a simple set of rules, others are capable of learning from experience and adjusting their behavior using machine learning and natural language processing. Siri, a virtual assistant application on Apple's iPhone and iPad, is an example. Siri uses natural language processing to answer questions, make recommendations, and perform actions. The software adapts to the user's individual preferences over time and personalizes results, performing tasks such as getting directions, scheduling appointments, and sending messages. Similar products include Microsoft's Cortana and Amazon's Alexa.

**Chatbots** (chatterbots) are software agents designed to simulate a conversation with one or more human users via textual or auditory methods. They try to understand what you type or say and respond by answering questions or executing tasks. They also provide automated conversations that allow users to do things like check the weather, manage personal finances, shop online, and receive help when they have questions for customer service. For example, the UK package delivery firm Evri (formerly Hermes) created a chatbot called Holly to help its call center handle customer service inquiries. The chatbot helps customers track shipments, change delivery orders, update account preferences, and handle other essential tasks quickly. Facebook has integrated chatbots into its Messenger messaging app so that an outside company with a Facebook brand page can interact with Facebook users through the chat program. Today's chatbots perform very basic functions, but are becoming more technologically advanced. Leading-edge tech firms including Google, Facebook and OpenAI have developed a new class of computer programs known as "large language models," with more sophisticated conversational capabilities. For example, Google engineer Blake Lemoine created controversy when he announced that he believed that Google's LaMDA conversational technology has the capability to be "sentient." People will increasingly use conversational agents as a major tool for interacting with system (Oremus, 2022).

## 11-5 Understand how MIS can help your career.

Here is how this chapter and this book can help you find an entry-level job as a sales coordinator for an AI company.

### THE COMPANY

SeeandHear AI Group is global advertising company serving 21 markets worldwide using artificial intelligence technology. The company is headquartered in Ann Arbor, Michigan, and is looking for a sales coordinator. SeeandHear specializes in computer vision and natural language processing technology for contextual advertising. (In contextual advertising, ads are placed on web pages based on the content of those pages.)

### POSITION DESCRIPTION

The Sales Coordinator plays a proactive support role for SeeandHear's sales teams. The Sales Coordinator creates presentations and keeps abreast of market research to develop more client demand for SeeandHear's products. Job responsibilities include:

- Conducting periodic LinkedIn meetings with the sales team to help identify new prospects and make connections
- Pulling research for upcoming new client meetings
- Assembling trend slides, research slides, and competitive slides to help sellers
- Sending follow-up emails and posting bi-weekly sales calls
- Creating bi-weekly Pipeline meeting recaps that are sent out to the sales team
- Updating the Master Account List

### JOB REQUIREMENTS

- Bachelor's degree preferred
- One year in a selling or marketing environment at a tech company or digital agency. Can include an internship.

- Strong writing and communication skills, with ability to summarize articles quickly and formulate short, concise emails
- Strong PowerPoint & Excel skills
- Proficiency in Salesforce
- Attention to detail, enthusiastic attitude, and the ability to thrive in a fast-paced environment
- Ability to multitask and work on multiple projects simultaneously
- Ability to work remotely

### INTERVIEW QUESTIONS

- What do you know about our company and about contextual advertising, computer vision systems, and natural language processing? Have you ever done any work with AI technology or contextual marketing?
- Have you worked with Salesforce.com? How have you used the software?
- What is your proficiency level with Microsoft Office tools? What work have you done with Excel and PowerPoint?
- Can you provide samples of your writing to demonstrate your communication skills and sense of detail?

### AUTHOR TIPS

- Review the section of this chapter on AI, and use the web to find out more about computer vision systems, natural language processing, and contextual advertising.
- Use the web and LinkedIn to find out more about this company—its products, services, and competitors and the way it operates. Think about what it needs to support its sales team and how you could specifically contribute.
- Learn what you can about Salesforce.com, with attention given to how it handles lead generation and account and contact management.
- Inquire exactly how you would be using Excel and PowerPoint in this job. Describe some of the Excel and PowerPoint work you have done, and perhaps bring samples with you to the interview.

## Review Summary

11-1 **Identify the different types of decisions and explain how the decision-making process works.** Decisions may be structured, semistructured, or unstructured, with structured decisions clustering at the operational level of the organization and unstructured decisions clustering at the strategic level. Decision making can be performed by individuals or groups and includes employees as well as operational, middle, and senior managers. There are four stages in decision making: intelligence, design, choice, and implementation.

11-2 **Describe how business intelligence and business analytics support decision making.** Business intelligence and analytics promise to deliver correct, nearly real-time information to decision makers, and the analytic tools help them quickly understand the information and take action. A business intelligence environment consists of data from the business environment, the BI infrastructure, a BA toolset, managerial users and methods, a BI delivery platform (MIS, DSS, or ESS), and the user interface. There are six analytic functionalities that BI systems deliver to achieve these ends: predefined production reports, parameterized reports, dashboards and scorecards, ad hoc queries and searches, the ability to drill down to detailed views of data, and the ability to model scenarios and create forecasts. BI analytics are starting to handle Big Data. Predictive analytics, location analytics, and operational intelligence are important analytic capabilities.

Management information systems (MIS) producing prepackaged production reports are typically used to support operational and middle management, whose decision making is fairly structured. For making unstructured decisions, analysts and power users employ decision support systems (DSS) with powerful analytics and modeling tools, including spreadsheets and pivot tables. Senior executives making unstructured decisions use dashboards and visual interfaces displaying key performance information affecting the overall profitability, success, and strategy of the firm. The balanced scorecard methodology, focusing on key performance indicators (KPIs), is used in designing executive support systems (ESS).

11-3 **Define artificial intelligence (AI) and explain how it differs from human intelligence.** The most ambitious vision of AI involves the attempt to build computer systems that try to think and act like humans. At present, artificial intelligence lacks the flexibility, breadth, and generality of human intelligence, but it can be used to capture, codify, and extend organizational knowledge in limited domains. AI systems today can perform many tasks that would be impossible for humans to accomplish and can equal or come close to human abilities in certain, well-defined tasks.

11-4 **Identify the major types of AI techniques and show how they benefit organizations.** Expert systems capture tacit knowledge from a limited domain of human expertise and express that knowledge in the form of rules. Machine learning software can learn from previous data and examples. It can identify patterns in very large databases without explicit programming, albeit with significant human training.

Neural networks consist of hardware and software that attempt to mimic the thought processes of the human brain. Neural networks are notable for their ability to learn on their own with some training and to recognize patterns that cannot be easily identified by humans. Deep learning neural networks use multiple layers of neural networks to reveal the underlying patterns in data and, in some limited cases, identify patterns without human training.

Genetic algorithms develop solutions to particular problems using genetically based processes such as fitness, crossover, and mutation. Genetic algorithms are useful for solving problems involving optimization where many alternatives or variables must be evaluated to generate an optimal solution.

Natural language processing technology makes it possible for a machine to understand human language and to process that information. Computer vision systems enable computers to emulate the human visual system to view and extract information from real-world images. Robotics deals with the design, construction, operation, and use of movable machines that can substitute for some human actions.

Intelligent agents are software programs with built-in or learned knowledge bases that carry out specific tasks for an individual user, business process, or software application. Intelligent agents can be programmed to navigate through large amounts of data to locate useful information and in some cases to act on that information on behalf of the user. Chatbots are software agents designed to simulate a conversation with one or more human users via textual or auditory methods.

## Key Terms

Artificial intelligence (AI), 407
Augmented reality (AR), 399
Balanced scorecard method, 405
Chatbot, 421
Choice, 395
Computer vision systems, 416
Data visualization, 398
"Deep learning," 413
Design, 395
Drill down, 406
Expert system, 408
Genetic algorithms, 415
Geographic information systems (GIS), 402
Implementation, 395
Inference engine, 409
Intelligence, 395
Intelligent agents, 420
Intelligent techniques, 407

Key performance indicators (KPIs), 405
Knowledge base, 409
Location analytics, 402
Machine learning (ML), 409
Natural language processing, 416
Neural networks, 412
Operational intelligence, 402
Pivot table, 404
Predictive analytics, 400
Robotics, 418
Semistructured decisions, 394
Sensitivity analysis, 404
Structured decisions, 394
Supervised learning, 410
Unstructured decisions, 394
Unsupervised learning, 411
Virtual reality (VR) systems, 398

## Review Questions

**11-1** Identify the different types of decisions and explain how the decision-making process works.

- List and describe the different decision-making levels and groups in organizations and their decision-making requirements.
- Distinguish among an unstructured, a semistructured, and a structured decision.
- List and describe the stages in decision making.

**11-2** Describe how business intelligence and business analytics support decision making.

- Define and describe business intelligence and business analytics.
- List and describe the elements of a BI environment.
- List and describe the analytic functionalities that BI systems provide.
- Describe how virtual reality and augmented reality enhance data visualization and decision making.
- Define predictive analytics and location analytics, and give two examples of each.
- List each of the types of BI users, and describe the kinds of systems that provide decision support for each type of user.
- Define and describe the balanced scorecard method.

**11-3** Define artificial intelligence (AI) and explain how it differs from human intelligence.

- Define artificial intelligence (AI).
- Explain how AI differs from human intelligence.

**11-4** Identify the major types of AI techniques and show how they benefit organizations.

- Define an expert system, describe how it works, and explain its value to business.
- Define machine learning, explain how it works, and give some examples of the kinds of problems it can solve.
- Compare supervised and unsupervised learning.
- Define neural networks and deep learning neural networks, describing how they work and how they benefit organizations.
- Define and describe genetic algorithms and intelligent agents. Explain how each works and the kinds of problems for which each is suited.
- Define and describe computer vision systems, natural language processing systems, and robotics, and give examples of their applications in organizations.

**MyLab MIS**
To complete these problems, go to EOC Discussion Questions in **MyLab MIS**.

## Discussion Questions

**11-5** MyLab MIS If businesses used DSS and ESS more widely, would they make better decisions? Why or why not?

**11-6** MyLab MIS How much can business intelligence and business analytics help companies refine their business strategy? Explain your answer.

**11-7** MyLab MIS How intelligent are AI techniques? Explain your answer.

## Hands-On MIS Projects

The projects in this section give you hands-on experience identifying opportunities for business intelligence, using a spreadsheet pivot table to analyze sales data, and using intelligent agents to research products for sale on the web. Visit MyLab MIS to access this chapter's Hands-On MIS Projects.

### MANAGEMENT DECISION PROBLEM

**11-8** Samsung sells around 80 million new mobile phones every quarter. Samsung has marketing and analytics teams with access to the massive quantities of complex customer data flowing into its systems from millions of active devices. The team was using standard business intelligence (BI) and dashboarding tools. However, these traditional business intelligence tools couldn't keep up with the volume and complexity of Samsung's data. It took too long to analyze these data with hundreds of variables in order to understand user preferences across demographics, location, mobile device profiles, carrier loyalty, past interactions with Samsung products, and customers to be targeted by a new product launch. Which customers are likely to upgrade to a new device? What factors influence an upgrade decision? It might take weeks to answer a single question. There was no way the team could reliably check every possible factor in the data. Using the four-step problem-solving methodology introduced in Chapter 1, identify Samsung's problem and suggest a solution. Identify the people, organizational, and technology issues to be addressed by the solution.

### IMPROVING DECISION MAKING: USING PIVOT TABLES TO ANALYZE SALES DATA

Software skills: Pivot tables
Business skills: Analyzing sales data

**11-9** This project gives you an opportunity to learn how to use Excel's PivotTable functionality to analyze a database or data list. Use the data file for Online Management Training Inc. described earlier in the chapter. This is a list of the sales transactions at OMT for one day. You can find this spreadsheet file at MyLab MIS. Use Excel's PivotTable to help you answer the following questions:

- Where are the average purchases higher? The answer might tell managers where to focus marketing and sales resources or pitch different messages to different regions.
- What form of payment is the most common? The answer could be used to emphasize in advertising the most preferred means of payment.
- Are there any times of day when purchases are most common? Do people buy products while at work (likely during the day) or at home (likely in the evening)?
- What's the relationship among region, type of product purchased, and average sales price?

### IMPROVING DECISION MAKING: USING INTELLIGENT AGENTS FOR COMPARISON SHOPPING

Software skills: Web browser and shopping bot software
Business skills: Product evaluation and selection

**11-10** This project will give you experience using shopping bots to search online for products, find product information, and find the best prices and vendors.

Select a digital camera you might want to purchase, such as the Canon PowerShot SX740 or the Olympus Tough TG-7. Visit PriceGrabber, BizRate, and Google Shopping to do price comparisons for you. Evaluate these shopping sites in terms of their ease of use, number of offerings, speed in obtaining information, thoroughness of information offered about the product and seller, and price selection. Which site or sites would you use and why? Which camera would you select and why? How helpful were these sites in making your decision?

## COLLABORATION AND TEAMWORK PROJECT

Investigating Data-Driven Analytics in Sports

**11-11** With three or four of your classmates, select a sport, such as football, baseball, basketball, or soccer. Use the web to research how the sport uses data and analytics to improve team performance or increase ticket sales to events. If possible, use Google Docs and Google Drive or Google Sites to brainstorm, organize, and develop a presentation of your findings for the class.

# BUSINESS PROBLEM-SOLVING CASE

## SHOULD AN ALGORITHM MAKE OUR DECISIONS?

Darnell Gates of Philadelphia had been jailed for running a car into a house in 2013 and later for violently threatening his former domestic partner. When he was released in 2018, he was initially required to visit a probation officer once a week because he had been identified as "high risk" by a computer algorithm. Gates's probation officer visits were eventually stretched to every two weeks and then once a month, but conversations with probation officers remained impersonal and perfunctory, with the officers rarely taking the time to understand Gates's rehabilitation. What Gates didn't know was that a computer algorithm developed by a University of Pennsylvania professor had made the "high-risk" determination that governed his treatment.

This algorithm is one of many now being used to make decisions about people's lives in the United States and Europe. Predictive algorithms are being used to determine prison sentences, probation rules, and police patrols. It is often not clear how these automated systems are making their decisions. Many countries and states have few rules requiring disclosure of algorithms' formulae. And even if governments provide explanation of how the systems arrive at their decisions, the algorithms are often too difficult for a layperson to understand.

Since 2019, the Dutch government has been embroiled in a scandal after the country's tax agency used a self-learning algorithm to create risk profiles for spotting fraud among people applying for childcare benefits. Families were penalized on suspicion of fraud based on the system's risk indicators. Tens of thousands of families became impoverished by exorbitant debts owed to the tax agency to pay back their claims. More than one thousand children were put into foster care due to the scandal. Some victims committed suicide.

The Dutch system had been launched in 2013 to create risk profiles of people in order to weed out benefits fraud at an early stage. The criteria for the risk profile were developed by the tax authority. Having dual nationality or low income were big risk indicators. The algorithm would first vet claims for signs of fraud, and then humans would scrutinize those claims that the algorithm had flagged as high risk. The algorithm falsely labeled claims as fraudulent, and civil servants rubber-stamped the fraud labels. The tax authorities then started clawing back benefits from families who were flagged by the system without proof that they had committed such fraud.

In December 2021, the Dutch data protection agency fined the Dutch tax administration €2.75 million (approximately US $3 million) for the unlawful, discriminatory manner in which the tax authority had processed data on the dual nationality of childcare benefit applicants. A parliamentary report on the childcare benefits scandal had found evidence of institutional bias and authorities hiding information or misleading the Parliament about the facts. Revelation of the scandal forced Dutch Prime Minister Mark Rutte's government to resign, regrouping 225 days later. The new government has pledged to create a new algorithm regulator under the country's data protection authority. The European Parliament is working on an AI Act that would put public-sector uses of AI under tighter scrutiny,

The city of Bristol, England is using an algorithm to compensate for tight budgets that reduced social services. A team that includes representatives from the police and children's services meets weekly to review results from an algorithm that tries to identify youths in the city who are most at risk for crime and children who are most in need. In 2019, Bristol introduced a software program that creates a risk score based on data extracted from police reports, social benefits, and other government records. The risk score takes into account crime, school attendance, and housing data; known ties to others with high-risk scores; and whether a youth's parents were involved in a domestic incident. The scores fluctuate, depending on how recently a youth had an incident such as a school suspension.

There is evidence that Bristol's algorithm is identifying the right people, and there are human decision makers governing its use. Charlene Richardson and Desmond Brown, two city workers who have been responsible for aiding young people flagged by the software, acknowledged that the computer doesn't always get it right, so they haven't relied on it entirely. The city government has been open with the public about the program. The government has posted some details online and staged community events. However, opponents believe the program isn't fully transparent. Young people and their parents do not know if they are on the list and have no way to contest their inclusion.

Studies of algorithms in credit scoring, hiring, policing, and healthcare have found that poorly designed algorithms can reinforce racial and gender biases. According to Solon Barocas, an assistant professor at Cornell University and principal researcher at Microsoft Research, algorithms can appear to be very data-driven, but there are subjective decisions that go into setting up the problem in the first place.

For example, a study by Ziad Obermeyer, a health policy researcher at the University of California, Berkeley, and colleagues published in October 2019 in the journal *Science*, examined an algorithm widely used in hospitals to establish priorities for patient care. The study found that black patients were less likely than white patients to receive extra medical help, despite being sicker. The algorithm has been used in the Impact Pro program from Optum, United Health Group's health services arm, to identify patients with heart disease, diabetes, and other chronic illnesses who could benefit from more attention from nurses, pharmacists, and case workers in managing their prescriptions, scheduling doctor visits, and monitoring their overall health.

The algorithm assigned healthier white patients the same ranking as black patients who had one more chronic illness as well as poorer laboratory results and vital signs. To identify people with the greatest medical needs, the algorithm looked at patients' medical histories and how much was spent treating them and then predicted who was likely to have the highest costs in the future. The algorithm used costs to rank patients.

The algorithm wasn't intentionally racist—in fact, it specifically excluded race. Instead, to identify patients who would benefit from more medical support, the algorithm used a seemingly race-blind metric: how much patients would cost the healthcare system in the future. But cost isn't a race-neutral measure of healthcare need. Black patients incurred about $1,800 less in medical costs per year than white patients with the same number of chronic conditions; thus, the algorithm scored white patients as equally at risk of future health problems as black patients who had many more diseases.

When the researchers searched for the source of the scores, they discovered that Impact Pro was using bills and insurance payouts as indicators of a person's overall health. However, healthcare costs tend to be lower for black patients, regardless of their actual well-being. Compared with white patients, many black patients live farther from their hospitals, making it harder to visit them regularly. They also tend to have less flexible job schedules and more childcare responsibilities.

As a result, black patients with the highest risk scores had higher numbers of serious chronic conditions, including cancer and diabetes, than white patients with the same scores, the researchers reported. Compared with white patients with the same risk scores, black patients also had higher blood pressure and cholesterol levels, more severe diabetes, and worse kidney function.

By using medical records, laboratory results, and vital signs of the same set of patients, the researchers found that black patients were sicker than white patients who had a similar predicted cost. By revising the algorithm to predict the number of chronic illnesses that a patient will likely experience in a given year—rather than the cost of treating those illnesses—the researchers were able to reduce the racial disparity by 84 percent. When the researchers sent their results to Optum, it replicated their findings and committed to correcting its model.

Civil rights lawyers, labor unions, community organizers, and some government agencies are trying to find ways to push back against the growing dependence on automated systems that remove humans from the decision-making process. Movement Alliance Project in Philadelphia and MediaJustice in Oakland, California, have compiled a nationwide database of prediction algorithms. Community Justice Exchange, a national organization supporting community organizers, has issued a guide advising organizers on how to confront the use of algorithms. Idaho passed a law in 2019 specifying that the methods and data used in bail algorithms should be publicly available so that the general public can understand how the algorithms work.

Concerns about biases have always been present whenever people make important decisions. What's new is the much larger scale at which we rely on algorithms in automated systems to help us decide, and even to take over the decision making for us. Algorithms are useful in making predictions that will help guide decision makers, but decision making requires much more. Good decision making requires bringing together and reconciling multiple points of view and the ability to explain why a particular path was chosen.

As automated systems increasingly shift from predictions to decisions, focusing on the fairness of algorithms is not enough because their output is just one of the inputs for a human decision maker. One must also look at how human decision makers interpret and integrate the output from algorithms and under what conditions they would deviate from an algorithmic recommendation. Which aspects of a decision process should be handled by an algorithm and which by a human decision maker to obtain fair and reliable outcomes?

Sources: Rahul Rao, "The Dutch Tax Authority Was Felled by AI—What Comes Next?" *IEEE Spectrum*, May 9. 2022; Melissa Heikkila, "AI Decoded: A Dutch Algorithm Scandal Serves a Warning to Europe—The AI Act Won't Save Us." *Politico*, March 30, 2022; Irving Wladawsky-Berger, "The Coming Era of Decision Machines," *Wall Street Journal*, March 27, 2020; Cade Metz and Adam Satariano, "An Algorithm that Grants Freedom, or Takes It Away," *New York Times*, February 6, 2020; Michael Price, "Hospital 'Risk Scores' Prioritize White Patients," *Science*, October 24, 2019.

## CASE STUDY QUESTIONS

**11-12** What are the problems in using algorithms and automated systems for decision making?

**11-13** What people, organizational, and technology factors have contributed to the problem?

**11-14** Should automated systems be used to make decisions? Explain your answer.

# Chapter 11 References


Alves, Carolina de Lima Salge, Elena Karahanna, and Jason Bennett Thatcher. "Algorithmic Processes of Social Alertness and Social Transmission: How Bots Disseminate Information on Twitter." *MIS Quarterly* 46, No 1 (March 2022).

Babic, Boris, I., Glenn Cohen, Theodoros Evgeniou, and Sara Gerke. "When Machine Learning Goes off the Rails." *Harvard Business Review* (January–February 2021).

Bean, Matt, and Erik Brynjolfsson. "Robots in a Post-Pandemic World." *MIT Sloan Management Review* 62, No. 2 (Winter 2021).

Bertini, Marco, and Oded Koenigsberg."The Pitfalls of Pricing Algorithns." *Harvard Business Review* (September–October 2021).

Burtka, Michael. "Genetic Algorithms." *The Stern Information Systems Review* 1, No. 1 (Spring 1993).

Chui, Michael, James Manyika, and Mehdi Miremadi. "What AI Can and Can't Do (Yet) for Your Business." *McKinsey Quarterly* (January 2018).

Davenport, Thomas H. "Big Data at Work: Dispelling the Myths, Uncovering the Opportunities." *Harvard Business Review* (March 2014).

Davenport, Thomas H., Abhijit Guha, and Dhruv Grewal. "How to Design an AI Marketing Strategy." *Harvard Business Review* (July–August 2022).

Fügener, Andreas, Jörn Grahl, Alok Gupta, and Wolfgang Ketterer. "Will Humans-in-the-Loop Become Borgs? Merits and Pitfalls of Working with AI (Open Access)." *MIS Quarterly* 45, No.3 (September 2021).

GumGum. "If the Eye Can See It, Valspar Can Colour Match It." Gumgum.com, accessed May 25, 2022.

Holland, John H. "Genetic Algorithms." *Scientific American* (July 1992).

IBM Cloud Education. "Unsupervised Learning." IBM.com, accessed May 14, 2022.

Joshi, Mayur P., Ning Su, Robert D. Austin, and Anand K. Sundaram. "Why So Many Data Science Projects Fail to Deliver." *MIT Sloan Management Review* 62, No. 3 (Spring 2021).

Kahneman, Daniel. *Thinking, Fast and Slow* (New York: Farrar, Straus and Giroux, 2011).

Kane, Gerald C., Amber G. Young, Ann Majchrzak, and Sam Ransbotham. "Avoiding an Oppressive Future of Machine Learning: A Design Theory for Emancipatory Assistants." *MIS Quarterly* 45, No.1 (March 2021).

Knight, Will. "Robots Won't Close the Warehouse-Worker Gap Any Time Soon." *Wired* (November 26, 2021).

Larson, Erik J. *The Myth of Artificial Intelligence: Why Computers Can't Think the Way We Do* (Cambridge, MA: Belknap Press, 2021).

Le, Quoc V., et al. "Building High-level Features Using Large Scale Unsupervised Learning." ArXiv. org:1112.6209, Machine Learning, Cornell University Library (November 2011).

Lebovitz, Sarah, Natalia Levina, and Hila Lifshitz-Assaf. "Is AI Ground Truth Really True? The Dangers of Training and Evaluating AI Tools Based on Experts' Know-What." *MIS Quarterly* 45, No. 3 (September 2021).

Lin, Yu-Kai, and Fang, Xiao, "First, Do No Harm: Predictive Analytics to Reduce In-Hospital Adverse Events." *Journal of Management Information Systems* 38, No. 4 (2021).

Lohr, Steve. “Is There a Smarter Path to Artificial Intelligence? Some Experts Hope So.” *New York Times* (June 20, 2018).

Maor, Itzakh, and T. A. Reddy. “Literature Review of Artificial Intelligence and Knowledge-Based Expert Systems in Buildings and HVAC&R System Design,” in M. Geshwiler, E. Howard, and C. Helms (Eds.), *ASHRAE Transactions* (2003).

Marcus, Gary. “Deep Learning: A Critical Appraisal.” (January 2, 2018).

Martens, David, Foster Provost, Jessica Clark, and Enric Junqué de Fortuny. “Mining Massive Fine-Grained Behavior Data to Improve Predictive Analytics.” *MIS Quarterly* 40, No. 4 (December 2016).

Maxfield, Max. “When Genetic Algorithms Meet Artificial Intelligence.” *Journal of Electronic Engineering* (July 9, 2020).

Möhlmann, Mareike, Lior Zalmanson, Ola Henfridsson, and Robert Wayne Gregory. “Algorithmic Management of Work on Online Labor Platforms: When Matching Meets Control.” *MIS Quarterly* 45, No. 4 (December 2021).

Moser, Christine, Frank den Hond, and Dirk Lindebaum. “When We Let AI Decide.” *MIT Sloan Management Review* 63, No.3 (Spring 2022).

Oremus, Will. “Google’s AI Passes a Famous Test, and Shows How the Test Is Broken.” *Washington Post* (June 17, 2022).

Pearl, Judea. “Theoretical Impediments of Machine Learning.” (November 2016).

Pearl, Judea, and Dana Mackenzie. “AI Can’t Reason Why.” *Wall Street Journal* (May 18, 2018).

Porter, Michael E., and James Heppelmann. “Why Every Organization Needs an Augmented Reality Strategy.” *Harvard Business Review* (November–December 2017).

Simon, H. A. *The New Science of Management Decision*. (New York: Harper & Row, 1960).

Tableau. “Sysco Labs Leverages Customer Behavior Insights to Help Field Reps Transition Customers to Self-Service.” Tableau.com, accessed May 28, 2022.

Teodorescu, Mike M., Lily Morse, Yazeed Awwad, and Gerald C. Kane. “Failures of Fairness in Automation Require a Deeper Understanding of Human–ML Augmentation.” *MIS Quarterly* 45, No. 3 (September 2021).

Tudorache, Adela. “How Apple Uses the Balanced Scorecard System.” *Performance Magazine* (September 10, 2021).

Vial, Gregory, Jinglu Jiang, Tanya Giannelia, and Ann-Frances Cameron. “The Data Problem Stallling AI.” *MIT Sloan Management Review* 62, No. 2 (Winter 2021).

Wilson, H. James, and Paul R. Daugherty. “Robots Need Us More than We Need Them.” *Harvard Business Review* (March–April 2022).

PART IV

**Chapter 12**
Making the Business Case for Information Systems and Managing Projects

# Building and Managing Systems

Part IV shows how to use the knowledge acquired in earlier chapters to analyze and design information system solutions to business problems. This part answers questions such as these: How can managers select systems projects and technologies that will deliver the greatest value to the firm? How can I develop a solution to an information system problem that provides genuine business benefits? How can the firm adjust to the changes introduced by the new system solution? What alternative approaches are available for building system solutions?

CHAPTER 12

# Making the Business Case for Information Systems and Managing Projects

## LEARNING OBJECTIVES

After completing this chapter, you will be able to:

12-1 Explain how managers should build a business case for the acquisition and development of a new information system.

12-2 Identify the core problem-solving steps for developing a new information system.

12-3 Compare the alternative methods for building information systems.

12-4 Describe how information systems projects should be managed.

12-5 Understand how MIS can help your career.

## CHAPTER CASES

- Angostura Builds a Mobile Sales System
- McAfee Turns to Automated Software Testing
- Sauder Woodworking Gets ERP Implementation Right
- JEDI and JWCC: A Cloud of Controversy

**MyLab MIS**

- **Video Case:** Software Design Is "Really on a Tear," Figma CEO Says
- Discussion Questions: 12-5, 12-6, 12-7
- Hands-On MIS Projects: 12-8, 12-9, 12-10, 12-11
- eText with Figure Videos

# ANGOSTURA BUILDS A MOBILE SALES SYSTEM

House of Angostura (also known as Angostura Limited), headquartered in Laventille, Trinidad, is one of the Caribbean's leading rum producers and the world market leader for bitters used in many cocktails. The company sells its iconic products in 170 markets, with its core rum market in Trinidad and Tobago.

Angostura has 347 full-time employees, with annual revenue exceeding $920 million in 2021. Despite the Covid-19 pandemic and worldwide closures of bars and restaurants, profit for the year increased by 8.7 percent compared to the prior year's profit, and Angostura expanded its reach into more international markets. Angostura still takes care of local distribution of its products in Trinidad and Tobago, with a team of 16 sales representatives taking orders out in the field. Although this arrangement worked well in the past, the process was heavily manual, tedious, and time consuming and sometimes produced inaccurate orders.

Each day, the 16 sales reps in the field had to copy the orders on paper and return to the office to hand off the order forms to a customer service representative, who would then manually input the order data into Angostura's SAP enterprise resource planning (ERP) system. Because the orders were handwritten, information might be read and entered incorrectly, which could result in the wrong goods being sent to a customer. Such inaccurate orders were often returned, creating more paperwork and higher costs. Angostura also used manual processes for reporting and tracking invoices and accounts receivable information, which could create additional delays and errors.

The sales representatives were also working with data on product availability that might be out of date. If the sales reps were away from the office, they would not be able to tell whether an order could actually be fulfilled. They would have to call Angostura's warehouse to find out if fulfilling an order was possible.

Angostura's management decided that the sales process needed to be more streamlined and efficient and that it should use mobile technology. The company identified a set of detailed information requirements for the improved sales process and spent more than a year evaluating system solutions from five mobile vendors. One important requirement was that the application be able to automatically update the availability of purchased products from the company's overall inventory and integrate with the firm's back-end SAP ERP system. Another requirement was that the mobile system be able to operate offline so that a sales representative could still input an order on a mobile device even if there was no online connectivity. Once online, the device could then send the order through to the ERP system.

© Prostock-studio/Shutterstock

The vendor selected was the one that could best develop the mobile application to the company's specifications and stay within the budget established by management. Angostura partnered with IDS Scheer and itCampus consultants to develop a mobile sales solution running on Apple iPads. The solution includes an offline customer database, product catalog, customer-specific pricing, order entry, order preview, and integration with Bluetooth wireless printers. A pilot application was quickly created using SAP NetWeaver Gateway technology to connect various devices and platforms to SAP software, and the entire application went live within six months.

Each of Angostura's 16 sales representatives was issued an iPad that includes not only the order application but other mobile apps, such as email, Google Maps, and a video and PDF document uploader to display the Angostura product line, to make the sales process more efficient. The sales application integrates with the corporate ERP system, providing the sales reps with up-to-date information on the availability of products in the warehouse.

With the Angostura Mobile Sales App, an order can be created in less than 30 seconds, depending on the size of the order, making the ordering process two times faster. There is a 20 percent time savings per salesperson because the sales reps now have the ability to send orders remotely as they place them rather than waiting until they return to the office. The amount of time customer service representatives would typically spend on data entry—which was considerable—has been reduced by 75 percent, freeing up time for more useful tasks. In addition, returned orders have been reduced by 30 percent.

Sources: "Inside a Multi-Phased SAP Digitization Journey II," Sapinsider.org, accessed April 2, 2022; Angostura.com, accessed March 24, 2022; "Angostura," Owler.com, accessed March 24, 2022; IDS Scheer Consulting Group, "Angostura's iPad-Based SAP Mobile Sales Solution," 2014.

---

Angostura's experience illustrates some of the steps required to plan, build, and implement new information systems. Building a new system for mobile sales orders entailed analyzing the organization's problems with existing systems, assessing information requirements, selecting appropriate technology, and redesigning business processes and jobs. Angostura's system builders had to make a convincing business case—namely, that it would solve the firm's order entry problems—for investing in a new sales system. Management had to oversee the systems-building effort and evaluate benefits and costs. The information requirements were incorporated into the design of the new system, which represented a process of planned organizational change. Angostura succeeded with this project because its management clearly understood that strong project management and attention to organizational change were essential to success.

The chapter-opening case calls attention to important points raised by this case and this chapter. Angostura's ability to handle sales orders was hampered by outdated and inefficient manual processes, which raised costs, slowed down work, and limited the company's ability to serve its customers.

The solution was to redesign the sales order process to use mobile devices and software and allow orders to be entered through iPads and transmitted to the firm's back-end ERP system. Angostura's information requirements were incorporated into the system design. The solution encompassed not just the application of new technology but also changes to corporate culture, business processes, and job functions. Angostura's sales operations have become much more efficient and far less error prone.

Here are some questions to think about: How did Angostura's Mobile Sales App meet its information requirements? How effective a solution was Angostura's Mobile Sales App? Why? How much did the new system change the way Angostura ran its business?

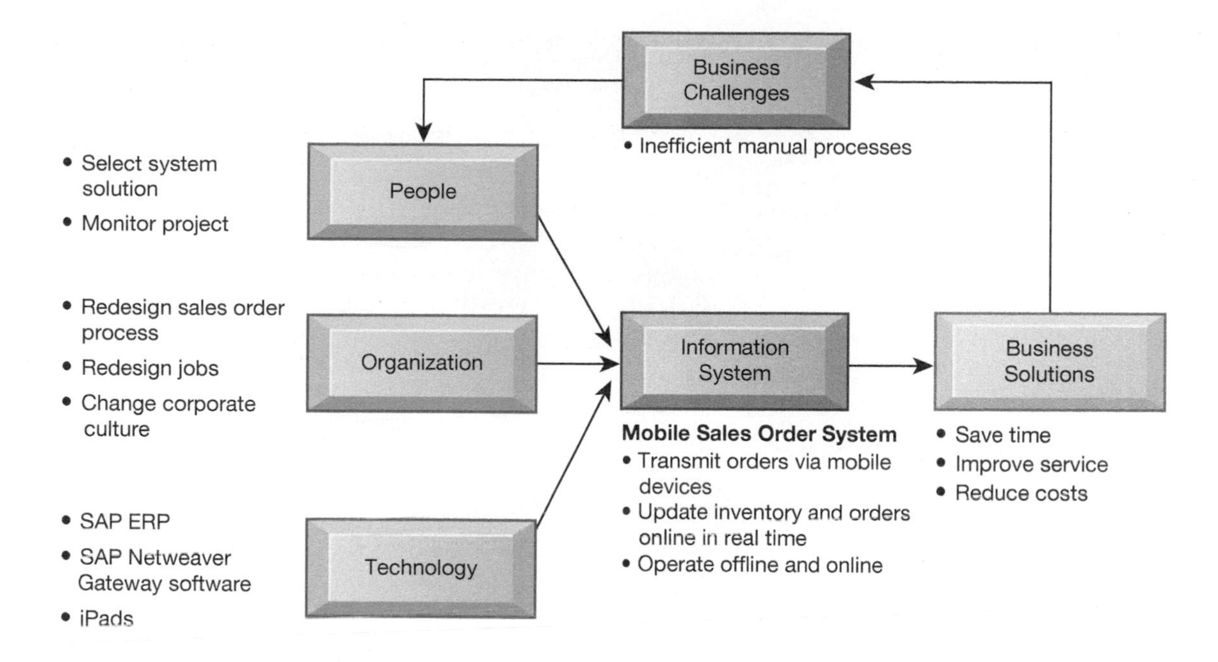


## 12-1 Explain how managers should build a business case for the acquisition and development of a new information system.

Companies typically are presented with many alternatives for solving problems and improving their performance, including the development of new information systems or the enhancement of existing ones. There are far more ideas for systems projects than there are resources. Your company will need to select the systems projects investments that promise the greatest benefit to the business. And you will need to make the business case for why the solution you select provides the greatest value to the firm when compared to other solutions.

A **business case** is a proposal to management seeking approval for an investment. The business case for an IT investment describes the problem facing the organization that can be solved by investing in a proposed system solution. It provides an analysis of all the costs, benefits, and risks associated with that investment and the justification for that proposed course of action. The business case describes the rationale for proceeding with an investment and shows how the investment supports the firm's strategic goals and business objectives and how it fits in with the overall information systems plan of the firm. It also provides the information necessary to make an informed decision about whether to proceed with the investment and in what form. The business case explains how this investment will provide value for the business and identifies any risks that could negatively affect outcomes. The business case identifies alternative solutions, along with the deciding factors for selecting the preferred option. A good business case will also describe how the proposed solution may require changes in organizational culture, systems, processes, and jobs.

Examine Figure 12.1, which summarizes the seven major factors that are used in making the business case for a specific new system. (Review the discussion of business drivers of information systems in Chapter 1.) These factors are: (1) long-term strategic (lowering production costs, differentiating products and services, increasing the scope of the firm [e.g., global expansion], and matching or exceeding competitor capabilities); (2) improved decision making; (3) customer and supplier relationships; (4) survival (required by the market and ESG goals); (5) new products and services; (6) financial rationale; and (7) fitting with the long-term IT plan of the firm. Smaller systems that focus on a single problem, like the Angostura mobile order entry system described in the opening case, will focus on just a few of these elements, such as "improved decision

**Figure 12.1**
Factors to Consider When Making the Business Case
*There are seven major factors that should be addressed when making the business case for a new information system.*

making," "customer relationships," and "lowering costs," whereas larger system projects may well include all of these factors when making the business case.

## THE INFORMATION SYSTEMS PLAN

In order to identify the information systems projects that will deliver the most business value, organizations need a corporate-wide **information systems plan** that supports their overall business plan, with strategic systems incorporated into top-level planning. The IS firm plan is developed by the Chief Information Officer and is approved annually by the CEO and often the Board of Directors. The plan serves as a roadmap indicating the direction of systems development (the purpose of the plan), the rationale, the state of current systems, new developments to consider, the management strategy, the implementation plan, and the budget (see Table 12.1). Without a comprehensive, firm-wide IS plan, it is difficult, if not impossible, to evaluate the worth of proposals for developing specific individual systems. You cannot make the case for a specific new system without understanding the larger context of all the many systems in the firm.

The plan contains a statement of corporate goals and specifies how information technology will support the attainment of those goals. It explains how general goals will be achieved by specific systems projects. It identifies specific target dates and milestones that can be used later to evaluate the plan's progress in terms of how many objectives were actually attained in the timeframe specified in the plan. The plan also indicates the key management decisions, technology, and required organizational change.

In order to plan effectively, firms will need to inventory and document all of their information system applications, IT infrastructure components, and long- and short-term information requirements. For projects in which benefits involve improved decision making, managers should try to identify the decision improvements that would provide the greatest additional value to the firm (see Chapter 11). They should then develop a set of metrics to quantify the value of more timely and precise information on the outcome of the decision.

The plan should describe organizational changes, including management and employee training requirements; changes in business processes; and changes in authority, structure, or management practice. When you are making the business case for a new information system project, you show how the proposed system fits into that plan.

**TABLE 12.1**

Information Systems Plan

1. **Purpose of the Plan**
   - Current business organization and future organization
   - Key business processes
   - Management strategy
2. **Strategic Business Plan Rationale**
   - Current situation
   - Current business organization
   - Changing environments
   - Major goals of the business plan
   - Firm's strategic plan
3. **Current Systems**
   - Major systems supporting business functions and processes
   - Current infrastructure capabilities
     - Hardware
     - Software
     - Database
     - Networking and Internet
     - Cloud services
     - Security
   - Difficulties meeting business requirements
   - Anticipated future demands
4. **New Developments**
   - New system projects
     - Project descriptions
     - Business rationale
     - Applications' role in strategy
   - New infrastructure capabilities required
     - Hardware
     - Software
     - Database
     - Networking and Internet
     - Cloud services
     - Security
5. **Management Strategy**
   - Acquisition plans
   - Organizational realignment
   - Management controls
   - Major training initiatives
   - Human Resources strategy
6. **Implementation of the Plan**
   - Anticipated difficulties in implementation
   - Progress reports and milestones
7. **Budget Requirements**
   - Resources
   - Potential savings
   - Financing
   - Acquisition cycle

## PORTFOLIO ANALYSIS AND SCORING MODELS

Once you have determined the overall direction of systems development in the firm by establishing a firm-wide IS Plan, **portfolio analysis** is one tool that can help you evaluate alternative system projects. Portfolio analysis inventories all of the firm's information systems projects and assets, including infrastructure, outsourcing contracts,

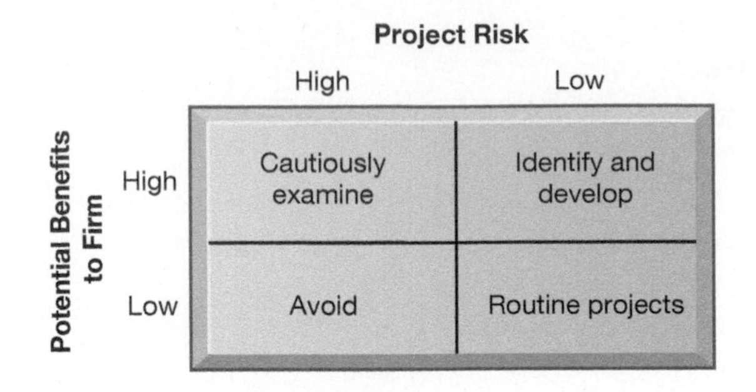


**Figure 12.2**
A System Portfolio
*Companies should examine their portfolio of projects in terms of potential benefits and likely risks. Certain kinds of projects should be avoided altogether and others developed rapidly. There is no ideal mix. Companies in different industries have different information systems needs.*

and licenses. This portfolio of information systems investments can be described as having a certain profile of risk and benefit to the firm (see Figure 12.2), similar to a financial portfolio. Each information systems project carries its own set of risks and benefits. Firms try to improve the return on their information system portfolios by balancing the risks and returns from their systems investments.

Examine Figure 12.2, which illustrates how firms can use a portfolio analysis matrix to evaluate whether the firm should undertake a project. Obviously, you begin first by focusing on systems of high benefit and low risk. These promise early returns and low risks. Second, high-benefit, high-risk systems should be examined; low-benefit, high-risk systems should be totally avoided; and low-benefit, low-risk systems should be re-examined for the possibility of rebuilding and replacing them with more desirable systems having higher benefits. By using portfolio analysis, management can determine the optimal mix of investment risk and reward for their firms, balancing riskier, high-reward projects with safer, lower-reward ones. View the Figure 12.2 video in the eText for an animated and more detailed discussion of this figure.

Another method for evaluating alternative system solutions is a **scoring model**. Scoring models give alternative systems a single score based on the extent to which they meet selected objectives. Table 12.2 shows part of a simple scoring model that Angostura could have used in evaluating proposed system solutions for improving the sales process. The first column lists the criteria that decision makers use to evaluate the systems. Table 12.2 shows that Angostura attaches the most importance to capabilities for sales order processing, ease of use, ability to support individual sales reps taking orders, and system access from mobile platforms. The second column in Table 12.2 lists the weights that decision makers attached to the decision criteria. Columns 3 and 5 show the percentage of requirements for each function that each alternative system solution meets. Each alternative's score is calculated by multiplying the percentage of requirements met for each function by the weight attached to that function. Solution alternative 2 has the highest total score.

## DETERMINING SOLUTION COSTS AND BENEFITS

As we pointed out earlier, the business case for a system solution includes an assessment of whether each solution represents a good investment for the company.

Even if a systems project supports a firm's strategic goals and meets user information requirements, it needs to be a good investment for the firm. The value of systems from a financial perspective essentially revolves around the issue of return on invested capital. Does a particular information system investment produce sufficient returns to justify its costs?

Table 12.3 lists some of the more common costs and benefits of systems. **Tangible benefits** can be quantified and assigned a monetary value. **Intangible benefits**, such as more efficient customer service or enhanced decision making, cannot be immediately quantified but may lead to quantifiable gains in the long run. Transaction and clerical systems that displace labor and save space always produce more measurable, tangible benefits than management information systems, decision-support systems, and systems for collaborative work (see Chapter 2). Some of the tangible benefits Angostura obtained were increased productivity and lower operational costs resulting from

**TABLE 12.2**

Example of a Scoring Model for the Angostura Mobile Sales System

| Criteria | Weight | Alternative 1 (%) | Alternative 1 Score | Alternative 2 (%) | Alternative 2 Score |
|---|---|---|---|---|---|
| **1.1 Order processing** | | | | | |
| 1.2 Online order entry | 5 | 67 | 335 | 83 | 415 |
| 1.3 Order tracking by sales rep | 5 | 81 | 405 | 75 | 375 |
| 1.4 Order tracking by customer | 5 | 30 | 150 | 80 | 400 |
| Total order processing | | | 890 | | 1,190 |
| **2.1 Ease of use** | | | | | |
| 2.2 System access from mobile platforms | 5 | 55 | 275 | 92 | 460 |
| 2.3 Short training time | 4 | 79 | 316 | 85 | 340 |
| 2.4 User-friendly online screens and data entry | 4 | 65 | 260 | 87 | 348 |
| Total ease of use | | | 851 | | 1,148 |
| **3.1 Costs** | | | | | |
| 3.2 Software costs | 3 | 51 | 153 | 65 | 195 |
| 3.3 Hardware (cloud services) costs | 4 | 57 | 228 | 90 | 360 |
| 3.4 Maintenance and support costs | 4 | 42 | 168 | 89 | 356 |
| Total costs | | | 549 | | 911 |
| Grand Total | | | 2,290 | | 3,249 |

streamlining the ordering process and reduced errors. Intangible benefits included customer satisfaction, more timely information, and improved operations.

Chapter 5 introduced the concept of total cost of ownership (TCO), which is designed to identify and measure the components of information technology expenditures beyond the initial cost of purchasing and installing hardware and software. TCO analysis, however, provides only part of the information needed to evaluate an information technology investment because it typically does not deal with benefits, cost categories such as complexity costs, and "soft" and strategic factors discussed later in this section.

## Capital Budgeting for Information Systems

To determine the benefits of a particular project, you'll need to calculate all of its costs and all of its benefits. Obviously, a project whose costs exceed its benefits should be rejected. But even if the benefits outweigh the costs, additional financial analysis is required to determine whether the project represents a good return on the firm's invested capital. **Capital budgeting** models are one of several techniques used to measure the value of investing in long-term capital investment projects.

Capital budgeting methods rely on measures of cash flows into and out of the firm; capital projects generate those cash flows. The investment cost for information systems projects is an immediate cash outflow caused by expenditures for hardware, software, and labor. In subsequent years, the investment may cause additional cash outflows that will be balanced by cash inflows resulting from the investment. Cash inflows take the form of increased sales of more products (for reasons such as new

**TABLE 12.3**

Costs and Benefits of Information Systems

| **Costs** |
| --- |
| Hardware |
| Networking |
| Software |
| Services |
| Personnel |
| **Tangible Benefits (Cost Savings)** |
| Increased productivity |
| Lower operational costs |
| Reduced workforce |
| Lower computer expenses |
| Lower outside vendor costs |
| Lower clerical and professional costs |
| Reduced rate of growth in expenses |
| Reduced facility costs |
| **Intangible Benefits** |
| Improved asset utilization |
| Improved resource control |
| Improved organizational planning |
| Increased organizational flexibility |
| More timely information |
| Improved customer experience |
| Increased organizational learning |
| Legal requirements attained |
| Enhanced employee goodwill |
| Increased job satisfaction |
| Improved decision making |
| Higher client satisfaction |
| Better corporate image |

products, higher quality, or increasing market share) or reduced costs in production and operations. The difference between cash outflows and cash inflows is used to calculate the financial worth of an investment. Once the cash flows have been established, several alternative methods are available for comparing different projects and deciding about the investment.

The principal capital budgeting models for evaluating IT projects are the payback method, the accounting rate of return on investment (ROI), net present value (NPV), and the internal rate of return (IRR). Examine Figure 12.3, which illustrates part of the capital budgeting analysis for an online ordering system similar to Angostura's.

### Limitations of Financial Models

The traditional focus on the financial and technical aspects of an information system tends to overlook the social and organizational dimensions of information systems that may affect the true costs and benefits of the investment. Many companies' information systems investment decisions do not adequately consider costs from organizational disruptions created by a new system, such as the cost to train end users, the impact that users' learning curves for a new system have on productivity, or the time managers need to spend overseeing new system-related changes. Intangible benefits such as more timely decisions from a new system or enhanced employee learning and expertise may also be overlooked in a traditional financial analysis.

| | A | B | C | D | E | F | G | H |
|---|---|---|---|---|---|---|---|---|
| 1 | Estimated Costs & Benefits - Mobile Online Ordering System | | | | | | | |
| 2 | Year | | 0 | 1 | 2 | 3 | 4 | 5 |
| 3 | | | 2022 | 2023 | 2024 | 2025 | 2026 | 2027 |
| 4 | Costs | | | | | | | |
| 5 | Hardware | | | | | | | |
| 6 | 50 iPads @$500 | | $25,000 | | | | | |
| 7 | Cloud IaaS | | $4,000 | $2,000 | $2,000 | $2,000 | $2,000 | $2,000 |
| 8 | Networking | | $1,500 | $1,500 | $1,500 | $1,500 | $1,500 | $1,500 |
| 9 | Software | | | | | | | |
| 10 | Mobile ordering app | | $35,000 | | | | | |
| 11 | Integration with ERP | | $25,000 | | | | | |
| 12 | Human Resources | | | | | | | |
| 13 | Business Staff | | $10,000 | | | | | |
| 14 | IT Staff + Consultants | | $45,000 | $2,000 | $2,000 | $2,000 | $2,000 | $2,000 |
| 15 | Training | | $7,000 | $1,000 | $1,000 | $1,000 | $1,000 | $1,000 |
| 16 | Maintenance and Support | | | $5,000 | $5,000 | $5,000 | $5,000 | $5,000 |
| 17 | | Annual Costs | $152,500 | $11,500 | $11,500 | $11,500 | $11,500 | $11,500 |
| 18 | | Total Costs | $210,000 | | | | | |
| 19 | Benefits | | | | | | | |
| 20 | Reduced labor costs | | | $52,000 | $52,000 | $52,000 | $52,000 | $52,000 |
| 21 | Reduced errors and returns | | | $70,000 | $70,000 | $70,000 | $70,000 | $70,000 |
| 22 | | Annual Net Cash Flow | -$152,500 | $110,500 | $110,500 | $110,500 | $110,500 | $110,500 |
| 23 | | Total Benefits | $400,000 | | | | | |
| 24 | | | | | | | | |
| 25 | | Net Present Value | $268,407 | | | | | |
| 26 | | ROI | 4.1% | | | | | |
| 27 | | Internal Rate of Return | 17.0% | | | | | |



**Figure 12.3**
Capital Budgeting for an Information System Investment

*This worksheet illustrates a simplified capital budgeting analysis for a mobile sales ordering system.*

## 12-2 Identify the core problem-solving steps for developing a new information system.

A new information system is built as a solution to a problem or set of problems that the organization perceives it is facing. The problem may be one in which managers and employees believe that the business is not performing as well as expected, or it may come from the realization that the organization should take advantage of new opportunities to perform more effectively.

The problem-solving process introduced in Chapter 1 provides the facts and findings needed to develop a strong business case and to implement the right solution. Examine Figure 12.4, which illustrates the four steps you would need to take: (1) define and understand the problem, (2) develop alternative solutions, (3) choose the best solution, and (4) implement the solution.

Before a problem can be solved, first it must be properly defined. Members of the organization must agree that a problem actually exists and that it is serious. The problem must be investigated so that it can be better understood. Next comes a period of devising alternative solutions, and then comes a period of evaluating each alternative and selecting the best solution. The final stage is one of implementing the solution, in which a detailed design for the solution is specified, translated into a physical system, tested, introduced to the organization, and further refined as it is used over time.

In the information systems world, we have a special name for these activities. As shown in Figure 12.4, the first three problem-solving steps, in which you identify the problem, gather information, devise alternative solutions, and make a decision about the best solution, are called **systems analysis**.

### DEFINING AND UNDERSTANDING THE PROBLEM

Defining the problem may take some work because various members of the company may have different ideas about the nature of the problem and its severity. What caused the problem? Why is it still around? Why wasn't it solved long ago? Systems analysts typically gather facts about existing systems and problems by examining documents, work papers, procedures, and system operations and by interviewing

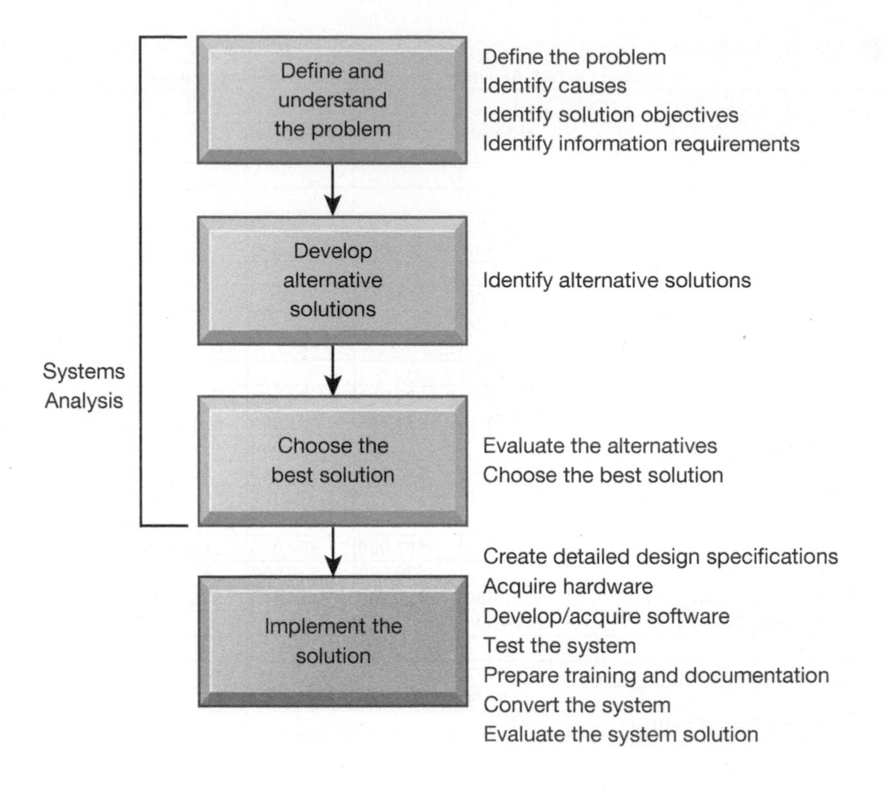


**Figure 12.4** Developing an Information System Solution

*Developing an information system solution is based on the problem-solving process.*

key users of the system, including employees, managers, and customers if they will be users of the system.

Information systems problems in the business world typically result from a combination of people, organization, and technology factors. When identifying a key issue or problem, ask what kind of problem it is: Is it a people problem, an organizational problem, a technology problem, or a combination of these? What people, organizational, and technological factors contributed to the problem?

Once the problem has been defined and analyzed, it is possible to make some decisions about what should and can be done. What are the objectives of a solution to the problem? Are they to reduce costs, increase sales, or improve relationships with customers, suppliers, or employees? Do managers have sufficient information for decision making? What information is required to achieve these objectives?

At the most basic level, the **information requirements** of a new system identify who needs what information and where, when, and how. Requirements analysis carefully defines the objectives of the new or modified system and develops a detailed description of the functions that the new system must perform. A system designed around the wrong set of requirements either will have to be discarded because of poor performance or will need to undergo major modifications. Section 12-2 describes alternative approaches to eliciting requirements that help minimize this problem. In the case of Angostura, the problem is that the traditional ordering process has been excessively manual and time consuming, with high error rates.

The objectives of a solution for Angostura would be to reduce the amount of time, effort, and errors in the ordering process while making it possible to submit orders online from any location. Information requirements for the solution include the ability to take orders instantly, the ability to track orders by type of product or account, the ability to track the status of orders, and the ability to interact with the company's ERP system.

## DEVELOPING ALTERNATIVE SOLUTIONS

What alternative solutions are possible for achieving these objectives and meeting these information requirements? The systems analysis lays out the most likely paths to follow given the nature of the problem. Some possible solutions do not require an information system solution but instead call for an adjustment in management, additional training, or refinement of existing organizational procedures. Some solutions, however, do require modifications of the firm's existing information systems or an entirely new information system.

## EVALUATING AND CHOOSING SOLUTIONS

The systems analysis includes a **feasibility study** to determine whether each proposed solution is feasible, or achievable, from financial, technical, and organizational standpoints. The feasibility study establishes whether each alternative solution is a good investment, whether the technology needed for the system is available and can be handled by the firm's information systems staff, and whether the organization is capable of accommodating the changes the system introduces.

A written systems proposal report describes the costs and benefits and advantages and disadvantages of each alternative solution. Which solution is best in a financial sense? Which works best for the organization? The systems analysis will detail the costs and benefits of each alternative and the changes that the organization will have to make to use the solution effectively. We provide a detailed discussion of how to manage change in the following section. On the basis of this report, management will select what it believes is the best solution for the company.

## IMPLEMENTING THE SOLUTION

The first step in implementing a system solution is to create detailed design specifications. **Systems design** shows how the chosen solution should be realized. The system design is the model or blueprint for an information system solution and consists of all the specifications that will deliver the functions identified during systems analysis. These specifications should address all the technical, organizational, and people components of the system solution.

Table 12.4 shows some of the design specifications for the online ordering system discussed earlier, which were based on information requirements for the solution that was selected. These design specifications apply to both the web and the mobile app platforms.

### Completing Implementation

In the final steps of implementing a system solution, the following activities would be performed:

- **Hardware selection and acquisition.** System builders select appropriate hardware for the application. They would either purchase the necessary computers and mobile devices, lease them from a technology provider, or lease processing services from a cloud computing vendor.
- **Software development and programming.** Software for ordering may be custom-programmed in house or purchased from an external source such as an outsourcing vendor, an application software package vendor, or an online software service provider. The core ordering system and databases are in corporate data centers or in remote servers accessed through the Internet.
- **Testing.** The system is thoroughly tested to ensure that it produces the right results. The **testing** process requires detailed testing of individual computer programs, called **unit testing**, as well as **system testing**, which tests the performance of the information system as a whole. **Acceptance testing** provides the final certification that the system is ready to be used in a production setting. Information systems tests are evaluated by users and reviewed by management. When all parties are satisfied that the new system meets their standards, the system is formally accepted for installation.

**TABLE 12.4**

Design Specifications for Online Ordering System

| | |
|---|---|
| Output | Online reports<br>Hard-copy reports<br>Online queries<br>Order transactions |
| Input | Online order entry<br>Order status request screen |
| Database | Database with order file, customer file |
| Processing | Calculate order totals by type of product.<br>Transmit orders to distribution centers.<br>Track orders by customer.<br>Track orders by sales rep.<br>Schedule deliveries.<br>Update customer data for account changes. |
| Manual procedures | Sales reps contact customers by phone, email, text message. |
| Security and controls | Online passwords<br>Only authorized sales reps and company employees can access the system.<br>Only company-owned mobile devices can be used for entering orders or accessing corporate data. |
| Conversion | Input<br>Input customer data<br>Test system |
| Training and documentation | System guide for users<br>Online training sessions and tutorials |

The systems development team works with users to devise a systematic test plan. The **test plan** includes all the preparations for the series of tests we have just described. Examine Figure 12.5, which shows a sample from a test plan that could be used for a mobile ordering system. The condition being tested is online access to the system by an authorized user.

- **Training and documentation.** End users and information system specialists require training so that they will be able to use the new system. Detailed **documentation** in the form of hard-copy training manuals or online tutorials showing how the system works from both a technical and end-user standpoint must be prepared.
- **Conversion** is the process of changing from the old to the new system. There are four main conversion strategies: the parallel strategy, the direct cutover strategy, the pilot study strategy, and the phased approach strategy.

  In a **parallel strategy**, both the old system and its potential replacement are run together for a time until everyone is assured that the new system functions correctly. The old system remains available as a backup in case of problems. The **direct cutover strategy** replaces the old system entirely with the new system on an appointed day, carrying the risk that there is no system to fall back on if problems arise. The **pilot study** strategy introduces a new system to only a limited area of the organization, such as a single department or operating unit. Once this pilot version is working smoothly, it is installed throughout the rest of the organization. A **phased approach** introduces the system in stages, such as first implementing payroll processing for hourly workers who are paid weekly and later for salaried employees paid monthly.

**Test Case Number:** GS02-010

| **Prepared by:** A. Patel | **Date:** February 19, 2023 |
|---|---|

**Objective:** This subtest checks for an authorized user accessing the system.

**Platform:** iOS

**Procedure Description:**
Select Sign In
Select Username
Enter Username
Select Password
Enter Password
Select Submit

**Expected Result:**
When user selects Sign In, the Sign In menu appears
When user selects User Name, the cursor moves to the User Name field
When user enters that person's system user name the user name appears on the screen
When user selects Password, the cursor moves to the Password field.
When user enters that person's password, the password appears on the screen as asterisks
When user selects Submit, the system verifies the entered data and allows the user to access the system
When user enters an incorrect (or unauthorized) username or password, the error message "Wrong User Name or Password" appears

**Test Results:**
All OK

**Figure 12.5**
A Sample Test Plan for a Mobile Ordering System
*When developing a test plan, it is imperative to include the various conditions to be tested, the requirements for each condition tested, and the expected results. Test plans require input from both end users and information systems specialists. Illustrated here is a test case for accessing the mobile ordering system for an authorized user.*

- **Production and maintenance.** After the new system is installed and conversion is complete, the system is said to be in **production**. During this stage, users and technical specialists review the solution to determine how well it has met its original objectives and to decide whether any revisions or modifications are in order. Changes in hardware, software, documentation, or procedures that are made to a production system to correct errors, meet new requirements, or improve processing efficiency are termed **maintenance**.

### Managing the Change

Developing a new information systems solution is not merely a matter of installing hardware and software. The business must also deal with the organizational changes that the new solution will bring about—new information, new business processes, and perhaps new reporting relationships and decision-making power. A well-designed solution may not work unless it is introduced to the organization very carefully. The process of planning change in an organization so that it is implemented in an orderly and effective manner is so critical to the success or failure of information system solutions that we devote Section 12-4 to a detailed discussion of this topic.

## 12-3 Compare the alternative methods for building information systems.

There are alternative methods for creating system solutions by using the basic problem-solving model we have just described. These alternative methods include the traditional systems development life cycle, prototyping, end-user development, application software packages, and outsourcing.

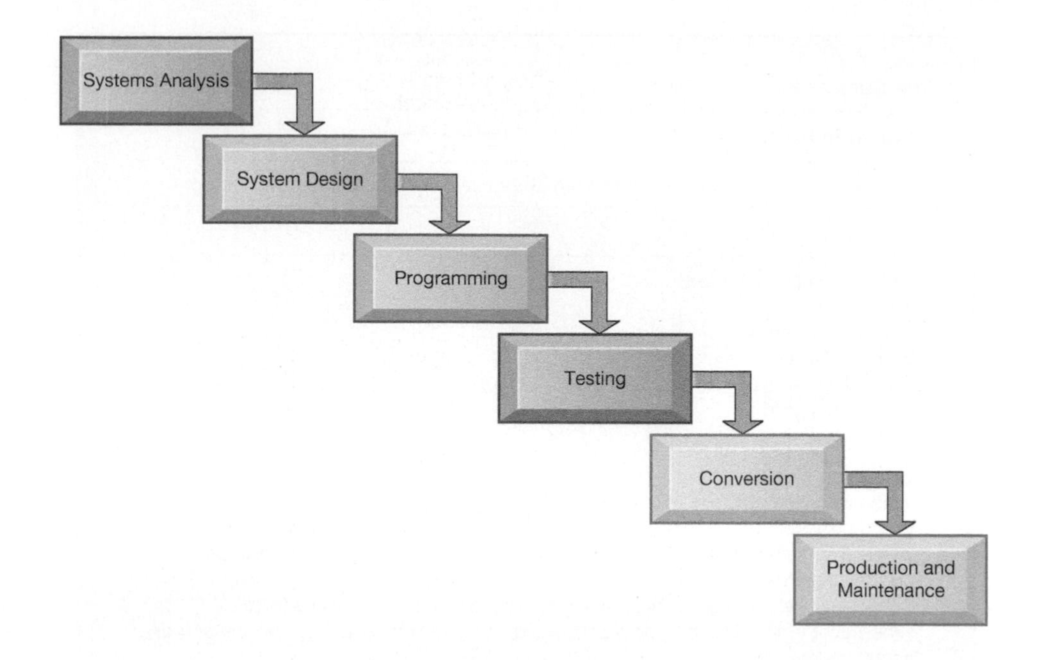


**Figure 12.6**
The Traditional Systems Development Life Cycle
*The systems development life cycle partitions systems development into formal stages, with each stage requiring completion before the next stage can begin.*

## TRADITIONAL SYSTEMS DEVELOPMENT LIFE CYCLE

The **systems development life cycle (SDLC)** is the oldest method for building information systems. Examine Figure 12.6, which illustrates the SDLC. The SDLC methodology is a phased approach to building a system, dividing systems development into a series of formal stages: systems analysis, system design, programming, testing, conversion, and production and maintenance. Although systems builders can go back and forth among stages in the life cycle, the systems life cycle is predominantly a waterfall approach in which tasks in one stage are completed before work for the next stage begins.

This approach maintains a formal division of labor between end users and information systems specialists. Technical specialists, such as systems analysts and programmers, are responsible for much of the systems analysis, design, and implementation work; end users are limited to providing information requirements and reviewing the technical staff's work. The life cycle also emphasizes formal specifications and paperwork, so many documents are generated during the course of a systems project.

The systems life cycle is still used for building large, complex systems that require rigorous and formal requirements analysis, predefined specifications, and tight controls over the systems-building process. However, this approach is also time consuming and expensive to use. Tasks in one stage are supposed to be completed before work for the next stage begins. Activities can be repeated, but volumes of new documents must be generated and steps retraced if requirements and specifications need to be revised. This encourages freezing of specifications relatively early in the development process. The life cycle approach is also not suitable for many small desktop systems and apps, which tend to be less structured and more individualized.

## PROTOTYPING

**Prototyping** consists of building an experimental system rapidly and inexpensively for end users to evaluate and then revising the prototype based on user feedback. Examine Figure 12.7, which shows a four-step model of the prototyping process. The first step is to identify basic requirements for the system. The second step is to develop a prototype: a working version of the information system or part of the system that is intended as a preliminary model. In the third step, users interact with the prototype to get a better idea of their information requirements, potentially refining the prototype multiple times. When the design is finalized, the prototype will be converted to a polished production system.

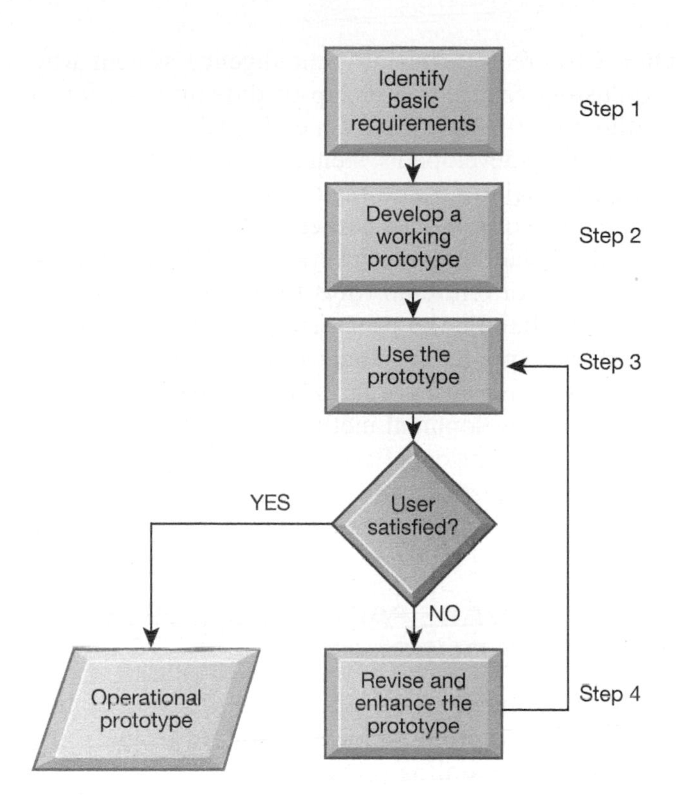


**Figure 12.7**
**The Prototyping Process**
*The process of developing a prototype consists of four steps. Because a prototype can be developed quickly and inexpensively, systems builders can go through several iterations, repeating Steps 3 and 4, to refine and enhance the prototype before arriving at the final, operational one.*

**Step 1:** *Identify users' basic requirements.* The system designer (usually an information systems specialist) works with users only long enough to capture users' basic information needs.

**Step 2:** *Develop an initial prototype.* The system designer creates a working prototype quickly, using tools for rapidly generating software.

**Step 3:** *Use the prototype.* Users are encouraged to work with the system to determine whether the prototype meets their needs and to suggest improvements for the prototype.

**Step 4:** *Revise and enhance the prototype.* The system builder notes all changes users request and refines the prototype accordingly. After the prototype has been revised, the cycle returns to Step 3. Steps 3 and 4 are repeated until users are satisfied.

Prototyping is especially useful for designing an information system's user interface. Because prototyping encourages intense end-user involvement throughout the systems development process, it is more likely to produce systems that fulfill user requirements.

However, rapid prototyping may gloss over essential steps in systems development, such as thorough testing and documentation. If the completed prototype works reasonably well, management may not see the need to build a polished production system. Some hastily constructed systems do not easily accommodate large quantities of data or a large number of users in a production environment.

## END-USER DEVELOPMENT

**End-user development** allows end users, with little or no formal assistance from technical specialists, to create simple information systems, reducing the time and steps required to produce a finished application. Using user-friendly query, reporting, website development, graphics, and PC software tools such as Excel or Access, end users can access data, create reports, and develop simple applications on their own with little or no help from professional systems analysts or programmers. For example Sound Credit Union, with 29 full-service branch locations and nearly 120,000 members, used

TIBCO WebFOCUS to create a business intelligence system where executives can work with easy-to-customize dashboards, up-to-date information, and dynamic visualizations—all without writing any program code (Tibco, 2022).

On the whole, end-user-developed systems are completed more rapidly than those developed with conventional programming tools. Allowing users to specify their own business needs improves requirements gathering and often leads to a higher level of user involvement and satisfaction with the system. End-user development tools, however, still cannot replace conventional tools for some business applications because the former cannot easily handle the processing of large numbers of transactions or applications with extensive procedural logic and updating requirements.

End-user development also poses organizational risks because systems are created rapidly, without a formal development methodology, testing, and documentation. To help organizations maximize the benefits of end-user applications development, management should require cost justification of end-user information system projects and establish hardware, software, and quality standards for user-developed applications.

## APPLICATION SOFTWARE PACKAGES, SOFTWARE SERVICES, AND OUTSOURCING

Chapter 5 points out that much of the software underlying contemporary information systems is not developed in-house but is acquired from external sources. Firms can rent the software from an online software service provider, purchase a software package from a commercial vendor to run in-house, or have an in-house application developed by an external outsourcing firm. Selection of the software or software service is often based on a **Request for Proposal (RFP)**, which is a detailed list of questions submitted to external vendors to see how well they meet the requirements for the proposed system.

### Application Software Packages and Cloud Software Services

Systems are increasingly based on commercially available application software packages or cloud software as a service (SaaS). For example, companies can choose to implement Oracle enterprise resource planning, supply chain management, or human capital management software in-house or pay to use this software running on the Oracle Cloud platform. Microsoft Office productivity software comes in both desktop and cloud (Microsoft 365) versions.

If a cloud software service or software package can fulfill most of an organization's requirements, the company does not have to write its own software. The company saves time and money by using prewritten, predesigned, pretested software programs from package and SaaS vendors, who also provide ongoing maintenance and upgrades for the system. Many packages include capabilities for customization to meet unique requirements not addressed by the prewritten software. **Customization** features allow prewritten software to be modified to meet an organization's unique requirements without destroying the integrity of the software. If extensive customization is required, however, additional programming and customization work may become so expensive and time consuming that it negates many of the advantages of software packages or services. If the software cannot be customized, the organization will have to adapt by changing its procedures.

### Outsourcing

If a firm does not want to use its internal resources to build or operate information systems, it can outsource the work to an external organization that specializes in providing these services. The outsourcing vendor might be domestic or in another country. Domestic outsourcing is driven primarily by the fact that outsourcing firms possess skills, resources, and assets that their clients do not have. Installing a new supply chain management system in a very large company might require hiring an additional 30 to 50 people with specific expertise in supply chain management software. Rather than hire new employees and then release them after the new system is built, it makes more sense, and is often less expensive, to outsource this work for a 12-month period.

In the case of offshore outsourcing, the decision tends to be driven by cost. Because of cost of living differences, a skilled programmer in India or Malaysia earns about US $10,000 to $20,000 per year, compared to $80,000 or more per year for a comparable programmer in the United States. The Internet and low-cost communications technology have drastically reduced the expense and difficulty of coordinating the work of global teams in faraway locations. In addition to cost savings, many offshore outsourcing firms offer world-class technology assets and skills. For example, leading companies such as Hilton, NBC, and Yahoo have outsourced website design and development work to India using tools that are not available internally in most companies. However, wage inflation outside the United States has eroded some of these advantages, and some jobs have moved back to the United States.

Your firm is most likely to benefit from outsourcing if it takes the time to evaluate all the risks and make sure outsourcing is appropriate for its particular needs. Any company that outsources its applications must thoroughly understand the project, including its requirements, method of implementation, source of expected benefits, cost components, and metrics for measuring performance.

Many firms underestimate costs for identifying and evaluating vendors of information technology services, for transitioning to a new vendor, for improving internal software development methods to match those of outsourcing vendors, and for monitoring vendors to make sure they are fulfilling their contractual obligations. Outsourcing offshore incurs additional costs for coping with cultural and language differences that drain productivity and dealing with human resources issues, such as terminating or relocating domestic employees. These hidden costs undercut some of the anticipated benefits from outsourcing. Firms should be especially cautious when using an outsourcer to develop or operate applications that give some type of competitive advantage.

Examine Figure 12.8, which shows best- and worst-case scenarios for the total cost of an offshore outsourcing project. It shows how much hidden costs affect the total project cost. The best case reflects the lowest estimates for additional costs, and the worst case reflects the highest estimates for these costs. As you can see, hidden costs increase the total cost of an offshore outsourcing project by an extra 15 to 57 percent. Even with these extra costs, many firms will benefit from offshore outsourcing if they manage the work well.

**TOTAL COST OF OFFSHORE OUTSOURCING**

| **Cost of outsourcing contract** | | | **$10,000,000** | |
|---|---|---|---|---|
| Hidden Costs | Best Case | Additional Cost ($) | Worst Case | Additional Cost ($) |
| 1. Vendor selection | 0.2% | 20,000 | 2% | 200,000 |
| 2. Transition costs | 2% | 200,000 | 3% | 300,000 |
| 3. Layoffs & retention | 3% | 300,000 | 5% | 500,000 |
| 4. Lost productivity/cultural issues | 3% | 300,000 | 27% | 2,700,000 |
| 5. Improving development processes | 1% | 100,000 | 10% | 1,000,000 |
| 6. Managing the contract | 6% | 600,000 | 10% | 1,000,000 |
| **Total additional costs** | | 1,520,000 | | 5,700,000 |
| | Outstanding Contract ($) | Additional Cost ($) | Total Cost ($) | Additional Cost |
| Total cost of outsourcing (TCO) best case | 10,000,000 | 1,520,000 | 11,520,000 | 15.2% |
| Total cost of outsourcing (TCO) worst case | 10,000,000 | 5,700,000 | 15,700,000 | 57.0% |

**Figure 12.8**
Total Cost of Offshore Outsourcing

*If a firm spends $10 million on offshore outsourcing contracts, that company will spend 15.2 percent in extra costs even in the best-case scenario. In the worst-case scenario, when there is a dramatic drop in productivity along with exceptionally high transition and layoff costs, a firm can expect to pay up to 57 percent in extra costs on top of the $10 million outlay for an offshore contract.*

## MOBILE APPLICATION DEVELOPMENT: DESIGNING FOR A MULTISCREEN WORLD

Today, employees and customers expect, and even demand, to be able to use a mobile device of their choice to obtain information or to be able to perform a transaction anywhere and at any time. To meet these needs, companies will need to develop mobile websites, mobile apps, and native apps as well as traditional information systems.

Once an organization decides to develop mobile apps, it has to make some important choices, including the technology it will use to implement these apps (whether to use software for a native app or mobile web app) and what to do about a mobile website. A **mobile website** is a version of a regular website that is scaled down in content and navigation for easy access and search on a small mobile screen. (Access Amazon's website from your computer and then from your smartphone to see the differences between the two types of websites.)

A **mobile web app** is an Internet-enabled app with specific functionality for mobile devices. Users access mobile web apps through their mobile device's web browser. The web app resides primarily on a server, is accessed through the Internet, and doesn't need to be installed on the device. The same application can be used by most devices that can surf the web, regardless of their brand.

A **native app** is a stand-alone application designed to run on a specific platform and device. The native app is installed directly on a mobile device. Native apps can connect to the Internet to download and upload data, and they can operate on these data even when not connected to the Internet. For example, an e-book reading app such as Kindle software can download a book from the Internet, disconnect from the Internet, and present the book for reading. Native mobile apps provide fast performance and a high degree of reliability. They can also take advantage of a mobile device's particular capabilities, such as its camera or touch features. However, native apps are expensive to develop because multiple versions of an app must be programmed for different mobile operating systems and hardware such as Android and Apple's iOS.

A **hybrid app** has many features of both a mobile web app and a native app. A hybrid app is built with web technologies such as HTML, CSS, and JavaScript, but like a native app, it runs within the mobile device itself and is able to access many of the device's features that are not accessible by a mobile web app.

Developing applications for mobile platforms is quite different from developing applications for PCs and their much larger screens. The reduced size of mobile devices makes using fingers and multitouch gestures much easier than typing and using keyboards. Mobile apps thus need to be optimized for the specific tasks they are to perform. They should not try to carry out too many tasks, and they should be designed for usability. The user experience for mobile interaction is fundamentally different from using a desktop or laptop PC. Saving resources—bandwidth, screen space, memory, processing, data entry, and user gestures—is a top priority. Mobile application developers also need to address accessibility issues that people with visual impairments experience when they interact with mobile devices.

When a full website created for the desktop shrinks to the size of a smartphone screen, it is difficult for the user to navigate through the site. The user must continually zoom in and out and scroll to find relevant material. Therefore, companies need to design websites specifically for mobile interfaces and create multiple mobile sites to meet the needs of smartphones, tablets, and desktop browsers. This equates to at least three sites with separate content, maintenance, and costs. Currently, websites know what device you are using because your browser will send this information to the server when you log on. Based on this information, the server will deliver the appropriate screen.

One solution to the problem of having multiple websites is to use **responsive web design**. Responsive web design enables websites to change layouts automatically according to the visitor's screen resolution, whether on a desktop, laptop, tablet, or smartphone. Responsive design uses tools such as flexible grid-based layouts, flexible images, and media queries to optimize the design for different viewing contexts. This eliminates the need for separate design and development work for each new device. HTML5, which we introduced in Chapter 5, is also used for mobile application development because it can support cross-platform mobile applications.

## RAPID APPLICATION DEVELOPMENT FOR E-BUSINESS

Technologies and business conditions are changing so rapidly that companies are adopting shorter, more informal systems development processes, including those for mobile applications. In addition to using software packages and online software services, businesses are relying more heavily on fast-cycle techniques such as rapid application development, agile development, automated software testing, and low-code/no-code development.

The term **rapid application development (RAD)** refers to the process of creating workable systems in a very short period of time with some flexibility to adapt as a project evolves. RAD includes the use of visual programming and other tools for building graphical user interfaces, iterative prototyping of key system elements, automation of program code generation, and close teamwork among end users and information systems specialists. Simple systems often can be assembled from prebuilt components. The process does not have to be sequential, and key parts of development can occur simultaneously.

**Agile development** focuses on rapid delivery of working software by breaking a large project into a series of small subprojects that are completed in short periods of time using iteration, continuous feedback, and ongoing user involvement. Iterations, called sprints, are short timeframes that typically last from one to four weeks. Each mini-project is worked on by a cross-functional team (consisting of programmers, testers, a user representative, and other people required by the project) and regularly released to the client. Improvement or addition of new functionality takes place within the next iteration as developers clarify requirements.

Features to be developed are assigned to the sprints on a priority basis, with the "critical path" through the program developed first. This makes it possible for a bare-bones but functioning version of a program to be demonstrated early on. As additional features are developed, the code is added to the code already developed, a process known as "continuous integration."

Testing occurs early and often throughout the entire development process. Agile methods emphasize face-to-face communication, encouraging people to collaborate and make decisions quickly and effectively, rapid and flexible response to change, and the production of working software rather than elaborate documentation. Instead of a separate testing phase, as in the traditional waterfall approach, testing occurs continuously in agile development during each sprint, and testing is encouraged. This often means that more development time will be spent on testing than when using traditional approaches.

Although essential, continuous testing can be very time consuming, especially since developing test scripts, executing test steps, and evaluating the results used to be largely manual. Automated testing tools are now available to meet this need. Automated testing tools perform examinations of the software, report outcomes, and compare results with earlier test runs. The Spotlight on Technology case shows how automated testing helped security software company McAfee use an agile methodology in its projects.

## McAfee Turns to Automated Software Testing

McAfee is a privately held company that sells security software to nearly 69,000 enterprise customers and more than 500 million individuals in 189 countries. You or your company may be using a McAfee product for combating malware, identity theft, and privacy invasion.

McAfee used a single global instance of SAP ERP 5.0 to run all of its back-end finance, controlling, accounting, materials management, and order fulfillment processes. It also has other non-SAP systems, which are integrated for processes involving sales, licensing, and customer service from an order-to-cash perspective. (Order-to-cash refers to the set of business processes for receiving, processing, and paying for customer orders.) The SAP ERP system, however, is the single source of information for McAfee's revenues and bookings. McAfee is a primarily partner-based organization that works through many resellers and distributors, and it needs to ensure that these partners can quickly and easily enter orders into the SAP system.

McAfee has been trying to juggle multiple systems projects. It has been trying to migrate to SAP S/4HANA (SAP's ERP business suite) and also to implement SAP Revenue Accounting and Reporting, which required updating McAfee's accounting codes to conform to new revenue recognition standards. This project had consumed most of McAfee's IT resources in 2018. The business had also been spun off from its parent company and had to separate its IT systems. Updates had to be released, tested, and confirmed to be working properly. McAfee's IT staff had to manage all of these projects without increasing headcount. At the same time, McAfee was adopting an agile methodology for all of its IT projects. In contrast to the traditional waterfall methodology, in which a project manager oversees individuals who are each dedicated to quality assurance, testing system functionality, and user acceptance testing, McAfee switched to an agile sprint cycle in which all members of a development team collaborate on incremental development of smaller pieces of software, which are released each sprint cycle every two weeks. With an agile methodology, software modules are constantly being created, tested, demonstrated for feedback, and revised with shorter timeframes than the waterfall approach. McAfee had to deal with approximately 40 systems for handling lead-to-order, order-to-cash, and source-to-pay processes that are actively worked on and enhanced. According to Mouli Subrahamanayan, IT director at McAfee India, "there are moving parts all over."

One way to do more, more quickly was to automate testing. McAfee had traditionally used manual processes for software testing, which could not easily handle an increased amount of testing. Whenever a system was changed or enhanced with a new feature, McAfee's IT staff had to ensure that the updated systems performed as expected and fix any problems before the system went into production.

It was very difficult to use manual processes to keep up with the testing because so many changes had to be made within a very short period of time. Manually creating test scenarios for end-to-end processes, such as order-to-cash, took a long time and was very costly.

McAfee's increased testing needs were also driven by other application changes that required testing as well. For example, McAfee replaced its custom-developed system for configure, price, and quote (CPQ) processes (for configuring product pricing and generating quotes) with a third-party non-SAP system. The company needed to test the end-to-end scenario to ensure that quotes created for sales were properly converted into orders in the SAP ERP system and then fulfilled seamlessly. The testing had to create and test end-of-quarter volumes of 60,000 orders and validate the loads in the supporting non-SAP applications, such as for invoicing, licensing, and analytics. The testing had to show that everything in that chain of events could handle the transaction volume and that the systems behaved as they should.

To complete a very large amount of testing in a short time period, McAfee opted for automated testing. Automated testing would allow time for agile development teams to focus on each scenario and ensure that back-end systems were working as expected to complete customer orders. McAfee's 13-member automation team consisted of automation engineers and business analysts, who were charged with selecting the automated testing product for the company. Automated testing software vendors had to use hands-on demonstrations to show that their tool could handle SAP testing and business process automation and that it was easy to use.

McAfee selected Worksoft Certify because it used a framework for SAP testing that came prebuilt within the software and because its testing framework could be applied to areas beyond the company's on-premises SAP system, including new custom applications. Worksoft Certify is an industry-leading test automation solution for enterprise applications including SAP, Workday, Salesforce, Oracle, and web apps. It is designed to test complex business processes that span multiple applications, and it is code-free. The tool models an application undergoing testing as a series of pages containing GUI (graphical user interface) objects and test steps, performing actions against those objects. It creates and stores automated test steps in a relational database without a single software script or program. Therefore, people who lack software coding skills can use the tool.

By the end of 2017, McAfee started putting its automated testing system into production. The company now uses Worksoft Certify for testing and for business process automation in its SAP environment. Manual tasks such as performing regular checks on the health of the SAP system are automated, as is the testing of changes to the system. That testing includes performance testing to test heavy processing loads. For example, McAfee was able to test how the SAP system processed a volume of 60,000 orders with 250,000 line items.

McAfee saved nearly 2,500 hours of manual effort through testing and business process automation, equivalent to $200,000, and the need for application maintenance is now at an all-time low.

Sources: Mcafee.com, accessed April 7, 2022; Worksoft.com, accessed April 7, 2022; Lauren Bonneau, "McAfee Saves 2500 Hours of Manual Effort with Test and Business Process Automation on Its Journey to SAP S/4HANA," *SAPInsider*, April 1, 2019.

## CASE STUDY QUESTIONS

1. Why would a company such as McAfee benefit from automated software testing?
2. What people, organization, and technology factors did McAfee address in moving to automated software testing?
3. Was Worksoft Certify a good solution for McAfee? Why or why not?
4. How did automated software testing change the way McAfee runs its business?

**Low-code development** is a software development approach that enables the delivery of applications more quickly and with minimal hand-coding using visual modeling in a graphical interface to assemble and configure applications. Such tools may produce entirely operational applications or require a small amount of additional coding. Low-code development platforms reduce the amount of hand-coding needed to create workable software so that business applications can be developed more rapidly and by a wider range of people (including business end users in some instances), not just those with formal programming skills.

**No-code development** tools are even easier for non-IT businesspeople to use. Everything the software vendor thinks the user needs to create an app is already built into the tool, with no coding required. For example, Goodgigs, a job board for "social impact" jobs that benefit people or the environment, used the no-code desktop and mobile web app development tool Bubble to build its web app (Low Code Agency, 2021).

However, many no-code tools are built to solve simple business problems and have limited functionality. No-code applications are also difficult to customize. Another downside is that these tools make it possible for business users to create their own applications without proper management oversight by the IT department, thus creating security, compliance, and integration problems as well as substandard systems.

## 12-4 Describe how information systems projects should be managed.

Your company might have developed what appears to be an excellent system solution. Yet when the system is in use, it does not work properly or it doesn't deliver the benefits that were promised. If this occurs, your firm is not alone. There is a high failure rate among information systems projects because they have not been properly managed. A joint study by McKinsey and Oxford University found that large software projects on average run 66 percent over budget and 33 percent over schedule. Another McKinsey study found that the average digital transformation project stands a 45 percent chance of delivering less profit than expected, with the chance of surpassing profit expectations only one in 10 (Bughin, Deakin, and O'Beirne, 2019).

Firms may have incorrectly assessed the business value of the new system or were unable to manage the organizational change the new technology required. That's why it's essential to know how to manage information systems projects and the reasons they succeed or fail.

## PROJECT MANAGEMENT OBJECTIVES

A **project** is a planned series of related activities for achieving a specific business objective. Information systems projects include the development of new information systems, the enhancement of existing systems, or projects for replacing or upgrading the firm's information technology (IT) infrastructure.

**Project management** refers to the applications of knowledge, skills, tools, and techniques to achieve specific targets within specified budget and time constraints. Project management activities include planning the work, assessing risks, estimating resources required to accomplish the work, organizing the work, acquiring human and material resources, assigning tasks, directing activities, controlling project execution, reporting progress, and analyzing the results. As in other areas of business, project management for information systems must deal with five major variables: scope, time, cost, quality, and risk.

**Scope** defines what work is or is not included in a project. For example, the scope of a project for a new order-processing system might include new modules for inputing orders and transmitting them to production and accounting but not any changes to related accounts receivable, manufacturing, distribution, or inventory control systems. Project management defines all the work required to complete a project successfully and should ensure that the scale of a project does not expand beyond what was originally intended.

Time is the amount of time required to complete the project. Project management typically establishes the amount of time required to complete major components of a project. Each of these components is further broken down into activities and tasks. Project management tries to determine the time required to complete each task and establish a schedule for completing the work.

Cost is based on the time to complete a project multiplied by the daily cost of human resources required to complete the project. Information systems project costs also include the cost of hardware, software, and workspace. Project management develops a budget for the project and monitors ongoing project expenses.

Quality is an indicator of how well the result of a project satisfies the objectives that management specified. The quality of information systems projects usually boils down to improved organizational performance and decision making. Quality also considers the accuracy and timeliness of information that the new system produces and the ease of use.

Risk refers to potential problems that would threaten the success of a project. These potential problems might prevent a project from achieving its objectives by increasing time and cost, lowering the quality of project outputs, or preventing the project from being completed altogether. We discuss the most important risk factors for information systems projects later in this section.

## MANAGING PROJECT RISK AND SYSTEM-RELATED CHANGE

Some systems development projects are more likely to run into problems or to suffer delays because they carry a much higher level of risk than others. The level of project risk is influenced by project size, project structure, and the level of technical expertise of the information systems staff and project team. The larger the project—as indicated by the dollars spent, the project team size, and how many parts of the organization will be affected by the new system—the greater the risk. Very large-scale systems projects have a failure rate that is 50 to 75 percent higher than that for other projects because such projects are complex and difficult to control. Risks are also higher for systems where information requirements are not clear and straightforward or when the project team must master new technology.

## Implementation and Change Management

Dealing with these project risks requires an understanding of the implementation process and change management. A broader definition of **implementation** refers to all the organizational activities working toward the adoption and management of an innovation, such as a new information system. Successful implementation requires a high level of user involvement in a project and management support.

If users are heavily involved in the development of a system, they have more opportunities to mold the system according to their priorities and business requirements and to control the outcome. They also are more likely to react positively to the completed system because they have been active participants in the change process.

The relationship between end users and information systems specialists has traditionally been a problem area for information systems implementation efforts because of differing backgrounds, interests, and priorities. These differences create a **user–designer communications gap**. Information systems specialists often have a highly technical orientation to problem solving, focusing on technical solutions in which hardware and software efficiency is optimized at the expense of ease of use or organizational effectiveness. End users prefer systems that are oriented toward solving business problems or facilitating organizational tasks. Often the orientations of both groups are so at odds that they appear to speak in different tongues. These differences are illustrated in Table 12.5.

If an information systems project has the backing and commitment of management at various levels, it is more likely to receive higher priority from both users and the technical information systems staff. Management backing also ensures that a systems project receives sufficient funding and resources to be successful. Furthermore, to be enforced effectively, all the changes in work habits and procedures and any organizational realignment associated with a new system depend on management backing.

## Controlling Risk Factors

There are strategies you can follow to deal with project risk and increase the chances of a successful system solution. If the new system involves challenging and complex technology, you can recruit project leaders with strong technical and administrative experience. Outsourcing or using external consultants are options if your firm does not have staff with the required technical skills or expertise.

**TABLE 12.5**
The User–Designer Communications Gap

| User Concerns | Designer Concerns |
|---|---|
| Will the system deliver the information we need for our work? | What demands will this system put on our servers? |
| Can we access the data on our mobile devices and PCs? | What kinds of programming demands will this place on our group? |
| What new procedures do we need to enter data into the system? | Where will the data be stored? What's the most efficient way to store the data? |
| How will the operation of the system change our daily routines? | What technologies should we use to secure the data? |

Large projects benefit from appropriate use of **formal planning and control tools** for documenting and monitoring project plans. The two most commonly used methods for documenting project plans are Gantt charts and PERT charts. Examine Figure 12.9, which illustrates a Gantt chart. A **Gantt chart** lists project activities and their corresponding start and completion dates. The Gantt chart visually represents the timing and duration of different tasks in a development project as well as their

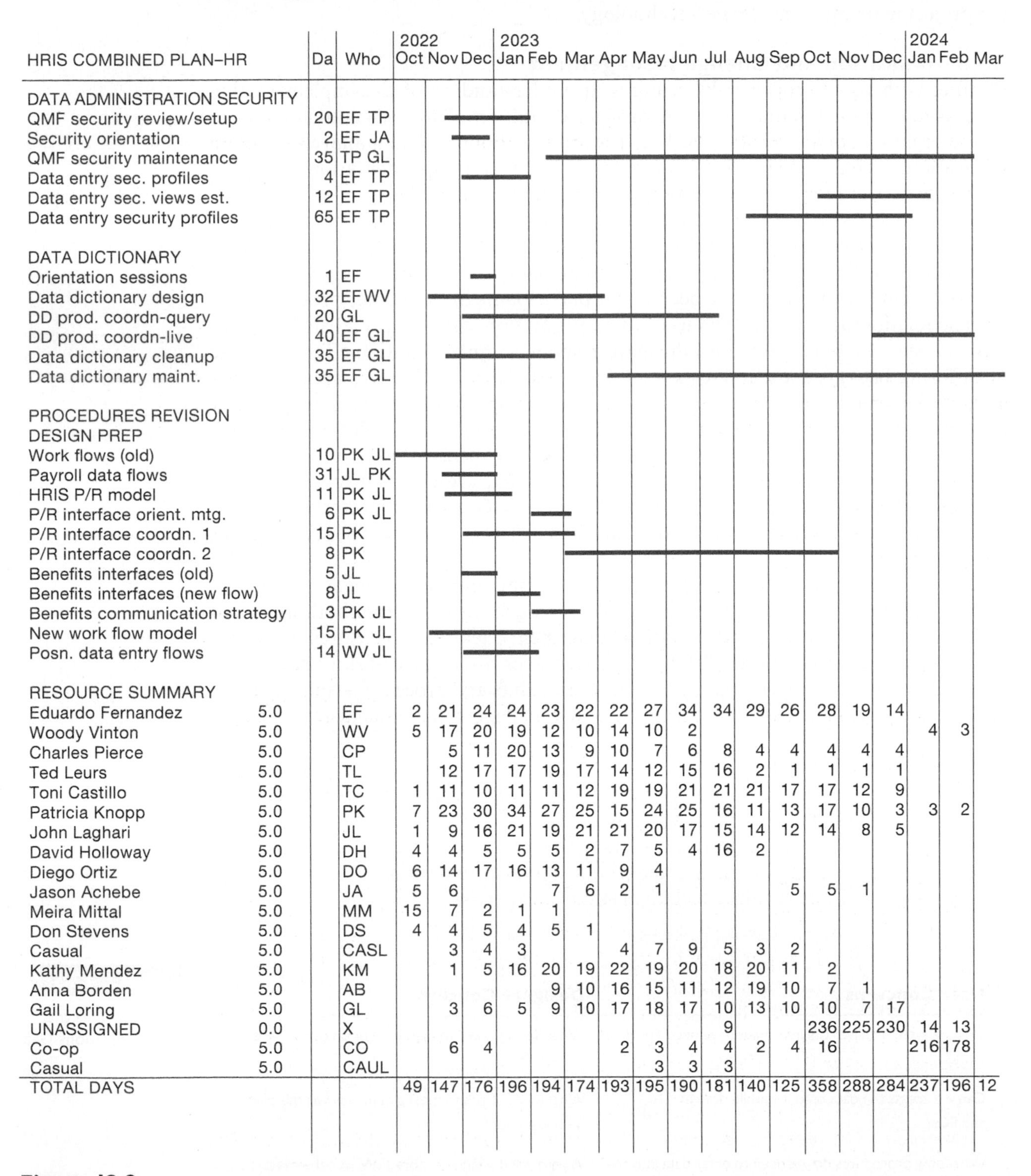

HRIS COMBINED PLAN–HR
Da Who
2022 Oct Nov Dec 2023 Jan Feb Mar Apr May Jun Jul Aug Sep Oct Nov Dec 2024 Jan Feb Mar
DATA ADMINISTRATION SECURITY
QMF security review/setup 20 EF TP
Security orientation 2 EF JA
QMF security maintenance 35 TP GL
Data entry sec. profiles 4 EF TP
Data entry sec. views est. 12 EF TP
Data entry security profiles 65 EF TP
DATA DICTIONARY
Orientation sessions 1 EF
Data dictionary design 32 EF WV
DD prod. coordn-query 20 GL
DD prod. coordn-live 40 EF GL
Data dictionary cleanup 35 EF GL
Data dictionary maint. 35 EF GL
PROCEDURES REVISION DESIGN PREP
Work flows (old) 10 PK JL
Payroll data flows 31 JL PK
HRIS P/R model 11 PK JL
P/R interface orient. mtg. 6 PK JL
P/R interface coordn. 1 15 PK
P/R interface coordn. 2 8 PK
Benefits interfaces (old) 5 JL
Benefits interfaces (new flow) 8 JL
Benefits communication strategy 3 PK JL
New work flow model 15 PK JL
Posn. data entry flows 14 WV JL
RESOURCE SUMMARY
Eduardo Fernandez 5.0 EF 2 21 24 24 23 22 22 27 34 34 29 26 28 19 14
Woody Vinton 5.0 WV 5 17 20 19 12 10 14 10 2 4 3
Charles Pierce 5.0 CP 5 11 20 13 9 10 7 6 8 4 4 4 4 4
Ted Leurs 5.0 TL 12 17 17 19 17 14 12 15 16 2 1 1 1 1
Toni Castillo 5.0 TC 1 11 10 11 11 12 19 19 21 21 21 17 17 12 9
Patricia Knopp 5.0 PK 7 23 30 34 27 25 15 24 25 16 11 13 17 10 3 3 2
John Laghari 5.0 JL 1 9 16 21 19 21 21 20 17 15 14 12 14 8 5
David Holloway 5.0 DH 4 4 5 5 5 2 7 5 4 16 2
Diego Ortiz 5.0 DO 6 14 17 16 13 11 9 4
Jason Achebe 5.0 JA 5 6 7 6 2 1 5 5 1
Meira Mittal 5.0 MM 15 7 2 1 1
Don Stevens 5.0 DS 4 4 5 4 5 1
Casual 5.0 CASL 3 4 3 4 7 9 5 3 2
Kathy Mendez 5.0 KM 1 5 16 20 19 22 19 20 18 20 11 2
Anna Borden 5.0 AB 9 10 16 15 11 12 19 10 7 1
Gail Loring 5.0 GL 3 6 5 9 10 17 18 17 10 13 10 10 7 17
UNASSIGNED 0.0 X 9 236 225 230 14 13
Co-op 5.0 CO 6 4 2 3 4 4 2 4 16 216 178
Casual 5.0 CAUL 3 3 3
TOTAL DAYS 49 147 176 196 194 174 193 195 190 181 140 125 358 288 284 237 196 12


**Figure 12.9**
A Gantt Chart

*The Gantt chart in this figure shows the task, person-days, and initials of each responsible person as well as the start and finish dates for each task. The resource summary provides a good manager with the total person-days for each month and for each person working on the project to manage the project successfully. The project described here is a data management project.*

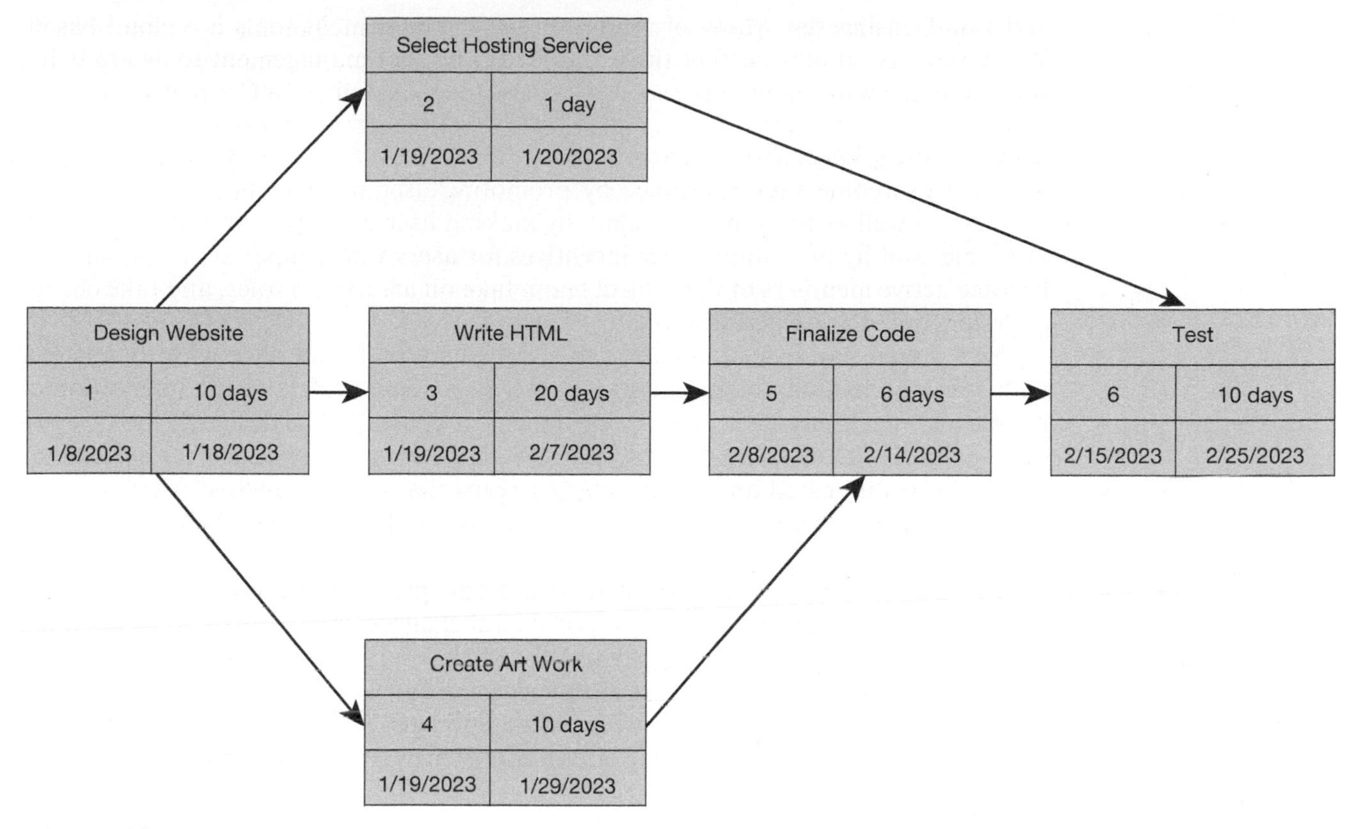


**Figure 12.10**
A PERT Chart
*This is a simplified PERT chart for creating a small website. It shows the ordering of project tasks and the relationship of a task with preceding and succeeding tasks.*

human resource requirements. It shows each task as a horizontal bar whose length is proportional to the time required to complete the task.

Although Gantt charts show when project activities begin and end, they don't depict task dependencies, how one task is affected if another is behind schedule, or how tasks should be ordered. That is when **PERT charts** are useful. PERT stands for Program Evaluation and Review Technique, a methodology the US Navy developed during the 1950s to manage the *Polaris* submarine missile program. A PERT chart graphically depicts project tasks and their interrelationships. The PERT chart lists the specific activities that make up a project and the activities that must be completed before a specific activity can start. Examine Figure 12.10, which shows a simplified PERT chart used for the creation of a small website. The PERT chart portrays a project as a network diagram consisting of numbered nodes (either rectangles, as shown in Figure 12.10, or circles) representing project tasks. Each node is numbered and shows the task, its duration, the starting date, and the completion date. The direction of the arrows on the lines indicates the sequence of tasks and shows which activities must be completed before the commencement of another activity. In Figure 12.10, the tasks in nodes 2, 3, and 4 do not depend on each other and can be undertaken simultaneously, but each depends on completion of the first task.

## Project Management Software

Commercial software tools are available to automate the creation of Gantt and PERT charts and facilitate the project management process. Project management software typically features capabilities for defining and ordering tasks, assigning resources to tasks, establishing starting and ending dates for tasks, tracking progress, and facilitating modifications to tasks and resources. The most widely used project management tool today is Microsoft Project, but there are also lower-cost tools for small projects and small businesses. Many of today's project management tools are cloud-based. We should also point out that these traditional project management tools are being supplemented with some of the social business tools described in Chapter 2.

### Overcoming User Resistance

You can overcome user resistance by promoting user participation (to elicit commitment as well as to improve design), by making user education and training easily available, and by providing better incentives for users who cooperate. End users can become active members of the project team, take on leadership roles, and take charge of system installation and training.

You should pay special attention to areas where users interface with the system, with special sensitivity to ergonomics issues. **Ergonomics** refers to the interaction of people and machines in the work environment. It considers the design of jobs, health issues, and the **end-user interface** of information systems. For instance, if a system has a series of complicated online data entry screens that are extremely difficult or time consuming to work with, users will reject the system if it increases their workload or level of job stress.

Users will be more cooperative if organizational problems are solved prior to introducing the new system. In addition to procedural changes, transformations in job functions, organizational structure, power relationships, and behavior should be identified during systems analysis, using an **organizational impact analysis**. You can see some of these project management strategies at work in the Spotlight on Organizations case, which describes how Sauder Woodworking implemented a new, state-of-the-art ERP system.

## 12-5 Understand how MIS can help your career.

Here is how this chapter and this book can help you find a job as an entry-level junior business systems analyst.

### THE COMPANY

Systems 100 Technology Consultants, a Chicago-based professional technology services firm, provides staffing and information technology consulting services to other US companies and has an open position for an entry-level junior business systems analyst. The company provides business and technology consultants to more than 150 firms in financial services, healthcare, communications, transportation, energy, consumer goods, and technology, helping them implement business and technology initiatives cost-effectively.

### POSITION DESCRIPTION

A junior business systems analyst is expected to work in project teams throughout all phases of the software development life cycle, including defining business requirements, developing detailed design specifications, and working with application developers to build or enhance systems and business processes. Before undertaking assignments, new business systems analysts receive the training they will need to succeed in their assignments.

### JOB REQUIREMENTS

- Upcoming or recent college graduate, with BA in Management Information Systems, Finance, Psychology, or related field
- Three to six or more months of corporate work or internship experience, including experience working with a project team
- Strong understanding of technology, systems, and business process improvement

## SPOTLIGHT ON: ORGANIZATIONS Sauder Woodworking Gets ERP Implementation Right

Sauder Woodworking is North America's leading producer of ready-to-assemble furniture and is one of the top-five U.S. residential furniture manufacturers. Ninety percent of Sauder furniture is manufactured in Archbold, Ohio, where the company was founded in 1934. (Sauder also sources furniture from a network of global suppliers.) Its Ohio facility has four million square feet of floor space housing high-tech furniture-making equipment from around the world and 2,000 employees. The company offers nearly 50 furniture collections. Furniture making requires skilled design, artistry, and attention to detail, and Sauder has historically excelled in these areas.

Sauder started using SAP ERP Central Component (ECC) software for enterprise resource planning (ERP) in 2004. ECC provides modules covering a full range of industry applications, including finance, logistics, HR, product planning, and customer service, all linked together into a single, customizable system run on a database of the user's choice. As a modular system, SAP ECC is designed so that organizations can use the pieces they need, configured in a way that makes sense for their business. Sauder used a phased implementation approach, starting with the modules for order-to-cash (business processes for receiving and processing customer orders). Sauder finished implementing all the main SAP ECC modules by 2015.

By that time, SAP had released SAP S/4HANA, a more leading-edge version of its ERP system that features in-memory computing (see Chapter 6). Sauder management had to decide whether to optimize the SAP modules it had already implemented on premises or to switch to S/4HANA to take advantage of its new functionality. Switching to the newest version of SAP ERP software would require more work than sticking with Sauder's existing system. Management opted for an on-premises version of SAP S/4HANA so that it could implement the new S/4HANA suite on its own timeline rather than wait and then be pressured to change software when SAP withdrew support for the older ECC system.

Another decision was whether to go greenfield (a completely new implementation) or brownfield (converting the existing system to the new one). A greenfield implementation would require installing a clean installation of the ERP system and then importing all the relevant data. A brownfield implementation would entail converting the existing system to the new one, using the data from the ECC system. Sauder opted for the brownfield approach. A key factor in the decision to go brownfield was the number of employees (2,000) who would be impacted by having to deal with an entirely new system. They might have trouble learning the system or resist working with the system because it was so unfamiliar. Sauder didn't want to have to deal with a lot of change management.

Sauder was quite familiar with SAP software. Sauder's SAP system supported more than 1,600 users conducting business transactions with the software, including those for finance, supply chain management, and shop floor interactions with SAP. In 2018 Sauder's SAP system processed 1,624,169 shipments, 1,661,225 sales orders, 14,438,089 warehouse moves, 894,664 production receipts, and 1,624,684 invoices.

There were challenges, however. To minimize business disruptions, management wanted the SAP ECC system converted to SAP S/4HANA within 72 hours, a very narrow timeframe for a project of this sort. There was no room for error. At the same time Sauder began transitioning to SAP S/4HANA, it made major changes to its IT infrastructure. The company migrated off legacy IBM iSeries systems to a new platform based on Dell servers and storage area networks (SANs) in a VMware virtual environment (see Chapter 5). (A *SAN* is a network of storage devices that can be accessed by multiple servers or computers, providing a shared pool of storage space.)

Sauder's IT staff had to learn how to support these new technologies but was fortunately assisted by Symmetry consultants. Symmetry helps firms manage complex SAP implementations on a global scale. Symmetry had provided services to Sauder for more than 12 years and understood the nature of Sauder's business and technology environment. Sauder also took advantage of SAP Enterprise Support, a premium support and maintenance plan for SAP customers geared toward ensuring that customers have access to the latest technologies and tools as well as access to SAP experts. Among the services Sauder utilized was advice from SAP experts about business process improvement and data volume management to formulate a strategy to reduce its data footprint and enhance system performace.

The system was successfully converted within the 72-hour downtime window, including conversion of the production planning and detailed scheduling functionality of the SAP Advanced Planning and Optimization component within SAP S/4HANA. The transition to the new S/4HANA system went very smoothly, with minimal disruption to the business, and the project was delivered on time and within budget.

The company plans to optimize its use of the S/4HANA system in phases, starting with finance functions. Running on SAP S/4HANA has better positioned Sauder Woodworking to achieve its strategic goals.

Sources: "How Does Assembling the Right Support Team Help a Business Run Precisely to Plan?" Sap.com, accessed April 3, 2022; "Case Study: Sauder Woodworking," Nttdata-solutions.com, accessed April 3, 2022; "Sauder Woodworking," Itelligence.com, accessed May 8, 2020; Eric Kavanah, "Measure Twice, Cut Once: Getting a Modern ERP Deployment Right," *eWeek*, February 11, 2020.

## CASE STUDY QUESTIONS

1. Why is an ERP system so important for Sauder Woodworking? Why did Sauder want to switch to a newer ERP system?
2. Were there any risks in undertaking this project? How did Sauder deal with them?
3. Was SAP S/4HANA a good choice for Sauder? Why or why not?
4. Evaluate the advantages and disadvantages of a brownfield implementation for Sauder. Was going brownfield the right approach for Sauder?

- Strong analytical, communication, and problem-solving skills
- Ability to work comfortably in a team environment and also to work remotely
- Knowledge and understanding of the software development life cycle and business process improvement
- Knowledge of Microsoft 365 applications
- Exposure to SQL desirable but not required

## INTERVIEW QUESTIONS

1. What information systems courses have you taken, including MIS, database, data analytics, and systems development? Can you write SQL queries?
2. Have you worked on any systems development projects? If so, what exactly did you do? What systems development practices did you use?
3. Have you worked on any other kinds of projects, and what role did you play? Do you have samples of the writing or output you produced for these projects?
4. Which Microsoft 365 tools have you used? What kinds of problems have you used these tools to solve?
5. Do you have any experience with agile software development?

## AUTHOR TIPS

1. Review the discussion of business processes in Chapters 2 and 3 along with this chapter's discussion of IT project management and implementation. Be prepared to talk about any systems development experience you have had, including analyzing or redesigning business processes. Also be prepared to discuss contemporary systems development practices.
2. Inquire about how you would be using SQL and Microsoft 365 tools for the job and what skills you would be expected to demonstrate. Bring samples of the work you have done with this software. Express interest in learning what you don't know about these tools to fulfill your job assignments.
3. Bring samples of your writing that demonstrate your analytical and business application skills and project experience.

## Review Summary

**12-1 Explain how managers should build a business case for the acquisition and development of a new information system.** The business case for an IT investment describes the problem facing the organization that can be solved by investing in a proposed system solution. It provides an analysis of whether an information system project is a good investment by calculating its costs and benefits. Organizations should develop an information systems plan that describes how information technology supports the company's overall business plan and strategy. Portfolio analysis and scoring models can be used to evaluate alternative information systems projects. Tangible benefits are quantifiable, and intangible benefits cannot be immediately quantified but may provide quantifiable benefits in the future. Benefits that exceed costs should then be analyzed using capital budgeting methods to make sure they represent a good return on the firm's invested capital.

**12-2 Identify the core problem-solving steps for developing a new information system.** The core problem-solving steps for developing new information systems are: (1) define and understand the problem, (2) develop alternative solutions, (3) evaluate and choose the solution, and (4) implement the solution. The third step includes an assessment of the technical, financial, and organizational feasibility of each alternative. The fourth step entails finalizing design specifications, acquiring hardware and software, testing, providing training and documentation, conversion, and evaluating the system solution once it is in production.

**12-3 Compare the alternative methods for building information systems.** The systems development life cycle requires information systems to be developed in formal stages. The stages must proceed sequentially and have defined outputs; each requires formal approval before the next stage can commence. The system life cycle is rigid and costly but useful for large projects.

Prototyping consists of building an experimental system rapidly and inexpensively for end users to interact with and evaluate. The prototype is refined and enhanced until users are satisfied that it includes all their requirements and can be used as a template to create the final system. End-user-developed systems can be created rapidly and informally using user-friendly software tools. End-user development can improve requirements determination and reduce application backlog.

Application software packages and cloud-based SaaS eliminate the need for writing software programs when developing an information system. Application software packages and SaaS are helpful if a firm does not have the internal information systems staff or financial resources to custom-develop a system.

Outsourcing consists of using an external vendor to build (or operate) a firm's information systems. If it is properly managed, outsourcing can save application development costs or enable firms to develop applications without an internal information systems staff.

Rapid application development (RAD) uses visual programming, prototyping, and tools for very rapid creation of systems. Agile development breaks a large project into a series of small subprojects that are completed in short periods of time using iteration and continuous feedback. Automated testing tools speed up testing and improve quality by automating tasks that were previously manual. Low-code and no-code development tools enable people with minimal or no programming skills to create workable systems in a short period of time. Mobile application development must pay attention to simplicity, usability, and the need to optimize tasks for tiny screens.

**12-4 Describe how information systems projects should be managed.** Information systems projects and the entire implementation process should be managed as planned organizational change using an organizational impact analysis. Management support and control of the implementation process are essential, as are mechanisms for dealing with the level of risk in each new systems project. Project risks are influenced by project size, project structure, and the level of technical expertise required of the information systems staff and project team. Formal planning and control tools (including Gantt and PERT charts) track resource allocations and specific project activities. Users can be encouraged to take active roles in systems development and become involved in installation and training.

## Key Terms

Acceptance testing, 443
Agile development, 451
Business case, 435
Capital budgeting, 439
Conversion, 444
Customization, 448
Direct cutover strategy, 444
Documentation, 444
End-user development, 447
End-user interface, 458
Ergonomics, 458
Feasibility study, 443
Formal planning and control tools, 456
Gantt chart, 456
Hybrid app, 450
Implementation, 455
Information requirements, 442
Information systems plan, 436
Intangible benefits, 438
Low-code development, 453
Maintenance, 445
Mobile web app, 450
Mobile website, 450
Native app, 450
No-code development, 453
Organizational impact analysis, 458
Parallel strategy, 444
PERT charts, 457
Phased approach, 444
Pilot study, 444
Portfolio analysis, 437
Production, 445
Project, 454
Project management, 454
Prototyping, 446
Rapid application development (RAD), 451
Request for Proposal (RFP), 448
Responsive web design, 451
Scope, 454
Scoring model, 438
System testing, 443
Systems analysis, 441
Systems design, 443
Systems development life cycle (SDLC), 446
Tangible benefits, 438
Test plan, 444
Testing, 443
Unit testing, 443
User-designer communications gap, 455

## Review Questions

**12-1** Explain how managers should build a business case for the acquisition and development of a new information system.

- Define and describe the components of a business case for a proposed systems investment.
- List and describe the major components of an information systems plan.
- Explain the difference between tangible and intangible benefits.
- List six tangible benefits and six intangible benefits of an IT investment.
- Describe how portfolio analysis and scoring models can be used to establish the worth of systems.

**12-2** Identify the core problem-solving steps for developing a new information system.

- List and describe the problem-solving steps for building a new system.
- Define information requirements, and explain why they are important for developing a system solution.
- List the various types of design specifications required for a new information system.
- Explain why the testing stage of systems development is so important. Name and describe the three stages of testing for an information system.
- Describe the roles of documentation, conversion, production, and maintenance in systems development.

**12-3** Compare the alternative methods for building information systems.

- Define the traditional systems development life cycle, and describe its advantages and disadvantages for systems building.

- Define prototyping, and describe its benefits and limitations. List and describe the steps in the prototyping process.
- Define end-user development and explain its advantages and disadvantages.
- Describe the advantages and disadvantages of developing information systems based on application software packages and cloud software services (SaaS).
- Define outsourcing. Describe the circumstances in which it should be used for building information systems. List and describe the hidden costs of offshore software outsourcing.
- Explain how businesses can rapidly develop e-business applications using rapid application development, agile development, automated software testing, and low-code/no-code development.
- Describe the issues that must be addressed when developing mobile applications.

**12-4** Describe how information systems projects should be managed.

- Explain the importance of implementation for managing the organizational change surrounding a new information system.
- Define the user-designer communications gap, and explain the kinds of implementation problems it creates.
- List and describe the factors that influence project risk, and describe strategies for minimizing project risks.

## Discussion Questions

**12-5** **MyLab MIS** It has been said that systems fail when systems builders ignore people problems. Why might this be so?

**12-6** **MyLab MIS** Why is building a system a form of organizational problem solving?

**12-7** **MyLab MIS** Discuss the roles of business end users and information system professionals in developing a system solution. How do both roles differ when the solution is developed using prototyping or end-user development?

**MyLab MIS**
To complete these problems, go to EOC Discussion Questions in **MyLab MIS**.

## Hands-On MIS Projects

The projects in this section give you hands-on experience evaluating a mobile application development project, using a database to clarify business strategy, and analyzing website information requirements. Visit MyLab MIS to access this chapter's Hands-On MIS Projects.

### MANAGEMENT DECISION PROBLEM

**12-8** Traditionally, sports teams have viewed apps as sideshows to their primary business. However, management for the NBA's Miami Heat basketball team wanted to create a mobile app that could serve as a one-stop shop for entertainment, digital ticketing, personalized content and marketing messages, and deepening fan engagement and loyalty. The app needed to be integrated with the Heat's customer relationship management (CRM) system as well as third-party systems such as TicketMaster SafeTix, which encrypts mobile tickets with technology to prevent fraud and counterfeit ticketing. Describe the mobile application development issues this mobile app project raised. What management decisions had to be made?

## IMPROVING DECISION MAKING: USING A DATABASE TO CLARIFY BUSINESS STRATEGY

Software skills: Database querying and reporting, querying a multi-table database
Business skills: Reservation systems, customer analysis

**12-9** In this exercise, you'll use database software to analyze the reservation transactions for a hotel and use that information to fine-tune the hotel's business strategy and marketing activities.

In MyLab MIS for Chapter 12, you'll find a database for hotel reservation transactions with information about Fair Winds Inn in Sanibel, Florida. At the Inn, 10 rooms overlook sidestreets, 10 rooms have windows with limited views of the ocean, and the remaining 10 rooms face the ocean. Room rates are based on the type of room.

The owners currently use a manual reservation and bookkeeping system, which cannot provide management with immediate data about the hotel's daily operations and revenue. They need to know which room and type of room has been assigned to each guest along with the guest's name and start and end dates for that guest's stay; the average number of visitors per room type; and the average length of visit per room type. Create queries and reports from the database EMIS15Chap12 Question File to answer these questions. Then write a brief report on how the new database system solves some of the problems created by the old, manual reservation system and how the database can help Fair Winds Inn refine its business strategy.

## ACHIEVING OPERATIONAL EXCELLENCE: ANALYZING WEBSITE DESIGN AND INFORMATION REQUIREMENTS

Software skills: Web browser software
Business skills: Information requirements analysis, website design

**12-10** Visit the website of your choice and explore it thoroughly. Prepare a report analyzing the various functions provided by that website and its information requirements. Your report should answer these questions: What functions does the website perform? What data does it use? What are its inputs, outputs, and processes? What are some of its other design specifications? Does the website link to any internal systems or systems of other organizations? What value does this website provide the firm?

## COLLABORATION AND TEAMWORK PROJECT

Identifying Implementation Problems

**12-11** With three or four of your classmates, select a system described in this text. Write a description of the implementation problems you might encounter with such a system and the steps you would take to solve or prevent these problems. If possible, use Google Docs and Google Drive or Google Sites to brainstorm, organize, and develop a presentation of your findings for the class.

# BUSINESS PROBLEM-SOLVING CASE

## JEDI AND JWCC: A CLOUD OF CONTROVERSY

Most major companies have moved some of their computing operations to the cloud, and the US military wants to follow suit. Unifying information in the cloud is more necessary than ever as the armed services deploy large numbers of remote sensors, semiautonomous weapons, and artificial intelligence applications to fight wars across multiple domains (land, sea, air, and cyberspace). All of these capabilities require instantaneous access to very large quantities of data gathered from many different locations.

Project JEDI was the US Department of Defense's plan to modernize its IT infrastructure so that employees, officers, and soldiers on the frontline could access and manipulate data at the speed of modern enterprises. Project JEDI sought to create a unified cloud infrastructure across the entire Department of Defense (DOD) that would speed the flow of data and analysis to combat troops. The new plan was part of a larger move toward replacing the military's branch-specific systems and networks with a more efficient and manageable enterprise model.

On July 26, 2018, the DOD issued a Request for Proposal (RFP) for a Joint Enterprise Defense Infrastructure (JEDI) Cloud Program, a cloud services solution that could support Unclassified, Secret, and Top Secret requirements with a focus on using commercially available technology services. The JEDI program presented a 10-year, $10 billion government contract to go to a single cloud computing vendor, which would serve as the exclusive cloud computing provider for the US Department of Defense.

The US Department of Defense maintains more than 500 public and private cloud infrastructures that support Unclassified and Secret requirements. The DOD's current cloud services are decentralized, creating an additional level of complexity for managing data and services at an enterprise-wide level. Current DOD systems are fragmented, slowing the decision-making process within the DOD both at home and abroad. Much of the US military operates on outdated computer systems built during the 1980s and 1990s. The Defense Department has spent billions of dollars trying to make these systems talk to one another. What the DOD wants and needs is an enterprise-wide cloud that supports rapid data-driven decision making and provides worldwide support for DOD operations. The JEDI contract was central to the Pentagon's efforts to modernize its technology.

The 10-year JEDI contract set off a showdown among Amazon, Microsoft, Oracle, IBM, and Google for the work to transform the military's cloud computing systems. (Google dropped out in October 2018.) Oracle, IBM, and Microsoft stated that the DOD shouldn't use a single cloud vendor for JEDI. A number of experts have backed them up. Justin Cappos, associate professor of computer science and engineering at New York University, said a single cloud solution is out of the norm: Many companies use multiple cloud vendors because doing so is safer. Leigh Madden, Microsoft's general manager for national security, stated that his company wanted to win the contract, but 80 percent of businesses use multiple cloud vendors.

Other experts have pointed out that deployment of a single cloud conflicts with established best practices and industry trends in the commercial marketplace. These experts feared that by awarding the JEDI contract to a single vendor, the DOD would be locked into that vendor long term and lose access to changes in technology. They also worried about security. Having a single cloud enterprise from a single vendor is a significant vulnerability risk. The size of this system would make it difficult to secure, and a single intrusion could result in catastrophic consequences. Having multiple clouds allows for each cloud to be more readily protected while also containing any security breaches. Those favoring JEDI noted, however, that using one provider would reduce complexity in military IT systems and streamline communications.

Oracle America and IBM both filed pre-award bid protests against the JEDI Cloud solicitation, claiming it favored Amazon and Microsoft. These protests were dismissed by the General Accounting Office (GAO) in late 2018. Amazon and Microsoft, the most experienced cloud service providers, became the two finalists, and Microsoft was awarded the contract in October 2019.

The Pentagon's Cloud Executive Steering Group described the acquisition process for the massive cloud migration that would stretch across the entire DOD IT infrastructure, focusing primarily on infrastructure-as-a-service (IaaS) and platform-as-a-service (PaaS). Instead of building and maintaining its own data centers and systems, the DOD wanted to take advantage of the existing strengths of commercially available cloud technologies and not saddle them with extensive customizations. The DOD wants to remain in pace with industry and be able to take advantage of new commercial software solutions. The Pentagon's acquisition regulations have in the past served as barriers to innovation. Internal acquisition policies needed to be revised to take full advantage of the commercial cloud platform.

The envisioned IaaS had to be more than a data center. Other requirements included vendor monitoring, identity, failover, scalability, and even artificial intelligence (AI) applications. The DOD wanted to be able to immediately take advantage of commercial innovations in these areas. If the DOD used commercially available cloud solutions, it would have the foundational technology in place to deliver better software to fighters, with better security, lower cost, and easier maintainability.

On February 13, 2020, a federal judge ordered the Pentagon to halt work on the JEDI contract. Amazon had been long considered the favorite to win the JEDI contract because of its dominance of cloud computing (it has 33 percent of the market) and its experience building cloud services for the Central Intelligence Agency. However, its bid was overshadowed by conflict-of-interest allegations: Amazon had filed suit in December 2019 to block the JEDI contract award going to Microsoft, contending that the selection of Microsoft had been improperly influenced by President Trump's public complaints about Amazon. Amazon's CEO, Jeff Bezos, owns *The Washington Post*, which has been a frequent critic of Trump and his policies.

On July 6, 2021, acting DOD CIO Johns Sherman announced that JEDI would be replaced entirely with a multicloud Joint Warfighting Cloud Capability (JWCC). Discontinuing JEDI became more likely when it appeared that litigation would persist for months or possibly years. In November 2021 the Pentagon issued formal solicitations to four major technology companies—Amazon and Microsoft as well as Google and Oracle—for pieces of the JWCC multibillion-dollar cloud-computing contract.

The Defense Department emphasized that the JWCC's multicloud approach would be more flexible than JEDI. The US military requires multicloud technology to ensure the best fit for each mission given the varying capabilities of cloud service providers. The switch to multicloud gives DOD IT staff the flexibility to pick the most appropriate cloud technology for various mission requirements. Cloud tools might have different capabilities in terms of features and processing speeds for various functions. Granting program offices and developers the authority to pick the most appropriate cloud delivers capability more quickly, with the best set features for each customer.

JWCC is a maximum-five-year project worth multiple billions of dollars. Each cloud provider will need to be able to offer services at the unclassified, secret, and top-secret security levels, with parity across each of those levels. In addition, the procurement's basic requirements call for integrated cross-domain solutions, global availability, including at the tactical edge, and enhanced cybersecurity controls. Rob Carey, former Navy CIO and Deputy DOD CIO and now president of Cloudera Government Solutions, stated that he expected the Joint Warfighting Cloud Capability to retain 90 percent of the requirements of the JEDI cloud, the fundamental difference being the multicloud architecture.

The JWCC project is somewhat open ended. It calls for indefinite-delivery, indefinite-quantity contracts that span a three-year base period and two optional years, with a ceiling of $9 billion. (An indefinite-delivery, indefinite-quantity contract provides for an indefinite quantity, within stated limits, of supplies or services during a specified period. It is utilized when the purchasing party cannot determine the precise quantities of services or supplies it requires during that period.) Price will be determined later in the procurement after some testing and tweaking of the new environment, Sherman said, emphasizing that people shouldn't get fixated on a certain figure. After the initial five-year period, the DOD hopes to move to a larger, multicloud environment.

The switch to a multicloud enterprise-wide capability could present the DOD with hurdles in both the acquisition of the cloud and its management. The department is striving to avoid the drawn-out litigation that plagued JEDI and allegations of unfair acquisition practices that favored different companies.

Another complicating factor is interoperability among the new systems. In order to create an enterprise cloud that is available to personnel worldwide and that connects sensors to shooters, the various clouds must stitch together seamlessly. Data in one cloud must be easily accessible to another cloud. According to Carey, there are now tools for managing a multicloud enterprise environment with multiple vendors that were not available when JEDI was being planned. These include more mature tools for cybersecurity, network maintenance automation and orchestration, and policy governance and compliance.

The DOD isn't mandating that everyone migrate to JWCC, especially the military services, all of which have their own cloud efforts underway. The Air Force has a Cloud One program, the Army maintains its own Enterprise Cloud Management Agency, and the Navy has consolidated its cloud activities within a new Cloud Service Management Organization. The JWCC is not intended to supplant the work the services have already done to incorporate commercial cloud services into their IT infrastructures, at least not in the near term. The hope is that as JWCC is proven, more of the services will transition to it. For the time being, JWCC is envisioned as the enterprise capability for the combatant commands and fourth-estate organizations. (The Defense Department has 11 combatant commands, each with a geographic or functional mission, such as the US European Command or the US Cyber Command, that provide command and control of military forces in peace and in war. Fourth-estate organizations are DOD agencies, such as the Defense Finance and Accounting Service, that are not part of the military, intelligence community or combatant commands,)

On March 29, 2022, top DOD officials announced that the JWCC project would be delayed until December of that year. Officials initially believed that they could award the final JWCC contracts in April 2022, but when they started working with the vendors, the DOD officials realized they needed more time. DOD CIO Sherman emphasized that there weren't any ongoing problems with the JWCC procurement effort and that DOD was simply conducting due diligence with the four selected cloud vendors. An enterprise cloud for the entire DOD is "unprecedented," he noted, and the department wanted to get it right. DOD officials also expressed confidence that the JWCC cloud won't be obsolete by the time it is awarded and deployed, as the department takes a modern approach to cloud acquisition.

Sources: Mark Pomerleau, "DOD's JWCC Enterprise Cloud Award Delayed Until December," *FedScoop*, March 29, 2022; Billy Mitchell, "2021 in Review: Out with JEDI, in with JWCC," *FedScoop*, December 27, 2021; Jared Serbu,"DOD Picks Amazon, Microsoft, Google, and Oracle for Multibillion Dollar Project to Replace JEDI Cloud," *Federal News Network*, November 19, 2021; Andrew Eversden, "With JEDI Cloud Scuttled, Pentagon Embraces Critics' Idea of Multicloud for Digital Warfare," *C4ISRNSET*, July 9, 2021; "Inside DOD's JEDI Replacement: The Joint Warfighting Cloud Capability," *FedScoop*, July 8, 2021; Lauren Williams, "DOD Moves on from JEDI," *Government Computer News*, July 6, 2021; Vikram Mittai, "The Next JEDI: The Joint Warfighting Cloud Capability," *Forbes*, July 1, 2021; and; and Sydney J. Freedberg Jr., "Experts Debate: Should JEDI Cloud Be Saved?" *Breaking Defense*, May 4, 2020.

## CASE STUDY QUESTIONS

**12-12** Describe the JEDI Project. Why was it so important? What problems was it meant to solve? Evaluate its pros and cons.

**12-13** How did JWCC differ from JEDI?

**12-14** Identify the key risk factors in the JWCC project. What people, organization, and technology problems could the project encounter?

**12-15** What could be done to mitigate the JWCC's risks?

## Chapter 12 References


Andriole, Steve. "3 Main Reasons Why Big Technology Projects Fail—& Why Many Companies Should Just Never Do Them." *Forbes* (March 24, 2021).

Benaroch, Michael, Yossi Lichtenstein, and Lior Fink. "Contract Design Choices and the Balance of Ex Ante and Ex Post Transaction Costs in Software Development Outsourcing." *MIS Quarterly* 40, No. 1 (March 2016).

Brock, Jon, Tamim Saleh, and Sesh Iyer. "Large-Scale IT Projects: From Nightmare to Value Creation." *Boston Consulting Group* (May 20, 2015).

Brown, Karen A., Nancy Lea Hyer, and Richard Ettenson. "Protect Your Project from Escalating Doubts." *MIT Sloan Management Review* 58, No. 3 (Spring 2017).

Browning, Tyson, R., and Ranga V. Ramases. "Reducing Unwelcome Surprises in Project Management." *MIT Sloan Management Review* 56, No. 3 (Spring 2015).

Bughin, Jacques, Jonathan Deakin, and Barbara O'Beirne. "Digital Transformation: Improving the Odds for Success." (October 2019).

Cecez-Kecmanovic, Dubravka, Karlheinz Kautz, and Rebecca Abrahall. "Reframing Success and Failure of Information Systems: A Performative Perspective." *MIS Quarterly* 38, No. 2 (June 2014).

Chandrasekaran, Sriram, Sauri Gudlavalleti, and Sanjay Kaniyar. "Achieving Success in Large Complex Software Projects." *McKinsey Quarterly* (July 2014).

Comella-Dorda, Santiago, Swati Lohiya, and Gerard Speksnijder. "An Operating Model for Company-Wide Agile Development." McKinsey & Company (May 2016).

Flyvbjerg, Bent. "Make Megaprojects More Modular." *Harvard Business Review* (November–December 2021).

Gnanasambandam, Chandra, Martin Harrysson, Rahul Mangla, and Shivam Srivastava. "An Executive's Guide to Software Development." McKinsey & Company (February 2017).

Hoehle, Hartmut, and Viswanath Venkatesh. "Mobile Application Usability: Conceptualization and Instrument Development." *MIS Quarterly* 39, No. 2 (June 2015).

Jenkin, Tracy A., Yolande E. Chan, and Rajiv Sabherwal."Mutual Understanding in Information Systems Development: Changes within and across Projects." *MIS Quarterly* 43, No. 2 (June 2019).

Keil, M., and J. S. Lee. "Judgment and Decision Making in Managing IT Project Risks: A Construal-Level Theory Perspective." PMI Sponsored Research (March 2019).

Kendall, Kenneth E., and Julie E. Kendall. *Systems Analysis and Design*, 10th ed. (Hoboken, NJ: Pearson Education, Inc., 2019).

Kloppenborg, Timothy J., and Debbie Tesch. "How Executive Sponsors Influence Project Success." *MIT Sloan Management Review* (Spring 2015).

Kudaravalli, Srinivas, Samer Faraj, and Steven L. Johnson. "A Configural Approach to Coordinating Expertise in Software Development Teams." *MIS Quarterly* 41, No. 1 (March 2017).

Langer, Nishtha, and Deepa Mani."Impact of Formal Controls on Client Satisfaction and Profitability in Strategic Outsourcing Contracts." *Journal of Management Information Systems* 35, No. 4 (2018).

Levina, Natalia, and Jeanne W. Ross. "From the Vendor's Perspective: Exploring the Value Proposition in Information Technology Outsourcing." *MIS Quarterly* 27, No. 3 (September 2003).

Low Code Agency. "7 Examples of No-Code Apps that Solve Real-World Problems." Blog.lowcode.agency (February 17, 2021).

Maruping, Likoebe M., and Sabine Matook. "The Multiplex Nature of the Customer Representative Role in Agile Information Systems Development." *MIS Quarterly* 44 No. 3 (September 2020).

Maruping, Likoebe M., ViswanathVenkatesh, James Y. L. Thong, and Xiaojun Zhang. "A Risk Mitigation Framework for Information Technology Projects: A Cultural Contingency Perspective." *Journal of Management Information Systems* 36, No. 1 (2019).

Moeini, Mohammad, and Suzanne Rivard. "Responding—or Not—to Information Technology Project Risks: An Integrative Model." *MIS Quarterly* 43, No. 2 (June 2019).

Overby, Stephanie. "9 Outsourcing Myths Debunked." *CIO* (April 13, 2022).

Ramasubbu, Narayan, and Indranil R. Bardhan. "Reconfiguring for Agility: Examining the Performance Implications for Project Team Autonomy through an Organizational Policy Experiment." *MIS Quarterly* 45 No. 4 (December 2021).

Ramasubbu, Narayan, and Chris F. Kemerer." Controlling Technical Debt Remediation in Outsourced Enterprise Systems Maintenance: An Empirical Analysis." *Journal of Management Information Systems* 38, No. 1 (2021).

Schwalbe, Kathy. *An Introduction to Project Management*, 7th ed. (Boston: Cengage, 2021).

Shivendu, Shivendu, David Zeng, and Vijay Gurbaxani. "Optimal Asset Transfer in IT Outsourcing Contracts."*MIS Quarterly* 44, No. 2 (June 2020).

Tibco. "Sound Credit Union Empowers Executives to Make Data-Driven Decisions." Tibco.com, accessed April 9, 2022.

Valacich, Joseph, and Joey George. *Modern Systems Analysis and Design*, 9th ed. (Hoboken, NJ: Pearson Education, Inc., 2020).

Wei He, J. J., Po-An Hsieh, Andreas Schroeder, and Yulin Fang. "Attaining Individual Creativity and Performance in Multi-Disciplinary and Geographically-Distributed IT Project Teams: The Role of Transactive Memory Systems." *MIS Quarterly* 46, No. 2 (June 2022).

Wiener, Martin, Magnus Mähring, Ulrich Remus, and Carol Saunders. "Control Configuration and Control Enactment in Information Systems Projects: Review and Expanded Theoretical Framework." *MIS Quarterly* 40, No. 3 (September 2016).

# Glossary

**3-D printing:** Uses machines to make solid objects, layer by layer, from specifications in a digital file. Also known as additive manufacturing.

**3G networks:** Cellular networks based on packet-switched technology, enabling users to transmit video, graphics, and other rich media in addition to voice; now being phased out in the United States.

**4G networks:** Wireless communication technology capable of up to 100 Mbps speeds for downloading.

**5G:** Newest wireless technology evolution, supporting transmission of very large amounts of data in the gigabit range, with fewer transmission delays and the ability to connect many more devices (such as sensors and smart devices) at once.

**acceptable use policy (AUP):** Defines acceptable uses of the firm's information resources and computing equipment and specifies consequences for noncompliance.

**acceptance testing:** Provides the final certification that the system is ready to be used in a production setting.

**accountability:** The mechanisms for assessing responsibility for decisions made and actions taken.

**advertising revenue model:** E-commerce revenue model based on generating revenue from online advertising.

**adware:** Software that can secretly install itself on an Internet user's computer by piggybacking on another application, and, once installed, can call out to websites to send ads and other unsolicited material to the user.

**affiliate revenue model:** E-commerce revenue model in which websites are paid for sending their visitors to other sites in return for a referral fee.

**agile development:** Rapid delivery of working software by breaking a large project into a series of small subprojects that are completed in short periods of time using iteration and continuous feedback.

**analytic platform:** Preconfigured hardware-software system that is specifically designed for high-speed analysis of large data sets.

**analytical CRM:** Customer relationship management applications dealing with the analysis of customer data to provide information for improving business performance.

**Android:** Open-source operating system for mobile devices developed by Google and the Open Handset Alliance; currently the most popular smartphone operating system worldwide.

**anti-malware software:** Software designed to detect, and often eliminate, malware from an information system.

**application controls:** Specific controls unique to each application that ensure that only authorized data are completely and accurately processed by that application.

**application server:** Software that handles all application operations between browser-based computers and a company's back-end business applications or databases.

**application software:** Program that applies a computer to a specific task for an end user.

**apps:** Software programs that can be accessed via the Internet.

**artificial intelligence (AI):** The effort to develop computer-based systems that can think and behave like humans.

**attributes:** Pieces of information describing a particular entity.

**augmented reality (AR):** Technology for enhancing visualization that provides a live view of a physical world environment whose elements are augmented by computer-generated imagery.

**authentication:** The ability of each party in a transaction to ascertain the identity of the other party.

**balanced scorecard method:** Framework for operationalizing a firm's strategic plan by focusing on measurable financial, business process, customer, and learning and growth outcomes of firm performance.

**bandwidth:** The capacity of a communications channel, as measured by the difference between the highest and the lowest frequencies that can be transmitted by that channel.

**behavioral targeting:** Tracking the clickstreams (history of clicking behavior) of individuals across multiple websites for the purpose of understanding their interests and intentions and exposing them to advertisements that are uniquely suited to their interests.

**benchmarking:** Involves comparing efficiency and effectiveness of business processes and measuring organizational performance against strict standards.

**best practices:** The most successful solutions or problem-solving methods for achieving a business objective that have been developed by a specific organization or industry.

**Big Data:** Data sets with volumes so huge that they are beyond the ability of typical relational DBMS to capture, store, and analyze. The data are often unstructured or semistructured.

**biometric authentication:** Technology for authenticating users that compares a person's unique characteristics, such as fingerprints, face, or retinal image, against a stored set profile of these characteristics.

**bit:** A binary digit representing the smallest unit of data in a computer system. It can have only one of two states, representing 0 or 1.

**blockchain:** Distributed ledger that stores a permanent and tamper-proof record of transactions and shares them among a distributed network of computers.

**blog:** Popular term for weblog, designating an informal yet structured website where individuals can publish stories, opinions, and links to other websites of interest.

**Bluetooth:** Standard for wireless personal area networks that can transmit up to 722 Kbps within a 10-meter area.

**botnet:** A group of computers that have been infected with bot malware without users' knowledge, enabling a hacker to use the amassed resources of the computers to launch distributed denial-of-service attacks, phishing campaigns, or spam.

**broadband:** High-speed transmission technology; also designates a single communications medium that can transmit multiple channels of data simultaneously.

**B2B e-commerce marketplace:** Provides a single digital marketplace based on Internet technology for many business buyers and sellers.

**bugs:** Software program code defects.

**bullwhip effect:** Distortion of information about the demand for a product as it passes from one entity to the next across the supply chain.

**business:** A formal organization whose aim is to produce products or provide services for a profit.

**business case:** Proposal to management seeking approval for an investment by describing its costs, benefits, and business value.

**business continuity planning:** Planning that focuses on how the company can restore business operations after a disaster strikes.

**business ecosystem:** Term used to describe loosely coupled interdependent networks of suppliers, distributors, outsourcing firms, transportation services firms, and technology manufacturers that provide related services and products that deliver value to the customer. Also referred to as industry sets.

**business intelligence (BI):** Applications and technologies to help users make more informed and thus better business decisions.

**business model:** Describes how an enterprise delivers a product or service, thus showing how the enterprise creates wealth.

**business process management (BPM):** Tools and methodologies for continuously improving and managing business processes.

**business process reengineering (BPR):** The radical redesign of business processes to maximize the benefits of information technology.

**business processes:** Logically related set of activities that define how specific business tasks are performed. Also refers to the unique ways in which organizations coordinate and organize work activities, information, and knowledge to produce a product or service.

**business-to-business (B2B) e-commerce:** Sales of goods and services among businesses via digital technology.

**business-to-consumer (B2C) e-commerce:** Selling products and services directly to individual consumers via digital technology.

**byte:** Group of bits that represents a single character, which can be a letter, number, or other symbol.

**BYOD:** Allowing employees to use their personal mobile devices in the workplace ("bring your own device").

**C:** A powerful programming language with tight control and efficient use of computer resources; used primarily by professional programmers to create operating systems and applications software.

**C++:** Newer version of C that has all the capabilities of C, plus additional features for working with software objects.

**cable Internet connections:** Using digital cable coaxial lines to deliver high-speed Internet access to homes and businesses.

**capacity planning:** The process of predicting when a computer hardware system will become saturated, in order to ensure that adequate computing resources are available for work of different priorities and that the firm has enough computing power for its current and future needs.

**capital budgeting:** The process of analyzing and selecting various proposals for capital expenditures.

**carpal tunnel syndrome (CTS):** Type of RSI in which pressure on the median nerve through the wrist's bony carpal tunnel structure produces pain.

**centralized processing:** Processing that is accomplished by one large central computer.

**change management:** Giving proper consideration to the impact of organizational change associated with a new system or alteration of an existing system.

**chatbot:** Software agent designed to simulate a conversation with one or more human users via textual or auditory methods.

**chief data officer (CDO):** Responsible for enterprise-wide governance and usage of information to maximize the value the organization can realize from its data.

**chief information officer (CIO):** Senior manager who oversees the use of information technology and information systems in the firm.

**chief knowledge officer (CKO):** Responsible for the firm's knowledge management program.

**chief privacy officer (CPO):** Responsible for ensuring that the company complies with data-privacy laws.

**chief security officer (CSO):** In charge of information systems security and responsible for enforcing the firm's information security policy.

**choice:** Simon's third stage of decision making, when the individual selects among the various solution alternatives.

**Chrome OS:** Google's lightweight, cloud-based operating system for Google's Chromebook laptop and tablet computers.

**churn rate:** Measurement of the number of customers who stop using or purchasing products or services from a company; used as an indicator of the growth or decline of a firm's customer base.

**client:** The user point of entry for the required function in client/server computing; normally a desktop computer, workstation, laptop computer or mobile device.

**client/server computing:** A model for computing that splits processing between clients and servers on a network, assigning functions to the machine that is most able to perform the function.

**cloud computing:** Model of computing that enables access to a shared pool of computing resources over a network, primarily the Internet.

**collaboration:** Working with others to achieve shared and explicit goals.

**community provider:** Online business model that involves creating a digital online environment in which people with similar interests can transact; share interests, photos, and videos; and receive interest-related information.

**competitive forces model:** Model used to describe the interaction of external influences—specifically, threats and opportunities—that affect an organization's strategy and ability to compete.

**computer abuse:** The commission of acts involving a computer that may not be illegal but are considered unethical.

**computer crime:** The commission of illegal acts through the use of a computer or against a computer system.

**computer forensics:** The scientific collection, examination, authentication, preservation, and analysis of electronically stored information (ESI) in such a way that the information can be used as evidence in a court of law.

**computer hardware:** Physical equipment used for input, processing, and output activities in an information system.

**computer literacy:** Knowledge about information technology, focusing on understanding of how computer-based technologies work.

**computer software:** Detailed, preprogrammed instructions that control and coordinate the work of computer hardware components in an information system.

**computer virus:** Rogue software program that attaches itself to other software programs or data files and activates, often causing hardware and software malfunctions.

**computer vision syndrome (CVS):** Eyestrain condition related to display screen use; symptoms include headaches, blurred vision, and dry and irritated eyes.

**computer vision systems:** Computer systems that try to emulate the human visual system to view and extract information from real-world images.

**computer-aided design (CAD) system:** Information system that automates the creation and revision of designs by using sophisticated graphics software.

**consumerization of IT:** New information technology originating in the consumer market that spreads to business organizations.

**consumer-to-consumer (C2C) e-commerce:** Consumers selling goods and services directly to other consumers.

**controls:** All of the methods, policies, and procedures that ensure protection of the organization's assets, accuracy and reliability of its records, and operational adherence to management standards.

**conversion:** The process of changing from the old system to the new system.

**cookies:** Small text files deposited on a computer or mobile device when a user visits certain websites; used to identify the visitor and track visits to the website.

**copyright:** Protects creators of intellectual property against copying by others for any purpose during the life of the author, plus an additional 70 years after the author's death; for corporate-owned works, extends for 95 years after initial creation.

**core competency:** Activity at which a firm is an industry leader.

**cost transparency:** The ability of consumers to discover the actual costs that merchants pay for products.

**critical thinking:** Sustained suspension of judgment with an awareness of multiple perspectives and alternatives.

**cross-selling:** Marketing complementary products to customers.

**crowdsourcing:** Using large Internet audiences for advice, market feedback, new ideas, and solutions to business problems; related to the wisdom-of-crowds theory.

**cryptocurrencies:** Currencies designed to work as a medium of exchange through a computer network that is not reliant on any central authority, such as a government or bank, to uphold or maintain it.

**culture:** Fundamental set of assumptions, values, and ways of doing things that has been accepted by most members of an organization.

**customer experience management:** Management of all the interactions between a customer and a company throughout their entire business relationship.

**customer lifetime value (CLTV):** Difference between revenues produced by a specific customer and the expenses for acquiring and servicing that customer minus the cost of promotional marketing over the lifetime of the customer relationship, expressed in today's dollars.

**customer relationship management (CRM) systems:** Information systems that track all the ways in which a company interacts with its customers and analyze these interactions to optimize revenue, profitability, customer satisfaction, and customer retention.

**customization:** The modification of a software package to meet an organization's unique requirements without destroying the package software's integrity. In e-commerce, changing a delivered product or service based on a user's preferences or prior behavior.

**cybervandalism:** Intentional disruption, defacement, or even destruction of a website or corporate information system.

**cyberwarfare:** State-sponsored activity designed to cripple and defeat another state or nation by damaging or disrupting its computers or networks.

**cycle time:** The total elapsed time from the beginning of a process to its end.

**dark web:** Portion of the deep web that has been intentionally hidden from search engines in order to preserve anonymity.

**data:** Streams of raw facts representing events occurring in organizations or the physical environment before they have been organized and arranged into a form that people can understand and use.

**data center:** Facility housing computer systems and associated components, such as telecommunications, storage and security systems, and backup power supplies.

**data cleansing:** Processes for detecting and correcting data in a database or file that are incorrect, incomplete, improperly formatted, or redundant. Also known as data scrubbing.

**data definition:** Specifies the structure of the content of a database.

**data dictionary:** Tool for storing and organizing information about the data maintained in a database.

**data governance:** Policies and procedures through which data can be managed as an organizational resource.

**data lake:** Repository for raw, unstructured data or structured data that for the most part have not yet been analyzed.

**data management software:** Software used for creating and manipulating lists, creating files and databases to store data, and combining information for reports.

**data management technology:** Software governing the organization of data on physical storage media.

**data manipulation language:** A language associated with a database management system that end users and programmers use to add, change, delete, and retrieve data in the database.

**data mart:** A subset of a data warehouse that contains only a portion of the organization's data for a specified function or population of users.

**data mining:** Analysis of large pools of data to find patterns and relationships that can be used to guide decision making and predict future behavior.

**data quality audit:** A structured survey and/or sample of files to determine accuracy and completeness of data in an information system.

**data visualization:** Technology for helping users see patterns and relationships in large amounts of data by presenting the data in graphical form.

**data warehouse:** A database, with reporting and query tools, that stores current and historical data extracted from various operational systems and consolidated for management reporting and analysis.

**data workers:** People such as clerks or bookkeepers who assist with administrative work at all levels of the firm.

**database:** A group of related files.

**database management system (DBMS):** Special software used to create and maintain a database and enable individual business applications to extract the data they need without having to create separate files or data definitions in their computer programs.

**database server:** A computer in a client/server environment that is responsible for running a DBMS to process SQL statements and perform database management tasks.

**decision-support systems (DSS):** Information systems at the organization's management level that combine data and sophisticated analytical models or data analysis tools to support semistructured and unstructured decision making.

**"deep learning":** Using multiple layers of neural networks to reveal the underlying patterns in data and, in some limited cases, to identify patterns without human training.

**demand planning:** Determining how much product a business needs to make to satisfy all its customers' demands.

**denial-of-service (DoS) attack:** Flooding a network server or web server with false communications or requests for services to crash the network.

**design:** Simon's second stage of decision making, when the individual conceives of possible alternative solutions to a problem.

**digital certificates:** Method of establishing the identity of users and digital assets.

**digital dashboard:** Displays all of a firm's key performance indicators as graphs and charts on a single screen to provide a one-page overview of all the critical measurements necessary to make key executive decisions.

**digital divide:** Disparities in access to computers and the Internet among different social groups and different locations.

**digital goods:** Goods that can be delivered over a digital network.

**Digital Millennium Copyright Act (DMCA):** Makes it illegal to make, distribute, or use devices that circumvent technology-based protections of copyrighted materials.

**digital resiliency:** An effort to maintain and increase the resilience of an organization and its business processes in an all-pervasive digital environment, not just the resiliency of the IT function.

**digital subscriber line (DSL):** A group of technologies providing high-capacity transmission over existing telephone lines.

**direct cutover strategy:** A risky conversion approach by which the new system completely replaces the old one on an appointed day.

**direct goods:** Goods used in a production process.

**disaster recovery planning:** Planning for the restoration of computing and communications services after they have been disrupted.

**disintermediation:** The removal of organizations or business process layers responsible for certain intermediary steps in a value chain.

**disruptive technologies:** Technologies with disruptive impact on industries and businesses, rendering existing products, services, and business models obsolete.

**distributed database:** Database stored in multiple physical locations.

**distributed denial-of-service (DDoS) attack:** Attack that uses numerous computers to inundate and overwhelm a network from many launch points.

**distributed processing:** The distribution of computer processing work among multiple computers linked by a communications network.

**documentation:** Descriptions of how an information system works from either a technical or an end-user standpoint.

**domain name:** Natural language name that corresponds to the unique numeric Internet Protocol (IP) address for each computer connected to the Internet.

**Domain Name System (DNS):** A hierarchical system of servers maintaining a database enabling the conversion of domain names to their numeric IP addresses.

**domestic exporter:** Form of business organization characterized by heavy centralization of corporate activities in the home country of origin.

**download:** involves transferring a file from a web server to the user's device, where it resides and can be accessed by the user at any time

**downtime:** Period of time in which an information system is not operational.

**drill-down:** The ability to move from summary data to increasingly granular levels of detail.

**drive-by download:** Malware that comes with a downloaded file that a user unintentionally opens.

**due process:** A process by which laws are well-known and understood and provide an ability to appeal to higher authorities to ensure that laws are applied correctly.

**dynamic pricing:** Pricing of items based on real-time interactions between buyers and sellers that determine what an item is worth at any particular moment.

**edge computing:** Method of optimizing cloud computing systems by performing some data processing on a set of linked servers at the edge of the network, near the source of the data.

**efficient customer response system:** System that directly links consumer behavior to distribution, production, and supply chains.

**e-government:** Use of the Internet and related technologies to enable government and public sector agencies' relationships with citizens, businesses, and other arms of government digitally.

**electronic business (e-business):** The use of the Internet and digital technology to execute all the business processes in the enterprise; includes e-commerce as well as processes for the internal management of the firm and coordination with suppliers and other business partners.

**electronic commerce (e-commerce):** The process of buying and selling goods and services by using the Internet, networks, and other digital technologies.

**Electronic Data Interchange (EDI):** The direct computer-to-computer exchange between two organizations of standard business transactions, such as orders, shipment instructions, or payments.

**email:** The computer-to-computer exchange of messages.

**employee relationship management (ERM):** Software dealing with employee issues that are closely related to CRM, such as setting objectives, employee performance management, performance-based compensation, and employee training.

**encryption:** The coding and scrambling of messages to prevent them from being read or accessed without authorization.

**end users:** Representatives of departments outside the information systems group for whom applications are developed.

**end-user development:** The development of information systems by end users with little or no formal assistance from technical specialists.

**end-user interface:** The part of an information system through which the end user interacts with the system, such as online screens and commands.

**enterprise applications:** Systems that can coordinate activities, decisions, and knowledge across many functions, levels, and business units in a firm; include enterprise systems, supply chain management systems, customer relationship management systems, and knowledge management systems.

**enterprise content management (ECM) systems:** Systems that help organizations manage structured, semistructured, and unstructured types of information, providing corporate repositories of documents, reports, presentations, and best practices and capabilities.

**enterprise software:** Software to support major business processes such as finance and accounting, human resources, manufacturing and production, and sales and marketing.

**enterprise systems:** Integrated, enterprise-wide information systems that coordinate key internal processes of the firm. Also known as enterprise resource planning (ERP) systems.

**entity:** A person, place, thing, or event about which information is kept.

**entity-relationship diagram:** A methodology for documenting databases that illustrates the relationship among various entities in the database.

**ergonomics:** The interaction of people and machines in the work environment, including the design of jobs, health issues, and the end-user interface of information systems.

**ethical no-free-lunch rule:** Assumption that all tangible and intangible objects are owned by someone else, unless there is a specific declaration otherwise, and that the creator wants compensation for this work.

**ethics:** Principles of right and wrong that can be used by individuals, acting as free moral agents, to make choices to guide their behavior.

**evil twins:** Wireless networks that pretend to be legitimate Wi-Fi networks to entice participants to log on and reveal passwords or credit card numbers.

**exchanges:** Third-party B2B e-commerce marketplaces that are primarily transaction-oriented and that connect many buyers and suppliers for spot purchasing.

**executive support systems (ESS):** Information systems at the organization's strategic level designed to address unstructured decision making, providing graphs and data from many sources through an interface that is easy for senior managers to use.

**expert systems:** Computer programs that capture the expertise of a human in limited domains of knowledge and represent knowledge as a set of rules.

**Extensible Markup Language (XML):** A more powerful and flexible markup language than hypertext markup language (HTML) for web pages, allowing data to be manipulated by the computer.

**extranets:** Private intranets that are accessible to authorized outsiders.

**Fair Information Practices (FIP):** A set of principles originally set forth in 1973 that governs the collection and use of information about individuals and forms the basis of most US and European privacy laws.

**fault-tolerant computer systems:** Systems that contain extra hardware, software, and power supply components that can back a system up and keep it running to prevent system failure.

**feasibility study:** As part of the systems analysis process, the way to determine whether the solution is achievable, given the organization's resources and constraints.

**feedback:** Output that is returned to the appropriate members of the organization to help them evaluate or correct input.

**field:** A grouping of characters into a word, a group of words, or a complete number, such as a person's name or age.

**file:** Group of records of the same type.

**File Transfer Protocol (FTP):** Tool for retrieving and transferring files from a remote computer.

**fintech:** Startup, innovative financial technology firms and services.

**firewalls:** Hardware and software placed between an organization's internal network and an external network to prevent outsiders from invading private networks.

**foreign key:** Field in a database table that enables users to find related information in another database table.

**formal planning and control tools:** Tools to improve project management by listing the specific activities that make up a project, their duration, and the sequence and timing of tasks.

**franchiser:** Form of business organization in which a product is created, designed, financed, and initially produced in the home country but, for product-specific reasons, relies heavily on foreign personnel for further production, marketing, and human resources.

**free/freemium revenue model:** E-commerce revenue model in which a firm offers free basic services or content while charging a premium for advanced or high-value features.

**Gantt chart:** Chart that visually represents the timing, duration, and human resource requirements of project tasks, with each task represented as a horizontal bar whose length is proportional to the time required to complete it.

**general controls:** Overall control environment governing the design, security, and use of computer programs and the security of data files in general throughout the organization's information technology infrastructure.

**General Data Protection Regulation (GDPR):** Updated framework adopted by the European Union that unifies data privacy laws across the European Union, focusing on making businesses more transparent and expanding the privacy rights of data subjects.

**genetic algorithms:** Form of machine learning that promotes the evolution of solutions to specified problems using the model of living organisms adapting to their environment.

**geoadvertising services:** Delivering ads to users based on their GPS location.

**geographic information systems (GIS):** Systems with software that can analyze and display data using digitized maps to enhance planning and decision making.

**geoinformation services:** Information on local places and things based on the GPS position of the user.

**geosocial services:** Social networking based on the GPS location of users.

**Golden Rule:** Putting oneself in the place of others as the object of a decision.

**Gramm-Leach-Bliley Act:** Requires financial institutions to ensure the security and confidentiality of customer data.

**graphical user interface:** The part of an operating system that users interact with that uses graphic icons and the computer mouse to issue commands and make selections.

**green computing (green IT):** Practices and technologies for producing, using, and disposing of computers and associated devices to minimize impact on the environment.

**grid computing:** Applying the resources of many computers in a network to a single problem.

**hacker:** A person who gains unauthorized access to a computer network for profit, criminal mischief, or personal pleasure.

**Hadoop:** Open-source software framework that enables distributed parallel processing of huge amounts of data across many inexpensive computers.

**hertz:** Measure of frequency of electrical impulses per second, with 1 hertz (Hz) equivalent to 1 cycle per second.

**HIPAA:** Law outlining medical security and privacy rules and procedures for simplifying the administration of healthcare billing and automating the transfer of healthcare data among healthcare providers, payers, and plans.

**hotspots:** Specific geographic locations in which an access point provides public Wi-Fi network service.

**HTML5:** Most recent version of HTML; makes it possible to embed images, video, and audio directly into a document without using add-on software.

**HTTPS:** Secure version of the HTTP protocol that uses TLS for encryption and authentication.

**hubs:** Very simple devices that connect network components, sending a packet of data to all other connected devices.

**hybrid app:** Mobile app that has features of both a mobile web app and a native app.

**hybrid cloud:** Cloud computing model in which firms use both their own IT infrastructure and public cloud computing services.

**Hypertext Markup Language (HTML):** Page description language for creating web pages.

**Hypertext Transfer Protocol (HTTP):** The communications standard that transfers pages on the web. Defines how messages are formatted and transmitted.

**identity and access management (IAM):** Business processes and software tools for identifying the valid users of a system and controlling their access to system resources.

**identity theft:** Theft of key pieces of personal information, such as credit card or social security numbers, to obtain merchandise and services in the name of the victim or to obtain false credentials.

**Immanuel Kant's categorical imperative:** A principle that states that if an action is not right for everyone to take, it is not right for anyone to take.

**implementation:** All the organizational activities surrounding the adoption, management, and regular reuse of an innovation, such as a new information system. When used with respect to Simon's model of the decision-making process, Simon's final stage of decision making, when the individual puts the decision into effect and reports on the progress of the solution.

**indirect goods:** Goods not directly used in the production process, such as office supplies.

**inference engine:** The strategy used to search through the rule base in an expert system; can be forward or backward chaining.

**information:** Data that have been shaped into a form that is meaningful and useful to human beings.

**information asymmetry:** Situation when the relative bargaining power of two parties in a transaction is determined by one party in the transaction possessing more information essential to the transaction than the other party.

**information density:** The total amount and quality of information available to all market participants, consumers, and merchants.

**information requirements:** A detailed statement of the information needs that a new system must satisfy; identifies who needs what information and when, where, and how the information is needed.

**information rights:** The rights that individuals and organizations have with respect to information that pertains to them.

**information system:** Interrelated components working together to collect, process, store, and distribute information to support decision making, coordination, control, analysis, and visualization in an organization.

**information systems audit:** Identifies all the controls that govern individual information systems and assesses their effectiveness.

**information systems department:** The formal organizational unit that is responsible for the information technology services in the organization.

**information systems literacy:** Broad-based understanding of information systems that includes behavioral knowledge about organizations and individuals as well as technical knowledge about computers.

**information systems managers:** Leaders of the various specialists in the information systems department.

**information systems plan:** A roadmap indicating the direction of systems development: the rationale, the current situation, the management strategy, the implementation plan, and the budget.

**information technology (IT):** All the hardware and software technologies that a firm needs to achieve its business objectives.

**information technology (IT) infrastructure:** Computer hardware, software, data, storage technology, and networks, along with the people required to run and manage them, providing a portfolio of shared IT resources for the organization.

**informed consent:** Consent given with knowledge of all the facts needed to make a rational decision.

**in-memory computing:** Technology for very rapid analysis and processing of large quantities of data by storing the data in the computer's main memory (RAM) rather than in secondary storage.

**input:** The capture or collection of raw data from within the organization or from its external environment for processing in an information system.

**input devices:** Devices that gather data and convert them into digital form for use by the computer.

**instant messaging:** Chat service that allows participants to create their own private chat channels so that they can be alerted whenever someone on their private list is online to initiate a chat session with that particular individual.

**intangible benefits:** Benefits that are not easily quantified, such as more efficient customer service or enhanced decision making.

**intellectual property:** Products of the human mind created by individuals or corporations that are subject to protections under trade secret, copyright, trademarks, and patent law.

**intelligence:** The first of Simon's four stages of decision making, when the individual collects information to identify problems occurring in the organization.

**intelligent agents:** Software programs that use a built-in or learned knowledge base to carry out specific, repetitive, and predictable tasks for an individual user, business process, or software application.

**intelligent techniques:** Another name for artificial intelligence.

**Internet:** Global network of networks using universal standards to connect millions of networks.

**Internet2:** Research network with new protocols and transmission speeds that provides an infrastructure for supporting high-bandwidth Internet applications.

**Internet of Things (IoT):** Network of physical objects (things) embedded with sensors, software, and other technologies enabling them to connect and exchange data with other devices and systems via the Internet.

**Internet Protocol (IP) address:** Four-part numeric address indicating a unique computer location on the Internet.

**Internet service provider (ISP):** A commercial organization with a permanent connection to the Internet that sells temporary connections to subscribers.

**interorganizational system:** Information system that automates the flow of information across organizational boundaries and links a company to its customers, distributors, or suppliers.

**intranets:** Internal networks based on Internet technology and standards.

**intrusion detection systems (IDS):** Tools to monitor the most vulnerable points in a network to detect and deter unauthorized intruders.

**intrusion prevention systems (IPS):** System that has all the functionality of an IDS, as well as the additional ability to take steps to prevent and block suspicious activities.

**iOS:** Operating system for the Apple iPhone, iPad, and Apple Watch.

**IPv6:** New IP addressing system using 128-bit IP addresses. Stands for Internet Protocol version 6.

**Java:** An operating system–independent, processor-independent, object-oriented programming language that has become a leading interactive programming environment for the web.

**JavaScript:** Core technology for making web pages more interactive. Used for implementing dynamic features on web pages.

**just-in-time strategy:** Scheduling system for minimizing inventory by having components arrive exactly at the moment they are needed and finished goods shipped as soon as they leave the assembly line.

**key field:** A field in a record that uniquely identifies instances of that record so that it can be retrieved, updated, or sorted.

**key performance indicators (KPIs):** Measures proposed by senior management for understanding how well the firm is performing along specified dimensions.

**keyloggers:** Spyware that records every keystroke made on a computer.

**knowledge base:** Model of human knowledge in the form of a set of rules that expert systems use.

**knowledge management systems (KMS):** Systems that support the creation, capture, storage, and dissemination of firm expertise and knowledge.

**knowledge workers:** People such as engineers, scientists, or architects who design products or services and create knowledge for the organization.

**legacy systems:** Systems that have been in existence for a long time and that continue to be used to avoid the high cost of replacing or redesigning them.

**liability:** Laws that permit individuals to recover the damages done to them by other actors, systems, or organizations.

**Linux:** Reliable and compactly designed operating system that is an open-source offshoot of UNIX, can run on many hardware platforms, and is available free or at very low cost.

**local area network (LAN):** A telecommunications network that requires its own dedicated channels and that encompasses a limited distance, usually one building or several buildings in close proximity.

**location analytics:** Ability to gain insight from the location (geographic) component of data, including location data from mobile phones, output data from sensors or scanning devices, and data from maps.

**location-based services:** Include services enabled by GPS map services available on smartphones, such as geosocial services, geoadvertisng services, and geoinformation services.

**long tail marketing:** Ability of firms to market goods profitably to very small online audiences, largely because of the lower costs of reaching very small market segments.

**low-code development:** Software development approach that enables the creation of workable software with minimal hand coding.

**machine learning (ML):** Software that can identify patterns and relationships in very large data sets without explicit programming, although with significant human training.

**mainframe:** Large-capacity, high-performance computer that can process large amounts of data very rapidly.

**maintenance:** Changes in hardware, software, documentation, or procedures that are made to a production system to correct errors, meet new requirements, or improve processing efficiency.

**malware:** Malicious software programs such as computer viruses, worms, and Trojan horses.

**managed security service providers (MSSPs):** Companies that provide security management services for subscribing clients.

**management information systems (MIS):** The study of information systems, focusing on their use in business and management. Also refers to the specific category of information system providing reports on organizational performance to help middle management monitor and control the business.

**market creator:** E-commerce business model in which firms provide a digital online environment where buyers and sellers can meet, search for products, and engage in transactions.

**market entry costs:** The cost merchants must pay simply to bring their goods to market.

**marketspace:** A marketplace extended beyond traditional boundaries and removed from a temporal and geographic location.

**mashup:** Software application that integrates functionality and data from multiple applications to create a new, customized application.

**mass customization:** The capacity to offer individually tailored products or services on a large scale.

**menu costs:** Merchants' costs of changing prices.

**metaverse:** Anticipated future iteration of the web featuring online 3-D or virtually integrated environments that provide users access to virtual reality and augmented reality experiences.

**metropolitan area network (MAN):** Network that spans a metropolitan area, usually a city and its major suburbs. Its geographic scope falls between a WAN and a LAN.

**Microsoft 365:** Hosted cloud version of productivity and collaboration tools, such as the Microsoft Office product line, available as a subscription service.

**middle management:** People in the middle of the organizational hierarchy who are responsible for carrying out the plans and goals of senior management.

**mobile device management (MDM):** Software that monitors, manages, and secures mobile devices that are deployed across multiple mobile service providers and multiple mobile operating systems used in the organization.

**mobile e-commerce (m-commerce):** The use of mobile devices to purchase goods and services from any location via the Internet.

**mobile web app:** Application residing on a server and accessed through the mobile web browser built into a smartphone or tablet computer.

**mobile website:** Version of a regular website that is scaled down in content and navigation for easy access and search on a small mobile screen.

**modem:** A device for translating a computer's digital signals into analog form or for translating analog signals back into digital form for reception by a computer.

**multicore processor:** Integrated circuit to which two or more processors have been attached for enhanced performance, reduced power consumption, and more efficient simultaneous processing of multiple tasks.

**multifactor authentication (MFA):** Tools that increase security by validating users via a multistep process.

**multinational:** Form of business organization that concentrates financial management and control out of a central home base while decentralizing production, sales, and marketing.

**multitouch:** Interface that features the use of one or more finger gestures to manipulate lists or objects on a screen without using a mouse or keyboard.

**nanotechnology:** Technology that builds structures and processes based on the manipulation of individual atoms and molecules.

**native advertising:** Placing ads within social network newsfeeds or traditional editorial content, such as a newspaper article.

**native app:** Standalone application specifically designed to run on a mobile platform.

**natural language processing:** Technology that makes it possible for a computer to understand spoken or written words expressed in human (natural) language and to process that information.

**near field communication (NFC):** Short-range wireless connectivity standard that uses electromagnetic radio fields to enable two compatible devices to exchange data when brought within a few centimeters of each other.

**network:** The linking of two or more computers to share data or resources such as a printer.

**network economics:** Model of strategic systems at the industry level that is based on the concept of a network when adding another participant entails zero marginal costs but can create much larger marginal gains.

**network operating system (NOS):** Special software that routes and manages communications on the network and coordinates network resources.

**networking and telecommunications technology:** Physical devices and software that link various pieces of hardware and transfer data from one physical location to another.

**neural network:** Algorithms loosely based on the processing patterns of the biological brain and that can be trained to classify objects into known categories based on data inputs.

**no-code development:** Platform for creating apps without writing any program code.

**nonobvious relationship awareness (NORA):** Technology that can find obscure connections among people or other entities by analyzing information from many sources to correlate relationships.

**nonrelational database management system:** Database management system for working with large quantities of structured and unstructured data that would be difficult to analyze with a relational model. Also referred to as NoSQL.

**normalization:** When designing a relational database, the process of streamlining complex groups of data to minimize redundant data elements and awkward many-to-many relationships.

**n-tier client/server architecture:** Client/server arrangement that balances the work of the entire network over multiple levels of servers.

**object:** Software building block that combines data and the procedures acting on the data.

**offshore software outsourcing:** Outsourcing systems development work or maintenance of existing systems to external vendors in another country.

**on-demand computing:** Firms off-loading peak demand for computing power to remote, large-scale data processing centers, investing just enough to handle average processing loads and paying for only as much additional computing power as they need. Also called utility computing.

**online analytical processing (OLAP):** Capability for manipulating and analyzing large volumes of data from multiple perspectives.

**online retailer:** Online retail stores, from the giant Amazon to tiny local stores, that have websites where retail goods are sold.

**online transaction processing:** Transaction processing mode in which transactions entered online are immediately processed by the computer.

**open-source software:** Software that provides free access to its program code, allowing users to modify the program code in order to make improvements or fix errors.

**operating system:** The system software that manages and controls the activities of the computer.

**operational CRM:** Customer-facing applications, such as sales force automation, call center and customer service support, and marketing automation.

**operational intelligence:** Business analytics that deliver insight into data, streaming events, and business operations.

**operational management:** People who monitor the day-to-day activities of the organization.

**opt-in:** Model of informed consent prohibiting an organization from collecting any personal information unless the individual specifically takes action to approve information collection and use.

**opt-out:** Model of informed consent permitting the collection of personal information until the consumer specifically requests that the data not be collected.

**organizational impact analysis:** Study of the way a proposed system will affect organizational structure, attitudes, decision making, and operations.

**output:** The distribution of processed information to the people who will use it or to the activities for which it will be used.

**output devices:** Devices that display data after they have been processed.

**outsourcing:** The practice of contracting computer center operations, telecommunications networks, applications development, or management of all components of an IT infrastructure, to external vendors.

**packet switching:** Technology that breaks messages into small, fixed bundles of data and routes them in the most economical way through any available communications channel.

**parallel strategy:** A safe and conservative conversion approach in which both the old system and its potential replacement are run together for a time until everyone is assured that the new one functions correctly.

**partner relationship management (PRM):** Automation of the firm's relationships with its selling partners using customer data and analytical tools to improve coordination and customer sales.

**password:** Secret word or string of characters for authenticating users so that they can access a resource such as a computer system.

**patches:** Software that repairs flaws in programs without disturbing the proper operation of the software.

**patent:** Grants the owner an exclusive monopoly on the ideas behind an invention for a certain period of years; designed to ensure that inventors of new machines or methods are rewarded for their labor while making available the widespread use of their inventions.

**peer-to-peer:** Network architecture that gives equal power to all computers on the network; used primarily in small networks.

**personal area networks (PANs):** Computer networks used for communication among digital devices that are located within close proximity to one another.

**personal computer (PC):** Desktop or laptop computer.

**personalization:** Ability of merchants to target their marketing messages to specific individuals by adjusting the message to a person's name, interests, and past purchases.

**PERT chart:** A chart that graphically depicts project tasks and their interrelationships, showing the specific activities that must be completed before others can start.

**pharming:** Phishing technique that redirects users to a bogus web page even when the users type the correct web page address into their browser.

**phased approach:** Introduces the new system in stages either by functions or by organizational units.

**phishing:** A form of spoofing involving setting up fake websites or sending email messages that look like those of legitimate businesses to ask users for confidential personal data.

**pilot study:** A strategy to introduce the new system to a limited area of the organization until it proves to be fully functional; only then can the conversion to the new system across the entire organization take place.

**pivot table:** Spreadsheet tool for reorganizing and summarizing two or more dimensions of data in a tabular format.

**platforms:** Businesses providing information systems, technologies, and services that thousands of other firms in different industries use to enhance their own capabilities.

**podcasting:** Method of publishing audio broadcasts through the Internet, allowing subscribing users to download audio files to their personal computers or mobile devices.

**portal:** Web interface for presenting integrated, personalized content from a variety of sources. Also refers to a website service that provides an initial point of entry to the web.

**portfolio analysis:** An analysis of the portfolio of potential applications within a firm to determine the risks and benefits and to select among alternatives for information systems.

**predictive analytics:** Use of data-mining techniques, historical data, and assumptions about future conditions to predict future trends and behavior patterns.

**predictive search:** Part of a search algorithm that predicts what a user query is looking for as it is entered, based on popular searches.

**presentation graphics:** Software to create professional-quality graphics presentations that can incorporate charts, sound, animation, photos, and video clips.

**price discrimination:** Selling the same goods, or nearly the same goods, to different targeted groups at different prices.

**price transparency:** The ease with which consumers can find out the variety of prices in a market.

**primary activities:** Activities most directly related to the production and distribution of a firm's products or services.

**primary key:** Unique identifier for all the information in any row of a database table.

**privacy:** The claim of individuals to be left alone, free from surveillance or interference from other individuals, organizations, or the state.

**private B2B networks:** Internet-enabled networks linking systems of multiple firms in an industry for the coordination of trans-organizational business processes.

**private cloud:** Cloud computing model in which the cloud infrastructure is operated solely for the benefit of a single organization.

**processing:** The conversion, manipulation, and analysis of raw input into a form that is more meaningful to humans.

**product differentiation:** Creating new products and services or enhancing customer convenience in using existing products and services.

**production:** The stage after the new system is installed and the conversion is complete; during this time, the system is reviewed by users and technical specialists to determine how well it has met its original goals.

**production or service workers:** People who actually produce the products or services of the organization.

**profiling:** The use of computers to combine data from multiple sources and create digital dossiers of detailed information on individuals.

**programmers:** Highly trained technical specialists who write computer software instructions.

**project:** A planned series of related activities for achieving a specific business objective.

**project management:** Application of knowledge, skills, tools, and techniques to achieve specific targets within specified budget and time constraints.

**protocol:** A set of rules and procedures that govern transmission among the components in a network.

**prototyping:** The process of building an experimental system quickly and inexpensively for demonstration and evaluation so that users can better determine information requirements.

**public cloud:** Cloud computing model in which the cloud infrastructure is maintained by a cloud service provider and made available to outside organizations and/or the general public.

**public key encryption:** Encryption using two keys: one shared (or public) and one private.

**public key infrastructure (PKI):** System for creating public and private keys by using a certificate authority (CA) and digital certificates for authentication.

**pull-based model:** Supply chain driven by actual customer orders or purchases so that members of the supply chain produce and deliver only what customers have ordered. Also known as demand-driven or build-to-order model.

**push-based model:** Supply chain driven by production schedules that are driven by forecasts or best guesses of demand for products; products are pushed to customers. Also known as build-to-stock model.

**quality:** From standpoint of a product, signifies a product's or service's conformance to specifications and standards.

**quantum computing:** Use of principles of quantum physics to represent data and perform operations on the data, with the ability to be in many states at once and to perform many computations simultaneously.

**query:** Request for data from a database.

**radio frequency identification (RFID):** Technology using tiny tags with embedded microchips containing data about an item and its location; the tags transmit short-distance radio signals to special RFID readers that then pass the data on to a computer for processing.

**ransomware:** Malware that extorts money from users by taking control of their computers, blocking access to files, or displaying annoying pop-up messages.

**rapid application development (RAD):** Process for developing systems in a very short time period by using prototyping, user-friendly tools, and close teamwork among users and systems specialists.

**record:** Group of related fields.

**referential integrity:** Rules to ensure that relationships among coupled database tables remain consistent.

**relational database:** A type of logical database model that organizes data into two-dimensional tables with columns and rows. It can relate data stored in one table to data in another table as long as the two tables share a common data element.

**repetitive stress injury (RSI):** Occupational disease that occurs when muscle groups are forced through repetitive actions with high-impact loads or thousands of repetitions with low-impact loads.

**report generator:** Computer program that can take data from a DBMS to generate reports that display data of interest in a more structured and polished format than would be possible just by querying.

**Request for Proposal (RFP):** A detailed list of questions submitted to vendors of software or other services to determine how well the vendor's product can meet the organization's specific requirements.

**responsibility:** Accepting the potential costs, duties, and obligations for the decisions one makes.

**responsive web design:** Ability of a website to change screen resolution and image size automatically as a user switches to devices of different sizes, such as a laptop, tablet computer, or smartphone. Eliminates the need for separate design and development work for each new device.

**revenue model:** Description of how a firm will earn revenue, generate profits, and produce a return on investment.

**richness:** Measurement of the depth and detail of information that a business can supply to the customer as well as information the business collects about the customer.

**risk assessment:** Determining the potential frequency of the occurrence of a problem and the potential damage if the problem were to occur. Used to determine the cost/benefit of a control.

**risk aversion principle:** Principle that one should take the action that produces the least harm or incurs the least cost.

**robotics:** Use of machines that can substitute for human movements as well as computer systems for their control, sensory feedback, and information processing.

**router:** Specialized communications processor that forwards packets of data from one network to another network.

**RSS:** Technology using aggregator software to pull content from websites and feed it automatically to subscribers' computers.

**safe harbor:** Private, self-regulating policy and enforcement mechanism that meets the objectives of government regulations but does not involve government regulation or enforcement.

**sales force automation (SFA):** CRM system module that helps staff increase productivity by focusing sales efforts on the most profitable customers.

**sales revenue model:** Selling goods, information, or services to customers as the main source of revenue for a company.

**Sarbanes-Oxley Act:** Law that imposes responsibility on companies and their management to protect investors by safeguarding the accuracy and integrity of financial information that is used internally and released externally.

**scalability:** The ability of a computer, product, or system to expand to serve a larger number of users without breaking down.

**scope:** Defines what work is or is not included in a project.

**scoring model:** A quick method for deciding among alternative systems based on a system of ratings for selected objectives.

**search costs:** The time and money spent locating a suitable product and determining the best price for that product.

**search engine:** Tool for locating specific sites or information on the Internet.

**search engine marketing:** Use of search engines to deliver sponsored links, for which advertisers have paid, in search engine results.

**search engine optimization (SEO):** Process of changing a website's content, layout, and format to increase the site's ranking on popular search engines and to generate more site visitors.

**security:** Policies, procedures, and technical measures used to prevent unauthorized access, alteration, theft, or physical damage to information systems.

**security policy:** Statements ranking information risks, identifying acceptable security goals, and identifying the mechanisms for achieving these goals.

**security token:** Physical device, similar to an identification card, designed to prove the identity of a single user.

**semantic search:** Search technology capable of understanding human language and behavior.

**semistructured decisions:** Decisions in which only part of the problem has a clear-cut answer provided by an accepted procedure.

**semistructured knowledge:** Information in the form of less structured objects such as email.

**senior management:** People occupying the top of an organization's hierarchy and who are responsible for making long-range decisions for the organization.

**sensitivity analysis:** Models that ask what-if questions repeatedly to determine the impact on the outcomes of changes in one or more factors.

**sentiment analysis:** Mining text comments in an email message, blog, or other social medium.

**server:** Computer specifically optimized to provide software and other resources to other computers over a network.

**service level agreement (SLA):** Formal contract between customers and their service providers that defines the specific responsibilities of the service provider and the level of service the customer expects.

**service-oriented architecture (SOA):** Software architecture of a firm built on a collection of software programs that communicate with each other to perform assigned tasks to create a working software application.

**shopping bots:** Software that uses intelligent agent software to help online shoppers locate and evaluate products or services they might wish to purchase.

**Six Sigma:** A specific measure of quality, representing 3.4 defects per million opportunities; used to designate a set of methodologies and techniques for improving quality and reducing costs.

**slippery slope rule:** Ethical principle that states that if an action cannot be taken repeatedly, it is not right to take it at all.

**smart card:** A credit-card-size plastic card that stores digital information and can be used for electronic payments in place of cash.

**smartphones:** Mobile devices that combine the functionality of a cell phone with that of a Wi-Fi-enabled laptop computer; provide voice, messaging, scheduling, email, and Internet capabilities.

**sniffer:** A type of eavesdropping program that monitors information traveling over a network.

**social business:** Use of social network platforms, such as Facebook, Twitter, and internal corporate social tools, to engage employees, customers, and suppliers.

**social CRM:** Tools enabling a business to link customer conversations, data, and relationships from social networks to CRM processes.

**social e-commerce:** Use of social networks to share knowledge about items of interest to other shoppers and, increasingly, to enable purchases directly via the social network.

**social engineering:** Tricking people into revealing their passwords by pretending to be legitimate users or members of a company in need of information.

**social graph:** Map of all significant online social relationships, comparable to a social network describing offline relationships.

**social listening:** Refers to the ability of marketers to understand what people are saying about a brand as well as its competitors.

**social networks:** Online community that enables users to connect with and expand their social or business contacts.

**software as a service (SaaS):** Software hosted by a vendor on the vendor's cloud infrastructure and delivered as a service on a subscription fee basis to end users who access them via a web browser.

**software localization:** Process of converting software so that it can operate in a second language.

**software package:** A prewritten, precoded, commercially available set of programs that eliminates the need to write software programs for certain functions.

**software-defined networking (SDN):** Using a central control program separate from network devices to manage the flow of data on a network.

**software-defined storage (SDS):** Software to manage provisioning and management of data storage independent of the underlying hardware.

**solid state drive (SSD):** Storage device that stores data on an array of semiconductor memory organized as a disk drive.

**spam:** Unsolicited commercial email.

**spoofing:** Tricking or deceiving computer systems or other computer users by hiding one's identity or faking the identity of another user on the Internet.

**spreadsheet software:** Software displaying data in a grid of columns and rows, with the capability of easily recalculating numerical data.

**spyware:** Technology that aids in gathering information about a person or organization without their knowledge.

**SQL injection attack:** Attack against a website that takes advantage of vulnerabilities in poorly coded SQL applications to introduce malicious program code into a company's systems and networks.

**streaming:** Method of distributing music and video as a continuous stream of content to a user's device without it having to be stored locally on the device.

**structured decisions:** Decisions that are repetitive and routine and that have a definite procedure for handling them.

**structured knowledge:** Knowledge in the form of formal text documents and reports.

**Structured Query Language (SQL):** The standard data manipulation language for relational database management systems.

**subscription revenue model:** E-commerce revenue model based on charging a subscription fee for access to some or all of its content or services on an ongoing basis.

**supercomputer:** Highly sophisticated and powerful computer that can perform very complex computations extremely rapidly.

**supervised learning:** Machine learning algorithm trained by providing specific examples of desired inputs and outputs classified by humans in advance.

**supply chain:** Network of organizations and business processes for procuring materials, transforming raw materials into intermediate and finished products, and distributing the finished products to customers.

**supply chain execution systems:** Systems to manage the flow of products through distribution centers and warehouses to ensure that products are delivered to the right locations in the most efficient manner.

**supply chain management (SCM) systems:** Information systems that automate the flow of information between a firm and its suppliers to optimize the planning, sourcing, manufacturing, and delivery of products and services.

**supply chain planning systems:** Systems that enable a firm to generate demand forecasts for a product and develop sourcing and manufacturing plans for that product.

**support activities:** Activities that make the delivery of a firm's primary activities possible; consist of the organization's infrastructure, human resources, technology, and procurement.

**switch:** Device to connect network components that has more intelligence than a hub and can filter and forward data to a specified destination.

**switching costs:** The expense a customer or company incurs in lost time and expenditure of resources when changing from one supplier or system to a competing supplier or system.

**system software:** Generalized programs that manage the computer's activities and resources, such as the central processor, communications links, and peripheral devices.

**system testing:** Tests the functioning of the information system as a whole to determine whether discrete modules will function together as planned.

**systems analysis:** The analysis of a problem that the organization will try to solve with an information system.

**systems analysts:** Specialists who translate business problems and requirements into information requirements and systems, acting as liaisons between the information systems department and the rest of the organization.

**systems design:** Details how a system will meet the information requirements as determined by the systems analysis.

**systems development life cycle (SDLC):** A traditional methodology for developing an information system that partitions the systems development process into formal stages that must be completed sequentially and that has a very formal division of labor between end users and information systems specialists.

**T1 lines:** High-speed data lines leased from communications providers that provide guaranteed transmission capacity at 1.5 Mbps.

**T3 lines:** High-speed data lines leased from communications providers that provide guaranteed transmission capacity at 45 Mbps.

**tablet computer:** Mobile handheld computer that is larger than a mobile phone and operated primarily by touching a flat screen.

**tacit knowledge:** Expertise and experience of organizational members that have not been formally documented.

**tangible benefits:** Benefits that can be quantified and assigned a monetary value, such as lower operational costs and increased cash flows.

**teams:** Formal groups whose members collaborate to achieve specific goals.

**telepresence:** Technology that allows a person to give the appearance of being present at a location other than his or her true physical location.

**Telnet:** Enables remote login on another computer.

**test plan:** Plan prepared by the development team in conjunction with the users; it includes all the preparations for the series of tests to be performed on the system.

**testing:** The exhaustive and thorough process that determines whether the system produces the desired results under known conditions.

**text mining:** Discovery of patterns and relationships from large sets of unstructured data.

**Total cost of ownership (TCO):** Designates the total cost of owning technology resources, including initial purchase costs, the cost of hardware and software upgrades, maintenance, technical support, and training.

**Total quality management (TQM):** A concept that makes quality control a responsibility to be shared by all people in an organization.

**touch point:** Method of firm interaction with a customer, such as telephone, email, customer service desk, conventional mail, or point of purchase. Also known as a contact point.

**trade secret:** Any intellectual work or product used for a business purpose that can be classified as belonging to that business, provided it is not based on information in the public domain.

**trademarks:** Marks, symbols, and images used to distinguish products in the marketplace.

**transaction costs:** The costs of participating in a market.

**transaction fee revenue model:** E-commerce revenue model in which the firm receives a fee for enabling or executing transactions.

**transaction processing systems (TPS):** Computerized systems that perform and record the daily, routine transactions necessary to conduct the business; they serve at the organization's operational level.

**Transmission Control Protocol/Internet Protocol (TCP/IP):** Dominant model for achieving connectivity among different networks. Provides a universally agreed-on method for breaking up digital messages into packets, routing them to the proper addresses, and then reassembling them into coherent messages.

**transnational:** Truly global form of business organization where value-added activities are managed from a global perspective without reference to national borders, thus optimizing sources of supply and demand and local competitive advantage.

**Transport Layer Security (TLS):** Enables client and server computers to manage encryption and decryption activities as they communicate with each other during a secure web session; successor to the Secure Sockets Layer (SSL) protocol.

**Trojan horse:** A software program that appears legitimate but contains hidden functionality that may cause damage.

**two-factor authentication:** Validating user identity with two means of identification, one of which is typically a physical token and the other of which is typically data.

**unified communications:** Integrates disparate channels for voice communications, data communications, instant messaging, email, and teleconferencing into a single experience by which users can seamlessly switch back and forth among different communication modes.

**unified threat management (UTM):** Comprehensive security management tool that combines multiple security tools, including firewalls, virtual private networks, intrusion detection systems, and web content filtering and anti-spam software.

**uniform resource locator (URL):** The address of a specific resource on the Internet.

**unit testing:** The process of separately testing each program in the system.

**UNIX:** Operating system that is machine independent and supports multiuser processing, multitasking, and networking; often used in high-end workstations and servers but can run on many kinds of computers.

**unstructured decisions:** Nonroutine decisions in which the decision maker must provide judgment, evaluation, and insights into the problem definition; there is no agreed-upon procedure for making such decisions.

**unsupervised learning:** Machine learning algorithm trained to use information that is neither classified nor labeled in advance and to find patterns in that information without explicit human guidance.

**user–designer communications gap:** The difference in backgrounds, interests, and priorities that impede communication and problem solving between end users and information systems specialists.

**utilitarian principle:** Principle that assumes one can put values in rank order and understand the consequences of various courses of action.

**value chain model:** Model that highlights the primary or support activities that add a margin of value to a firm's products or services and where information systems can best be applied to achieve a competitive advantage.

**value web:** Customer-driven network of independent firms who use information technology to coordinate their value chains to produce a product or service collectively for a market.

**virtual company:** A company that uses networks to link people, assets, and ideas, enabling it to ally with other companies to create and distribute products and services without being limited by traditional organizational boundaries or physical locations.

**virtual private network (VPN):** A secure connection between two points across the Internet to transmit corporate data. Provides a low-cost alternative to a private network.

**virtual reality (VR) systems:** Interactive graphics software and hardware that create computer-generated simulations that provide sensations that emulate real-world activities.

**virtual worlds:** Online 3-D environments populated by users who have built graphical representations of themselves called avatars.

**virtualization:** Presenting a set of computing resources so that they can all be accessed in ways that are not restricted by physical configuration or geographic location.

**Visual Basic:** Widely used visual programming tool and environment for creating applications that run on Microsoft Windows operating systems.

**visual programming language:** Allows users to manipulate graphic or iconic elements to create programs.

**visual web:** Refers to websites and apps, such as Pinterest and Instagram, that focus on images.

**Voice over IP (VoIP):** Technology for managing the delivery of voice information using the Internet Protocol (IP).

**war driving:** An eavesdropping technique in which eavesdroppers drive by buildings or park outside and try to intercept wireless network traffic.

**web beacons:** Tiny images invisibly embedded in email messages and web pages that are designed to monitor the behavior of the user visiting a website or sending an email.

**web browsers:** Easy-to-use software tools for accessing the web and the Internet.

**web hosting service:** Company with web server computers on which they maintain the websites of fee-paying subscribers.

**web mining:** Discovery and analysis of useful patterns and information from the web.

**web server:** Software that manages requests for web pages on the computer where they are stored and delivers the page to the user's computer.

**web services:** Set of universal standards using Internet technology for integrating different applications from different sources without time-consuming custom coding. Used for linking systems of different organizations or for linking disparate systems within the same organization.

**website:** A collection of web pages linked to a home page.

**wide area networks (WANs):** Telecommunications networks that span a large geographical distance. May consist of a variety of cable, satellite, and microwave technologies.

**Wi-Fi:** Refers to the 802.11 family of wireless networking standards.

**wiki:** Collaborative website where visitors can add, delete, or modify content on the site, including the work of previous authors.

**WiMax:** Popular term for IEEE Standard 802.16 for wireless networking over a range of up to 31 miles with a data transfer rate of up to 75 Mbps.

**Windows 11:** Most recent Microsoft client operating system.

**wireless sensor networks (WSNs):** Networks of interconnected wireless devices with built-in processing, storage, and radio frequency sensors and antennas that are embedded in the physical environment to provide measurements of many points over large spaces.

**wisdom of crowds:** Belief that large numbers of people can make better decisions about a wide range of topics and products than a single person or even a small committee of experts can.

**word processing software:** Software for digitally creating, editing, formatting, and printing documents.

**workstation:** Desktop computer with powerful graphics and mathematical capabilities.

**World Wide Web:** A service the Internet provides that uses universally accepted standards for storing, retrieving, formatting, and displaying information in a page format; usually referred to as the web.

**worms:** Independent software programs that propagate themselves to disrupt the operation of computer networks or to destroy data and other programs.

**zero-day vulnerabilities:** Security vulnerabilities in software, unknown to the creator, that hackers can exploit before the vendor becomes aware of the problem.

**zero trust:** Cybersecurity framework based on the principle of maintaining strict access controls and not trusting anyone or anything by default, even those behind a corporate firewall.

# Index